Essential Procedures for Clinical Microbiology

Essential Procedures for Clinical Microbiology

Henry D. Isenberg
EDITOR IN CHIEF
Long Island Jewish Medical Center
Long Island Campus for Albert Einstein College of Medicine
New Hyde Park, New York

ASM
PRESS
Washington, D.C.

1325 Massachusetts Avenue, N.W.
Washington, DC 20005-4171

Library of Congress Cataloging-in-Publication Data

Essential procedures for clinical microbiology / editor in chief, Henry D. Isenberg.
p. cm.
Includes bibliographical references and index.
ISBN 1-55581-125-6 (pbk., comb binding)
1. Medical microbiology—Laboratory manuals. I. Isenberg, Henry D.
[DNLM: 1. Microbiological Techniques—laboratory manuals. 2. Communicable Diseases—microbiology—laboratory manuals. QW 25 E78 1998]
QR46.E87 1998
616′.01—dc21
DNLM/DLC
for Library of Congress 97-25189
CIP

Contents

Editorial Board

Abbreviations

Abbreviations

In this handbook (EPCM), most abbreviations have been introduced in parentheses after the terms they abbreviate on their first occurrence, e.g., "a central nervous system (CNS) specimen." Some exceptions to this rule are explained below and given in Tables 1 to 4.

Because of their frequent use in EPCM and/or their familiarity to readers, the terms listed in Table 1 have been abbreviated in the procedures; i.e., they have not been spelled out or introduced. Based on the editorial style for books and journals published by the American Society for Microbiology (ASM), the abbreviations listed in Table 2 have also been used without introduction in EPCM. Table 3 lists abbreviations that have been used without introduction in the bodies of tables. Abbreviations for commonly accepted units of measurement have been used without definition if they appeared with numerical values. Table 4 lists some common units of measurement appearing in EPCM. These last two items are also based on ASM style.

Finally, throughout EPCM, there are many references to the *Clinical Microbiology Procedures Handbook* (H. D. Isenberg, ed., American Society for Microbiology, Washington, D.C., 1992), which has been abbreviated "CMPH."

As readers use the various procedures in this handbook and see unfamiliar abbreviations that are not defined in the procedures themselves, they should refer to these tables for definitions.

Table 1 Common abbreviations used without introduction in EPCM

Abbreviation	Definition
ATCC	American Type Culture Collection
BAP (*not* SBA)	5% Sheep blood agar plate
BHI	Brain heart infusion
BSL	Biosafety level
CAP	College of American Pathologists
CDC	Centers for Disease Control and Prevention
CHOC	Chocolate agar
CMPH	*Clinical Microbiology Procedures Handbook*
CSF	Cerebrospinal fluid
EIA	Enzyme immunoassay
ELISA	Enzyme-linked immunosorbent assay
EMB	Eosin-methylene blue
GLC	Gas-liquid chromatography
JCAHO	Joint Commission on Accreditation of Healthcare Organizations

(continued)

Table 1 Common abbreviations used without introduction in EPCM *(continued)*

Abbreviation	Definition
MAC	MacConkey agar
MSDS	Material safety data sheet
N.A.	Numerical aperture
NBS	National Bureau of Standards (pertaining to a special calibrated thermometer)
NCCLS	National Committee for Clinical Laboratory Standards
NIH	National Institutes of Health
OSHA	Occupational Safety and Health Administration
PMNs	Polymorphonuclear leukocytes
PPE	Personal protective equipment
QA	Quality assurance
QC	Quality control
RBCs	Red blood cells *or* erythrocytes
TCBS	Thiosulfate citrate bile salt sucrose agar
THIO	Thioglycolate broth
TSA	Trypticase soy agar *or* tryptic soy agar
TSB	Trypticase soy broth *or* tryptic soy broth
WBCs	White blood cells *or* leukocytes

Table 2 Additional abbreviations used without introduction (according to ASM style)

Abbreviation	Definition
AIDS	Acquired immunodeficiency syndrome
AMP, ADP, ATP, GTP, dCMP, ddGTP, etc.	Adenosine 5′-monophosphate, adenosine 5′-diphosphate, adenosine 5′-triphosphate, guanosine 5′-triphosphate, deoxycytidine 5′-monophosphate, dideoxyguanosine triphosphate, etc.
ATPase, dGTPase, etc.	Adenosine triphosphatase, deoxyguanosine triphosphatase, etc.
cDNA	Complementary deoxyribonucleic acid
CFU	Colony-forming unit(s)
cRNA	Complementary ribonucleic acid
DEAE	Diethylaminoethyl
DNA	Deoxyribonucleic acid
DNase	Deoxyribonuclease
EDTA	Ethylenediaminetetraacetate, ethylenediaminetetraacetic acid
EGTA	Ethylene glycol-bis(β-aminoethyl ethyl)-*N,N,N′,N′*-tetraacetic acid
HEPES	*N*-2-hydroxyethylpiperazine-*N′*-2-ethanesulfonic acid
MIC	Minimal inhibitory concentration
mRNA	Messenger ribonucleic acid
NAD	Nicotinamide adenine dinucleotide
NAD^+	Oxidized nicotinamide adenine dinucleotide
NADH	Reduced nicotinamide adenine dinucleotide
NADP	Nicotinamide adenine dinucleotide phosphate
NADPH	Reduced nicotinamide adenine dinucleotide phosphate
oligo(dT), etc.	Oligodeoxythymidylic acid, etc.
PCR	Polymerase chain reaction
PFU	Plaque-forming unit(s)
poly(A), poly(dT), etc.	Polyadenylic acid, polydeoxythymidylic acid, etc.
RNA	Ribonucleic acid
RNase	Ribonuclease
rRNA	Ribosomal ribonucleic acid
Tris	Tris(hydroxymethyl)aminomethane
tRNA	Transfer ribonucleic acid
UV	Ultraviolet

Table 3 Abbreviations used without introduction in the bodies of tables

Abbreviation	Definition	Abbreviation	Definition
amt	Amount	SD	Standard deviation
approx	Approximately	SE	Standard error
avg	Average	SEM	Standard error of the mean
concn	Concentration	sp act	Specific activity
diam	Diameter	sp gr	Specific gravity
expt	Experiment	temp	Temperature
exptl	Experimental	tr	Trace
ht	Height	vol	Volume
mo	Month	vs	Versus
mol wt	Molecular weight	wk	Week
no.	Number(s)	wt	Weight
prepn	Preparation	yr	Year

Table 4 Some common units of measurement used in EPCM

Abbreviation	Definition
°C	Degree Celsius
h	Hour
μg	Microgram
mg	Milligram
min	Minute
ml	Milliliter
mm	Millimeter
mM	Millimolar
s	Second

Icons

The three icons listed below are used throughout this handbook. The icons direct the readers to follow important directions as they carry out the procedures. As a reminder, an explanation of the icon appears next to it at each appearance in the text.

It is imperative that these cultures be handled in a biosafety hood.

Include QC information on reagent container and in QC records.

Observe universal (standard) precautions.

Preface

The object of *Essential Procedures for Clinical Microbiology* is to provide a concise approach to the performance of laboratory analyses aimed at identifying whenever possible the etiological agents of infectious diseases with presently available modalities. The methods included here are not exhaustive; rather, they represent the choices of a group of experts in the various divisions that make up the discipline of clinical microbiology. Their choices are invariably supplemented with references to additional approaches to the same tasks, to detailed explanations of reagent preparations, and to other pertinent information occasionally needed to explain and interpret microbiological analyses.

This effort by the editors of the *Clinical Microbiology Procedures Handbook* is a response to numerous requests for a guide to those examinations that constitute the core of clinical microbiology. To this end, we have kept choices to a minimum and eliminated detailed descriptions but maintained compliance with the intent of the National Committee for Clinical Laboratory Standards manual of approved guidelines for laboratory procedures (GHP2-A2-1992).

The format of this volume enables the reader to follow a specimen through all of the steps from proper collection to a final report issued to the clinician. We provide numerous tables, flow sheets, and illustrations. Descriptions are kept to a minimum. Suggestions for commercial reagents and kits are provided with the clear understanding that these suggestions do not constitute an endorsement and the caveat that manufacturer's instructions must be followed to the letter.

This book is intended as a laboratory workbook to be used at the bench whenever it is needed. As with the *Clinical Microbiology Procedures Handbook*, we invite all users to communicate the errors that have escaped us and to provide suggestions and any advice that will help improve this guide.

Henry D. Isenberg

Acknowledgments

In preparing the *Essential Procedures for Clinical Microbiology* (EPCM), the section editors selected and modifed procedures from the *Clinical Microbiology Procedures Handbook* (CMPH) to conform to the format of EPCM. In some cases, changes to the original CMPH procedures were minimal and at other times the changes were more extensive. Although the names of the original CMPH authors have not been listed in this volume, the editors and publisher do want to acknowedge again the contributions of these authors. Readers should refer to CMPH for these authors' names and affiliations.

In the tables of contents for the various sections, some authors' names are listed along with the titles of the procedures. When names are listed, this means that the material is new to EPCM. When items within this new material have been picked up from CMPH, an appropriate credit will be given in the text. For the most part, the names that are listed are those of the section editors themselves since they wrote most of the new material for this volume.

For space considerations, when a general reference has been made in the text to CMPH, there will not be corresponding citation given in the References or Supplemental Reading section. Readers should be aware that the reference is to the *Clinical Microbiology Procedures Handbook* (H. D. Isenberg, ed., American Society for Microbiology, Washington, D.C., 1992). In those instances in which a particular portion of CMPH is being referred to, the appropriate citation will be given in the References or Supplemental Reading section.

SECTION 1

Collection, Transport, and Manipulation of Clinical Specimens and Initial Laboratory Concerns

SECTION EDITOR: *Henry D. Isenberg*

1.1 Collection, Transport, and Manipulation of Clinical Specimens and Initial Laboratory Concerns

The tables in this section address the initial events that lead to the accurate, rapid identification of microorganisms and viruses. Remember that results can be only as good as the original specimens. All the long-established precautions must be observed, keeping in mind that many specimens are obtained from anatomic sites that encourage specimen contamination with indigenous microbiota. Laboratorians in clinical microbiology laboratories must be aware of these quasi-autochthonous organisms and evaluate their results accordingly (6).

Table 1.1–1 summarizes the microorganisms, including viruses, involved in disease production at various anatomic sites. The term "clinical impression" is used to indicate the most likely diagnosis for the patient conditions listed. The corresponding organisms are those most frequently isolated and that demonstrate some involvement with the disease manifestations in the patient (5). This table should be useful in counseling individuals preparing to collect specimens and in alerting laboratory personnel to choose the most appropriate means to ensure the isolation of the most likely etiological agents.

Table 1.1–2 summarizes instructions for the collection of specimens for transport to the laboratory and bacteriological analyses. These instructions are intended to help health care personnel charged with the responsibility of collecting specimens and laboratory personnel in their effort to ensure that only specimens properly obtained, placed into appropriate containers, and of adequate volume are analyzed. In this time of cost containment efforts, specimens that do not meet all requirements cannot be examined cost-effectively (*see* Table 1.1–7). The excuse most often proffered for inadequate or mishandled specimens is the inability to secure a correct specimen for repeat examination. Decisions about proceeding with analyses under these circumstances are laboratory directors' responsibilities and should be referred to these individuals whenever possible. A policy reflecting the laboratory director's opinion on such claims and the consequent action to be taken should be made part of the instruction manual of the laboratory and that provided to the clinical staff. In Table 1.1–2, the column headed "helpful clinical information" is intended to heighten the awareness of all personnel involved that pertinent clinical information and the patient's history are indispensable in the interpretation of microbiological results. Similarly, the "comment" column provides additional information and caveats where needed.

Table 1.1–3 provides guidelines for the collection of fungal specimens. Pityriasis versicolor cannot be cultured; therefore, scrapings are best submitted in sterile tubes or small sterile containers for direct examinations in the laboratory.

Table 1.1–4 lists the requirements for collecting and transporting clinical specimens for parasite examinations. Blood smears for the detection of plasmodia are frequently submitted to the Hematology Division. Arrangements between the two laboratories are advisable, enabling both to provide input concerning the presence and identity of these protozoa. For blood parasites other than the etiological agents of malaria and babesiosis, note Table 1.1–4 footnote *b* and the appropriate references provided.

A great variety of skin preparations for obtaining blood, CSF, and other specimens have been advocated. Remember, the purpose of such preparations is the elimination of contaminants capable of obfuscating your analyses and confusing the clinician.

Table 1.1–5 represents one of the simplest approaches to the problem. Microbiology personnel may not be involved in securing many specimens requiring skin disinfection, although phlebotomy for blood cultures may fall within their purview in some institutions. However, microbiology staff must be fully cognizant of these requirements, and the laboratory information manual must highlight the need for these precautions. Instructions in aseptic techniques for all occasions should be a quality management task for microbiologists, infectious disease specialists, epidemiologists, phlebotomists, surgical staff, neurologists, emergency department personnel, and all other individuals charged with providing specimens that require invasive steps. Periodic review of these requirements is advisable.

Table 1.1–6 is an abbreviated listing of commercially available transport media that should be used to ensure the integrity of the specimen from patient to laboratory. Most of those preparations do not require refrigeration if the specimens can be delivered to the laboratory within 24 h (or in some cases, 48 h). Plastic meshes (polyurethane), rather than cotton, on collection swabs provide an effective and nontoxic means for adsorbing, rather than absorbing, microorganisms from clinical material and can be used for the isolation of anaerobic bacteria in casual specimens. For sinus tracts, biopsy specimens, deep surgical or traumatic wounds, and body fluids, a variety of excellent transport media are available. Microbiologists who still manufacture their own transport media can consult appropriate sections of CMPH for details. While this reference also addresses the requirements for transporting parasites and ova, even more detailed advice is provided by Garcia and Bruckner (4). Detailed instructions concerning the transport of viral specimens and the requirements for shipping specimens to reference laboratories can be found in CMPH and in the virology section of this volume.

Once a specimen arrives in the microbiology laboratory, personnel must ascertain that all pertinent information has been provided, that the specimen has been collected in the proper transport device, and that all other conditions for an acceptable specimen have been met. The previously mentioned need for a policy of rejecting specimens that do not meet these requirements is expanded in Table 1.1–7 in an effort to provide a sample policy that should be modified by each individual laboratory to address its particular problems. Footnote *b* is a special reminder to all laboratory personnel to be acutely aware that a record of any and all actions concerning the lack of acceptibility of a clinical specimen, including the name and location of the person notified by telephone, must be maintained. Similarly, footnote *c* may seem redundant but is meant to emphasize that only appropriate specimens are suitable for anaerobic cultures. To attempt using unsuitable specimens for this purpose is a waste of material and labor and will likely provide meaningless results.

Table 1.1–8 outlines an example of how laboratory personnel can proceed to process specimens. Accessioning varies greatly from institution to institution, but the essential steps enumerated in section I of Table 1.1–8 must be met. Note that each specimen receives an identification number reserved for this particular specimen alone. Specimens obtained from the same patient but of different anatomic origins must be designated separately. Table 1.1–8 does not mention that smears should be prepared on many of the specimens. While this step is not needed universally, smears remain one of the best and most cost-effective means of providing important information almost immediately. The smear also serves as a control for culture results and permits an informed guess as to the cellular responses of immunocompetent patients. The suggestions for media to be used are just that: suggestions. Many laboratory workers prefer to use the media they encountered when first introduced to microbiology. It does not matter which particular medium is used as long as the microorganisms presumed to be present as pathogens or members of the normal microbiota can be separated from the clinical specimen (*see* Table 1.1–9).

The procedure for streaking agar plates is intended to make the experienced laboratorian pause to consider whether shortcuts have been introduced into this procedure with time and to guide the neophyte microbiologist to the easiest way to dilute an inoculum to obtain isolated colonies. Urine is singled out, since some quantitation of the microbial bioload is desirable. Laboratories that use dipstick technology as a prescreen for culture of urine must remember that immunocompromised patients may not meet a microbial challenge with increased segmented neutrophils and thus will not react positively for leukocyte esterase, while many of the microorganisms other than members of the family *Enterobacteriaceae* do not reduce nitrate to nitrite. Microbiologists should keep in mind that a Gram stain of well-mixed, uncentrifuged urine showing two or more bacteria per oil immersion field reflects approximately 100,000 CFU/ml.

Table 1.1–9 lists the various media that are among the most frequently used in clinical microbiology laboratories, atmosphere requirements for their incubation, inhibitor and/or indicator contents, and the rationale for their use. As mentioned above and in footnote *a*, substitutions that reflect individual preferences are perfectly acceptable as long as the spectrum of microorganisms expected in a specimen by virtue of the clinical impression, history, or body site can be met. The media listed here are intended primarily for bacteria. Media for fungi are listed in Table 1.1–3, while the section on virology provides appropriate guidelines for tissue culture use. Consult CMPH for cultivation of select protozoa.

Time and again, health care personnel and even an occasional administrator will inquire about "normal values" for microbiology. Confusion and disbelief usually meet the microbiologist's assertion that "normal" in microbiology can be defined only by the isolation or demonstration of an "abnormal" representation of the microbial world, that is, an organism that should not be found in a particular specimen. However, all clinical microbiology personnel must be aware of the urgent need to report the observations and isolations listed in Table 1.1–10. This list may be expanded, shortened, or modified in accordance with existing conditions and the emergence or reemergence of potentially pathogenic microorganisms. For example, vancomycin-resistant enterococci and penicillin-tolerant and -resistant pneumococci deserve such special attention. In addition, the organisms listed in Table 1.1–11 might be missed unless the laboratory staff is alerted to the clinician's suspicion. This "Alert Request" should be part of the manual that provides health care personnel with the requirements of the clinical pathology service.

Table 1.1–12 lists the serodiagnostic tests presently available to be performed in-house or to be sent to reference laboratories. Unless a laboratory performs a test routinely (at least once a week on several specimens), it is advisable to enlist the services of an established reference laboratory. It is important that the reference laboratory provide not only reports of expeditiously performed tests but also the values obtained with appropriate controls. These reports should be made part of the patient's record, with clear indications that the test was not performed in the institutional laboratory. Copies of reference laboratory reports should be kept in the microbiology laboratory files. Table 1.1–12 also lists "normal values" that represent the cutoff points generally accepted for these examinations. However, these interpretations are subject to the instructions on the package insert, the variations reported by the reference laboratory, or the results obtained when the population in a particular geographic locale is studied. Instructions provided by reference laboratories concerning the collection and transport for these examinations must be followed to the letter. For some tests, refrigerated storage space for acute- and convalescent-stage specimens should be available, unless the reference laboratory indicates that the acute-stage specimen will be held in its facility. If the tests are performed in-house, it is advisable to store serum rather than clotted blood.

Acknowledgments. The tables in this chapter have been gleaned from the laboratory manuals used by several of the editors and authors who participated in preparing this volume and CMPH. I am indebted to each and every one of them for their generosity and assistance.

REFERENCES

1. **Amies, C. R.** 1967. A modified formula for the preparation of Stuart's transport medium. *Am. J. Public Health* **58:**296–299.
2. **Baron, E. J., L. R. Peterson, and S. M. Finegold.** 1994. *Bailey and Scott's Diagnostic Microbiology,* 9th ed. The C. V. Mosby Co., St. Louis.
3. **Carey, S. G., and E. B. Blair.** 1964. New transport medium for shipment of clinical specimens. I. Fecal specimens. *J. Bacteriol.* **88:** 96–98.
4. **Garcia, L. S., and D. A. Bruckner.** 1993. *Diagnostic Medical Parasitology.* American Society for Microbiology, Washington, D.C.
5. **Gorbach, S. L., J. G. Bartlett, and N. R. Blacklow.** 1992. *Infectious Diseases.* The W. B. Saunders Co., Philadelphia.
6. **Isenberg, H. D., and R. F. D'Amato.** 1994. Indigenous and pathogenic microorganisms of humans, p. 5–18. *In* P. R. Murray, E. J. Baron, M. A. Pfaller, F. C. Tenover, and R. H. Yolken (ed.), *Manual of Clinical Microbiology,* 6th ed. ASM Press, Washington, D.C.
7. **Isenberg, H. D., F. D. Schoenknecht, and A. von Graevenitz.** 1979. *Cumitech 9, Collection and Processing of Bacteriological Specimens.* Coordinating ed., S. J. Rubin. American Society for Microbiology, Washington, D.C.
8. **Stuart, R. D., S. R. Tosach, and T. M. Patsula.** 1954. The problem of transport for gonococci. *Am. J. Public Health* **45:**73–77.
9. **Summanen, P., E. J. Baron, D. M. Citron, C. A. Strong, H. M. Wexler, and S. M. Finegold.** 1993. *Wadsworth Anaerobic Bacteriology Manual,* 5th ed. Star Publishing Co., Belmont, Calif.

Table 1.1–1 ''Rule-out'' clinical impressions and potential etiological agents

Clinical impression	Potential etiological agent
Head and neck infections	
Gingivitis	Spirochetes, *Prevotella intermedia, Prevotella oralis*
Peridontitis	*Prevotella melaninogenica*, spirochetes
Juvenile	*Actinobacillus actinomycetemcomitans*
Early onset	Spirochetes, *Porphyromonas gingivalis, Actinobacillus actinomycetemcomitans*
Adult	Spirochetes, *Prevotella melaninogenica, Actinobacillus actinomycetemcomitans*
Progressive	*Bacteroides forsythus, Campylobacter rectus*
Peritonsillar abscess (Quinsy)	Polymicrobic: *Streptococcus pyogenes*, aerobic bacteria plus normal microbiota of the oral cavity
Ludwig's angina	Viridans streptococci, peptostreptococci, *Prevotella melaninogenica, Fusobacterium nucleatum*
Pterygopalantine, infratemporal	Anaerobes, aerobic and facultatively anaerobic gram-negative rods
Para- and retropharyngeal abscesses, temporal fossea infections	Polymicrobic with anaerobes, streptococci, staphylococci, *Enterobacteriaceae*
Septic jugular thrombophlebitis (postanginal sepsis)	Anaerobic streptococci, *Prevotella* spp., *Porphyromonas* spp., viridans streptococci, *Streptococcus pyogenes, Streptococcus pneumoniae, Eickenella corrodens*
Suppurative parotitis	*Staphylococcus aureus*, anaerobes, *Enterobacteriaceae, Pseudomonas aeruginosa*, oral microbiota, *Eickenella corrodens*
Parotitis (viral)	Mumps, coxsackievirus, influenza virus, parainfluenza virus types 1 and 3, lymphocytic choriomeningitis virus, cytomegalovirus
Sinusitis	*Streptococcus pneumoniae, Haemophilus influenzae, Prevotella* spp., *Porphyromonas* spp., *Fusobacterium* spp., *Peptostreptococcus* spp., *Staphylococcus aureus, Veillonella* spp., *Streptococcus pyogenes, Moraxella catarrhalis*, aerobic and facultatively anaerobic gram-negative rods, microsporidia, free-living amebae, rhinovirus, influenza virus, parainfluenza virus, adenovirus (children)
Cranial epidural abscess	*Peptostreptococcus* spp., viridans streptococci, *Streptococcus milleri* group, *Bacteroides* spp., (*Prevotella* spp.); *Enterobacteriaceae, Pseudomonas aeruginosa, Staphylococcus aureus*
Otitis media	*Streptococcus pneumoniae, Haemophilus influenzae, Moraxella catarrhalis, Streptococcus pyogenes, Staphylococcus aureus, Pseudomonas aeruginosa*
Chronic suppurative	As above plus *Staphylococcus epidermidis, Candida* spp., *Corynebacterium* spp., *Bacteroides* spp., *Peptostreptococcus* spp.
Pharyngitis	*Streptococcus pyogenes*, group C and G, β-hemolytic streptococci; (*Corynebacterium diphtheriae, Corynebacterium ulcerans, Arcanobacterium haemolyticum, Neisseria gonorrhoeae*, mixed anaerobes [Vincent's angina], *Yersinia enterocolitica, Treponema pallidum*—rare), rhinovirus, coronavirus, adenovirus, influenza viruses A and B, parainfluenza viruses (less common herpes simplex virus [types 1 and 2], coxsackie A [types 2, 4, 5, 6, 8, 10], Epstein-Barr virus, cytomegalovirus, HIV)
Laryngitis	*Moraxella catarrhalis, Bordetella pertussis, B. parapertussis, Haemophilus influenzae* (may be secondary invaders), *Mycobacterium hominum* (rare—isolated from granulomatous laryngitis, also *Histoplasma capsulatum, Coccidioides immitis, Blastomyces dermatitidis, Candida* spp., *Treponema pallidum*, herpes simplex virus, varicella), influenza virus, parainfluenza virus, rhinovirus, adenovirus

Table 1.1–1 ''Rule-out'' clinical impressions and potential etiological agents *(continued)*

Clinical impression	Potential etiological agent
Epiglotitis	*Haemophilus influenzae* serogroup b, *Haemophilus parainfluenzae, Streptococcus pneumoniae, Staphylococcus aureus, Streptococcus* spp., non-b *Haemophilus influenzae, Pasteurella multocida*
Tracheitis	*Staphylococcus aureus, Streptococcus pyogenes, Haemophilus influenzae* b
Thyroiditis	
Acute suppurative	Oral microbiota, *Staphylococcus* spp., *Streptococcus pneumoniae*, anaerobes (needle aspiration); in immunocompromised patients: *Pseudoallescheria boydii, Candida* spp., *Aspergillus* spp., *Coccidioides immitis, Actinomyces* spp.
Subacute granulomatous	Mumps, rubeola virus, influenza virus, adenovirus; Epstein-Barr virus, coxsackievirus; cytomegalovirus, *Yersinia* spp.
Common cold	>200 different viruses, including different serotypes of adenoviruses, coronaviruses, influenza virus, parainfluenza virus, respiratory syncytial virus, rhinovirus, enterovirus
Pleuropulmonary infections	
Acute community-acquired pneumonia (adults)	*Streptococcus pneumoniae, Haemophilus influenzae, Staphylococcus aureus, Legionella pneumophila, Legionella* spp., oral anaerobes (aspiration), *Moraxella catarrhalis, Mycoplasma pneumoniae*, (very rare = *Yersinia pestis, Bacillus anthracis, Francisella tularensis, Pseudomonas pseudomallei*), *Histoplasma capsulatum, Blastomyces dermatitidis, Coccidioides immitis, Actinomyces* spp., *Cryptococcus neoformans, Chlamydia pneumoniae*, influenza virus, adenovirus, *Nocardia asteroides, Sporothrix schenckii, Penicillium marneffei, Geotrichum* spp.
Pneumonia in immunocompromised host	Members of *Enterobacteriaceae, Pseudomonas* spp., and related obligately aerobic gram-negative rod-shaped bacteria, *Staphylococcus aureus, Streptococcus pneumoniae, Nocardia* spp., *Legionella* spp., *Mycobacterium tuberculosis, Mycobacterium avium-intracellulare, Mycobacterium haemophilum, Aspergillus* spp., zygomycetes, *Cryptococcus neoformans, Curvularia lunata, Fusarium* spp., *Paecilomyces varioti, Candida* spp., *Trichosporon* spp., *Torulopsis glabrata, Pneumocystis carinii, Toxoplasma gondii, Strongyloides stercoralis, Cryptosporidium parvum*, microsporidia, cytomegalovirus, varicella-zoster, herpes simplex virus
Bronchitis	*Haemophilus influenzae, Haemophilus parainfluenzae, Streptococcus pneumoniae, Moraxella catarrhalis, Neisseria* spp., *Klebsiella* spp., *Pseudomonas* spp., obligately aerobic gram-negative rod-shaped bacteria
Aspiration pneumonia (caused by microorganisms)	*Peptostreptococcus* spp., *Fusobacterium nucleatum, Porphyromonas asaccharolytica, Prevotella melaninogenica*, (*Enterobacteriaceae, Pseudomonas* spp., and related bacteria in nosocomial aspiration pneumonia)
Lung abscess	*Peptostreptococcus* spp., *Fusobacterium nucleatum, Prevotella melaninogenica, Bacteroides fragilis* group, *Staphylococcus aureus, Escherichia coli, Klebsiella pneumoniae, Pseudomonas aeruginosa, Streptococcus pneumoniae*
Cardiovascular infections	
Bloodstream infections	*Streptococcus pneumoniae, Streptococcus pyogenes, Streptococcus agalactiae, Enterococcus* spp., especially *E. faecalis* and *E. faecium, Streptococcus bovis, Streptococcus* serogroup G, viridans streptococci, *Staphylococcus aureus, Staphylococcus epidermidis*, other *Staphylococcus* spp., *Corynebacterium jeikeium, Enterobacteriaceae, Pseudomonas* spp. and other obligately aerobic gram-negative rod-shaped bacteria, *Stenotrophomonas maltophilia, Burkholderia cepacia, Acinetobacter* spp., *Flavobacterium* spp., *Haemophilus* spp., *Neisseria* spp., (*Moraxella catarrhalis*), *Bacteroides* spp., *Fusobacterium* spp., *Prevotella* spp.,

(continued)

Table 1.1–1 "Rule-out" clinical impressions and potential etiological agents *(continued)*

Clinical impression	Potential etiological agent
	Porphyromonas spp., *Peptostreptococcus* spp., other anaerobic bacteria, *Mycobacterium* spp., *Candida* spp., *Hansenula* spp., *Malasezzia furfur* and *Plasmodium* spp., *Babesia* spp., *Trypanosoma* spp. (rare), *Leishmania donovani* (rare), microfilariae (rare), *Toxoplasma gondii* (rare), and other microorganisms on rare occasions
Septic shock	Most gram-negative bacteria, *Neisseria meningitidis, Haemophilus influenzae, Salmonella* spp., other *Enterobacteriaceae* including *Yersinia* spp., *Vibrio* spp., *Pseudomonas* spp., *Streptococcus pneumoniae, Mycobacterium* spp. (steroids; T-cell abnormalities), zygomycetes (diabetes, neutropenia), *Aspergillus* spp., *Candida* spp., *Listeria* spp. (T-cell abnormality), herpes simplex (T-cell abnormalities), cytomegalovirus (T-cell abnormalities), varicella-zoster virus (T-cell abnormalities)
Endocarditis	
Native valve	Viridans streptococci (*Streptococcus sanguis, Streptococcus salivarius, Streptococcus mutans, Streptococcus mitior*, etc.), *Enterococcus faecalis, Enterococcus faecium, Enterococcus durans, Streptococcus bovis*, (caveat = colon malignancy), *Streptococcus equinus, Streptococcus pyogenes, Streptococcus agalactiae, Staphylococcus aureus*, coagulase-negative staphylococci, *Streptococcus pneumoniae, Neisseria gonorrhoeae, Haemophilus* spp., *Pseudomonas* spp., *Listeria* spp., *Corynebacterium* spp., (*Candida* spp., *Torulopsis glabrata*, and *Aspergillus* spp. can produce native valve endocarditis in severe underlying disease, corticosteroid therapy, prolonged antibiotic use, or cytotoxic therapy)
Intravenous drug abuser	*Staphylococcus aureus, Streptococcus* spp., *Enterococcus* spp., gram-negative rod-shaped bacteria (mostly *Pseudomonas* spp. and *Serratia* spp.), *Candida* spp., *Plasmodium* spp., *Leishmania* spp., anaerobic oral bacteria
Prosthetic valve	
Early	*Staphylococcus epidermidis, Staphylococcus aureus*, aerobic gram-negative rods, fungi (usually *Candida* spp. and/or *Aspergillus* spp.), *Streptococcus* spp., *Enterococcus* spp., *Corynebacterium* spp.
Late	Viridans streptococci, *Staphylococcus* spp., *Enterococcus* spp., (staphylococci, gram-negative rods, fungi, and corynebacteria are isolated in earlier infections occurring in <18 months)
Vascular graft infections	*Staphylococcus aureus, Staphylococcus epidermidis, Streptococcus* spp., *Escherichia coli, Proteus mirabilis, Pseudomonas aeruginosa*, anaerobic bacteria, *Candida* spp.
Pericarditis, myocarditis	*Staphylococcus aureus, Streptococcus pneumoniae, Streptococcus pyogenes, Escherichia coli, Proteus* spp., *Pseudomonas aeruginosa, Salmonella* spp., *Shigella* spp., *Neisseria meningitidis, Histoplasma capsulatum, Candida* spp., *Aspergillus* spp., *Trypanosoma cruzi, Aspergillus* spp., coxsackievirus (especially B), echoviruses, *Mycobacterium tuberculosis, Echinococcus granulosum, Entamoeba histolytica, Toxoplasma gondii, Trichinella spiralis, Mycoplasma pneumoniae*
Intestinal tract infections	
Diarrhea	*Shigella* spp., *Salmonella* spp., *Vibrio cholerae*, enterotoxigenic *Escherichia of coli, Giardia lamblia, Campylobacter* spp., *Entamoeba histolytica, Dientamoeba fragilis*, rotavirus, Norwalk agent, toxins of *Clostridium difficile, Staphylococcus aureus, Bacillus cereus, Clostridium perfringens, Yersinia enterocolitica, Plesiomonas shigelloides, Cryptosporidium* spp., *Isospora belli, Sarcocystis hominis, Cyclospora cayetanensis, Encephalitozoon* spp., *Nosema* spp., *Enterocytozoon* spp., *Pleistophora* spp., *Microsporidium* spp.
Salmonellosis	
Gastroenteritis	*Salmonella typhimurium, Salmonella enteritidis, Salmonella newport, Salmonella aviatum*, other *Salmonella* spp.

Table 1.1–1 ''Rule-out'' clinical impressions and potential etiological agents *(continued)*

Clinical impression	Potential etiological agent
Enteric fever	*Salmonella typhi, Salmonella paratyphi* A, *Salmonella schottmuelleri*, other *Salmonella* spp.
Bacteremia	*Salmonella typhimurium, Salmonella choleraesuis, Salmonella heidelberg*, other *Salmonella* spp.
Carrier state	*Salmonella* spp.
Escherichia coli	
Diarrheagenic infections	Enterotoxigenic *Escherichia coli*, enteropathogenic *E. coli*, enteroadherent *E. coli*, enteroinvasive *E. coli*, enterohemorrhagic *E. coli*
Intestinal campylobacteriosis (diarrhea)	*Campylobacter jejuni, Campylobacter coli, Campylobacter lari, Campylobacter upsalensis, Campylobacter fetus*
Gastric/duodenal ulcers	*Helicobacter pylori*
Viral gastroenteritis	Rotavirus, Norwalk agent, adenovirus (enteric), astrovirus, calicivirus, Norwalk-like viruses
Food poisoning (food as vehicle and/or contains toxins)	*Bacillus cereus, Campylobacter* spp., *Clostridium perfringens, Salmonella* spp., *Shigella* spp., *Staphylococcus aureus, Clostridium botulinum, Escherichia coli, Vibrio parahaemolyticus, Yersinia enterocolitica, Listeria monocytogenes, Giardia lamblia*, microsporidia, *Entamoeba histolytica, Anisakis* spp., (raw saltwater fish), *Taenia saginata, Taenia solium, Diphyllobothrium latum*, lung and liver trematodes, *Trichinella spiralis*, hepatitis A virus, Norwalk virus
Intra-abdominal infections	*Escherichia coli, Klebsiella* spp., *Proteus* spp., *Enterobacter* spp., *Pseudomonas aeruginosa, Staphylococcus aureus, Enterococcus* spp., *Bacteroides* spp., *Fusobacterium* spp., *Veillonella* spp., *Peptostreptococcus* spp., *Propionibacterium* spp., *Staphylococcus* spp.
Intra-abdominal abscess, including appendicitis, diverticulitis	*Escherichia coli, Bacteroides fragilis, Enterococcus* spp., *Peptostreptococcus* spp., *Clostridium* spp., *Proteus* spp., *Fusobacterium* spp., *Klebsiella* spp., *Pseudomonas* spp., aerobic gram-negative rod-shaped bacteria, *Staphylococcus* spp., *Eubacterium* spp., *Streptococcus pyogenes, Streptococcus* spp.
Liver abscess/infections	*Streptococcus* spp., *Escherichia coli, Proteus* spp., *Peptostreptococcus* spp., *Fusobacterium* spp., *Bacteroides* spp., *Entamoeba histolytica, Leishmania donovani*, microsporidia
Liver granulomata	*Mycobacterium tuberculosis, Mycobacterium* spp., *Brucella* spp., *Francisella tularensis, Histoplasma capsulatum, Coccidioides immitis, Coxiella burnetii, Treponema pallidum* (secondary syphilis), *Echinococcus* spp., *Schistosoma* spp., cytomegalovirus, Epstein-Barr virus
Pancreatic infections	*Escherichia coli, Enterococcus* spp., *Staphylococcus* spp., *Klebsiella* spp., *Proteus* spp., *Candida* spp., *Pseudomonas* spp., *Streptococcus* spp., *Torulopsis glabrata, Haemophilus* spp., *Corynebacterium* spp., *Serratia marcescens*
Splenic abscess	*Staphylococcus* spp., *Salmonella* spp., *Escherichia coli, Enterococcus* spp., *Streptococcus* spp., *Klebsiella* spp., *Enterobacter* spp., *Proteus* spp., *Pseudomonas* spp., *Corynebacterium* spp., *Shigella* spp., *Bacteroides* spp., *Propionibacterium* spp., *Clostridium* spp., *Fusobacterium* spp., *Candida* spp., *Aspergillus* spp., *Leishmania donovani*, microsporidia, *Blastomyces dermatitidis*
Genitourinary tract infections	
Urinary tract infections	*Escherichia coli, Staphylococcus saphrophyticus, Proteus* spp., *Klebsiella* spp., *Enterococcus* spp., *Pseudomonas* spp., *Candida* spp., *Staphylococcus* spp.

(continued)

Table 1.1–1 "Rule-out" clinical impressions and potential etiological agents *(continued)*

Clinical impression	Potential etiological agent
Urethritis	
Male	*Neisseria gonorrhoeae, Chlamydia trachomatis, Ureaplasma urealyticum*
Female	*Escherichia coli, Staphylococcus saprophyticus, Chlamydia trachomatis, Ureaplasma urealyticum*
Prostatitis	*Escherichia coli*, other *Enterobacteriaceae, Pseudomonas* spp., *Staphylococcus aureus, Enterococcus* spp., *Trichomonas vaginalis*
Epididymitis	*Chlamydia trachomatis, Neisseria gonorrhoeae, Enterobacteriaceae, Pseudomonas* spp.
Renal abscess	
Cortical	*Staphylococcus aureus*
Corticomedullary	*Escherichia coli, Klebsiella* spp., *Proteus* spp.
Perinephric	*Staphylococcus aureus, Escherichia coli, Proteus* spp., *Klebsiella* spp., *Pseudomonas* spp., *Streptococcus* spp., *Enterococcus* spp., *Mycobacterium* spp., *Bacteroides* spp., *Candida* spp.
Sexually transmitted diseases	
Acute pelvic inflammatory disease	*Neisseria gonorrhoeae, Chlamydia trachomatis, Mycoplasma hominis, Peptostreptococcus* spp., *Bacteroides* spp., *Enterobacteriaceae*
Neonatal/perinatal complications of STD	*Neisseria gonorrhoeae, Chlamydia trachomatis*, cytomegalovirus, herpes simplex virus, *Ureaplasma urealyticum, Mycoplasma hominis*
Potential factors in neoplasia	Human papillomavirus, hepatitis B virus, HIV, herpes simplex virus
Mucopurulent cervicitis	*Chlamydia trachomatis, Neisseria gonorrhoeae*, herpes simplex virus
Vaginitis, vulvovaginitis	*Trichomonas vaginalis, Candida* spp., *Bacteroides* spp., *Prevotella bivia, Prevotella disiens, Prevotella* spp., *Actinomyces* spp., *Peptostreptococcus* spp., *Eubacterium nodatum, Mobiluncus* spp.
Vaginosis	*Gardnerella vaginalis, Mycoplasma hominis, Mobiluncus* spp.
Genital ulcers with lymphadenopathy	*Treponema pallidum, Haemophilus ducreyi, Chlamydia trachomatis* (lymphogranuloma venereum strains), *Calymmatobacterium granulomatis*, herpes simplex virus
Endometritis	*Chlamydia trachomatis, Neisseria gonorrhoeae*, (*Streptococcus agalactiae, Mycoplasma hominis*, isolated from cervix)
Amniotic fluid infections	*Ureaplasma urealyticum, Mycoplasma hominis, Bacteroides* spp., *Gardnerella vaginalis, Streptococcus agalactiae, Peptostreptococcus* spp., *Escherichia coli, Enterococcus* spp., *Fusobacterium* spp., *Bacteroides* spp.
Skin and soft tissue infections	
Impetigo	*Streptococcus pyogenes, Staphylococcus aureus*
Echthyma	*Streptococcus pyogenes, Staphylococcus aureus*
Blistering distal dactylitis	*Streptococcus pyogenes, Staphylococcus aureus*
Foliculitis	*Staphylococcus aureus*
Furuncles and carbuncles	*Staphylococcus aureus*
Erysipelas	*Streptococcus pyogenes*, (rare: serogroup G, C, and B streptococcus)
Cellulitis, acute	*Streptococcus pyogenes, Staphylococcus aureus*, (rare: *Haemophilus influenzae, Enterobacteriaceae, Clostridium* spp., *Bacillus anthracis, Pasteurella multocida, Erysipelothrix* spp., *Aeromonas hydrophila, Vibrio vulnificus, Mycobacterium* spp.)
Fasciitis, necrotizing	*Streptococcus pyogenes* or synergistic infection of facultative and anaerobic bacteria

Table 1.1–1 "Rule-out" clinical impressions and potential etiological agents *(continued)*

Clinical impression	Potential etiological agent
Fournier's gangrene	*Escherichia coli, Pseudomonas aeruginosa, Proteus mirabilis, Enterococcus* spp., *Staphylococcus* spp., *Peptostreptococcus* spp., *Bacteroides* spp.
Abscess, cutaneous	
Axilla, paronychia, breast, hands, head, neck, and trunk	*Staphylococcus aureus, Staphylococcus epidermidis, Propionibacterium* spp., *Peptostreptococcus* spp., *Leishmania tropica* complex, *Leishmania braziliense* complex
Perineal, vulvovaginal, scrotal, perianal, and buttocks	*Staphylococcus aureus, Staphylococcus epidermidis, Propionibacterium* spp., *Peptostreptococcus* spp., *Streptococcus* spp.
Myositis	
Bacterial	*Staphylococcus aureus, Streptococcus* spp., influenza virus, coxsackievirus, Epstein-Barr virus, herpes simplex virus type 2, parainfluenza virus type 3, adenovirus type 21, echovirus type 9
Protozoan	*Toxoplasma gondii, Trypanosoma* spp., microsporidia
Tinea favosa	*Trichophyton schoenleinii* (rare: *Trichophyton mentagrophytes, Trichophyton verrucosum, Microsporum canis*)
Piedra	
White	*Trichosporum beigelii* (Indonesia, South America, Far East)
Black	*Piedraia hortae* (tropical South America, Africa, Pacific Islands, Far East)
Tinea corporis (ringworm)	*Trichophyton* spp., *Microsporum* spp., *Epidermophyton* spp.
Tinea imbricata	*Trichophyton rubrum, Trichophyton mentagrophytes, Epidermophyton floccosum*
Onychomycosis	
Dermatophytic	*Trichophyton rubrum, Trichophyton mentagrophytes*
Nondermatophytic	*Candida albicans, Geotrichum candidium, Scopulariopsis brevicaulis, Cephalosporium* spp., *Acremonium* spp.
Viral skin infections	Rubeola, rubella, varicella, herpes zoster, papillomavirus (warts), parvovirus (erythema infectiosum), human herpesvirus 6 (roseola), enteroviral exanthems: hand-foot-mouth syndrome—coxsackievirus A (select types); other types of rashes: echovirus, coxsackievirus—maculopapular; coxsackievirus and echovirus—petechial; caveat: hard to differentiate from meningococcemia rash
Bone and joint infections	
Osteomyelitis	
Hematogenous	*Staphylococcus aureus, Staphylococcus* spp., *Streptococcus agalactiae, Candida* spp., *Pseudomonas aeruginosa, Salmonella* spp.
Trauma associated	*Streptococcus* spp., *Propionibacterium* spp., *Enterobacteriaceae, Pseudomonas* spp., *Staphylococcus* spp., anaerobic bacteria
Vascular insufficiency	*Enterobacteriaceae*, anaerobic bacteria
Septic arthritis	
Nongonococcal	*Neisseria gonorrhoeae, Staphylococcus aureus, Streptococcus pyogenes, Streptococcus pneumoniae, Streptococcus* spp., *Enterococcus* spp., *Enterobacteriaceae*, obligately aerobic gram-negative rod-shaped bacteria
Rare arthritides	*Brucella* spp., *Salmonella* spp., rubella, mumps, coxsackievirus, echovirus (transient), parvovirus B 19, hepatitis B virus (immune response), *Mycobacterium* spp.

(continued)

Table 1.1–1 "Rule-out" clinical impressions and potential etiological agents *(continued)*

Clinical impression	Potential etiological agent
Infections of the eye	
Eyelid	*Staphylococcus aureus*, herpes simplex virus, varicella virus, papillomavirus, *Trichophyton* spp., *Microsporum* spp., *Trichosporon* spp.
Styes	*Staphylococcus aureus*
Chalazion	*Staphylococcus aureus, Moraxella lacunata*
Conjunctivitis	
Purulent	*Neisseria gonorrhoeae, Neisseria meningitidis* (newborn); *Chlamydia trachomatis, Staphylococcus aureus, Streptococcus pyogenes, Streptococcus pneumoniae, Haemophilus influenzae, (Haemophilus aegyptius), Pseudomonas aeruginosa, Escherichia coli*
Chronic	*Moraxella lacunata, Staphylococcus* spp.
Parinaud's oculoglandular	*Bartonella henselae* (or possibly *Afipia felis* [both possibly agents of cat scratch disease], *Lymphogranuloma venereum, Mycobacterium tuberculosis, Treponema pallidum, Haemophilus ducreyi, Francisella tularensis*, Epstein-Barr virus, mumps
Viral	Adenovirus type 8 or 19, causing keratoconjunctivitis, serotypes 3, 7, and 4—pharyngoconjunctival fever, herpes simplex viruses, varicella-zoster virus, Epstein-Barr virus, cytomegalovirus, rubeola virus, mumps, influenza virus, Newcastle disease virus (paramyxovirus)
Corneal infections	
Bacterial keratitis	*Staphylococcus aureus, Staphylococcus* spp., *Streptococcus pneumoniae, Streptococcus* spp., *Pseudomonas aeruginosa, Bacillus cereus, Enterobacteriaceae, Neisseria* spp., *Moraxella lacunata, Mycobacterium fortuitum, Mycobacterium chelonei*, anaerobes
Fungal keratosis	*Fusarium solani, Candida albicans, Aspergillus fumigatus, Alternaria* spp., *Curvularia* spp., *Acremonium* spp.
Viral keratosis	Herpes simplex virus, varicella-zoster virus
Protozoan keratitis	*Acanthamoeba* spp.
Lacrimal system infections	
Dacryoadenitis	*Staphylococcus aureus, Streptococcus* spp., *Neisseria gonorrhoeae, Mycobacterium tuberculosis, Treponema pallidum*, mumps virus, Epstein-Barr virus
Canaliculitis	*Actinomyces israelii, Streptococcus* spp., *Candida* spp., *Aspergillus* spp., herpes simplex virus, varicella-zoster virus
Dacryocystitis	*Streptococcus pneumoniae, Staphylococcus aureus, Haemophilus influenzae, Pseudomonas aeruginosa, Proteus mirabilis, Candida albicans, Aspergillus* spp.
Retina and choroid infections	Cytomegalovirus, *Toxoplasma gondii*
Endophthalmitis	
Exogenous (surgical or nonsurgical trauma)	*Staphylococcus epidermidis, Staphylococcus aureus, Streptococcus* spp., *Bacillus* spp., *Pseudomonas* spp., *Enterobacteriaceae, Haemophilus influenzae, Propionibacterium* spp.
Endogenous	*Staphylococcus aureus, Neisseria meningitidis, Streptococcus* spp., *Bacillus cereus, Enterobacteriaceae, Pseudomonas aeruginosa, Nocardia asteroides*
Fungal (endogenous)	*Candida albicans, Aspergillus fumigatus, Aspergillus flavus*
Fungal (exogenous)	*Candida* spp., *Aspergillus* spp., *Cephalosporium* spp., *Penicillium* spp., *Curvularia* spp.
Orbital infections	*Staphylococcus aureus, Streptococcus* spp., *Peptostreptococcus* spp., *Pseudomonas aeruginosa, Haemophilus influenzae, Mycobacterium* spp., zygomycetes, *Aspergillus* spp., *Echinococcus* spp., *Taenia solium*

Table 1.1–1 ''Rule-out'' clinical impressions and potential etiological agents *(continued)*

Clinical impression	Potential etiological agent
Nervous system infections	
Acute bacterial meningitis	*Neisseria meningitidis, Streptococcus pneumoniae, Streptococcus agalactiae, Listeria monocytogenes, Escherichia coli, Enterobacteriaceae, Haemophilus influenzae, Staphylococcus aureus*; less commonly, *Peptostreptococcus* spp., *Fusobacterium necrophorum, Prevotella melaninogenica, Bacteroides fragilis, Clostridium perfringens*; rare (zoonotic): *Brucella* spp., *Francisella tularensis, Streptococcus suis*
Acute viral meningitis	Coxsackievirus, echovirus, poliovirus, herpes simplex viruses 1 and 2, varicella-zoster virus, flaviviruses (St. Louis encephalitis), mumps virus, bunyaviruses (California group, La Crosse), rubeola, lymphocytic choriomeningitis virus, adenoviruses
Chronic meningitis	*Mycobacterium tuberculosis, Brucella* spp., *Francisella tularensis, Listeria monocytogenes* (rare), *Neisseria meningitidis* (rare), *Tropheryma whippelli* (rare), *Borrelia burgdorferi, Leptospira* spp. (rare), *Treponema pallidum, Actinomyces* spp., *Nocardia* spp., *Cryptococcus neoformans, Coccidioides immitis* (rare), *Histoplasma capsulatum* (rare), *Candida* spp., *Aspergillus* spp. (rare), zygomycetes (rare), lymphocytic choriomeningitis virus, mumps, herpes simplex virus, varicella-zoster virus, arbovirus, flavivirus, echovirus, parasites (rare)
Brain abscess	*Streptococcus* spp., *Peptostreptococcus* spp., *Porphyromonas* spp., *Bacteroides* spp., *Prevotella* spp., *Fusobacterium* spp., *Staphylococcus aureus, Enterobacteriaceae, Burkholderia cepacia, Streptococcus pneumoniae, Neisseria meningitidis, Haemophilus influenzae, Listeria monocytogenes, Haemophilus aphrophilus, Actinomyces* spp., *Nocardia* spp., zygomycetes, *Mycobacterium* spp., *Naegleria* spp. (primary meningoencephalitis), *Acanthamoeba* spp., *Balamuthia mandrillaris* (granulomatous encephalitis)
Spinal cord, peripheral and cranial nerves	Poliomyelitis virus, herpesvirus simiae, HIV-1, human T-lymphotrophic virus 1; myelitis associated: cytomegalovirus, herpes simplex virus; following infection with rubeola virus, varicella-zoster virus, influenza virus, mumps virus; *Borrelia burgdorferi, Borrelia recurrentis, Chlamydia* spp.
Epidural abscess	*Staphylococcus aureus, Escherichia coli, Pseudomonas aeruginosa, Streptococcus* spp., *Peptostreptococcus* spp., *Salmonella* spp., *Staphylococcus* spp., *Nocardia* spp., *Actinomyces* spp., *Fusobacterium* spp., *Mycobacterium* spp., *Aspergillus* spp., *Brucella* spp., *Treponema pallidum*
Infections caused by rather recently recognized etiological agents that may or may not be cultured in routine microbiology laboratories	
Bartonellosis	*Bartonella bactiformis*
Chromobacterium bacteremia	*Chromobacterium violacium*
Strawberry foot rot (exudative, scabbing dermatitis)	*Dermatophilus congolensis*
Bacterial vaginosis	*Gardnerella vaginalis, Mobiluncus* spp., *Mycoplasma* spp.
Legionellosis and related diseases	*Legionella* spp.
Chronic nodular skin lesions (joint infections, wound infections)	*Prototheca* spp.
Occasional endocarditis, dialysis	*Stomatococcus mucilaginosus*
Rat bite fever (Haverhill fever)	*Streptobacillus moniliformis*
Rat bite fever (Sodoku)	*Spirillum minus* (Wright stain/dark field)
Cat scratch fever	*Bartonella henselae* (*Afipia felis*)
Rhinosporidiosis	*Rhinosporidium seeberi* (epithelial cell tissue culture)
Lobomycosis	*Loboa loboi* (histological diagnosis)
Whipple's disease	*Tropheryma whippelii*
Erythema infectiosum (fifth disease)	Parvovirus B 19

Table 1.1–2 Collection of specimens for bacteriological analysis[a,b]

Specimen	Preparation	No., type, volume	Container	Helpful clinical information	Comments
Anaerobic cultures					
Actinomycosis	Decontaminate skin	Aspirated pus	Anaerobic transport medium	History of "Lumpy Jaw"	Fistulating chronic infection often in the neck, jaw, and upper chest area
Body fluids					
Secretions, pus	Decontaminate skin	>1 ml[c]	Anaerobic TM	Foul-smelling discharge, abdominal surgery, abscess aspirate	Do not refrigerate; immediately transport to lab.
Respiratory tract	Transtracheal aspirate, pleural or empyema fluid only	>1 ml	Anaerobic TM	Foul-smelling sputum, history of aspiration	Do not submit sputum.
Tissues	Surgery	1 cm if possible	Anaerobic TM	Clinical impression	Do not add fluid. Larger specimens (>1 cm) tolerate short exposure to air.
Autopsy material					
Blood	Best collected before body is handled too much or opened. Decontaminate skin or sear surface of heart or other organ before inserting needle or cutting out tissue block.	10 ml of right-heart blood	Sterile tube, vacutainer with anticoagulant, Isolator tube	Clinical diagnosis, postmortem interval, autopsy impression, previous positive cultures, suspected infection	Autopsy cultultures are often contaminated with bacteria from the water faucet and with enteric bacteria.
Tissue	Same as above	6 cm^3 (if possible) with one serosal or other surface; this large size is preferred because aseptic collection is difficult.	Sterile container		A block of spleen tissue may be submitted in lieu of a blood culture. Note: coccidiomycosis and tuberculosis are discovered often at autopsy only.[d,e,f,g]
Blood					
Peripheral	Skin decontamination with 70% alcohol followed by povidone iodine	10 ml (adults and older child); 1–2 ml (infants); 3 samples per 24 h or 4–6 for FUO	Culture bottle for direct inoculation; vacutainer with SPS; Isolator tube; designated containers for automated instruments		Clinical diagnosis, antimicrobics and chemotherapy, immune status Inoculate and incubate as soon as possible.

Table 1.1–2 Collection of specimens for bacteriological analysis[a,b] *(continued)*

Specimen	Preparation	No., type, volume	Container	Helpful clinical information	Comments
Bone marrow	Same as above	≥1 ml	Same as above	Same as above	Direct smears should be made. Recommended by many authorities for diagnosis of systemic histoplasmosis and for other fungus infections, also for diagnosis of miliary TB, brucellosis
Body fluids (other than blood, urine, CSF)					
Bile	Surgery or decontaminate skin area before aspiration	Several ml (first ml from post-op drain site often contains contaminants.)	Sterile container, TM	Consider viruses, fungi, and parasites for analysis	Sample may contain gallstones, which should be examined. Duodenal aspirates sometimes submitted for special tests
Hematomas	Skin decontamination	Several ml	Sterile tube, vacutainer, TM	Suspected abscess	May clot; when in doubt use anticoagulant (SPS)
Joint fluid	Same	Several ml	Same	History of trauma, previous surgery, or infection: consider GC chlamydia, etc., in analysis.	Often proteinaceous; may clot. Do not add acetic acid or other fluid which may precipitate protein; this makes cell evaluation impossible. Distilled sterile water is acceptable.
Pericardial fluid	Same as above	Several ml	Sterile jar, tube, TM	History of TB or previous surgery	Consider viral etiology, especially coxsackievirus.
Peritoneal fluid	Same	Several ml or more	Sterile jar, tube, TM	History of TB, surgery, or cancer	Same as for joint fluids; consider also GC; specimen may be peritoneal dialysis fluid.
Pleural fluid	Same	Several ml or more	Sterile tube, jar, TM	As above	
Breast milk	Skin decontamination of nipple	Several ml; first few may be contaminated.	Sterile tube, jar, TM	Suspected abscess	Often submitted for presence of *Staphylococcus aureus* and/or hemolytic streptococci
Catheter tips					
Foley catheter		Not recommended			Foley catheters or tips should not be cultured except unused, as sterility check.

(continued)

Table 1.1–2 Collection of specimens for bacteriological analysis[a,b] *(continued)*

Specimen	Preparation	No., type, volume	Container	Helpful clinical information	Comments
Vascular cannulae venous access devices, arterial lines	Skin decontamination, careful asceptic removal mandatory	Segment near skin and tip end should be used; use sterile scissors to cut.	Sterile jar or tube	History of local infection, signs and symptoms	Peripheral blood cultures help with interpretation.
Central nervous system, brain biopsy	Surgery	See tissues	As for aerobic and anerobic cultures	Suspected abscess, cryptococcosis, toxoplasmosis, etc.	Needs coordination with Pathology. Include viruses in diagnostic analysis.
CSF	Skin decontamination	Several ml if possible	Sterile, clean, screw-cap tube	Tentative clinical diagnosis and/or suspicion	Since culture yields more from a larger volume, an alternate method is to aseptically pool all tubes collected (after cell count). The supernatant goes to Chemistry and Serology, with aliquot reserved for antigen detection when indicated. Obtain smears from precipitate.
Eye					
Internal	Surgery	Volume of specimen often suboptimal	Sterile tube	History of trauma or postop infection	Since specimen is usually small and obtained under great difficulty, speed in transport and care in handling are very important.
External	Cleanse skin around eye with mild antiseptic. Gently remove makeup and ointment with sterile cotton and saline.	For most cases moistened swabs are used. For diagnosis of viral or chlamydial infections and for cytology, conjunctival and/or corneal scrapings are necessary. Make two slides per lesion.	Moist sterile swabs in sterile tubes with a small amount of nutrient broth; alcohol-cleaned glass slides for scrapings; sterile tube for scrapings to be cultured (AFB, fungi)	History and suspected agent, e.g., bacterial, fungal, AFB, inclusion bodies (viral or chlamydial), GC only, allergic	Handle carefully; inoculate at bedside (or office) culture for *Acanthameba* on special media. Transport to laboratory immediately. Often only a few microorganisms present. Scraping should be done by ophthalmologist. Consult with physician about use of terms or the handling of the specimen (CD-RIGHT EYE, OS-LEFT EYE). Giemsa and Gram stains are frequently requested.

Table 1.1–2 Collection of specimens for bacteriological analysis[a,b] *(continued)*

Specimen	Preparation	No., type, volume	Container	Helpful clinical information	Comments
Genital tract—female					
Amniotic fluid		Uncontaminated fluid	Sterile tube	Premature rupture of membranes >24 h	Treat as any other normally sterile body fluid; may contain *Neisseria gonorrhoeae*.
Cervix (endocervix)	Wipe cervix clean of vaginal secretion and mucus. Use speculum and no lubricant.	Uncontaminated endocervical secretions; take two swabs.	Sterile container with TM; appropriate collection containers for demonstration of specific organisms such as *N. gonorrhoeae*, *Chlamydia*, HSV, etc.	Venereal disease, postpartum infection	Viability of *N. gonorrhoeae* held in Amies or modified Stuart transport medium decreases substantially after several hours.
Cul de sac (culdocentesis)	Surgical procedure	Fluid, secretions	See Anaerobic cultures, STD diagnosis	Venereal disease, pelvic inflammatory disease	Pelvic inflammatory disease
Endometrium	As for cervix	Curettings or aspiration	Sterile container, anaerobic TM	Postpartum fever, venereal disease	Likelihood of external contamination is high for cultures obtained through the vagina.
Intrauterine device	Surgical	Entire device plus secretion, pus	Sterile container	History of bleeding	Unusual organisms may be isolated, e.g., *Actinomyces*, *Torulopsis*, and other yeasts.
Lymph nodes (inguinal)	Skin decontamination	Biopsy or needle aspirate	Sterile container	History of venereal disease	May require sending to reference lab.
Products of conception (fetal tissue, placenta, membranes, lochia)	Surgical	Tissue or aspirates	Sterile container		Occasionally this type of specimen is expelled into toilet and is grossly contaminated.
Urethra	Wipe clean with sterile gauze or swab	Swab with urethral secretion or free discharge	Transport medium	History of discharge	Discharge may be stimulated by gently stripping and massaging the urethra against pubic symphysis through the vagina.
Tubes, ovaries	Surgical	Tissue, aspirates, or swabs	Sterile container; see anaerobic cultures	Salpingo-oophoritis	Consider venereal, fungal, anaerobic, and AFB infection.

(continued)

Table 1.1–2 Collection of specimens for bacteriological analysis[a,b] *(continued)*

Specimen	Preparation	No., type, volume	Container	Helpful clinical information	Comments
Vagina	Use of speculum without lubricant	Aspirate or swab, Gram stains, and wet mounts	Swab with TM, Transgrow	History of discharge	Ulcerations should be checked for syphilis, soft chancre, or genital herpes. Yeast common. For GC, cervical specimen is preferred. Wet mount is for yeast and *Trichomonas*, "clue cells," *Gardnerella*, and organisms of vaginosis.
Vaginal cuff		Aspirate of abscess	See Anaerobic cultures	Postsurgery	
Vulva (including labia, Bartholin glands)	Do not use alcohol for mucous membranes. Skin prep for regular skin sites.	Swab or aspirate (Bartholin gland abscess)	Swab with TM. See anaerobic culture for aspirate.	Discharge	Same as above
Lesion (dark field, for *Treponema pallidum*)	1- to 2-h soaking with sterile saline on gauze	Several slide preparations or aspirate fluid into capillary tube	Slide and coverslip or capillary tube		Characteristic motility is seen only on warm material. Seal coverslip or capillary tube with lanolin or petrolatum.
Genital tract—male					
Lymph nodes		Same as for female genital tract			
Penile lesion	Do not use alcohol for mucous membranes. Skin prep for regular skin sites.			Duration of lesion, pain, discomfort	Special culture/microscopic techniques are required for chancroid and granuloma inguinale.
Culture		Swab	Swab with TM if pus		
Dark field (for *T. pallidum*)			Same as for female genital tract		
Prostatic fluid		Secretion for smear and culture	Sterile tube or swab with TM	History of chronic UTI	Not recommended for GC cultures, but helpful in some chronic UTI, *Trichomonas* spp.
Urethra		Same as for female	Secretion, slide, and/or swab	History and duration of painful discharge	In males the diagnosis of gonorrhea can often be made by microscopic examination of a Gram-stained smear.

Table 1.1–2 Collection of specimens for bacteriological analysis[a,b] *(continued)*

Specimen	Preparation	No., type, volume	Container	Helpful clinical information	Comments
Pus/abscess			See Anaerobic cultures and Skin (deep suppurative lesion)		May be labeled "incision and drainage" (I & D)
Intestinal					
Duodenal contents	Through tube	Several ml	Sterile tube	Travel, food	Examine for bacterial overgrowth, *Salmonella typhi*, and parasites.
Feces		At 1 g, 3 consecutive specimens	Stool preservative, Culturette; for bacteria, special transport for parasites and ova	Travel, food, suspected etiology	See section 8 for further information.
Rectal swab		3 consecutive specimens	Swab with TM; GN broth	As above	Not useful for the detection of carriers
Gastric aspirate, neonate		Enough for smear and culture	Sterile container	History of ruptured membranes	May visualize and isolate causative agent of septicemia before blood cultures become positive
Respiratory tract					
Throat/pharynx		Swab	Swab with TM; commercial kits for *Streptococcus*	Agent suspected (e.g., group A streptococci, *N. gonorrhoeae*)	Do not touch oral mucosa or tongue with swab.
Epiglottis	Swab		TM	Suspected bacteria include *Haemophilus influenzae*	Do not swab throat in cases of acute epiglottitis unless prepared for tracheostomy.
Nasal sinuses			See Anaerobic cultures: body fluids		
Nasopharynx		Swab	Thin wire or flexible swab with TM	Agent suspected (e.g., *Bordetella pertussis*)	Transport to laboratory immediately or inoculate at bedside. Prepare smear for DFA.
Nose			Swab with TM	Mainly for staphylococcal carriers	
Oral cavity					
Mucosal surface of gums or teeth	Rinse mouth	Scraping, swab	Swab, tongue depressor, or slide in sterile container	Duration, agent suspected	Culture for yeast, smear for yeast or organisms of Vincent's angina

(continued)

Table 1.1–2 Collection of specimens for bacteriological analysis[a,b] *(continued)*

Specimen	Preparation	No., type, volume	Container	Helpful clinical information	Comments
Dental abscess, root abscess	Rinse mouth; prep with dry sterile gauze.	Exudate	See Anaerobic cultures		Predominant pathogens are anaerobes, including *Actinomyces* and various streptococci and gram-negative rods.
Bronchoscopy		Brushings, transbronchial biopsies, bronchial secretions	Sterile container	Agent(s) suspected on clinical impression	Gram stains may help direct culture effort. Patient's history important. May require culture for mycobacteria, legionellae in addition to DFA, stains for *Pneumocystis*, culture for fungi, viruses, and parasites.
Expectorated sputum	May require ultrasonic nebulization, hydration, physiotherapy, or postural drainage	Sputum, not saliva	Sterile container	Pneumonia	Culture processing must be preceded by Gram stain demonstrating >25 PMNs and <10 epithelial cells/lpf. See section on acid-fast organisms if submitted for mycobacteria. May be refrigerated overnight.
Trachael aspirate		Sputum	Sterile container	Pneumonia	Cellular composition may be misleading due to inflammatory reaction caused by endotracheal tube. May be refrigerated.
Transtracheal aspirate (infralaryngeal aspirate)	Skin is cleansed, anesthetized, and decontaminated.		Sterile container. See Anaerobic cultures	Pneumonia, aspiration, TB	Process immediately. See bronchoscopy.
Skin					
Superficial wound	Clean wound surface with 70% alcohol.	Pus, biopsy	Aspirate or swab with transport medium	Animal bite or trauma, duration, travel	

Table 1.1–2 Collection of specimens for bacteriological analysis[a,b] *(continued)*

Specimen	Preparation	No., type, volume	Container	Helpful clinical information	Comments
Extensive burns,[a] decubitus ulcer	Clean wound surface with 70% alcohol.	For quantitative culture 3- to 4-mm dermal punch	Sterile container		Consider quantitative culture. See CMPH. Decubitus ulcers should not be cultured casually; appropriate debridement and punch biopsy are required for quantitative evaluation of decubitus.
Deep suppurative lesion, closed abscess	Clean and decontaminate.	Pus, >1 ml if possible	Syringe or anaerobic container	Duration, location	See Anaerobic cultures
Fistula, sinus tract	Clean surface; decontaminate.	Pus, >1 ml if possible	Syringe or swab with TM	Duration, location	Examine stained smear for guidance of isolation media.
Rash	Clean surface with 70% alcohol.	Pus, fluid	Syringe		
Tissue, surgical or biopsy	Surgical	5 to 10 mm^3 or aspirate	Sterile container with TM		See also Anaerobic cultures. Do not discard leftover tissue. Freeze in sterile broth until culture and pathology are completed.

[a] NOTE: Universal precautions must be observed at all times. When the ''Container'' column indicates specimen may be submitted in syringe, the caveats of universal precautions must be followed; i.e., the syringe must be capped with a sterile closure; syringes with needles in place are unacceptable. Many specimens may contain important yeasts, molds, or viruses. Follow the instructions of the director of the laboratory in concert with Infectious Disease Advisory Committee which designate such examinations as routine procedures with select specimens or follow the request of the physician of record based on patient's history and clinical impression. Perform smears whenever possible.

[b] Abbreviations: TM, transport medium; TB, tuberculosis; SPS, sodium polyanethole sulfonate; GC, gonorrhea; AFB, acid-fast bacillus; STD, sexually transmitted disease; UTI, urinary tract infection; GN, gram negative; DFA, direct fluorescent-antibody assay.

[c] If copious amounts are available, fill sterile tube.

[d] Preparation of smears for Gram (and other) stains at time of autopsy is helpful.

[e] Culture for viruses based on clinical impression; inoculate virus transport vial if intended for reference laboratory.

[f] Culture for yeasts and molds if suspected from clinical history or premortem findings.

[g] Cultures for mycobacteria and fungus require special attention. See appropriate sections.

Table 1.1–3 Laboratory approaches to suspected fungal infections

Disease	Specimen type	Culture medium[a]
Superficial mycoses		
Pityriasis versicolor	Scrapings	Not necessary
Tinea nigra	Skin scrapings	SAB-SPEC
Piedra	Cut hair	SAB
Cutaneous mycoses		
Tinea capitis	Epilated hair	SAB-SPEC, DTM
Tinea corporis	Skin scrapings	SAB-SPEC
Onychomycosis	Nail scrapings	SAB-SPEC C
Candidiasis	Skin, nail scrapings, mucocutaneous scrapings, vaginal swab	SAB-SPEC
Subcutaneous mycoses		
Chromoblastomycoses	Scrapings, crust exudate from lesions	SAB-SPEC
Mycetoma	Pus from draining sinuses, aspirated fluids, biopsy	SAB, BHI, BHIA with blood
Phaeohyphomycosis	Sputum, BW, body fluids, pus, corneal scrapings	SAB-SPEC C
Sporotrichosis	Pus from lesions, aspirated fluids	SAB-SPEC
Systemic mycoses		
Yeastlike fungi		
Candidiasis	Sputum, BW, biopsy, CSF, urine, stool, blood	SAB-SPEC C, BHIA, blood culture (isolation)
Cryptococcus	CSF, sputum, blood, bone marrow, urine, scrapings from skin lesions, pus from abscesses, sinus tracts	SAB-SPEC C, birdseed agar
Geotrichosis	Sputum, BW, stools	SAB-SPEC C
Dimorphic fungi		
Blastomycosis	Scrapings from edge of lesions	SAB-SPEC
	Pus from abscesses, sinus tracts, urine, sputum, BW	BHIA, yeast extract phosphate medium
Paracoccidioidomycosis	Scrapings from edge of lesions, mucous membranes, biopsied lymph nodes, sputum, BW	SAB-SPEC, BHIA yeast extract, phosphate agar
Coccidioidomycosis	Sputum, BW, CSF, urine scrapings from lesions, pus from sinuses, abscesses	SAB-SPEC, yeast extract, phosphate agar
Histoplasmosis	Blood, bone marrow, sputum, BW, CSF, pus from sinus tract or ulcer, scrapings from lesion	Yeast extract, phosphate agar, blood agar, immerse modified SAB with blood
Miscellaneous mycoses		
Aspergillosis	Sputum, BW	SAB-SPEC C
Zygomycosis	Sputum, BW, biopsy specimen	SAB-SPEC C
Hyalohyphomycosis	Sputum, BW, nail scrapings, blood, body fluids, pus, wound scrapings	SAB-SPEC C
External otitis	Epithelial scales and detritus	SAB-SPEC C

[a] SAB, Sabouraud's agar, SAB-SPEC, Sabouraud's agar with chloramphenicol and cycloheximide; SAB-SPEC C, Sabouraud's agar with chloramphenicol only; DTM, dermatophyte test medium (presumptive); BHIA, BHI agar; blood culture, any approach extant in laboratory is acceptable—however, lysis-centrifugation (Isolator-Wampole) is preferred. All media required are available commercially (*see* Appendix A). For selected commercial suppliers, see Appendix A.

Table 1.1–4 Collection of specimens to detect parasites[a]

Site	Specimen options	Collection method
Blood	Smears of whole blood	A minimum of four each (separate slides) of thick and thin blood films (first choice)
	Anticoagulated blood[b,c]	Anticoagulant (second choice) EDTA vacutainer tube (purple top)
Bone marrow	Aspirate[b,c]	Sterile tube
Central nervous system	Spinal fluid[b,c]	Sterile tube
Cutaneous	Aspirates from below surface	Sterile tube plus air-dried smears
Ulcers	Biopsy	Sterile tube, nonsterile to Histopathology (formalin acceptable)
Eye	Biopsy	Sterile tube (in saline), nonsterile tube to Histopathology
	Scrapings	Sterile tube (in saline)
	Contact lens	Sterile tube (in saline)
	Lens solution	Sterile tube
Intestinal tract	Feces[d]	PVA (polyvinyl alcohol or non-mercury base), 5 or 10% buffered formalin. SAF (sodium acetate, acetic acid, formalin), single-vial collection system (zinc-based, proprietary formulas)
	Sigmoidoscopy	Fresh, PVA, or Schaudinn's smears
	Duodenal contents	Entero-test or aspirates
	Anal impression smear	Cellulose tape (pinworm exam)
	Adult worm/worm segments	Saline, 70% alcohol
Liver, spleen	Aspirates	Sterile tube, collected in four separate aliquots (liver)
	Biopsy	Sterile tube
Lung	Sputum	True sputum (not saliva)
	Induced sputum	No preservative (10% formalin if time delay)
	BAL	Sterile tube (container)
	Transbronchial aspirate[b,c]	Air-dried smears
	Tracheobronchial aspirate[b,c]	Same as above
	Brush biopsy[b,c]	Same as above
	Open lung biopsy[b,c]	Same as above
	Aspirate	Sterile
Muscle	Biopsy	Fresh, squash preparation, nonsterile, to Histopathology (formalin acceptable)
Skin	Scrapings	Aseptic, smear or vial
	Skin snip	No preservative
	Biopsy	Sterile (in saline)
Urogenital system	Vaginal discharge	Saline swab, transport swab (no charcoal), culture medium, plastic envelope culture, air-dried smear for FA
	Urethral discharge	Same as above
	Prostatic secretions	Same as above
	Urine	Single unpreserved specimen, 24-h unpreserved specimen, early morning

[a] Abbreviations: FA, fluorescent antibody; BAL, bronchoalveolar lavage.

[b] Immediate delivery to laboratory desirable; if request is for organisms other than *Plasmodium* spp., request it be indicated on label and on requisition for parasites such as filariae (2,4).

[c] Immediate delivery to laboratory desirable; requires prompt processing.

[d] Examinations of fresh stool specimens require special attention. Liquid stool specimens for protozoan trophozites must be examined within 30 min of passage (not 30 min after arriving at laboratory); soft stools should be examined within 1 h of being passed, although protozoan cysts survive and can be detected in firm stools within 24 h of being passed. The preserved stool specimens are suitable for examination for *Cryptosporidium* spp., *Isospora* spp., and related coccidia. NOTE: Stool examinations for ova, parasites, and enteropathogenic bacteria should not be requested if patients have been hospitalized for 3 days or more.

Table 1.1–5 Skin disinfection

Select area
↓
70% alcohol
↓
Povidone iodine
Let dry 1 min
↓
70% alcohol

Table 1.1–6 Commercially available transport media[a]

Medium	Purpose
Stuart's medium (8)	Most aerobic and facultatively anaerobic bacteria
Amies medium (1)	As above
Amies medium with charcoal	Use for transport of gonococci.
Carey-Blair medium (3)	Transport of pathogenic stool bacteria, e.g., *Salmonella, Shigella, Vibrio, Campylobacter, Yersinia* spp.
Buffered glycerol	Transport of potentially pathogenic stool bacteria
Anaerobic transport medium	Numerous variations are available; objective is to preserve appropriate atmosphere. Use of swabs should be discouraged; aspirates or tissue in proper transport devices is preferable.
Synthetic mesh	Recently made available on plastic shaft in a protective tube; mesh entraps organisms and preserves them without buffer or medium; can be used for recovery of aerobic and most anaerobic organisms. Suitable for smear preparation and/or antigen detection when two shafts are provided.
PVA (4)	For transport of intestinal parasites; suitable for preparation of slides to be stained; specimens may be concentrated.
SAF (4)	Sodium acetate-acetic acid-formaldehyde as PVA substitute
Buffered formalins (4)	Especially for ova, larvae and for concentrations
Viral transport	Contains salt solutions to stabilize viral agents

[a] Various commercial suppliers have these transport vials, tubes, and devices available, often packaged with polyester tips and plastic shanks (to avoid toxic effects of cotton and wood). Consult Appendix A for names of suppliers.

Table 1.1–7 Rejection criteria for microbiological specimens[a,b]

Criteria[c]	Procedure
Clerical errors	
Discrepancy between patient identification on requisition and specimen container label	Notify physician or nurse in charge. Request new specimen or have physician or nurse correct error in person in the laboratory.
No identification on container	Notify physician or charge nurse. Proper label must be brought to laboratory and specimen identity verified.
Specimen source or type not noted	Call physician or charge nurse to ascertain missing information.
Test not indicated on requisition	Call physician or charge nurse to ascertain missing information.
General microbiology	
Specimen received in fixative (formalin). Exception, stool for parasites and ova.	Notify physician or charge nurse and request new specimen; indicate ''received in fixative'' on requisition and return.
Foley catheter tip	Notify physician or charge nurse that specimen is not suitable for microbiological analysis. Note rejection on requisition and return.
Containers Unpreserved urines held in refrigerator >24 h Improper or nonsterile container Leaking container	Notify physician or nurse in charge and request new specimen. If contact insists specimen be processed, refer to supervisory personnel. Return requisition with appropriate comment.
Dry swab	Notify physician or charge nurse and request new specimen, properly submitted in appropriate transport device. If physician insists dry swab be cultured, note on laboratory record and report with caveat: ''microorganisms recovered may not reflect actual microbiota.''
More than one specimen of urine, stool, sputum, wound, routine throat specimens submitted on the same day from the same source	Notify physician or charge nurse that as stated in laboratory manual only one specimen will be processed per day.
Only one swab submitted with multiple requests for various organisms (bacteria, AFB, fungi, virus, ureoplasmas, etc.)	Notify physician or charge nurse and request additional material. If additional material cannot be procured, ask physician to prioritize.
Anaerobes	
Specimen for anaerobes not received in appropriate container	Notify physician or charge nurse and request properly handled specimen. If the physician insists specimen be processed, refer to supervisory personnel or alternatively comment in laboratory record and in report form that inappropriate transport may have influenced recovery of significant anaerobic bacteria.
Anaerobic cultures requested on autopsy material, bronch wash, decubitus ulcer material (not punch biopsy of tissue beneath eschar) drain, drain site, environment, exudate, feces, gastric washing (other than newborn), midstream or catheterized urine, mouth, nose, prostatic secretions, sputum material on swabs from ileostomy, colostomy, fistual, intestinal contents, throat, vaginal secretions	Inform physician or charge nurse that, as detailed in laboratory manual, these specimens are not cultured for anaerobic bacteria since these anatomic sites harbor anaerobes normally and usually. If physician insists, refer him/her to supervisory personnel.
Aerobic bacteriology	
Gram stain for *Neisseria gonorrhoeae* on specimens from cervix, vagina, and anal crypts	Notify physician or charge nurse that these smears are not examined for GC since these anatomic loci may harbor nongonococcal neisseriae.

(continued)

Table 1.1–7 Rejection criteria for microbiological specimens[a,b] *(continued)*

Criteria[c]	Procedure
Specimens for GC and/or *Chlamydia* culture received in Gen Probe transport media	Notify physician or charge nurse that the specimens are in a fixative that kills bacteria; only molecular probe results may be offered with this fixed specimen.
Respiratory culture requested on throat specimens of patients older than 10 years	Notify physician or charge nurse that a screening culture for group A streptococci only will be performed. Exceptions are transplant, cystic fibrosis, and head and neck clinic patients.
Sputum specimen with <25 WBC per >10 epithelial cells/lpf	Inform physician or charge nurse that specimen is mostly saliva and is not appropriate for culture. Request and repeat specimen.
Sputum with or without WBC and numerous and varying tinctorial bacterial morphotypes for *Legionella* DFA	Notify physician or charge nurse that specimen is unsuitable for DFA.
Mycobacteriology/mycology	
24-h collection of urine or sputum for AFB or fungus culture	Inform physician or nurse that as per laboratory manual, three separate first morning specimens of sputum or of urine are the best samples for analysis; reject 24-h specimens.
Sputum swabs for AFB or fungi	Notify physician or charge nurse that specimen is inadequate in quantity for the isolation of these microorganisms. Request properly collected specimens.
Parasitology	
Ova and parasites examination received in only formalin or only PVA	Inform physician or charge nurse that stool in both transport media required for adequate parasitological analysis; request repeat with both transport media used.
Stool for *Cryptosporidium* and/or *Isospora* spp. received in PVA only	Notify physician or charge nurse that both PVA and formalin are required for accurate performance of test; request repeat submission with both transport media.
Excess barium or oil noted in stool submitted for parasitological examination	Notify physician or charge nurse that the barium and/or oil will obscure the exam and that a specimen should be resubmitted in 10 days.
Virology	
Stool specimens received in preservative for viral culture for detection of *C. difficile* toxins, rotavirus, adenovirus 40/41	Notify physician or charge nurse to request new proper specimen; record on requisition ''specimen unsatisfactory; received in preservative.''
Clotted blood for viral culture (serum acceptable when enterovirus suspected)	Inform physician or charge nurse that blood has clotted; request new specimen and admonish to mix blood well to preserve clotting.

[a] Abbreviations: GC, gonococcus; AFB, acid-fast bacillus.

[b] A record for all rejections or discrepancies must be maintained (book, card file, computer). Note demographic patient information; physician of record; date and time specimen received; type and source of specimen; examination requested; reason for rejection or discrepancy noted; person contacted by phone; date, time, and manner of contact (phone, computer, FAX); final disposition/resolution. Records should be reviewed by supervisory personnel at regular intervals.

[c] Specimens unsuitable for routine anaerobic cultures: throat swabs; nasopharyngeal swabs; gingival and other internal mouth surface swabs; expectorated sputum; sputum obtained by nasotracheal or orotracheal suction; bronchial washings or other specimens obtained with bronchoscopy unless procured with a protected; double-lumen catheter or properly executed bronchoalveolar lavage; gastric and small-bowel contents; large-bowel contents except for *Clostridium difficile, Clostridium botulinum, Anaerobiospirillum succiniciproducens*, and other specific causative agents; ileostomy, colostomy effleunts; feces, except for large-bowl contents (as directed above); voided or catheterized urine; vaginal or cervical swabs; female genital tract cultures collected via vagina, except for suction curettings or other specimens collected with a double-lumen catheter; surface swabs from decubitus ulcers, perirectal abscesses, foot ucers, exposed wounds, eschars, pilonidal and other sinus tracts; any materal adjacent to a mucous membrane that has not been adequately decontaminated.

Table 1.1–8 Procedure for processing clinical specimens in microbiology[a]

I. Evaluate the specimen for adequacy
   A. Specimen must be properly labeled.
   B. The transmittal must be submitted with the specimen, and the information on the transmittal must match the information on the specimen label.
   C. The specimen must be submitted in the proper transport container.
   D. The specimen volume must be adequate to perform all tests requested.
   E. The specimen sent must be appropriate for the test ordered.
   NOTE: If any of these requirements are not met, refer to list of rejection criteria in Table 1.1–7.

II. Media selection and labeling
   A. Select appropriate media for the tests ordered.
   B. Examine all media for expiration date and contamination before the media are inoculated.
   C. Individually label all media with an accession number and date. Do not obscure the names and expiration dates of the media.
   D. All media made in Microbiology must be properly labeled with the name of medium and expiration date.

III. Order of inoculation media
   Inoculate the least selective media first. This prevents any carry over of an inhibitory substance to another medium. Arrange labeled plates in order from least to most selective.
   A. Routine aerobic culture
      1. BAP
      2. CHOC
      3. MS
      4. MAC
      5. SEA
      6. Nutrient broth
   B. Genital culture
      1. BAP
      2. CHOC
      3. MS
      4. TM
      5. MAC or EMB
   C. Stool culture
      1. PEA
      2. CAMPY
      3. MAC or EMB
      4. HE
      5. GN broth
   D. Fungal culture
      1. SAB-SEL
      2. BHI blood agar with gentamicin and chloramphenicol
      3. SAB
   E. Anaerobe culture
      1. BRU
      2. PEA
      3. BBE/LKV (optional)
      4. Chopped meat broth or THIO
   F. Urine culture[b]
      1. BAP
      2. MAC or EMB
      3. SEA

IV. Procedure for streaking plates for primary isolation
   A. Sterilize the inoculating loop in the incinerator for 5 to 10 s. Allow the loop to cool thoroughly before streaking media.
   B. Pass the cooled loop back and forth through the inoculum in the first quadrant several times.

(continued)

Table 1.1–8 Procedure for processing clinical specimens in microbiology[a] *(continued)*

C. Do not flame between the quadrants unless necessary.
D. Turn the plate a quarter turn, and pass the loop through the edge of the first quadrant approximately four times while streaking into the second quadrant. Continue streaking in the second quadrant without going back to the first squadrant, approximately four times.
E. Turn the plate another quarter turn, and pass the loop through the edge of the second quadrant approximately four times while streaking into the third quadrant. Continue streaking in the third quadrant without going back into the second quadrant, approximately four times.
F. Flame the loop between plates to prevent carrying over a possible contaminant from the previous plate.
G. Continue streaking the rest of the culture media in the same manner.
H. Inoculate suitable nutrient broth with 1 to 2 loops full of specimen or swab.

[a] Abbreviations: MS, mitis salivarius agar; SEA, selective enterococcus agar; TM, transport medium; PEA, phenyl ethanol agar; CAMPY, campylobacter agar; HE, Hektoen enteric agar; GN, gram-negative broth; BRU, brucella agar; BBE/LKV, bacteroides bile esculin agar/laked blood-kanamycin-vancomycin agar.

[b] Special instructions for processing urine specimens:
1. Do not centrifuge urine before culturing.
2. Mix urine well before culturing.
3. Use 0.001 calibrated loop for inoculating media for routine urine.
4. Transfer one loop full of urine to BAP.
5. Pull the loop down the surface of the agar to form a single streak in the center of the first quadrant.
6. Spread the inoculum over the first quadrant by streaking the loop back and forth.
7. Streak for isolation in the other quadrants.
8. Transfer one loop full of urine to MAC or EMB plate. Repeat procedure using SEA or CNA agar.
9. Pull loop down surface of agar to form a single streak that crosses the center of the plate.
10. Cross streak through the initial inoculum streak by moving the loop back and forth at perpendicular angles to initial streak.
11. To inoculate media for special collection urines, i.e., suprapubic tap, high and low counts or nephrostomy, inoculate two sets of plates:
    a. 0.001 loop: BAP and MAC (EMB) and SEA (CNA) agars
    b. 0.01 loop: BAP and MAC (EMB) and SEA (CNA) agars
    c. Use the same inoculation procedure outlined above for both the 0.001 loop and the 0.01 loop.
12. If a urine Gram stain is ordered, use a ringed microscope slide.

NOTE: Biplates, commercially available, consisting of MAC and CNA or MAC and SEA may be used instead of single-agar plates of each.
NOTE: bioMérieux-Vitek AMS has urine cards that can be substituted for culture analysis.

Table 1.1–9 Commonly used media[a]

Medium	Atmosphere[b]	Type[c]	Inhibition/indicator	Purpose
BAP (usually 5% defibrinated sheep blood in a nutrient base)	O_2, CO_2, ANO_2	N		Most bacteria and fungi grow in this medium; determination of hemolysis depends on source of RBC; sheep blood is especially suited to classify streptococcal hemolysis; BAP usually will not permit growth of *Haemophilus* spp., *Neisseria gonorrhoeae, Calymmatobacterium granulomatis*, mycobacteria, leptospires, mycoplasmas, borreliae, treponemes, rickettsiae, *Bordetella pertussis, Francisella tularensis*, and legionellae.
Enteric agars				Selective for *Enterobacteriaceae* and supports other gram-negative rods; inhibits most gram-positive bacteria.
MAC	O_2	S, D	Bile salts, crystal violet, lactose, neutral red	Sorbitol substitution helps in detection of *Escherichia coli* O157:H7.
EMB	O_2	S, D	Eosin Y, methylene blue, lactose, sucrose	

Table 1.1–9 Commonly used media[a] *(continued)*

Medium	Atmosphere[b]	Type[c]	Inhibition/indicator	Purpose
Phenylethyl alcohol	CO_2	S	Phenylethyl alcohol	Isolation of streptococci and staphylococci; inhibits most gram-negative bacteria
Mannitol salt agar	O_2	S, D	NaCl, mannitol, phenol red	Isolation of streptococci and staphylococci; inhibits most gram-negative bacteria
Selective enterococcus agar	O_2	S, D	Oxgall, esculin, sodium, azide	Isolation of enterococcus
Selective agars for "enteric pathogens"				Isolation of *Salmonella* spp., *Shigella* spp. Use two types, one with moderate selectivity and one with high selectivity.
Brilliant green agar	O_2	S, D	Brilliant green, lactose, sucrose, phenol red	Isolation of *Salmonella* spp.; inhibits most other *Enterobacteriaceae*, including *Shigella*. May be used especially after enrichment broth initial incubation.
Hektoen enteric agar	O_2	S, D	Bile salts, ferric ammonium citrate, sodium thiosulfate, lactose, sucrose, salicin bromthymol blue, fuchsin	Isolation of most *Enterobacteriaceae*; differentiates lactose/sucrose fermentors and nonfermentors; has H_2S producers and nonproducers.
GN broth	O_2	E	Deoxycholate, citrate	Enrichment for *Salmonella* spp. and *Shigella* spp.
Campylobacter agar (enriched)	5% O_2, 10% CO_2	S	Amphotericin (2 μg/ml), cephalothin (15 μg/ml), trimethoprim (5 μg/ml), vancomycin (10 μg/ml), polymyxin B (2.5 μg/ml)	Isolation of *Campylobacter* spp.
CHOC (enriched)	CO_2	N, S		Growth of most bacteria, especially more fastidious ones such as *Haemophilus* spp. and *Neisseria gonorrhoeae*
Selective agars for pathogenic neisseriae				
Thayer Martin agar	CO_2	S	Vancomycin (3 μg/ml), colistin (7.5 μg/ml), nystatin (12.5 μg/ml), trimethoprim lactate (15 μg/ml)	Isolation of *Neisseria meningitidis, N. gonorrhoeae*, and *N. Lactamica*; most other bacteria inhibited
Anaerobic agars				
Anaerobic blood	AN	N		Growth of all anaerobic bacteria should contain yeast extract, menadione, hemin.
Kanamycin	AN	S	Kanamycin (100 μg/ml), vancomycin (7.5 μg/ml), menadione (0.5 μg/ml)	Isolation of *Bacteriodes* spp. and *Prevobacter* spp.
THIO	O_2	E		Support growth of most bacteria
Broths such as brain heart infusion and/or tryptic digest broth	O_2	E	Add inhibitors and indicators as needed.	Support growth of most bacteria
Legionella medium	CO_2	S	Vancomycin (2 μg/ml), colistin (3.75 μg/ml), anisomycin 10 μg/ml)	Available in various combinations

[a] NOTE: the media listed are examples and suggestions. Many other media serving the same purpose are available and may be preferred without any effect on obtaining proper isolation. All media, their manufacturers, their availability as prepared, and ready-for-use reagents are listed in Appendix A.

[b] O_2, aerobic incubation; CO_2, incubation with 3 to 10% CO_2; AN, incubation under anaerobic conditions.

[c] N, nutrient; D, differential; S, selective; E, enrichment.

Table 1.1–10 "Panic values" in microbiology[a]

Panic value
Organisms seen in CSF
Organisms seen in joint fluids
Positive India ink preparations
Positive cryptococcal antigen detection
Positive CSF antigen detection for pneumococci, *Streptococcus agalactiae, Neisseria meningitidis*, and *Haemophilus influenzae* (now rare)
Positive AFB smear
Positive blood cultures (not contaminated)
Positive CSF cultures
Positive wound cultures
Isolation of *Streptococcus pyogenes*
Isolation of *Mycobacterium tuberculosis*
Isolation of *Salmonella* spp.
Isolation of *Shigella* spp.
Isolation of *Escherichia coli* O157:H7
Isolation of pathogenic neisseriae
Isolation of any of the reportable etiological agents that follow:

Lymphogranuloma venereum	
Malaria	Rubella
Meningitis	Salmonellosis
Aseptic	Shigellosis
Haemophilus spp.	Syphilis (specify stage)
Meningococcal	Tetanus
Other (specify type)	Toxic shock syndrome
Meningococcemia	Trichinosis
Mumps	Tuberculosis
Pertussis	Tularemia
Plague	Typhoid
Poliomyelitis	Typhus
Psittacosis	Varicella
Rabies	Yellow fever
Rocky Mountain spotted fever	Yersiniosis

Table 1.1–10 ''Panic values'' in microbiology[a] *(continued)*

Panic value	
Examples of communicable diseases (requiring immediate notification of referring physician and, for some, governmental agencies)	
Amebiasis	Giardiasis
Anthrax	Gonococcal infection
Babesiosis	Granuloma inguinale
Botulism	Hemolytic uremic syndrome
Brucellosis	Hemophilus influenzae (invasive disease)
Campylobacteriosis	Hepatitis (A, B, C)
Chancroid	Histoplasmosis
Cholera	Legionellosis
Cryptosporidiosis	Leprosy
Diphtheria	Leptospirosis
Encephalitis	Listeriosis
Escherichia coli O157:H7 infections	Lyme disease

[a] ''Panic values'' in microbiology encompass the detection of clinically important microorganisms and viruses that require notification, immediate action by the physician of record or his/her designate, action by hospital personnel and visitors, and notification of governmental agencies (this may differ from state to state).

[b] AFB, acid-fast bacteria.

Table 1.1–11 Alert request

Request all physicians to notify Microbiology Laboratory if they suspect the following etiological agents in specimens to be sent to the laboratory:
Bacillus anthracis
Bordetella pertussis
Brucella spp.
Francisella tularensis
Pseudomonas pseudomallei
Yersinia pestis
Blastomyces dermatitidis
Coccidioides immitis
Histoplasma capsulatum
Paracoccidioides brasiliensis

(continued)

Table 1.1–12 Serodiagnostic tests[a]

Test name	Method(s)[b]	Normal value(s)	Specimen requirement(s)[c]
Adenovirus antibody (IgG)	IFA	<1:40	Blood (5 ml)
Amoeba antibody	IHA	<1:64	Blood (5 ml)
Amoeba antibody panel	IHA, ID	<1:64 negative	Blood (5 ml)
Antideoxyribonuclease B	EN	<1:170	Blood (5 ml)
Antihyaluronidase titer	EN	<1:32	Blood (5 ml)
Antistreptolysin-O titer	LPA	<200 Todd units	Blood (5 ml)
Arbovirus antibody panel Western equine encephalitis Eastern equine encephalitis St. Louis encephalitis California encephalitis	IFA	<1:16 (IgG) <1:10 (IgM)	Blood (5 ml) CSF (0.5 ml)
Aspergillus antibody (*A. fumigatus, A. flavis, A. niger*)	ID	Negative	Blood (5 ml)
Bacterial antigen	LPA	Negative	CSF (2 ml)
Detection panel (group B streptococci, *Streptococcus pneumoniae, Neisseria meningitidis, Haemophilus influenzae* type b)			
Blastomyces antibody	CF	<1:8 (serum); <1:2 (CSF)	Blood (5 ml)
Blastomyces antibody panel	ID CF	Negative <1:8	Blood (5 ml)
Bordetella pertussis antibody	MAT	<1:2	Blood (5 ml)
Borrelia (relapsing fever)	IFA	<1:16	Blood (5 ml)
Borrelia burgdorferi antibody	(*see* Lyme disease)		
Brucella antibody	SA	<1:20	Blood (5 ml)
C-reactive protein	LPA	<6 mg/liter	Blood (5 ml)
California encephalitis antibody	IFA	<1:16 (IgG) <1:10 (IgM)	Blood (5 ml) CSF (0.5 ml)
Candida antibody	ID	Negative	Blood (5 ml)
Candida antigen	LPA	Negative	Blood (5 ml)
Chagas disease antibody panel (*Trypanosoma cruzi*)	IFA	<1:16	Blood (5 ml)
Chlamydia antibody *C. psittici* *C. pneumoniae* *C. trachomatis*	IFA	<1:64	Blood (5 ml)
Coccidioides antibody	CF	<1:2	Blood (5 ml) or CSF (0.5 ml)
Coccidioides antibody panel Latex agglutination Immunodiffusion Complement fixation	 LPA (IDTP) IDCF CF	 Negative Negative <1:2	 CSF (0.5 ml)
Cold agglutinin titer	TA	<1:4	Blood (5 ml) (Do not refrigerate.)
Coxsackie A antibody panel Includes types 7, 9, 10, 21	CF	<1:8 (serum) <1:2	Blood (5 ml)

Table 1.1–12 Serodiagnostic tests[a] *(continued)*

Test name	Method(s)[b]	Normal value(s)	Specimen requirement(s)[c]
Coxsackie B antibody panel Includes types 1 through 6	CF	<1:8 (serum) <1:2 (CSF)	Blood (5 ml) CSF (0.5 ml)
Cryptococcus antigen	LPA	Negative	Blood (5 ml) or CSF (1 ml)
Cysticercosis antibody	EIA	<1:10 (serum) <1:2 (CSF)	Blood (5 ml) CSF (1 ml)
Cytomegalovirus antibody (IgG)	IFA	<1:10	Blood (5 ml)
Cytomegalovirus antibody panel includes IgG and IgM	IFA	<1:10 (IgG)	Blood (7 ml)
Diphtheria antitoxin antibody	EIA	>0.01 IU/ml	Blood (5 ml)
Eastern equine encephalitis antibody	IFA	<1:16 (IgG) <1:10 (IgM)	Blood (5 ml) or CSF (0.5 ml)
Echinococcosus antibody	IHA	<1:10	Blood (5 ml)
Echovirus antibody panel Includes types 4, 9, 11, 16	CF	<1:8 (serum) <1:2 (CSF)	Blood (7 ml) CSF (0.5 ml)
Epstein-Barr virus antibody panel			
Viral capsid IgG	IFA	>1:10	Blood (10 ml)
Viral capsid IgM	IFA	<1:10	
EBNA IgG	ACIF	>1:2	
Early Ag IgG	IFA	<1:40	
Febrile agglutinins panel *S. typhi* *S. paratyphi* A, B *Brucella* *Proteus* OX19	SA, TA	<1:20	Blood (5 ml)
Fluorescent treponemal antibody absorption (FTA-ABS)	IFA	Nonreactive	Blood (5 ml)
Filariasis antibody	EIA	<1:160	Blood (5 ml)
Francisella tularensis antibody	EIA	<1:20	Blood (5 ml)
Fungal diseases antibody panel *Aspergillus* *Blastomyces* *Coccidioides* *Cryptococcus* *Histoplasma*	LPA, ID, CF	By report	Blood (10 ml) or CSF (2 ml)
Haemophilus influenzae b vaccine response (polyribose phosphate antibody)	EIA	By report	Blood (5 ml)
Herpes simplex virus antibody type 1 (IgG)	IFA	>1:10	Blood (5 ml)
Herpes simplex virus antibody Antibody panel Includes types 1 and 2 IgG and IgM	IFA	>1:10 (IgG) <1:10 (IgM)	Blood (7 ml)
Heterophile antibody screen	LPA	Negative	Blood (5 ml)
Histoplasma antibody by CF	CF	<1:8 (serum) <1:2 (CSF)	Blood (5 ml) CSF (0.5 ml)
Histoplasma antibody panel	ID, CF	Negative <1:0 (serum) <1:2 (CSF)	Blood (5 ml) CSF (0.5 ml)

(continued)

Table 1.1–12 Serodiagnostic tests[a] *(continued)*

Test name	Method(s)[b]	Normal value(s)	Specimen requirement(s)[c]
HIV-1 antibody screen	EIA	Negative	Blood (5 ml)
HIV-1 antibody by Western blot	IB	No bands detected (IB)	Blood (5 ml)
HIV-1 antigen assay	EIA	Negative	Blood (5 ml)
HTLV-1 antibody	EIA	Negative	Blood (5 ml)
Hypersensitivity pneumonitis Antibody panel Includes 9 antigens: 5 thermophilic *Actinomycetes*, 3 *Aspergillus*, and avian serum proteins	ID	None detected	Blood (7 ml)
Influenza virus A and B IgG antibody panel	IFA	<1:40	Blood (5 ml)
Influenza virus A and B antibody	IFA	<1:40 (IgG)	Blood (5 ml)
Legionella pneumophila antibody	IFA	<1:64	Blood (5 ml)
Leishmaniasis (visceral) antibody includes *L. donovani, L. mexicana, L. tropica, L. braziliensis*	IFA	<1:16	Blood (5 ml)
Leptospira antibody	IHA	<1:50	Blood (5 ml)
Lyme disease antibody	IFA	<1:16	Blood (5 ml)
Lymphocytic choriomeningitis antibody	CF	<1:8 (serum) <1:2 (CSF)	Blood (5 ml)
LGV antibody	IFA	<1:64	Blood (5 ml)
Measles (rubeola) IgG antibody	IFA	>1:10	Blood (5 ml)
Measles (rubeola) IgG and IgM antibody panel	IFA	>1:10 (IgG)	Blood (5 ml)
Melioidosis (*Pseudomonas pseudomallei*)	IFA	<1:16	Blood (5 ml)
Mumps virus IgG antibody	IFA	>1:10	Blood (5 ml)
Mumps virus IgG and IgM antibody panel	IFA	>1:10 (IgG) <1:10 (IgM)	Blood (5 ml)
Mycoplasma IgG antibody	IFA	<1:40	Blood (5 ml)
Mycoplasma IgG and IgM antibody panel	IFA	<1:40 (IgG) <1:20 (IgM)	Blood (5 ml)
Paragonimus antibody	EIA	None detected	Blood (5 ml)
Parainfluenza virus IgG antibody panel includes types 1, 2, 3	IFA	<1:40	Blood (5 ml)
Parainfluenza virus IgG and IgM antibody panel includes types 1, 2, 3	IFA	<1:40 (IgG) <1:10 (IgM)	Blood (5 ml)
Poliovirus antibody by neutralization	NT	By report	Blood (5 ml)
Poliovirus antibody panel (1, 2, 3)	CF	<1.8	Blood (5 ml)
Premarital testing, male	RPR	Nonreactive	Blood (5 ml)
Premarital testing, female RPR Rubella	 RPR LPA	 Nonreactive Immune	 Blood (5 ml)
Psittacosis antibody	IFA	<1:64	Blood (5 ml)

Table 1.1–12 Serodiagnostic tests[a] *(continued)*

Test name	Method(s)[b]	Normal value(s)	Specimen requirement(s)[c]
Q-fever (*Coxiella burnetii*) total antibody	IFA	<1:16	Blood (5 ml)
Q-fever (*Coxiella burnetii*) IgG and IgM antibody panel	IFA	<1:16 (IgG) <1:10 (IgM)	Blood (5 ml)
Respiratory virus IgG and IgM antibody panel, adult influenza A and B, mycoplasma	IFA	<1:40 (IgG) <1:10 (IgM)	Blood (7 ml)
Respiratory syncytial virus IgG antibody	IFA	<1:40	Blood (3 ml)
Rickettsial diseases total antibody Rocky Mountain spotted fever Typhus Q-fever	IFA	<1:16	Blood (5 ml)
Rickettsial disease IgG and IgM antibody panel Rocky Mountain spotted fever Typhus Q-fever	IFA	<1:16 (IgG) <1:10 (IgM)	Blood (5 ml)
Rocky Mountain spotted fever total antibody	IFA	<1:10	Blood (5 ml)
Rocky Mountain spotted fever IgG and IgM antibody panel	IFA	<1:16 (IgG) <1:10 (IgM)	Blood (5 ml)
Rotavirus antigen detection	EIA	None detected	Stool (1 g)
Rubella virus IgG antibody	LPA	>1:10	Blood (5 ml)
Rubella virus IgG and IgM antibody panel	LPA, EIA	>1:10 (IgG) Negative (IgM)	Blood (5 ml)
Streptococcal antibody panel			Blood (7 ml)
Antistreptolysin-O	LPA	<1:200	Peds: blood (3 ml)
Antideoxyribonuclease B	EN	<1:170	
Syphilis serology Nontreponemal antibody Rapid plasma reagin (RPR)	RPR	Nonreactive	Blood (5 ml)
Syphilis serology Nontreponemal antibody VDRL on CSF	VDRL	Nonreactive	CSF (1 ml)
Syphilis serology Treponemal antibody MHA-TP	IHA	Nonreactive	Blood (5 ml)
Syphilis serology Fluorescent treponemal antibody Absorption (FTA) FTA-ABS	IFA	Nonreactive	Blood (5 ml)
Tetanus antitoxin antibody (referred to outside lab)	EIA	0.01–0.05 IU/ml	Blood (5 ml)
TORCH IgG antibody panel			
Toxoplasma	IFA	<1:16	Peds: blood (3 ml)
Rubella	LPA	<1:10	
CMV	IFA	<1:10	
Herpes simplex	IFA	<1:10	

(continued)

Table 1.1–12 Serodiagnostic tests[a] *(continued)*

Test name	Method(s)[b]	Normal value(s)	Specimen requirement(s)[c]
TORCH IgG and IgM			
Antibody panel		IgG IgM	Peds: blood (3 ml)
Toxoplasma	IFA	<1:16 <1:10	
Rubella	LPA, EIA	<1:10 Negative	
CMV	IFA	<1:10 <1:10	
Herpes simplex	IFA	<1:10 <1:10	
Toxocara antibody	EIA	<1:4	Blood (5 ml)
Toxoplasma total antibody	IFA	<1:16	Blood (5 ml)
Toxoplasma IgG antibody	IFA	<1:16	Blood (5 ml)
Toxoplasma IgG and IgM	IFA	<1:16 (IgG)	Blood (5 ml)
Antibody panel		<1:10 (IgM)	
Trichinosis antibody	LPA	<1:5	Blood (5 ml)
Typhus (epidemic, endemic) total antibody	IFA	<1:16	Blood (5 ml)
Typhus (epidemic and endemic) IgG and IgM antibody panel	IFA	<1:16 (IgG) <1:10 (IgM)	Blood (5 ml)
Varicella-zoster virus IgG	EIA	Seropositive	Blood (5 ml)
Antibody (immune status)		Index >1.0	
Varicella-zoster virus IgG antibody titer	IFA	>1:10	Blood (5 ml)
Varicella-zoster IgG and IgM	IFA	>1:10 (IgG)	Blood (5 ml)
Antibody panel		<1:10 (IgM)	
Western equine encephalitis antibody	IFA	<1:16 (IgG)	Blood (5 ml)
Yersinia antibody	MAT	<1:20	Blood (5 ml)

[a] Tests can be performed in the laboratory or sent to a reference laboratory. Abbreviations: IgG, immunoglobulin G; LGV, lymphogranuloma venereum; MHA-TP, microhemagglutination; CMV, cytomegalovirus.

[b] Methods: ACIF, anticomplement immunofluorescence; CF, complement fixation; DA, direct agglutination; IB, immunoblot; EIA, enzyme immunoassay; EN, enzyme neutralization; FTA-ABS, fluorescent treponemal antibody absorption; ID, immunodiffusion; IFA, indirect fluorescent antibody; IHA, indirect hemagglutination; LPA, latex particle agglutination; MAT, microagglutination titer; MIF, micro indirect fluorescence; MC, mucin clot technique; NT, neutralization; PHA, passive hemagglutination; RPR, rapid plasma reagin test; SA, slide agglutination; VDRL, venereal disease research lab.

[c] Draw blood in red-top tube except when indicated otherwise.

SECTION 2 Aerobic Bacteriology

SECTION EDITOR: *Marie Pezzlo*

(continued)

2.1 Gram Stain

I. PRINCIPLE

The Gram stain is used to classify bacteria on the basis of their forms, sizes, cellular morphologies, and Gram reactions; in a clinical microbiology laboratory, it is additionally a critical test for the rapid presumptive diagnosis of infectious agents and serves to assess the quality of clinical specimens. The test was originally developed by Christian Gram in 1884. The modification currently used for general bacteriology was developed by Hucker in 1921; it provides greater reagent stability and better differentiation of organisms. Other modifications have been specifically developed for staining anaerobes (Kopeloff's modification) and for weakly staining gram-negative bacilli (*Legionella* spp., *Campylobacter* spp., *Bacteroides* spp., *Fusobacterium* spp., *Brucella* spp.) by using a carbol-fuchsin or basic fuchsin counterstain.

Interpretation of Gram-stained smears involves consideration of staining characteristics and cell size, shape, and arrangement. These characteristics may be influenced by many factors, including culture age, medium, incubation atmosphere, staining methods, and presence of inhibitory substances. Similar considerations apply to the interpretation of smears from clinical specimens, but additional factors include the presence of particular host cell types and phagocytosis.

II. SPECIMENS

Smears for Gram stain may be prepared from clinical specimens, broth cultures, or colonies growing on solid medium. Young cultures (<24 h) from noninhibitory media and fresh clinical specimens yield the most accurate results; for certain morphological considerations, broth culture smears are required.

III. MATERIALS

Include QC information on reagent container and in QC records.

A. Reagents

Reagents may be purchased commercially or prepared in the laboratory. The various Gram stain modifications and uses are summarized in Table 2.1–1.

1. Hucker's modification
   - **a.** Crystal violet
   - **b.** Gram's iodine
     Caution: *Iodine is corrosive. Avoid inhalation, ingestion, or skin contact.*
   - **c.** Decolorizers
     - (1) Slowest: ethanol, 95%
     - (2) Intermediate: acetone-alcohol; mixture of
       ethanol, 95% .... 100 ml
       acetone (reagent grade) .... 100 ml
       Combine in brown-glass bottle, label with 1-year expiration date, and store at room temperature.
     - (3) Fastest: acetone (reagent grade)
     Caution: *Ethanol and acetone are flammable.*
   - **d.** Counterstains
     - (1) Safranin
     - (2) Alternatively: basic fuchsin, 0.1 or 0.2% (wt/vol)
2. Kopeloff's modification for anaerobes (for details, *see* CMPH 1.5.3)
3. Carbol-fuchsin–basic fuchsin counterstain (for details, *see* CMPH 1.5.4)

B. Supplies

1. Glass slides (25 by 75 mm), frosted ends desirable
2. 0.85% NaCl, sterile
3. Pasteur pipettes and wood applicator sticks, sterile
4. Microbiological loops, inoculating needles
5. Supplies for disposal of biological waste, including "sharps"

Table 2.1–1 Gram stain modifications, recommended reagents, timing, and uses[a]

Stain and use	Hucker's		Carbol-fuchsin		Kopeloff's	
	Reagent	Time	Reagent	Time	Reagent	Time
Initial stain	Crystal violet	30 s	Crystal violet	30 s	Alkaline crystal violet: flood with solution A; add 5 drops of solution B	2–3 min
Iodine	Gram's iodine	30 s	Gram's iodine	30 s	Kopeloff's iodine	≥2 min
Decolorizer	Acetone-alcohol	~1–5 s	95% ethanol and reagent grade acetone	~30 s	3:7 acetone-alcohol: rinse immediately after applying	
Counterstain	Safranin[b]	30 s	Carbol-fuchsin or 0.8% basic fuchsin	≥1 min	Kopeloff's safranin	10–30 s
Recommended use	General bacteriology		*Bacteroides* spp. *Fusobacterium* spp. *Legionella* spp. *Campylobacter* spp. *Brucella* spp. and other faintly staining gram-negative organisms		Anaerobes	

[a] Adapted from CMPH 1.5.2.
[b] Some laboratories may prefer to use 0.1 to 0.2% basic fuchsin as a counterstain.

III. MATERIALS *(continued)*

6. Microincinerator or Bunsen burner
7. Immersion oil

C. **Equipment**

Optional materials, depending on specimen source or laboratory protocol

1. Electric slide warmer, 60°C
2. Centrifuge
3. Cytospin centrifuge
4. Vortex mixer
5. Sterile tubes, screw cap
6. Sterile scissors, scalpels, forceps
7. Tissue grinder
8. Methanol, absolute

   NOTE: Store methanol in brown screw-cap bottles. A working supply may be stored in plastic containers if replenished every 2 weeks.

IV. QUALITY CONTROL

A. Check appearance of reagents daily.
   1. If crystal violet has precipitate or crystal sediment, refilter before use even when purchased commercially.

      NOTE: Some stains, especially basic fuchsin and safranin, can become contaminated. When suspected, either culture or start with fresh material in a clean bottle.
   2. Evaporation may alter effectiveness of reagents. Working solutions should be changed regularly if not depleted with normal use.

B. Daily and when a new lot is used, prepare a smear of *Escherichia coli* (ATCC 25922) and *Staphylococcus epidermidis* (ATCC 12228) or *Staphylococcus aureus* (ATCC 25923). Fix, and stain as described above. Expected results:
   1. Gram-negative bacilli, pink
   2. Gram-positive cocci, deep violet

   NOTE: An alternative quality control source is material scraped from between teeth with a wooden applicator stick; both gram-positive and -negative organisms will be present.

C. The following are some common causes of poor Gram stain results.
   1. Use of glass slides that have not been precleaned or degreased

      NOTE: Storing slides in a jar with 95% ethanol will ensure clean slides. Drain excess alcohol or flame slide before use.
   2. Smear preparations that are too thick
   3. Overheating of smears when heat fixation is used
   4. Excessive rinsing during the staining procedure

D. Additionally, to ensure accuracy of interpretation, establish a system for reviewing Gram stain reports.

1. A daily review of selected Gram stains by supervisory personnel may help determine training needs and aid in correlating relevant clinical information.

2. Compare final culture results with Gram stain reports.

◩ **NOTE:** Not all organisms discerned on a smear can be cultivated.

3. A set of reference slides is invaluable for training and comparison.

V. PROCEDURE

A. Smear preparation

Proper preparation should produce a monolayer of organisms sufficiently dense for easy visualization but sparse enough to reveal characteristic arrangements. Use clean, new glass slides.

◩ **NOTE:** Always inoculate culture media first before preparing smear when using the same pipette or swab.

1. Clinical specimens

Observe universal (standard) precautions.

◩ **NOTE:** Wear latex gloves and other protection commensurate with universal precautions when handling clinical specimens. Observe other BSL 2 recommendations.

a. Specimens received on swabs

A separate swab should preferably be submitted.

(1) Roll the swab gently across the slide to avoid destruction of cellular elements and disruption of bacterial arrangements.

(2) Alternatively, when only one swab is received, place the swab in a small amount of saline and vortex it. Squeeze the swab against the side of the tube, and use it to prepare a smear. Use the remaining suspension to inoculate culture media.

b. Specimens not received on swabs: aspirates, exudates, sputa, stools

(1) If the specimen is received in a syringe, first transfer entire amount to a sterile tube. Vortex specimen if appropriate.

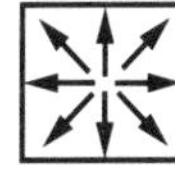

Observe universal (standard) precautions.

◩ **NOTE:** Do not accept syringes with attached needles. Establish such a policy and educate medical personnel regarding removal of needles before transport. Observe universal precautions.

(2) Select purulent or blood-tinged portions by using a sterile applicator stick, pipette, or wire loop. Extremely thick or purulent specimens may be diluted in a drop of saline on the slide for easier smear preparation.

(3) Spread the sample over a large area of the slide to form a thin film.

c. CSF and other body fluids requiring centrifugation

Some laboratories may choose to use a cytospin slide centrifuge to concentrate body fluids for smear preparation. This method has been used to increase Gram stain sensitivity and to decrease centrifugation and examination time for more rapid results.

◩ **NOTE:** Using slides with etched rings helps locate the inoculated area.

(1) After centrifugation, use a sterile pipette to remove supernatant to a sterile tube, leaving approximately 0.5 ml as sediment.

(2) Vortex or forcefully aspirate the sediment in and out of a sterile Pasteur pipette several times.

(3) Use the Pasteur pipette to transfer a small drop of the sediment to a clean slide.

(4) Do not spread the drop out. Allow it to air dry.

V. **PROCEDURE** *(continued)*

d. Urine specimens
(1) Do not centrifuge. Mix specimen well.
(2) Use a sterile Pasteur pipette to transfer 1 drop to a slide. Do not spread the drop.
(3) Allow the drop to dry.

e. Dried material or very small amounts of clinical specimen
(1) Emulsify specimen in 0.5 ml of sterile saline. Vortex if necessary.
(2) Use a sterile Pasteur pipette to transfer 1 drop to a slide.
(3) Use the pipette tip to spread the drop into an even thin film.

f. Biopsies and tissue sections
A ''touch prep'' and/or ground specimen preparation
(1) Mince with sterile scissors or scalpel.
(2) Prepare a ''touch prep.'' Use sterile forceps to hold pieces, and touch the sides of one or more of the minced fragments to a sterile glass slide, grouping the touches together for easier examination.
(3) For homogenized ground specimens, place 1 drop onto the slide and spread within a 1-in. circle.

2. Broth cultures
One smear per slide is recommended to avoid wash-off from one area to another during staining.
a. Use a sterile Pasteur pipette (or a venting needle for containers with septa, such as blood culture bottles, to avoid manipulating a needle and syringe) to transfer 1 or 2 drops to the slide.
b. Spread drop into an even thin film.

3. Colonies from solid media
a. Place a drop of sterile saline or water on slide.
b. Transfer a small portion of colony with a sterile applicator stick, wire needle, or loop.
c. Gently mix to emulsify.
NOTE: Never mix vigorously. Avoid creation of aerosols.

B. Smear fixation

Smears may be fixed with heat or methanol.

1. Heat
a. Air dry smears on a flat surface, or place them on a 60°C electric slide warmer until dry.
b. If smears are air dried, pass them two or three times through a flame, or hold the slide against the front of a microincinerator for 5 to 10 s. To avoid distortions, do not overheat.
c. Allow slide to cool before staining.

2. Methanol fixation prevents the lysis of RBCs, avoids damage to all host cells, and results in a cleaner background. It is strongly recommended for all clinical material, especially urines, and it also prevents wash-off of urine specimens.
a. Air dry the smear on a flat surface.
b. Place a few drops of methanol on a slide for 1 min, drain off remaining methanol without rinsing, and allow slide to air dry.
c. Do not use heat before staining.

C. Staining procedures

1. Hucker's modification
a. Flood the fixed smear with the crystal violet solution. Allow the stain to remain for 30 s.
b. Decant crystal violet, and rinse slide gently with running tap water.
Caution: *Excessive rinsing in this step could cause crystal violet to be washed from gram-positive cells. The flow of water may be applied to*

the underside of the angled slide to ensure a gentle flow across the smeared side.

c. Rinse off excess water with iodine solution, and then flood the slide with fresh iodine solution. Allow iodine to remain for 30 s.
d. Rinse gently with flowing tap water.
e. Decolorize by letting the reagent flow over the smear while the slide is held at an angle. Stop when the runoff becomes clear. Adjust decolorization time to thickness of smear and type of decolorizer used.
f. Remove excess decolorizer with gentle flow of tap water.

◩ **NOTE:** A properly decolorized smear appears with an almost olive-green hue and without observable evidence of crystal/gentian violet.

g. Flood the slide with safranin and allow counterstain to remain for 30 s.
h. Remove excess counterstain with a gentle flow of tap water.
i. Drain slide, and air dry it in an upright position. Clean off the bottom of the slide by wiping with alcohol or acetone on a tissue or paper towel.
j. Examine the smear microscopically.

2. Basic/carbol-fuchsin counterstain (for details, *see* CMPH 1.5.8)
Basic/carbol-fuchsin is recommended for detecting faintly staining gram-negative organisms. See Table 2.1–1 for timing.
3. Kopeloff's modification (for procedure, *see* CMPH 1.5.9)
Kopeloff's modification is recommended for better visualization and differentiation of anaerobes, which may easily overdecolorize and stain faintly with Hucker's modification.

VI. REPORTING RESULTS

A. Results and interpretation

1. Evaluate the general nature of the smear under low-power magnification (10×).
   a. Determine if smear has been properly decolorized. Depending on the source of the specimen, the background should be generally clear or gram negative. If WBCs are present, they should appear completely gram negative. Do not mistake thin crystal/gentian violet precipitate needles for gram-positive bacillus-shaped bacteria.
   b. Determine if thickness of smear is appropriate. For proper interpretation, areas must be no more than one cell thick with no overlapping of cells.
2. Examine smears prepared from clinical specimens under low power for evidence of inflammation. If appropriate for culture source, note the following.
   a. Relative amounts of PMNs, mononuclear cells, and RBCs
   b. Relative amounts of squamous epithelial cells, bacteria consistent with normal microbiota, which may indicate an improperly collected specimen
   c. Location and arrangements of microorganisms
3. Examine several areas of the smear under oil immersion for presence of microorganisms.
   a. If no microorganisms are detected in a smear of a clinical specimen, report "No microorganisms seen."
   b. If microorganisms are observed, report relative numbers and describe morphology (Fig. 2.1–1 and –2).
4. Observe predominant shapes of microorganisms.
   a. Overall shape: coccus, coccoid, coccobacillary, bacilli, filament, yeastlike
   b. Appearance of ends: rounded, tapered, flattened, clubbed (swollen), concave; swelling of sides can suggest the presence of spores but can also be caused by vacuoles, marked pleomorphism, or irregular staining.

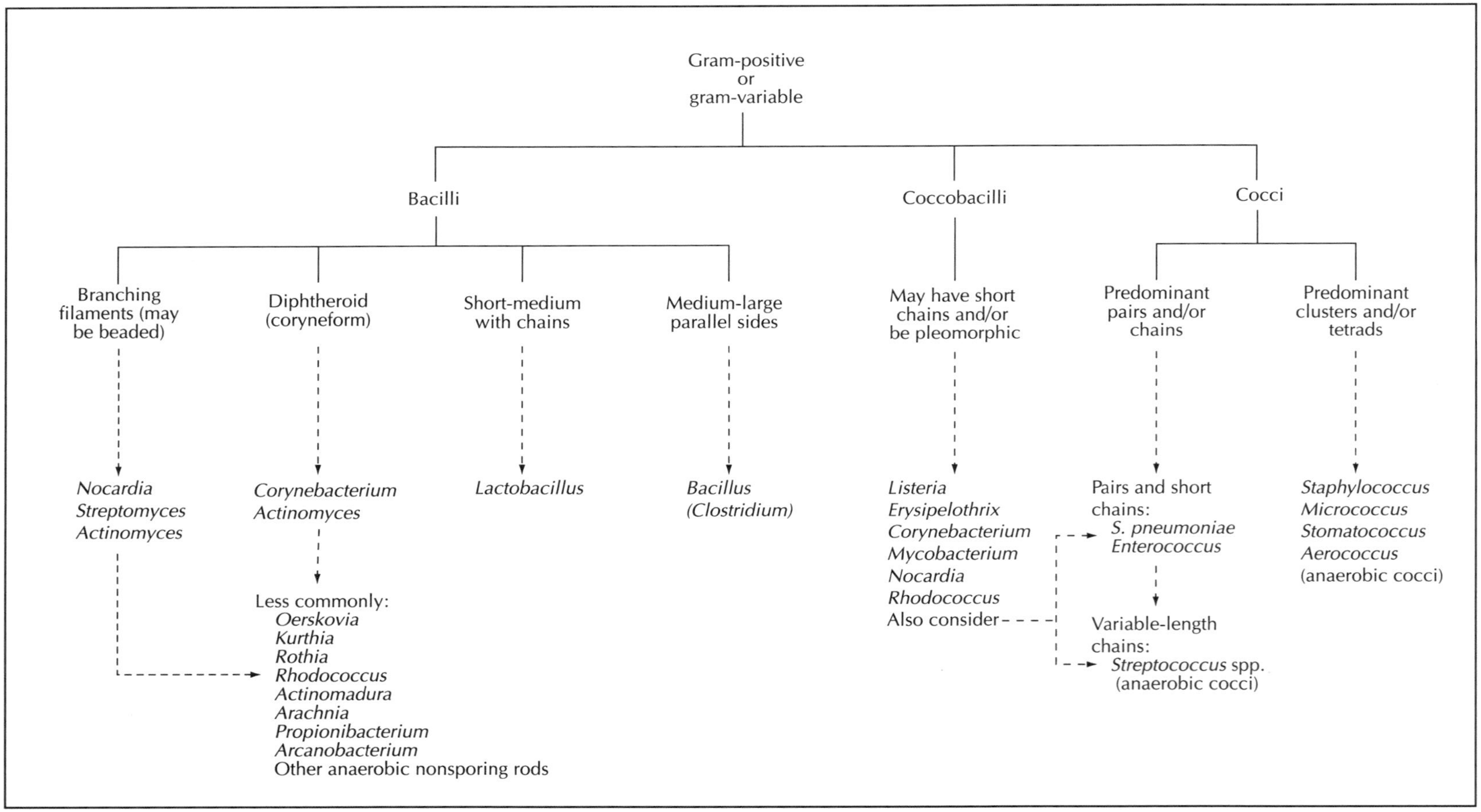

- Gram-positive or gram-variable
  - Bacilli
    - Branching filaments (may be beaded) → *Nocardia*, *Streptomyces*, *Actinomyces* → (see Less commonly list)
    - Diphtheroid (coryneform) → *Corynebacterium*, *Actinomyces* → Less commonly: *Oerskovia*, *Kurthia*, *Rothia*, *Rhodococcus*, *Actinomadura*, *Arachnia*, *Propionibacterium*, *Arcanobacterium*, Other anaerobic nonsporing rods
    - Short-medium with chains → *Lactobacillus*
    - Medium-large parallel sides → *Bacillus* (*Clostridium*)
  - Coccobacilli
    - May have short chains and/or be pleomorphic → *Listeria*, *Erysipelothrix*, *Corynebacterium*, *Mycobacterium*, *Nocardia*, *Rhodococcus*; Also consider → (Pairs and short chains; Variable-length chains)
  - Cocci
    - Predominant pairs and/or chains → Pairs and short chains: *S. pneumoniae*, *Enterococcus* → Variable-length chains: *Streptococcus* spp. (anaerobic cocci)
    - Predominant clusters and/or tetrads → *Staphylococcus*, *Micrococcus*, *Stomatococcus*, *Aerococcus* (anaerobic cocci)

Figure 2.1–1 Typical Gram stain morphologies of gram-positive and -variable genera (from CMPH 1.5.12). Most gram-positive species can appear gram variable or even negative owing to overdecolorization, phagocytosis, effect of antimicrobial agents, age, etc. *Mycobacterium* species may stain gram neutral and appear as ''ghost'' forms; they may also appear beaded, resembling chains of streptococci. Individual cells in the chains of *Enterococcus* species vary in size.

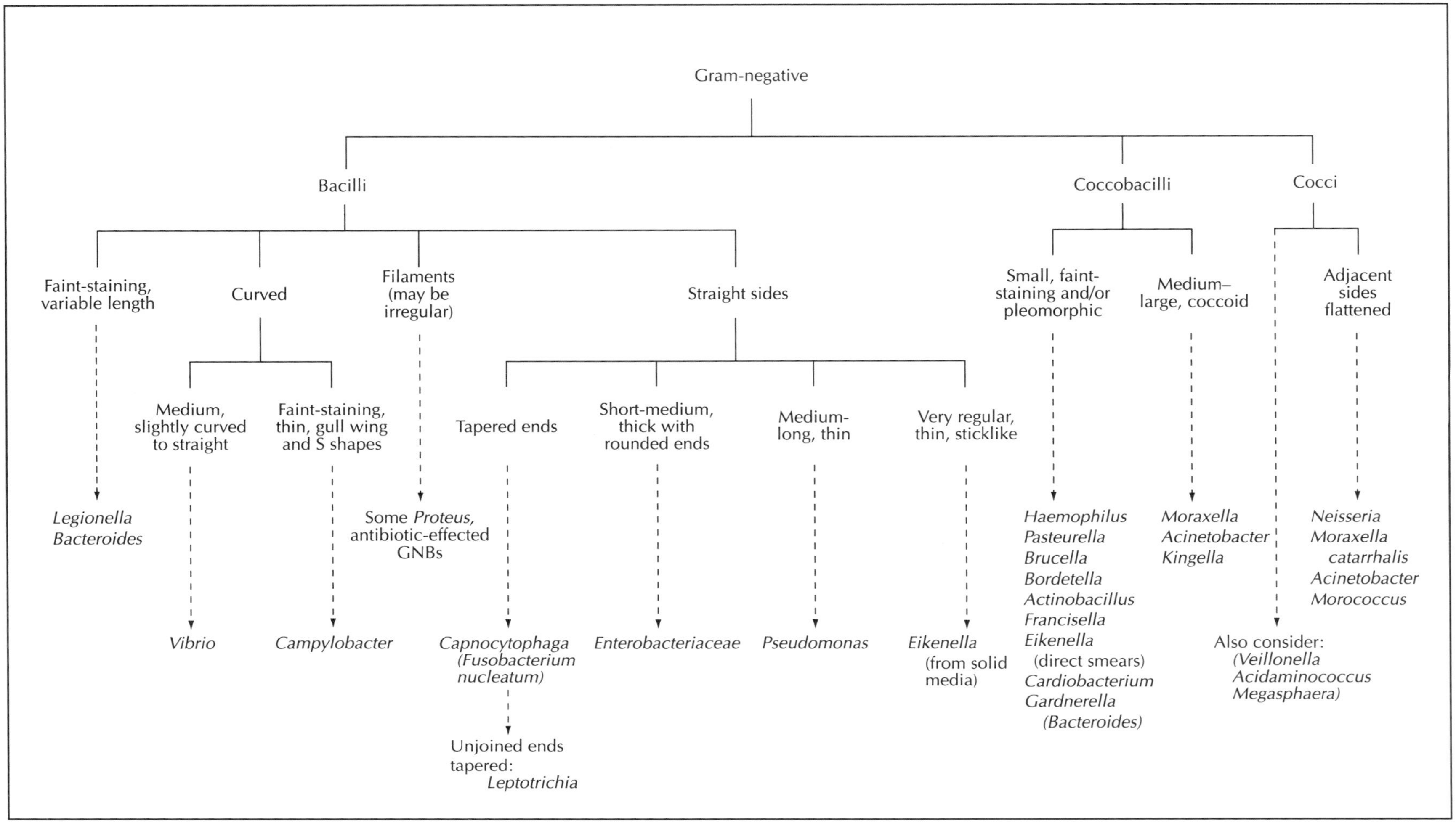

- Gram-negative
  - Bacilli
    - Faint-staining, variable length → *Legionella*, *Bacteroides*
    - Curved
      - Medium, slightly curved to straight → *Vibrio*
      - Faint-staining, thin, gull wing and S shapes → *Campylobacter*
    - Filaments (may be irregular) → Some *Proteus*, antibiotic-effected GNBs
    - Straight sides
      - Tapered ends → *Capnocytophaga* (*Fusobacterium nucleatum*) → Unjoined ends tapered: *Leptotrichia*
      - Short-medium, thick with rounded ends → *Enterobacteriaceae*
      - Medium-long, thin → *Pseudomonas*
      - Very regular, thin, sticklike → *Eikenella* (from solid media)
  - Coccobacilli
    - Small, faint-staining and/or pleomorphic → *Haemophilus*, *Pasteurella*, *Brucella*, *Bordetella*, *Actinobacillus*, *Francisella*, *Eikenella* (direct smears), *Cardiobacterium*, *Gardnerella* (*Bacteroides*)
    - Medium–large, coccoid → *Moraxella*, *Acinetobacter*, *Kingella*
  - Cocci
    - Adjacent sides flattened → *Neisseria*, *Moraxella catarrhalis*, *Acinetobacter*, *Morococcus*
    - Also consider: (*Veillonella*, *Acidaminococcus*, *Megasphaera*)

Figure 2.1–2 Typical Gram stain morphologies of gram-negative genera (from CMPH 1.5.13). *Gardnerella* species may stain gram variable.

VI. REPORTING RESULTS *(continued)*

- **c.** Appearance of sides: parallel, ovoid (bulging), concave, irregular
- **d.** Nature of axis: straight, curved, spiral
- **e.** Pleomorphism (variation in shape): the descriptive term "diphtheroid" or "coryneform" is used to describe gram-positive bacilli that are pleomorphic, club shaped, or irregularly staining or that have palisading and/or angular arrangements (V and L shapes).
- **f.** Branching or cellular extensions

B. Recording observations

Each laboratory must standardize specific reporting policies. Clinically significant findings should be called to the attention of the attending physician.

1. Smears of clinical specimens

For urine cultures, examine 20 or more fields. Report as positive if an average of one or more organisms are seen per oil immersion field. This correlates with a colony count of $\geq 10^5$ CFU/ml.

- **a.** Record relative amounts of observed cells and microorganisms. Commonly used quantitation systems include the following.
  - (1) Numerical
    - (a) 1+ (<1 per oil immersion field [100×])
    - (b) 2+ (1 per oil immersion field)
    - (c) 3+ (2 to 10 per oil immersion field)
    - (d) 4+ (predominant or >10 per oil immersion field)
  - (2) Descriptive
    - (a) Rare (>1 per oil immersion field)
    - (b) Few (1 to 5 per oil immersion field)
    - (c) Moderate (5 to 10 per oil immersion field)
    - (d) Many (>10 per oil immersion field)
- **b.** Record the morphology of observed bacteria.

2. Refer to Tables 2.1–1 to 2.1–4 for characteristic morphologies that may be seen in smears from clinical specimens.

C. Gram stain review

1. After interpretation of smears, hold slides long enough to allow a confirmatory review, especially if culture or other laboratory test results are inconsistent.

- **a.** Drain or gently blot excess oil.
- **b.** For slide libraries and teaching collections that will be stored for longer periods, immersion oil can be removed with xylene solution and the slides can be overlaid with a sealer, such as permount solution (used to coverslip surgical tissue specimens), to prevent fading.
- **c.** When discarding stained smears, handle as biological waste and use biohazard containers as required by local/state regulations. Slides may puncture biohazard bags and should be treated as "sharps." Cardboard boxes or other protective containers may be used.

2. If a repeat Gram stain or a special stain to confirm findings suggested by a Gram smear interpretation is needed when extra unstained smears are not available, a Gram-stained smear may be destained.

- **a.** Remove immersion oil on the slide with xylene solution.
- **b.** Flood slide with acetone-alcohol until smear appears colorless.
- **c.** Restain.

Table 2.1–2 Gram-positive microorganisms found in direct smears of some clinical sources[a]

Organism	Gram stain morphology	Frequent sources	Comments and additional tests or media that may be included
Actinomyces spp.	Thin, beaded, branched gram-positive filaments; may be within sulfur granules with peripheral clubs	Cervicofacial, thoracic, abdominal, and pelvic abscesses and drainages; pleural fluid; bronchial washings	Modified acid-fast stain; if sulfur granules present, wash and crush in thioglycolate medium
Nocardia spp.	Long, thin, branching, beaded, gram-positive or irregularly staining bacilli; in culture smears, may be pleomorphic with branching and coccoid forms	Sputum, bronchial washings, biopsy material, purulent exudates, CSF, blood	Modified acid-fast stain; if mixed microbiota, mycobacterial decontamination procedures may be used; plate incubated at 45°C may enhance recovery; use Sabouraud dextrose agar without chloramphenicol
Mycobacterium spp.	Gram-positive beaded or gram-neutral bacilli; often found inside macrophages; bacilli may be short to long, banded or beaded, and/or slightly curved; some species appear pleomorphic and coccoid	Respiratory tract, urine, blood, biopsy material, CSF	Confirm with acid-fast stain
Corynebacterium spp.	Gram-positive pleomorphic, club-shaped, irregularly staining bacilli or coccobacilli with palisading and/or angular arrangements	Blood, tissue aspirates, skin lesions, wounds, indwelling catheters, prosthetic heart valves, upper and lower respiratory tracts	Add selective and differential media for *C. diphtheriae*, if suspected
C. jeikeium	Often small coccobacilli resembling streptococci		
Propionibacterium spp.	Gram-positive, very pleomorphic "diphtheroid" forms that may branch	Blood, CSF, other body fluids, skin lesions	GLC for major peak of propionic acid; common skin contaminant during needle aspiration
Listeria monocytogenes	Gram-positive small to medium coccobacilli that may be pleomorphic; occur in short chains or palisades; may be confused with corynebacteria or enterococci	CSF, blood, amniotic fluid, placental or fetal tissue	Direct wet mount for tumbling motility; if mixed microbiota, cold enrichment may be used
Erysipelothrix rhusiopathiae	In tissue, long, slender, gram-positive bacilli; in blood, small "coryneforms"	Skin lesions, biopsy material, tissue aspirates, blood	Associated with occupational or animal contact
Lactobacillus spp.	Medium, straight, uniform gram-positive bacilli with rounded ends; may form chains or spirals; sometimes short and coccobacillary	Usually involved in mixed infections; rarely from blood, CSF	Recovery may be improved by anaerobic incubation; normal vaginal, mouth, and gastrointestinal tract microbiota
Bacillus spp.	Medium to large square-ended gram-positive bacilli with parallel sides with or without spores; some species have spores that swell sides; may stain gram variable or gram negative with age	May be clinically relevant from any source in compromised patient or intravenous-drug abuser; also intraocular	Frequent culture contaminants; may cause ocular infections
Clostridium perfringens	Gram-positive large "boxcars" with no spores; may stain gram negative	Blood, wounds, intra-abdominal	Add egg yolk agar incubated anaerobically; absence of inflammatory cells may indicate clostridial myonecrosis; normal gastrointestinal tract microbiota

(continued)

Table 2.1–2 Gram-positive microorganisms found in direct smears of some clinical sources[a] *(continued)*

Organism	Gram stain morphology	Frequent sources	Comments and additional tests or media that may be included
Clostridium spp.	Gram-positive, -variable, or -negative bacilli with or without spores; bacilli may be large, slender and short, or long; sometimes form coils; often smaller than *Bacillus* spp.	Blood, intra-abdominal, wounds, abscesses	Normal gastrointestinal and genital tract microbiota
Streptococcus pneumoniae	Gram-positive cocci in pairs, lancet shapes, short chains	Lower respiratory tract, blood, CSF, sterile fluids	Quellung test and other antigen detection procedures may be used on selected clinical specimens
Enterococcus spp.	Gram-positive cocci in pairs, short chains; may resemble pneumococci	Urine, wounds, blood, intra-abdominal abscesses	Normal gastrointestinal tract microbiota; common cause of superinfections in patients treated with expanded-spectrum cephalosporins
Streptococcus spp.	Round to oval gram-positive cocci, occasionally elongated; in pairs and/or short to long chains; nutritionally variant streptococci often seen as highly pleomorphic, gram-variable to gram-negative coccobacilli with pointed ends and spindle shapes	Blood, CSF, respiratory tract, multiple other sources	May be difficult to distinguish from corynebacteria and lactobacilli; antigen detection procedures may be used on selected clinical specimens for group A and B streptococci; other than group A and B, normal microbiota, which may be opportunistic; *S. bovis* from blood cultures has strong correlation with colon cancer; if nutritionally variant streptococci are suspected, inoculate BAP with *S. aureus* cross-streak
Aerococcus viridans	Gram-positive cocci in pairs, tetrads, clusters	Blood, CSF	
Staphylococcus spp.	Gram-positive cocci in pairs, tetrads, clusters	Abscesses, drainages, wounds, respiratory tract, blood, tissue, sterile fluids, indwelling catheters	Normal microbiota, especially skin, nares
S. aureus	May often be characterized by very uniform, geometric clusters of small cocci, whereas coagulase-negative species are often irregular, more pleomorphic, with greater size variation		
Stomatococcus mucilaginosus	Large gram-positive cocci in pairs, tetrads	Blood in compromised patients, peritoneal dialysates	Normal oral microbiota

[a] Reprinted from CMPH 1.5.14–15.

Table 2.1–3 Gram-negative microorganisms seen in direct smears of some clinical sources[a]

Organism	Gram stain morphology	Frequent sources	Comments
Enterobacteriaceae	Straight thick bacilli; short to medium length with rounded ends; antimicrobial-agent-affected microorganisms may be pleomorphic, filamentous, and/or irregularly staining	Urinary tract, multiple other sources	Includes microorganisms that cause gastroenteritis and bacterial dysentery; also normal gastrointestinal tract microbiota; nosocomial strains may be multiply resistant to antimicrobial agents
Pseudomonas spp.	Thin bacilli; medium length to long with rounded to pointy ends; often in pairs; antimicrobial-agent-affected microorganisms may appear filamentous, coiled, and/or pleomorphic	Urine, lower respiratory tract, wounds, eyes, multiple sites in compromised patients	Selective agar for *P. aeruginosa* may be used for recovery from mixed microbiota; nosocomial strains may be multiply resistant to antimicrobial agents
Haemophilus spp., *Pasteurella* spp., fastidious gram-negative bacilli	Small coccoid to bacillary forms; pleomorphic; often with filamentous forms; may be faintly staining	Blood, sterile fluid including CSF, lower respiratory tract, abscesses, wounds, eyes, genital tract	Direct antigen detection procedures may be used for *H. influenzae* type b on selected clinical specimens
Legionella spp.	Pleomorphic slender bacilli of variable lengths that may stain pale; may not take stain in clinical specimens	Lower respiratory tract	Add special growth media; direct fluorescent-antibody stains and molecular probes available; Dieterle silver or Gimenez stains may be used; acid-wash method may be used to enhance recovery from specimens with mixed microbiota
Fusobacterium nucleatum, Capnocytophaga spp.	Long slender bacilli with tapered to pointed ends; "needlelike"; may be in pairs, end to end, or filamentous	Respiratory tract, wounds, blood, abscesses	Endogenous microbiota
Fusobacterium necrophorum, F. mortiferum, or *F. varium*	Highly pleomorphic bacilli with rounded to tapered ends; pale and irregularly staining, with bizarre forms and round bodies	Wounds, blood, abscesses	Endogenous microbiota
Bacteroides spp.	Pleomorphic straight bacilli with possible irregular to bipolar staining	Wounds, blood, abscesses	Direct fluorescent-antibody stain available; endogenous microbiota
Vibrio spp.	Slightly curved to straight bacilli	Stool, wounds	If mixed microbiota, selective medium (TCBS) recommended
Campylobacter spp. (*Helicobacter* spp.)	Thin, curved bacilli including S shapes, gull wings, long spiral forms	Stool, blood, gastric biopsy	Microaerophilic or capnophilic atmosphere required; if mixed microbiota, 42°C incubation recommended for recovery of thermophilic species
Acinetobacter spp.	Medium to large cocci in pairs; occasionally coccoid, bacillary, and filamentous forms; often resistant to decolorization	Urine, lower respiratory tract, blood, sterile fluids, wounds, abscesses, tissues, stool	Mutliple sites in compromised patients
Neisseria spp., *Moraxella catarrhalis, Veillonella* spp.	Medium to large cocci in pairs and tetrads, coffee bean shaped; no bacilli seen	Genital tract, urine, lower respiratory tract, blood, sterile fluids, wounds, abscesses	Direct antigen detection procedures may be used for some *N. meningitidis* serotypes in selected clinical specimens; if mixed microbiota, selective medium may be used to enhance recovery of *N. gonorrhoeae*

[a] Adapted from CMPH 1.5.16.

Table 2.1–4 Other microorganisms seen in direct smears of some clinical sources[a]

Organism	Gram stain morphology	Frequent sources	Comments
Pneumocystis carinii	Gram-negative spherical cysts (5–7 μm) often containing rosette of eight gram-negative intracystic bodies; or cluster of gram-negative cysts surrounded by halos in background of amorphous gram-negative material	Open lung and transtracheal biopsies, bronchial washings and lavages, sputum	Confirm with Grocott methenamine-silver, toluidine blue O, Gram-Weigert, or direct or indirect fluorescent-antibody stain
Blastomyces dermatitidis	Gram-variable, broad-based, thick-walled yeast cell with figure-eight appearance	Bronchial washings, sputum, purulent exudates, skin lesions	
Cryptococcus neoformans	Partially or completely gram-positive round yeast cell with clear or red-orange halo; yeast cells may appear stippled or gram neutral	CSF, blood, biopsy material, sputum, bronchial washings, cutaneous lesions	Confirm capsule with India ink; direct antigen detection procedures may be used on selected clinical specimens
Candida spp.	Gram-positive budding yeast cell with or without pseudohyphae; may also appear stippled or gram neutral	Sputum, urine, blood, biopsy material, vaginal discharge, upper respiratory tract	Endogenous microbiota
Malassezia furfur	Bottle-shaped yeast cells in compact clusters, usually with short hyphal elements	Skin scrapings, blood drawn through catheter lines, hyperalimentation fluids	Inoculate lipid-enriched medium

[a] Reprinted from CMPH 1.5.17.

VII. LIMITATIONS OF THE PROCEDURE

A. Use results of Gram stains in conjunction with other clinical and laboratory findings. Use additional procedures (e.g., special stains, direct antigen tests, inclusion of selective media, etc.) to confirm findings suggested by Gram-stained smears.

B. Careful adherence to procedure and interpretive criteria is required for accurate results. Accuracy is highly dependent on the training and skill of microscopists.

C. Additional staining procedures are recommended for purulent clinical specimens in which no organisms are observed by the Gram stain method.

D. Gram stain-positive, culture-negative specimens may be the result of contamination of reagents and other supplies, presence of antimicrobial agents, or failure of organisms to grow under usual culture conditions (medium, atmosphere, etc.).

E. False Gram stain results may be related to inadequately collected specimens.

SUPPLEMENTAL READING

Baron, E. J., L. R. Peterson, and S. M. Finegold. 1994. *Bailey and Scott's Diagnostic Microbiology*, 9th ed. The C. V. Mosby Co., St. Louis.

Bartholomew, J. W. 1962. Variables influencing results, and the precise definition of steps in gram staining as a means of standardizing the results obtained. *Stain Technol.* **37:**139–155.

Clarridge, J. E., and J. M. Mullins. 1987. Microscopy and staining, p. 87–103. *In* B. J. Howard (ed.), *Clinical and Pathogenic Microbiology*. The C. V. Mosby Co., St. Louis.

Mangels, J. I., M. Cox, and L. Lindberg. 1984. Methanol fixation: an alternative to heat fixation of smears before staining. *Diagn. Microbiol. Infect. Dis.* **2:**129–137.

Murray, P. R., and J. A. Washington. 1975. Microscopic and bacteriologic analysis of sputum. *Mayo Clin. Proc.* **50:**339–344.

Shanholtzer, C. J., P. Schaper, and L. Peterson. 1982. Concentrated Gram stain smears prepared with a cytospin centrifuge. *J. Clin. Microbiol.* **16:**1052–1056.

Washington, J. A. 1986. Rapid diagnosis by microscopy. *Clin. Microbiol. Newsl.* **8:**135–137.

2.2 Interpretation of Aerobic Bacterial Growth on Primary Culture Media

I. PRINCIPLE

The initial interpretation of bacterial growth on primary culture media, which usually follows the first 24 to 48 h of incubation, is an opportunity for the skilled microbiologist to make a preliminary identification and to decide what additional tests and procedures must be performed to arrive at a definitive identification.

II. DESCRIPTION OF COLONIAL MORPHOLOGY

A. Initial examination of primary plates

1. Note the different types of colonial morphology appearing on each agar plate and the number of each morphotype present. It is very important when making these initial assessments to look at the culture plates from different angles and to use direct illumination of the plate.
2. Describe the gross colonial morphology of each colony. Refer to Table 2.2–1 for terms to describe colonial morphology.

B. Interpretation of primary culture media

1. Note the appearance of microorganisms on general-purpose primary media. Some common microorganisms and their characteristics on primary media are listed in Table 2.2–2.
2. Be aware of microorganisms that will grow or are inhibited on various selective primary media. Commonly used primary culture media and the purpose of each are described in Table 2.2–3.
3. Describe changes that occur in the medium surrounding the bacterial colonies. These changes can provide information on various characteristics of the bacteria. A variety of dyes, indicators, and other ingredients can be incorporated into the primary medium to help detect enzymatic activities and end products of bacteria. These activities may be indicated by a color change in the medium surrounding the bacteria or in the colony.
4. Hemolysis is helpful in characterizing microorganisms.
   a. Alpha-hemolysis is the reduction of hemoglobin to methemoglobin in the medium surrounding the colony. This causes a greenish discolorization of the medium. Microscopic inspection of alpha-hemolyzed RBCs shows that the cell membrane is intact.
   b. Beta-hemolysis is the lysis of RBCs, resulting in a clear zone surrounding the colony.
   c. Gamma-hemolysis indicates no hemolysis. No destruction of RBCs occurs, and there is no change in the medium.

Table 2.2–1 Terms to describe gross colonial morphology (reprinted from CMPH 1.6.2)

SIZE (diameter in mm)
- Large = greater than 1mm in diameter
- Medium = 1mm in diameter
- Small = less than 1mm in diameter

SHAPE
- Circular
- Filamentous
- Irregular
- Punctiform
- Rhizoid
- Spindle

ELEVATION 
- Flat
- Raised
- Convex
- Dome shaped
- Umbonate
- Umbilicate

MARGIN (edge of colony)
- Entire
- Undulate
- Lobate
- Erose
- Filamentous
- Curled

COLOR White, Black, Cream, Orange etc.

SURFACE APPEARANCE
- Glistening
- Smooth
- Granular
- Dull
- Rough
- Creamy

DENSITY (ability to see through the colony)
- Opaque = can not see through the colony
- Transparent = can see through the colony
- Translucent = only with light shining through

CONSISTENCY (best observed by picking up a colony with a loop or needle)
- Butyrous (buttery)
- Viscid (sticky)
- Friable (crumbles easily)
- Brittle
- Membranous (pliable)

II. DESCRIPTION OF COLONIAL MORPHOLOGY *(continued)*

d. Alpha-prime-hemolysis is a small zone of complete hemolysis that is surrounded by an area of partial lysis. Alpha-prime-hemolysis is best seen when magnification is used to observe the colonies.

5. Note pigment production when primary culture medium is examined. Pigment can be water soluble and produce a discoloration of the surrounding medium. Pigment may be nondiffusible and confined to the colonies themselves.

6. Include odors in the evaluation of colonies. Many bacteria have distinct odors that are strong clues to their presence and/or identification. *Pseudomonas aeruginosa* is described as having a fruity, grapelike odor. *Eikenella corrodens* often smells like bleach. *Proteus* species have been described as smelling like devil's food cake or burnt chocolate.

◪ **NOTE:** The descriptions and tables address aerobic and facultatively anaerobic bacteria only.

III. PRIMARY PLATING MEDIA

Additional information on primary plating media is provided in Table 2.2–3.

A. Bile-esculin (enterococcal selective) agar

Enterococci are able to grow and hydrolyze esculin in the presence of bile. Bile-esculin agar can also be made more selective for the recovery of vancomycin-resistant enterococci by adding 6 μg of vancomycin per ml. In this case, vancomycin-resistant strains produce black colonies on this agar, but susceptible strains of enterococci will not grow.

B. Blood agar plate with 5% sheep blood (BAP)

Blood agar with 5% sheep blood is an enriched, primary isolation medium that should be incorporated in the plating of most clinical specimens since most microorganisms grow on this medium.

C. Campylobacter blood agar (Campy agar)

Campy agar is the basic medium for the isolation of *Campylobacter jejuni*. It can be inoculated directly with rectal swabs and fecal specimens.

D. Chocolate agar (CHOC)

CHOC is a medium to which blood, hemoglobin, or heme has been added. Enriched CHOC supplies the special growth requirements needed for the isolation of such fastidious organisms as *Haemophilus influenzae* and, when incubated in CO_2, *Neisseria meningitidis* and *Neisseria gonorrhoeae*. It has a low concentration of agar, which provides the higher moisture content required by some organisms. Additional enrichments, such as IsoVitaleX (BBL Microbiology Systems, Cockeysville, Md.) or enrichments provided by other manufacturers, are required for the more fastidious bacteria.

E. Columbia colistin-nalidixic acid agar (CNA)

CNA is selective for the growth of gram-positive cocci. The colistin in the medium inhibits the growth of most gram-negative bacilli, while the nalidixic acid inhibits the growth of most strains of *Proteus*.

F. Eosin-methylene blue (EMB) agar

EMB is a differential medium used for the detection and isolation of gram-negative enteric bacilli. EMB contains a combination of eosin and methylene blue as an indicator, resulting in a sharp distinction between colonies that ferment lactose and/or sucrose and those that do not. Differences occur as a result of the pH of end products. Colonies of some microorganisms that ferment one or both of the sugars are black (or dark purple) or possess dark centers with colorless borders; others appear pinkish violet. Some microorganisms, especially *Escherichia coli*, also have a metallic sheen when viewed by reflected light. Colonies of microorganisms that do not ferment lactose or sucrose are

Table 2.2–2 Colonial morphology on primary isolation media[a]

Organism	Morphology on[b]: BAP	CNA	CHOC
Escherichia coli	Gray, mucoid, flat or dome shaped, may be beta-hemolytic	I	Same as BAP
Proteus spp.	Flat, gray, spreading	I	Same as BAP
Pseudomonas aeruginosa, P. fluorescens	Flat, gray-green, may have spreading margins, metallic sheen, grape odor	I	Same as BAP
Neisseria gonorrhoeae	I	I	Small, gray, entire, sticky
N. meningitidis	Medium to large, creamy and gray, alpha-hemolytic	I	Same as BAP, no hemolysis
Haemophilus spp.	I	I	Gray, raised, smooth, may be mucoid
Branhamella spp.	Whitish, medium to large, raised or dome shaped	I	Same as BAP
Staphylococcus aureus	Large, convex, white-yellow, creamy opaque, may be beta-hemolytic	Same as BAP	Same as BAP, no hemolysis
Coagulase-negative staphylococci	White-gray, raised, creamy	Same as BAP	Same as BAP
Beta-hemolytic streptococci	Pinpoint to medium, zone of beta-hemolysis (clear zone) translucent, dull, gray	Same as BAP	Same as BAP, no hemolysis
Viridans streptococci	Pinpoint to medium; white-gray; caramel odor; alpha-hemolysis (greenish color) surrounding colony	Same as BAP	Same as BAP
Enterococcus spp.	Gray, medium, usually no hemolysis	Same as BAP	Same as BAP
Streptococcus pneumoniae	Umbilicate, alpha (greening)-hemolysis, may be mucoid or teardrop shaped	Same as BAP	Same as BAP
Listeria monocytogenes	Whitish gray similar to group B streptococcus, flat, narrow zones of beta-hemolysis	Same as BAP	Same as BAP
Corynebacterium spp.	White, dry, may be sticky	Same as BAP	Same as BAP
Yeast cells	White, creamy, bread odor	Same as BAP	Same as BAP

[a] Adapted from CMPH 1.6.3–4.
[b] Abbreviations: I, inhibited; NA, not applicable.
[c] Some strains of *S. aureus* and *Enterococcus* spp. grow on EMB or MAC.

Table 2.2–3 Commonly used primary plating media[a,b]

Medium	Form	Type	Expected isolates	Differential reactions	Comments
Campy agar	P	E, S	*Campylobacter jejuni*	NA	Direct or indirect inoculation may be performed.
CHOC	P, Sl	E	*Haemophilus influenzae, Neisseria gonorrhoeae, N. meningitidis*	NA	Low agar content provides increased moisture required by some organisms.
CNA	P	E, S	Gram-positive microorganisms	NA	Colistin inhibits gram-negative organisms, and nalidixic acid inhibits *Proteus* spp. Hemolysis may be observed.
EMB	P	D, S	Gram-negative enteric bacilli	L^+ or Su^+: black, purple, metallic sheen L^- or Su^-: colorless	Lactose and/or sucrose fermenters are generally normal enteric bacteria. Gram-positive organisms are inhibited by EMB.
HE	P	D, S	*Salmonella* and *Shigella* species	S/S: blue, green NF: orange, pink	Contains lactose, sucrose, and salicin. Most *Enterobacteriaceae* usually ferment one of these. Detects H_2S.
MAC	P	D, S	Gram-negative enteric bacilli	L^+: pink L^-: colorless	Gram-positive organisms are inhibited by bile salts.

Table 2.2–2 Colonial morphology on primary isolation media[a] *(continued)*

Morphology on[b]:				
MAC	EMB	MTM	HE	XLD
Pink or colorless	Dark center, green sheen	I	Orange-salmon pink	Opaque yellow
Colorless	Colorless	I	I or green with dark center	Yellow or red
Colorless	Colorless	I	I or green	Red
I	I	Same as CHOC	NA	NA
I	I	Same as BAP	NA	NA
I	I	I	NA	NA
I	I	Same as BAP	NA	NA
I[c]	I[c]	I		
I	I	I		
I	I	I		
I	I	I		
I	I[c]	I		
I	I	I		
I	I	I		
I	I	I		
I	I	I		

Table 2.2–3 Commonly used primary plating media[a,b] *(continued)*

Medium	Form	Type	Expected isolates	Differential reactions	Comments
Salmonella-shigella	P	D, S	*Salmonella* and *Shigella* species	L^+: pink L^-: colorless	Bile salts inhibit gram-positive microorganisms. Salmonella-shigella agar may inhibit some *Shigella* species.
TM or MTM agar	P	E, S	*N. gonorrhoeae*, *N. meningitidis*	NA	Vancomycin inhibits gram-positive microorganisms, colistin inhibits gram-negative microorganisms, and nystatin inhibits yeast cells.
BAP	P	E	Gram-positive and gram-negative organisms	NA	General-purpose medium. Hemolysis can be observed.
XLD agar	P	D, S	EP	NF: yellow EP: red	Contains xylose, lactose, and sucrose. NF usually ferment one of these. Detects H_2S.

[a] Reprinted from CMPH 1.6.5.
[b] Abbreviations: D, differential; E, enriched; EP, enteric pathogen; L, lactose; NA, not applicable; NF, normal microorganisms; P, plate; S, selective; Sl, slant; S/S, *Salmonella* or *Shigella* spp.; Su, sucrose.

III. PRIMARY PLATING MEDIA *(continued)*

transparent and/or colorless. Lactose and/or sucrose fermenters are, for the most part, normal enteric bacteria. Gram-positive microorganisms are inhibited by the eosin-methylene blue mixture. The agar concentration may be increased to 5% to inhibit swarming of *Proteus* spp. Certain strains of staphylococci and enterococci will form small colonies; *Pseudomonas aeruginosa* will appear as an umbonate colony with a filamentous margin, usually violet in appearance.

G. Hektoen enteric agar (HE)

1. Use

   HE is recommended for the isolation of *Salmonella* and *Shigella* spp. from fecal specimens. HE contains lactose, sucrose, and salicin. *Salmonella* or *Shigella* spp. generally do not ferment lactose, sucrose, or salicin and therefore will produce blue or green colonies on this medium. Most members of the family *Enterobacteriaceae* will ferment at least one of these sugars, resulting in acid production. In the presence of acid, bromthymol blue in HE produces a yellow color, and the acid fuchsin produces a red color. The combination of red and yellow results in the *Enterobacteriaceae* appearing orange to pink.

2. Expected reactions
   - a. *Salmonella* spp.: Green to blue-green, usually with black centers
   - b. *Salmonella* and *Shigella* spp.: When surrounded by many bright fermenting *Enterobacteriaceae*, may appear to be faint pink with a green tinge, but there is usually a clear halo around the colonies in the area of precipitated bile. This halo is most apparent when the plates are held up to a light. The colonies are more translucent than those of other members of the family *Enterobacteriaceae*.
   - c. *Shigella* spp.: Usually green rather than blue
   - d. *Citrobacter* spp.: Usually inhibited; if present, small and blue-green
   - e. *Proteus* spp.: Usually inhibited; if present, H_2S producers may produce yellow or green colonies with black centers. Non-H_2S producers resemble *Shigella* spp. but are smaller.
   - f. *E. coli, Klebsiella* spp., and *Enterobacter* spp.: Brightly colored orange to salmon-pink; usually heavy pink precipitate surrounding the colonies

H. MAC

MAC is used for the differentiation and isolation of enteric bacilli. Lactose fermenters produce pink colonies on this medium, while the colonies of microorganisms that do not ferment lactose are colorless. Gram-positive microorganisms are inhibited by the bile salts in the medium. The agar concentration may be increased to 5% to inhibit the swarming of *Proteus* species.

I. Thayer-Martin (TM) or modified Thayer-Martin (MTM) agar

TM and MTM are modifications of CHOC and are used for the isolation of *N. gonorrhoeae* and *N. meningitidis*, particularly from cultures that may contain other organisms. The VCN inhibitor contains vancomycin (which inhibits most gram-positive organisms), colistin (which inhibits most gram-negative organisms except *Proteus* spp.), and nystatin (which inhibits yeast cells). Trimethoprim lactate is added to inhibit *Proteus* species. *N. lactamica* may grow on this medium.

J. Trypticase soy agar (TSA)

TSA is a general-purpose medium used with or without blood for isolating and cultivating a number of microorganisms. The addition of blood makes this agar suitable for determining hemolytic activity. Several stabs should be made when this medium is inoculated to detect the beta-hemolysis of group A streptococci. Stabbing creates an area of reduced oxygen tension, which is necessary for the demonstration of hemolysis by oxygen-labile hemolysin O.

K. Xylose-lysine deoxycholate (XLD) agar

1. Use

XLD agar is a differential, selective medium used for the isolation of enteric pathogens, especially *Shigella* species. The selectivity of the medium is increased through the addition of sodium deoxycholate, which inhibits some of the normal enteric bacteria. When normal enteric bacteria do appear, yellow colonies are produced because of the fermentation of xylose, sucrose, and lactose. Organisms such as *Shigella, Providencia*, and some *Proteus* spp. do not ferment any of these sugars and therefore produce red (alkaline) colonies. Species of *Edwardsiella* and *Salmonella* ferment xylose but not sucrose or lactose. To offset the acid production, lysine is incorporated into the medium. Species of *Edwardsiella* and *Salmonella* decarboxylate lysine. The xylose-lysine ratio in XLD permits these microorganisms to exhaust the xylose and then attack the lysine, causing a reversion to an alkaline pH, and thus produce red colonies. To prevent other lysine decarboxylase-positive enteric bacteria from causing this reversion, double portions of lactose and sucrose are added to produce acid in excess. Phenol red serves as the pH indicator. Ferric ammonium citrate reacts with hydrogen sulfide. Specimens suspected of containing species of *Salmonella, Edwardsiella*, or *Shigella* are inoculated onto XLD agar and incubated overnight at 35°C.

2. Expected reactions

a. Yellow colony: Species of *Escherichia, Enterobacter, Klebsiella*, and *Serratia; Citrobacter diversus; Providencia rettgeri; Morganella morganii*; and *Yersinia enterocolitica*

b. Yellow colony with black center: *Citrobacter freundii, Proteus vulgaris, Proteus mirabilis*

c. Red colony: Species of *Shigella, Providencia*, H_2S-negative *Salmonella, Pseudomonas*, and some *Proteus rettgeri*

d. Red colony with black center: Species of *Salmonella* and *Edwardsiella*

SUPPLEMENTAL READING

Difco Laboratories. 1984. *Difco Manual*, 10th ed., p. 546–551, 1025–1026. Difco Laboratories, Detroit, Mich.

Dowell, V. R., Jr., and T. M. Hawkins. 1974. *Laboratory Methods in Anaerobic Bacteriology.* CDC manual DHEW publication (CDC) 74-8272. U.S. Government Printing Office, Washington, D.C.

Holt-Harris, J. E., and O. Teague. 1916. A new culture medium for the isolation of *Bacillus typhosus* from stools. *J. Infect. Dis.* **18:**596–600.

Horvath, R. S., and M. E. Ropp. 1974. Mechanism of action of eosin-methylene blue agar in the differentiation of *Escherichia coli* and *Enterobacter aerogenes. Int. J. Syst. Bacteriol.* 24:221–224.

King, S., and W. Metzger. 1968. A new plating medium for the isolation of enteric pathogens. *Appl. Microbiol.* **16:**577–578.

Koneman, E. W., S. D. Allen, V. R. Dowell, Jr., W. M. Janda, H. M. Sommers, and W. C. Winn, Jr. 1988. Introduction to medical microbiology, part 2. Definitive laboratory diagnosis of infectious diseases, p. 70–73. *In* E. W. Koneman (ed.), *Color Atlas and Textbook of Diagnostic Microbiology,* 3rd ed. J. B. Lippincott, Philadelphia.

Miller, J. M., and H. T. Holmes. 1995. Specimen collection, transport, and storage, p. 19–32. *In* P. R. Murray, E. J. Baron, M. A. Pfaller, F. C. Tenover, and R. H. Yolken (ed.), *Manual of Clinical Microbiology,* 6th ed. American Society for Microbiology, Washington, D.C.

Taylor, W. I., and D. Schellart. 1968. Isolation of shigellae. *Appl. Microbiol.* **16:**1383–1386.

Thayer, J. D., and J. E. Martin. 1966. Improved medium selective for cultivation of *N. gonorrhoeae* and *N. meningitidis.* Public Health Rep. **81:**559–562.

2.3 Processing and Interpretation of Blood Cultures

I. PRINCIPLE

When bacteria or fungi overcome the host's normal reticuloendothelial cell barriers and enter the bloodstream through the lymphatics or from extravascular sites, they can quickly disseminate throughout the body, causing severe illness. In addition, the by-products of their metabolism can lead to sepsis and shock, among the most serious complications of infectious diseases. Rapid recognition and immediate institution of appropriate treatment are essential. Laboratory diagnosis of bacteremia and fungemia is currently dependent on blood cultures. However, because the culture methods are so sensitive, the procedure must be carefully controlled beginning at the preanalytical stage (collection), to avoid the misinterpretation of a procurement-associated skin commensal microorganism as an agent of infection.

II. SPECIMEN COLLECTION AND INITIAL HANDLING

(For details, see CMPH 1.1.)

A. Skin antisepsis and venipuncture

1. Select a different venipuncture site for each blood culture bottle. If poor access requires that blood for culture must be drawn through a port in an indwelling catheter, the second culture must be from a peripheral site because cultures drawn through catheters are more likely to be contaminated. Do not draw blood from a vein into which an intravenous solution is running. Draw the two blood cultures in succession.

   NOTE: Comparison of quantitative cultures drawn through an indwelling intravenous catheter and from a peripheral site may be useful for diagnosis of catheter-related infection.
2. Prepare the site (skin or port) with 70% isopropyl or ethyl alcohol to remove surface dirt and oils.
3. Swab or wipe concentric circles of tincture of iodine, moving outward from the center of the site. Allow the iodine to dry for 1 min and avoid touching the site.

   NOTE: If povidone iodine is used, it must be allowed to dry for 2 min.
4. Prepare the septum of the blood culture bottle and the rubber stopper on bottles/tubes by vigorously wiping with 70% alcohol and allowing the septum to dry.

   NOTE: Do not use iodine on BACTEC bottles (Becton Dickinson [B-D] Instrument Systems, Sparks, Md.) to be read by analysis of headspace gas for carbon dioxide production.

   One Isolator tube may be used for aerobic and anaerobic bacteria in addition to recovery of yeast cells and mycobacteria.
5. Withdraw the blood. Use a new needle if the first attempt is not successful.
6. If using a needle and syringe, inoculate the blood into the bottles or Isolator tubes as stipulated by laboratory protocol. Inoculate the aerobic bottle first. There is no need to change needles.

   NOTE: Best recovery is achieved when blood is inoculated directly into blood culture media at the bedside. If the blood must be transported to the laboratory or a distant site before being inoculated into the blood culture media, it should be collected with 0.05% sodium polyanetholesulfonate (SPS) anticoagulant-containing Vacutainer tubes (Becton Dickinson and Co., Paramus, N.J.).

7. After phlebotomy, remove residual iodine from the patient's skin by cleansing with alcohol to avoid development of irritation.
8. Label the blood culture bottles or tubes with patient information and the initials of the phlebotomist, gently mix the contents, and deliver them to the laboratory.

B. Recommended total volume and numbers of cultures

◩ **NOTE:** Adequate volume is the single most important factor in the laboratory detection of microbes in the bloodstream; thus collection methods must be chosen to maximize the volume collected. Total volume of blood drawn should be divided equally between two separate venipunctures, to minimize the chance of a false-positive result caused by recovery of a skin contaminant.

1. Neonates to 1 year: 0.5 to 1.5 ml/tube, although at least 1.0 ml is preferred

   ◩ **NOTE:** Two separate venipunctures may not always be possible.
2. Children 1 to 6 years old: 1 ml per year of age, divided between two blood cultures

   Example: For a 3 year old, draw 1.5 ml from each of two sites, for a total of 3.0 ml of blood.
3. Children weighing 30 to 80 lb: 10 to 20 ml, divided between two blood cultures
4. Children and adults weighing >80 lb: 30 to 40 ml divided between two blood cultures.

   ◩ **NOTE:** At least 20 to 30 ml of blood in two to three draws is the minimal requirement.

C. Timing of blood cultures

◩ **NOTE:** Although drawing blood cultures before the fever spike is optimal for recovery, volume is more important than timing in the detection of agents of septicemia.

1. When acute sepsis or another condition (osteomyelitis, meningitis, pneumonia, or pyelonephritis) requires immediate institution of antimicrobial-agent therapy, two blood cultures of maximum volume should be drawn consecutively before starting therapy.
2. For fever of unknown origin, subacute bacterial endocarditis, or other continuous bacteremia/fungemia, three blood cultures of maximum volume should be drawn on day 1. Two sets should be drawn consecutively, and the third set can be drawn an hour or more later.
3. For patients on antimicrobial therapy or for whom initial cultures are negative, a maximum of an additional three blood cultures may be obtained on day 2 or day 3. There is no added recovery with additional blood beyond 120-ml total volume cultured.

   ◩ **NOTE:** Blood cultures from patients on antimicrobial therapy should be drawn when antimicrobial agents are at their lowest concentration. Use of a resin-containing medium may enhance recovery. However, appropriate dilution and action of some anticoagulants and antifoaming agents may also diminish the effect of antimicrobial agents.

III. GENERAL CONSIDERATIONS REGARDING BLOOD CULTURES

(For details, see CMPH 1.7.)

Include QC information on reagent container and in QC records.

A. Media

◩ **NOTE:** Follow manufacturer's instructions for inoculation, special handling, and incubation of blood culture media.

1. SPS acts as an anticoagulant, antiphagocytic agent that inactivates complement and neutralizes many antimicrobial agents and antibacterial factors in blood. Media should contain 0.025 to 0.05% SPS.

   ◩ **NOTE:** SPS also inhibits some bacteria, such as *Neisseria* spp. and *Gardnerella vaginalis*. Addition of 1.2% gelatin may counteract this effect.
2. One blood culture usually consists of blood from a single venipuncture

III. GENERAL CONSIDERATIONS REGARDING BLOOD CULTURES *(continued)*

inoculated into two separate bottles to accommodate the volume of blood removed, since dilution ratios cannot exceed 1:5 to 1:10 blood to medium and bottles with total volume >100 ml are impractical. The use of more than one formulation of medium for each blood culture (aerobic and anaerobic; aerobic lytic and aerobic resin containing; etc.) may help maximize recovery of all possible pathogens.

3. Alternatives to broth media include biphasic media (agar and broth in one bottle such as Septi-Chek [B-D Microbiology Systems, Cockeysville, Md.]) and the lysis-centrifugation system of ISOLATOR (Wampole Laboratories, Cranbury, N.J.).
4. Automated systems have their own media, formulated to maximize detection based on the indicator system (for more information on automated blood culture instruments, *see* CMPH 12.5).
5. Special media for isolation of fungi and mycobacteria should be used if those agents are suspected. Although yeast cells are usually detected in most aerobic broth systems, specialized fungal media will allow faster and more extensive recovery of all fungal agents of sepsis.

B. Organisms that have special requirements (*see* CMPH 1.7 and specific protocols based on organisms sought)

It is imperative that these cultures be handled in a biosafety hood.

1. *Brucella* spp.

   ◪ **NOTE:** Handle specimens and cultures in a class II biosafety cabinet; these organisms are considered to be class III pathogens.

   a. Extended incubation up to 4 weeks is usually necessary for recovery of *Brucella* spp. if standard broth media are used. Biphasic cultures or lysis-centrifugation may yield better recovery.

   b. Growth in broth may not be visible on visual inspection and may not be detected by the indicator systems of all automated instruments. Therefore, routine subcultures of samples from blood culture bottles to brucella or BHI agar with sheep blood must be performed.
2. *Leptospira* spp.

   ◪ **NOTE:** Special media, inoculation, incubation, and detection methods are required.
3. Nutritionally variant streptococci
4. Cell wall-deficient bacteria, including *Streptobacillus moniliformis*
5. *Bartonella* (formerly *Rochalimeae*) *henselae*

   ◪ **NOTE:** Best recovery has been in the lysis-centrifugation system.
6. *Malassezia furfur* (usually seen in catheter-related infections in infants receiving lipid-rich total parenteral nutrition)

   a. Blood in the culture provides enough lipids to allow growth of *M. furfur* in broth.

   b. Subcultures onto blood agar require addition of a small drop of pure virgin olive oil (never autoclaved) on the agar surface to provide sufficient nutrients. Addition of a drop of lipid solution used in hyperalimentation also suffices for the isolation of *M. furfur*.
7. Fungi other than *M. furfur*
8. Mycobacteria
9. *Legionella* spp.

   ◪ **NOTE:** Refer to index to CMPH for complete description and identification of the above-mentioned microorganisms.

C. Removal of antimicrobial agents

◪ **NOTE:** Most systems contain enough SPS, resins, and/or dematiaceous earth and use a dilution factor sufficient to neutralize the antimicrobial agents in blood of patients on therapy. If high-dosage penicillins are being used, however, additional measures may be taken.

III. LABORATORY PROCESSING OF BLOOD CULTURES

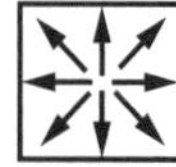

Observe universal (standard) precautions.

◪ **NOTE:** Working with blood cultures, keep the cultures within a biosafety cabinet or behind a shield, or wear a face mask and always wear gloves, because blood cultures contain material from patients that may harbor blood-borne pathogens. Follow all precautions necessary for prevention of blood-borne disease (standard or universal precautions).

A. Incubate blood cultures for the predetermined time period (usually 5 to 7 days) unless special circumstances require longer incubation. Maintain an atmosphere that allows recovery of the microorganisms sought (follow manufacturer's instructions) and maintain rotation or agitation of the media if at all possible.

B. Examine the cultures at least daily, whether detection of positives is by visual inspection or by an automated system.

C. At least one subculture to solid agar is necessary from negative bottles for manual broth systems. An early subculture after overnight incubation may detect most pathogens in a timely manner.

D. For the lysis-centrifugation system, follow manufacturer's instructions for centrifugation and subculturing of the specimen. Plate onto media that will allow recovery of the organism sought. Incubate the media under the correct conditions and for sufficient time to recover all pathogens sought. For example, recovery of *Bartonella henselae* may require up to 4 weeks of incubation in a moist, carbon dioxide-enriched atmosphere. If the lysis-centrifugation system is used alone, inoculation of a portion of the sediment into thioglycolate broth or onto anaerobic plates is necessary for recovery of anaerobes. Since positive results will be detected by the presence of colonies, subculture protocols detailed below are not applicable to this system.

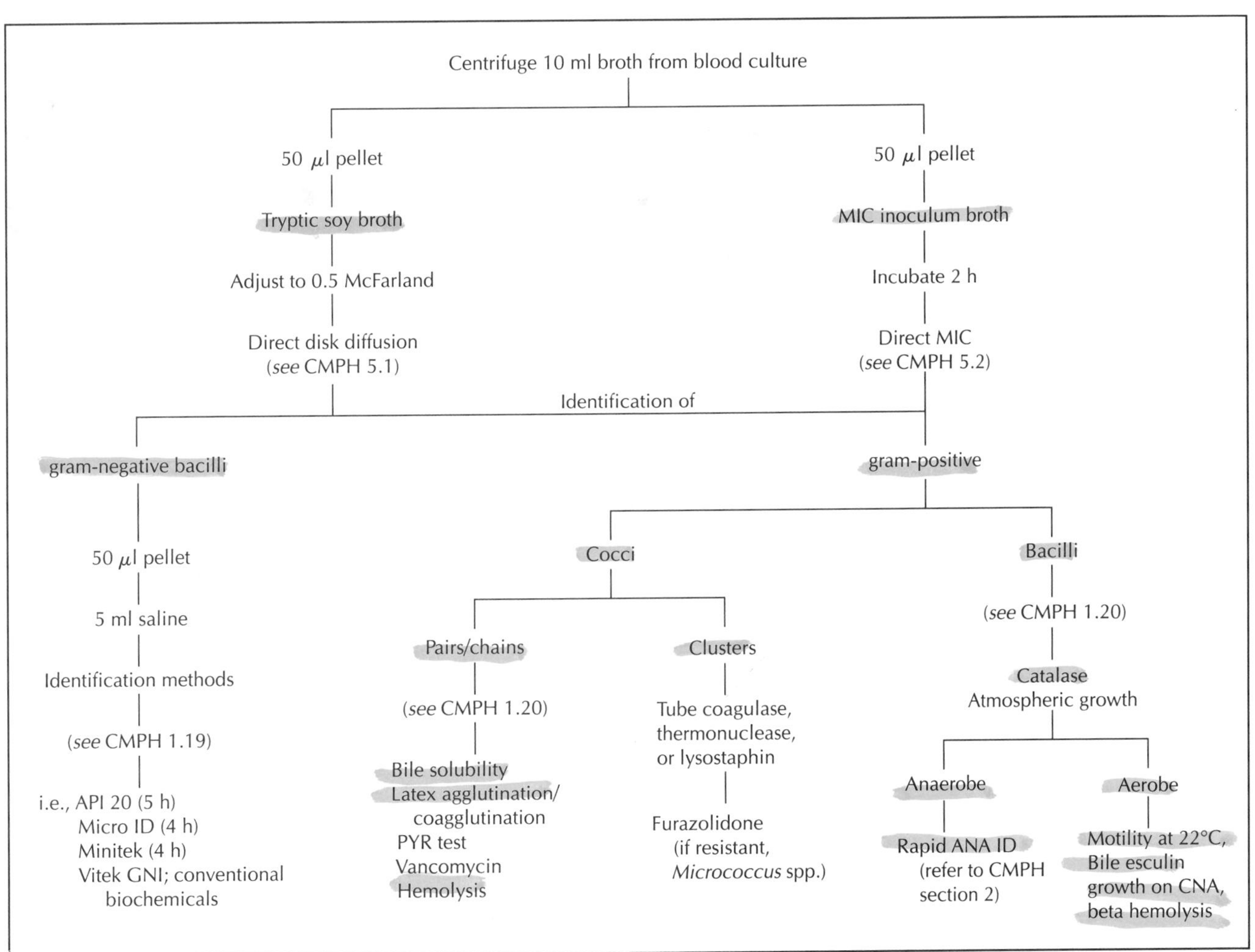

Figure 2.3–1 Identification of microorganisms isolated from positive blood cultures. Adapted from CMPH 1.7.8. Procedure numbers refer to CMPH.

IV. WORKING WITH POSITIVE BLOOD CULTURE RESULTS

A. Broth bottles suggestive of growth (automated or manual detection) must be examined immediately. Optimal patient care requires 24-h processing of positive blood cultures.

B. Initial examination of positive blood cultures

1. Gram stain of the broth should be made immediately. A thin smear of the medium should be air dried, methanol fixed, and stained and examined.
2. Subcultures to agar media are made depending on the Gram stain indications and examination of the bottles in which the microorganism grew.
3. Gram stain results are called immediately to the physician responsible for the patient, with as much interpretive information as possible.

   Example: Gram-positive cocci in chains and pairs, suggestive of streptococci. If only one bottle among two or more yields gram-positive cocci in clusters, then the interpretation that this isolate may be a procurement-associated skin organism can also be stated. The number of positive bottles compared with the total number of specimens collected, the hours or days from collection until positive results, similar organisms isolated from other specimens, and any other relevant information should be reported. Laboratories can develop their own criteria for verbal reports to physicians. Documentation of all telephone communications is essential.
4. Written or computer-generated reports should follow the verbal reports.
5. Based on laboratory protocol, the broth may be inoculated directly into an identification system and onto media for direct antimicrobial susceptibility testing (Fig. 2.3–1).

   ◩ **NOTE:** RBCs may interfere with some tests. Therefore, centrifuge broth from blood cultures slowly to remove RBCs; then centrifuge at 1,500 × *g* for 15 min to obtain pellet of microorganisms for testing.

C. Identification of isolates

◩ **NOTE:** When more than one blood culture obtained from the same patient during a 24-h period yield microorganisms of identical morphology, it is necessary to perform complete identification and antimicrobial susceptibility testing more than once. The time period after which similar isolates should again be fully examined in the laboratory should be determined in advance. The extent of identification of isolates considered to be procurement-associated skin microorganisms should be determined in consultation with the institution's infectious disease specialists.

1. Follow protocols in standard references, such as CMPH.
2. Report results to physicians as soon as the results are available.

SUPPLEMENTAL READING

Aronson, M. D., and D. H. Bor. 1987. Diagnostic decision: blood cultures. *Ann. Intern. Med.* **106:** 246–253.

Kellogg, J. A., F. L. Ferrentino, J. Liss, S. L. Shapiro, and D. A. Bankert. 1994. Justification and implementation of a policy requiring two blood cultures when one is ordered. *Lab. Med.* **25:** 323–330.

Li, J., J. J. Plorde, and L. G. Carlson. 1994. Effects of volume and periodicity on blood cultures. *J. Clin. Microbiol.* **32:**2829–2831.

Pfaller, M. A. 1995. Laboratory diagnosis of catheter-related bacteremia. *Infect. Dis. Clin. Pract.* **4:** 206–210.

2.4 Processing and Interpretation of Sterile Body Fluids (Excluding Blood, Cerebrospinal Fluid, Dialysate, and Urine)

I. PRINCIPLE

Infection of normally sterile body fluids often results in severe morbidity and mortality; therefore, rapid and accurate microbiological assessment of these samples is important for successful patient management. Most microorganisms infecting these sites are not difficult to culture, but sometimes occur in low numbers. With the increased use of prostheses, immunosuppressive therapeutic regimens, and long-term care of individuals with chronic debilitating disease, the likelihood of true infection with commensal microorganisms has increased. Care must be taken during specimen collection and transport to ensure that the specimen is not contaminated. Any microorganism found in a normally sterile site must be considered significant, and all isolates must be reported.

II. SPECIMEN

A. Joint fluid, or synovial fluid

B. Pleural fluid

1. Thoracentesis fluid
2. Empyema fluid

C. Peritoneal fluid

1. Ascites fluid
2. Paracentesis fluid

D. Pericardial fluid

E. Cul-de-centesis fluid

III. MEDIA AND REAGENTS

Include QC information on reagent container and in QC records.

A. Media

1. CHOC
2. BAP
3. THIO; chopped meat-carbohydrate or chopped meat-glucose may be used as an alternative to THIO. See CMPH 2.3.
4. Blood culture media (optional), i.e., BACTEC (B-D Microbiology Instrument Systems, Sparks, Md.) or BacT/Alert Media (Organon Teknika, Durham, N.C.) with added supplements as recommended by the manufacturers

B. Reagents

1. Gram stain reagents
2. Refer to the specific microorganism identification procedure for the reagents needed for identification confirmation (*see* CMPH 1.19 and 1.20).

C. Supplies

1. Single-ring slide
2. Sterile pipettes
3. Sterile tubes

IV. PROCESSING PROCEDURE

A. Blood culture bottles received for culture

1. Label bottles.
2. Incubate bottles based on specific procedure.
3. See EPCM 2.3 and CMPH 1.7 for additional information.

B. Fluid received in sterile tube

1. Aliquot the appropriate amount of fluid for other requests. If the specimen is scant, notify the physician and request prioritization of test requests.

IV. PROCESSING PROCEDURE *(continued)*

2. Less than 1 ml of fluid

a. Place ≤1 drop (depending on the amount of specimen) onto a ring slide for Gram stain.

b. Inoculate 1 drop to CHOC, BAP, and THIO.

c. If the specimen is very scant

(1) Centrifuge/vortex (quick spin) to concentrate all fluid at the bottom of the tube.

(2) Prepare the slide.

(3) Inoculate the CHOC only.

(4) Inoculate the THIO by rinsing the tube or syringe with a small amount of THIO. Consult supervisor or specialist as needed.

NOTE: Centrifugation at 3,000 × g for 20 min is not recommended because of the small volume of fluid; use entire volume during processing.

Table 2.4–1 Media selection

Volume	Centrifuge	Smear	Media	Instructions
Very scant	Yes, quick spin	≤1/2 drop	CHOC, THIO	1. Place ≤1/2 drop onto a slide. 2. Inoculate CHOC. 3. Inoculate THIO by rinsing tube or syringe with small amount of THIO.
1 ml	No	1 drop direct	CHOC, BAP, THIO	1. Place 1 drop onto a slide for staining. 2. Inoculate 1 drop onto each plate. 3. Inoculate 2–3 drops to THIO, depending on specimen volume. 4. Save remaining fluid.
<4 ml	Yes	1 drop sediment	CHOC, BAP, THIO	1. Centrifuge. 2. Decant supernatant into sterile tube. 3. Leave 0.5 ml to suspend sediment and vortex. 4. Place 1 drop onto a slide for staining. 5. Inoculate 1 drop onto plates. 6. Inoculate the remaining sediment into the THIO. 7. Save the supernatant tube with the sediment tube.
4–10 ml	Yes	1 drop sediment	CHOC, BAP, THIO, BACTEC 26 (aer)	1. Centrifuge. 2. Decant supernatant into sterile tube. 3. Leave 0.5 ml to suspend sediment. 4. Place 1 drop onto a slide for staining. 5. Inoculate 1 drop onto each plate. 6. Add supernatant back to the sediment, and vortex. 7. Save an aliquot (0.5–1 ml) and inoculate the remainder into a blood culture bottle.
>10 ml	Yes	1 drop sediment	CHOC, BAP, blood culture, media-aerobic, anaerobic	1. Centrifuge entire volume (use 50-ml conical tube as needed). 2. Decant supernatant into sterile tube. 3. Leave 0.5 ml to suspend sediment. 4. Place 1 drop onto a slide for staining. 5. Inoculate 1 drop onto each plate. 6. Add supernatant back to the sediment, and vortex. 7. Save an aliquot (0.5–1 ml) and inoculate an equal volume (up to 10 ml) into blood culture media. Add supplement if indicated.

3. Greater than 1 ml of fluid
   a. Centrifuge the specimen for 20 min at 3,000 × *g* (3,750 rpm).
   b. Decant supernatant into a sterile tube, leave approximately 0.5 ml, and resuspend the pellet by vortexing for 5 s.
   c. Place a drop of sediment onto a ring slide for Gram stain. If the specimen is viscous, bloody, or has pus, spread the specimen on the slide to ensure that the Gram stain is readable.
   d. See Table 2.4–1 for media selection.
   e. Hold any remaining fluid in the designated rack in the refrigerator for 14 days.

C. For an anaerobic request, include anaerobic media in the primary media, if the specimen has been transported anaerobically.

D. Media are selected, labeled, inoculated, and incubated at 35°C in 5 to 10% CO_2.

E. Prepare a smear, air dry, methanol fix, Gram stain, and interpret.

V. CULTURE EVALUATION

A. Examine aerobic plates after 24-h incubation, reincubate, and reexamine after 48 and 72 h.

B. Correlate varieties and quantities of colonial morphotypes with the direct Gram stain results.

C. Broth cultures

1. Visually inspect THIO daily for 7 days.
   a. Smear all positive tubes.
   b. Correlate the culture result with the Gram stain of the THIO and with the direct Gram stain result. Do not subculture the THIO if the smear correlates with the growth on the plates.
2. Blood culture bottles
   a. Smear all positive bottles.
   b. If direct plates were not inoculated, select and inoculate media based on the smear result.
   c. If direct plates were inoculated, correlate the culture result with the Gram stain of the bottle and with the direct Gram stain result. Do not subculture the bottles if the smear correlates with the growth on the plates.

D. For culture positive with one to two species of microorganisms

1. Semiquantitate the growth of each microorganism when solid media have been inoculated. Record the approximate distribution of isolates by morphotype.
2. Establish a definitive identification (ID) and perform antimicrobial susceptibility testing (AST).

E. For culture positive with greater than three microorganisms

1. Quantitate the growth of each organism when solid media have been inoculated.
2. Definitively identify (ID) and perform antimicrobial susceptibility testing (AST) if one or two microorganisms are predominant.
3. Descriptively identify if no microorganism is predominant.
4. Consult the physician for further workup.

VI. RESULTS

A. Negative cultures

1. Report "No growth after 7 days."
2. If the specimen was transported anaerobically, anaerobes were requested, and anaerobic solid media were inoculated, report "No aerobic or anaerobic growth after 7 days."

VI. RESULTS *(continued)*

B. Positive cultures

1. Report ID and AST if a complete workup was performed.
2. Report descriptive ID on cultures with multiple (more than three) organisms isolated.
3. If an anaerobic workup was performed and no anaerobes were isolated, report "No anaerobes isolated."
4. Save representative plates for 5 days if the microorganism(s) has not been worked up completely.

SUPPLEMENTAL READING

Baron, E. J., M. A. Pfaller, F. C. Tenover, R. H. Yolken, and P. R. Murray (ed.). 1995. *Manual of Clinical Microbiology,* 6th ed. American Society for Microbiology, Washington, D.C.

Bobadilla, M., J. Sifuentes, and G. Garcia-Tsao. 1989. Improved method for the bacteriological diagnosis of spontaneous bacterial peritonitis. *J. Clin. Microbiol.* **27:**2145–2147.

Glenn, S., and S. Vincent. 1992. Processing and interpretation of sterile body fluids (excluding blood, cerebrospinal fluid, dialysate, and urine), p. 1.8.1–1.8.8. *In* H. D. Isenberg (ed.), *Clinical Microbiology Procedures Handbook,* vol. 1. American Society for Microbiology, Washington, D.C.

Reinhold, C. E., D. J. Nickolai, T. E. Piccinini, B. A. Byford, M. K. York, and G. F. Brooks. 1988. Evaluation of broth media for routine culture of cerebrospinal and joint fluid specimens. *Am. J. Clin. Pathol.* **89:**671–674.

Runyon, B. A., E. T. Umland, and T. Merlin. 1987. Inoculation of blood culture bottles with ascitic fluid: improved detection of spontaneous bacterial peritonitis. 1987. *Arch. Intern. Med.* **147:** 73–75.

Von Essen, R., and A. Holtta. 1986. Improved method of isolating bacteria from joint fluids by the use of blood culture bottles. *Ann. Rheum. Dis.* **45:**454–457.

2.5 Processing and Interpretation of Cerebrospinal Fluid

I. PRINCIPLE

Bacterial meningitis is the result of infection of the meninges. Identification of the infecting agents is one of the most important functions of the diagnostic microbiology laboratory because acute meningitis is a most serious infection. CSF from a patient suspected of having meningitis is an emergency specimen that requires immediate processing to determine the etiologic agent.

Aerobic bacteria commonly cause bacterial meningitis; however, anaerobes may be present in CSF only when meningeal abscess or another similar infectious process occurs adjacent to the meninges. Inoculation of anaerobic media is not routinely recommended. Anaerobes must be ruled out in the presence of traumatic head injury or prostheses, such as metal cranial plates and drains.

CSF may contain very few microorganisms per milliliter of fluid; therefore, concentration of the specimen is recommended. CSF is obtained by transcutaneous aspiration, thus all organisms recovered from the culture are potential pathogens. Correlation with Gram stain findings is important. Any positive finding on Gram stain or culture must be reported to the physician immediately.

II. SPECIMENS

A. Specimen collection

1. CSF obtained by lumbar puncture (for details, *see* CMPH 1.1.2)
   a. CSF specimens must be collected prior to antimicrobial therapy.
   b. Place CSF into sterile leak-proof tubes available in lumbar puncture trays.
   c. Submit at least three separate tubes for CSF analysis in chemistry, hematology, and microbiology.
   d. Submit tube 2 or the most turbid tube to the microbiology laboratory.
   e. Submit a sufficient volume of fluid. Suggested volumes are
      (1) 1 ml for bacterial culture
      (2) 2 ml for fungal culture
      (3) 2 ml for mycobacterial culture
2. CSF collected from Ommaya reservoir
   a. Disinfect reservoir collection site before collection of CSF.
   b. Submit specimen in sterile tube with appropriate volumes as described in item A.1.b–e above.
3. Material from brain abscess
   a. Aspirate material as described in CMPH 1.1.2.
   b. Place aspirated material in anaerobic container. Alternatively, transport material immediately to the laboratory in the collection syringe after the needle has been removed by the physician collecting the specimen. (For details, *see* CMPH 1.1.2.).
   c. Refer to CMPH 2.2 and 2.3 for procedures related to collection, transport, and processing of specimens for anaerobic culture.

B. Specimen volume

The sensitivity of laboratory methods for detecting infectious agents in CSF is dependent on the volume of fluid received.

1. The chance of organism recovery increases with the volume of the specimen.
2. Call the physician to prioritize the tests to be performed if multiple tests have been ordered and an insufficient volume of CSF has been submitted.

II. SPECIMENS *(continued)*

C. Specimen transport

1. Transport CSF specimens to the laboratory immediately.
2. Do not refrigerate CSF specimens unless viral studies alone are requested.
3. Only CSF for viral studies should be refrigerated or frozen. (For details, *see* CMPH section 8.)

D. Specimen handling

1. Handle a CSF specimen as a stat specimen.
2. Perform all the initial processing of a CSF specimen in a biological safety cabinet to avoid contamination of the specimen and/or the primary inoculation medium.
3. For detailed information regarding initial processing, inoculation, and incubation of CSF specimens, see CMPH 1.4 and EPCM section 1.

III. MEDIA, REAGENTS, AND SUPPLIES

Include QC information on reagent container and in QC records.

A. Media

1. BAP
2. CHOC
3. Broth medium, e.g., THIO without indicator or TSB

NOTE: There are very few organisms that commonly cause bacterial meningitis (Table 2.5–1). For this reason, few media are routinely required for primary inoculation of CSF. If viruses or fungi are suspected, include appropriate media for isolation as described in CMPH sections 6 and 8.

B. Reagents

1. Gram stain

   For detailed reagents, supplies, and staining procedure, see CMPH 1.5.
2. Bacterial antigen test
   a. For detailed reagents and supplies, see CMPH 9.2.
   b. Refer to the appendix of this procedure for a general bacterial antigen protocol.

C. Supplies

1. Centrifuge (at least 3000 × *g*)
2. Vortex mixer
3. Sterile screw-cap tubes
4. Alcohol-soaked and air-dried glass slides
5. Sterile Pasteur pipettes
6. Inoculating loop

IV. PROCEDURE

A. Initial processing of CSF (Fig. 2.5–1)

1. Record volume of CSF.
2. Record gross appearance of CSF, i.e., clear, bloody, cloudy, xanthochromic.
3. Centrifuge the specimen if >1 ml is received.
   a. If the volume of fluid is >1 ml, centrifuge for 20 min at 1,500 to 3,000 × *g*. The higher speed is preferred to sediment the bacteria.
   b. If ≤1 ml is received, vortex.
4. Allow the centrifuge to stop. *Do not brake.*

Table 2.5–1 Common organisms causing acute meningitis[a]

Patient group (age)	Organisms causing acute meningitis
Neonates	*Escherichia coli, Streptococcus agalactiae* (group B streptococci), *Listeria monocytogenes*, herpes simplex 2 virus
Infants (<2 mo)	*S. agalactiae, L. monocytogenes, E. coli*
Children (2 mo–10 yr)	Viruses, *Haemophilus influenzae, Streptococcus pneumoniae, Neisseria meningitidis*
Young adults (>10–18 yr)	Viruses, *N. meningitidis*
Adults (>18–70 yr)	*S. pneumoniae, N. meningitidis*
Elderly (>70 yr)	*S. pneumoniae*, gram-negative bacilli, *L. monocytogenes*
AIDS patients	*Cryptococcus* neoformans

[a] Adapted from CMPH 1.9.2.

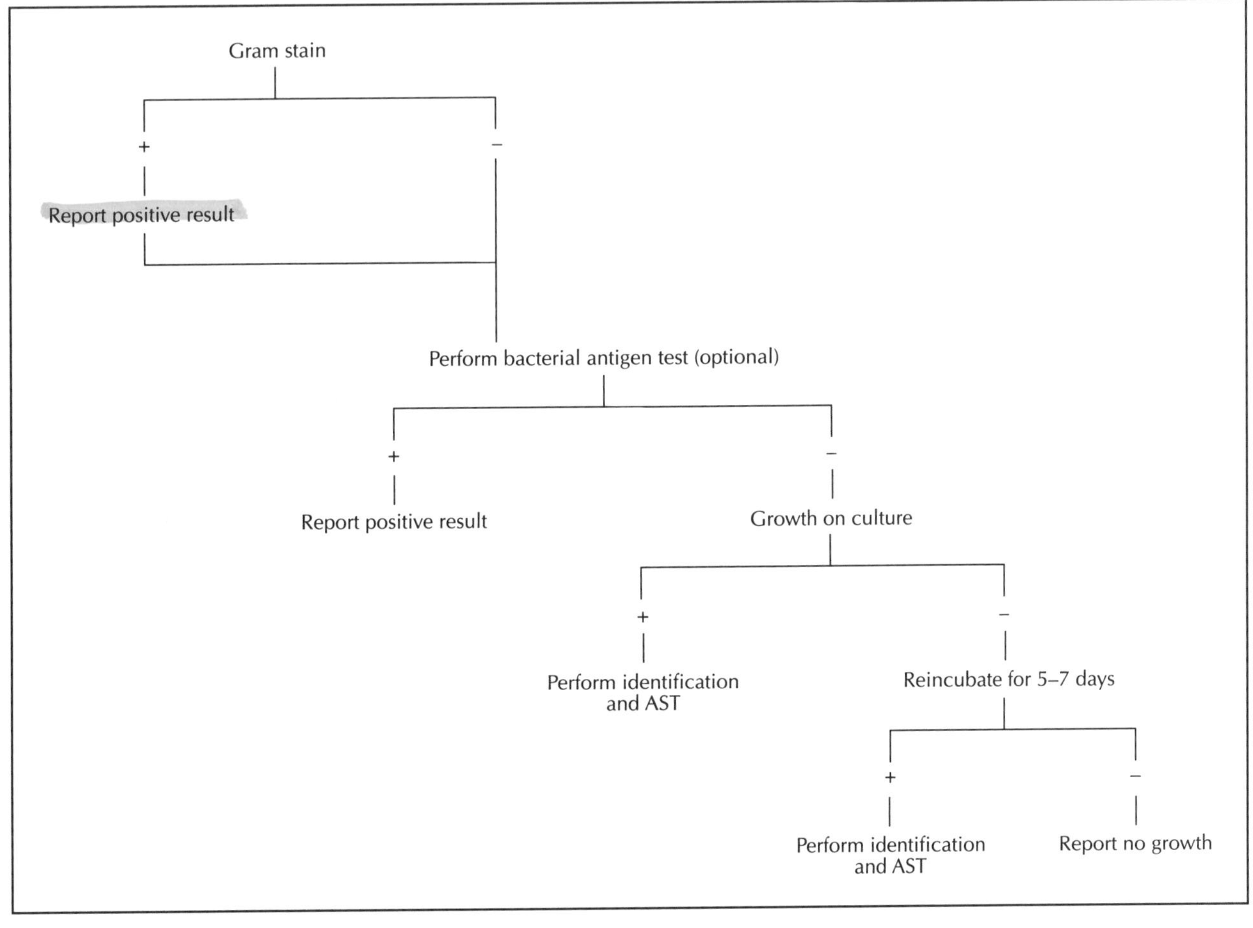

Figure 2.5.–1 Processing of CSF for common bacterial isolates. AST, Antimicrobial susceptibility testing. Adapted from CMPH 1.9.3.

5. Aspirate the supernatant with a sterile pipette, leaving approximately 0.5 to 1.0 ml of fluid in the specimen tube. Reserve the supernatant for additional studies. Hold for 1 week.
6. Vortex the sediment vigorously for at least 30 s to resuspend the pellet. This step is critical. Do not use a pipette to mix the sediment because the bacteria and cells may adhere to the sides of the tube, resulting in false-negative findings.
7. Using a sterile pipette, inoculate media by placing 1 or 2 drops of the vortexed sediment onto each plate and into the broth.
8. Prepare the smear by placing 1 or 2 drops of sediment on an alcohol-rinsed slide, allowing the drop(s) to form a large heap. Do not spread the fluid.
9. Air dry the slide in a biosafety cabinet. Alternatively, allow the slide to dry on a slide warmer.
10. Methanol fix and Gram stain the slide.
11. An alternate smear preparation method is to cytospin.
    a. Place $\geq$0.2 ml uncentrifuged CSF into a slide centrifuge (cytospin).
    b. Methanol fix and Gram stain the preparation.

B. Incubation

1. Incubate inoculated plated media at 35°C in 5 to 10% CO_2 for 48 to 72 h.
2. Alternatively, a candle jar or CO_2-generating Bio-Bag will provide the proper atmosphere if a CO_2 incubator is unavailable.
3. Incubate broth media in CO_2 or aerobically at 35°C.

IV. PROCEDURE *(continued)*

C. Microscopic examination

1. Interpret all CSF Gram stains immediately.
2. Report all positive CSF Gram stains immediately to the physician by telephone and to the nursing unit.
3. Document the telephone notification.

D. Bacterial antigen

1. See Appendix 2.5–1 for general procedure.
2. Refer to Table 2.5–2 for testing guidelines.

E. Culture examination

1. Examine all plated and broth media for macroscopic evidence of growth.
2. If no visible growth is observed on the culture media, reincubate.
   a. Examine negative plates daily for 72 h before discarding.
   b. Examine broth media daily for 5 to 7 days before discarding.
3. Culture with growth
   a. Gram stain the broth, if it is positive.
   b. Semiquantitate growth on plated media.
   c. Gram stain each morphotype.
   d. Notify physician of culture findings.
   e. Identify all microorganisms.
4. Organism identification
   a. Perform a definitive identification on all CSF isolates.
   b. For detailed identification procedures, see CMPH 1.19 and 1.20.
   c. If a microorganism is determined by the patient's physician to be a probable contaminant, a complete identification may not be required.
   d. Isolates of coagulase-negative staphylococci and corynebacteria may cause infection in patients with CNS shunts or who are otherwise compromised with head injuries.
5. Antimicrobial susceptibility testing
   a. For detailed procedures, refer to CMPH section 5 or EPCM section 5.
   b. Perform beta-lactamase tests (cefinase) on isolates of *Haemophilus influenzae* and *Neisseria* spp. For detailed procedure, see CMPH 5.3 or EPCM section 5.
   c. Perform antimicrobial susceptibility testing on other isolates as described in CMPH section 5 or EPCM section 5.

F. Reporting results

1. Establish a protocol for immediate, preliminary, and final reporting of CSF microscopic and culture results.
2. Telephone all positive results.

Table 2.5–2 Guidelines for bacterial antigen testing

Patient group	Gram stain result	Antigen tested
Neonates (<6 wk)	Gram-positive cocci (pairs and chains)	Group B streptococcus
	Gram-negative bacilli	*E. coli* K1 (react with *N. meningitidis* serogroup B)
All other patients	Gram-negative bacilli (small)	*H. influenzae*
	Gram-negative diplococci	*N. meningitidis*
	Gram-positive cocci (pairs and chains)	*S. pneumoniae*
Immunocompromised patients	Yeast cells	*C. neoformans*

V. PROCEDURE NOTES

A. Refer to appropriate sections of this manual or to the following sections of CMPH for details on isolation and identification of other microorganisms causing meningitis: fungi (CMPH section 6), mycobacteria (CMPH section 3), parasites (CMPH section 7), and viruses (CMPH section 8).

B. Rapid bacterial antigen tests of CSF may be helpful in partially treated patients to distinguish between bacterial and nonbacterial meningitis. Some laboratories use the test routinely, especially in infants and children. A general procedure has been included as an appendix to this procedure.

VI. QUALITY CONTROL

A. To check for dead microorganisms in the collection tubes or on the slides used for staining procedures, filter sterilize (through a 0.22-μm-pore-size membrane filter) CSF from saved supernatants, and periodically process along with patient samples.

B. No microorganisms should be detected on these slides, and cultures from these slides should yield no growth.

SUPPLEMENTAL READING

Baron, E. J., and S. M. Finegold. 1994. Microorganisms encountered in the cerebrospinal fluid, p. 210–218. *In* E. J. Baron and S. M. Finegold (ed.), *Bailey and Scott's Diagnostic Microbiology,* 9th ed. The C. V. Mosby Co., St. Louis.

Daly, J., K. C. Seskin, and M. Pezzlo. 1992. Processing and interpretation of cerebrospinal fluid, p. 1.9.1–1.9.5. *In* H. D. Isenberg (ed.), *Clinical Mircrobiology Procedures Handbook,* vol. 1. American Society for Microbiology, Washington, D.C.

Forbes, B. A., and P. A. Granato. 1995. Processing specimens for bacteria, p. 265–281. *In* P. R. Murray, E. J. Baron, M. A. Pfaller, F. C. Tenover, and R. H. Yolken (ed.), *Manual of Clinical Microbiology,* 6th ed. American Society for Microbiology, Washington, D.C.

Ray, C. G., J. A. Smith, B. L. Wasilauskas, and R. J. Zabransky. 1993. *Cumitech 14A. Laboratory Diagnosis of Central Nervous System Infections.* Coordinating ed., A. J. Smith. American Society for Microbiology, Washington, D.C.

APPENDIX 2.5–1

Rapid Immunologic Screen for Antigens Associated with Bacterial Meningitis

I. PRINCIPLE

Bacterial meningitis is an acute infection of the meninges characterized by the presence of inflammatory cells (WBCs) and bacteria. Clinical diagnosis requires the detection of these microorganisms as rapidly as possible. Use of a bacterial meningitis panel latex agglutination test provides rapid detection and also allows for detection of low-level antigen even in those cases in which prior antimicrobial therapy has been administered.

The common bacterial microorganisms causing meningitis carry specific polysaccharide surface antigens that can be detected by "antibody-coated" latex "particles" agglutination. The most common agents associated with bacterial meningitis include *Haemophilus influenzae* type b (patients 2 months to 10 years old), *Neisseria meningitidis* (patients >5 years old), *Streptococcus pneumoniae* (patients >18 years old), and *Streptococcus agalactiae* group B streptococci (neonatal patients), although both *N. meningitidis* and *S. pneumoniae* have been recovered from children <5 years old.

II. SPECIMEN PREPARATION

A. Depending upon the commercial kit used, it may be necessary to heat the CSF specimen in boiling water for 5 min.

B. Cool the specimen to room temperature, if heated.

C. Centrifuge at 3,000 × *g* for 10 min.

III. REAGENTS AND SUPPLIES

A. Commercial bacterial antigen kits

1. Bactigen (Wampole Laboratories)
2. Directogen (Hynson, Westcott and Dunning)
3. Wellcogen (Wellcome Diagnostics)

NOTE: For detailed information on commercial suppliers, *see* CMPH volume 2, Appendix 2, p. A.2.1.

B. Reagents

Each manufacturer provides reagents to detect bacterial antigens for:

1. *H. influenzae* type b

APPENDIX 2.5–1 *(continued)*

2. *N. meningitidis*, several serotypes
3. *S. pneumoniae*
4. *S. agalactiae*

C. Supplies
1. Boiling-water bath; heating block at 80°C
2. Centrifuge
3. Sterile pipettes

IV. PROCEDURE

A. General test procedure (varies with each kit)
1. Using a reaction card (slide), dispense latex reagents onto separate circles. Use all control suspensions of organisms included in the test kit before adding the clinical sample(s).
2. Dispense the prepared specimen next to the latex reagent.
3. Mix specimen and latex reagent.
4. Rotate the card (slide) as described by the manufacturer and observe for agglutination.
5. Read results macroscopically under a strong light.

B. General test interpretation
1. A positive test is indicated by the development of an agglutinated clumping of latex particles.
2. With some commercial kits, a positive result with *N. meningitidis/E. coli* K1 against a neonatal specimen suggests *E. coli* K1 infection; with older patients, *N. meningitidis* is more likely. In neonates <3 weeks old, the galactosyl moiety in the neonate's brain disappears and *E. coli* K1 no longer poses an additional problem.
3. A negative test is indicated when latex does not agglutinate and the appearance of the mixture remains unchanged after the designated reaction time. A negative test indicates the absence of a detectable antigen level.
4. In some cases, a noninterpretable test is indicated by visible agglutination in more than one of the reaction wells.

V. QUALITY CONTROL

A positive and negative control must be tested each time the test is performed. The control suspensions of organisms must be tested before testing the patients' specimens.

VI. LIMITATIONS OF THE TEST

A. *N. meningitidis* group B antigen is more difficult to detect, as well as being structurally and immunologically related to *E. coli* K1 antigen.

B. A positive result depends on the level of detectable antigen. This is a screening test and does not replace culture.

C. The sensitivity of the latex agglutination tests for the antigens ranges from approximately 65% for *S. pneumoniae* to 95% for *H. influenzae* type b (1).

NOTE: Since children are now immunized against *H. influenzae* type b, the incidence of meningitis caused by this bacterium has declined precipitously. Therefore, a positive result for this microorganism must be interpreted with great care.

D. Each laboratory must evaluate the sensitivity, specificity and predictive values for the protocol and the patient population.

Reference

1. **Ballard, T. L., M. H. Roe, R. C. Wheeler, J. K. Todd, and M. P. Goode.** 1987. Comparison of three latex agglutination kits and counterimmunoelectrophoresis for the detection of bacterial antigens in a pediatric population. *Pediatr. Infect. Dis. J.* **6:**630–634.

Supplemental Reading

Hoban, D. J., E. Witivicki, and G. W. Hammond. 1985. Bacterial antigen detection in cerebrospinal fluid of patients with meningitis. *Diagn. Microbiol. Infect. Dis.* **3:**373–379.

Tilton, R. C., F. Dias, and R. W. Ryan. 1984. Comparative evaluation of three commercial products and counterimmunoelectrophoresis for detection of antigens in cerebrospinal fluid. *J. Clin. Microbiol.* **20:**231–234.

2.6 Processing and Interpretation of Respiratory Tract Cultures

PART 1 Laboratory Diagnosis of Streptococcal Pharyngitis

I. PRINCIPLE

The most common cause of bacterial pharyngitis in North America is *Streptococcus pyogenes* (group A streptococci [GAS]). Routine laboratory diagnosis of bacterial pharyngitis should consist of procedures which are able to detect low numbers of GAS.

II. SPECIMEN

A. Throat swab of the tonsillar area and/or posterior pharynx

NOTE: Avoid the tongue and uvula.

B. For additional information on specimen collection, see CMPH 1.1.

III. MATERIALS

Include QC information on reagent container and in QC records.

A. Media

1. BAP
2. Selective BAP containing sulfamethoxazole and trimethoprim (SXT)

   NOTE: Selective BAP is an option. If screening for *S. pyogenes* only, it may be used alone. For details, see CMPH 1.14.1.

B. Reagents

1. 3% hydrogen peroxide for catalase test
2. L-Pyrrolidonyl β-naphthylamide hydrolysis (PYR)
3. Latex agglutination, using antibody-coated latex particles for antigen detection
4. Bacitracin susceptibility—0.04-U taxos disk
5. Refer to Table 2.6–1 for additional tests and reagents and to CMPH 1.19 and 1.20 for detailed testing methods.

IV. PROCEDURE

A. Direct antigen detection

1. Numerous commercial kits for the rapid detection of GAS directly from throat swabs are now available. Most procedures are based on extraction of group A antigen from the swab by enzymatic or chemical means followed by an antigen detection step using coagglutination, enzyme immunoassay, or latex particle agglutination. For performance advice, see CMPH 1.14.
2. Because a common problem is the low sensitivity of the current tests, all negative results should be followed by culture.

IV. PROCEDURE *(continued)*

B. Group A streptococci screening culture

1. Inoculate medium and incubate under the selected atmospheric condition at 35°C for 18 to 24 h.

   NOTE: An alternative approach is to train health care personnel to inoculate BAP immediately after collection and then transport to the laboratory for appropriate streaking. The inoculum should be spread over an area no larger than 1.5 to 2.0 cm.
2. Examine for the presence of beta-hemolytic colonies.
3. Reincubate negative cultures for an additional 18 to 24 h.

C. Identification of group A streptococci

1. Catalase negative
2. PYR positive

   NOTE: Beta-hemolytic *Enterococcus* spp., although rare, will give a positive PYR result.
3. Antigen detection (for details, *see* CMPH 1.14)
4. Bacitracin susceptibility—0.04-U taxos disk

D. Reporting results

1. *S. pyogenes* group A streptococci isolated.
2. Beta-hemolytic streptococci, not group A, isolated.
3. No *S. pyogenes* or beta-hemolytic streptococci isolated.

PART 2

Laboratory Diagnosis of Diphtheria

I. PRINCIPLE

Diphtheria is an acute infectious disease primarily of the upper respiratory tract but occasionally of the skin. It is caused by toxigenic strains of *Corynebacterium diphtheriae*. The disease is characterized by a classic pseudomembrane of the pharynx; however, this symptom may be lacking in mild cases. To confirm a clinical diagnosis of diphtheria, the strain isolated must be shown to produce toxin.

II. SPECIMEN

A. Respiratory: Include both throat and nasopharyngeal specimens.

B. Cutaneous: Include skin, throat, and nasopharyngeal specimens.

III. MATERIALS

Include QC information on reagent container and in QC records.

A. Media

For complete media selection, see CMPH 1.6.

1. Cystine tellurite blood agar (CTBA)
2. BAP
3. Loeffler agar slant (LAS)

B. Reagents

1. Gram stain reagents (for procedures, *see* EPCM 2.1)
2. Methylene blue stain
   - **a.** Dissolve 0.3 g of methylene blue in 30 ml of 95% ethanol.
   - **b.** Add 100 ml of 0.01% potassium hydroxide.
3. Conventional biochemical media (*see* CMPH 1.14 and 1.19)

IV. SPECIMEN PROCESSING

A. Inoculate CTBA, BAP, and LAS. Leave swab on the slant during incubation.

B. Incubate aerobically at 35°C.

C. LAS

1. Prepare a smear after 2 to 4 h of incubation and stain with Gram stain and Loeffler's methylene blue (LMB [optional]). A Gram stain will ascertain typical corynebacterial morphology. If an LMB stain is performed, observe for metachromatic granules.

   NOTE: LMB stain is not reliable when performed directly from clinical material since metachromatic granules can be observed in other bacteria.

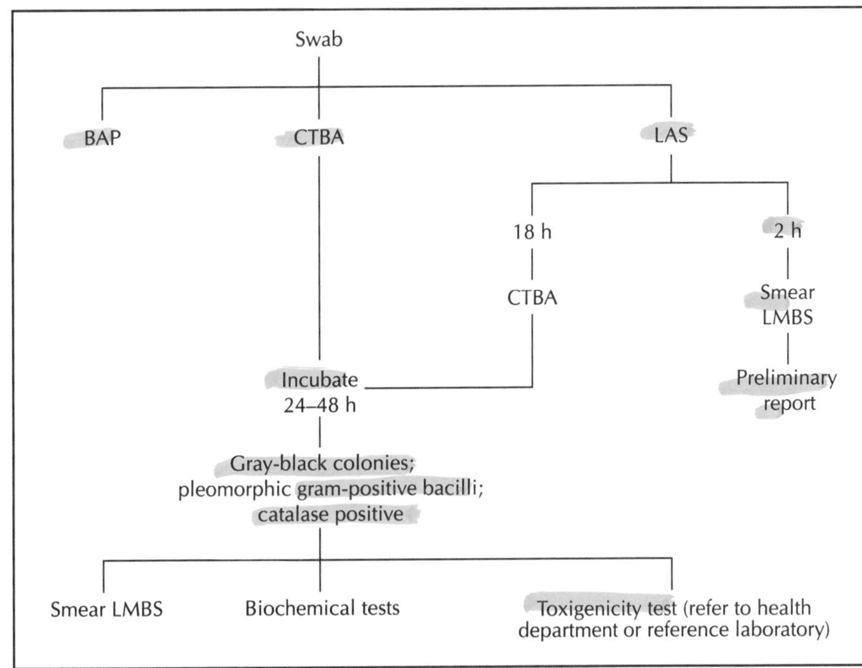

Figure 2.6–1 Evaluation of culture for *C. diptheriae.* CTBA, cystine-tellurite-blood agar; LAS, Loeffler agar slant; LMBS, Loeffler's methylene blue stain. Adapted from CMPH 1.14.16.

2. Provide a preliminary report.
3. Subculture to CTBA after 18 to 24 h of incubation.

D. CTBA

1. Examine at 24 and 48 h.
2. Gram stain suspicious black colonies and perform catalase test.

   ◪ **NOTE:** *C. diphtheriae* produces a garlic-like odor on CTBA. Also other *Corynebacterium* spp., staphylococci, and some streptococci can also reduce tellurite to tellurium, therefore producing black colonies on CTBA.
3. Subculture suspicious colonies to BAP for identification.

E. For processing and culture evaluation, see Fig. 2.6–1.

V. IDENTIFICATION OF *C. DIPHTHERIAE* (Fig. 2.6–1)

A. Cellular morphology

1. Pleomorphic gram-positive bacilli
2. Metachromatic granules present on LAS slant

B. Colonial morphology on CTBA

1. Grayish black (gunmetal gray) colonies approximately 1 to 3 mm in diameter
2. Black colonies with or without gray rims

C. Perform biochemical tests as needed to confirm identification. At minimum include glucose, sucrose, starch, glycogen, nitrate, and urea (for details, *see* CMPH 1.14).

D. All isolates of *C. diphtheriae* must be sent to a public health department for toxigenicity testing.

◪ **NOTE:** If reagents are available in the laboratory, a modified Elek test may be performed. The test requires skill and experience in interpretation of results, and for this reason, it is recommended that it be referred to a reference laboratory or public health laboratory.

E. The isolate must be toxigenic to confirm a clinical diagnosis of diphtheria.

PART 3

Processing Lower Respiratory Tract Specimens

I. PRINCIPLE

Although lower respiratory tract infections are a major cause of morbidity and mortality, diagnosis of these infections is often complicated by the contamination of specimens with upper respiratory tract secretions during collection. Because the upper respiratory tract may be colonized with potential pathogens not involved in infection of the lower respiratory tract, the laboratory must ensure that an appropriate specimen is processed. The specimen should be examined microscopically to assess its quality and to look for microorganisms associated with an inflammatory cell response.

II. SPECIMENS

A. Acceptable specimens

1. Sputum
2. Trachael and transtracheal aspirates
3. Bronchial washings, bronchial alveolar lavage, bronchial brushes, and bronchial biopsy
4. Lung aspirate and lung biopsy

B. Unacceptable specimens

1. Saliva submitted as sputum
2. Twenty-four-hour sputum collection
3. Swabs

III. MATERIALS

Include QC information on reagent container and in QC records.

A. Media

1. BAP
2. MAC or EMB
3. CHOC
4. Broth—BHI or THIO for aspirates and biopsy specimens obtained by invasive technique (transtracheal aspiration; for details, *see* CMPH 1.15)

B. Reagents

Refer to CMPH 1.19 and 1.20 for reagents.

IV. GENERAL CONSIDERATIONS

A. If there is a delay in processing, refrigerate specimens at 2 to 8°C.

NOTE: A delay in processing of more than 1 to 2 h may result in loss of recovery of fastidious pathogens and overgrowth of gram-negative bacilli.

B. All specimens should be processed in a biological safety cabinet, or a face mask or face shield must be worn when processing lower respiratory specimens.

C. Specimens collected by an invasive procedure should be handled as "precious" specimens and processed as rapidly as possible. For steps in processing bronchoscopy specimens, see Fig. 2.6–2 and –3.

NOTE: Any invasive procedure puts the patient at risk, and therefore every effort should be made to handle the specimen properly and evaluate it as completely as possible to avoid a repeat procedure.

D. Appropriate specimens with possible anaerobes that are collected and transported anaerobically should be cultured for anaerobes. These include specimens collected by invasive techniques, e.g., transtracheal aspiration, bronchial biopsy, and bronchial brushes.

NOTE: Protected bronchial brushes may be cultured for anaerobes when requested. Bronchial washings, lavages, and unprotected brushes are not acceptable specimens for anaerobic culture.

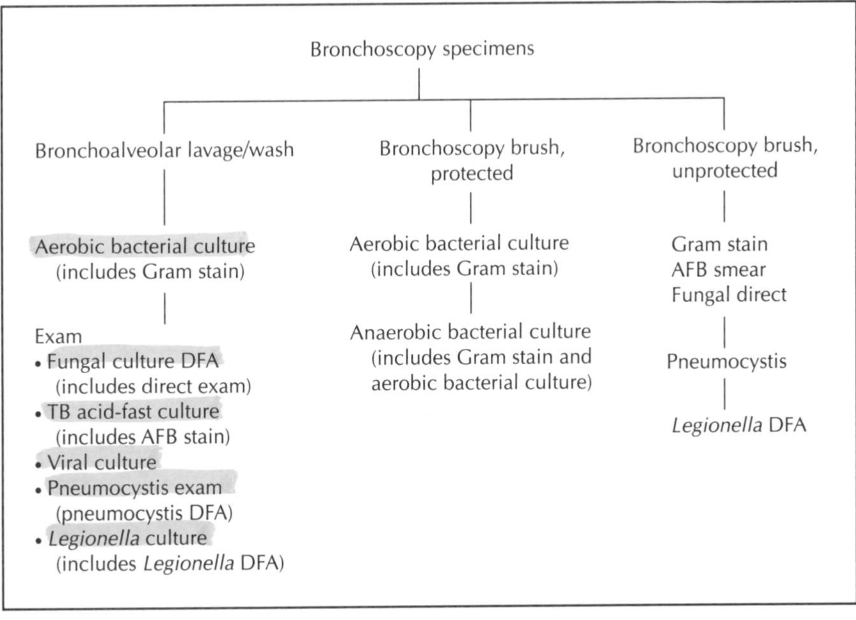

Figure 2.6–2 Acceptable requests for bronchoscopy specimens. DFA, Direct fluorescent-antibody assay. Not all *Legionella* spp. or variants can be detected by DFA.

E. Protected bronchoscopy brush sent for quantitative culture

1. Quantitative culture: Plate 10 μl.
Vortex the brush in 1 ml of BHI or steril saline (Fig. 2.6–3).

2. Colony count: number of colonies per plate × 100 (dilution factor)

Colonies on plate	Colony count (CFU/ml)
<10	10^3
10 to 100	10^3 to 10^4
100 to 1,000	10^4 to 10^5
>1,000	$>10^5$

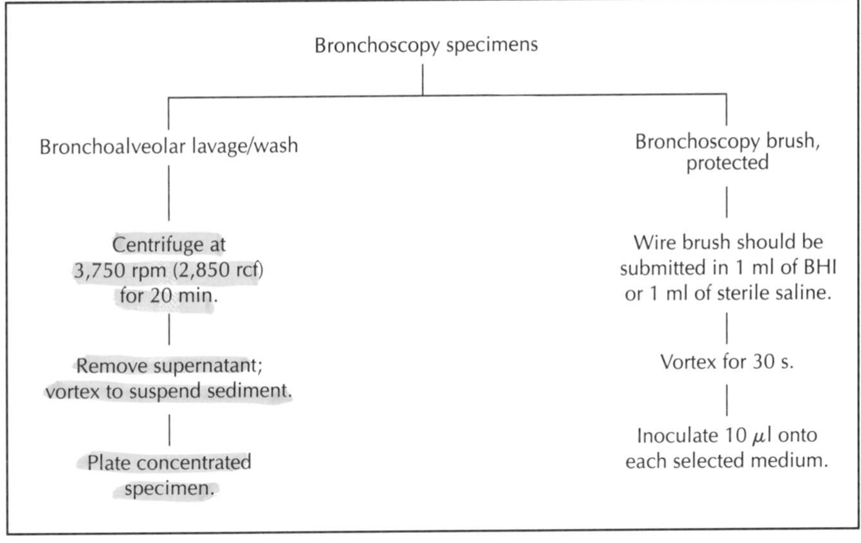

Figure 2.6–3 Processing bronchoscopy specimens.

V. MICROSCOPIC EXAMINATION

A. Smear preparation

1. Select purulent portions of the specimen.
2. Gently roll the material on the slide with a sterile swab.
3. For details, see EPCM 2.1.
4. Scan the entire slide under low power (10× objective) to detect the presence of fungal structures or any other unusual findings.

   ◩ **NOTE:** It is possible to observe faintly staining fungal structures, other than yeast cells, e.g., *Aspergillus* spp.

B. Microscopic screening (sputum specimens only)

1. Screen slide of sputum to assess specimen quality.
2. Examine the slide under low-power magnification (10× objective) for the presence of squamous epithelial cells (SECs) and PMNs.
3. Examine a minimum of 10 representative low-power (10× objective) fields.
   a. Concentrate on areas where WBCs are seen.
   b. Avoid areas with oropharyngeal contamination.
4. For criteria used to assess the quality of sputum specimens, see CMPH 1.3. In general, there should be a ratio of greater than 2:1 WBCs to SECs.
5. Unacceptable sputum specimens
   a. Include a comment indicating the presence of many SECs and few or no PMNs, which is consistent with oropharyngeal contamination (specimen contaminated with saliva).
   b. Notify the appropriate person and request a new specimen.

C. Interpretation of Gram stain

1. Examine the slide under oil immersion (100× objective).
2. Semiquantitate the cells (SECs and WBCs) as few, moderate, or many.
3. Semiquantitate the organisms as few, moderate, or many.
   a. Predominant organism
      (1) If one organism predominates in areas of inflammation, semiquantitate and report the microorganism.
      (2) If other types of bacteria are present, semiquantitate and add "and mixed bacterial morphotypes."
   b. No predominant microorganism: Semiquantitate and report "mixed bacterial morphotypes."
   c. No microorganisms seen: Report "no microorganisms seen."
   d. Fungal structures and yeast cells
      (1) Semiquantitate and report hyphal structures present. If possible, describe the appearance of the hyphae, e.g., segmented, pseudohyphae, etc.
      (2) Do not semiquantitate and report yeast separately unless it is a predominant microorganism or it is present in large amounts (many).

VI. CULTURE EVALUATION

A. Workup of probable pathogens (PP) from acceptable-quality sputum and other noninvasive specimens

PP may be defined as any microorganisms that have been associated with infections of the lower respiratory tract. These include both hospital-acquired pathogens, such as members of the family *Enterobacteriaceae, P. aeruginosa*, other gram-negative bacilli, and fungi, and non-hospital-acquired *S. pneumoniae* and *H. influenzae* to mention a few (Table 2.6–1).

1. No PP present: Report "Isolates are consistent with microorganisms encountered in the upper respiratory tract."
2. One or two PP present
   a. Semiquantitate, identify (ID), and perform antimicrobial susceptibility testing (AST) on the PP.
   b. Semiquantitate and report the amount of normal microbiota.

Table 2.6–1 Identification of respiratory pathogens and potential pathogens

Organism	Gram stain	Rapid test	Alternative test
S. aureus	Gram-positive cocci (clusters)	Slide coagulase positive	Tube coagulase positive
Coagulase-negative staphylococcus		Slide coagulase negative	Tube coagulase negative
S. pneumoniae	Gram-positive cocci (pairs/chains)	Bile solubility positive	Optochin sensitive
S. pyogenes		PYR positive	Streptococcus typing
Viridans streptococcal group		Bile solubility negative	
Enterococcus spp.		PYR positive	Bile esculin positive, 6.5% salt broth positive
H. influenzae	Small, gram-negative bacilli	PPT positive (porphyrin production)	X and V dependent
Moraxella catarrhalis	Gram-negative diplococci	Oxidase positive	Glucose alkaline Maltose alkaline Lactose alkaline Sucrose alkaline DNase positive
N. meningitidis	Gram-negative diplococci	Oxidase positive	Glucose acid Maltose acid Lactose alkaline Sucrose alkaline
Mycobacteria spp.	Gram-positive rods (beaded)	Acid-fast stain positive	
Legionella spp.		Direct detection: direct fluorescent antibody or DNA probe	Use BCYE agar; refer to CMPH 1.12 for detailed information
Nocardia spp.	Gram-positive rods (branching)	Modified acid-fast stain positive	
C. neoformans	Yeast cells	Urea positive	
Candida albicans	Germ tube positive		

3. More than two PP present
   a. Semiquantitate, ID, and perform AST on up to two predominant PP.
   b. Semiquantitate and report a descriptive ID of the PP if no PP are predominant.
   c. Quantitate and report the amount of normal microbiota.
4. More PP than normal microbiota
   a. Quantitate, identify (ID), and perform AST on the PP if growth is moderate to heavy.
   b. Report a descriptive ID of the PP if growth is light.
   c. Quantitate and report the amount of normal microbiota.
5. Pure culture of an organism that is part of the normal microbiota
   a. Any organism, bacteria or fungi, in pure culture from the lower respiratory tract
   b. Semiquantitate, identify, and report.
6. No organisms isolated: See EPCM sections 4, 6, and 8 and CMPH sections 3, 4, 6, and 8 for isolation of mycobacteria, actinomycetes, fungi, viruses, rickettsiae, chlamydiae, and mycoplasmas.

   ◪ **NOTES:** Ascertain that all appropriate media have been used, including media for the isolation of *Legionella* spp., *Haemophilus* spp., and other fastidious organisms. If not, request a repeat specimen. Repeatedly negative specimens in the presence of inflammatory cells require consultation with an infectious disease specialist or pulmonary specialist.

VI. CULTURE EVALUATION *(continued)*

B. Bronchial washes/lavage and protected bronchoscopy brush (Fig. 2.6–2 and –3)

1. If PP are present, identify all microorganisms (Table 2.6–1).

a. Semiquantitate and ID all PP regardless of amount of growth.

b. Perform AST on up to two PP if the microorganism is predominant or growth is moderate to heavy.

2. Normal oropharyngeal microbiota

a. Pure growth of a single microorganism: Identify.

b. Growth of multiple organism types: Report ''multiple microorganisms indicative of normal upper respiratory microbiota.''

C. Lung aspirates and biopsies

1. Quantitate and identify all colony types isolated.

2. One or two PP

a. Definitive ID and AST on PP

b. Descriptive ID of microorganisms considered normal microbiota

3. More than two PP

Descriptive ID of all microorganisms present

SUPPLEMENTAL READING

Bannatyne, R. M., C. Clausen, and L. R. McCarthy. 1979. *Cumitech 10, Laboratory Diagnosis of Upper Respiratory Tract Infections.* Coordinating ed., I. B. R. Duncan. American Society for Microbiology, Washington, D.C.

Bartlett, J. G., K. J. Ryan, T. F. Smith, and W. R. Wilson. 1987. *Cumitech 7A, Laboratory Diagnosis of Lower Respiratory Tract Infections.* Coordinating ed., J. A. Washington. American Society for Microbiology, Washington, D.C.

James, L., and J. E. Hoppe-Bauer. 1992. Processing and interpretation of lower respiratory tract specimens, p. 1.15. *In* H. D. Isenberg (ed.), *Clinical Microbiology Procedures Handbook,* vol. 1. American Society for Microbiology, Washington, D.C.

Sneed, J. O. 1992. Processing and interpretation of upper respiratory tract specimens, p. 1.14. *In* H. D. Isenberg (ed.), *Clinical Microbiology Procedures Handbook,* vol. 1. American Society for Microbiology, Washington, D.C.

2.7 Processing and Interpretation of Genital Cultures

I. PRINCIPLE

The normal human genital tract is lined with a mucosal layer composed of transitional columnar and squamous epithelial cells. Many indigenous microorganisms colonize these surfaces. The microorganisms colonizing the female genital tract have been studied extensively and include lactobacilli, *Corynebacterium* spp., *Gardnerella vaginalis*, coagulase-negative staphylococci, *Staphylococcus aureus*, *Streptococcus agalactiae*, *Enterococcus* spp., *Escherichia coli*, anaerobes, and yeasts. The male urethra normally contains relatively few microorganisms found on the skin, such as coagulase-negative staphylococci, micrococci, and corynebacteria.

Many female genital tract infections arise from endogenous microorganisms, the pathogenicity of which has been activated by host factors and other microorganisms. The role of various viruses, including herpes simplex virus (HSV), human papillomavirus (HPV), and human immunodeficiency virus (HIV), may also influence the receptivity of the host surface to microbial attack.

If specific pathogens such as *Neisseria gonorrhoeae* and *S. agalactiae* (group B streptococci [GBS]) are suspected, it is both efficacious and cost-effective to have screening culture methods available. If common sexually transmitted bacterial pathogens such as *Chlamydia trachomatis*, *G. vaginalis*, *N. gonorrhoeae*, *Haemophilus ducreyi*, *Treponema pallidum*, *Ureaplasma urealyticum*, and *Mycoplasma hominis* are suspected, special media and/or techniques are required for recovery of these microorganisms. If the laboratory is unable to isolate these pathogens, a reference laboratory must be employed.

II. SPECIMENS

A. General considerations

1. If the culture site allows, aspirated material is preferred.
2. Swabs should be submitted in a suitable transport medium such as Stuart or Amies medium. *Dry swabs are unacceptable.*
3. Specimens must be transported anaerobically if anaerobes are suspected. If an abscess is present, transport material anaerobically.

   NOTE: Do not process vaginal specimens for anaerobes.
4. An appropriate transport system should be used for the isolation of *N. gonorrhoeae*, such as a charcoal media. *Do not refrigerate.*
5. All specimens should be labeled appropriately and transported promptly to the laboratory (For details on various transport media, *see* EPCM section 1).

B. Appropriate specimens from the female and male genital tracts are listed by site in Tables 2.7–1 and –2, respectively.

Table 2.7–1 Female genital cultures: sites and appropriate specimens[a]

Site	Specimen
Amniotic fluid	Catheter aspirate, amniocentesis fluid, swab from cesarean section
Bartholin gland	Duct aspirate (syringe or swab)
Cervix	Swab of endocervical canal
Endometrium	Transcervical aspirate
Fallopian tubes	Abscess aspirate, culdocentesate swab
Genital ulcer	Swab of ulcer base, bubo aspirate
IUD	IUD, associated secretions and pus
Urethra	Discharge on swab
Vagina	Discharge on swab, swab from posterior vaginal vault or vaginal orifice
Vesicle	Vesicular material (syringe or swab)

[a] Reprinted from CMPH 1.11.5.

Table 2.7–2 Male genital specimens: sites and appropriate specimens[a]

Site	Specimen
Epididymis	Swab of urethra
Genital ulcer	Swab of ulcer base, biopsy sample or curettings of ulcer base, bubo aspirate
Penile discharge	Discharge on swab
Prostate	Expressed prostatic secretions
Urethra	Expressed exudate or swab from distal urethra
Vesicle	Vesicular material (syringe or swab)
Testicle	Abcess aspirate (syringe or swab)

[a] Reprinted from CMPH 1.11.5.

III. MATERIALS

Include QC information on reagent container and in QC records.

A. Media (For details, *see* CMPH 1.11.6, Table 6.)

1. BAP
2. CHOC with IsoVitaleX or other enriched CHOC
3. Modified Thayer-Martin (MTM) or other selective gonoccal agar
4. MAC or other differential gram-negative agar
5. Selective gram-positive agar
6. Enriched and selective anaerobic agar

B. Reagents (Fig. 2.7–1; for details, *see* CMPH 1.20.)

1. Gram-positive organisms
   - **a.** L-Pyrrolidonyl-β-naphthylamide
   - **b.** CAMP
   - **c.** Sodium hippurate
   - **d.** Catalase
   - **e.** Coagulase
2. Gram-negative organisms (Fig. 2.7–2 and –3; for details, *see* CMPH 1.19.)

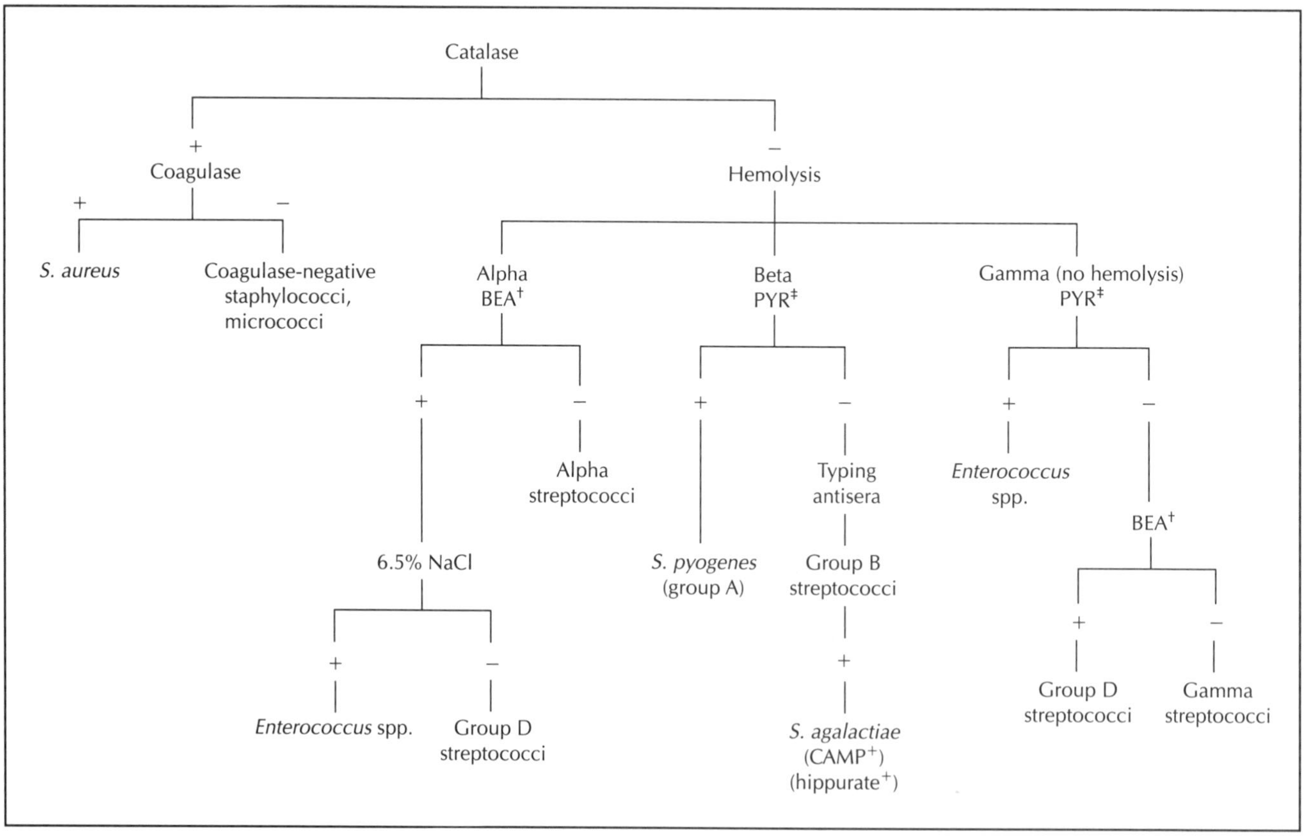

Figure 2.7–1 Identification of gram-positive cocci. † BEA, bile esculin agur; ‡ PYR, L-pyrrolidonyl-β-naphthylamide. Adapted from CMPH 1.11.8.

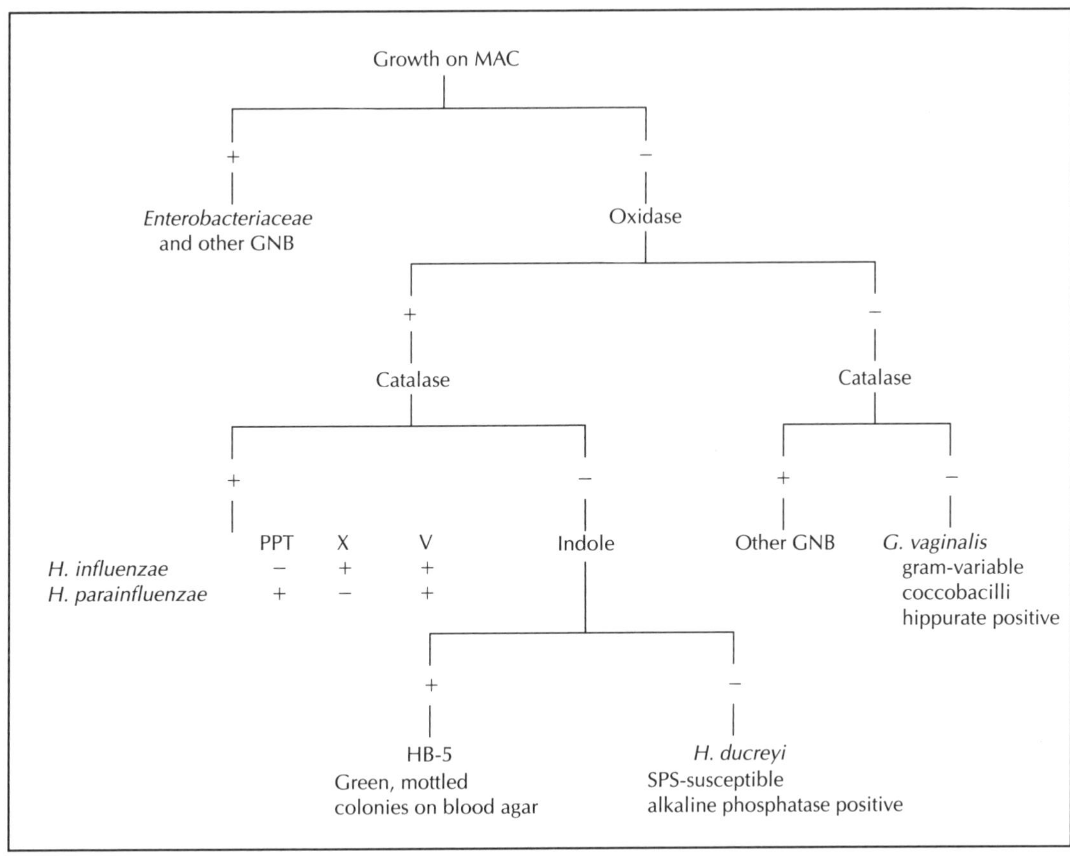

Figure 2.7–2 Identification of aerobic gram-negative bacilli. GNB, gram-negative bacilli. Adapted from CMPH 1.11.10.

IV. SPECIMEN PROCESSING

A. General guidelines

1. Proper specimen transport and processing, use of appropriate selective and enriched media, and appropriate environmental conditions will facilitate the recovery of fastidious genital tract pathogens.
2. Aerobic media should be incubated in ambient air at 35°C and CHOC should be placed in 5 to 10% CO_2 at 35°C.
3. Anaerobic media should be incubated under anaerobic conditions.
4. For organisms associated with various female and male infectious diseases, see Tables 2.7–3 and –4.

B. Gram stain

1. Prepare Gram-stained smears on all specimens as described previously (*see* EPCM 2.1).
2. The information provided by the direct smear may suggest the inclusion of special media for primary plating and/or provide useful interpretative information when growth is evaluated.

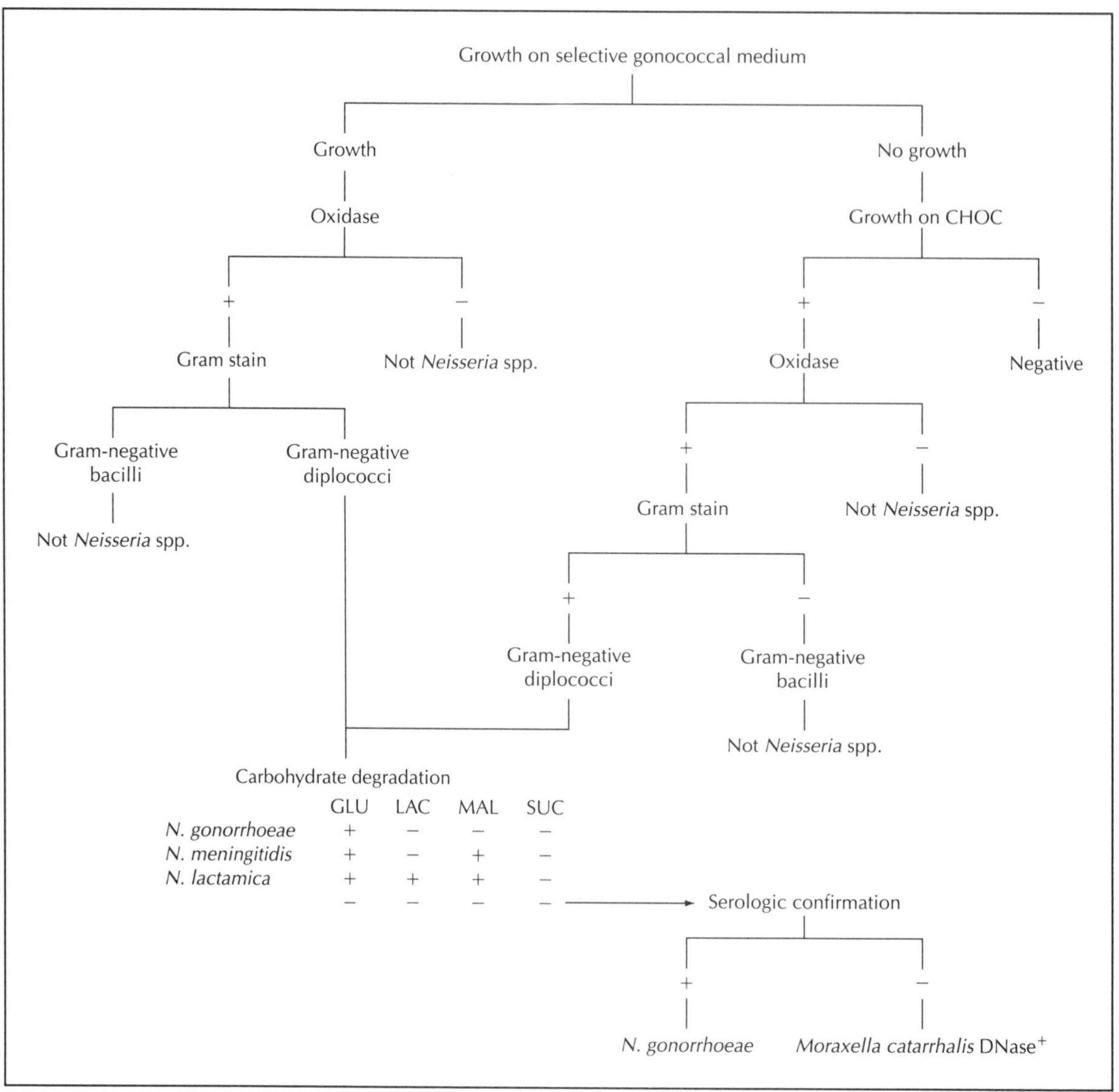

Figure 2.7–3 Identification of aerobic gram-negative cocci. *Kingella* spp., *Moraxella* spp., *Acinetobacter* spp., or *Capnocytophaga* spp. may grow on MTM medium. *N. gonorrhoeae* and *N. meningitidis* may confirm serologically. GC serology should be used to confirm *N. gonorrhoeae*. Adapted from CMPH 1.11.11.

V. PROCEDURE FOR ROUTINE CULTURE EVALUATION

A. General guidelines

1. Examine cultures for growth at 24 and 48 h.
2. MTM plates should be held for 72 h.
3. Extend the incubation of cultures with requests for fastidious microorganisms to allow for the recovery of the specific microorganism.
4. Correlate quantity and types of colonial morphotypes with direct smear results. Consider the presence of epithelial cells and or WBCs to determine the significance of growth.
5. Evaluate for the presence of pathogens. Perform identification tests (Fig. 2.7–1 to 2.7–3) for the following:
   a. *N. gonorrhoeae*
   b. *S. agalactiae*

Table 2.7–3 Female genital infectious disease syndromes and associated microbial pathogens[a]

Syndrome	Microbial pathogens[b]
Amnionitis	*Capnocytophaga* spp. *Escherichia coli* *Gardnerella vaginalis*[c] HB-5 *Haemophilus influenzae* *H. parainfluenzae* *Listeria monocytogenes* *Neisseria gonorrhoeae* **Streptococcus agalactiae* *S. pyogenes*
Bartholinitis	Anaerobes *Chlamydia trachomatis* **E. coli* HB-5 *H. influenzae* **N. gonorrhoeae* **Proteus mirabilis* *Staphylococcus aureus* Streptococci **Ureaplasma urealyticum*
Cervicitis	Actinomycetes[c] *Capnocytophaga* spp. **C. trachomatis* HB-5 *Herpes simplex virus *Mycobacterium tuberculosis* **N. gonorrhoeae* *S. agalactiae*
Chancroid	**H. ducreyi* *Calymmatobacterium granulomatis*
Endometritis	Actinomycetes[d] **Bacteroides* spp. **C. trachomatis* *Enterococcus* spp. *E. coli* *G. vaginalis* *H. influenzae* *Klebsiella* spp. *L. monocytogenes* *Prevotella bivia* *S. agalactiae* *S. pyogenes*
Postpartum endomyometritis	*Bacteroides fragilis* **B. bivius* **B. disiens* **Enterococcus* spp. **E. coli* **Klebsiella* spp. *Peptostreptococcus* spp. **S. agalactiae*

(continued)

Table 2.7–3 Female genital infectious disease syndromes and associated microbial pathogens[a] *(continued)*

Syndrome	Microbial pathogens[b]
Salpingitis	Actinomycetes[d]
	**Bacteroides* spp.
	**C. trachomatis*
	Enterococcus spp.
	E. coli
	HB-5
	H. influenzae
	**N. gonorrhoeae*
	Prevotella bivia
	Staphylococcus aureus
	S. epidermidis
	S. agalactiae
	Streptococci[c]
Skenitis	**N. gonorrhoeae*
Urethritis, urethral syndrome	**C. trachomatis*
	**E. coli*
	**N. gonorrhoeae*
Vulvovaginitis	Actinomycetes[d]
	Capnocytophaga spp.
	**Candida* spp.
	C. trachomatis
	Enterobacteriaceae
	HB-5
	Herpes simplex virus
	M. tuberculosis
	N. gonorrhoeae
	Salmonella spp.
	Shigella spp.
	Staphylococcus aureus[e]
	Streptococcus pyogenes
	**Trichomonas vaginalis*

[a] Reprinted from CMPH 1.11.2.
[b] *, major pathogens.
[c] Found in 50 to 80% of asymptomatic females.
[d] Associated with use of IUD.
[e] In wounds, toxic shock syndrome, and tampon-associated ulcerations.

V. PROCEDURE FOR ROUTINE CULTURE EVALUATION *(continued)*

c. *S. pyogenes*
d. *L. monocytogenes*
e. *H. ducreyi*

For details, see CMPH 1.11.

6. Evaluate for the presence of probable pathogens (PPs; Tables 2.7–3 and 2.7–4). If there is heavy growth, if isolated as the predominant microorganism, or if isolated in quantities greater than or equal to the normal genital microbiota, perform identification tests for the following PPs.

a. Gram-negative bacilli
b. *S. aureus*
c. *Streptococcus pneumoniae*
d. *Haemophilus* spp.
e. *N. meningitidis*
f. Anaerobes, including *Actinomyces* spp.

NOTE: *Actinomyces* should be selectively identified and reported in women with intrauterine devices (IUDs).

7. *Gardnerella vaginalis*

a. On routine genital cultures, laboratories should not selectively culture for *G. vaginalis* with human bilayer Tween agar or vaginalis agar.

Table 2.7–4 Male genital infectious disease syndromes and associated microbial pathogens[a]

Syndrome	Microbial pathogens[b]
Epididymitis	**Chlamydia trachomatis* *Enterobacteriaceae*[b] *Mycobacterium tuberculosis* **Neisseria gonorrhoeae* *Pseudomonas* spp.[c]
Orchitis	**Escherichia coli* **Klebsiella pneumoniae* **Pseudomonas aeruginosa* Staphylococci Streptococci
Prostatic abscess	**Staphylococcus aureus*
Prostatitis	**Escherichia coli* **Enterobacter* spp. *Enterococcus* spp. **Klebsiella* spp. *N. gonorrhoeae* **Proteus* spp. **Pseudomonas* spp. *Serratia* spp.
Urethritis	**C. trachomatis* *Haemophilus influenzae* *Haemophilus parainfluenzae* HB-5 **N. gonorrhoeae* *Unreaplasma urealyticum*

[a] Reprinted from CMPH 1.11.4.
[b] *, major pathogens.
[c] Most commonly seen in men over 35 years of age.

b. *G. vaginalis* grows well on both colistin-nalidixic agar (CNA) and CHOC.
- (1) When *G. vaginalis* is isolated in quantities less than the mixed bacterial morphotypes, it should be included as part of the normal vaginal microbiota.
- (2) If *G. vaginalis* is the predominant microorganism isolated, it should be identified and reported.

c. The presence of *G. vaginalis* (even in mixed culture) in children should be reported because it may indicate sexual abuse.

d. Examine the direct smear from the clinical specimen for the presence of clue cells, a finding suggestive of *G. vaginalis*.

e. The physician may request a KOH preparation if *G. vaginalis* is suspected.

8. Semiquantitate growth (few, moderate, many) and provide a descriptive identification when multiple PPs are isolated and none are predominant.

9. Report the presence of indigenous normal genital microbiota.
- **a.** In other than vaginal or cervical/endocervical sites, it is appropriate to give a descriptive ID of each isolate.
- **b.** In vaginal, cervical, or endocervix cultures, the mixed indigenous microbiota may be reported as ''normal vaginal microbiota.''

10. Yeasts
- **a.** The presence of yeasts, even in mixed culture, may be significant.
- **b.** Diagnosis of vulvovaginal candidasis is made from clinical findings, direct fungal exam, and/or culture results.

V. PROCEDURE FOR ROUTINE CULTURE EVALUATION *(continued)*

c. Semiquantitate and identify yeasts if there is heavy growth or yeasts are isolated as the predominant organism.

d. If there is light growth, semiquantitate and report "yeast isolated."

B. Identification methods

For details, see CMPH 1.19 and 1.20. Flow charts of the processes for identification of organisms by Gram reaction and morphology are provided in Fig. 2.7–1 through 2.7–3.

VI. SCREENING CULTURES

A. *N. gonorrhoeae*

1. Specimen
   a. Genital
   b. Anorectal
   c. Oropharyngeal
2. Materials

   Use of a self-contained nutritive system consisting of selective gonococcal medium that is held and transported in an environment containing CO_2 is appropriate and cost-effective, e.g., Transgrow.
3. Procedure
   a. Inoculate medium and activate CO_2 generation system as needed.

      ◩ **NOTE:** Transgrow must be inoculated in an upright position to prevent loss of CO_2.

   b. Incubate at 35°C and examine daily for 72 h.
4. Identification
   a. Gram-negative diplococci
   b. Small, translucent, grayish, convex colonies
   c. Oxidase positive
   d. Catalase positive
   e. Confirmatory tests (for details, *see* CMPH 1.19)
      (1) Carbohydrate degradation: glucose, maltose, sucrose, and lactose
      (2) Commercial identification systems

      ◩ **NOTES:** Some strains of *Neisseria cinerea* may give a weak reaction in glucose in some rapid tests. Identification of *N. gonorrhoeae* by using oxidase, Gram stain morphology, growth on a selective medium, and glucose fermentation is sufficient.

B. *S. agalactiae* (group B streptococci)

1. Specimen (Cervical cultures are not acceptable.)
   a. Genital
   b. Anorectal
2. Materials
   a. BAP
   b. Todd-Hewitt broth supplemented with either colistin (10 g/ml) and nalidixic acid (15 g/ml) or gentamicin (8 g/ml) and naladixic acid (15 g/ml).
3. Procedure
   a. Inoculate both swabs together into the selective broth medium.
   b. Incubate at 35°C for 18 to 24 h.
   c. Subculture broth to BAP and incubate at 35°C for 18 to 24 h in 5% CO_2.
   d. Reincubate negative plates for an additional 24 h.
4. Identification
   a. Beta-hemolytic or nonhemolytic
   b. Gram-positive cocci
   c. Catalase negative
   d. Presumptive identification: CAMP positive
   e. Confirmatory identification
      (1) Agglutination tests
      (2) Antigen detection tests

SUPPLEMENTAL READING

Eschenbach, D., H. M. Pollack, and J. Schachter. 1983. *Cumitech 17, Laboratory Diagnosis of Female Genital Tract Infections.* Coordinating ed., S. J. Rubin. American Society for Microbiology, Washington, D.C.

Kellogg, D. S., Jr., K. K. Holmes, and G. A. Hill. 1976. *Cumitech 4, Laboratory diagnosis of gonorrhea.* Coordinating ed., S. Marcus and J. C. Sherris. American Society for Microbiology, Washington, D.C.

Lewis, B. 1992. Identification of aerobic bacteria from genital specimens, p. 1.11.1–1.11.22. *In* H. D. Isenberg (ed.), *Clinical Microbiology Procedures Handbook,* vol. 1. American Society for Microbiology, Washington, D.C.

Mazzulli, T., A. E. Simor, and D. E. Low. 1990. Reproductibility of interpretation of Gram-stained vaginal smears for the diagnosis of bacterial vaginosis. *J. Clin. Microbiol.* **28:**1506–1508.

Wentworth, B. B., and F. N. Judson. 1984. *Laboratory Methods for the Diagnosis of Sexually Transmitted Diseases.* American Public Health Association, Washington, D.C.

2.8 Processing and Interpretation of Bacterial Fecal Cultures

I. PRINCIPLE

Gastroenteritis can be caused by bacteria, parasites, or viruses. Physician input can help the laboratory determine which tests are appropriate for detecting the etiological agent. When only routine fecal culture is requested, however, the laboratory should search for the most common or the most easily detected bacterial agents of diarrhea, as outlined here. Although certain microorganisms, including *Pseudomonas aeruginosa, Candida* spp., *Staphylococcus aureus*, and *Clostridium difficile*, may be seen in fecal cultures from patients treated with broad-spectrum antibiotics, their role in disease is not always clear. Their presence may be reported, along with a statement indicating that the microorganism was the predominant microorganism recovered and that the number of expected enteric microorganisms was decreased.

II. SPECIMEN

(For rejection criteria, additional details, and materials, see CMPH 1.10.)

A. Fresh specimens in waxed cardboard or other similar clean container

1. Should be received within 30 min of passage to allow for isolation of *Shigella* spp., which are extremely fragile in feces. If *Shigella* spp. are suspected, a rectal swab, submitted in a broth medium as described below, is preferred.

   NOTE: The fatty acids in cotton are inhibitory to shigellae.

   Specimens received within 2 h are generally acceptable, but recovery of pathogens will be diminished.
2. At least 1 g of material or a walnut-sized portion should be submitted. The specimen should not be mixed with urine, but good material can be scooped out of semisolid to solid feces in urine, if necessary.
3. Routine bacterial fecal cultures should not be performed on patients hospitalized for >3 days.
4. Direct examination should be used to determine the presence of overt blood or mucus and the consistency or fluidity of the feces. These observations should be documented.
5. The presence of WBCs is one factor suggestive of invasive infection.
   a. A Gram or Wright's stain of a very thin smear of fecal matter on a glass slide can be used to determine the presence of WBCs. Predominance of one bacterial morphotype, presence of yeast cells, or absence of enteric-appearing gram-negative bacilli can also be detected.
   b. Mucus-appearing portions of feces can be mixed with a drop of Loeffler's methylene blue on a slide, coverslipped, and examined for a predominance of WBCs.
   c. A latex agglutination test for lactoferrin, a breakdown product of WBCs, is also useful for detection of WBCs in feces, particularly when the feces has sat for several hours before processing, when intact WBCs are often no longer visible.

B. Preserved specimens or specimens in transport media

NOTE: Transport medium usually destroys the laboratory's ability to determine relative numbers of WBCs or presence of blood in the feces.

1. Modified Cary-Blair is the most versatile preservative, although it does not support *Shigella* spp. viability as well as buffered glycerol saline.
2. Special transport media, such as alkaline peptone water, are recommended for isolation of *Vibrio* spp.

III. INOCULATION AND INCUBATION OF MEDIA

(For more details, see CMPH 1.10.)

Include QC information on reagent container and in QC records.

A. Routine media should include a selective agar for *Campylobacter* spp., a differential agar that supports growth of all enteric gram-negative bacilli, and a moderately selective and differential agar for isolation of *Salmonella* and *Shigella* spp. Use of additional selective agars and enrichment broths is optional, based on the laboratory's patient population and the needs of the physicians served. For example, if *Staphylococcus aureus, Aeromonas* spp., *Plesiomonas shigelloides*, or *Vibrio* spp. are suspected, supportive media should be available. A basic battery of media might include some type of campylobacter agar, Trypticase soy agar with BAP, MAC or EMB agar, and Hektoen enteric or xylose-lysine-deoxycholate (XLD) agar. Although public health authorities recommend inoculation of MAC-sorbitol (SMAC) agar for detection of *Escherichia coli* O157:H7, this medium should be used only if the feces is grossly bloody and/or upon a physician's request. Routine screening is not cost-effective, and the recovery is low.

B. Enrichment broths are used for detecting small numbers of *Salmonella* or *Campylobacter* spp. in carriers and occasionally for culturing feces of workers in sensitive occupations, such as day care and food service employees. Laboratories that have historical data showing very poor recovery of additional isolates of pathogens not seen on initial plates can make a case for abandoning routine use of enrichment broths.

C. Special media have been developed to enhance recovery of other fecal pathogens. They should be used based on prevalence of disease, public health requirements, and physician requests.

1. SMAC agar for *E. coli* O157:H7
2. Cefsulodin-irgasan-novobiocin (CIN) agar for *Yersinia enterocolitica*
3. TCBS agar for *Vibrio* spp.
4. Blood agar with 10 μg of ampicillin per ml for *Aeromonas* spp.
5. Salmonella-shigella agar, brilliant green agar, and bismuth sulfite agar for more selective isolation of *Salmonella* spp.
6. Inositol-brilliant green-bile salts agar for *P. shigelloides*
7. Cycloserine-cefoxitin-fructose-egg yolk agar (CCFA) for *Clostridium difficile*

   NOTE: Since as many as 20% of asymptomatic hospitalized patients may be colonized with this microorganism, tests for presence of toxin in feces are more specific for diagnosis of *C. difficile*-associated diarrhea. (For details, *see* CMPH 1.10.) Epidemiological studies require isolation of the microorganism.

D. The filter method for isolation of *Campylobacter* spp. removes the need for selective agar and increased temperature of incubation and probably increases the yield compared with that from conventional methods.

1. Place a 0.65- or 0.8-μm-pore-size cellulose-acetate filter disk on the surface of a TSA plate with BAP.
2. Place a pea-sized portion of feces on top of the filter and allow to sit at room temperature for 30 min. *Campylobacter* spp., if present, will migrate through the filter onto the agar surface.
3. Remove the filter and the specimen from the agar surface and incubate the plate in a microaerobic atmosphere at 35 to 37°C for up to 72 h.

E. Inoculation of media

A swab of the specimen should be rolled over one small area of the plate, and the material should be streaked for isolated colonies. Larger amounts of the specimen should be placed onto the more selective media, and streaking should be heavier handed to compensate for the inhibitory activity of the media.

F. Incubation

1. Routine agars should be incubated in the air at 35 to 37°C.

III. INOCULATION AND INCUBATION OF MEDIA *(continued)*

2. Media for isolation of *Clostridium difficile* should be incubated under strict anaerobic conditions for 48 to 72 h.
3. Media for isolation of *Campylobacter* spp. should be incubated in a microaerobic atmosphere (5% oxygen) for at least 72 h. Incubation at 42°C will suppress growth of other bacteria, although pathogenic *Campylobacter* spp. also grow at 35 to 37°C.
4. Incubate CIN agar at 25°C or room temperature for 48 h.

IV. ROUTINE CULTURE EXAMINATION

A. Blood agar (BAP)

1. Examine for a predominance of *Staphylococcus aureus* or yeast spp. Report if present, but do not perform antimicrobial susceptibility testing.
2. Use a sweep of colonies from the plate to test for oxidase production. If oxidase-positive colonies are found, not resembling *Pseudomonas aeruginosa*, they should be isolated and identified further to rule out *Aeromonas*, *Plesiomonas*, and *Vibrio* spp. (for more information, *see* CMPH 1.10).

B. Campy agar

1. Examine plates for pinkish-beige colonies that often tail along the line of streaking. Colony morphology varies, however, from flat spreading colonies to tiny convex colonies. The color of colonies ranges from gray to a yellowish or pinkish tint.
2. Suspicious colonies should be Gram stained for visualization of the typical curved and spiral-shaped thin gram-negative bacilli. Counterstaining with 0.1% basic fuchsin helps in differentiation of the microorganism. Alternatively, a wet preparation in a nutrient broth medium should reveal the characteristic darting or tumbling motility. (Distilled water and some salines inhibit motility.)
3. Isolates are identified presumptively by typical cellular morphology on Gram stain or wet preparation and a positive oxidase reaction.
4. For more information on identification of *Campylobacter* spp., see CMPH 1.10 and 1.19.

C. Differential and selective gram-negative enteric agars (*see* Appendix 2.8–2 for suspicious colony morphologies)

1. Subculture suspicious colonies to blood agar for oxidase testing and to the laboratory's choice of a limited battery of tests useful for identifying potential stool pathogens.
2. Alternatively, perform preliminary serological typing by using commercial colored latex agglutination particles (Wellcolex; Murex Diagnostics, Norcross, Ga.) or limited enzymatic or biochemical characterization (*see* CMPH 1.19).
3. Complete identification, both biochemically and serologically, should follow for any isolate still suspicious as an enteric pathogen after the initial screening tests. If required, send isolates identified biochemically as pathogens to a public health or reference laboratory for definitive serological studies.

D. Enrichment broth may be tested directly for the presence of *Salmonella* and *Shigella* spp. by using commercial latex agglutination reagents. The usual procedure, however, is to subculture the broth to one or two selective and differential agars after 8 h of incubation at 35 to 37°C. If enrichment broths are incubated for too long, the nonpathogenic enteric bacteria can overgrow the pathogens and negate the value of the procedure. Subcultured agar plates are incubated and then examined in the same manner as are original plates.

E. Recommended approach to antimicrobial susceptibility testing of fecal pathogens

NOTE: In most cases it should not be done.

1. Mild gastroenteritis is usually treated clinically without antimicrobial agents, since treatment can prolong the carrier state or lead to development of resistance to these agents.
2. If necessary, only ampicillin, a quinolone, and trimethoprim/sulfamethoxazole should be tested against isolates of *Salmonella* and *Shigella* spp. Several in vitro results do not correlate with the in vivo response. For example, *Salmonella* gastroenteritis is not treated effectively by cephalosporins regardless of the antimicrobial susceptibility test results.
3. Standardized methods for testing the susceptibility of *Campylobacter* isolates are not available.

V. REPORTING RESULTS

(For specific common reporting protocols, see CMPH 1.10.)

NOTE: Patients shedding pathogens in their feces are placed on both standard and contact isolation. Nursing units and epidemiologists should be notified as soon as possible of the presence of an enteric pathogen.

A. Report presumptive identification of microorganisms suspicious for enteric pathogens to the epidemiologist and the nursing care unit immediately by telephone so that proper patient isolation protocols can be instated. Follow the verbal report with a written or computer-generated report stating that further results are pending. (For suggested report statements, *see* CMPH procedure 1.10.)

B. Report definitive identifications as soon as they are available. For patients who have been discharged from the hospital or whose specimens were received as outpatient specimens, the physician responsible should be telephoned or faxed directly with the final result.

C. Report all reportable enteric pathogens to the appropriate public health agency as required by the local regulatory and government jurisdictions.

D. Report the presence of *C. difficile* toxin immediately to the physician, the epidemiologist, and the patient care unit by telephone and follow with a written report.

E. Report negative culture results by listing the microorganisms sought specifically.

Example: ''No *Salmonella, Shigella, Campylobacter*, or *E. coli* O157:H7 isolated.''

F. In the absence of expected enteric gram-negative microorganisms, the predominant microorganism (if one is seen) can be reported along with a statement such as ''Decreased or absent usual enteric gram-negative microbiota.''

SUPPLEMENTAL READING

Baron, E. J., and S. M. Finegold. 1990. Microorganisms encountered in the gastrointestinal tract, p. 238–252. *In Bailey and Scott's Diagnostic Microbiology*, 8th ed. C. V. Mosby Co., St. Louis.

Ewing, W. H. 1986. Differentiation of Enterobacteriaceae by biochemical reactions, p. 47–72. *In Edwards and Ewing's Identification of Enterobacteriaceae*, 4th ed. Elsevier Science Publishing Co., Inc., New York.

Grasmick, A. 1992. Processing and interpretation of bacterial fecal cultures, p. 1.10.1–1.10.25. *In* H. D. Isenberg (ed.), *Clinical Microbiology Procedures Handbook*, vol. 1. American Society for Microbiology, Washington, D.C.

Koneman, E. W., S. D. Allen, V. R. Dowell, Jr., W. M. Janda, H, M. Sommers, and W. C. Winn, Jr. (ed.). 1988. The Enterobacteriaceae, p. 124–126. *In Diagnostic Microbiology*, 3rd ed. J. B. Lippincott Co., Philadelphia.

Sack, R. B., R. C. Tilton, and A. S. Weissfeld. 1980. *Cumitech 12, Laboratory Diagnosis of Bacterial Diarrhea.* Coordinating ed., S. J. Rubin. American Society for Microbiology, Washington, D.C.

APPENDIX 2.8–1

Symptoms associated with pathogenic mechanisms of diarrheal disease[a]

Parameter	Symptom[b] with:	
	Toxin production	Tissue invasion
Stool consistency	Watery	Loose
Stool volume	Voluminous	Small
Vomiting	Yes	No
Fever	No	Yes
Dehydration	Yes	Mild
Time of onset	A few h–2 days p.i.	1–3 days p.i.
PMNs in stool	No	Yes
Blood in stool	No	Yes
Mucus in stool	No	Yes
Primary infection site	Small bowel	Large bowel

[a] Data obtained from F. L. Bryan. 1985. Procedures to use during outbreaks of food-borne disease, p. 36–51. *In* E. H. Lennette, A. Balows, W. J. Hausler, Jr., and H. J. Shadomy (ed.), *Manual of Clinical Microbiology*, 4th ed. American Society for Microbiology, Washington, D.C. Table reprinted from CMPH 1.10.22.
[b] p.i., postinoculation.

APPENDIX 2.8–2

Interpretation of growth on culture media for stool samples[a]

Medium	Colony morphology	Identification procedure
Differential		
MAC	Colorless or transparent	Enteric screening procedure[b]
EMB	Colorless or transparent amber (light purple)	Enteric screening procedure
Moderately selective		
HE	Blue or blue-green, with or without black centers	Enteric screening procedure
XLD	Red, with or without dark centers; *S. typhi* can be orange to light pink	Enteric screening procedure
SS	Colorless, with or without black centers	Enteric screening procedure
Deoxycholate	Transparent; colorless to light pink or tan, with or without black centers	Enteric screening procedure
Highly selective		
Brilliant green	Red, pink, or white surrounded by red zones	Enteric screening procedure
Bismuth sulfite	Black, with or without brownish-black zones, or green, with no zones around colonies	Enteric screening procedure
Other		
BAP	Predominating numbers of *Staphylococcus* spp.	Coagulase
	Predominating numbers of *Pseudomonas*-like colonies	Identify nonfermenters
	Predominating numbers of yeast cells	Identify predominating yeast cells
BAP-AMP or BAP	Hemolytic and nonhemolytic non-*Pseudomonas* colonies	Oxidase production
Campy plate	Gray to pinkish, flat to mucoid to convex to spreading all over plate	Gram stain

[a] Data extracted from J. F. McFadden. 1985. *Media for Isolation-Cultivation-Identification-Maintenance of Medical Bacteria*, vol. 1. The Williams & Wilkins Co., Baltimore. Reprinted from CMPH 1.10.23.
[b] Method used in the laboratory to detect potential pathogens.

2.9 Processing and Interpretation of Urine Cultures

I. PRINCIPLE

Urine specimens are submitted for culture from patients with symptoms of urinary tract infections and from asymptomatic patients with a high risk of infection. The etiologic agents of urinary tract infection are limited to a few rapidly growing microorganisms. *Escherichia coli, Enterococcus* spp., *Klebsiella-Enterobacter* spp., *Proteus* spp., and *Pseudomonas* spp. represent a majority of isolates from both hospitalized patients and outpatients.

II. SPECIMEN

Urine is normally a sterile body fluid. However, unless it is collected properly, it can become contaminated with microbiota from the perineum, prostate, urethra, or vagina. The microbiologist must provide detailed instructions to ensure proper specimen collection.

A. Specimen collection

1. Clean-voided midstream urine

2. Catheter urine

a. Indwelling catheters increase the risk of urinary tract infections; only closed systems should be used. Avoid contamination during urine collection.

b. A straight catheter (non-indwelling) is used by a physician or trained practitioner to obtain urine directly from the bladder. This procedure is not routinely recommended because there is a risk of introducing microorganisms into the bladder.

3. Urine collection by suprapubic needle aspiration of the bladder avoids contamination associated with the collection of voided urine. This is the preferred method for infants and for patients for whom the interpretation of results of voided urine is difficult and is also preferred when anaerobic bacteria are the suspected cause of infection.

◩ Foley catheter tips are unacceptable for culture.

B. Timing and number of specimens

1. Obtain early-morning specimens whenever possible because bacterial counts increase after overnight incubation in the bladder.

2. Do not force fluids to have the patient void urine. Excessive fluid intake will dilute the urine and may decrease the colony count to $<10^5$ CFU/ml.

3. Collect three consecutive early-morning specimens from asymptomatic patients.

C. Specimen transport

1. Transport urine to the laboratory as soon as possible after collection.

2. Culture urine specimens within 2 h after collection, or refrigerate and culture them within 8 h whenever possible.

II. SPECIMEN *(continued)*

◪ **NOTE:** If refrigeration is not possible, transport tubes containing a preservative, such as boric acid-glycerol, may be used.

3. Refrigerated urine specimens may be held for ≤24 h because bacterial counts usually remain stable for 24 h at 4°C.
4. Request a repeat urine specimen when there is no evidence of refrigeration and the specimen is >2 h old.
5. Request a repeat specimen when the collection time and method of collection have not been provided.
6. If an improperly collected, transported, or handled specimen cannot be replaced, document in the final report that specimen quality may have been compromised.

III. MATERIALS

A. Media
1. BAP
2. MAC or EMB
3. Colistin-nalidixic acid blood agar (CNA) or selective enterococcus agar (SEA)

B. Gram stain reagents (*see* EPCM 2.1)

C. Supplies (streak plate method)
1. Platinum loops or plastic disposable calibrated loops
2. Platinum-rhodium or disposable plastic 0.001-ml loop for colony counts >1,000 CFU/ml
3. 0.01-ml loop for colony counts between 100 and 1,000 CFU/ml.

◪ **NOTE:** Refer to item IV of this procedure and CMPH 1.10 for quality control methods.

IV. QUALITY CONTROL

Inspect calibrated loops regularly to confirm that they remain round and are free of bends, dents, corrosion, or incinerated material. Additionally, at least monthly, check the loops to ensure that the delivery volume is accurate.

Drill bit method

A. Calibration of 0.001-ml loop
1. Obtain two twist drill bits (no. 53 and 54).
2. Carefully slip the 0.001-ml loop over the end of the no. 54 bit. If the loop is calibrated, it will fit over the bit.
3. Repeat the procedure with the no. 53 bit. If the loop is calibrated, it will not fit over the end.
4. Discard the loop if it fits over the no. 53 bit.

B. Calibration of 0.01-ml loop
1. Obtain two twist drill bits (no. 21 and 22).
2. Repeat the procedure described above for the 0.001-ml loop.
3. A calibrated 0.01-ml loop will fit over the end of a no. 22 drill bit and will not fit a no. 21 bit.
4. Drill bits can be purchased at a hardware store.

V. GRAM STAIN

The Gram stain method can detect the presence of both bacteria and WBCs in urine specimens.

A. Method
1. Place 0.01 ml of well-mixed, uncentrifuged urine onto a glass slide, and allow it to air dry without spreading. (For details on Gram stain procedure, including fixing, staining, and organism identification, *see* EPCM 2.1.)
2. Determine the number of organisms per oil immersion field (OIF).

B. Interpretation

1. Report the number of microorganisms per OIF.
2. The presence of one or more microorganisms per OIF correlates with a colony count of $\geq 10^5$ CFU/ml.
3. The presence of many squamous epithelial cells and different microbial morphotypes indicates contamination. Request a repeat specimen.

VI. CULTURE METHOD

A. Surface streak method

1. Hold the flamed and cooled calibrated loop (0.001 or 0.01 ml) vertically, and immerse it just below the surface of a well-mixed, uncentrifuged urine specimen.
2. Spread the loopful of urine over the surface of the agar plate as illustrated in CMPH 1.4.
3. Without reflaming the loop, repeat the procedure with an additional loopful of urine for each agar plate medium to be inoculated.
4. Incubate aerobically overnight at 35 to 37°C.

   ◩ **NOTE:** When anaerobic cultures are requested, they should be performed only on suprapubic bladder aspirates.
5. Prolong the incubation when specimens are obtained by invasive techniques, such as suprapubic bladder aspiration or cystoscopy, and when culture results do not correlate with Gram stain findings or clinical conditions.

B. Examination of culture media

1. Examine cultures that have been incubated overnight.
2. If there is no visible growth and the specimen was collected by voiding or with a catheter, report as follows.
   - **a.** "No growth at less than 1,000 CFU/ml" (0.001-ml inoculum).
   - **b.** Or "No growth at less than 100 CFU/ml" (0.01-ml inoculum).
3. If the specimen was collected by an invasive technique, e.g., suprapubic bladder aspiration, reincubate the culture media for an additional 24 h.
4. Reincubate culture plates with tiny or scant colonies that are not discernible.
5. For positive cultures, examine culture media for the quantity and morphological type of organisms present.
   - **a.** With 0.001-ml loop, one colony equals 1,000 CFU/ml.
   - **b.** With 0.01-ml loop, one colony equals 100 CFU/ml.
6. Perform additional testing based on the colony count method of urine collection.

C. Interpreting culture results

1. Voided urine specimens (Fig. 2.9–1)

   ◩ **NOTE:** When voided urine specimens are submitted without specific clinical information suggesting that colony counts of $<10^5$ CFU/ml may cause infection, only a single microorganism (pathogen) at $>10^5$ CFU/ml should be identified to the species level. Probable contaminant is defined as diphtheroids, viridans streptococci, lactobacilli, and coagulase-negative staphylococci other than *Staphylococcus saprophyticus.*
   - **a.** Colony count of $>10^5$ CFU/ml
     - (1) One probable pathogen at $>10^5$ CFU/ml
       - (a) Definitively identify to the species level.
       - (b) Perform antimicrobial susceptibility testing (AST), if indicated, as described in section 5 of CMPH.
       - (c) Enumerate all species present at $<10^4$ CFU/ml; i.e., report "1 (2, 3, 4) microorganism(s) present at $<10^4$ CFU/ml."
     - (2) One probable contaminant at $>10^5$ CFU/ml
       - (a) Perform limited identification; i.e., distinguish between *S. saprophyticus* and other coagulase-negative staphylococci.
       - (b) Enumerate all species present at $<10^4$ CFU/ml.

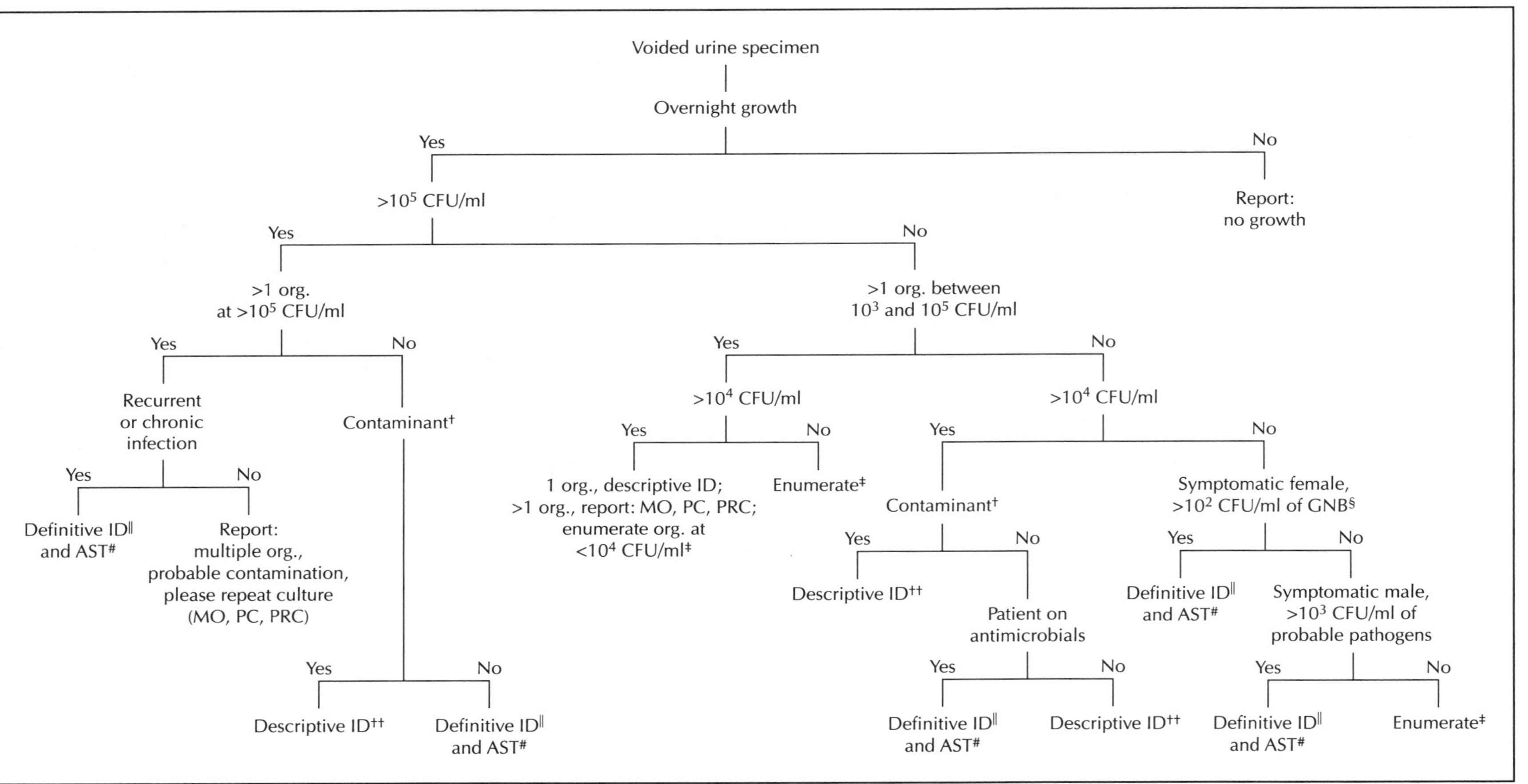

Figure 2.9–1 Interpretation of cultures of voided urine specimen. † Contaminant: diptheroids, lactobacilli, viridans group streptococci, not group D, and mixed cultures; ‡ enumerate, list the number of species present at $<10^4$ CFU/ml; § GNB, gram-negative bacilli; ‖ definitive ID, complete identification to species level; #AST, antimicrobial susceptibility testing; †† descriptive ID, identification from colonial morphology on plates or Gram stain. Adapted from CMPH 1.17.9.

VI. CULTURE METHOD *(continued)*

(3) Two probable pathogens, with a diagnosis of recurrent or chronic urinary tract infection
(a) Definitively identify.
(b) Perform AST as described in section 5 of CMPH.
(4) Two probable pathogens and no clinical information provided
(a) Describe each microorganism on the basis of its morphology.
(b) Also, report "Multiple microorganisms present; probable contamination; please repeat culture."
(5) More than two microorganisms at $>10^4$ CFU/ml: report "Multiple microorganisms present; probable contamination; please repeat culture."

b. Colony count of $\leq 10^5$ CFU/ml
(1) One probable pathogen at $\leq 10^5$ CFU/ml
(a) For patients on antimicrobial agents, female patients with urethritis, and symptomatic males, perform definitive identification and AST.
(b) If no clinical information is provided
i. Provide a descriptive identification of colony morphology for any microorganism present at 10^4 to 10^5 CFU/ml.
ii. Enumerate all species present at $<10^4$ CFU/ml.
iii. Hold the culture at room temperature for 3 days for possible further workup if requested by the patient's physician.
(2) One probable contaminant at $\leq 10^5$ CFU/ml: descriptively identify the isolate.
(3) Two or more microorganisms present at $<10^4$ CFU/ml: report "Multiple microorganisms present; probable contamination; please repeat culture."
(4) One or more microorganisms at $<10^4$ CFU/ml: enumerate, and report the number of morphological types present; i.e., "1 (2, 3, 4) microorganism(s) present at $<10^4$ CFU/ml."

2. Catheterized urine specimens (Fig. 2.9–2)

a. Colony count of $>10^4$ CFU/ml
(1) Two or more probable pathogens present at $>10^4$ CFU/ml
(a) Perform definitive identification and AST on both isolates.
(b) Descriptively identify the species present at $<10^4$ CFU/ml.
(2) One or two probable contaminants present at $>10^4$ CFU/ml
(a) Report the microorganism(s) present by providing a description of the morphological type(s), e.g., diphtheroids, viridans streptococcus group.
(b) Report "Multiple microorganisms present; probable contamination; please repeat culture."
(3) One probable pathogen and one probable contaminant
(a) Definitively identify the probable pathogen.
(b) Provide a description of the morphological type of the probable contaminant.
(4) Three or more microorganisms
(a) Provide a description of the morphological types.
(b) Report "Multiple microorganisms present; probable contamination; please repeat culture."

b. Colony count of $<10^4$ CFU/ml
(1) For patients on antimicrobial agents, female patients with urethritis, and symptomatic males, perform definitive identification and AST.
(2) For all other patients, provide a description of the morphological type(s) present and request a repeat specimen.
(3) Hold the culture at room temperature for 3 days for possible further workup if requested by the patient's physician

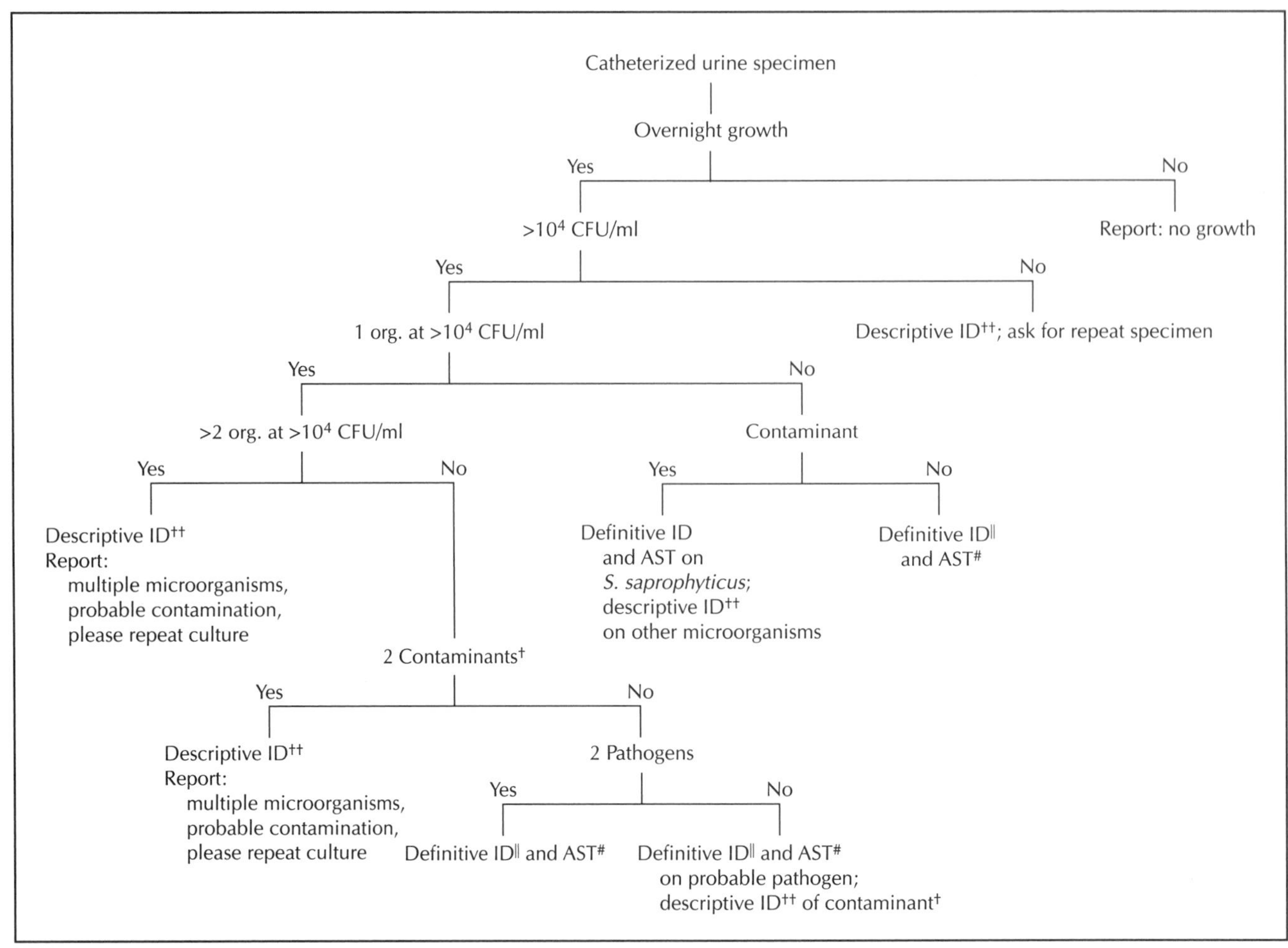

Figure 2.9–2 Interpretation of cultures of catheterized urine specimens. For explanation of terms, see legend to Fig. 2.9–1. Adapted from CMPH 1.17.11.

VI. CULTURE METHOD *(continued)*

3. Specimen obtained by invasive technique (Fig. 2.9–3)
   a. One or two microorganisms present
      (1) Definitively identify.
      (2) Perform AST on probable pathogen only.
   b. Three or more microorganisms present
      (1) Definitively identify.
      (2) Hold culture for 3 days for possible consultation.
   c. No growth
      (1) Examine culture at 24 h.
      (2) Reincubate for an additional 24 h.
      (3) Report "No growth at $<10^2$ CFU/ml" at 48 h.

D. Culture procedure notes.

1. Do not culture Foley catheter tip.
2. The criterion of $>10^5$ CFU/ml can be applied to a majority of specimens submitted for culture.
3. Perform anaerobic cultures only on suprapubic bladder aspirates when requested.
4. Do not perform AST directly from the urine specimen. Refer to section 5 of CMPH for an explanation.

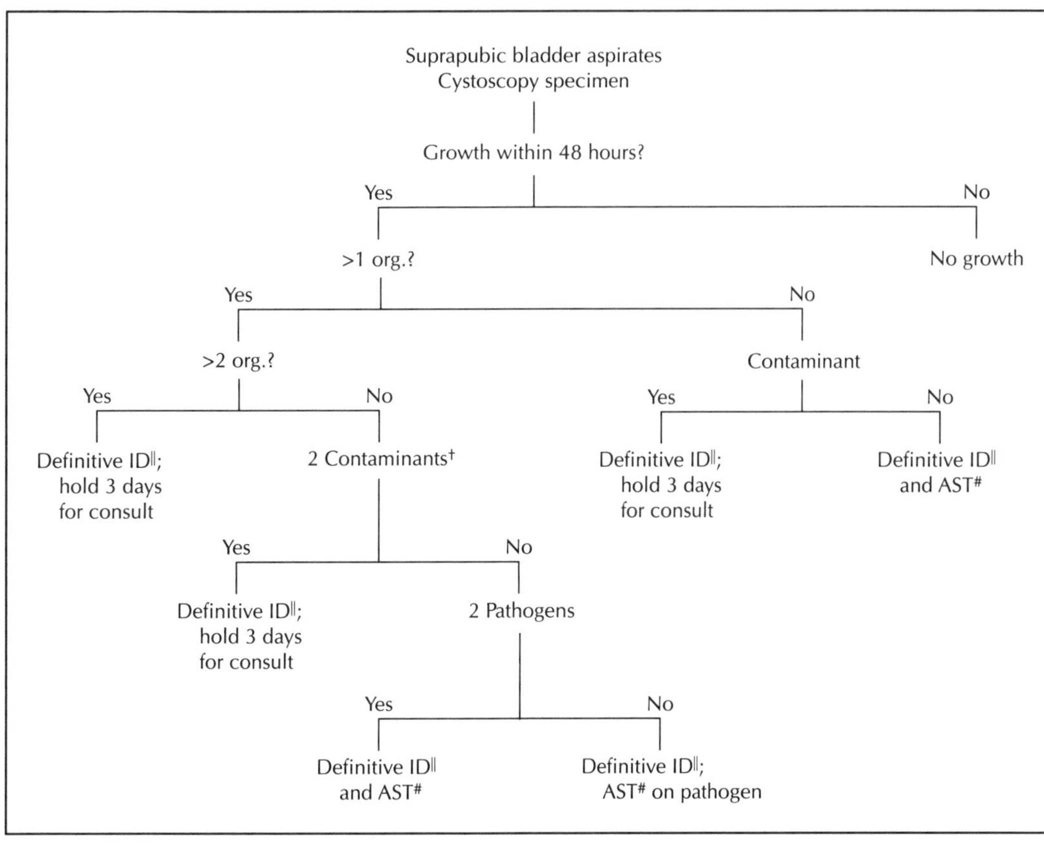

Figure 2.9–3 Interpretation of cultures of specimens obtained by invasive techniques. For explanation of terms, see legend to Fig. 2.9–1. Adapted from CMPH 1.17.12.

SUPPLEMENTAL READING

Albers, A. C., and R. D. Fletcher. 1983. Accuracy of calibrated-loop transfer. *J. Clin. Microbiol.* **18:**40–42.

Claridge, J. E., M. T. Pezzlo, and K. L. Vosti. 1987. *Cumitech 2A, Laboratory Diagnosis of Urinary Tract Infections.* Coordinating ed., A. L. Weissfeld. American Society for Microbiology, Washington, D.C.

Kass, E. H. 1956. Asymptomatic infections of the urinary tract. *Trans. Assoc. Am. Phys.* **69:**56–63.

Kunin, C. M. 1987. *Detection, Prevention and Management of Urinary Tract Infections*, 4th ed. Lea & Febiger, Philadelphia.

Lipsky, B. A., R. C. Ireton, S. D. Fihn, R. Hackett, and R. E. Berger. 1987. Diagnosis of bacteriuria in men: specimen collection and culture interpretation. *J. Infect. Dis.* **155:**847–854.

Pezzlo, M. T. 1990. Significance of low count bacteriuria. *Clin. Microbiol. Newsl.* **12:**60–61.

Stark, R. P., and D. G. Maki. 1984. Bacteriuria in the catheterized patient: what quantitative level of bacteriuria is relevant? *N. Engl. J. Med.* **311:** 560–564.

Washington, J. A., II, C. M. White, M. Laganiere, and L. H. Smith. 1981. Detection of significant bacteriuria by microscopic examination of urine. *Lab. Med.* **12:**294–296.

2.10 Processing and Interpretation of Skin and Subcutaneous-Tissue Specimens

I. PRINCIPLE

Microorganisms resident on the skin and mucous membranes of humans, as well as microorganisms in the environment, can cause infections if they enter normally sterile tissue through breaks in the skin or normally intact mucous membranes. Because virulence mechanisms are not always necessary for such microorganisms to cause infection, virtually any species can be involved. Interpretation of culture results should be based on Gram stain criteria, and extensive laboratory testing should be done only after consultation with the clinician. Procedures for performing bacterial cultures are presented below. Refer to the appropriate sections of CMPH and EPCM for analytical details when viruses (CMPH section 8), parasites (CMPH section 7), or fungi (CMPH section 6) are considered among the potential etiological agents.

II. SPECIMEN COLLECTION AND TRANSPORT

(See CMPH 1.1 and 1.16 for rejection criteria, additional details, and materials.)

A. Types of specimens

1. Tissue obtained during surgical procedures is the best type of specimen because organisms present are usually the cause of infection. The skin surface and surgical area are disinfected before collection, and the specimen is obtained by an invasive technique. Depending upon the source of the tissue, in most cases, most normal microbiota are absent.
2. Aspirated material from an abscess or deep wound, obtained to prevent contamination by surface or mucosal microorganisms, is the next best specimen type. Uninvolved skin or mucous membrane should be thoroughly disinfected with alcohol followed by betadine before aspiration.
3. Pus and exudates obtained from an infected site during surgery can be aspirated into a syringe directly through a plastic catheter or without a needle to avoid puncturing an organ. This method can be used during surgery when there is sufficient volume to allow the fluid to flow into the container.
4. Swab specimens are the least desirable because they hold the least volume and are most subject to contamination. Before a swab is applied, skin surrounding the infected site should be disinfected as thoroughly as for an aspiration.
5. Fluid from the peritoneal dialysate of patients undergoing chronic ambulatory peritoneal dialysis (CAPD) is a special case. Either this fluid should be inoculated at the patient's bedside directly into resin-containing blood culture broth media or the total volume should be sent to the laboratory for processing as described in EPCM 2.4, covering sterile body fluids.

B. Transport

1. Specimens should remain moist and at room temperature (to preserve anaerobic organism viability) during transport. Tissue can be placed into a sterile container or petri dish with a sterile, moistened gauze pad. Refer to CMPH 1.1 and EPCM section 1 for detailed descriptions of available specimen collection and transport containers.

2. Aspirated material should be injected into an anaerobic transport vial that excludes oxygen. Syringes containing specimen material, especially with needles attached, should never be used for transport.
3. A transport medium that prevents overgrowth of rapidly growing microorganisms, such as modified Stuart's or Amies, is recommended for swabs. Unless swabs are the only available specimen material, they should not be processed for recovery of anaerobes. (Special foam or synthetic swabs said to retain anaerobe viability are available, but their use should not be encouraged because the volume of specimen is not optimal.)

III. INOCULATION AND INCUBATION OF MEDIA

(See CMPH procedures 1.4 and 1.16 for more details.)

Include QC information on reagent container and in QC records.

A. Routine media should include a supportive blood agar to reveal hemolytic reactions, a CHOC agar plate, a selective and differential medium for gram-negative organisms (MAC or EMB), and other media depending on the specimen type and organisms sought.
   1. A blood-containing medium selective for gram-positive microorganisms, such as colistin-nalidixic acid agar (CNA) or phenyl-ethyl alcohol agar (PEA) may be useful when overgrowth by swarming *Proteus* spp. or other enteric flora is possible.
   2. CHOC may be added for specimens from normally sterile genital sites or other sites from which *Neisseria* spp., *Haemophilus* spp., and other fastidious-type microorganisms are suspected.
   3. For appropriate specimens, supplemented (vitamin K and hemin) anaerobic agars including a supportive blood agar, bacteroides bile esculin agar, and a differential agar for pigmenting gram-negative bacilli, such as laked sheep blood agar, should be inoculated. (*See* CMPH 2.1 to 2.4a for more details.)
   4. A broth enrichment, such as thioglycolate or chopped meat medium, may be added for certain sensitive specimens, but is not recommended for routine specimens received on swabs. Specimens suitable for enrichment broths include biopsied tissue, cerebrospinal fluid from patients with shunts, and perhaps joint fluids when original Gram stains show numerous PMNs.
   5. B culture broths have been recommended for CAPD fluids. They should be handled in the same manner as blood cultures. Refer to EPCM 2.3 and 2.4 covering procedures for blood cultures and sterile body fluids, respectively.
   6. Recovery of certain fastidious or unusual microorganisms requires special media and procedures, as outlined in various sections of CMPH and EPCM. Refer to the index for specific microorganism requirements.

B. Inoculation of media and preparation of Gram stain
   1. Although all specimens should be emulsified in broth (thioglycolate, Trypticase soy, or brucella) to create a homogeneous suspension for inoculation onto media and for preparation of a Gram-stained smear, a small piece of intact tissue should be saved to press onto the slide surface to create an impression smear. The homogenized material can be spread onto the slide next to the impression smear and both areas on the slide can be stained together.

      ◪ **NOTE:** Other stains, such as acid-fast, calcofluor, and Wright's, should be prepared if there is the least suspicion that organisms detected with these stains may be the causative agents of infection.

      Commercially available, disposable plastic grinding devices (Becton Dickinson; Sage Products Inc., Crystal Lake, Ill.) are superior to glass hand-held grinders for safety reasons. Mortars and pestles can be used, but only in an anaerobic chamber if anaerobes are suspected, because these tools aerate the material too much. An automated pummeling instrument (stomacher; Tekmar Co., Cincinnati, Ohio) is also very efficient for homogenizing tissues for culture. Refer to CMPH 1.4 for details.

III. INOCULATION AND INCUBATION OF MEDIA *(continued)*

2. Tissues that may harbor or are suspected of infection with anaerobic organisms should be ground or homogenized in thioglycolate or other reduced broth to avoid oxygenation, which may affect the recovery of anaerobic bacteria.

   ◩ **NOTE:** A small portion of intact tissue should be inoculated onto fungal media by pushing the tissue partway into the agar. Also, a portion of the tissue and/or homogenate should be refrigerated or frozen for future examination for a variety of other microorganisms, e.g., viruses and mycobacteria.
3. In a biological safety cabinet, purulent-appearing aspirated material should be mixed well, and drops should be placed directly onto media, into broth (if applicable), and onto a slide for Gram stain. Alternatively, a second slide can be placed over the drop and the two slides carefully pulled apart with minimal pressure, providing a thin layer on both slides for staining.
4. Clear-appearing aspirated material should be concentrated by centrifugation. The supernatant should be aspirated, leaving approximately 0.5 to 1.0 ml of sediment, which should be mixed gently by aspirating up and down with a sterile pipette so as not to add air. Drops of the concentrate should be inoculated to media, into broth, and placed onto a slide for the Gram stain.

   ◩ **NOTE:** The use of a cytocentrifuge for preparation of the Gram stain from clear aspirates or body fluids is highly recommended (*see* CMPH 12.6.1).
5. Swabs should be vortexed in 0.5 to 1.0 ml of broth to extract as much specimen as possible. Each swab should be squeezed against the side of the broth tube to express remaining fluid and then discarded. Drops of the remaining suspension should be inoculated to media and placed onto a slide for the Gram stain.
6. After air drying, the smear should be fixed by flooding with 90 to 100% methanol and allowing the smear to air dry again before staining. Smears from bloody or very purulent specimens should be spread thinly on the slide so that they do not stain too darkly and obscure recognition of microorganisms.

C. Incubation

1. Routine agars should be incubated in a moist atmosphere of 5 to 10% CO_2 at 35 to 37°C for a minimum of 48 h before discarding. Alternatively, although not necessary, MAC can be incubated in air.
2. Anaerobic media should be incubated in an anaerobic atmosphere for a minimum total of 72 h. Plates should not be exposed to air during the first 48 h. For isolation of some fastidious anaerobes, such as *Actinomyces* spp., prolonged incubation is indicated, up to 10 days.
3. All broths should be incubated in the air and held for at least 10 days before discarding, in the event that growth of an organism, aerobe, anaerobe, or facultative anaerobe is delayed.

IV. GRAM STAIN EXAMINATION

(For details, see EPCM 2.1.)

A. The Gram stain helps to determine the extent of workup required by the culture. Record the relative numbers of PMNs, histiocytes or macrophages, epithelial cells, and bacterial and fungal morphotypes. If abundant epithelial cells are seen, surface contamination is likely, and the isolates on culture should be minimally processed. Presence of numerous PMNs or other phagocytic cells indicates an infectious process.

B. If clinically important microorganisms are recognized or suspected on Gram stain, immediately telephone or report results to the appropriate physician. Some examples of such organisms are *Clostridium*-like gram-positive bacilli seen on a specimen from a soft-tissue infection or aspirate, even in the absence of intact PMNs; numerous PMNs and gram-positive cocci in clusters or gram-negative cocci from joint fluid; mixed morphologies suggestive of anaerobic bacteria on a specimen from a brain abscess; or gram-positive lancet-shaped diplococci suggestive of *Streptococcus pneumoniae* on peritoneal fluid.

V. ROUTINE CULTURE EXAMINATION

(See CMPH 1.16.)

A. Day 1 (*see* Fig. 2.10–1)

1. Examine all plates and broths. Gram stain and subculture broths if there is no growth on plates but evident growth in broth.
2. Describe colony types, Gram stain all colony types, and perform initial rapid identification procedures for clinically relevant colony types as correlated with direct-specimen Gram stain performed previously. Many specimen types will yield mixed cultures of more than three morphotypes of bacteria.

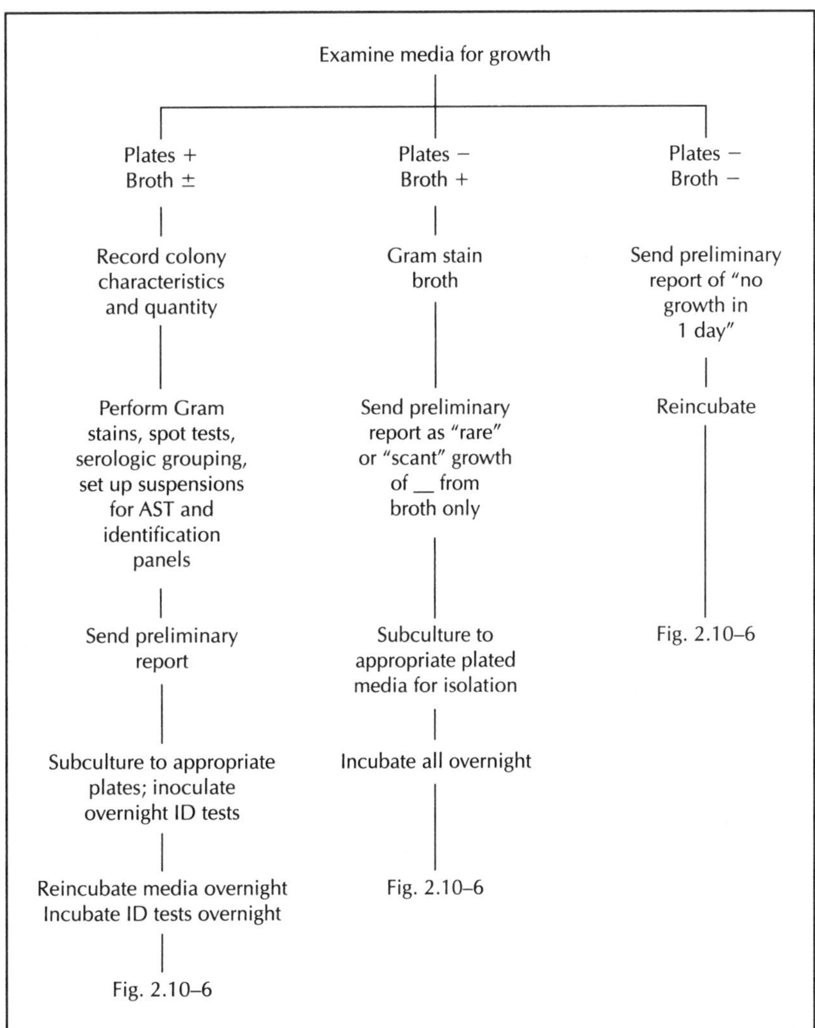

Figure 2.10–1 Examination of cultures—day 1. AST, antimicrobial susceptibility testing; ID, identification; +, growth; −, no growth; _, space for microorganism name or type. Adapted from CMPH 1.16.5.

V. ROUTINE CULTURE EXAMINATION *(continued)*

Depending on the source, such cultures can often be reported with morphologic descriptions and minimal workup.

Examples: ''Culture yields abundant growth of >3 colony types of enteric gram-negative bacilli. Please consult microbiology laboratory if more definitive studies are necessary.''

''Culture yields >3 colony types of gram-positive coryneform bacilli and cocci suggestive of skin flora. Please consult microbiology laboratory if more definitive studies are necessary.''

3. Report preliminary results on all cultures. Telephone clinically or epidemiologically critical results to the appropriate persons.
4. Set up antimicrobial susceptibility tests (rapid or conventional) on clinically relevant isolates if sufficient pure culture material is available. Probable pathogens include *Staphylococcus aureus, Streptococcus pyogenes, Pseudomonas aeruginosa,* and isolates representing the predominant cellular morphotype associated with PMNs on the initial Gram stain.
5. Set up definitive biochemical or other identification protocols on clinically relevant isolates if sufficient pure culture material is available. (For details, *see* Fig. 2.10–2 through 2.10–5.)
6. Make subcultures of nonisolated colonies to appropriate media for later workup.
7. Reincubate all primary and subculture media for an additional 8 to 12 h.

B. Day 2 (*see* Fig. 2.10–6.)

1. Read and record reactions and test results from previous day.
2. Set up additional tests as needed.
3. Send updated or final report, notifying appropriate persons by telephone or fax of clinically or epidemiologically important results, if warranted.
4. If no growth is detected on plates, send final report (unless cultures are to be held longer for special reasons) as ''No growth in 2 days.'' If incubation is continuing, add the comment ''Cultures will be held for ___ days.''

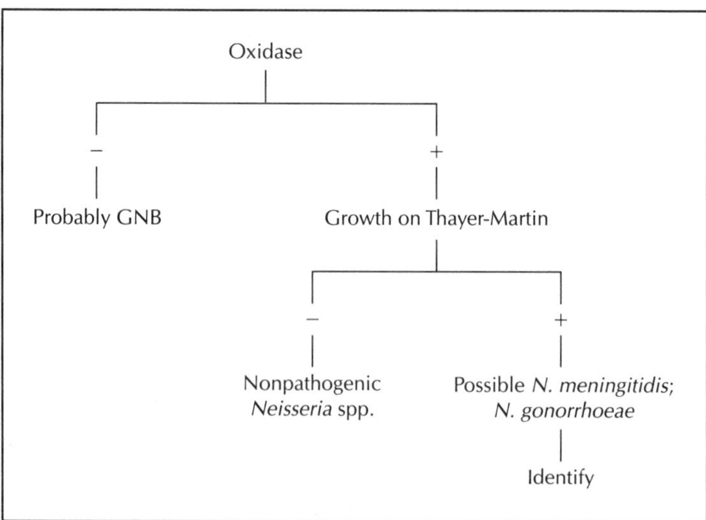

Figure 2.10–2 Identification of gram-negative cocci from skin and subcutaneous tissue. GNB, gram-negative bacilli; +, growth; −, no growth. For details on identification, see CMPH 1.19. Adapted from CMPH 1.16.9.

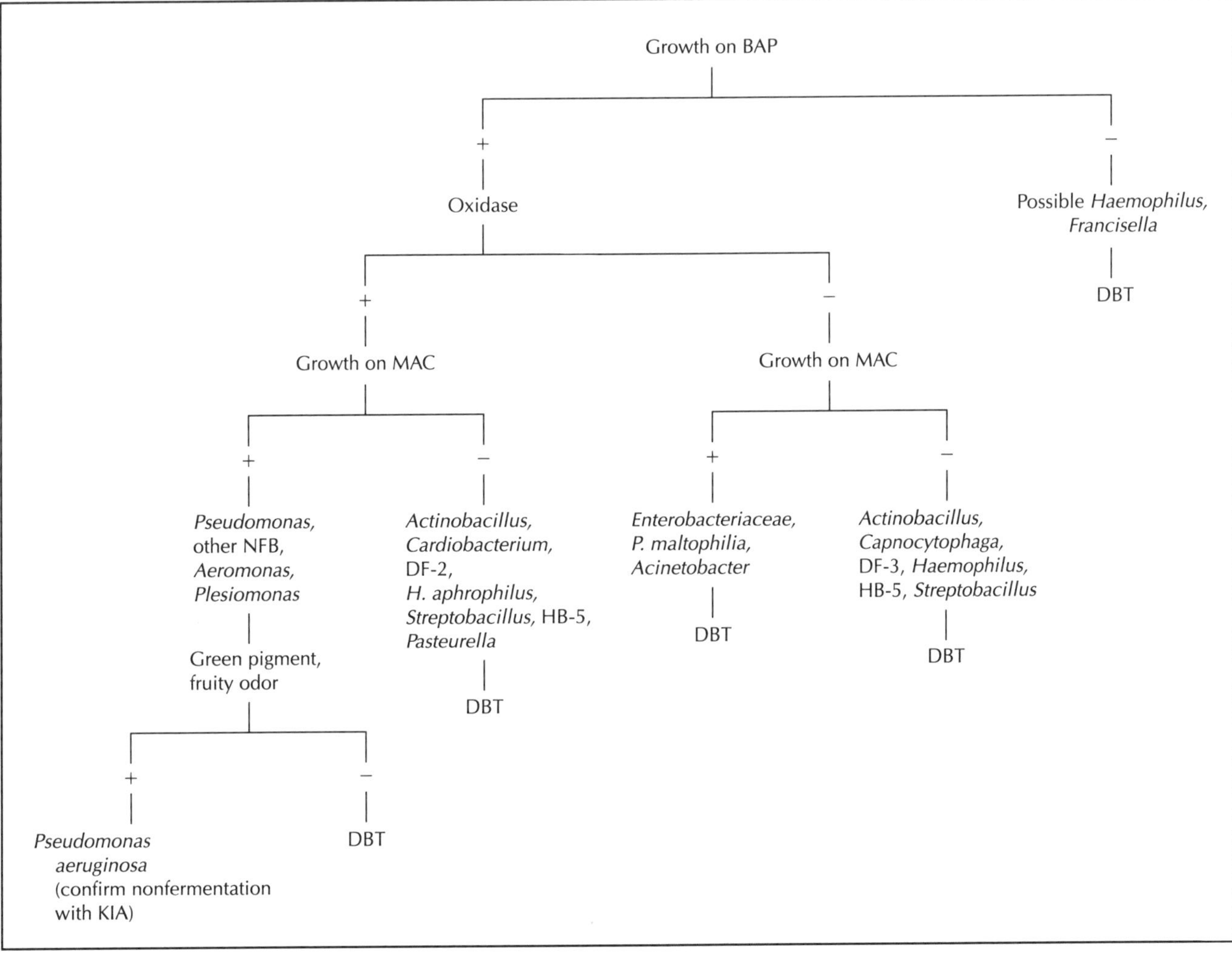

Figure 2.10–3 Identification of gram-negative bacilli from skin and subcutaneous tissue. NFB, Nonfermentative bacteria; KIA, Kligler iron agar; DBT, definitive biochemical testing. For details on DBT, see CMPH 1.18, 1.19, and 1.16 (Table 2). Adapted from CMPH 1.16.10.

5. For positive cultures or isolates, whether complete workup was performed or not, hold a representative plate at room temperature for 5 days in case the physician calls for further studies.

C. Additional days

Perform follow-up identification and antimicrobial susceptibility testing procedures until all relevant isolates have been finalized, and then send a final updated report.

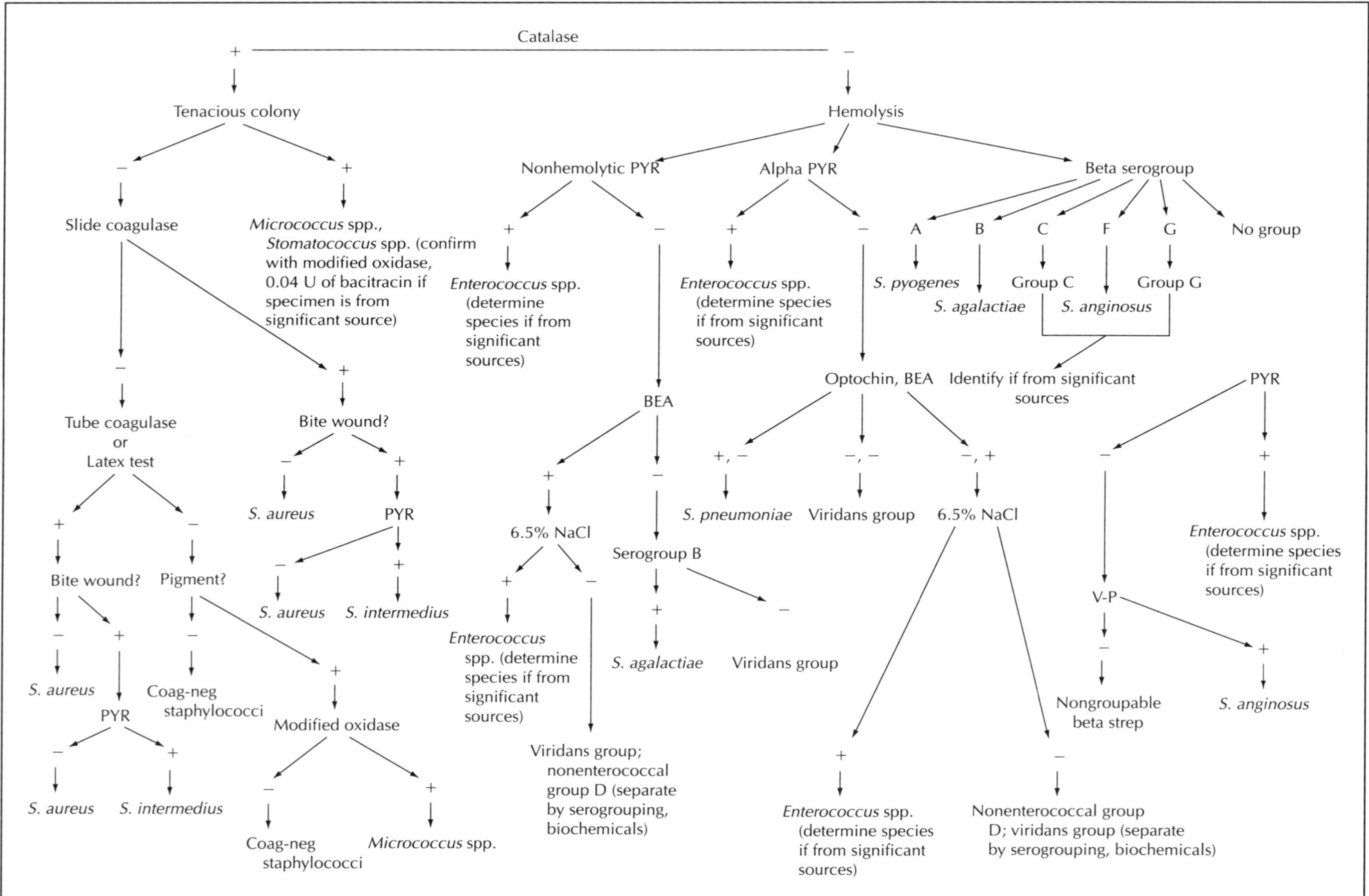

Figure 2.10–4 Identification of gram-positive cocci from skin and subcutaneous tissue. BEA, bile esculin agar; V-P, Voges-Proskauer; PYR, L-pyrrolidonyl-β-naphthylamide; coag-neg, coagulase negative; +, growth; −, no growth. For details on separating microorganisms by serogrouping or biochemical procedures, see CMPH 1.20. Adapted from CMPH 1.16.11.

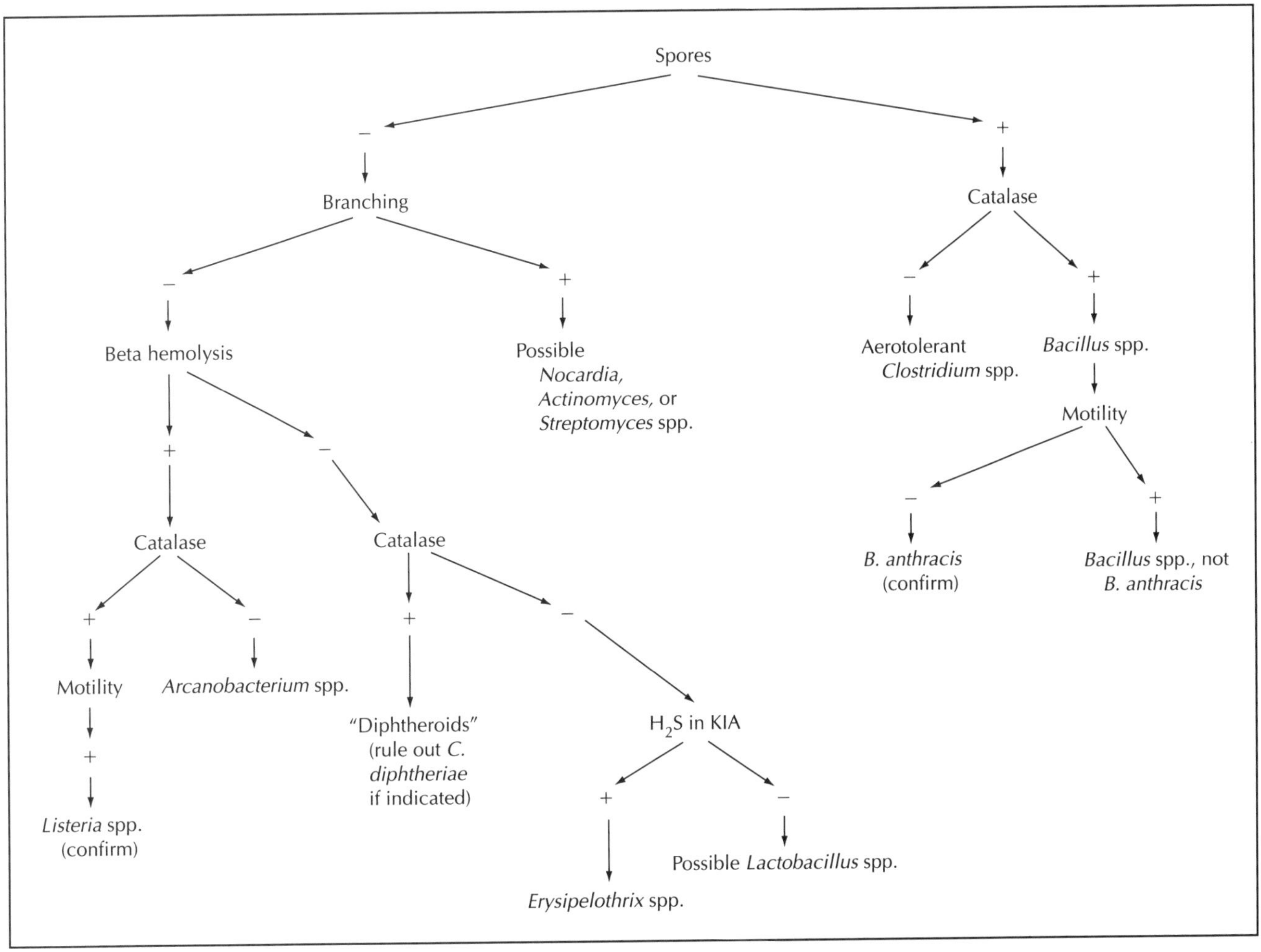

Figure 2.10–5 Identification of gram-positive bacilli from skin and subcutaneous tissues. KIA, Kligler iron agar; +, growth; −, no growth. Adapted from CMPH 1.16.12.

VI. ANTIMICROBIAL SUSCEPTIBILITY TESTS

(For details, see EPCM section 5.)

A. Perform standard antimicrobial susceptibility tests on clinically relevant isolates (*see* CMPH section 5).

B. Perform screening tests for epidemiologically important, resistant microorganisms, such as vancomycin-resistant enterococci and methicillin-resistant staphylococci, even if routine antimicrobial susceptibility testing is not indicated.

C. Report antimicrobial susceptibility testing results as soon as they are available. Rapid reporting leads to decreased morbidity and decreased hospital costs.

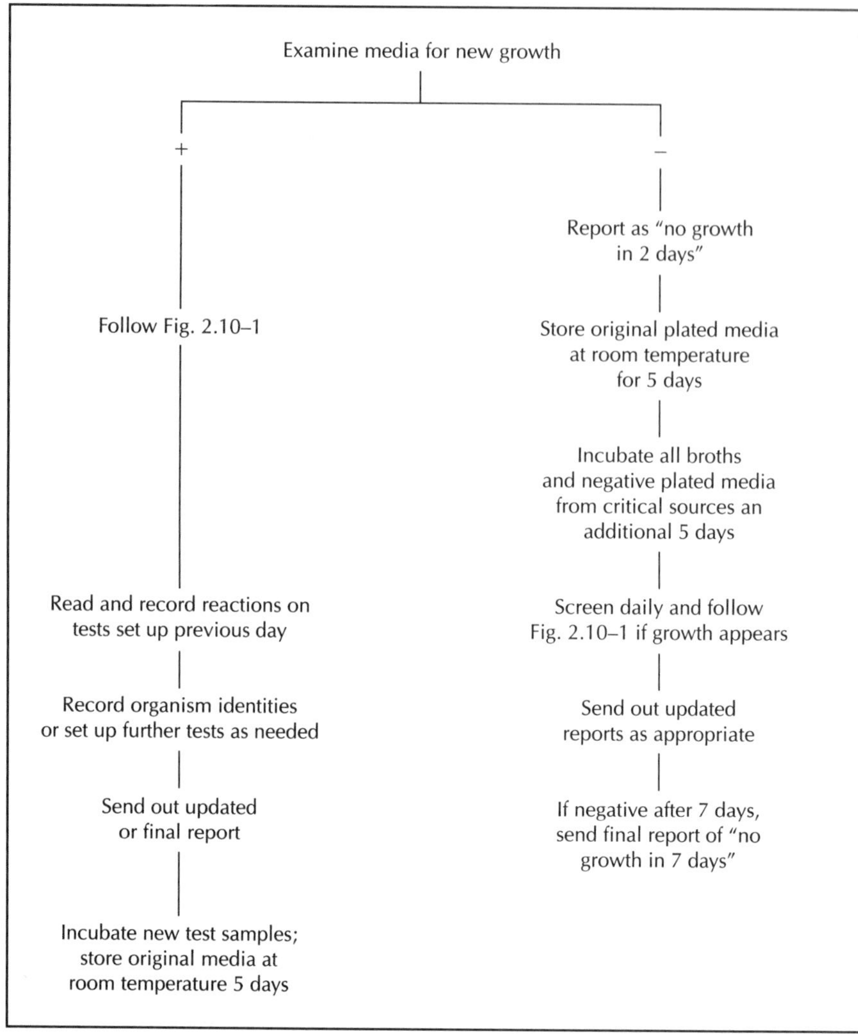

Figure 2.10–6 Examination of cultures—day 2. +, growth; −, no growth. Adapted from CMPH 1.16.6.

SUPPLEMENTAL READING

(For other references and supplemental reading, refer to CMPH 1.16.14–15.)

Blondeau, J. M., G. B. Pylypchuk, J. E. Kappel, R. B. Baltzan, Y. Yaschuk, and A. J. Adolph. 1995. Evaluation of aerobic Bactec 6A non-resin- and 16A resin-containing media for the recovery of microorganisms causing peritonitis. *Diagn. Microbiol. Infect. Dis.* **22:**361–368.

Morris, A. J., S. J. Wilson, C. E. Marx, M. L. Wilson, S. Mirrett, and L. B. Reller. 1995. Clinical impact of bacteria and fungi recovered only from broth cultures. *J. Clin. Microbiol.* **33:**161–165.

Runyon, B. A., M. R. Antillon, E. A. Akriviadis, and J. G. McHutchison. 1990. Bedside inoculation of blood culture bottles with ascitic fluid is superior to delayed inoculation in the detection of spontaneous bacterial peritonitis. *J. Clin. Microbiol.* **28:**2811–2812.

APPENDIXES

Aerobic Bacteriology

APPENDIX 2–1

Identification of Aerobic Gram-Negative Bacteria; Figures and Tables

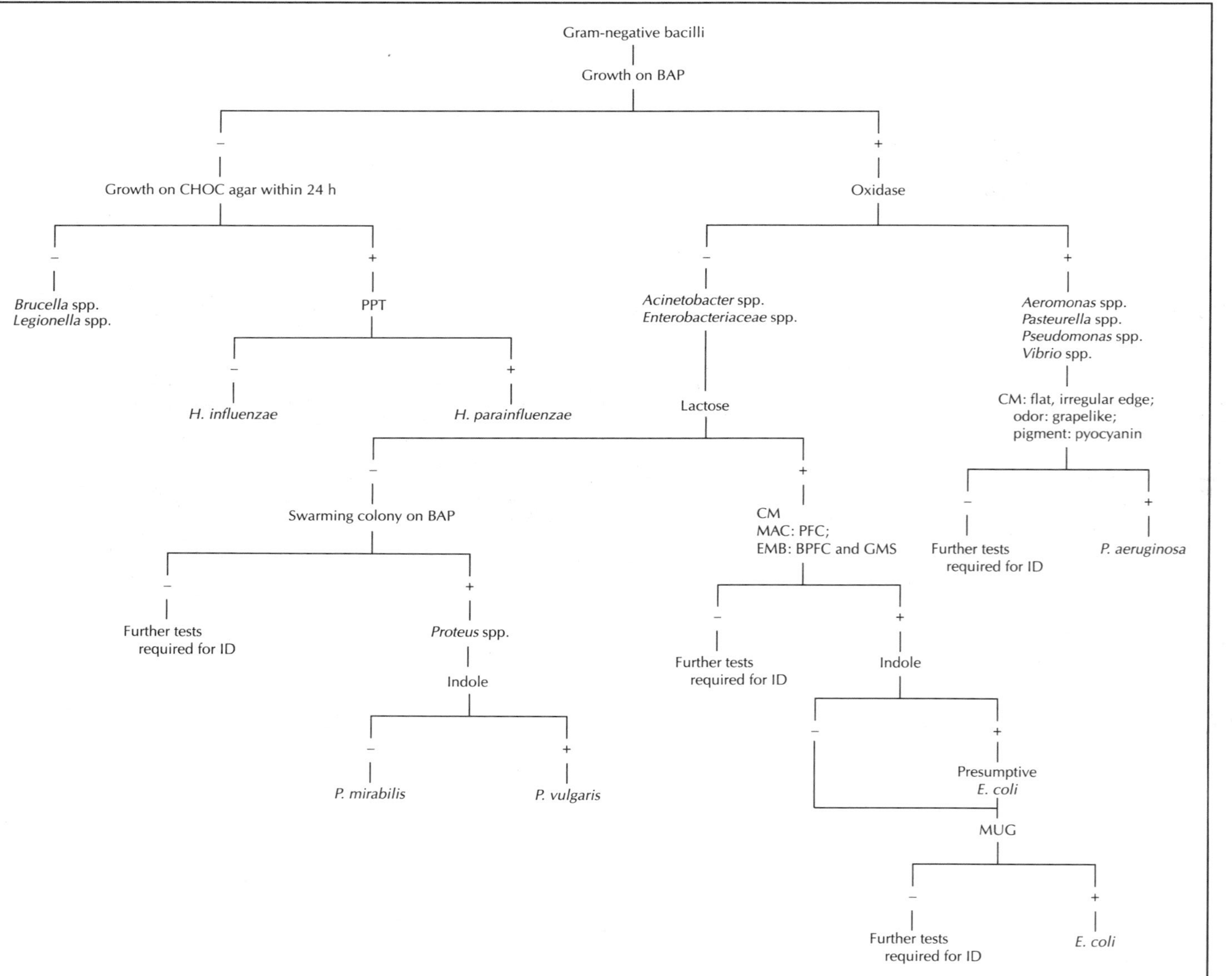

Figure A2–1–1 Flow diagram illustrating the use of rapid tests in the identification of gram-negative bacilli. Abbreviations: BPFC, blue-purple, flat colony; CM, colony morphology; GMS, green, metallic sheen; ID, identification; MUG, β-glucuronidase test; PFC, pink, flat colony; PPT, porphyrin production test. Organisms listed are examples of common clinical isolates. This chart is not all-inclusive for members of the family *Enterobacteriaceae*. Adapted from CMPH 1.19.2.

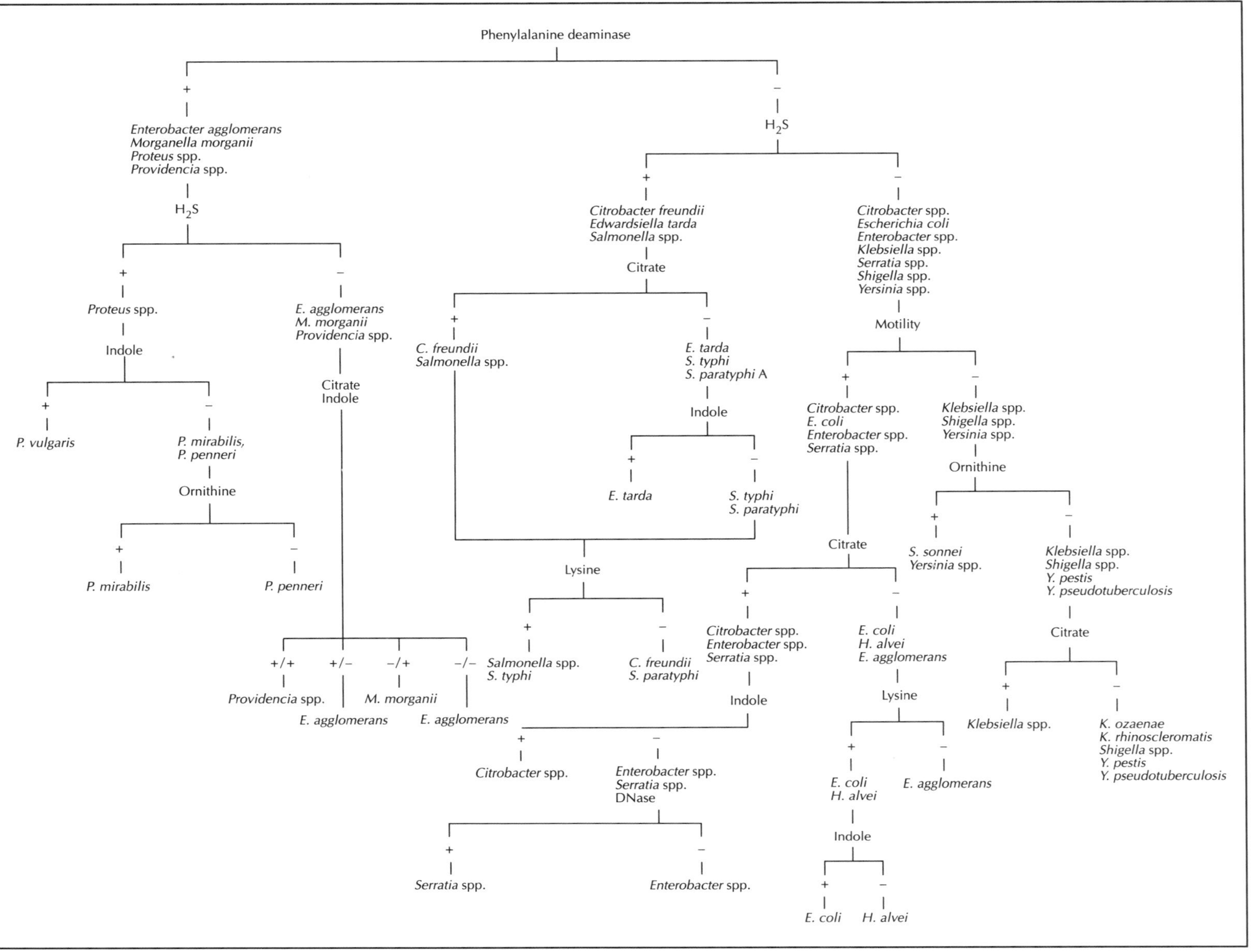

Figure A2–1–2 Flow diagram for identification of members of the family *Enterobacteriaceae*. Organisms listed are examples of common clinical isolates. This chart is not all-inclusive for the *Enterobacteriaceae*. Confirm biochemical identifications of *Salmonella* and *Shigella* by serology. Adapted from CMPH 1.19.3.

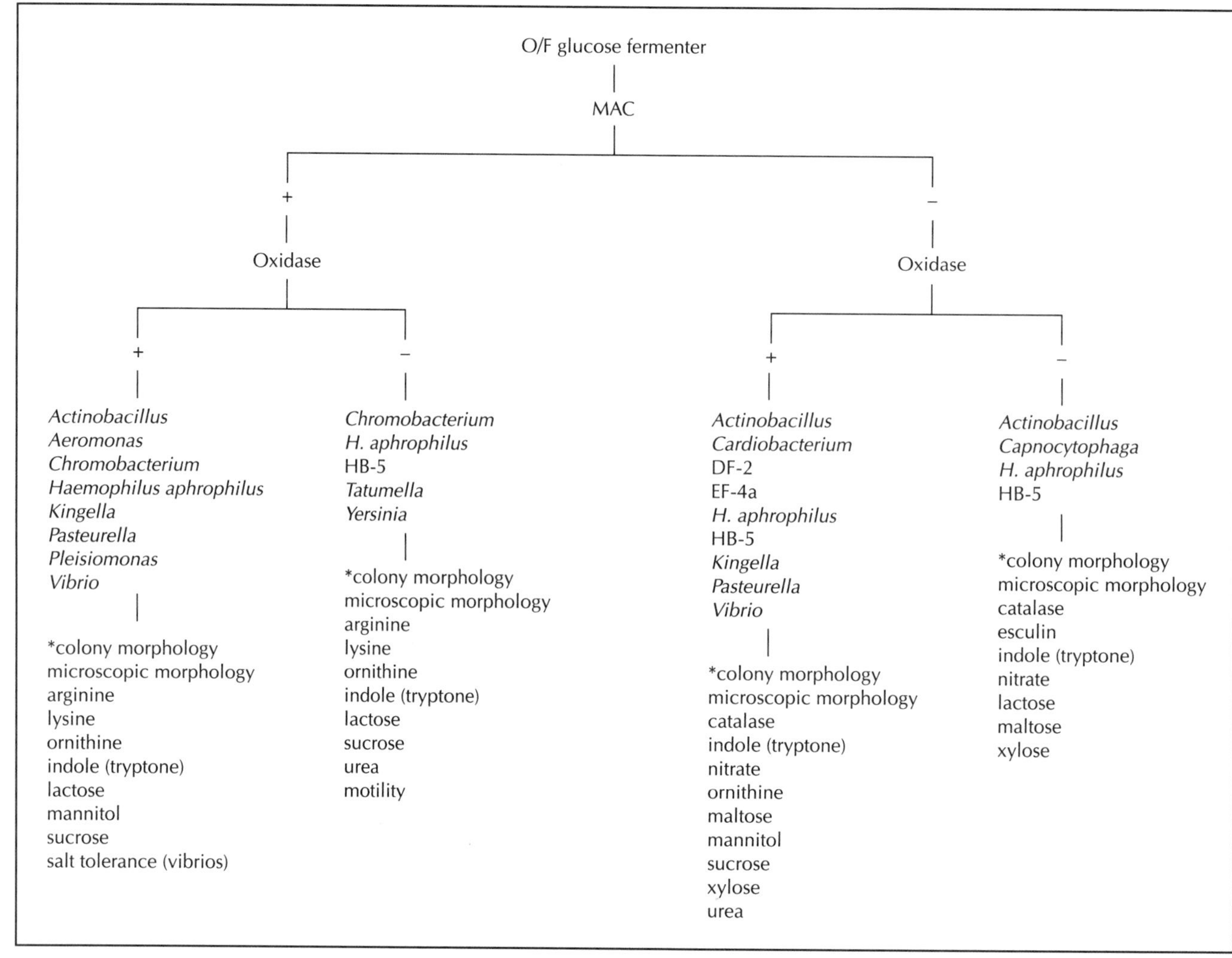

Figure A2–1–3 Flow diagram for presumptive identification of glucose-fermenting gram-negative bacilli other than members of the family *Enterobacteriaceae*. *, additional biochemical tests required for identification of most microorganisms; O/F, oxidative-fermentative. Use Andrade's carbohydrates. Adapted from CMPH 1.19.4.

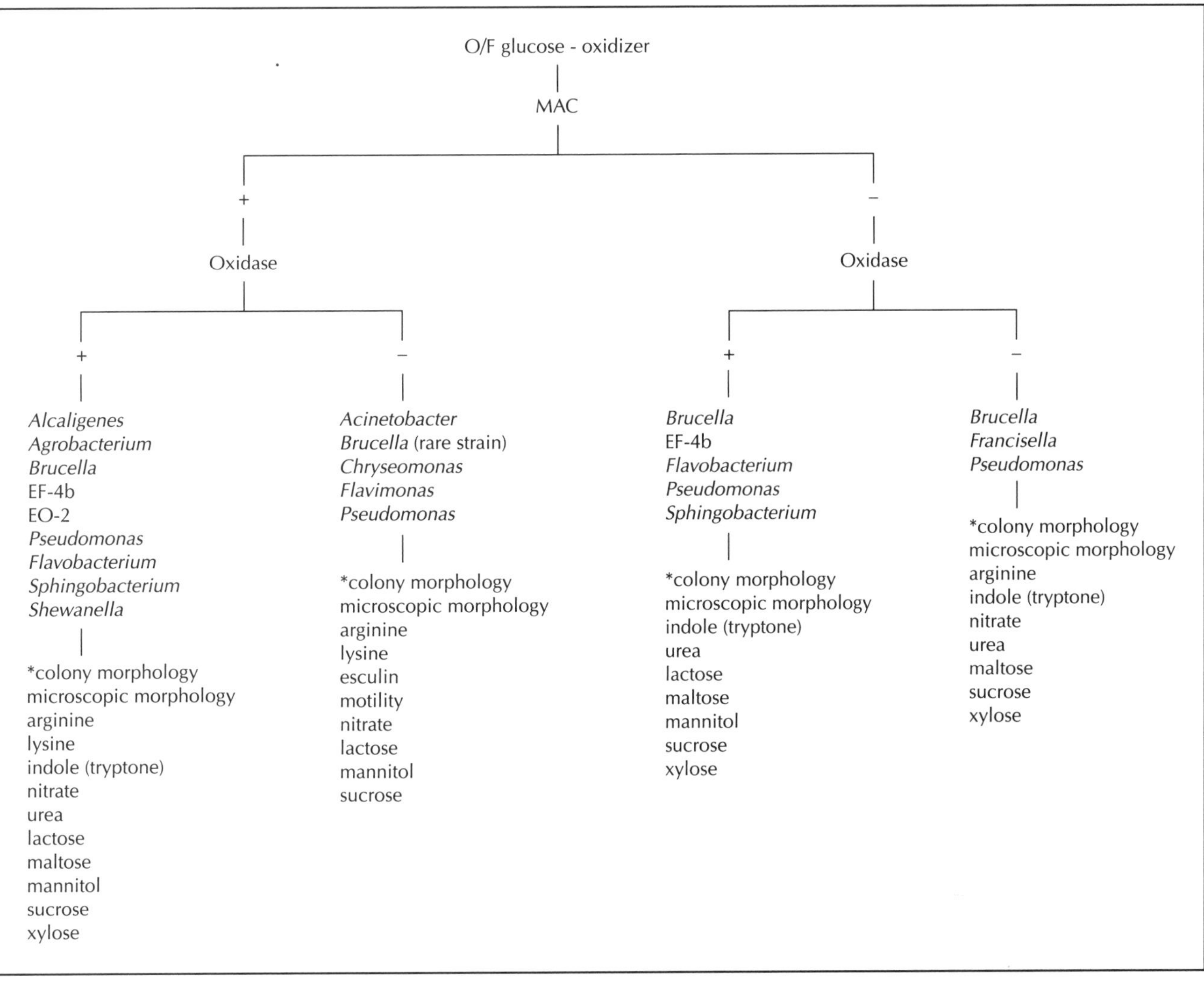

Figure A2–1–4 Flow diagram for presumptive identification of glucose-oxidizing gram-negative bacilli. O/F, oxidative-fermentative; *, additional biochemical tests required for identification of most microorganisms. Use oxidative-fermentative carbohydrates. Adapted from CMPH 1.19.5.

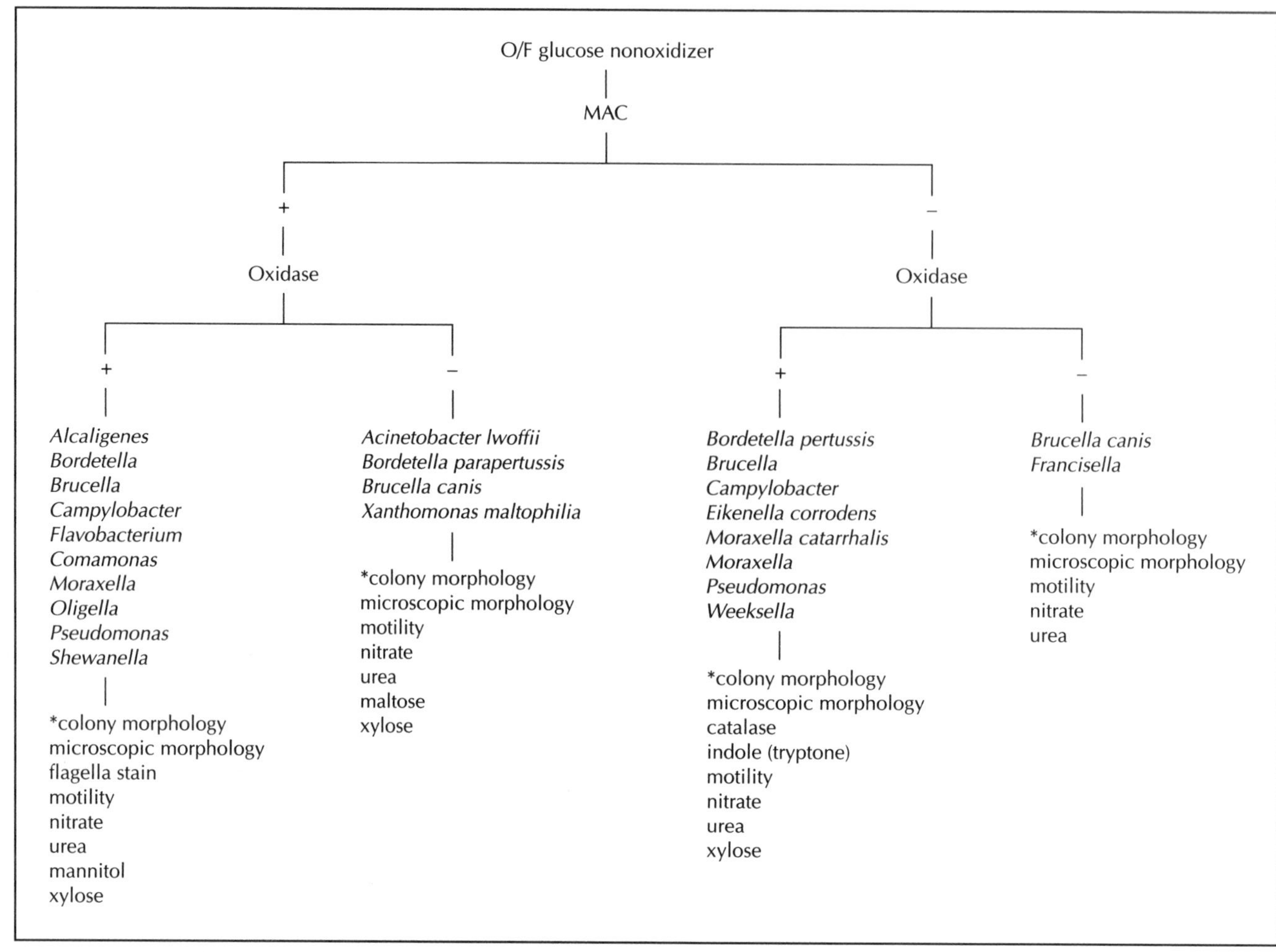

Figure A2–1–5 Flow diagram for presumptive identification of non-glucose-oxidizing gram-negative bacilli. O/F, oxidative-fermentative; *, additional biochemical tests required for identification of most microorganisms. Use oxidative-fermentative carbohydrates. Adapted from CMPH 1.19.6.

Table A2–1–1 Biochemical tests and their uses in differentiating gram-negative bacilli[a]

Test	Manual commercial ID system component	Comments
Carbohydrate fermentation	All commercial ID systems	Aids in characterization of bacterial groups or species
Citrate	API Rapid E, API 20E, Enterotube II, Key Scientific, Minitek, OXI/FERM tube, Rapid NFT, r/b System	Aids in differentiation between genera

Positive	Negative
Salmonella spp. (usually +)	*Edwardsiella tarda*
Klebsiella-Enterobacter group (usually +)	*Escherichia coli*
Providencia rettgeri	*Morganella morganii*

Aids in species differentiation

Positive	Negative
Bordetella spp.	*Bordetella pertussis*
Pseudomonas cepacia	*Pseudomonas maltophilia*
Klebsiella spp.	*Klebsiella rhinoscleromatis*
Salmonella spp.	*Salmonella paratyphi* A

Test	Manual commercial ID system component	Comments
CTA carbohydrates	Minitek, quadFERM+, NEISSERIA-KWIK	Aid in differentiation of *Neisseria* spp. and *M. catarrhalis*

	Glucose	Maltose	Lactose	Sucrose
N. gonorrhoeae	+	−	−	−
N. meningitidis	+	+	−	−
N. lactamica	+	+	+	−
N. sicca	+	+	−	+
N. subflava	+	+	−	+/−
N. mucosa	+	+	−	+
N. flavescens	−	−	−	−
N. cinerea	−	−	−	−
N. elongata	−	−	−	−
M. catarrhalis	−	−	−	−

Decarboxylase and dihydrolase — Aids in species differentiation

Test	Manual commercial ID system component		Arginine	Lysine	Ornithine
Arginine	API 20E, Key Scientific, Minitek, OXI/FERM tube, Rapid NFT	*Enterobacter* spp.			
		E. aerogenes	−	+	+
		E. agglomerans	−	−	−
		E. cloacae	+	−	+
		E. gergoviae	−	+	+
		E. sakazakii	+	−	+
Lysine	API Rapid E, API 20E, Enterotube II, Key Scientific, MICRO-ID, r/b System	*Proteus* spp.			
		P. mirabilis			+
		P. penneri			−
		P. vulgaris			−
Ornithine	API Rapid E, API 20E, Enterotube II, Key Scientific, MICRO-ID, Minitek, r/b System	*Salmonella* spp.			
		S. choleraesuis		+	+
		S. paratyphi A		−	+
		S. typhi		+	−
		Salmonella spp.		+	+
		Serratia spp.			
		S. liquefaciens group			+
		S. marcescens			+
		S. odorifera			+/−
		S. rubidaea			−
		Shigella spp.			
		S. boydii			−
		S. dysenteriae			−
		S. flexneri			−
		S. sonnei			+

(continued)

Table A2–1–1 Biochemical tests and their uses in differentiating gram-negative bacilli[a] *(continued)*

Test	Manual commercial ID system component	Comments
		Yersinia spp. *Y. enterocolitica* + *Y. frederiksenii* + *Y. kristensenii* + *Y. pestis* – *Y. pseudotuberculosis* – *Y. intermedia* +
Esculin	API Rapid E, API 20E, MICRO-ID, Minitek, N/F System, Rapid NFT	Aids in differentiation of nonfermentative gram-negative bacilli; Aids in species differentiation Positive: *Aeromonas hydrophilia*, *Aeromonas caviae* Negative: *Aeromonas sobria*
Fluorescence identification (FN)	Key Scientific, N/F System	Aids in differentiation of nonfermentative gram-negative bacilli
Gelatin	API 20E, Key Scientific, Rapid NFT	Aids in identification of some nonfermentative gram-negative bacilli: *Pseudomonas aeruginosa* and *Flavobacterium* spp. (+) Not routinely used for identification of *Enterobacteriaceae*: *Serratia* spp., *Proteus* spp., and some *Yersinia* spp. (+)
Hydrogen sulfide (H_2S)	API 20E, Enterotube II, MICRO-ID, Minitek, N/F System, OXI/FERM tube, r/b System	All members of *Enterobacteriaceae* can produce H_2S. Detection depends on sensitivity of method. Lead acetate paper is most sensitive detection method. Aids in differentiation between genera (TSI method): *Proteus* spp. (+) from *Morganella* and *Providencia* spp. (–) and other genera that produce H_2S (*Edwardsiella, Citrobacter, Salmonella*) Aids in species differentiation (TSI method) (Positive / Negative) *Citrobacter* spp. *C. freundii* + *C. diversus* – *C. amalonaticus* – *Campylobacter* spp. "*C. faecalis*" + "*C. hyointestinalis*" + *C. sputorum* + Other *Campylobacter* spp. –
Indole	API Rapid E, API 20E, Enterotube II, Micro-ID, Minitek, N/F System, OXI/FERM tube, Rapid NFT, r/b System	Conventional tube method is more sensitive than rapid spot test. Extraction with xylene is most sensitive detection method (often required with nonfermenters). Aids in differentiation between genera: *Edwardsiella* spp. (+) from *Salmonella* spp. (–); *Escherichia coli* (usually +) from *Klebsiella-Enterobacter* group (usually –) Aids in species differentiation Positive: *Proteus vulgaris*, *Klebsiella oxytoca*, *Citrobacter diversus*, *Citrobacter amalonaticus*, *Shigella* serogroups A, B, C (50% of strains) Negative: *Proteus mirabilis*, *Klebsiella pneumoniae*, *Citrobacter freundii*, *Shigella sonnei*
KIA		Directs initial characterization of gram-negative bacilli Aids in identification of *Erysipelothrix rhusiopathiae* and *Bacillus* spp.

Table A2–1–1 Biochemical tests and their uses in differentiating gram-negative bacilli[a] *(continued)*

Test	Manual commercial ID system component	Comments	
Malonate	API Rapid E, MICRO-ID, Minitek	Aids in differentiation between genera	
		Positive	Negative
		Alcaligenes faecalis	*Acinetobacter* spp.
		Salmonella arizonae	*Salmonella* spp. (usually −)
		Klebsiella-Enterobacter group (usually +)	*Escherichia coli, Serratia* spp.
		Aids in species differentiation	
		Positive	Negative
		Citrobacter diversus	*Citrobacter amalonaticus*
		Klebsiella spp.	*Klebsiella ozaenae*
		Serratia rubidaea	*Serratia* spp.
Methyl red		Component of IMViC set of reactions formerly used in initial characterization of *Enterobacteriaceae*	
		Aid in differentiation between genera	
		Positive	Negative
		Escherichia spp.	*Enterobacter* spp.
		Yersinia spp.	*Klebsiella* spp.
			Other gram-negative nonenteric bacilli
		Same substrate is used for both methyl red and Voges-Proskauer tests.	
MUG		Aids in rapid identification of *Escherichia coli*	
OF medium		Aids in differentiation of nonenteric, gram-negative bacilli from *Enterobacteriaceae*	
		Directs initial characterization of nonenteric gram-negative bacilli	
		Glucose fermenters	
		All members of *Enterobacteriaceae*	HB-5
		Actinobacillus spp.	*Haemophilus aphrophilus*
		Aeromonas spp.	*Kingella* spp.
		Capnocytophaga spp.	*Neisseria* spp. (some)
		Cardiobacterium spp.	*Pasteurella* spp.
		Chromobacterium spp.	*Plesiomonas* spp.
		EF-4a	*Streptobacillus* spp.
		Gardnerella spp.	*Vibrio* spp.
		Glucose oxidizers	*Flavobacterium* spp.
		Acinetobacter calcoaceticus	*Francisella* spp. (some)
		Alcaligenes spp.	*Neisseria* spp. (some)
		Agrobacterium spp.	*Pseudomonas* spp. (some)
		Brucella spp.	*Shewanella* spp. (some)
		Chryseomonas spp.	*Sphingobacterium* spp.
		EF-4b	
		Glucose nonoxidizers	*Francisella* spp. (some)
		Alcaligenes spp.	*Moraxella* spp.
		Acinetobacter lwoffii	*Neisseria* spp. (some)
		Bordetella spp.	*Oligella* spp.
		Brucella spp. (some)	*Pseudomonas* spp. (some)
		Campylobacter spp.	*Shewanella* spp. (some)
		Comamonas spp.	*Weeksella* spp.
		Eikenella spp.	
		Flavobacterium odoratum	

(continued)

Table A2–1–1 Biochemical tests and their uses in differentiating gram-negative bacilli[a] *(continued)*

Test	Manual commercial ID system component	Comments
ONPG	API Rapid E, API 20E, Key Scientific, MICRO-ID, N/F System	Aids in differentiating lactose-delayed organisms from lactose-negative organisms Aids in species differentiation Positive / Negative *Pseudomonas cepacia* / Other *Pseudomonas* spp. *Xanthomonas maltophilia* *Neisseria lactamica* / *Neisseria* spp. *Klebsiella* spp. / *Klebsiella rhinoscleromatis* Aids in differentiation of some nonfermentative gram-negative bacilli ONPG-negative organisms in *Enterobacteriaceae* include: *Edwardsiella tarda*; *Salmonella* spp. *Proteus* spp. (*P. vulgaris*, 10% +); *Shigella* spp. *Providencia* spp. (*P. stuartii*, 10% +); *Tatumella ptyseos*
Oxidase		Aids in differentiation of *Enterobacteriaceae* (−) from other glucose-fermenting gram-negative rods (+) Aids in identification of *Aeromonas, Campylobacter, Neisseria, Plesiomonas, Pasteurella, Pseudomonas*, and *Vibrio* spp. (+)
Phenylalanine deaminase	API Rapid E, API 20E, Enterotube II, Key Scientific, MICRO-ID, Minitek, r/b System	Aids in differentiation between genera Positive *Enterobacter agglomerans* (20% of strains) *Enterobacter sakazakii* (50% of strains) *Morganella morganii* *Proteus* spp. *Providencia* spp. *Rhanella aquatilis* *Tatumella ptyseos* Aids in species differentiation of certain *Pseudomonas* spp. Aids in species differentiation of *Moraxella* spp. Positive / Negative *M. phenylpyruvica* / Other *Moraxella* spp. *M. urethralis*
Porphyrin production		Aids in differentiation of *Haemophilus parainfluenzae* (+) from *Haemophilus influenzae* (−)
Tributyrin		Rapid identification of *Moraxella catarrhalis*
TSI agar		See KIA
Voges-Proskauer	API Rapid E, API 20E, Enterotube II, MICRO-ID	Component of IMViC reactions formerly used in initial characterization of *Enterobacteriaceae* Aids in differentiation Positive / Negative *Klebsiella* spp. / *Klebsiella ozaenae*, *Klebsiella rhinoscleromatis* *Aeromonas hydrophila* / *Aeromonas caviae* *Vibrio alginolyticus* / *Vibrio parahaemolyticus*, *Vibrio vulnificus*

[a] Abbreviations: ID, identification; CTA, cystine tryptic agar; FN, fluorescence-denitrification; OF, oxidation-fermentation; TSI, triple sugar iron agar; KIA, Kligler's iron agar; MUG, 4-methylumbelliferyl-β-D-glucoronide; ONPG, *O*-nitrophenyl-β-D-galactopyranoside; IMViC, indole–methyl red–Voges-Proskauer–citrate. Reprinted from CMPH 1.19.7–1.19.10.

Table A2–1–2 Characteristics of *Neisseria* species and *Moraxella catarrhalis*[a]

Organism	Colony morphology	Pigment	Growth on:			Acid production from[b]:				DNase	NO_3
			Selective medium[c]	CHOC or BAP (22°C)	Nutrient agar (35°C)	GLU	MAL	LAC (ONPG)	SUC		
N. gonorrhoeae	Smooth; multiple colony types	–	+	–	–	+	–	–	–	–	–
N. meningitidis	Smooth, transparent	–	+	–	–	+	+	–	–	–	–
N. lactamica	Smooth, transparent	–	+	V	+	+	+	+	–	–	–
N. cinerea	Slightly granular	–	–[d]	–	+	–[e]	–	–	–	–	–
N. flavescens	Opaque, smooth	+ (yellow)	–	+	+	–	–	–	–	–	–
N. sicca	Wrinkled, coarse, dry, adherent	–	–	+	+	+	+	–	+	–	–
N. subflava	Smooth, often adherent	+ (greenish yellow)	–	+	V	+	+	–	V	–	–
N. mucosa	Mucoid because of capsule production	+ (sometimes yellow)	–	+	+	+	+	–	+	–	+
M. catarrhalis	Opaque, smooth	–	V	V	+	–	–	–	–	+	+

[a] Adapted from CMPH 1.19.31.
[b] GLU, glucose; MAL, maltose; LAC, lactose; ONPG, *O*-nitrophenyl-*β*-D-galactopyranoside; SUC, sucrose.
[c] Modified Thayer-Martin, Martin-Lewis, or New York City medium. V, variable.
[d] Some strains isolated.
[e] Some strains may give a weak reaction.

Table A2–1–3 Sample selection of biochemical tests for identification of gram-negative rods in the family *Enterobacteriaceae*[a]

Test	Reaction and interpretation			
	Positive		Negative	
	Color	Microorganism	Color	Microorganism
Citrate	Blue	*K. pneumoniae*	No change	*E. coli*
Decarboxylase				
Arginine	Purple	*E. cloacae*	Yellow	*K. pneumoniae*
Lysine	Purple	*K. pneumoniae*	Yellow	*E. cloacae*
Ornithine	Purple	*E. cloacae*	Yellow	*K. pneumoniae*
Hydrogen sulfide	Brown to black	*P. mirabilis*	No change	*K. pneumoniae*
Indole (spot test)	Blue to green	*E. coli*	No change	*P. aeruginosa*
ONPG	Yellow	*K. pneumoniae*	No change	*P. mirabilis*
Phenylalanine	Green	*P. mirabilis*	No change	*K. pneumoniae*
Urea	Pink to red	*P. mirabilis*	No change	*E. coli*
Voges-Proskauer	Pink to red	*E. cloacae*	None	*E. coli*
Adonitol	Yellow	*K. pneumoniae*	Red	*P. mirabilis*
Arabinose	Yellow	*K. pneumoniae*	Red	*P. mirabilis*
Sorbitol	Yellow	*K. pneumoniae*	Red	*P. mirabilis*

[a] QC test organisms: *Enterobacter cloacae* ATCC 13047, *Escherichia coli* ATCC 25922, *Klebsiella pneumoniae* ATCC 13883, *Proteus mirabilis* ATCC 29245, and *Pseudomonas aeruginosa* ATCC 27853. ONPG, *O*-nitrophenyl-*β*-D-galactopyranoside. Reprinted from CMPH 1.19.31.

Table A2–1–4 Sample selection of biochemical tests for identification of aerobic, nonfastidious gram-negative bacilli other than members of the family *Enterobacteriaceae*[a]

Test	Result and interpretation			
	Positive		Negative	
	Color	Organism	Color	Organism
Oxidase (spot test)	Blue to purple	*P. aeruginosa*	No change	*E. coli*
Citrate	Blue	*P. aeruginosa*	Green	*P. putrefaciens*
Decarboxylase				
Arginine	Purple	*P. aeruginosa*	Yellow	*P. putrefaciens*
Lysine	Purple	*P. cepacia*	Yellow	*P. putrefaciens*
Ornithine	Purple	*P. putrefaciens*	Yellow	*P. aeruginosa*
Esculin	Brown to black	*A. hydrophila*	No change	*P. putrefaciens*
Indole (tryptophan)	Pink to red	*A. hydrophila*	No change	*P. putrefaciens*
Nitrate	Pink to red	*P. putrefaciens*	None	*P. aeruginosa*
Urea	Pink to red	*P. aeruginosa*	No change	*P. putrefaciens*
OF carbohydrates				
Glucose	Yellow	*P. cepacia*	Red	*P. putrefaciens*
Lactose	Yellow	*P. cepacia*	Red	*P. putrefaciens*
Maltose	Yellow	*P. cepacia*	Red	*P. putrefaciens*
Mannitol	Yellow	*P. cepacia*	Red	*P. putrefaciens*
Sucrose	Yellow	*P. cepacia*	Red	*P. putrefaciens*
Xylose	Yellow	*P. cepacia*	Red	*P. putrefaciens*

[a] QC test organisms: *Aeromonas hydrophila* ATCC 7965, *Escherichia coli* ATCC 25922, *Pseudomonas aeruginosa* ATCC 27853, *Pseudomonas cepacia* ATCC 17765, and *Pseudomonas putrefaciens* ATCC 8071. OF, oxidation-fermentation. Reprinted from CMPH 1.19.32.

APPENDIX 2–2 Identification of Aerobic Gram-Positive Bacteria; Figures and Tables

Table A2–2–1 Differentiation of catalase-negative aerobic gram-positive cocci and coccobacilli[a] (extracted from references 11 and 12 in CMPH 1.20)

Genus	VA	GG	PYR	LAP	NaCl	Growth at:	
						45°C	10°C
Streptococcus	S	–	–[b]	+	–	V	–
Enterococcus	S(R)	–	+	+	+	+	+
Lactococcus	S	–	–[c]	+	V	–[d]	+
Aerococcus	S	–	+	–	+	–	–
Gemella	S	–	+	V	–	–	–
Pediococcus	R	–	–	+	V	+	–
Leuconostoc	R	+	–	–	V	–	+
Lactobacillus	R(S)	V	–	V	V	V	V

[a] Abbreviations: VA, vancomycin; GG, gas from glucose; LAP, leucine aminopeptidase; NaCl, 6.5% saline; S, susceptible; R, resistant; V, variable. Reprinted from CMPH 1.20.16.

[b] *S. pyogenes* is positive.

[c] *L. garviae* is positive.

[d] Occasionally positive.

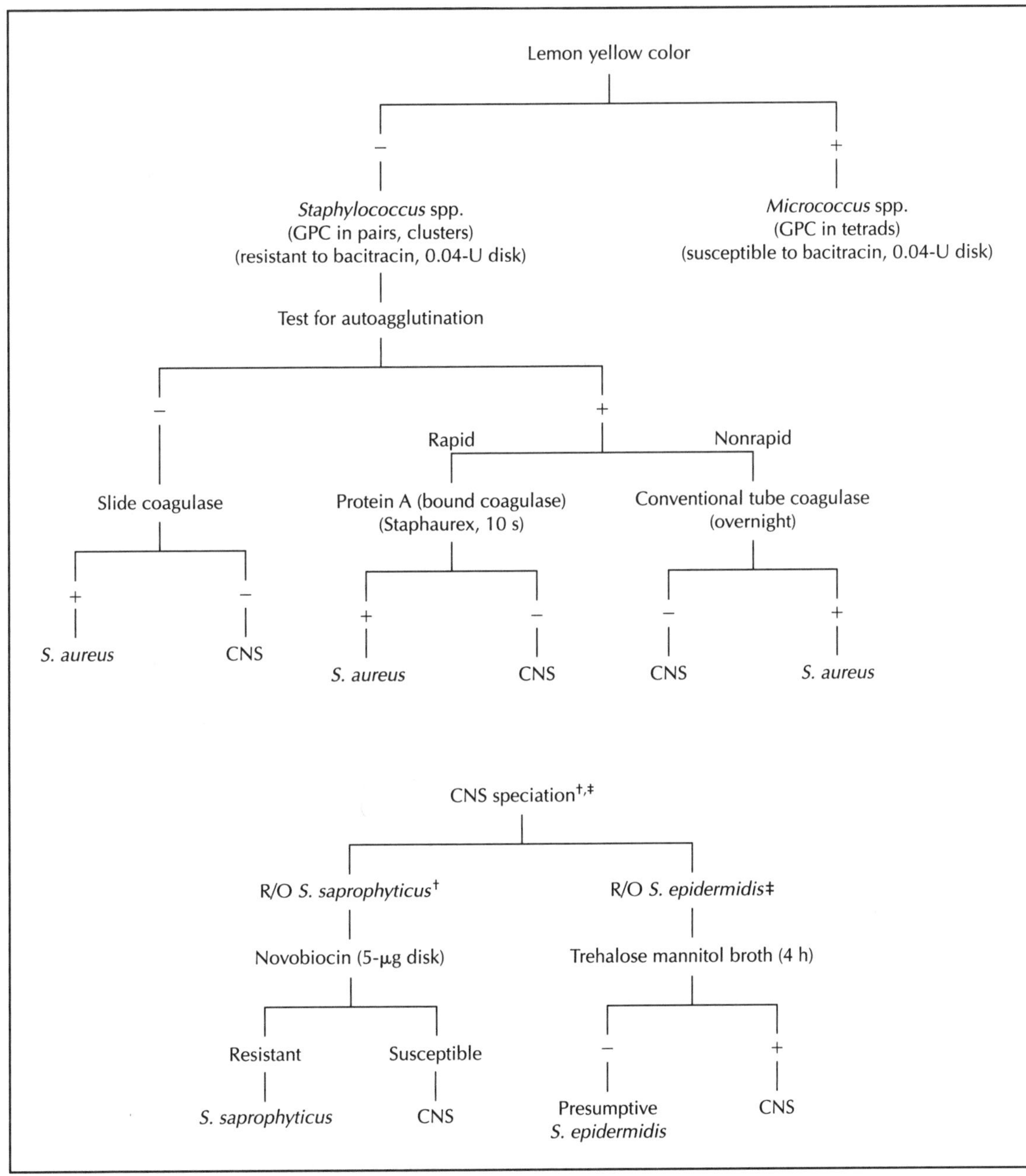

Figure A2–2–1 Differentiation of *Staphylococcus* and *Micrococcus* spp. (adapted from CMPH 1.20.3). Abbreviations: GPC, gram-positive cocci; CNS, coagulase-negative staphylococci; R/O, rule out. †, pure culture of CNS isolated from female urine specimen. ‡, isolates from blood, catheter tips, invasive hardware (e.g., prosthetic devices for valves and joints, pacemaker wires, epidural plates), CSF, or ventricular fluid collected through a shunt.

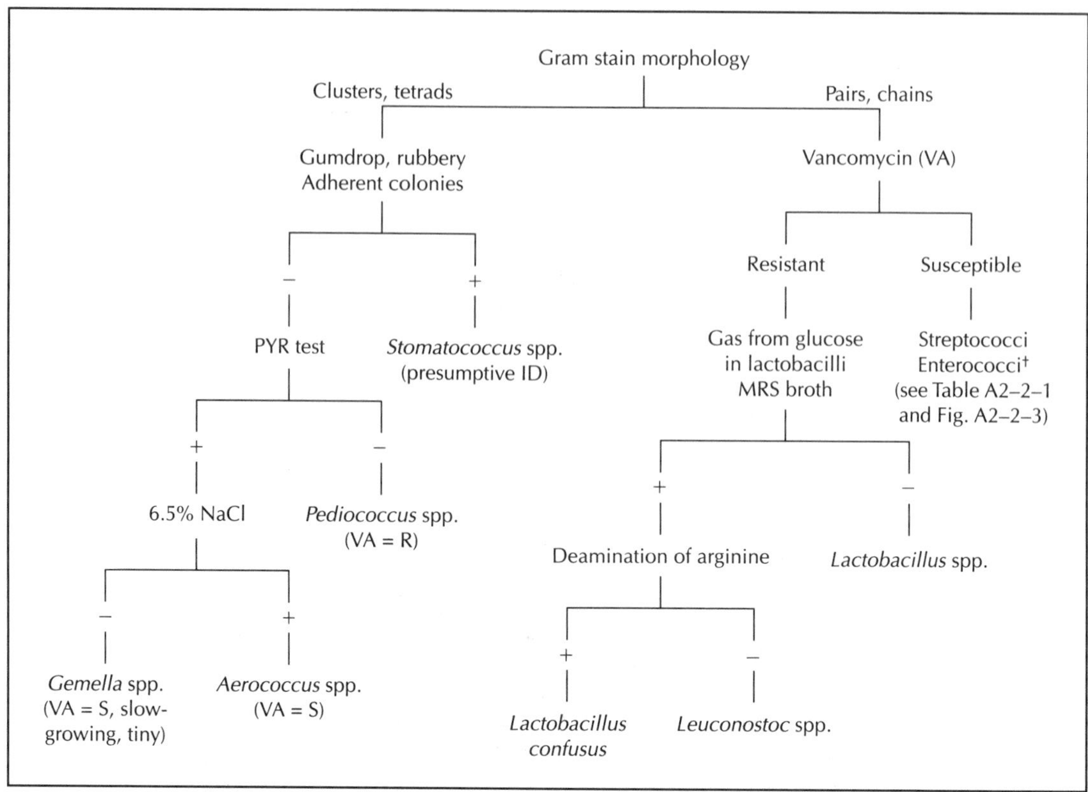

Figure A2–2–2 Presumptive differentiation of catalase-negative gram-positive cocci and gram-positive coccobacilli (adapted from CMPH 1.20.13). Abbreviations: VA, vancomycin; PYR, L-pyrrolidonyl-β-naphthylamide; R, resistant; S, susceptible; ID, identification. †, Vancomycin-resistant enterococci may also be isolated from clinical specimens.

Table A2–2–2 Differentiation of *Bacillus cereus* and related species[a]

Bacillus sp.	Motility	Hemolysis	Penicillin susceptibility	7% NaCl	Virulence in mice
B. cereus	+	+	−	+	−
B. anthracis	−	−	+	+	+
B. thuringiensis	+	+	−	+	−
B. cereus subsp. *mycoides*[b]	V	V	−	V	−

[a] V, variable. Reprinted from CMPH 1.20.31.
[b] Rhizoid growth.

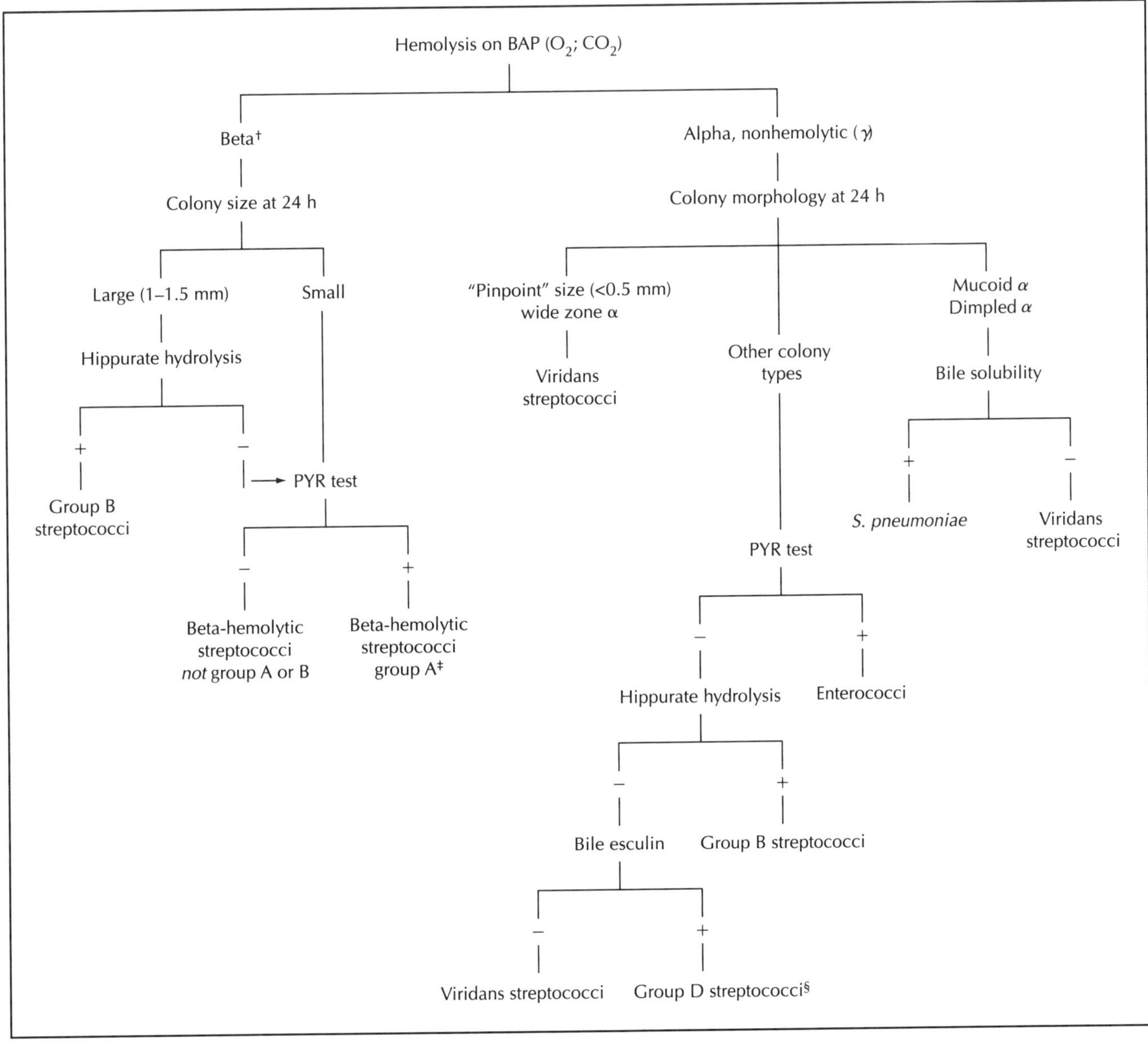

Figure A2–2–3 Differentiation of streptococci and enterococci (adapted from CMPH 1.20.14). PYR, L-pyrrolidonyl-β-naphthylamide. †, If isolated from sterile body fluids such as blood or CSF, perform a serologic test to identify groups A, B, C, D, F, and G. ‡, This flow diagram is based on the premise that beta-hemolytic enterococci are rarely encountered in clinical specimens. §, Identify to the species level if isolated from blood culture.

Table A2–2–3 Differentiation of *Listeria monocytogenes* from related gram-positive bacilli

Organism(s)	Catalase	Esculin	Motility	H_2S on TSI[a]
L. monocytogenes	+	+	+	–
Corynebacterium spp.	+	–[b]	–/+	–
E. rhusiopathiae	–	–	–	+
Lactobacillus spp.	–	V[c]	–	–

[a] TSI, triple sugar iron agar. Reprinted from CMPH 1.20.35.
[b] *C. aquaticum* hydrolyzes esculin and is yellow pigmented.
[c] V, variable reaction.

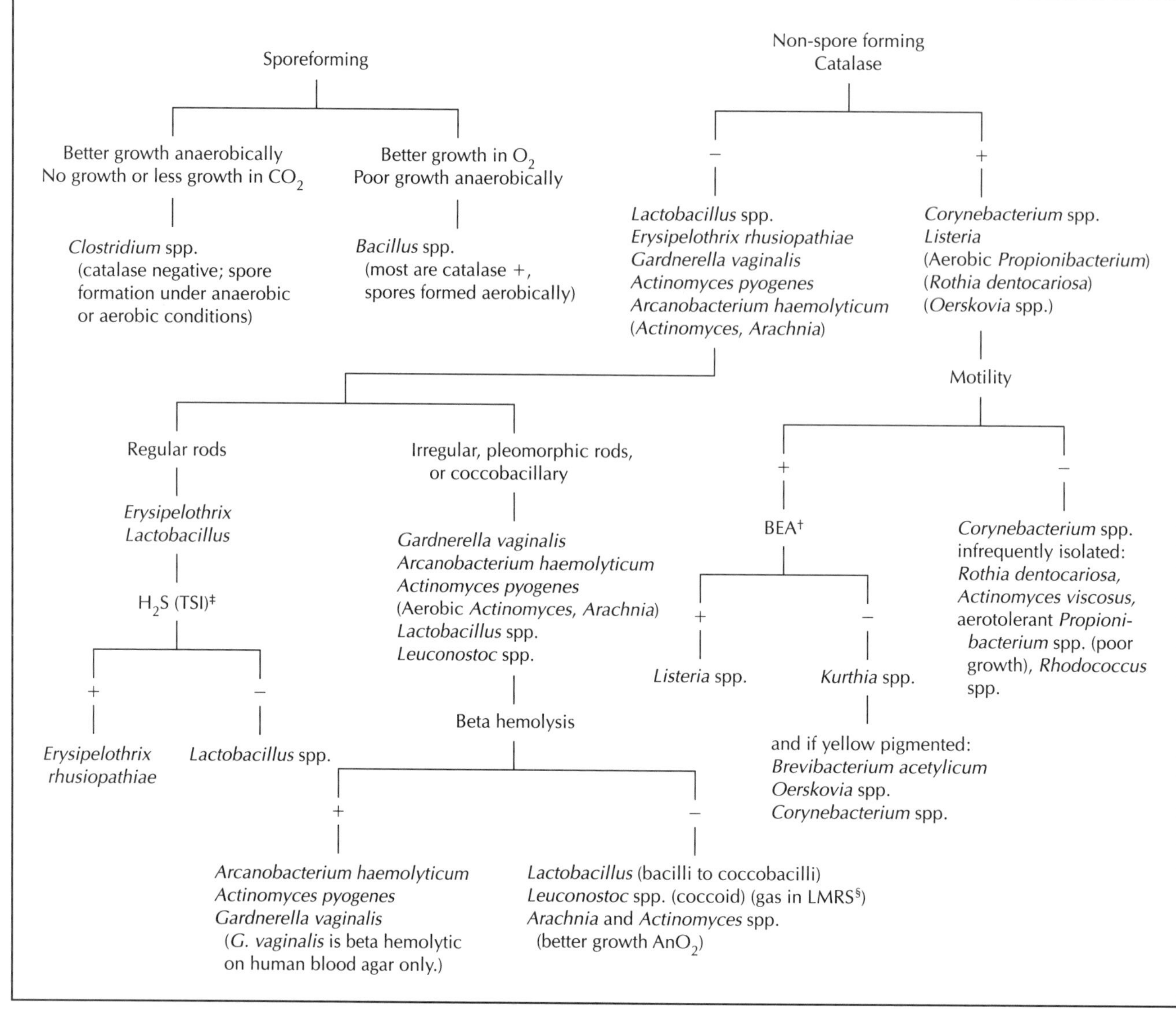

Figure A2–2–4 Differentiation of common aerobic gram-positive bacilli (adapted from CMPH 1.20.29). †, BEA, bile esculin agar (growth and hydrolysis of esculin); ‡, hydrogen sulfide formation along the stab line in triple sugar iron agar; §, lactobacilli MRS broth (gas trapped in Durham tube).

Table A2–2–4 Differentiation of corynebacteria and related microorganisms[a]

Microorganism(s)	Catalase	Nitrate	Gelatin hydrolysis	Glucose[b]	Xylose
Corynebacterium spp.	+	+/−	−	+ (OF)	−
Rhodococcus spp.[c]	+	V	−	+ (O)	V
Rhodococcus equi[c]	+	V	−	V (O)	−
Arcanobacterium haemolyticum	−	−	−	+ (OF)	+
Oerskovia spp.	+	+	+	+ (F)	+
Kurthia spp.	+	−	−	−	−

[a] V, variable reaction. Reprinted from CMPH 1.20.36.
[b] O, oxidative; F, fermentative.
[c] Salmon-colored colonies.

SECTION 3 Anaerobic Bacteriology

SECTION EDITOR: *James I. Mangels*

3.1 Introduction

Anaerobic bacteria cause a variety of infections in humans, including appendicitis, cholecystitis, otitis media, dental and oral infections, endocarditis, endometritis, myonecrosis, osteomyelitis, peritonitis, empyema, salpingitis, septic arthritis, sinusitis, wound infections following bowel or gynecological surgery or trauma, perirectal and tuboovarian abscess, and bacteremia (1, 2, 4). Many reports associate 50 to 60% of important infections with anaerobic bacteria (1–4). Anaerobes are also important because they can resist many commonly used antimicrobial agents, and they exhibit many virulence factors (2, 3).

Anaerobic infections are generally endogenous (from our own normal anaerobic microbiota). However, despite the variety of normal human anaerobic microbiota, anaerobic infections are generally limited to a few organisms: one-third are *Bacillus fragilis* group, 25% are members of the peptostreptococci, 11% are pigmented gram-negative rods, and 8% are fusobacteria (2, 3).

Anaerobic bacteria are often overlooked or missed unless the specimen is properly collected and transported to the laboratory and then isolated properly. Anaerobes vary in their susceptibility to oxygen; a brief exposure (10 min) to atmospheric oxygen is enough to kill some organisms. They also vary in their nutritional requirements, but most isolates require vitamin K (menadione) and hemin for growth.

This section outlines practical, cost-effective procedures a laboratory can use to establish the presence of clinically important anaerobes and to rapidly identify them.

REFERENCES

1. **Baron, E. J., L. R. Peterson, and S. M. Finegold.** 1994. *Bailey and Scott's Diagnostic Microbiology,* 9th ed., p. 474–550. The C. V. Mosby Co., St. Louis.
2. **Finegold, S. M., and W. L. George.** 1989. *Anaerobic Infections in Humans.* Academic Press, Inc., New York.
3. **Murray, P. R., E. J. Baron, M. A. Pfaller, F. C. Tenover, and R. H. Yolken (ed.).** 1995. *Manual of Clinical Microbiology,* 6th ed., p. 574–619. American Society for Microbiology, Washington, D.C.
4. **Smith, L. D. S.** 1975. *The Pathogenic Anaerobic Bacteria,* 2nd ed. Charles C Thomas, Publisher, Springfield, Ill.

3.2 Processing, Media Selection, and Anaerobic Incubation

I. PRINCIPLE

This section describes methods of processing; the use and importance of primary, selective, and differential media for the rapid workup of anaerobes; and techniques for providing proper anaerobic incubation conditions. The laboratory's goal is to rapidly isolate significant anaerobic organisms present in the original specimen for identification and susceptibility testing if needed.

II. SPECIMENS

(See Tables 3.2–1, 3.2–2, and 3.2–3.)

Proper collection of specimens to avoid contamination with indigenous microbiota and prompt transport to the laboratory for processing are extremely important for the isolation of anaerobes (1, 3, 7). The laboratory must provide the clinical staff with guidelines for the optimal amount and type of specimen required for anaerobic culture and must stress the need to transport the properly collected specimen to the laboratory without delay. Good communication between the laboratory and the clinical staff is essential to ensure specimen quality (4).

III. MATERIALS

(See also CMPH 2.3.1 for more details.)

Include QC information on reagent container and in QC records.

The following media may be used. These media are available in dehydrated form and prepared based on the manufacturers' directions. Commercially prepared, enriched primary agar media are available from suppliers.

A. Enriched all-purpose medium for primary growth

B. Kanamycin-vancomycin-laked blood agar (KVLB or LKV)

C. Phenylethyl alcohol agar plates (PEA)

D. *Bacteroides* bile esculin agar (BBE)

E. Cycloserine-cefoxitin fructose agar (CCFA)

F. Egg yolk agar (EYA)

G. Broths

1. THIO supplemented with vitamin K_1 (menadione) (0.1 μg/ml), hemin (5 μg/ml), and a marble chip or sodium bicarbonate (1 μg/ml)
2. Chopped meat-carbohydrate or chopped meat-glucose supplemented with vitamin K_1 and hemin

IV. PROCEDURE: PROCESSING SPECIMENS FOR ANAEROBIC CULTURE

(See CMPH 2.4 for more details.)

A. Grind bone or tissue with approximately 1 ml of THIO or chopped meat. Wring out swabs in 0.5 ml of THIO or chopped meat, and then treat them as a liquid specimen.

B. Centrifuge large volumes of nonpurulent material. Use the sediment to inoculate the media and to prepare the Gram stain.

C. The primary Gram stain of the specimen is critical. Carefully describe and record all morphotypes. The Gram stain provides preliminary information to the physician as well as quality control information to the microbiologist.

D. The use of methanol fixation has been shown to improve the quality of the Gram stain smear because bacterial and host cell morphologies remain intact (5).

Table 3.2–1 Acceptable specimens for anaerobic culture[a]

Site	Acceptable specimens	Unacceptable specimens
Head and neck	Abscess aspirate obtained by needle and syringe after surface decontamination Biopsy material surgically obtained Anaerobic swab surgically obtained when aspiration is not feasible	Throat or nasopharyngeal swabs Gingival swabs Superficial material collected with swabs
Lungs	Transtracheal aspirate Material from percutaneous lung puncture Biopsy material surgically obtained Bronchoscopic specimen obtained by protected brush Thoracotomy specimen Anaerobic swab surgically obtained	Expectorated sputum Induced sputum Endotracheal aspirate Bronchoscopic specimens not specially collected
Central nervous system	Abscess aspirate obtained by needle and syringe Biopsy material surgically obtained Anaerobic swabs surgically obtained	Aerobic swabs
Abdomen	Peritoneal fluid obtained by needle and syringe Abscess aspirate obtained by needle and syringe Bile Biopsy material surgically obtained Anaerobic swab surgically obtained	Aerobic swabs
Urinary tract	Suprapubic aspirate	Voided urine Catheterized urine
Female genital tract	Culdoscopy specimens Endometrial aspirate obtained by suction or protected collector Abscess aspirate obtained by needle and syringe Biopsy material surgically obtained Anaerobic swabs surgically obtained IUD[b] for *Actinomyces* species or *Eubacterium nodatum*	Vaginal or cervical swabs
Bone and joint	Aspirate obtained by needle and syringe Biopsy material surgically obtained Anaerobic swab surgically obtained	Superficial material collected with swabs
Soft tissue	Aspirate obtained by needle and syringe Biopsy material surgically obtained Aspirate from sinus tract obtained by needle and small plastic catheter Deep aspirate of open-wound margin obtained through decontaminated skin Deep aspirate of surface ulcer obtained through decontaminated skin	Superficial material collected from skin surface or edges of wound
Stomach and small bowel	Only for workup of blind-loop or malabsorption syndrome	
Large bowel	Only for culture or toxin assay for suspected involvement of *Clostridium difficile* or *Clostridium botulinium, Anaerobiospirillum succiniciproducens*, and other etiologic agents	

[a] Adapted from CMPH 2.2.2.
[b] IUD, intrauterine device.

Table 3.2–2 Anaerobic specimen transport devices[a]

Specimen type	Transport system	Commercially available system (manufacturer)
Aspirated material	Vial or tube with anaerobic atmosphere and agar base with indicator system	Port-A-Cul (BD Microbiology Systems) Anaerobic Transport Medium (Anaerobe Systems) Anaport and Anatube (Adams Scientific)
Tissue, biopsy material, or curettings	Bag or tube systems that act by removing molecular oxygen	Bio-Bag (Type A) (BD Microbiology Systems) Anaerobic Pouch (Difco Laboratories) Anaerobic Specimen Collector (BD Vacutainer Systems)
Collected on swabs	Tube with anaerobic atmosphere and agar base with indicator system	Anatube (Adams Scientific)
	Tube with catalytic system to generate anaerobic atmosphere after swabs are inserted	Anaerobic Specimen Collector (BD Vacutainer System) Anaerobic Culturette System (BD Microbiology Systems)
	Tube with anaerobic atmosphere and reduced transport medium	Anaerobic Transport Medium (Anaerobe Systems) Port-A-Cul (BD Microbiology Systems)

[a] Adapted from CMPH 2.2.3.

V. MEDIA SELECTION AND INOCULATION METHODS

A. Principle

Successful isolation of anaerobes depends on choosing the correct, primary growth media and environmental conditions to culture clinically significant anaerobes. Use a combination of enriched, nonselective, selective, and differential media for the isolation and presumptive identification of anaerobic bacteria from clinical specimens. Most anaerobes require hemin and vitamin K. Some studies suggest that freshly prepared, properly stored, highly enriched media are essential for recovery of anaerobes (2, 8), while another study has shown that prereduced, anaerobically sterilized media (PRAS) best support the growth of anaerobes (6).

Table 3.2–3 Suggested transport time for certain specimen volumes and collection methods[a]

Specimen type	Optimal time for transport to laboratory	Additional comments
Aspirated material		
Very small vol (<1.0 ml)	≤10 min	Transport small vol of aspirated material in anaerobic transport vial whenever possible for best possible results.
Small vol (~1.0 ml)	≤30 min	
Large vol (>2.0 ml)	≤2–3 h	Transport large vol of purulent material; large pieces of tissue; or aspirated material, tissue, biopsy material, or curettings in an anaerobic transport medium or container.
In anaerobic transport device	≤2–3 h	These specimens can generally be accepted for anaerobic culture with good results even after a delay of 8–24 h. Include comment regarding transport delay in report when these cultures are processed.
Tissue or biopsy material		
In sterile container	≤30 min	
In anaerobic bag or transport device	≤2–3 h	
Anaerobic swabs		
In tube with moist anaerobic atmosphere	≤1 h	
In anaerobic transport medium	≤2–3 h	

[a] Adapted from CMPH 2.2.4.

Table 3.2–4 Recommended primary media setup[a]

Medium[b]	Organisms inhibited	Organisms that grow
Brucella blood agar	None	All
PEA	Faculative gram-negative organisms	All anaerobes
BBE	Most except *B. fragilis*	*B. fragilis* group, some *Fusobacterium* spp.
LKV	All gram-positive and facultative gram-negative (*Fusobacterium* spp., *Bacteroides ureolyticus* group, and *Porphyromonas* spp.) organisms	*Bacteroides* spp., some *Fusobacterium* spp., *Prevotella* spp.
Broth		
Enriched THIO	None	All
Chopped meat Carbohydrate	None	All

[a] Adapted from CMPH 2.3.4.
[b] These and all media for isolation of anaerobes should contain vitamin K_1 and hemin.

B. Media for anaerobic culture (*see* Tables 3.2–4 and 3.2–5)

1. Brucella blood agar, CDC anaerobe agar, Columbia, Schaedler blood agar, or other enriched primary media with 5% sheep blood supplemented with vitamin K_1 (10 μg/ml) and hemin (5 μg/ml) for the isolation of most organisms
2. Phenylethyl alcohol-BAP for the selection of anaerobic organisms
3. KVLB (LKV) for the selection of pigmented and nonpigmented *Prevotella* and other *Bacteroides* spp.
4. BBE for the selection and presumptive identification of *Bacteroides fragilis* group organisms and *Bilophilia wadsworthia*
5. Chopped meat broth or THIO (supplemented with vitamin K and hemin). The liquid broth used for backup purposes must contain vitamin K_1 and hemin.

Table 3.2–5 Anaerobic media and their uses[a]

Medium	Purpose	Interpretation	Limitations	QC organisms
BHI agar Brucella blood agar CDC anaerobe agar Columbia blood agar TSA Schaedler blood agar	General all-purpose enriched primary media that allow growth of all clinically significant anaerobes and facultative anaerobes	Observed initially in 24 h. Determine if organism is anaerobe by aerotolerance testing. Hold medium 5–7 days before completing report.	You must perform Gram stain and aerotolerance test. Aerobes will grow on medium.	*Clostridium perfringens* ATCC 13124 *Bacteroides fragilis* ATCC 25285 *Fusobacterium nucleatum* ATCC 25586 *Peptostreptococcus anaerobius* ATCC 27337
KVLB or LKV	Rapid isolation and selection of *Bacteroides* species and earlier detection of pigmented anaerobic gram-negative rods	Growth is presumptive; *Bacteroides* spp., pigmented anaerobic gram-negative rods, *Prevotella* spp., or *Fusobacterium mortiferum*	You must perform Gram stain and aerotolerance test. Yeast cells and other kanamycin-resistant organisms may grow on this medium.	*Bacteroides levii* ATCC 29147 (growth, pigment) *Bacteroides fragilis* ATCC 25285 (growth) *Clostridium perfringens* ATCC 13124 (no growth) *Escherichia coli* ATCC 25922 (no growth)

(continued)

Table 3.2–5 Anaerobic media and their uses[a] *(continued)*

Medium	Purpose	Interpretation	Limitations	QC organisms
PEA	Inhibition of growth of facultative gram-negative rods; prevents *Proteus* spp. from swarming	Most gram-positive and gram-negative anaerobes will grow on PEA. Growth may be considered presumptive evidence of anaerobic organism, but further testing is required.	Examine at 24–48 h. Additional time may be necessary for some slower-growing anaerobes. You must perform Gram stain and aerotolerance test.	*Bacteroides fragilis* ATCC 25285 (growth) *Proteus mirabilis* ATCC 12453 (inhibition of swarming and growth) *Escherichia coli* ATCC 25922 (no growth)
BBE	Rapid selection, isolation, and presumptive identification of *B. fragilis* group	Supports growth of bile-resistant *Bacteroides* species, which grow and form brown to black colonies, but not bile-sensitive members of *B. fragilis* group.	Some strains of *F. mortiferum, Klebsiella pneumoniae*, enterococci, and yeasts may grow to limited extent on this medium. Some anaerobic organisms that should grow on BBE may be inhibited, so inoculate nonselective medium also. You must perform Gram stain and aerotolerance test. *B. vulgatus* does not form black colonies or discolor medium.	*Bacteroides fragilis* ATCC 25285 (growth, black colonies) *Bacteroides levii* ATCC 29147 (no growth) *Escherichia coli* ATCC 25923 (no growth)
CCFA	Selective and differential medium for recovery and presumptive identification of *Clostridium difficile*	*C. difficile* produces yellow ground-glass colonies, and original pink-colored agar turns yellow in vicinity of colonies.	Colonial morphology by Gram stain should be consistent with that of *C. difficile*. Additional identification methods should be used.	*Clostridium difficile* ATCC 9689 (growth) *Escherichia coli* ATCC 25922 (no growth) *Bacteroides fragilis* ATCC 25285 (no growth)
EYA	For use when *Clostridium* spp. are suspected or when proteolytic enzyme (lecithinase or lipase) may be useful for identification of anaerobic isolate	Positive lecithinase reaction is indicated by opaque zone (white) in medium around bacterial growth. Positive lipase reaction is indicated by iridescent sheen on surface of bacterial growth of agar. Proteolysis is indicated by total clearing around bacterial growth.	Lipase is observed under oblique light. It is best not to use all the plate so that lecithinase activity can be compared with portion of EYA acting as negative control.	*Clostridium perfringens* ATCC 13124 (positive lecithinase) *Fusobacterium necrophorum* ATCC 25286 (positive lipase) or *Clostridium sporogenes* ATCC 3584 (positive lipase)
THIO supplemented with vitamin K_1, hemin, marble chip, or sodium bicarbonate Chopped meat-carbohydrate Chopped meat-glucose	General enrichment broth media that support growth of most anaerobes; provide backup source of culture material if anaerobic jar or chamber fails, for enrichment for small numbers, or when growth is inhibited	Incubate at 35°C until there is growth on primary plates. If plates have no growth, incubate liquid medium for at least 7 days.	Anaerobes can be overgrown by more rapidly growing facultative organisms. Examine and Gram stain broth only if plating medium reveals no growth or if chamber, jar, or pouch is not functioning. Never rely on broth cultures exclusively for isolation of anaerobes. Some anaerobes may be inhibited by metabolic products or acids produced from more rapidly growing facultative anaerobes.	*Staphylococcus aureus* ATCC 25923 (growth) *Clostridium perfringens* ATCC 13124 (growth) *Bacteroides vulgatus* ATCC 8482 (growth)

[a] Adapted from CMPH 2.3.3–2.3.4.

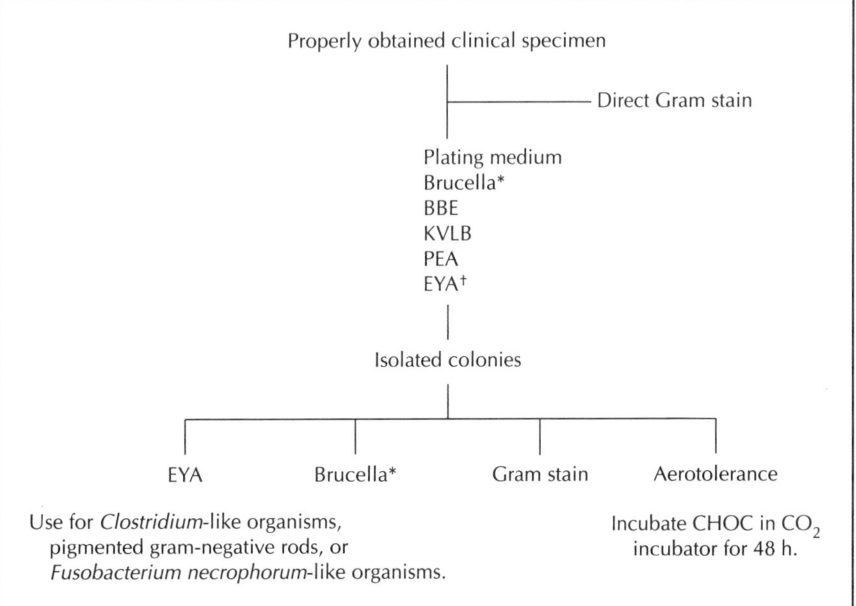

Figure 3.2–1 Procedure for the processing of clinical specimens for anaerobes. *, Use any other suitable enriched primary media that contain vitamin K and hemin and that allow good growth of anaerobes.[†], Use EYA if clostridia are suspected from Gram stain of the specimen or from nature of clinical specimen.

V. MEDIA SELECTION AND INOCULATION METHODS *(continued)*

Include QC information on reagent container and in QC records.

C. Inoculation procedures (*see* Fig. 3.2–1)

D. Microscopic examination of clinical specimens

1. Materials

a. Reagents (*See* EPCM procedure 2.1 for a complete list of reagents.)

(1) Crystal violet solution
(2) Gram's iodine
(3) Sodium bicarbonate, 5% (wt/vol)
(4) Decolorizers
(5) Counterstain: safranin (alternatively, basic fuchsin, 0.1 or 0.2% [wt/vol])

b. Supplies (*See* EPCM procedure 2.1 for a complete list of supplies.)

(1) Glass slides (25 by 75 mm), frosted ends desirable
(2) NaCl (0.85%), sterile
(3) Pasteur pipettes and wood applicator sticks, sterile
(4) Microbiological loops, inoculating needles
(5) Supplies for disposal of biological waste, including ''sharps''
(6) Microincinerator or Bunsen burner
(7) Immersion oil

2. Gram stain. (*See* EPCM 2.1 for details.)

3. Correlation of specimen type with bacterial morphology on the Gram stain can provide the clinician with rapid presumptive information about the identity of the bacteria present.

E. Gram stain clues

1. Large gram-positive rods with boxcar-shaped cells and no spores usually indicate *Clostridium perfringens.*

2. Gram-negative coccobacillary forms suggest pigmenting *Prevotella* or *Porphyromonas* species.

3. Thin gram-negative bacilli with tapered ends suggest *Fusobacterium nucleatum.*

V. MEDIA SELECTION AND INOCULATION METHODS *(continued)*

4. Pleomorphic palely staining gram-negative bacilli suggest *Bacteroides* spp.
5. Very small gram-negative cocci suggest *Veillonella* spp.
6. Results of the Gram stain may indicate the need for additional media or special stains. Aspirated material from a lung nodule, for example, may reveal long, thin, branching gram-positive bacilli. These bacilli suggest the possibility of *Actinomyces* or *Nocardia* spp. The addition of a modified Kinyoun acid-fast stain may separate these two organisms.

F. Incubation

1. One option is to batch the setup and incubation of anaerobic cultures by holding the specimens in the anaerobic transport medium until ready. Inoculate the specimens on plating media and incubate the multiple specimens together. This allows convenient and cost-efficient use of equipment and the microbiologist's time. Do this twice a day (once in the morning and again towards the end of the microbiologist's shift).
2. Plates incubated in the chamber can be inspected for growth after 24 h or earlier, since they do not need to be removed from the anaerobic environment. Hold plates incubated in an anaerobic jar or pouch 48 h before opening. Plates incubated in anaerobic pouches can sometimes be visually inspected without opening the pouch by observing the plates through the clear plastic pouch.
3. Hold negative culture plates for a minimum of 5 days before final examination. Hold liquid backup medium for a minimum of 7 days before final examination and report.

VI. ANAEROBIC INCUBATION TECHNIQUES

Incubation systems

The choice of incubation system is influenced by cost, the number of anaerobic cultures performed, and space limitations. The most common choices of anaerobic incubation systems are listed below. Regardless of which incubation system you use, monitor the anaerobic environment with a methylene blue strip or a resazurin indicator. Monitor the indicator strip daily. Also, monitor the incubation temperature daily. (*See* CMPH 2.4a for more details on the systems and the QC requirements.)

A. Anaerobic chamber

B. Anaerobic bag or pouch

C. Anaerobic jars

REFERENCES

1. **Baron, E. J., L. R. Peterson, and S. M. Finegold.** 1994. *Bailey and Scott's Diagnostic Microbiology*, 9th ed. The C. V. Mosby Co., St. Louis.
2. **Hanson, C. W., and W. J. Martin.** 1976. Evaluation of enrichment, storage, and age of blood agar medium in relation to its ability to support growth of anaerobic bacteria. *J. Clin. Microbiol.* **4:**394–399.
3. **Isenberg, H. (ed.).** 1992. *Clinical Microbiology Procedures Handbook*, p. 2.0.1–2.12.6. American Society for Microbiology, Washington, D.C.
4. **Mangels, J. I.** 1994. Anaerobic transport systems: are they necessary? *Clin. Microbiol. Newsl.* **16:**101–104.
5. **Mangels, J. I., M. E. Cox, and L. H. Lindberg.** 1984. Methanol fixation—an alternative to heat fixation of smears before staining. *Diagn. Microbiol. Infect. Dis.* **2:**129–137.
6. **Mangels, J. I., and B. P. Douglas.** 1989. Comparison of four commercial brucella agar media for growth of anaerobic organisms. *J. Clin. Microbiol.* **27:**2268–2271.
7. **Miller, J. M., and H. T. Holmes.** 1995. Specimen collection, transport, and storage, p. 19–32. *In* P. R. Murray, E. J. Baron, M. A. Pfaller, F. C. Tenover, and R. H. Yolken (ed.), *Manual of Clinical Microbiology*, 6th ed. American Society for Microbiology, Washington, D.C.
8. **Murray, P. R.** 1978. Growth of clinical isolates of anaerobic bacteria on agar media: effects of media composition, storage conditions, and reduction under anaerobic conditions. *J. Clin. Microbiol.* **8:**708–714.

3.3 Isolation, Examination, and Workflow of Primary Plates

I. PRINCIPLE

Anaerobes are usually present in mixed culture with other anaerobes and/or with aerobic and facultative bacteria. Therefore, it is important to use selective and differential media to facilitate the recovery of anaerobic organisms. It is important to investigate and rule out all colonial morphotypes from the primary anaerobic culture media since many facultative and anaerobic bacteria may have similar colonial morphotypes. Special precautions must be taken to avoid exposure of culture plates to oxygen during examination. A 10-min exposure will kill some oxygen-sensitive anaerobes (*Fusobacterium* spp., *Porphyromonas* spp., *Prevotella* spp., and anaerobic cocci). Clues to the presence of anaerobes during examination may include a foul odor upon opening anaerobic jar, more colonial morphotypes present from anaerobic media than from aerobically incubated media, growth on bacteroides bile esculin agar (BBE), phenylethyl alcohol agar (PEA), or laked kanamyan-vancomycin agar (LKV), and red-fluorescing or black-pigmenting colonies.

II. MATERIALS

(See also CMPH 2.4 for additional details.)

Equipment (*See* CMPH 2.3.1 for more details.)

A. Anaerobic environment

A mechanism to protect culture plates from exposure to oxygen is essential. Suitable devices include holding jars and anaerobic chambers. (Holding jars are available from BBL. Manufacturers of anaerobic chambers include Anaerobe Systems, Coy, and Forma.) Individual anaerobic bags may be used if a few cultures are processed at one time and if the plates are quickly returned to an anaerobic environment.

B. Stereoscopic microscope (7× to 15×) or hand lens (8×)

C. UV light, 366-nm wavelength

III. QUALITY CONTROL

See Table 3.2–5 for QC of media.

IV. PROCEDURE

Figure 3.2–1 is a flow for the processing of clinical specimens for anaerobes. Figure 3.3–1 is a flow chart for the processing of primary culture plates.

A. Perform the initial plate examination 24 h after plate inoculation if an anaerobic chamber is used. Delay this first examination until 48 h after inoculation if anaerobic jars or bags are used.

B. Carefully examine the anaerobic blood agar plates (anaBAP) with a stereoscopic microscope or hand lens. Notice size, shape, edge, profile, color, opacity, and other characteristics such as pigment fluorescence, pitting, and/or hemolysis. Record a detailed description of each colony type, noting such characteristics as pitting, swarming, hemolysis, or "greening" of the medium, etc. These colony characteristics can provide valuable clues to the identity of the isolates when used in conjunction with rapid identification tests and Gram stain. (*See* Table 3.3–1 for colony characteristics on supplemental media.)

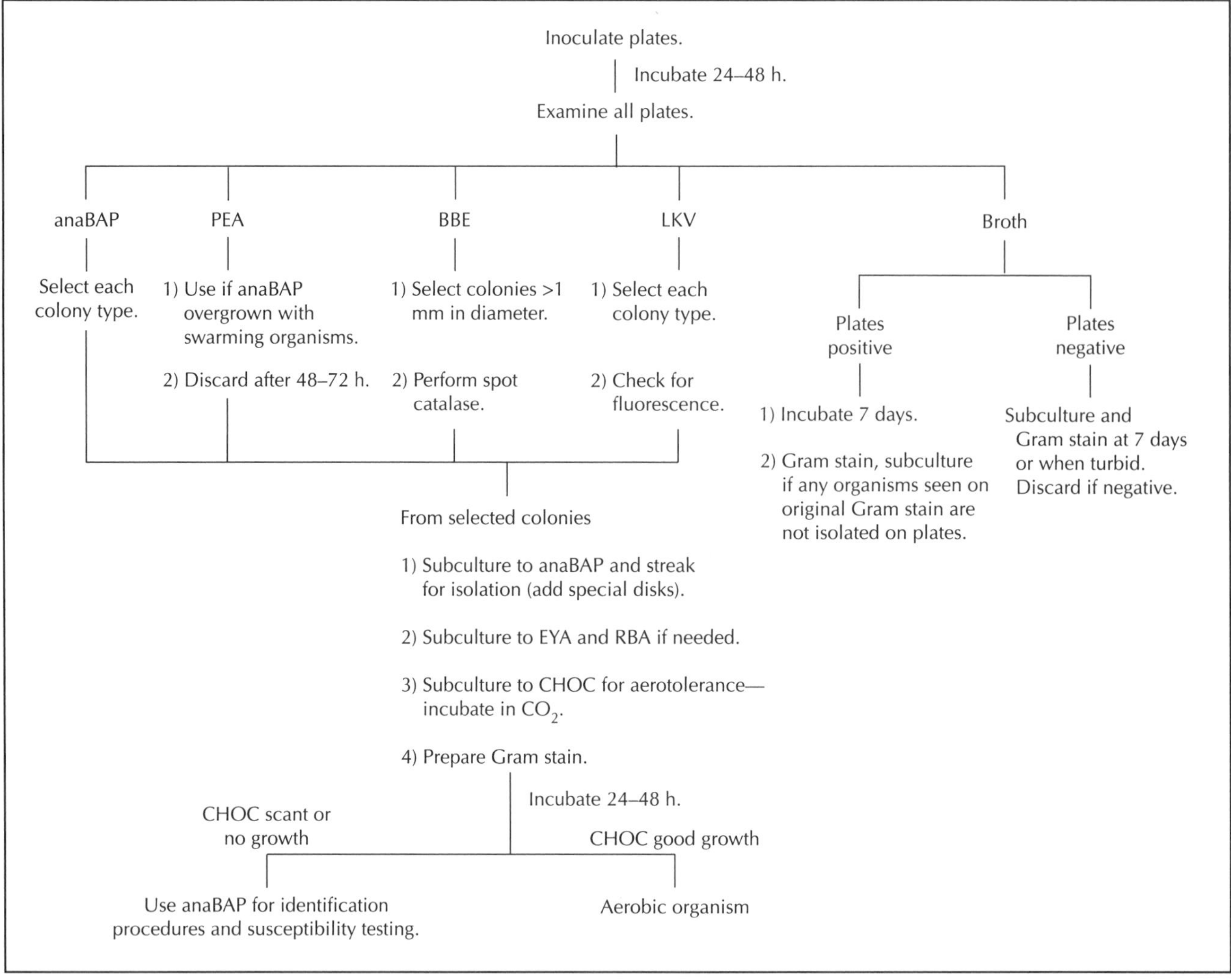

Figure 3.3–1 Flow chart for processing primary anaerobic culture plates. Chart adapted from CMPH 2.4.3.

Table 3.3–1 Supplemental media[a]

Colony morphology	Possible identification	Supplemental medium
Agar pitting	*Bacteroides ureolyticus* group	
Black pigmentation	*Porphyromonas* and *Prevotella* spp.	EYA for lipase (*Prevotella intermedia*)
Brick-red fluorescence	*Prevotella intermedia, Prevotella loeschii*	EYA for lipase, RBA for pigment
Chartreuse fluorescence	*Fusobacterium* spp.	EYA for lipase
Double-zone beta hemolysis	*Clostridium perfringens*	EYA for lecithinase
"Fried egg"	*Fusobacterium necrophorum, Fusobacterium varium*	EYA for lipase, BBE for bile growth
"Greening" of medium	*Fusobacterium* spp.	EYA for lipase
Large with irregular margin	*Clostridium* spp.	EYA for lecithinase
"Medusa-head"	*Clostridium septicum*	
"Molar tooth"	*Actinomyces* spp.	
Speckled	*Fusobacterium nucleatum*	
Swarming growth	*Clostridium septicum, Clostridium sordelii, Clostridium tetani*	

IV. PROCEDURE *(continued)*

C. Select a single, well-isolated colony of each morphological type. Touch each colony with a loop or sterile stick, subculture it to an anaerobic blood agar plate (anaBAP) and CHOC, and make a smear for Gram stain.

1. anaBAP

a. Use the following special potency disks, commercially available from Anaerobe Systems, BBL, Difco, Oxoid, and Remel. (*See* Table 3.3–2 for a list of tests.)

(1) Vancomycin, 5 μg/ml

(2) Kanamycin, 1,000 μg/ml

(3) Colistin, 10 μg/ml

b. Add special-potency kanamycin, vancomycin, and colistin disks to the first quadrant of the anaBAP plate. Nitrate and/or sodium polyanethol sulfonate disks may also be added to the second or third quadrant if needed for identification of isolate.

c. Incubate anaerobically at 35 to 37°C for 24 to 48 h. Record special-potency disk pattern. (*See* Table 3.2–1 for presumptive identification data from special-potency disks.)

2. CHOC

a. Incubate at 35 to 37°C in a 5% CO_2 environment for 48 to 72 h to detect slow-growing aerobic organisms such as *Capnocytophaga, Actinobacillus,* and *Eikenella* spp.

b. Use only CHOC for aerotolerance testing. *Haemophilus* spp. will grow anaerobically on BAP. If blood agar is used for aerotolerance testing, *Haemophilus* isolates will be mistaken for anaerobic gram-negative rods.

D. Perform Gram stain on each colony picked for subculture.

1. Gram stain

a. Air dry the smear. Fix smears by flooding the slide with absolute methanol. (*See* section 2 for details.) Methanol fixation is preferred because bacterial and host cell morphologies remain intact.

b. Perform Gram stain by standard procedures.

E. Examine selective media such as PEA, BBE, and LKV.

F. Pick any colonies on PEA that are different from the colonies isolated on the primary medium (brucella and others). Subculture all colonies from PEA if the original primary medium is overgrown with swarming *Clostridium* spp., *Proteus* spp., or other organisms.

G. From BBE, subculture *Bacillus fragilis* group, which are generally >1 mm in diameter surrounded by a dark-gray zone (*B. vulgatus* is esculin negative). Record the esculin hydrolysis reaction (black equals a positive reaction), and perform a spot catalase test on each colony type. A granular precipitate around the colonies is characteristic of *B. fragilis* species. Small translucent colonies (<1 mm) appearing on BBE in 3 to 4 days with black centers are suggestive of *Bilophila* sp.

H. Subculture all colony types from the LKV and check them for pigment or fluorescence by exposing the plate to UV light. Subculture organisms to rapid blood agar (RBA) to enhance pigment production (optional). (*See* Table 3.3–1.)

I. Process organisms isolated from selective media in the manner described for isolates from the anaBAP. Subculture to anaBAP or CHOC, and prepare a smear for Gram stain.

J. Use of supplemental media (EYA or BBE) can aid in the rapid identification of anaerobic organisms. These media provide evidence of lipase and/or lecithinase production (EYA) and ability to grow in the presence of 20% bile (BBE). *See* Table 3.3–2.

Table 3.3–2 Summary of tests used for the rapid identification of anaerobes[a]

Test	Principle	Reagents	Results	Control organisms (ATCC no.)	Expected results
Special-potency disks	Special-potency disks are used as an aid in determining the Gram reaction of anaerobes as well as in preliminary categorization of some genera and species. Susceptible zones, ≥10 mm; resistant, ≤10 mm	Vancomycin, 5 μg Kanaymcin, 1,000 μg Colistin, 10 μg	See Table 3.4–1 for details. In general, gram-positive organisms are resistant to colistin and susceptible to vancomycin, while most gram-negative organisms are resistant to vancomycin, except for *Porphyromonas* sp.	*B. fragilis* ATCC 25285 *F. necrophorum* ATCC 25286 *C. perfringens* ATCC 13124	Resistant to all three antibiotics Resistant to vancomycin; susceptible to kanamycin and colistin Resistant to colistin; susceptible to vancomycin and kanamycin
Spot indole test	Indole is split from tryptophan by certain organisms. It is used in grouping and identifying many anaerobic bacteria.	*p*-Dimethyl-aminocin-namaldehyde. To perform test, ensure that media contain tryptophan, such as blood agar, or egg yolk medium. Use a small piece of filter paper.	Positive indole, development of a blue or green color on filter paper within 30 s. A negative test is no color change or pinkish color.	*E. coli* ATCC 25922 *P. aeruginosa* ATCC 27853	Indole positive Indole negative
Nitrate disk reduction test	Nitrate can be reduced to nitrite and other reduction products by organisms possessing the enzyme nitrate reductase. The nitrate test is useful for separating *Bacillus ureolyticus* grp from *Fusobacterium* grp which have similar special-potency disk results. *B. ureolyticus* grp are nitrate positive and the *Fusobacterium* grp are nitrate negative.	Nitrate disks are commercially available. Nitrate A and B reagents, and zinc	See Fig. 3.4–1 for details. A positive reaction is indicated by the development of a red or pink color or no color after the addition of zinc. A negative reaction is indicated by no color development after the reagents are added and development of a red color after zinc is added.	*E. coli* ATCC 25922 *A. lwoffii* ATCC 43498	Nitrate positive, red color Nitrate negative, no color change

Table 3.3–2 Summary of tests used for the rapid identification of anaerobes[a] *(continued)*

Test	Principle	Reagents	Results	Control organisms (ATCC no.)	Expected results
Catalase test	Some anaerobic bacteria possess catalase, an enzyme that decomposes hydrogen peroxide into oxygen and water.	A 15% solution of hydrogen peroxide is preferred.	A positive reaction is indicated by immediate bubbling; a negative reaction is indicated by no bubbling. Formation of bubbles after 20 s is considered a negative test.	*S. aureus* ATCC 25923 *S. pyogenes* ATCC 19615	Catalase positive Catalase negative
SPS disk test	SPS is used for the differentiation of anaerobic cocci.	SPS disks are commercially available.	*P. anaerobius* produces a large zone ≥12 mm around the SPS disk. *P. micros* may produce small zones around the SPS disk ≤10 mm.	*P. anaerobius* ATCC 27337 *P. asaccharolyticus* ATCC 29745	Susceptible to SPS Resistant to SPS
Bile test	*B. fragilis, F. mortiferum, F. varium*, and *Bilophila wadsworthia* are capable of growing in the presence of bile (bile resistant). This is a key reaction in separating *B. fragilis* grp from many other anaerobic gram-negative rods.	Bile disks are commercially available, or use BBE agar plates, which are also commercially available.	A positive test (bile resistance) is indicated by growth around bile disks or growth on BBE agar plates.	*B. fragilis* ATCC 25285 *P. melaninogenica* ATCC 25845	Resistant to bile Susceptible to bile
Fluorescence	Some anaerobic organisms are capable of fluorescing when exposed to UV light. The presence and color of fluorescing colonies can aid in the rapid detection and identification of certain anaerobic bacteria.	Use any isolated colony from plates containing blood. Prolonged incubation of >72 h is necessary. Use a long-wave UV light source (366 nm).	A positive test is a distinct color detected with UV light from blood agar plates. See Table 3.4–2 for use as an aid in the presumptive identification of certain anaerobic bacteria.	*P. melaninogenica* ATCC 25845 *B. fragilis* ATCC 25285	Brick red fluorescence No fluorescence
Esculin	To determine the ability of an organism to hydrolyze the glycoside esculin to esculetin. Esculetin reacts with an iron salt to form a dark-brown or black complex.	Use commercially prepared BBE agar plates, or alternatively use esculin broth.	A positive esculin test is the presence of black to brown color. A positive test on BBE agar is growth.	*B. fragilis* ATCC 25285 *B. vulgatus* ATCC 29327	Black, positive esculin No black, esculin negative

(continued)

Table 3.3–2 Summary of tests used for the rapid identification of anaerobes[a] *(continued)*

Test	Principle	Reagents	Results	Control organisms (ATCC no.)	Expected results
Lipase	Free fats in EYA are broken down by the enzyme lipase to produce glycerol and fatty acids. The fatty acids appear as a surface iridescent layer that covers the colony and may extend beyond edge of colony.	Use commercially prepared EYA plates, or alternatively prepare media by using EYA media.	Examine egg yolk plates for an iridescent and multicolored layer on top of the colonies. This is a positive test; it may take 48 h.	*F. necrophorum* ATCC 25286 *B. fragilis* ATCC 25285	Lipase positive, multicolored layer Lipase negative, no color on top
Lecithinase	Bacterial lecithinase splits lecithin to insoluble diglycerides, resulting in an opaque halo surrounding a colony on a medium containing egg yolk.	Use commercially prepared EYA plates, or alternatively prepare media with EYA media.	Examine egg yolk plates for a white opacity in the medium that surrounds the colony and extends beyond the edge of growth. This is a positive test.	*C. perfringens* ATCC 13124 *B. fragilis* ATCC 25285	Lecithinase positive Lecithinase negative
Pigment production	Some anaerobic gram-negative rods, namely, *Porphyromonas* spp. and some *Prevotella* spp., produce a dark pigment that causes their colonies to become brown to black.	Use commercially prepared anaerobic BAP; alternatively use laked BAP plates or rabbit blood agar plates.	Examine colonies from anaerobic plates for brown to black pigment. Some strains may produce pigment in 4 to 6 days; others may take up to 2 weeks. A brown to black pigment is a positive test.	*P. melaninogenica* ATCC 25845 *B. fragilis* ATCC 25285	Black pigment No black pigment
Urease	To determine the ability of an organism to split urea. Hydrolysis of urea by the enzyme urease releases ammonia, the alkalinity of which causes the indicator phenol to change from yellow to red.	Use commercially available urea broth, or alternatively use commercially available rapid urea disks.	A color change from pale yellow to dark bright-pink represents a positive test for urea hydrolysis.	*B. ureolyticus* ATCC 33387 *B. fragilis* ATCC 25285	Urease positive, pink color Urease negative, no color change

[a] Abbreviations: grp, group; SPS, sodium polyanethol sulfonate.

IV. PROCEDURE *(continued)*

K. Broth culture

1. Use as backup only. Prepare a Gram stain, and subculture to anaBAB and CHOC ONLY if primary plates are negative or if *Acintomyces* spp. are expected.
2. Incubate negative broth cultures for 7 days, examine visually, and discard.

L. Subsequent plate examination

1. Incubate primary anaBAP for 5 to 7 days. Examine the primary plates at 24- to 48-h intervals, depending on the type of anaerobic environment. Isolate and perform aerotolerance tests on any new colony types that appear.
2. Pigmented *Prevotella* spp., *Porphyromonas* spp., and *Actinomyces* spp. commonly appear after 2 to 3 days of incubation. *Bilophilia wadsworthia* commonly appears after 3 to 4 days of incubation. Examine the primary anaBAP for fluorescent and pigmented organisms. Use the stereoscope to check for the characteristic ''molar tooth'' colonies of *Actinomyces* spp.
3. Discard PEA and BBE plates after 2 to 3 days of incubation. These media lose their selective properties upon incubation because of evaporation and antibiotic degradation. Secondary growth will occur as these media lose selectivity.

SUPPLEMENTAL READING

Baron, E. J., L. R. Peterson, and S. M. Finegold. 1994. *Bailey and Scott's Diagnostic Microbiology,* 9th ed., p. 474–550. The C. V. Mosby Co., St. Louis.

Isenberg, H. (ed.). 1992. *Clinical Microbiology Procedures Handbook.*, p. 2.0.1–2.12.6. American Society for Microbiology, Washington, D.C.

Murray, P. R., E. J. Baron, M. A. Pfaller, F. C. Tenover, and R. H. Yolken (ed.). 1995. *Manual of Clinical Microbiology,* 6th ed., p. 574–619. American Society for Microbiology, Washington, D.C.

Summanen, P., E. J. Baron, D. M. Citron, C. Strong, H. M. Wexler, and S. M. Finegold. 1993. *Wadsworth Anaerobic Bacteriology Manual,* 5th ed. Star Publishing Co., Belmont, Calif.

3.4 Identification Methods for Anaerobic Bacteria

I. PRINCIPLE

This procedure describes various methods for the rapid identification of anaerobic organisms. The extent of identification required varies by the type of isolate, the source of the specimen, the needs of the physician, and the resources of the laboratory. In general, there are three different methods that enable rapid and cost-efficient identification of anaerobic isolates: method 1, presumptive and preliminary grouping with Gram stain information, plate morphology, and various spot and disk tests; method 2, use of a variety of preformed-enzyme tests; and method 3, use of commercial systems.

Identification of anaerobes by either of the first two methods is less expensive (about $0.50 per isolate) than by the third method (commercial systems [about $5.00 each]).

Any size of laboratory should be able to identify anaerobic organisms presumptively to a group level from primary plates and to isolate and maintain an anaerobe in pure culture so that it can be sent to a reference laboratory as needed. The identification of anaerobic isolates to group level by either method 1 or 2 may be all that is necessary for many laboratories to provide clinically relevant information to allow initiation of appropriate therapy. Most laboratories should also be able to use simple tests for further grouping of the anaerobes and to identify certain anaerobes to species level if required.

II. SPECIMEN

See procedure 3.3 for processing, isolating, and determining the presence of anaerobic isolates.

III. IDENTIFICATION METHODS

A. Identification method 1, Presumptive and preliminary grouping by using plate and cell morphology, Gram stain information, and various spot and disk tests

1. See Tables 3.3–1 and 3.4–1 and Appendix 3.4–1 for plate and cell morphology and Gram stain information.
2. Various rapid spot and disk tests allow preliminary or presumptive identification of many anaerobes to genus and species level.
   a. See Table 3.3–2 for a list of rapid identification tests for anaerobes.
   b. See Table 3.4–2 for special-potency antibiotic disk reactions.
   c. See Table 3.4–3 for fluorescence patterns.
   d. See Fig. 3.4–1 for nitrate disk procedure.

B. Identification method 2, Use of a variety of preformed-enzyme tests

In some instances, identification is necessary beyond that which may be obtained by using rapid spot tests. Presumptive or definitive identification of anaerobes is possible by using individual biochemical tests that detect the presence of preformed enzymes in anaerobic bacteria.

1. See Table 3.4–4 for a list of rapid preformed-enzyme tests. Many of these rapid biochemical tests are commercially available.
2. See CMPH 2.9a for more details.

Table 3.4–1 Gram stain and colony morphology of common anaerobic gram-negative rods[a]

Organism	Gram stain morphology	Colony morphology
Bacteroides fragilis group	Gram-negative coccobacilli or straight rods with variable length. Some cells are pleomorphic or contain vacuoles.	Circular, entire, gray to white, 2- to 3-mm-diameter colony that is shiny and smooth on primary blood agar media. Good growth on BBE >1.0-mm-diameter colonies that are circular, entire, and convex, usually surrounded by a dark-gray zone or brown to blackening of medium caused by esculin hydrolysis. *B. vulgatus* grows on BBE but is esculin negative.
Bacteroides ureolyticus group	Gram-negative coccobacilli or short rods. Some cells are in filaments.	Circular to slightly umbonate; some are gray-white; others produce spreading or swarming growth that forms a depression in the agar. Pitting is best observed if the plate surface is at a 45° angle.
Bilophilia wadworthia	Gram-negative, pleomorphic to straight, short rods.	Small gray colonies within 3 to 4 days on BAP. Growth on BBE, clear colonies with black centers, ''fish-eye.''
Fusobacterium spp.	Gram-negative uneven staining, pleomorphic, coccoid, and rod-shaped cells. Some cells have rounded ends.	Circular, flat to convex colonies >1-mm diameter. Usually gray-white translucent to shiny colony.
Fusobacterium necrophorum	Gram-negative fairly large cells, usually pleomorphic. Rounded-end cells.	Circular, convex colonies >1- to 2-mm diameter. Gray-white to shiny translucent colonies.
Fusobacterium nucleatum	Pale staining, thin gram-negative cells with sharply pointed or tapered ends; spindle-shaped rods; some may have swellings.	Small colonies, usually <1-mm diameter. Circular to slightly irregular; some strains produce rough ''breadcrumb'' colonies. Some strains have ''flecked'' or ''ground-glass'' appearance. Most strains when exposed to oxygen produce greenish discoloration of the blood agar under the colony.
Porphyromonas spp.	Short gram-negative rods; some shorter spherical cells are seen.	Small colony, circular, convex, light gray after 48 h, 6 to 10 days is required for black color. No growth on LKV.
Prevotella spp. (pigmented or nonpigmented)	Gram-negative rods, some short; some coccobacilli forms.	Circular, convex colony about 1- to 2-mm diameter. Gray to slightly shiny. Growth on LKV. Some species form a brown-tan to black pigment in 5 to 10 days.

[a] Abbreviations: BBE, *Bacteroides* bile esculin agar; LKV, laked kanamycin-vancomycin agar.

Table 3.4–2 Identification of anaerobic organisms by using special-potency disks

Organism	Response[a] to disk of		
	Vancomycin (5 g)	Kanamycin (1,000 g)	Colistin (10 g)
Gram negative	R	V	V
Gram positive	S	V	R
B. fragilis group	R	R	R
B. ureolyticus group	R	S	S
Fusobacterium spp.	R	S	S
Porphyromonas spp.	S	R	R
Prevotella spp.	R	R	V
Veillonella spp.	R	S	S

[a] R, resistant; S, susceptible; V, variable.

Table 3.4–3 Fluorescence of anaerobic bacteria[a]

Organism(s)	Color of fluorescence
Porphyromonas asaccharolytica, P. endodontalis	Red[b]
P. gingivalis	None
Pigmented *Prevotella* spp.	Red[b]
Nonpigmented gram-negative bacilli	No fluorescence or pink, orange, or yellow
Fusobacterium spp.	Chartreuse
Veillonella spp.	Red
Eubacterium lentum	Red
Clostridium difficile	Chartreuse

[a] Adapted from CMPH 2.5.10.
[b] Fluorescence disappears when black pigment has developed.

Table 3.4–4 Summary of single or combination rapid enzymatic tests for anaerobes

Test	Principle	Reagents	Results
Alkaline phosphatase	Hydrolysis of 4-nitrophenyl phosphate by alkaline phosphatase releases free 4-nitrophenol, which is yellow. Used to help identify anaerobic gram-positive cocci.	Alkaline phosphatase in kits or in individual tablets	A yellow color within 4 h is positive; no color is negative.
Glutamic acid decarboxylase	Enzymatic action on glutamic acid by a specific decarboxylase releases an amine indicated by a dark blue. Used for the presumptive identification of *B. fragilis* group and some *Clostridium* spp.	Glutamic acid tube, commercially available	A positive test for glutamic acid decarboxylate is indicated by a color shift from green to dark blue; no change is negative.
L-Alanyl-alanylaminopeptidase	Hydrolysis of L-alanyl-alanylaminopeptide releases β-naphythylamine, which complexes with cinnamaldehyde to produce a pink to purple color. Used to separate *Fusobacterium* spp. from *Bacteroides* spp.	ALN Disk, commercially available	A positive test is indicated by a pink to purple color in 30 s. A negative test is indicated by no color change.
L-Proline-aminopeptidase	Hydrolysis of L-proline-β-naphythyl-amine releases β-naphythylamine, which complexes with cinnamaldehyde to produce a pink to purple color. Used to presumptively identify *Clostridium difficile.*	PRO Disk, commercially available	A positive test is indicated by a pink to purple color in 30 s. A negative test is indicated by no color change.
4-Methylumbelliferone derivative substrates	A glucosidase is linked to a fluorescent substrate, 4-methylumbelliferone, to yield rapid determination of enzymatic activity as indicated by fluorescence when viewed with a Wood's lamp.	4-Methylumbelliferone glucoside substrate, commercially available and 360-nm UV light	An immediate light-blue fluorescence when viewed with a Wood's lamp is positive. A negative test is no color.
Combination glycosidase/ naphthylamide tablets	Combination tablets in which two or more enzymatic tests can be performed in a single tube. Some substrates may also be linked to 4-methylumbelliferone to permit detection of fluorescent end products.	Glycosidase/naphyl-amine/methyumberriferyl commercially available tablets	A positive glycosidase is indicated by bright yellow within 2 h; a positive peptidase is indicated by red within 15 min; 4-methylumbelliferone shows a blue-green fluorescence immediately.

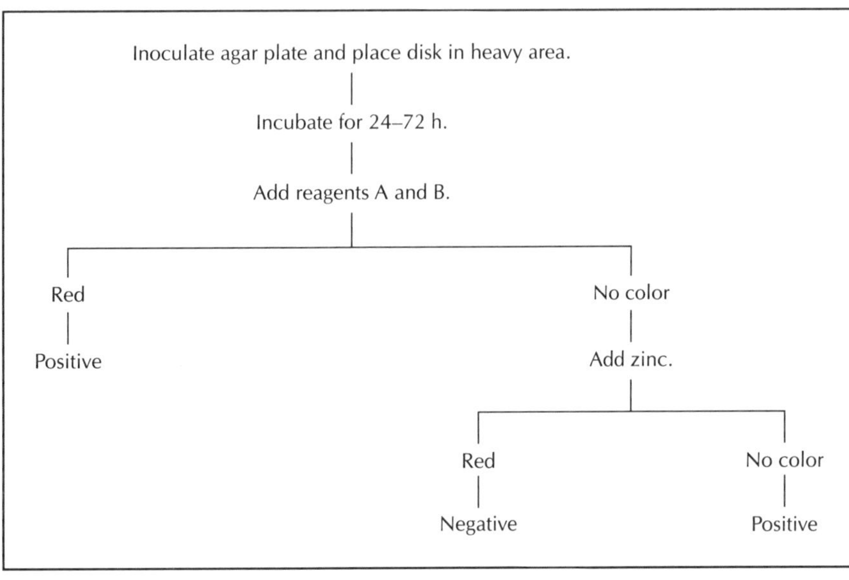

Figure 3.4–1 Flow chart for nitrate disk reduction test.

III. IDENTIFICATION METHODS *(continued)*

C. Identification method 3, Use of commercial systems

1. Principle
   Rapid identification of anaerobes can be accomplished with commercially available microsystems for the detection of preformed enzymes within a few hours following inoculation (1–6). These systems allow the identification and differentiation of many species not identified by conventional microbiochemical tests. The systems and their characteristics are listed in Table 3.4–5. All systems require only 4 h of aerobic incubation after inoculation.
2. Specimen
   See procedure 3.3.
3. Materials

Include QC information on reagent container and in QC records.

   a. AN-Ident (Analytab Products), BBL Crystal System, RapID-ANA II (Innovative Diagnostics Systems), Anaerobe Panel (MicroScan), or Anaerobe ANI Card (Vitek)
   b. See Table 3.4–5 for details on required materials.
   c. Some systems have their own equipment and supplies (BBL Crystal, Anaerobe Panel, and Anaerobe ANI Card).
4. Quality control
   Each manufacturer has its own recommended specific QC procedures (*see* Table 3.4–5).
5. Procedure
   Suspend enough colonies from a 24- to 48-h culture from a nonselective medium to make a turbid suspension. Each test system has its own medium and McFarland requirements (*see* Table 3.4–5).
6. Limitations of the procedure
   a. Specific medium requirements differ from manufacturer to manufacturer; however, systems generally do not use media containing glucose because the sugar suppresses glycolytic activity (Schaedler's). The BBL Crystal system, however, uses different data bases for its identification schema depending upon choice of primary media, one of which is Schaedler's medium.

Table 3.4–5 Characteristics of rapid identification systems[a]

Parameter	An-Ident[b]	RapID-ANA II[c]	Anaerobe Panel[d]	Anaerobe ANI Card[e]	BBL CRYSTAL ANR System[f]
No. of tests	20	18	24	28	29
Inoculum source	Plate	Plate	Plate	Plate	Plate
Inoculum age	24 h	24–72 h	24–48 h	24–48 h	24–72 h
McFarland turbidity	No. 5 in 3.0 ml	No. 3 in 1.0 ml	No. 3 in 3.0 ml	No. 3 in 1.5 ml	No. 4 in 2.3 ml
Diluent	Sterile distilled H_2O	Provided	Sterile deionized H_2O	Sterile saline	Provided
Size of inoculum	85 μl/cupule	0.1 ml/well	50 μl/well	Semiautomatic filling	Semiautomatic filling
Data base	Codebook, computer assisted	Codebook, computer assisted	Codebook, computer assisted	Computer program	Electronic codebook. The database used depends upon the type of primary media used to prepare the inoculum.
Inoculation time	4 h	4 h	4 h	4 h	4 h
Additional required reagents and apparatus	Kovacs or *p*-dimethyl-aminocin-namalde-hyde Cinnamaldehyde reagent 30% H_2O_2 Pasteur pipettes Codebook Sterile saline	Spot indole *p*-dimethylamino-cinnamaldehyde ANA II reagent (cinnamaldehyde) RapID inoculation fluid Pipette Codebook	Mineral oil Peptidase reagent 0.8% sulfanilic acid 3% H_2O_2 Xylene 0.5% *N*,*N*-dimethyl-α-naphthylamine Inoculation fluid Ehrlich's reagent Cover panel 50-μl pipette Codebook	Spot indole *p*-dimethyla-minocin-namaldehyde Vitek System computer Off-line or hand-held viewer Printer Sterile saline Filling stand Sealer module	Spot indole *p*-dimethylaminocinnamaldehyde BBL CRYSTAL panel viewer BBL CRYSTAL electronic computer codebook 15% H_2O_2 Pipette Sterile cotton swabs
QC organisms	*Bacteroides ovatus* ATCC 8483 *Clostridium histolyticum* ATCC 19401 *Actinomyces odontolyticus* ATCC 17929 *Bacteroides fragilis* ATCC 23745 *Clostridium perfringens* ATCC 13124	*Clostridium sordelli* ATCC 9714 *Bacteroides distasonis* ATCC 8503 *Bacteroides uniformis* ATCC 8492 *Peptostreptococcus magnus* ATCC 29328	*Clostridium perfringens* ATCC 13124 *Clostridium sordellii* ATCC 9714 *Bacteroides fragilis* ATCC 25285	*Bacteroides ureolyticus* ATCC 33387 *Bacteroides vulgatus* ATCC 8482 *Propionibacterium acnes* ATCC 11827 *Porphyromonas gingivalis* ATCC 33277 *Bacteroides distasonis* ATCC 8503	*Bacteroides fragilis* ATCC 25285 *Bacteroides distasonis* ATCC 8503 *Peptostreptococcus asaccharolyticus* ATCC 29743 *Lactobacillus acidophilus* ATCC 314 *Fusobacterium varium* ATCC 27725

[a] Adapted from CMPH 2.9.2.
[b] Analytab Products, Plainview, N.Y.
[c] Innovative Diagnostics Systems, Atlanta, Ga.
[d] Microscan, Baxter Healthcare Corp., West Sacramento, Calif.
[e] Vitek, St. Louis, Mo.
[f] Becton Dickinson Microbiology Systems, Cockeysville, Md.

III. IDENTIFICATION METHODS *(continued)*

b. Rapid enzymatic systems should be used in conjunction with other conventional information, such as Gram stain, colonial morphology, and organism growth requirements. Special-potency antimicrobial-agent disks and other presumptive tests can be extremely useful in verifying identification and Gram stain reaction. All aggregate reactions must be considered.

c. Interpretation of colors produced can be difficult but critical for obtaining accurate, reproducible results.

d. Organisms that have been sequentially transferred for long periods may demonstrate aberrant reactions and incorrect identification.

e. For advantages and disadvantages of systems, see CMPH 2.9.3 (2).

REFERENCES

1. **Burlage, R. S., and P. D. Ellner.** 1985. Comparison of the PRAS II, An-Ident, and RapID-ANA systems of identification of anaerobic bacteria. *J. Clin. Microbiol.* **22:**32–35.
2. **Isenberg, H. (ed.).** 1992. *Clinical Microbiology Procedures Handbook,* p. 2.0.1–2.12.6. American Society for Microbiology. Washington, D.C.
3. **Karachewski, N. O., E. L. Busch, and C. L. Wells.** 1985. Comparison of PRAS II, RapID ANA, and API 20A systems for identification of anaerobic bacteria. *J. Clin. Microbiol.* **21:** 122–126.
4. **Murray, P. R., C. J. Weber, and A. C. Niles.** 1985. Comparative evaluation of three identification systems for anaerobes. *J. Clin. Microbiol.* **22:**52–55.
5. **Schrekenberger, P. C., D. M. Celig, and W. M. Janda.** 1988. Clinical evaluation of the Vitek ANI card for identification of anaerobic bacteria. *J. Clin. Microbiol.* **26:**225–230.
6. **Stoakes, L., T. Kelly, K. Manarian, B. Schieven, R. Lannigan, D. Groves, and Z. Hussain.** 1990. Accuracy and reproducibility of the MicroScan Rapid Anaerobe Identification system with an automated reader. *J. Clin. Microbiol.* **28:**1135–1138.

SUPPLEMENTAL READING

Carr-Scarborough Microbiologicals, Inc. 1990. Technical Bulletins for ALA and PRO disk. Stone Mountain, Ga.

Jilly, B. J. 1984. Rapid glutamic acid decarboxylase test for identification of *Bacteroides* and *Clostridium* spp. *J. Clin. Microbiol.* **19:**592–593.

Key Scientific Products. Technical Inserts: Wee-Tabs. Key Scientific Products, Round Rock, Tex.

Lombard, G. L., and V. R. Dowell. 1983. Comparison of three reagents for detecting indole production by anaerobic bacteria in microtest systems. *J. Clin. Microbiol.* **18:**609–613.

MacFaddin, J. F. 1980. *Biochemical Tests for Identification of Medical Bacteria,* 2nd ed. The Williams & Wilkins Co., Baltimore.

Mangels, J., I. Edvalson, and M. Cox. 1993. Rapid presumptive identification of *Bacteroides fragilis* group organisms with use of 4-methylumbelliferone-derivative substrates. *Clin. Infect. Dis.* **16**(Suppl. 4)**:**S319–321.

Moncla, B. J., P. Braham. L. K. Rabe, and S. L. Hillier. 1991. Rapid presumptive identification of black-pigmented gram-negative anaerobic bacteria by using 4-methylumbellifereone derivatives. *J. Clin. Microbiol.* **29:**1955–1958.

Shah, H. N., R. Bonnett, B. Mateen, and R. A. D. Williams. 1979. The porphyrin pigmentation of subspecies of *Bacteroides melaninogenicus*. *Biochem. J.* **180:**45–50.

Slots, J., and H. S. Reynolds. 1982. Long-wave UV light fluorescence for identification of black-pigmented *Bacteroides* sp. *J. Clin. Microbiol.* **16:** 1148–1151.

APPENDIX 3.4–1

Gram stain and colonial characteristics of anaerobic gram-positive bacilli[a]

Organism	Gram stain characteristics[b]	Colonial characteristics
Actinomyces israelii	Long, thin; some branching, some club shaped	Rough, ''molar tooth'' after 5–7 days; can be smooth, white; slow growth (note: white, crumblike molar tooth best seen on BHI agar)
A. meyeri	Diphtheroidal; may be branching	Smooth, white (note: strict anaerobe)
A. naeslundii	Long, thin; many short branches	White, smooth or rough, raised irregular; tan pigment on older colonies; rapid growth
A. odonolyticus	Diphtheroidal, branching	Smooth; may have pink-red pigment
A. viscosus	Diphtheroidal, branching	Smooth; rapid growth
P. propionicum	Diphtheroidal, branching	Rough; slow growth
Bifidobacterium dentium (formerly *B. eriksonii*)	Short, thick, with clubbed or bifurcated ends	White, smooth, glistening, convex with irregular edge; rapid growth; aerotolerant
Clostridium barati	Large, with blunt ends; nonmotile; spores (ST) rarely seen	No hemolysis
C. bifermentans	Large, motile, oval (ST) spores in chains	Gray, irregular edge; narrow zone of hemolysis (note: chalk white on egg yolk agar)
C. botulinum	Large, motile, spores (ST)	Variable hemolysis
C. butyricum	Round or blunt ends, motile, large oval (ST) spores	Nondescript
C. cadaveris	Motile, oval (T) spores	
C. clostridioforme	Stains gram negative; elongated with tapered ends; football-shaped cells; spores rarely seen	Small, convex, translucent; mottled or mosaic surface
C. difficile	Relatively long, thin, motile, oval (T) spores readily seen; horse barn or stable odor	Slightly raised; umbonate with filamentous edge; translucent, with crystalline internal speckling; chartreuse fluorescence
C. histolyticum	Pleomorphic, motile, oval (ST) spores	Smooth and rough colonies; rough have flat edges with rhizoids; aerotolerant
C. innocuum	Small, nonmotile spores (T)	White, glossy, raised; chartreuse fluorescence
C. novyi	Medium, motile, oval (ST) spores	Gray, translucent; irregular surface; may swarm; double zone of hemolysis
C. perfringens	Large, blunt square ends; boxcar appearance; spores rarely seen, nonmotile	Gray, opaque; low, flat, somewhat rhizoid; tend to spread but not swarm; double zone of hemolysis
C. ramosum	Frequently gram negative; thin, pleomorphic, in chains with bulges; nonmotile; spores (T) round or oval, rarely seen	Translucent; circular or slightly irregular; entire; low, convex; red fluorescence
C. septicum	Long, thin, some oval; tend to be pleomorphic, sometimes producing long thin filaments; chain formation common; motile, oval (ST) spores	Medusa head-like that becomes heavy film that covers plate; flat, gray, glistening, semitranslucent; markedly irregular to rhizoid margins
C. sporogenes	Oval (ST) spores, filamentous in older cultures, motile	Raised gray-yellow center, rhizoid edge; swarms; colonies adhere firmly to agar
C. sordellii	Straight, in singles and pairs; spores central to ST, cause slight swelling of cell; free spores often seen, motile	Translucent to opaque; flat or raised; can have mottled internal structure; swarms or spreads
C. tertium	Large oval (T) spores; sporulates only anaerobically; motile	Small, low, translucent, glossy; aerotolerant
C. tetani	Slender, motile, round (T) spores; tennis racket appearance	Translucent, gray; irregular edge; narrow zone of hemolysis
Eubacterium lentum	Short, coccoidal or diphtheroidal, pleomorphic; in short chains	Smooth, opaque; slightly irregular edge; aerotolerant; red fluorescence
E. limosum	Pleomorphic; in pairs and short chains	Translucent to white; entire edge; aerotolerant
Lactobacillus catenaforme	Pleomorphic; sometimes long, straight, and slender; often in long chains, some streptococcuslike	Slightly translucent; entire edge; aerotolerant
Propionibacterium spp.	Pleomorphic; club shapes, pointed ends	White to pink, shiny, opaque; entire edge; aerotolerant

[a] Adapted from CMPH 2.11.8.
[b] Spore location: ST, subterminal; T, terminal.

3.5 Identification of Anaerobic Gram-Negative Bacilli

I. PRINCIPLE

This section describes characteristics of and identification methods for the clinically important members of the *Bacteroidaceae* family, including *Bacteroides* spp., *Fusobacterium* spp., *Porphyromonas* spp., *Prevotella* spp., *Wolinella* spp., and *Bilophila wadsworthia*, and will help differentiate them from other gram-negative bacilli such as *Capnocytophaga* spp. Anaerobic gram-negative bacilli are the most commonly encountered anaerobes in clinical specimens, with *Bacteroides fragilis* isolated more frequently than any other anaerobe (1, 3, 5).

Initial differentiation of anaerobic gram-negative bacilli is based on cell and colony morphology, pigment production, fluorescence under long-wave UV light, susceptibility to special-potency antibiotic disks (*see* procedure 3.4), and rapid enzyme testing (*see* procedure 3.4). Definitive species identification currently requires additional biochemical testing as outlined in this procedure and in procedure 3.4.

II. ISOLATES

Follow the directions in procedures 3.2 and 3.3 to obtain a pure culture. It is essential to confirm that the isolate is an anaerobe and a gram-negative rod, since an appreciable number of gram-negative rods are facultatively anaerobic instead of true anaerobes. The following steps will aid in identification.

A. Based on the results obtained with special-potency antibiotic disks (*see* Table 3.4–2), refer to Fig. 3.5–1 of this procedure for direction to the appropriate follow-up flow chart to determine the tests needed for further identification.

B. The *Bacteroides fragilis* group is resistant to all three special-potency antimicrobial-agent disks. (*See* Table 3.4–2 and Fig. 3.5–2.) Both the *B. ureolyticus* group and *Fusobacterium* sp. are sensitive to kanamycin and colistin, but resistant to vancomycin. They may be separated by nitrate (the *B. ureolyticus* group is nitrate positive). (*See* Table 3.5–1 and Fig. 3.5–3 for identification of the *B. ureolyticus* group and Table 3.5–2 and Fig. 3.5–4 for identification of *Fusobacterium* sp.)

C. *Porphyromonas* spp. are sensitive to vancomycin and resistant to the other two antimicrobial agents. *Prevotella* spp. are resistant to kanamycin and vancomycin and may be either resistant or sensitive to colistin. See Tables 3.5–3 and 3.5–4 and Fig. 3.5–5 for further identification.

III. CELL AND COLONY MORPHOLOGY

A. Gram stain morphology (*See* Table 3.4–1.)

B. Colony morphology (*See* Table 3.4–1.)

1. BBE will select for those species able to grow in the presence of 20% bile and to hydrolyze esculin (Tables 3.5–5 and 3.5–6; Fig. 3.5–2). See procedure 3.4 for more details.
    a. Some non-*B. fragilis* group organisms are resistant to bile (*F. mortiferum*, *Enterococcus* spp., and some members of the family *Enterobacteriaceae*). Carefully evaluate growth on BBE. Perform Gram stain if necessary.
    b. *B. wadsworthia* grows on BBE at 3 to 5 days, has black-centered colonies, and has a strong positive catalase reaction (1, 3) (*See* Table 3.5–5.)

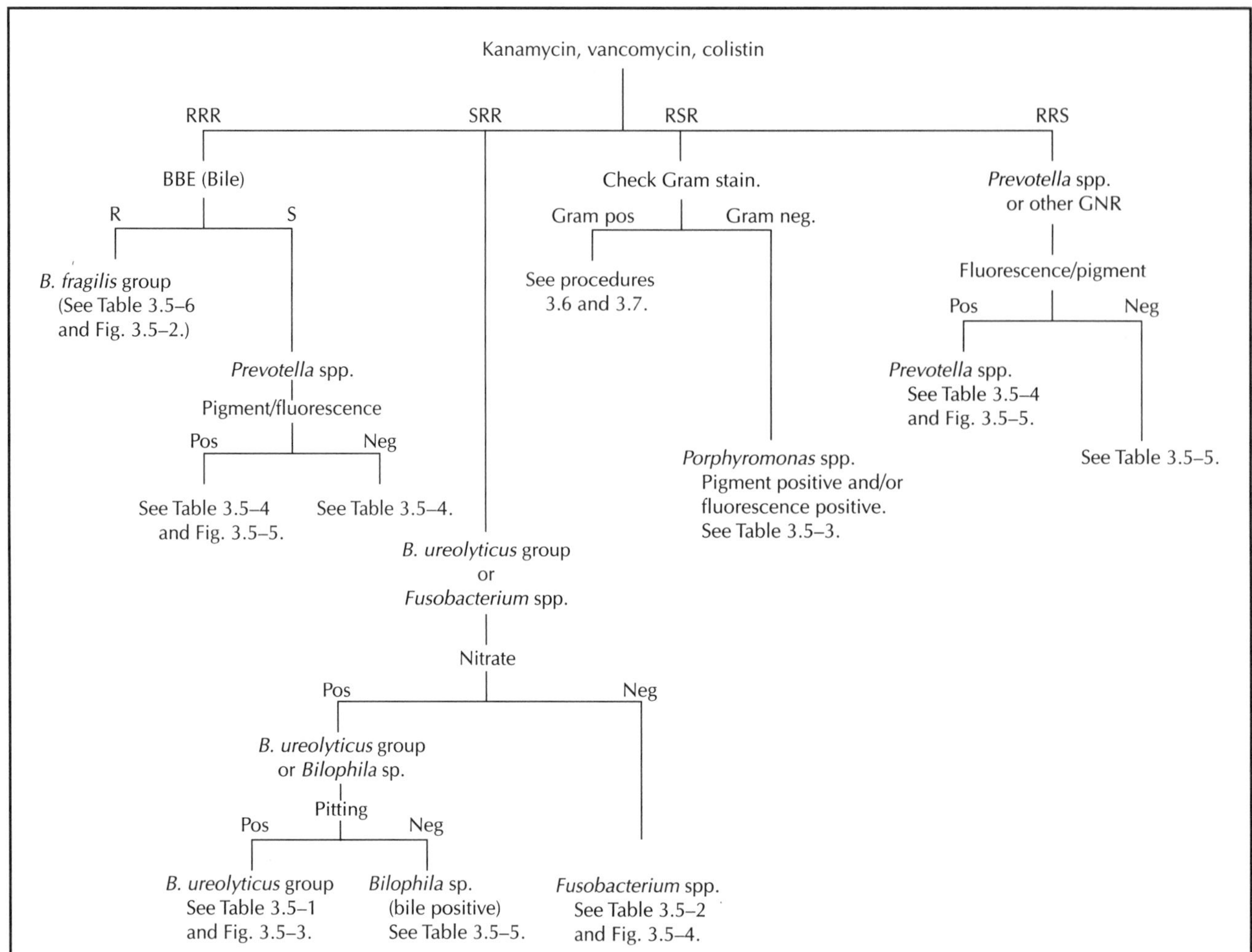

Figure 3.5–1 Kanamycin, vancomycin, colistin special antimicrobial-agent disk profile. R, resistant; S, sensitive; GNR, gram-negative rods; Pos, positive, Neg, negative.

Table 3.5–1 Identification of *B. ureolyticus* group[a,b]

Species	Kan (1 mg)	Nitrate reduced	Motile	Urease	Strong oxidase	Agar pitting
B. gracilis	S	+	−	−	−	+(−)
B. ureolyticus	S	+	−	+	V	+(−)
Wolinella spp.	S	+	+	−	−	−(+)
Campylobacter concisus	S	+	+	−	+	−(+)

[a] Adapted from CMPH 2.10.4.
[b] Kan, kanamycin; R, resistant; S, susceptible; −, negative; −(+), rare strain positive; +, most strains positive; +(−), rare strain negative; V, variable reaction.

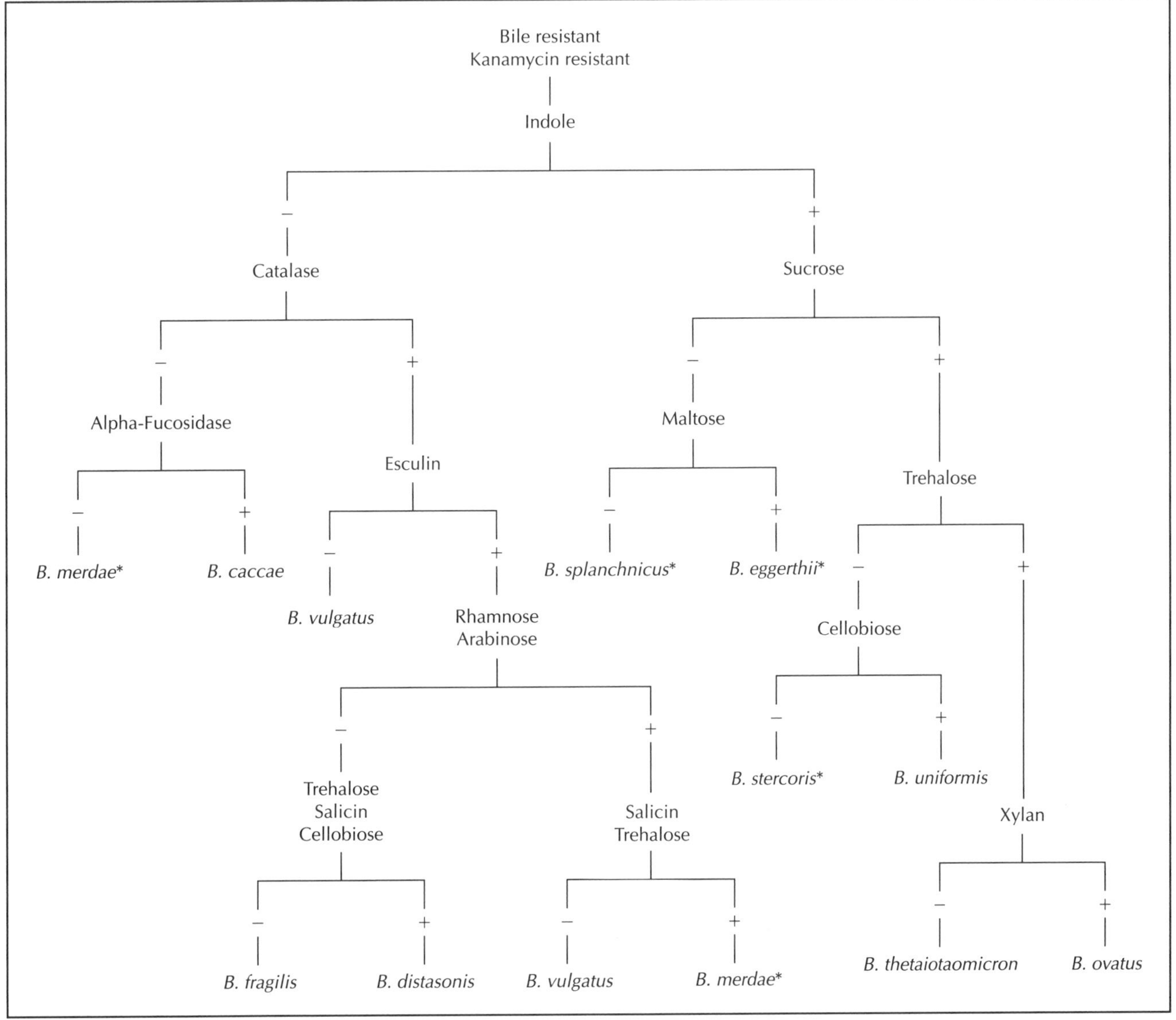

Figure 3.5–2 Identification of *Bacteroides fragilis* group (adapted from CMPH 2.10.11).*, rarely isolated.

Table 3.5–2 *Fusobacterium* spp.[a]

Species	Cell morphology	Colony morphology	Kan (1 mg)	Indole	20% Bile	Esculin	Lipase	Lactate to propionate	Threonine to propionate
F. nucleatum	Tapered ends	Breadcrumblike, speckled	S	+	−	−	−	−	+
F. gonidiaformans	Gonidial forms	Smooth	S	+	−	−	−	−	+
F. necrophorum	Round ends (bizarre)	Umbonate (greening)	S	+	−(+)	−	+(−)	+	+
F. naviforme	Boat shaped	Mottled	S	+	−	−	−	−	−
F. varium	Round ends	Smooth (''fried egg'')	S	+(−)	+	−	−	−	+
F. mortiferum	Bizarre	''Fried egg''	S	−	+	+	−	−	+
F. russi	Round ends	Smooth	S	−	−	−	−	−	−

[a] Adapted from CMPH 2.10.4. Kan, kanamycin; S, susceptible; −, negative; −(+), rare strain positive; +, most strains positive; +(−), rare strain negative; V, variable reaction.

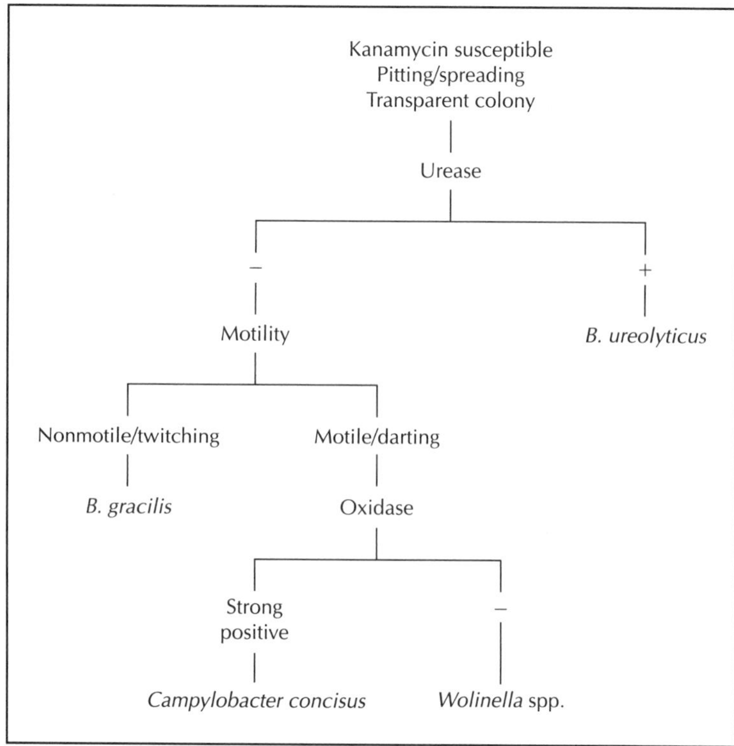

Figure 3.5–3 Identification of *B. ureolyticus* group (adapted from CMPH 2.10.9) See also Table 3.5–1.

Table 3.5–3 *Porphyromonas* spp.[a,b]

Species	Colony morphology	Pigment	Col	Van	Kan	Indole	Arginine	α-Fucosidase	Trypsin	NAG	Fluorescence under UV
P. asaccharolytica	Mucoid, smooth	Dark brown to black	R	S	R	+	+(−)	+	−	−	+
P. endodontalis	Mucoid, smooth	Dark brown to black	R	S	R	+	−(+)	−	−	−	+
P. gingivalis	Mucoid, smooth	Dark brown to black	R	S	R	+	+	−	+	+	−

[a] Adapted from CMPH 2.10.6.
[b] Col, colistin; Van, vancomycin; Kan, kanamycin; R, resistant; S, susceptible; −, negative; −(+), rare strain positive; +, most strains positive; +(−), rare strain negative; NAG, *N*-acetyl-β-glucosaminidase.

Table 3.5–4 Pigmented *Prevotella* spp.[a]

Species	Colony morphology	Pigment	Van (5 μg)	Col (10 μg)	Indole	Lipase	Esculin	Lactose	Sucrose	Cellibiose	Arginine
P. melaninogenica	Smooth	Tan to buff	R	R^s	−	−	−(+)	+	+	−	+
P. denticola	Smooth	Tan to buff	R	R^s	−	−	+	+	+	−	−
P. loescheii	Smooth	Tan to buff	R	R^s	−	−(w)	−	+	+(−)	+	−
P. corporis	Dry	Black	R	S	−	−	−	−	−	−	−
P. intermedia	Dry	Black	R	S	+	+(−)	−	−	+	−	+
P. bivia[b]	V	V	R	S	−	−	−	+	−	−	V

[a] Adapted from CMPH 2.10.7. Van, vancomycin; Col, colistin; R, resistant; S, susceptible; R^s, some strains susceptible: −, negative; −(+), rare strain positive; +, most strains positive; +(−), rare strain negative; V, variable reaction; w, weak reaction.
[b] May produce pigment upon prolonged incubation.

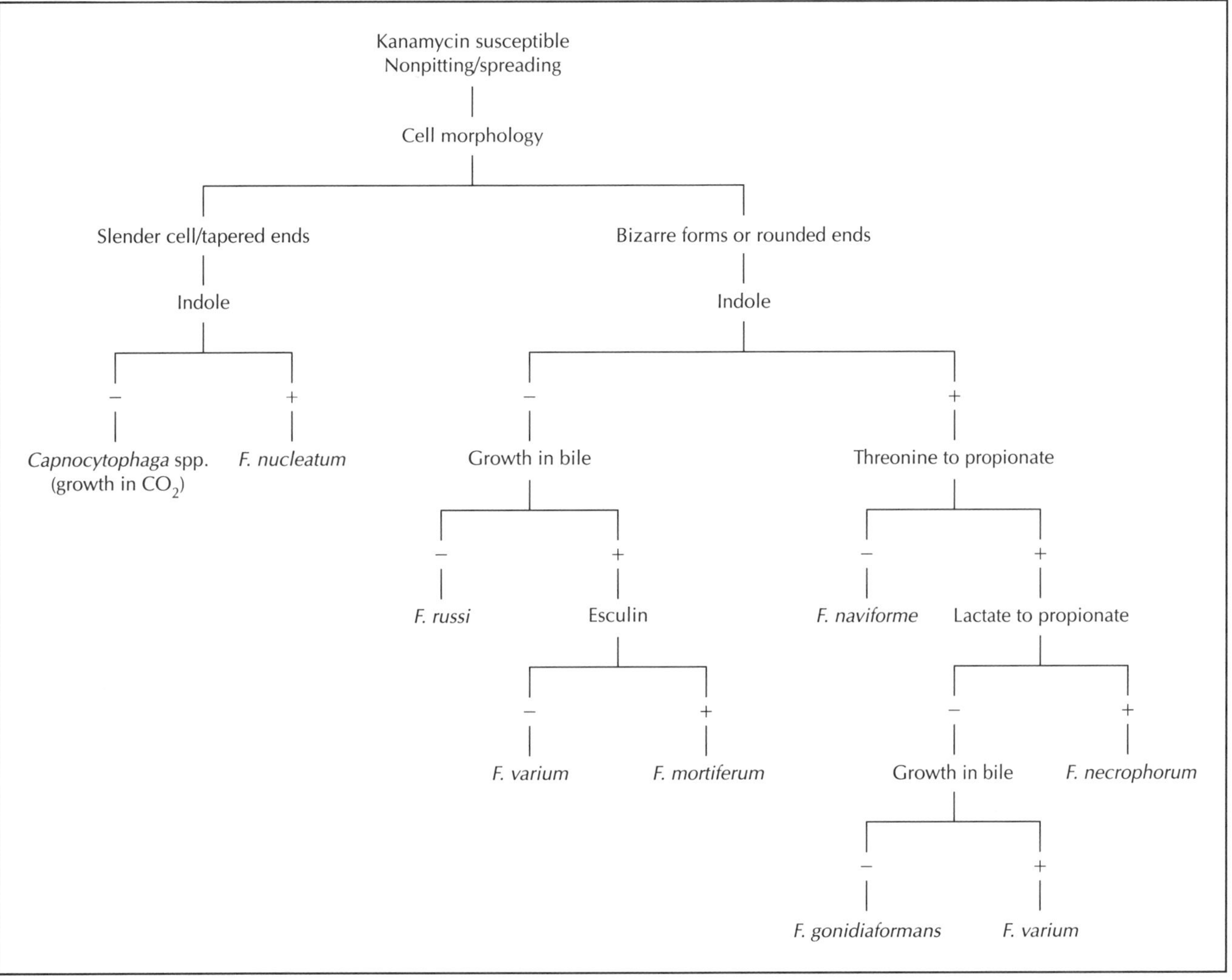

Figure 3.5–4 Identification of *Fusobacterium* spp. Adapted from CMPH 2.10.3.

III. CELL AND COLONY MORPHOLOGY *(continued)*

2. *Bacteroides ureolyticus* group members can form three different colony types. (*See* CMPH 2.10 [2] for more details.)
   a. Pitting (corroding) is usually best detected if the surface of the agar is inspected at a 30° to 45° angle.
   b. Ability to corrode agar and spread can be medium dependent and may be enhanced by using homemade fresh medium or PRAS media (2).
3. *Fusobacterium* species (Table 3.5–2 and Fig. 3.5–4)
   a. Along with its distinctive cell morphology, *F. nucleatum* possesses three different characteristic colony forms: speckled, breadcrumb, and smooth with greening of the agar (1, 3, 5).
   b. *F. necrophorum* produces umbonate colonies with greening of the agar.
   c. *F. mortiferum* produces ''fried-egg'' colonies (translucent with opaque centers and irregular borders). *Fusobacterium varium* may also produce this ''fried-egg'' colony.

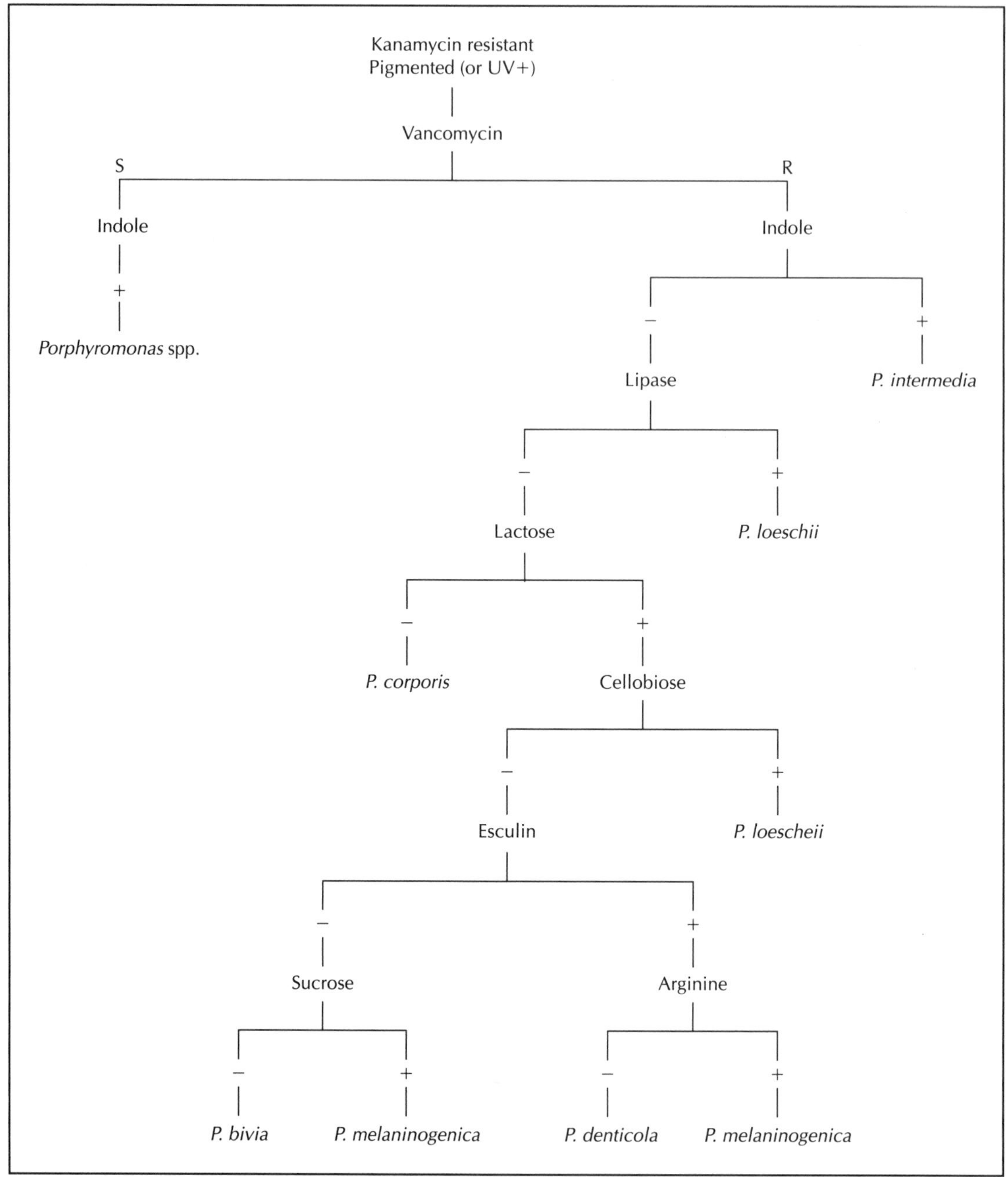

Figure 3.5–5 Identification of pigmented gram-negative bacilli (for *Porphyromonas* spp., *see* Table 3.5–3). S, susceptible; R, resistant. Adapted from CMPH 2.10.7.

Table 3.5–5 Nonpigmented *Bacteroides, Prevotella, Bilophia*, and *Tissierella* spp.[a]

Subgroup and species	Growth in 20% bile	Indole	Esculin	Catalase	Arabinose	Lactose	Sucrose	Salicin	Xylose	α-Fucosidase	Urease
B. splanchnicus	+	+	+	−	+	+	−	−	−	+	−
P. buccae	−	−	+	−	+	+	+	+	+	−	−
P. oris	−	−	+	NA	+(−)	+	+	+	+	+	−
P. heparinolytica	−	+	+	NA	+	+	+	+	+	+	−
P. zoogleoformans	−	−	+	NA	+(−)	+	+	V	+	+	−
P. buccalis	−	−	+	−	−	+	+	−	−	+	−
P. oralis	−	−	+	−	−	+	+	+	−	+	−
P. oulorum	−	−	−	NA	−	+	−	−	−	−	−
P. veroralis	−	−	+	NA	−	+	+	−	−	+	−
P. bivia	−	−	−	−	−	+	−	−	−	+	−
P. disiens	−	−	−	−	−	−	−	−	−	−	−
Bilophila wadsworthia	+	−	−	+	NA	NA	NA	NA	NA	NA	+(−)
B. capillosus	−(+)	−	+	−	−	−	−	−	−	NA	−
B. putredinis	+(−)	+	−	+(−)	NA	−	−	−	−	NA	−
B. gracilis	−+	−	−	−	NA	NA	NA	NA	NA	NA	−
B. forsythus	−	−	+	−	NA	NA	NA	NA	NA	NA	−
Tissierella praeacuta	+	−	−	−	NA	−	−	−	−	NA	−

[a] NA, not tested; V, variable reaction; −, negative; +, most strains positive; +(−), rare strain negative; −(+), rare strain positive.

Table 3.5–6 Bile-resistant *Bacteroides* spp.[a]

Species	Kan (1 mg)	20% Bile	Indole	Catalase	Esculin	Sucrose	Maltose	Arabinose	Cellobiose	Rhamnose	Salicin	Trehalose	Xylan	α-Fucosidase
B. fragilis	R	+	−	+	+	+	+	−	−(w)	−	−	−	−	NA
B. vulgatus	R	+	−	+(−)	−(+)	+	+	+	−	+	−	−	V	NA
B. distasonis	R	+	−	+(−)	+	+	+	−(+)	+	V	+	+	−	NA
B. merdae	R	+	−	−(+)	+	+	+	−(+)	−(+)	+	+	+	−	−
B. caccae	R	+	−	−	+	+	+	+	−(+)	+(−)	V	+	−	+
B. thetaiotaomicron	R	+	+	+	+	+	+	+	+	+	−(+)	+	−	NA
B. uniformis	R	+(w)	+	−(+)	+	+	+	+	+(w)	−(+)	+(−)	−		NA
B. ovatus	R	+	+	−(+)	+	+	+	+	+	−(+)	+	+	+	NA
B. stercoris	R	+	+	−	+	+	+	−(+)	−	+	−	−	V	V
B. eggerthii	R	+	+		+	−	+	+	−(w)	w(−)	−	−		NA

[a] Adapted from CMPH 2.10.5. Kan, kanamycin; R, resistant; −, negative; −(+), rare strain positive; +, most strains positive; +(−), rare strain negative; V, variable reaction; w, weak reaction; NA, not tested.

IV. PIGMENTATION AND FLUORESCENCE

A. **Pigmentation** (*see* procedure 3.4)
   1. Pigmentation of *Prevotella* spp. and *Porphyromonas* spp. usually occurs within 3 to 14 days but may be delayed, taking as long as 21 days for some isolates (1, 3–5).
   2. Note the degree of pigmentation; it may be useful in differentiating this group from other groups. *Porphyromonas* spp. tend to produce dark brown to black colonies (4) (Table 3.5–3), while some pigmenting *Prevotella* spp. (*Prevotella loescheii*, *Prevotella denticola*, and *Prevotella melaninogenica*) do not develop as dark a pigmentation (Table 3.5–4).

B. **Long-wave UV fluorescence** (*see* procedure 3.4)
   1. Test any suspicious colonies for fluorescence by exposure to a long-wave UV light (Wood's lamp) (*see* procedure 3.4) to recognize black-pigmented *Prevotella* spp. and *Porphyromonas* spp. before a distinct pigment develops.
   2. Only those isolates that fluoresce brick red can be presumptively identified as pigmented *Prevotella* spp. or *Porphyromonas* spp. Although *Veillonella* spp. may fluoresce red (but not brick red) (*see* procedure 3.4), differentiation is easily achieved by Gram stain.
   3. *Porphyromonas gingivalis* lacks the ability to fluoresce but will produce black colonies, so note both pigmentation and fluorescence (Table 3.5–3).
      NOTE: *F. nucleatum* and *F. necrophorum* can fluoresce chartreuse (yellow-green).
   4. If colonies are too old, they lose their ability to fluoresce by direct exposure to UV light (4, 5).

V. RAPID PRESUMPTIVE IDENTIFICATION

Use rapid tests to rule out or confirm presumptive identifications based on Gram stain and colony morphology (*see* procedure 3.4 for additional information). When typical morphology (cell and colony) is apparent and is combined with rapid tests, a presumptive identification can usually be made. Use more exhaustive tests if found necessary (*see* procedure 3.4 for more details).

VI. SPOT INDOLE

(See procedure 3.4.)

A. Some species of the *B. fragilis* group (Table 3.5–6 and Fig. 3.5–2) can be differentiated by spot indole (1, 2, 5). See procedure 3.4 for method.
   Caution: *Indole is diffusible, so the spot indole test should be performed only on plates with pure cultures.*

B. A small coccobacillary, gram-negative bacillus that produces pigmented colonies and indole and is vancomycin resistant can be identified as *Prevotella intermedia* (Table 3.5–4; Fig. 3.5–5).

C. Vancomycin-susceptible pigmented isolates are identified as *Porphyromonas* spp., which are indole positive, whereas vancomycin-resistant pigmented isolates are identified as *Prevotella* spp. (some may be indole positive) (Tables 3.5–3 and 3.5–4; Fig. 3.5–5).

VII. NITRATE AND UREASE

(See procedure 3.4 for use of nitrate and urea disk tests.)

A. A thin, gram-negative bacillus with rounded ends that is resistant to vancomycin and susceptible to both kanamycin and colistin should be tested for nitrate reduction. (*See* Tables 3.5–1 and 3.5–2; *also see* procedure 3.4.)

B. The agar-pitting group (*B. ureolyticus, B. gracilis, Wolinella* spp., and *Campylobacter concisus*) resembles some of the less frequently isolated *Fusobacterium* spp. in cell morphology and special-potency disk pattern, but is easily differentiated by its ability to reduce nitrates.

C. Nitrate-positive, corroding isolates that are urease positive and have the proper patterns of susceptibility to special-potency disks are identified as *B. ureolyticus* (Table 3.5–1 and Fig. 3.5–3). A rapid urea test may also assist in the identification (*see* procedure 3.4).

VIII. BILE

(See procedure 3.4.)

A. The *B. fragilis* group is resistant to bile (Table 3.5–6 and Fig. 3.5–2), while pigmented, pitting, and other anaerobic gram-negative rods are generally sensitive to it. The use of a primary isolation BBE plate provides a quick means of separating bile-resistant organisms.

B. Some *Fusobacterium* spp. are resistant to bile (Table 3.5–2).

C. *Bilophilia wadsworthia* is resistant to bile and will grow on BBE medium in 3 to 4 days with clear colonies with black centers (fish-eye appearance) (1, 5) (Table 3.5–5).

IX. LIPASE

(See procedure 3.4.)

A. Lipase (EYA plate) can be combined with rapid tests to aid in identification of pigmented *Prevotella* spp., *Porphyromonas* spp., and *Fusobacterium* spp. (*See* procedure 3.4 for description of lipase test.)

B. An indole- and lipase-positive coccobacillus that produces black-pigmented or brick red fluorescent colonies is identified as *P. intermedia* (Table 3.5–4).

C. An indole- and lipase-positive isolate, nitrate negative with a fusobacterium special-potency antibiotic-disk pattern, is identified as *F. necrophorum* (Table 3.5–2). Lipase-positive strains of *F. necrophorum* are often beta-hemolytic (1, 5).

X. BIOCHEMICALS

Commercially prepared single and multiple spot tests and rapid enzymatic biochemical tests may be useful in addition to the mentioned spot tests and biochemicals for identifying isolates (*see* procedure 3.4).

REFERENCES

1. **Baron, E. J., L. R. Peterson, and S. M. Finegold.** 1994. *Bailey and Scott's Diagnostic Microbiology,* 9th ed. The C. V. Mosby Co., St. Louis.
2. **Isenberg, H. (ed.).** 1992. *Clinical Microbiology Procedures Handbook,* p. 2.0.1–2.12.6. American Society for Microbiology, Washington, D.C.
3. **Murray, P. R., E. J. Baron, M. A. Pfaller, F. C. Tenover, and R. H. Yolken (ed.).** 1995. *Manual of Clinical Microbiology,* 6th ed. American Society for Microbiology, Washington, D.C.
4. **Shah, H. N., and M. D. Collins.** 1988. Proposal for the reclassification of *Bacteroides asaccharolyticus, Bacteroides gingivalis,* and *Bacteroides endodontalis* in a new genus, *Porphyromonas. Int. J. Syst. Bacteriol.* **38:** 128–131.
5. **Summanen, P., E. J. Baron, D. M. Citron, C. Strong, H. M. Wexler, and S. M. Finegold.** 1993. *Wadsworth Anaerobic Bacteriology Manual,* 5th ed. Star Publishing Co., Belmont, Calif.

3.6 Identification of Anaerobic Gram-Positive Bacilli

I. PRINCIPLE

Anaerobic gram-positive bacilli of human clinical relevance are divided into two distinct groups: one genus of endospore formers (*Clostridium* spp.) and five genera of nonsporeformers (*Actinomyces, Bifidobacterium, Eubacterium, Lactobacillus,* and *Propionibacterium*). These bacilli are part of the normal microbiota of the oral cavity, gastrointestinal and genitourinary tracts, and skin. *Clostridium* species can cause acute, severe, or chronic infections (2). The nonsporeformers are infrequently clinically significant and usually cause chronic disease. *Many of these nonsporeformers are resistant to metronidazole.*

The use of spot tests may help in the identification of these organisms. If further identification is required, many of the commercial rapid enzymatic systems are capable of accurately identifying this group of organisms. For some organisms or some groups of organisms, however, the commercial systems may not satisfactorily identify the isolate; therefore, other tests such as GLC may be necessary.

II. SPECIMEN

See procedures 3.2 and 3.3 for specimen and workup requirements.

III. MATERIALS

See procedures 3.3 and 3.4 for use of media, and Gram stain techniques, spot tests, and other tests for the identification of anaerobic gram-positive bacilli.

IV. PROCEDURE

A. Day 1

1. Pick colony of anaerobic gram-positive rod from brucella BAP or other suitably enriched media and subculture to the following.
   - **a.** Use egg yolk agar for detection of lipase and lecithinase.
   - **b.** Brucella BAP with nitrate
   - **c.** CHOC in CO_2 for aerotolerance
2. Gram stain colony for specific characteristics and record results. (*See* Appendix 3.6–1 and Fig. 3.6–1.)
   - **a.** If pleomorphic and diphtheroidal, do indole and catalase tests. If both tests are positive, report *Propionibacterium acnes.*
   - **b.** If cells are large, positive bacilli are arranged in pairs as "boxcars," and the colony produces a double zone of hemolysis on blood agar, report as *Clostridium perfringens* (1 to 2 h of refrigeration will develop the zone of hemolysis).
   - **c.** If spores are seen and the cells are not the large boxcar type, report *Clostridium* sp. but not *C. perfringens.*

B. Day 2 (or when adequate growth is achieved)

1. Examine brucella BAP.
   - **a.** Record results with disks.
   - **b.** Record colonial morphology.
   - **c.** Record odor. *Clostridium difficile* smells of horse barn, and *P. acnes* has a strong odor of tryptophan.

2. Examine egg yolk plate. (*See* procedure 3.4 for more information.) Record lecithinase and/or lipase reactions (*see* Fig. 3.6–2).
3. Stain chopped meat-carbohydrate.
   a. Gram stain, and look for spores. If spores are present, record their presence and locations.
   b. Spore stain can also be used (1, 4).

C. **Days 3 to 5**
Perform spore test (1, 4).

V. IDENTIFICATION

A. ***Clostridium* spp.**
See Appendixes 3.4–1 and 3.6–1 and Fig. 3.6–1 for Gram stain, colonial, and biochemical characteristics of species.
1. Vegetative cells rod shaped, can vary from coccoid to filamentous
2. Obligate anaerobes (majority)
   Exceptions (will grow on CHOC in air) are *Clostridium carnis, Clostridium histolyticum*, and *Clostridium tertium*.

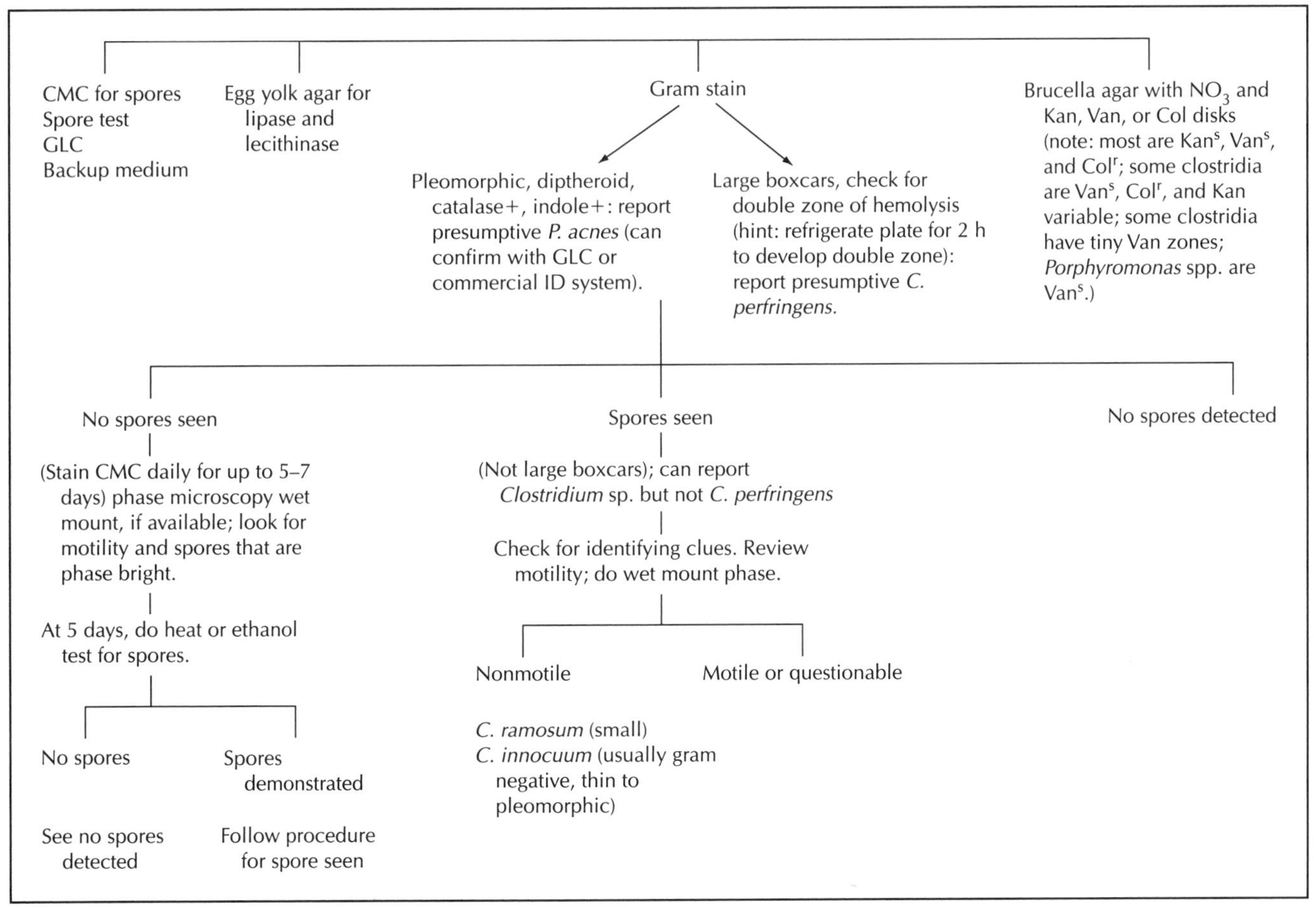

Figure 3.6–1 Procedure for identification of pure-colony anaerobic gram-positive bacillus from brucella or other agar. Adapted from CMPH 2.11.5. Refer to previous procedures for details on collection, isolation, culture, and examination of plates and on obtaining pure colonies of anaerobic gram-positive rods. CMC, chopped meat-carbohydrate; ID, identification; Kan, kanamycin; Van, vancomycin; Col, colistin: r, resistant; s, susceptible.

V. IDENTIFICATION *(continued)*

3. Gram positive (most)
Exceptions (sometimes appear as gram negative) are *Clostridium clostridiiforme, Clostridium ramosum*, and *Clostridium tetani* (by time of spore formation). A study (3) has found that many of these anaerobic gram-positive organisms when stained under an anaerobic environment stain true, e.g., gram positive, as compared with gram negative when stained under aerobic conditions.

4. Motile
Exceptions are *C. perfringens, C. ramosum*, and *C. innocuum.*

5. Catalase not produced, but if produced, will be weak and in small amounts.

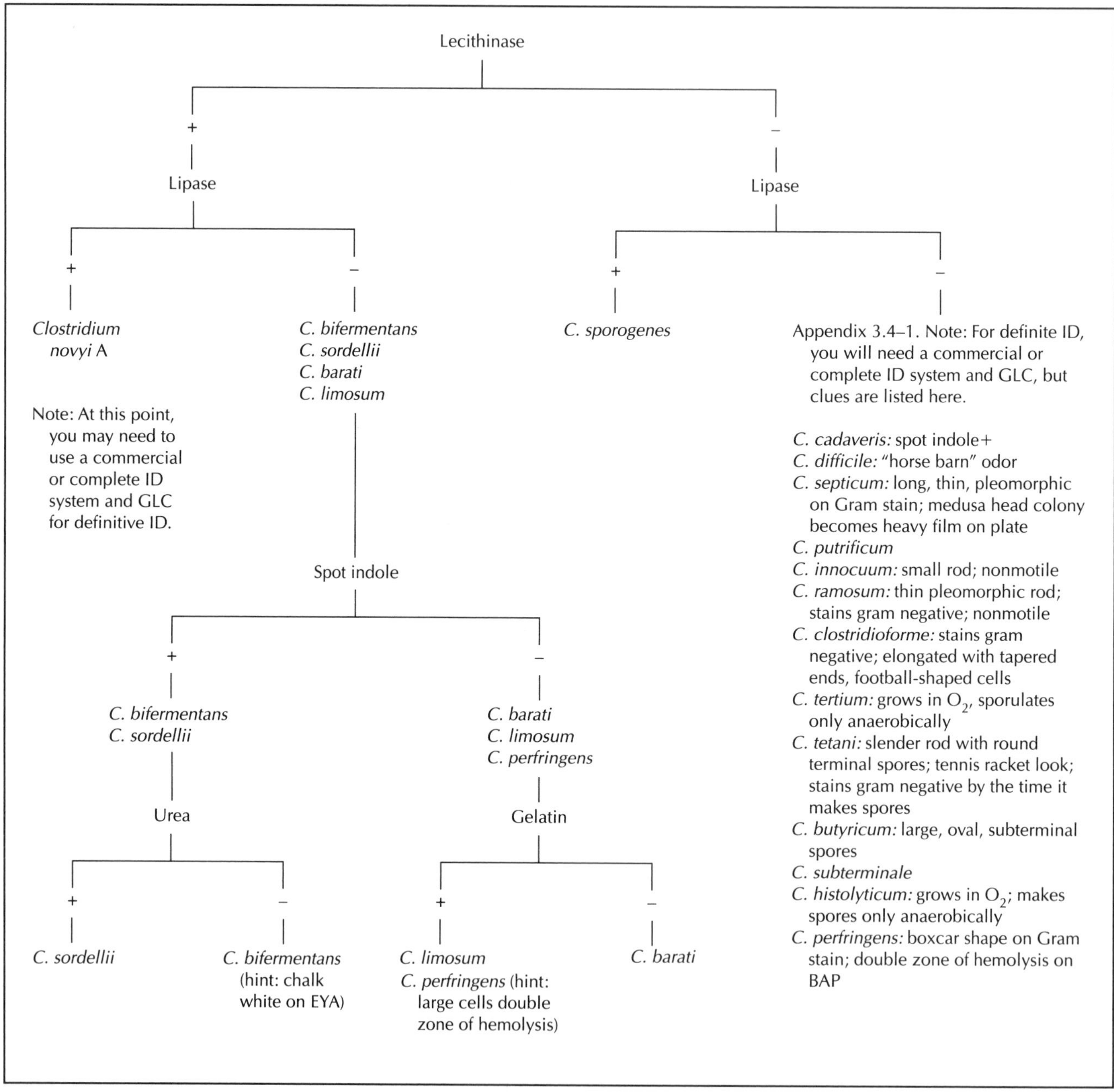

Figure 3.6–2 Examine egg yolk agar. Adapted from CMPH 2.11.4. Abbreviations: ID, identification; EYA, egg yolk agar.

V. IDENTIFICATION *(continued)*

6. Most commonly encountered in infection: *C. perfringens* (hint: boxcar-shaped cells that produce a double zone of hemolysis on BAP and are lecithinase positive) (*see* Fig. 3.6–1).
7. Lecithinase and lipase reactions may help differentiate among the clostridia (*see* Fig. 3.6–2).

B. Gram-positive nonsporeforming bacilli. (*See* Appendixes 3.4–1 and 3.6–2 and Fig. 3.6–1). These bacilli include *Actinomyces, Bifidobacterium, Eubacterium, Lactobacillus*, and *Propionibacterium* spp.

1. Distinctively variable morphology
   a. Vary from branched, pleomorphic, diphtheroidal, or streptococcus-like to long and slender
   b. Gram positive to gram variable
2. Considerable variation in relationship to oxygen
   a. Most are obligate anaerobes.
   b. Some facultative anaerobes or microaerophilic *Actinomyces* spp. (exception *A. meyeri*) are microaerophilic.
   c. *Bifidobacterium* and *Propionibacterium* spp. grow aerobically in the presence of CO_2.
3. Nonmotile
   Exceptions are rare *Lactobacillus* spp. and rare *Eubacterium* spp.
4. Catalase not produced
   Exceptions are *Propionibacterium* sp. and *A. viscosus.*
5. Most commonly encountered in infection
   a. *Actinomyces* spp.: actinomycosis
   b. *Bifidobacterium dentium*: pulmonary infections
   c. *Propionibacterium* sp.: endocarditis and infections of implanted devices such as ventricular-arterial shunts or artificial joints
6. At times other nonsporeforming bacilli may be isolated. *See* Appendix 3.6–2 for identification of gram-positive nonsporeforming bacilli.

VI. PROCEDURE NOTES

A. Confirm purity of all isolates.
B. Confirm Gram reaction. (The use of a vancomycin disk may be helpful.)
C. Confirm presence of spores.
   Do not confuse spores with vacuoles. There will be one spore per cell in the same location in every cell. Vacuoles (one or more) appear in various locations within the cell.
D. Confirm motility. Don't confuse Brownian movement with motility.
E. Confirm that you have an anerobe. Some anaerobic gram-positive bacilli can be aerotolerant.
F. Confirm that you have a bacillus. Look at a gram-stained sample taken from around a penicillin disk on a plate where your organism is growing. A rod elongates and sometimes takes on long, bizarre shapes; cocci remain coccal.

REFERENCES

1. **Baron, E. J., L. R. Peterson, and S. M. Finegold.** 1994. *Bailey and Scott's Diagnostic Microbiology,* 9th ed. The C. V. Mosby Co., St. Louis.
2. **Finegold, S. M., and W. L. George.** 1989. *Anaerobic Infections in Humans.* Academic Press, Inc., San Diego, Calif.
3. **Murray, P. R., E. J. Baron, M. A. Pfaller, F. C. Tenover, and R. H. Yolken (ed.).** 1995. *Manual of Clinical Microbiology,* 6th ed. American Society for Microbiology, Washington, D.C.
4. **Summanen, P., E. J. Baron, D. M. Citron, C. Strong, H. M. Wexler, and S. M. Finegold.** 1993. *Wadsworth Anaerobic Bacteriology Manual,* 5th ed. Star Publishing Co., Belmont, Calif.

APPENDIX 3.6–1

Characteristics of gram-positive sporeforming bacilli[a,b]

Type and species	Gelatin hydrolysis	Glucose fermentation	Lecithinase	Lipase	Indole	Aerobic growth	Urea	Nitrate	Motility	Spore shape and location[c]	Esculin
Saccharolytic proteolytic											
C. bifermentans	+	+	+	−	+	−	−	−		OS	+
C. sordellii	+	+	+	−	+	−	$+^{-}$	−		OS	−
C. perfringens	+	+	+	−	−	−	−	$+^{-}$	−	RS	−
C. novyi A	+	+	+	+	−	−	−	−		OS	−
C. sporogenes	+	+	−	+	−	−	−	−		OS	+
C. cadaveris	+	+	−	−	+	−	−	−		OT	−
C. septicum	+	+	−	−	−	−	−	V		OS	+
C. difficile	+	+	−	−	−	−	−	−		OS	+
C. putrificum	+	+	−	−	−	−	−	−		T	$-^{+}$
Saccharolytic nonproteolytic											
C. barati	−	+	+	−	−	−	−	−		S, RS	+
C. tertium	−	+	−	−	−	+	−	−		OT	+
C. butyricum	−	+	−	−	−	−	−	−	−	OA	+
C. innocuum	−	+	−	−	−	−	−	−	−	OT	+
C. ramosum	−	+	−	−	−	−	−	−		R, OT, RS	+
C. clostridioforme	−	+	−	−	$-^{+}$	−	−	$+^{-}$		OS, RS	+
Asaccharytotic proteolytic											
C. tetani	+	−	−	−	V	−	−	−		RT	−
C. hastiforme	+	−	−	−	−	−	−	$-^{+}$		S	−
C. subterminale	+	−	$-^{+}$	−	−	−	−	−		OS, RS	$-^{+}$
C. histolyticum	+	−	−	−	−	$+^{-}$	−	−		OS	−
C. limosum	+	−	+	−	−	−	−	−		S	−

[a] Adapted from CMPH 2.11.9.
[b] Key: +, positive reactions for 90 to 100% of strains; −, negative reactions for 90 to 100% of strains; $+^{-}$, most strains positive, some strains negative; $-^{+}$, most strains negative, some strains positive; V, variable (strains may be either + or −).
[c] OA, only anaerobic; RS, rarely seen; O, oval; R, round; S, subterminal; T, terminal.

APPENDIX 3.6–2

Identification of gram-positive nonsporeforming bacilli[a,b]

Organism	Nitrate reduction	Catalase	Indole production	Esculin hydrolysis	Urease	Red colony	Oxygen tolerance[c]
Actinomyces spp.		+	$-^{+}$	−	$+^{-}$	V	
A. israelli	$+^{-}$	−	−	+	−	−	A, M
A. odontolyticus	+	−	−	$+^{-}$	−	$+^{-}$	A, M
A. naeslundii	$+^{-}$	−	−	$+^{-}$	+	−	M, F
A. viscosus	$+^{-}$	+	−	V	+	−	A, M
A. meyeri	$-^{+}$	−	−	$-^{+}$	−	−	A
Propionibacterium spp.	V	V	$-^{+}$	V			A, M
P. acnes	$+^{-}$	$+^{-}$	$+^{-}$	−			A, M
P. granulosum	−	+	−	−			
P. avidum	−	+	−	+			
P. propionicus							
Bifidobacterium spp.	−	$-^{+}$	−	$+^{-}$			A, M
B. dentium	−	−	−	+			A
Lactobacillus spp.	$-^{+}$	−	−	V			A, M
Eubacterium spp.	V	−	$-^{+}$	$+^{-}$			A
E. lentum	+	$-^{+}$	−	−			A

[a] Adapted from CMPH 2.11.10.
[b] Key: +, positive reactions for 90 to 100% of strains; −, negative reactions for 90 to 100% of strains; $+^{-}$, most strains positive, some strains negative; $-^{+}$, most strains negative, some strains positive; V, variable (strains may be either + or −).
[c] Oxygen tolerance: A, anaerobic; M, microaerophilic; F, facultative.

3.7 Identification of Anaerobic Cocci

I. PRINCIPLE

The anaerobic cocci are a prominent part of the normal human microbiota in the bowel, oral cavity, upper respiratory tract, and female genital tract. The anaerobic gram-positive cocci are important human pathogens; next to the anaerobic gram-negative bacilli, they are the most commonly isolated anaerobes in clinically significant infections. Anaerobic gram-negative cocci account for a very small percentage of the anaerobic cocci isolated from human specimens (1–3). Anaerobic cocci can be identified by Gram stain, colony morphology, and various biochemical reactions. However, in most clinical situations, identification of the isolate to *Peptostreptococcus* spp. may provide information about treatment and clinical impact.

II. SPECIMEN

See procedures 3.2 and 3.3 for specimen requirements.

III. MATERIALS

A. See procedure 3.3 for isolation requirements.

B. See procedure 3.4 for identification tests.

IV. PROCEDURE

A. Day 1

The first day that workable colonies are seen on the primary brucella agar plate is considered day 1.

1. See procedures 3.2 and 3.3 for isolation, Gram stain, and aerotolerance testing.
2. See procedure 3.4 for identification procedures.
3. Document the description of colony morphology, Gram reaction, and all work performed on the anaerobe worksheet.
4. If the original colony is too small, the aerotolerance testing and Gram stain can be done on day 3 by using growth from the brucella subculture plate.

B. Day 2

1. Examine the CHOC aerotolerance plate. Growth on CHOC indicates that the organism is not an anaerobe. There is no need to proceed with its anaerobic identification.
2. If there is no growth on the CHOC aerotolerance plate, incubate the plate for another 24 h.

C. Day 3

1. Examine the CHOC again to confirm that the organism is an anaerobe.
2. Examine the brucella plate inoculated on day 1. If there is good growth, proceed to read the potency disk results.
3. If the organism is a gram-negative coccus, perform the nitrate test and see Fig. 3.7–1 for presumptive identification (*see* procedure 3.4).

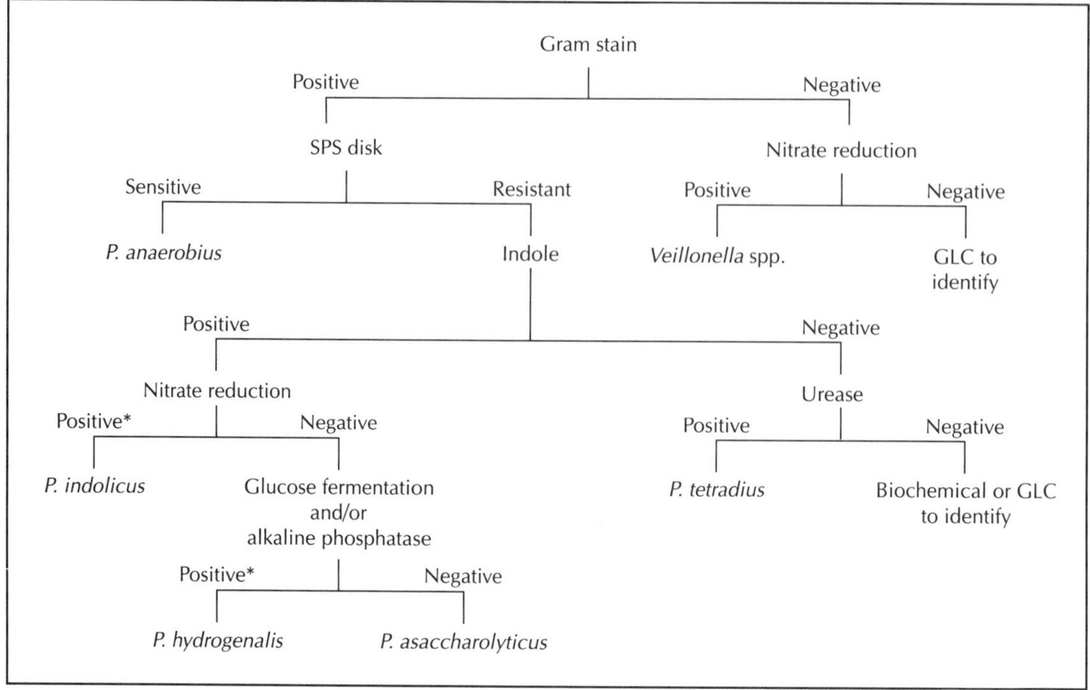

Figure 3.7–1 Flow chart for anaerobic cocci identification. Use this chart only as a guide; unusual strain variations may give anomalous results. *, All pathogenic strains of the genus *Peptococcus* (except *P. niger*) have been transferred to the genus *Peptostreptococcus*. Adapted from CMPH 2.12.3.

IV. PROCEDURE *(continued)*

4. If the organism is a gram-positive coccus, read sodium polyanethanolsulfonate (SPS) result and proceed according to Fig. 3.7–1.
5. Perform spot indole and/or rapid urease tests if necessary (*see* procedure 3.4).
6. If the organism cannot be identified by rapid tests, proceed to set up rapid enzymatic biochemical tests if required.

V. RESULTS

A. See Fig. 3.7–1 for anaerobic cocci identification.

B. Colony morphology

1. Growth of most anaerobic cocci is usually slower than that of *Bacteroides* or *Clostridium* spp. Small colonies are not apparent on brucella plates until after 48 h of incubation.
2. Colonies of gram-positive cocci are small, convex, grayish white, and opaque. The edge of the colony is entire, and the surface may appear stippled or pockmarked. Colony diameter is usually <0.5 to 2 mm. *Peptococcus niger*, very rarely isolated clinically, is weakly catalase positive and produces black-pigmented colonies (3).
3. *Veillonella* spp., the most commonly isolated gram-negative cocci, produce small, convex, translucent to transparent colonies with entire edges. These colonies may show red fluorescence under long-wave UV light (Wood's lamp).

C. Gram stain morphology

1. There are no unique microscopic characteristics to differentiate anaerobic cocci from facultative cocci. Some peptostreptococci may resemble staphylococci microscopically. Therefore, the presence of staphylococcal forms on direct Gram stain with no staphylococci recovered aerobically may suggest a *Peptostreptococcus* sp. (2, 3).

2. Microscopically, the gram-positive cocci are usually more consistent, although coccobacillary forms are seen occasionally. *Peptostreptococcus anaerobius* and *Peptostreptococcus productus* are large coccobacilli and usually appear in pairs or chains. *Peptostreptococcus magnus* and *Peptostreptococcus micros* are differentiated primarily by size. *P. magnus* cells are >0.7 mm in diameter and are usually found singly or in masses, and *P. micros* cells are <0.7 mm in diameter and usually form short chains.
3. *Veillonella* spp. are <0.5 mm in diameter and are usually seen in clusters or as diplococci.
4. Unusually large cocci, especially when found in groups or packets, suggest *Peptostreptococcus tetradius*, *Peptostreptococcus magnus*, *Sarcina* spp., or *Megasphaera* sp. *Megasphaera* sp. commonly stains gram positive but is really gram negative on the basis of cell wall composition (1, 2).

D. Rapid spot tests

1. *Peptostreptococcus asaccharolyticus* is the most commonly isolated indole-positive anaerobic gram-positive coccus. Another indole-positive coccus is *P. hydrogenalis*, which can be separated by its positive reaction to alkaline phosphatase (*see* procedure 3.4).
2. *Peptostreptococcus prevotii* and *Peptostreptococcus indolicus* are also nonfermentative, but *P. prevotii* is indole negative and *P. indolicus* reduces nitrate and produces indole.
3. Susceptibility to sodium polyanetholsulfonate (SPS) is a unique characteristic of *P. anaerobius*. Some strains of *P. micros* are susceptible to SPS; generally they produce smaller zones, and the Gram stain would exhibit very small cell size.

E. Rapid nitrate and urease tests (*see* procedure 3.4)

1. *P. indolicus*, which is rarely isolated from humans, reduces nitrate, but *P. asaccharolyticus* does not. *Veillonella parvula*, the only anaerobic gram-negative coccus of clinical significance, also reduces nitrate (3).
2. *P. tetradius* is the only clinically encountered anaerobic, urea-positive coccus (2, 3).

F. Rapid enzymatic biochemical identification

If the tests described above fail to identify the organism, use a commercially prepared rapid enzymatic biochemical identification system to complete the identification.

1. Generally, anaerobic gram-positive cocci can be identified fairly easily with the above-mentioned tests.
2. For anaerobic gram-negative cocci, the nitrate disk can be useful (3).

VI. REPORTING RESULTS

Anaerobic isolates that cannot be identified further by spot rapid tests and commercial rapid enzymatic identification systems may be reported as *Peptostreptococcus* sp. or *Veillonella* sp. Consult with the requesting physician to ensure that results are clinically significant and to determine whether to pursue definitive identification and susceptibility testing.

REFERENCES

1. **Baron, E. J., L. R. Peterson, and S. M. Finegold.** 1994. *Bailey and Scott's Diagnostic Microbiology,* 9th ed. The C. V. Mosby Co., St. Louis.
2. **Murray, P. R., E. J. Baron, M. A. Pfaller, F. C. Tenover, and R. H. Yolken (ed.).** 1995. *Manual of Clinical Microbiology,* 6th ed. American Society for Microbiology, Washington, D. C.
3. **Summanen, P., E. J. Baron, D. M. Citron, C. Strong, H. M. Wexler, and S. M. Finegold.** 1993. *Wadsworth Anaerobic Bacteriology Manual,* 5th ed., Star Publishing Co., Belmont, Calif.

SECTION 4

Mycobacteriology

SECTION EDITORS AND AUTHORS: *Phyllis Della-Latta and Irene Weitzman*

4.1 Levels of Laboratory Service for Mycobacteriology

I. PRINCIPLE

The levels of laboratory service promulgated by the CDC, American Thoracic Society, and CAP (1, 2) are based on workload, expertise, and cost effectiveness.

To maintain proficiency to qualify for Level I status, a laboratory must prepare at least 10 to 15 specimens per week and examine them for acid-fast bacilli; to qualify for Levels II and III, a laboratory must culture and identify at least 20 specimens per week.

II. PROCEDURE

A laboratory qualifies for Levels I through III by performing the procedures listed below.

Procedure	Levels		
	I	II	III
Acid-fast smears	Yes	Yes	Yes
Culture	No	Yes	Yes
Identification of MTB complex	No	Yes	Yes
Identification of all mycobacteria	No	No	Yes
Antituberculosis susceptibility	No	Yes	Yes

Procedures not performed by Level I or II laboratories are referred to a nationally recognized reference facility or a higher-level laboratory.

REFERENCES

1. **Hawkins, J. E., R. C. Good, G. P. Kubica, P. R. J. Gangadharam, H. M. Guft, D. D. Stottmeier, H. M. Sommers, and L. G. Wayne.** 1983. Levels of laboratory services for mycobacterial diseases: official statement of the American Thoracic Society. *Am Rev. Respir. Dis.* **128:**213.
2. **Salfinger, M.** 1995. Role of the laboratory in evaluating patients with mycobacterial disease. *Clin. Microbiol. News.* **17**(14)**:**108–109.

SUPPLEMENTAL READING

Salfinger, M., and A. J. Morris. 1994. The role of the microbiology laboratory in diagnosing mycobacterial diseases. *Am. J. Clin. Pathol.* **101**(Suppl.)**:**56.

4.2 Specimen Collection

I. PRINCIPLE

Accurate laboratory results require adequate collection and transport of appropriate specimens for mycobacterial detection. Specimens submitted for mycobacterial culture may be pulmonary or extrapulmonary, from a contaminated or a noncontaminated source. Contaminated specimens require digestion-decontamination before culture, whereas aseptically collected specimens from sterile sites do not.

II. SAFETY PRECAUTIONS

BSL 2 practices, facilities, containment equipment (e.g., a class I or II biological safety cabinet and centrifuges with safety carriers), and personal protective clothing (gowns, gloves, respirators) are required for processing, smear preparation, and culturing for mycobacteria (2).

III. MATERIALS

Sterile sputum collection kit for mycobacteria, such as the Sage Collection Kit (Becton Dickinson Microbiology Systems)

IV. PROCEDURE

Detailed instructions for specimen collection and handling of individual specimens are provided elsewhere (1). General guidelines are summarized below.

A. Recommendations for collection

1. Use a sterile, leakproof, disposable plastic container (Sage Collection Kit).
2. Collect specimens before antimicrobial therapy is started.
3. Perform a gastric lavage process within 4 h or add 100 mg of sodium carbonate.

B. Common mistakes

1. Insufficient volume
2. Excessive or inadequate number of specimens
3. Saliva, not sputum collected
4. Refrigerated blood

C. Unacceptable specimens

Immediately request another specimen and discard original unsuitable specimen after 48 h (4°C).

1. Dry swabs
2. 24-h urine or sputum
3. Wax containers
4. Nonsterile containers
5. Broken tubes
6. Leaking specimens
7. Unlabeled specimens

V. SPECIMEN STORAGE

A. Specimens not processed within 1 h of collection should be stored at 2 to 8°C (except blood).

B. If processing of gastric lavage specimens is delayed by more than 4 h, add approximately 100 mg of sodium carbonate to the collection container to neutralize the acidity.

C. Store any remaining specimens at 2 to 8°C until after processing is completed and the specimen has been stained and examined.

D. Hold normally sterile specimens (body fluids, etc.) at 2 to 8°C for several days in case of media contamination.

REFERENCES

1. **Gullans, C. R., Sr.** 1992. Preparation of specimens for mycobacterial culture, p. 3.3.1–3.3.6. *In* H. D. Isenberg (ed.), *Clinical Microbiology Procedures Handbook*, vol. 1. American Society for Microbiology, Washington, D.C.

2. **Richmond, J. Y., R. C. Knadsen, and R. C. Good.** 1996. Biosafety in the clinical microbiology laboratory. *Clin. Lab. Med.* **16**(3): 527–550.

SUPPLEMENTAL READING

Nolte, F. S., and B. Metchok. 1995. *Mycobacterium*, p. 400–437. *In* P. R. Murray, E. J. Baron, M. A. Pfaller, F. C. Tenover, and R. H. Yolken (ed.), *Manual of Clinical Microbiology*, 6th ed. American Society for Microbiology, Washington, D.C.

4.3 Digestion-Decontamination Procedure

I. PRINCIPLE

In Levels II and III laboratories, optimal recovery of mycobacteria from clinical specimens requires liquefying the organic debris (digestion) and eliminating contaminating organisms (decontamination). Decontaminating procedures are toxic to mycobacteria; therefore, the procedure must be followed precisely to ensure maximal survival. The most widely used method in the U.S. is *N*-acetyl-L-cysteine–sodium hydroxide (NALC-NaOH), whereas others include Zephiran-trisodium phosphate (ZTP) and sodium hydroxide (NaOH). *N*-Acetyl-L-cysteine (NALC) is a mucolytic agent (the digestant), and sodium hydroxide is the decontaminating agent.

II. SPECIMENS

Specimens from contaminated sources require digestion-decontamination, whereas normally sterile specimens do not require this procedure.

A. Contaminated

1. Sputum
2. Bronchial wash
3. Skin
4. Soft tissue
5. Gastric lavage
6. Stool and urine

B. Sterile

1. CSF
2. Bone marrow
3. Tissue
4. Biopsy
5. Blood

III. MATERIALS

Include QC information on reagent container and in QC records.

A. Reagents

1. *N*-Acetyl-L-cysteine powder
2. Sodium hydroxide solution (4%)
3. Prepackaged reagents for the NALC-NaOH procedures (BBL-Mycoprep may be purchased commercially from Becton Dickinson Microbiology Systems, Cockeysville, Md.)
4. Sodium citrate (0.1 M)
5. 0.067 M phosphate buffer, pH 6.8
6. Bovine albumin 0.2% fraction V solution (not recommended for use with BACTEC)

   Label all reagents with the date of preparation or receipt, contents, concentration, storage requirements, and expiration date. The NaOH and sodium citrate solutions can be mixed (Table 4.3–1), sterilized, and stored in sterile screw-cap flasks. Once NALC is added, use the solutions within 24 h. Follow manufacturer's instructions for commercial preparations.

B. Supplies

1. Discard containers
2. Sterile 50-ml conical polypropylene screw-cap tubes (aerosol free and graduated)

C. Equipment

1. Centrifuge with aerosol-free sealed centrifuge cups
2. Vortex mixer

Table 4.3–1 Preparation of NALC-NaOH digestant-decontaminant[a]

Volume (ml)	NaOH-sodium citrate (ml)[b]	NALC (g)
50	50	0.25
100	100	0.50
200	200	1.00
500	500	2.50
1,000	1,000	5.00

[a] Adapted from references 1 and 2.
[b] 1:1 mixture of 4% NaOH to 2.9% sodium citrate.

IV. PROCEDURE

(Adapted from references 1 and 2)

A. Add an equal volume of the NALC-NaOH solution to the specimen in a sterile 50-ml conical polypropylene screw-cap centrifuge tube (aerosol free and graduated). If the volume of specimen is more than 10 ml, transfer 10 ml of the most purulent, mucoid, or bloody portion of the specimen with a sterile pipette. Tightly close the cap.

B. Mix the contents of the tube on a vortex mixer for no more than 30 s and invert.

C. Incubate each tube for 15 min at room temperature (20 to 25°C) to decontaminate the specimen.

D. Dilute the contents with either sterile distilled water or sterile 0.067 M phosphate buffer (pH 6.8) up to the 50-ml mark on the tube to stop the decontamination procedure. Recap the tube tightly.

E. Invert the tube contents several times to mix.

F. Centrifuge the tube at ≥3,000 × *g* for 15 to 20 min, using aerosol-free sealed centrifuge cups.

G. Gently pour off the supernatant into a splash-proof discard container filled with a suitable disinfectant (e.g., Ampyl).

H. Add 1 to 2 ml of sterile 0.2% bovine albumin fraction V (pH 6.8) to the sediment. Bovine albumin is not recommended for use with BACTEC; use the phosphate buffer instead. Sterile saline (0.85%), sterile distilled, or deionized water may be used if media will be inoculated immediately. Swirl gently to suspend evenly.

I. Using a separate, sterile disposable capillary pipette per specimen, deliver 3 drops onto each solid and liquid medium and up to 0.5 ml per mycobacteria growth indicator tube (Becton Dickinson). For the BACTEC 12B, disinfect the rubber septum with 70% alcohol and inject 0.5 ml with a tuberculin syringe.

J. Prepare a smear for the acid-fast stain with one of the following:

1. Bacteriological loop (3-mm diameter)
2. Sterile wooden applicator stick
3. One drop from a sterile capillary pipette

K. Cover an area approximately 1 by 2 cm on a new, clean, unscratched glass slide.

V. QUALITY CONTROL

Monitor the percentage of contaminated specimens. The acceptable range is 3 to 5% with media without antimicrobial agents. A contamination rate significantly less than 3% in the non-antimicrobial-agent-containing media suggests overly harsh decontamination, whereas greater than 5% suggests inadequate decontamination or incomplete digestion (2).

REFERENCES

1. **Gullans, C. R., Sr.** 1992. Digestion-decontamination procedures, p. 3.4.1–3.4.14. *In* H. D. Isenberg (ed.), *Clinical Microbiology Procedures Handbook*, vol. 1. American Society for Microbiology, Washington, D.C.

2. **Kent, P. T., and G. P. Kubica.** 1985. *Public Health Mycobacteriology. A Guide for the Level III Laboratory.* U.S. Public Health Service publication no. 86-21654. U.S. Government Printing Office, Washington, D.C.

4.4 Acid-Fast Stain Procedures

I. PRINCIPLE

Acid-fast mycobacteria resist decolorization by acid-alcohol after primary staining owing to the high lipid (mycolic acid) content in their cell walls. The acid-fast stain is the most rapid method for detection of mycobacteria. It can monitor the progress of patients on antimycobacterial therapy, dictate appropriate dilution of sediments for direct drug susceptibility testing, and confirm the presence of acid-fast bacilli (AFB) in culture. The most widely used methods to determine acid-fastness are the Kinyoun (carbolfuchsin) and the fluorochrome (e.g., auramine O or auramine-rhodamine) stains. The fluorochrome stains are recommended for specimen examination because of their increased sensitivity and speed. They may be examined at a lower magnification (2–4) and confirmed by the Kinyoun stain on the same slide. The Kinyoun is used to confirm the presence of AFB in cultures. Only new, clean slides must be used for acid-fast staining.

II. SAFETY PRECAUTIONS

A. Level I laboratories

Before acid-fast staining, Level I laboratories must render the mycobacteria nonviable by treating the specimen with an equal volume of 5% sodium hypochloride solution (undiluted household bleach) and waiting exactly 15 min before pouring this into the 50-ml sterile centrifuge tube. After centrifuging and decanting, staining can be conducted on the open bench (1, 2).

B. Level II Laboratories

BSL 2 practices, facilities, containment equipment (e.g., a class I or II biological safety cabinet [BSC] and centrifuges with safety carriers), and personal protective clothing are required for processing and preparation of smears and culturing for mycobacteria (2).

III. MATERIALS

Include QC information on reagent container and in QC records.

A. Carbolfuchsin acid-fast stains

1. Carbolfuchsin (Ziehl-Neelson or Kinyoun)
2. 3% Acid-alcohol
3. Counterstain (methylene blue or brilliant green)

B. Fluorochrome acid-fast stains

1. Auramine-O (with or without rhodamine)
2. 0.5% Acid-alcohol
3. Counterstain (potassium permanganate or acridine orange)

IV. SPECIMENS

Clinical specimens or pure cultures suspected of harboring mycobacteria are stained for acid fastness. However, the value of staining gastric aspiration and urine is controversial. A smear directly from a clinical specimen (not concentrated) is discouraged because it lacks the sensitivity of a concentrated (centrifuged) smear. A negative result from a direct smear must be followed by a concentrated smear.

Preparation of smears for staining

For details, see references 1 and 2.

A. Save sediments from each specimen and refrigerate as a backup for stain failure, a doubtful smear result, or contamination of culture.

B. Air dry smears in the BSC. Fix the smeared material by allowing it to remain on an electric slide warmer at 65 to 75°C for at least 2 h. Smears may be fixed by passing through flames, but this procedure is not recommended in recent papers from the CDC.

V. STAINING PROCEDURES

A. Fluorochrome

1. Flood the slide with fluorochrome stain.
2. Stain for 15 min.
3. Rinse the slide with water; drain excess water from the slide.
4. Flood with 0.5% acid-alcohol.
5. Decolorize for 2 min.
6. Rinse the slide with water; drain excess water from the slide.
7. Flood the slide with counterstain (potassium permanganate or acridine orange).
8. Counterstain 2 min.

   NOTE: Timing is critical during the counterstaining step with potassium permaganate. Counterstaining for a longer time may quench the fluorescence of acid-fast organisms.
9. Rinse the slide with water. Drain excess water from the slide.
10. Air dry; do not blot.
11. Examine the smear with a fluorescent microscope.

    NOTE: Fluorochrome-stained slides may be directly restained with the carbolfuchsin staining procedure after immersion oil is removed with xylene to confirm a positive fluorochrome slide. Retain positive acid-fast smears for 1 year.

B. Kinyoun's carbolfuchsin

1. Flood the entire slide with Kinyoun's carbolfuchsin.
2. Allow the smear to stain for 2 to 5 min.
3. Rinse the slide with water.
4. Flood the slide with 3% acid-alcohol, until no more color drains from the slide.
5. Rinse the slide with water; drain excess water from the slide.
6. Flood the slide with counterstain (methylene blue or brilliant green).
7. Counterstain for 20 to 30 s.
8. Rinse the slide thoroughly with water; drain excess water from the slide.
9. Air dry. Do not blot.
10. Examine with a 100× oil immersion objective.

VI. SMEAR EXAMINATION AND RESULTS

A. Examination

1. Fluorochrome-stained smears are screened with a 25× or 40× objective.
2. Kinyoun (carbolfuchsin)-stained smears are examined with the 100× oil immersion objective.
3. Mycobacteria are approximately 1 to 10 μm long and typically appear as slender rods. However, they may also appear curved or bent, coccobacillary, or even filamentous. Some may be beaded or banded.

B. Result

1. Negative—Report ''No acid-fast bacilli'' seen.
2. Positive—Currently recommended reporting (2, 3) is listed below for both fuchsin and fluorochrome staining techniques. Data are numbers of AFB as per indicated number of microscopic fields (F's). This takes into account the larger number of fields covered with the lower magnification examined under the fluorescent microscope.

Fluorochrome		Carbolfuchsin	Report
250×	450×	1,000×	
0	0	0	No AFB seen
1–2/30F	1–2/70F	1–2/300F	Doubtful, repeat
1–9/10F	2–18/50F	1–9/100F	1+
1–9/F	4–36/10F	1–9/10F	2+
10–90/F	4–36/F	1–9/F	3+
>90/F	>36/F	>9/F	4+

VII. LIMITATIONS

A. Specificity

The acid-fast stain is nonspecific. Slow-growing mycobacteria (not just *M. tuberculosis*) are consistently acid-fast. Microorganisms other than mycobacteria that demonstrate various degrees of acid-fastness include *Nocardia, Rhodococcus, Legionella micdadei*, cysts of *Cryptosporidium* spp., and *Cyclospora* spp.

B. Sensitivity

The acid-fast smear is insensitive. Sensitivities ranging from 22 to 81% have been reported (4). A negative smear does not rule out tuberculosis because detection levels are only 5,000 to 10,000 bacilli per ml of sputum.

C. Safety

Since heat fixing and staining may not kill all the mycobacteria, discard slides in a sharps receptacle and wear gloves.

VIII. QUALITY CONTROL

A. Purpose

A positive and negative control slide must be included whenever the acid-fast stain is performed and upon receipt of each new lot of materials, reagents, and media to verify the correct performance of the procedure. *Escherichia coli* is commonly used for the negative control and *M. tuberculosis* H37Ra ATCC 25177 is the positive control.

B. Results

1. Kinyoun carbolfuchsin
   - **a.** Positive—*Mycobacterium* spp.: red or magenta against a blue or green background, depending on the counterstain used
   - **b.** Negative—*E. coli*: blue or green, depending on the counterstain used
2. Fluorochrome (auramine O or auramine-rhodamine)
   - **a.** Positive—*Mycobacterium* spp.: yellow to orange fluorescence (color may vary with the filter system used)
   - **b.** Negative—*E. coli*: no fluorescence
3. Record the results of the control slide before reviewing the smears from clinical specimens. If the control slide is unacceptable, review procedures and reagent preparations.

REFERENCES

1. **Ebersole, L. L.** 1992. Acid-fast stain procedures, p. 3.5.1–3.5.10. *In* H. D. Isenberg (ed.), *Clinical Microbiology Procedures Handbook*, vol. 1. American Society for Microbiology, Washington, D.C.
2. **Kent, P. T., and G. P. Kubica.** 1985. *Public Health Mycobacteriology: a Guide for the Level III Laboratory*. U.S. Department of Health and Human Services publication no. (CDC) 86-21654 6, p. 57–68. U.S. Government Printing Office, Washington, D.C.
3. **Nolte, F. S., and B. Metchok.** 1995. *Mycobacterium*, p. 400–437. *In* P. R. Murray, E. J. Baron, M. A. Pfaller, F. C. Tenover, and R. H. Yolken (ed.), *Manual of Clinical Microbiology*, 6th ed. American Society for Microbiology, Washington, D.C.
4. **Salfinger, M., and H. E. Pfyffer.** 1994. The new diagnostic mycobacteriology laboratory. *Eur. J. Clin. Microbiol. Infect. Dis.* **13:** 961–979.

SUPPLEMENTAL READING

Federal Register, April 28, 1997. *Goals for Working Safely with* Mycobacterium tuberculosis *in Clinical, Public Health and Research Laboratories*. Federal Register **62:**23066–23079. U.S. Government Printing Office, Washington, D.C.

Klotz, S. A., and R. L. Penn. 1987. Acid-fast staining of urine and gastric contents is an excellent indicator of mycobacterial disease. *Am. Rev. Respir. Dis.* **136:**1197–1198.

Richmond, J. Y., R. C. Knudsen, and R. C. Good. 1996. Biosafety in the clinical mycobacteriology laboratory. Clin. Lab. Med. **16**(3)**:**527–550.

4.5 Solid Media: Selection, Inoculation, Incubation, and Examination

I. PRINCIPLE

The principal advantage of solid media (tubed or in plates) is that they enable detection of mixed cultures and contaminants. Colonial morphology may also aid in the identification of mycobacteria. A combination of egg-based and agar-based media (in addition to a liquid medium) is optimal for the primary isolation of mycobacteria (1).

II. MEDIA SELECTION

Selective media, containing antibacterial and antifungal agents, are recommended for contaminated specimens, whereas less inhibitory media should be inoculated from normally sterile sites. Specimens, especially skin, bone, joint, and synovial fluid, from immunocompromised patients should be inoculated onto media supplemented with hemin, hemoglobin, or ferric ammonium citrate (e.g., CHOC, Middlebrook 7H10 or 7H11 with X factor disk, or Lowenstein-Jensen [LJ] with 1% ferric ammonium citrate) and incubated at 30 to 32°C to enhance the isolation of *M. haemophilum* (2, 4). Either plates or tubes may be used for agar-based media. Plates with agar media have the advantage of permitting early observation of colonies and detection of mixed cultures.

Some *M. tuberculosis* isolates prefer growth on egg-based rather than agar media.

III. MATERIALS

(For details, see reference 2.)

Include QC information on reagent container and in QC records.

NOTE: Malachite green, contained in egg-based media, is a photosensitive dye. Therefore, media should not be stored in exposed light.

A. Nonselective media for normally sterile sites

1. Egg based
   LJ
2. Agar based
   Middlebrook 7H10, 7H11
3. Comments
   Excessive exposure to light results in the formation of toxic compounds in the 7H10 or 7H11 medium.

B. Selective media for nonsterile sites

1. Egg based
   - **a.** LJ-Gruft
   - **b.** Mycobactosel-LJ
2. Agar based
   - **a.** Mitchensen's
   - **b.** Selective Middlebrook 7H11 (7H11S)
   - **c.** Mycobactosel-7H11

IV. INOCULATION AND INCUBATION

A. Inoculation

Add 3 drops of sediment to media with a sterile transfer pipette.

B. Incubation

1. Incubate all cultures in the dark at 35 to 37°C with the exception of skin or soft tissue suspected to be infected with *M. marinum, M. ulcerans*, or any specimen suspected with *M. haemophilum* for which an additional separate set of media is incubated at 30°C.

IV. INOCULATION AND INCUBATION *(continued)*

2. Incubate in 5 to 10% CO_2; monitor CO_2 levels daily. Screw caps should be loose for the first 2 weeks of incubation, and tubes should be slanted during this period. Afterwards, tighten caps and set the tubes upright. Plates should be incubated with the medium side down and placed in CO_2-permeable bags.
3. Incubate all cultures a minimum of 8 weeks.

V. EXAMINATION OF CULTURE

A. Schedule

1. Examine solid media within 3 to 5 days after incubation, twice a week up to Week 4, and weekly thereafter until Week 8 (4).
2. Invert clear media (Middlebrook 7H10 and/or 7H11) in petri dishes, and examine these plates microscopically to facilitate the early observation of microcolonies.

B. Contamination

Discard cultures completely overgrown with contaminants, if the patient's specimen is still available. Decontaminate by another method or request another specimen. Specimens repeatedly contaminated with *Pseudomonas* species should be decontaminated by the oxalic acid method (1) or NaLC-NaOH–oxalic acid (6). Contaminated urine may be decontaminated by using the sulfuric acid method (1).

C. Morphological features

1. Colonies on solid media should be examined with respect to growth rate, pigmentation, and colony morphology (5).
2. Colonies on clear agar media in petri dishes should be inverted and examined under a microscope for cording and presence of filaments. Slow-growing, dry, corded colonies are suggestive of *M. tuberculosis* or *M. kansasii* (1).

VI. QUALITY CONTROL

(For details, see references 2 and 3.)

A. Prepared media

1. Laboratories purchasing commercially prepared media that are quality controlled by the manufacturers in accordance with NCCLS proposed standards need not perform QC checks for sterility, growth, selectivity, or biochemical response on LJ and Middlebrook media.
2. Documentation of the manufacturers' QC procedures should be obtained, and the lot number of each medium must be placed in a log book.
3. All other media, including user-prepared media, must be quality controlled with the appropriate positive- and negative-control organisms to ensure satisfactory performance and examined for signs of deterioration (color, dehydration, presence of contamination, bubbles, etc.).
4. Do not use media beyond the expiration date.
5. User-prepared media may be checked for sterility by incubating 1 to 3% of each batch and must be quality controlled and documented for performance. The microorganisms chosen for QC depend on the media. Generally, ATCC stock cultures of *M. tuberculosis, M. kansasii, M. scrofulaceum, M. intracellulare*, and *M. fortuitum* are used. *Escherichia coli* can be the test organism to determine the inhibitory effectiveness of selective antimicrobial media.
6. All QC results must be documented and reviewed by a designated supervisor. Documentation of corrective action, resolution, and follow-up of the situation is required.

B. Inoculation

1. Inoculate a few colonies of test organisms into soybean-casein digest broth. Incubate for several hours to achieve a suspension equal to a 0.5 McFarland standard.

2. For tubed media, use a calibrated loop or pipette to inoculate each tube to be tested with 10 μl of suspension. Incubate at 35 to 37°C in 5 to 10% CO_2 for up to 21 days.
3. For testing selective properties of media, dilute the suspension 1:10 in sterile 0.85% NaCl and inoculate the media with 10 μl of the suspension.

C. Interpretation of results

Expected results of QC testing of mycobacterial media are shown below.

Control type and recommended ATCC strains	Result
Positive controls	
M. tuberculosis ATCC 25177	Growth on all media
M. kansasii ATCC 12478	Growth on all media
M. scrofulaceum ATCC 19981	Growth on all media
M. intracellulare ATCC 13950	Growth on all media
M. fortuitum ATCC 2841	Growth on all media
Negative control	
E. coli ATCC 25922	Partial or total inhibition

REFERENCES

1. **Kent, P. T., and G. P. Kubica.** 1985. *Public Health Mycobacteriology. A Guide for the Level III Laboratory.* U.S. Public Health Service publication no. 86-21654. U.S. Government Printing Office, Washington, D.C.
2. **Lambi, E. A.** 1992. Medium selection and incubation for the isolation of mycobacteria, p. 3.6.1–3.6.8. *In* H. D. Isenberg (ed.), *Clinical Microbiology Procedures Handbook*, vol. 1. American Society for Microbiology, Washington, D.C.
3. **NCCLS.** 1990. *Quality Assurance for Commercially Prepared Microbiological Culture Media.* NCCLS document M-22A. NCCLS, Wayne, Pa.
4. **Nolte, F. S., and B. Metchock.** 1995. *Mycobacterium*, p. 400–437. *In* P. R. Murray, E. J. Baron, M. A. Pfaller, F. C. Tenover, and R. H. Yolken (ed.), *Manual of Clinical Microbiology*, 6th ed. American Society for Microbiology, Washington, D.C.
5. **Silcox, V.** 1992. Identification of mycobacteria, p. 3.11.1–3.11.11. *In* H. D. Isenberg (ed.), *Clinical Microbiology Procedures Handbook*, vol. 1. American Society for Microbiology, Washington, D.C.
6. **Whittier, S., R. Hopfer, M. R. Knowles, and P. H. Gilligan.** 1993. Improved recovery of mycobacteria from respiratory secretions of patient with cystic fibrosis. *J. Clin. Microbiol.* **31:** 861–864.

4.6 Septi-Chek Biphasic Medium

I. PRINCIPLE

Septi-Chek AFB (Becton-Dickinson Microbiology Systems, Cockeysville, Md.) is a biphasic medium with a self-contained CO_2 environment. It consists of Middlebrook 7H9 broth and a slide with Middlebrook 7H11, modified egg, and CHOC. The system offers comparable sensitivity to conventional agar and broth methods (1, 3, 5).

II. SPECIMENS

The Septi-Chek System (Becton-Dickinson) is used for the detection and isolation of mycobacteria from sputum, bronchial washings or aspirate, body fluids (pleural, CSF, ascites, or synovial), urine, stool, biopsy tissues, or wounds and skin (4) when mycobacteriosis is suspected. For blood specimens, Septi-Chek has been successfully used with Isolator-procured blood concentrated by centrifugation and the addition of 0.75-ml volumes to two sets of Septi-Chek AFB broth (2).

III. MATERIALS

Include QC information on reagent container and in QC records.

A. Broth

B. Supplement

C. Slide with:

1. Side 1: nonselective Middlebrook 7H11 agar, modified, suitable for growth of most mycobacteria
2. Side 2: egg-based medium, suitable for growth of most mycobacteria
3. Side 3: chocolate agar

IV. PROCEDURE

Follow the manufacturer's instructions.

◪ **NOTE:** Identification of growth on solid or liquid medium may proceed by nucleic acid probes, HPLC, or biochemicals.

REFERENCES

1. **D'Amato, R. F., H. D. Isenberg, L. Hochstein, A. J. Mastellone, and P. J. Alperstein.** 1991. Evaluation of the Roche Septi-Chek AFB system for recovery of mycobacteria. *J. Clin. Microbial.* **29:**2906–2908.
2. **Isenberg, H. D.** Personal communication.
3. **Isenberg, H. D., R. F. D'Amato, L. Heifets, P. R. Murray, M. Scardamaglia, M. C. Jacobs, P. Alperstein, and A. Miles.** 1991. Collaborative feasibility study of a biphasic system (Roche Septi-Chek) for rapid detection and isolation of mycobacteria. *J. Clin. Microbiol.* **29:** 1719–1722.
4. **Lederman, C., J. C. Spitz, B. Scully, L. L. Schulman, P. Della-Latta, I. Weitzman, and M. E. Grossman.** 1994. *Mycobacterium haemophilum* cellulitis in a heart transplant recipient in New York City. *J. Am. Acad. Dermatol.* **30:**804–806.
5. **Whittier, P. S., K. Westfall, S. Setterquist, and R. Hofer.** 1992. Comparison of Septi-Chek AFB to BACTEC and Lowenstein-Jensen agar for detection and isolation of mycobacteria. *Eur. J. Clin. Microbiol. Infect. Dis.* **11:** 915–917.

4.7 Liquid Media

I. PRINCIPLE

Broth-based systems, radiometric or nonradiometric, permit faster growth of mycobacteria than systems that use solid media. It is recommended that both liquid and solid media be inoculated with all patient specimens (6, 7). The use of the liquid media described in this section should adhere to the following recommendations:

A. Perform all procedures in a biological safety cabinet (BSC) with appropriate precautions and protective clothing.
B. Examine all vials containing culture medium for signs of contamination and cracks before inoculation.
C. Properly label all vials before inoculation.
D. Carefully read manufacturer's instructions before implementing each of the following procedures.
E. Follow manufacturer's recommendations for digesting and decontaminating specimens from nonsterile sites.

II. SPECIMENS

All liquid-medium systems are devised for respiratory specimens; some include gastric aspirates, tissue, stool, and specimens from sterile body sites. Blood and bone marrow specimens can be used only in the ESP Myco System (Difco) and the BACTEC 13A (Becton-Dickinson).

III. RADIOMETRIC SYSTEMS

Include QC information on reagent container and in QC records.

The BACTEC system is based on Middlebrook 7H9 broth with the addition of ^{14}C-labelled substrate, BSA, casein hydrolysate, and catalase (5). An antimicrobial mixture (PANTA) is added for nonsterile specimens.

A. Materials

1. Media
   a. BACTEC 12B (all specimens except blood and bone marrow)
      Contains 4 ml of Middlebrook 7H12 broth
   b. BACTEC 13A (for blood and bone marrow)
      Contains 30 ml of modified Middlebrook 7H12 broth
2. Supplements
   a. BACTEC 13A enrichment supplement
      Antimicrobial supplement (PANTA)
   b. Mixture contains polymyxin B, amphotericin B, nalidixic acid, trimethoprim, and azlocillin.

B. Procedure

1. BACTEC 12B (5) setup
   a. Before inoculation, test all 12B vials in the 460TB system to eliminate vials with a high background reading and to establish the required 5 to 10% CO_2 atmosphere. Eliminate vials with a growth index (GI) of $\geq$20.
   b. Inoculate each BACTEC 12B vial with 0.1 ml of PANTA plus solution after PANTA is reconstituted with 5 ml of reconstituting fluid. Noncontaminated specimens do not need PANTA.
   c. Digest-decontaminate nonsterile specimens with NALC-NaOH, and add 1 ml of phosphate buffer (pH 6.8) to the processed, concentrated specimen.

III. RADIOMETRIC SYSTEMS *(continued)*

d. Inoculate each BACTEC 12B vial with up to 0.5 ml of the specimen with a tuberculin syringe that has a permanently attached needle.
e. Clean the rubber septum of each vial before inoculation with 70% alcohol, followed after inoculation with a suitable tuberculocidal disinfectant and 70% alcohol.
f. Incubate vials, without shaking, at 37 ± 1°C. A CO_2 incubator is not necessary. Specimens suspected of containing *M. marinum* or *M. chelonae* should be incubated at 30°C.
g. Check for growth in the 12B vials on the BACTEC 460 instrument every 3 to 4 days for the first 2 weeks (for a high-volume lab) or every 2 to 3 days for the first 3 weeks (for a low-volume lab) and weekly thereafter for a total of 6 to 8 weeks.
h. If a vial becomes positive (GI of 10), pull the vial from routine testing and read daily until the GI reaches 100 or more. Prepare a smear (GI, 100) and stain for acid-fast bacilli (AFB).

2. BACTEC 13A (5) setup
   a. Do not test 13A vials on the 460 instrument before inoculation for blood cultures.
   b. Clean the septum of each 13A vial with 70% alcohol before and after inoculation.
   c. Enrichment for the 13A is provided in separate vials and enhances mycobacterial growth. Add 0.5 ml of the enrichment solution to each vial either before or after inoculation of the blood. Use a single syringe for each vial if enrichment medium is added after inoculation.
   d. Inoculate each 13A vial with 5 ml of blood.
   e. Incubate at 37 ± 1°C without shaking.
   f. Test the inoculated vials on the BACTEC 460 TB system every 2 to 3 days for the first 2 weeks and weekly thereafter for a total of 6 weeks.
   g. A culture is considered presumptively positive when the GI reaches 20 or more.
   h. Prepare an AFB smear after the GI reaches >20.
3. BACTEC 12B and 13A—working up positive vials
   a. If the smear is positive for AFB, subculture to solid media to check for contamination, a mixed culture, for additional tests, and to save for 1 year. Continue to incubate until the GI reaches 300 to 500.
   b. Centrifuge and probe the sediment with the Accuprobe for the *Mycobacterium tuberculosis* complex and for the *M. avium* complex. Nucleic acid probes identify mycobacteria more rapidly because a sample can be removed from a positive BACTEC vial and tested directly with specific nucleic acid probes (3). This eliminates the additional time required for isolating the organism on a solid medium before identification procedures can be started. This is also faster than performing the NAP test.

C. QC

1. Instrument
   a. Follow the manufacturer's recommendations for QC of the BACTEC 460 TB instrument. These recommendations include a daily performance test to check the delivery of CO_2 and a maintenance schedule for the needle heater, filter, medium trap, and UV light to prevent cross contamination. Maintenance of the instrument needles is especially critical.
   b. It is recommended to keep two uninoculated 12B vials at the end of each run to observe for cross contamination. These blank vials should be changed every 3 to 7 days, depending upon workload. Also, the vials should be incubated at 37°C for 6 weeks and tested weekly to observe for sterility.

2. Growth in 12B medium
   Cultures of several mycobacteria species (preferably from ATCC) are recommended to become familiar with the growth patterns in 12B. These may include *M. tuberculosis, M. bovis, M. kansasii, M. scrofulaceum*, the *M. avium* complex, and *M. fortuitum*. The manufacturer recommends the following procedure:
   a. Make a suspension from the growth of any of the above cultures on solid medium and adjust to McFarland no. 1 turbidity.
   b. Prepare 2 dilutions, 1:100 and 1:500, and inoculate 0.1 ml of each into a 12B vial (containing 0.1 ml of reconstituting fluid).
   c. Incubate the vials and test daily on the 460 TB system until a peak GI value is achieved.
   d. Compare the growth patterns of each culture as indicated by the daily GI and observe the growth visually. Make a smear when the GI reaches 100 to 500, stain for AFB, and record the morphology of the mycobacteria.

4 *Mycobacteriology*

IV. NONRADIOMETRIC SYSTEMS

The features of four nonradiometric systems are summarized in Table 4.6–1.

A. BBL mycobacteria growth indicator tube

1. Medium
   The BBL mycobacteria growth indicator tube (MGIT) contains modified Middlebrook 7H9 broth base. The complete medium requires the addition of OADC (oleic acid-albumin-dextrose-catalase) broth enrichment and PANTA, an antimicrobial mixture. A fluorescent compound sensitive to the presence of oxygen is embedded in silicone on the bottom of the round-bottom tubes.
2. Specimen
   This medium can be used for the detection and recovery of mycobacteria from all clinical specimens except urine and blood. The MGIT system compares favorably with the radiometric system (1, 4).
3. Procedure
   Follow the manufacturer's instructions.

B. BACTEC 9000 MB/MYCO/F

1. Medium
   The BACTEC MYCO/F sputa medium vials contain a modified 7H9 broth. The complete medium requires supplementation with BACTEC PANTA reconstituted with BACTEC supplement F. Each vial contains a sensor which can detect decreases in disolved oxygen in the medium.
2. Specimen
   This medium is to be used for the recovery of mycobacteria from processed sputum and other respiratory specimens.
3. Procedure
   Follow the manufacturer's instructions.

Table 4.6–1 Features of nonradiometric systems

Features	MGIT (Becton-Dickinson)	BACTEC 9000MB (Becton-Dickinson)	ESP-AFB (Difco)	MB/BacT (Organon-Teknika)
Format	Tubes (manual)	Bottles (continuous monitoring)	Bottles (continuous monitoring)	Bottles (continuous monitoring)
Blood specimens	No	No	Yes	No
Detection	O_2 consumption and fluorescence	O_2 consumption and fluorescence	Manometric measurement of gas consumption and production	CO_2 production
Specimen volume	0.5 ml	0.5–1.5 ml	0.5–10 ml	0.5 ml

IV. NONRADIOMETRIC SYSTEMS *(continued)*

C. ESP Myco system (Difco)

1. Medium

The ESP instrument system, which monitors for changes in gas pressure, utilizes ESP Myco bottles each containing Middlebrook 7H9 broth and a cellulose sponge as a growth platform. An OADC supplement (ESP Myco GS) and an antimicrobial-agent solution (MycoPVNA) are added. The ESP system is accurate and effective in detecting mycobacteria (2, 8).

2. Specimen

This system can use all specimens submitted for mycobacterial culture.

3. Procedure

Follow manufacturer's instructions.

D. MB/BacT System (Organon Technika)

1. Medium

The MB/Bact T microbial detection system uses a colorimetric sensor in the bottom of each tube to detect the presence of CO_2. The isolation medium is composed of Middlebrook broth base, pancreatic digest of casein, BSA, and catalase. An MB/BacT reconstitution fluid is added to the isolation tubes for isolation of mycobacteria from sterile specimens or a MB/BacT antimicrobial-agent supplement for the culture of nonsterile mycobacteria.

2. Specimens

This system can be used to recover mycobacteria from all clinical specimens except blood.

3. Procedure

Follow manufacturer's instructions.

REFERENCES

1. **Badek, F. Z., D. L. Kiska, S. Setterquist, C. Hartley, M. A. O'Connell, and R. L. Hopfer.** 1996. Comparison of mycobacteria growth indicator tube with BACTEC 460 for detection and recovery of mycobacteria from clinical specimens. *J. Clin. Microbiol.* **34:** 2236–2239.
2. **LaBombardi, V. J., L. Carter, and S. Massarella.** 1997. Use of nucleic acid probes to identify mycobacteria directly from Difco ESP-Myco bottles. *J. Clin. Microbiol.* **35:** 1002–1004.
3. **Peterson, E. M., R. Lu, C. Floyd, A. Nakasone, G. Friedly, and L. M. De la Maza.** 1989. Direct identification of *Mycobacterium tuberculosis, Mycobacterium avium*, and *Mycobacterium intracellulare* from amplified primary culture in BACTEC media using DNA probes. *J. Clin. Microbiol.* **27:**1543–1547.
4. **Pfyffer, G. E., H. M. Welscher, P. Kissling, C. Cieslak, M. J. Casal, J. Gutierrez, and S. Rusch-Gerdes.** 1997. Comparison of the mycobacteria growth indicator tube (MGIT) with radiometric and solid culture for recovery of acid-fast bacilli. *J. Clin. Microbiol.* **35:** 364–368.
5. **Siddiqi, S. H.** 1992. Primary isolation of mycobacteria: BACTEC Method, p. 3.7.1–3.8.4. *In* H. D. Isenberg (ed.), *Clinical Microbiology Procedures Handbook*, vol. 1. American Society for Microbiology, Washington, D.C.
6. **Stager, C. E., J. P. Libonati, S. H. Siddiqui, J. R. David, N. M. Hooper, J. F. Baker, and M. E. Carter.** 1991. Role of solid media when used in conjunction with the BACTEC system for mycobacteria isolation and identification. *J. Clin. Microbiol.* **29:**154–157.
7. **Tenover, F. C., J. T. Crawford, R. E. Huebner, L. J. Geiter, C. R. Horsburgh, Jr., and R. C. Good.** 1993. The resurgence of tuberculosis: is your laboratory ready? *J. Clin. Microbiol.* **31:**767–770.
8. **Woods, G. L., G. Fish, M. Plaunt, and T. Murphy.** 1997. Clinical evaluation of Difco ESP Culture System II for growth and detection of mycobacteria. *J. Clin. Microbiol.* **35:** 121–124.

4.8 Identification Procedures from Culture

I. PRINCIPLE

Whenever possible, mycobacteria should be identified to the species level. Cultures should be referred to a reference laboratory if the staff of the clinical microbiology laboratory cannot perform identification or drug susceptibility procedures. In addition to colony morphology (1) and acid fastness, the identification of mycobacteria is largely based on rapid DNA probes (AccuProbe) and conventional methods (Table 4.8–1). Table 4.8–2 is a detailed identification chart that has been reproduced from the *Manual of Clinical Microbiology*, 6th ed. (6). In addition, an algorithm routinely used by clinical laboratories is depicted in Fig. 4.8–1 (1). Alternative methods available for the detection and identification of mycobacteria from culture include chromatographic analysis gas-liquid chromatography (GLC) and reverse-phase high performance chromatography (HPLC) as shown in Fig. 4.8–2.

Table 4.8–1 Principles and procedures of conventional tests to identify mycobacteria

Test	Principle	Procedure
Arylsulfatase	Arylsulfatase hydrolyzes the bond between the sulfate and the aromatic rings of tripotassium phenolphthalein. Free phenolphthalein can be recognized by the red color formed when an alkali is added. The 3-day test is used to identify and distinguish some rapid growers (*M. fortuitum, M. peregrinum, M. chelonae, M. abscessus*, and *M. mucogenicum*), which give a positive reaction, from other rapid growers. The 14-day test identifies slower growing species (*M. marinum, M. szulgai, M. xenopi*, and *M. triviale*) and some rapid growers (*M. flavescens* and *M. smegmatis*).	1. Inoculate labelled tubes of 3- and 14-day test media with 0.1 ml of a 7-day broth culture or 0.1 ml of a slightly turbid suspension, or a loopful of growth from a culture actively growing on solid medium. 2. Incubate at 35 to 37°C. 3. After 3 days of incubation, add 6 drops of 2 N sodium carbonate to the 3-day test tube. 4. After 14 days, add 6 drops of 2 N sodium carbonate to the 14-day test tube. Results Positive—Immediate color change to pink or red after the addition of the carbonate solution Negative—No color change
Catalase	The enzyme catalase splits hydrogen peroxide into water and oxygen, which appears as bubbles. Two measurements of catalase activity test (i) heat stability of catalase at 68°C and (ii) amount of catalase produced (semiquantitative test), i.e., those species producing <45 mm of bubbles and those producing >45 mm of bubbles. Most mycobacteria produce catalase, with the exception of *M. gastri* and INH-resistant *M. tuberculosis* and *M. bovis*. At 68°C, *M. tuberculosis, M. bovis, M. gastri*, and *M. haemophilum* lose their catalase activity. In the semiquantitative test, *M. tuberculosis, M. bovis, M. avium* complex, *M. gastri, M. malmoense, M. xenopi*, and *M. haemophilum* produce <45 mm of bubbles, whereas most *M. kansasii* and the other photochromogens, the scotochromogens, *M. terrae* complex, *M. triviale*, and the rapid growers produce >45 mm of bubbles.	68°C Test 1. Make a heavy suspension of the organism in 0.5 ml of 0.067 M phosphate buffer (pH 7) in a screw-cap tube. 2. Incubate in a 68°C water bath or heating block for 20 min. 3. Cool to room temperature. 4. Add 0.5 ml of the catalase reagent. Recap tubes loosely. 5. Read after 5 min. Hold negative tubes for 20 min before discarding. Results Positive—Formation of bubbles. Rarely, bubbles may be seen rising from bottom in such small quantity that foam does not form at the surface. Record this as a positive reaction. Negative—No bubbles Note: Do not shake tubes! Tween 80 may form bubbles when shaken, giving a false-positive result.

(continued)

Table 4.8–1 Principles and procedures of conventional tests to identify mycobacteria *(continued)*

Test	Principle	Procedure
		Semiquantitative test 1. Inoculate surface of LJ deep (butt) tubes with 0.1 ml of 7-day-old liquid culture or a loopful of growth from an actively growing slant. 2. Incubate at 37°C for 2 weeks (caps loose). 3. Add 1 ml of catalase reagent. Observe for 5 min. 4. Measure the height of the column of bubbles (mm) above the surface of the medium. Results Weak positive—<45 mm; strong positive—>45 mm Negative—No bubbles
Growth rate and pigment production	Mycobacteria may be separated into two main groups based on growth rate. Those that form visible colonies within 7 days are called rapid growers, whereas those requiring longer periods are the slow growers. The rapid growers encompass the Runyoun group IV mycobacteria, e.g., *M. fortuitum*; the slower growers include the *M. tuberculosis* complex and groups I through III. The photochromogens (group I) are slow-growing, photoreactive mycobacteria. Some mycobacteria tolerate higher temperatures, e.g., *M. xenopi* with an optimum growth rate at 35–45°C; others may be inhibited at higher temperatures, e.g., *M. marinum*, whose optimum growth is at 30–32°C and may not grow at 37°C.	Growth rate 1. Prepare a McFarland 0.5 suspension of the test organism in 7H9 broth. 2. Make a 10^2 dilution of this in 7H9 broth. 3. Inoculate an LJ or 7H11 slant with 1 to 2 drops of the 10^2 dilution. 4. Incubate at 35–37°C or at the optimum growth temperature. 5. Examine after 5 to 7 days and weekly thereafter. Results Record the growth rate as the number of days required for the appearance of discrete colonies. Rapid grower, <7 days; slow grower, >7 days Pigment production 1. Inoculate five LJ slants with a barely turbid broth suspension. Wrap four in aluminum foil. 2. Incubate one set of two tubes at 37°C and the other set of two at 25°C (room temperature). Incubate the slant without foil at 37°C. 3. When growth is seen (about 10 days), remove foil from one tube at each temperature and note growth and pigmentation. The second shielded slant in each incubator is the control for photoactive pigment production and should not be exposed to light. 4. If no pigment is observed, expose culture to strong natural or artificial (fluorescent or incandescent) light for 3 to 5 h with caps loosened. 5. Replace aluminum foil and reincubate. 6. Examine after 24, 48, and 72 h. 7. Compare the light-exposed culture with those maintained in the dark (wrapped in foil). Results Photochromogens produce a lemon-yellow pigment only if exposed to light. Bright-orange crystals may often be observed in *M. kansasii* colonies exposed continuously to light. In general, the scotochromogenic mycobacteria, when grown in the dark, produce a bright-yellow pigment that may deepen to orange and rarely to brick red if exposed to light continuously for 2 weeks. Note: *Mycobacterium szulgai* may be scotochromogenic at 37°C and photochromogenic at 25°C.
Growth on MAC (without crystal-violet) plate	Most isolates of the *M. fortuitum* and *M. chelonae* complexes will grow on MAC without crystal violet, whereas most other rapid growers will not. The incorporation of bile salts results in some inhibition of gram-positive organisms. However, owing to the absence of crystal violet, enterococci and some staphylococci may grow on this medium.	1. Inoculate rapid growers to be tested onto MAC by placing a 3-mm loopful of 7-day liquid culture, and streak for isolation. 2. Incubate without CO_2 (28°C). 3. *M. fortuitum* and *M. chelonae* complexes show growth in <7 days.

Table 4.8–1 Principles and procedures of conventional tests to identify mycobacteria *(continued)*

Test	Principle	Procedure
Iron uptake	Some mycobacteria, i.e., the *M. fortuitum* complex and most rapid growers (exception, the *M. chelonae* complex), are capable of converting FAC[a] to iron oxide. This is visible as a rust color in the colonies grown in media containing FAC. The test is most useful in differentiating the *M. fortuitum* complex (positive) from the *M. chelonae* complex (negative).	1. Inoculate LJ slant with barely turbid liquid suspension of the organism. 2. Incubate until there is visible growth. 3. Add 1 drop of citrate solution for each ml of Lowenstein medium (usually about 8 drops). 4. Reincubate up to 21 days and read weekly. Results Positive—Rusty brown color in the colonies and tan discoloration of the media Negative—No change in color
NaCl tolerance	Few mycobacteria are able to grow in culture media containing 5% sodium chloride, the exceptions including *M. triviale* and most of the rapid growers except the *M. chelonae* complex (including *M. mucogenicum*).	1. Add 0.1 ml of a barely turbid suspension of a liquid culture or growth from a slant suspended in sterile water. 2. Inoculate a slant of LJ with 5% NaCl and a control slant without NaCl with 0.1 ml of inoculum (4 drops). 3. Incubate at 35–37°C. 4. Read once a week for growth. 5. Culture may be discarded after 4 weeks of incubation. Results Positive—Growth on NaCl Negative—No growth
Niacin	Certain mycobacteria, e.g., most isolates of *M. tuberculosis* and *M. simiae*, accumulate niacin and excrete it into the culture media. The niacin test in conjunction with others is a key test for identification of these species. For example, niacin in conjunction with nitrate and the TCH test will differentiate *M. tuberculosis* from *M. bovis*. These tests differentiate to species level, the *M. tuberculosis* complex.	1. Add 1–1.5 ml of sterile distilled water or saline to a 4-week-old LJ culture slant. 2. Cut and/or stab medium surface several times with a sterile needle or loop. 3. Slant the tube so that the liquid covers the surface of the medium. 4. Allow 30 min to 2 h for niacin extraction. 5. Remove 0.6 ml of extract to a sterile test tube. 6. With forceps, insert a niacin strip into the extract at the bottom of the tube with the arrows on the strip pointing downward. 7. Stopper the tube tightly. 8. Leave at room temperature for 15 to 20 min with occasional gentle agitation. 9. Observe the color of the liquid at the bottom of the tube against a white background. Results Positive—Yellow color in extract fluid Negative—Colorless fluid
Nitrate	Mycobacteria differ quantitatively in their ability to reduce nitrate. The reduction of nitrate to nitrite is denoted by a color development when nitrite reacts with the three reagents. For a negative test with zinc dust, the reduction of the diazonium salt by the zinc in the presence of acetic acid produces a colored compound. The nitrate reduction test is useful to distinguish between organisms with similar characteristics such as growth rate, colony morphology, and pigmentation. Species that reduce nitrate include *M. tuberculosis, M. kansasii, M. szulgai, M. terrae* complex, and *M. flavescens*.	1. Inoculate nitrate broth with a loopful of growth from an actively growing culture. 2. Shake by hand to mix and incubate at 35–37°C for 2 h. 3. Add 1 drop of reagent 1 and shake by hand to mix. 4. Add 2 drops of reagent 2. 5. Add 2 drops of reagent 3. Results Positive—Pink to deep red color Negative—No color change; confirm by adding small amount of zinc dust.

(continued)

Table 4.8–1 Principles and procedures of conventional tests to identify mycobacteria *(continued)*

Test	Principle	Procedure
Pyrazin-amidase	Certain mycobacteria possess the enzyme pyrazinamidase and are thus able to hydrolyze PZA[a] to pyrazinoic acid, which can be detected by the addition of ferrous ammonium sulfate. This test is useful in differentiating *M. bovis* (−) from *M. tuberculosis* (+) and *M. marinum* (+) from *M. kansasii* (−). This test is also useful for detecting PZA-resistant *M. tuberculosis* which reportedly loses the ability to produce pyrazinovic acid, thereby becoming pyrazinamidase negative.	1. Include an uninoculated control tube and place a heavy inoculum from a fresh culture slant on the surface of each of two tubes of medium. Use enough inoculum so that it may be readily seen with the naked eye. 2. Incubate at 35–37°C in a non-CO_2 atmosphere. 3. After 4 days, remove one tube from the incubator, and add 1.0 ml of freshly prepared 1% ferrous ammonium sulfate to the tube. 4. Place the tubes at room temperature for 30 min, then refrigerate at 2–8°C to prevent growth of contaminants. 5. After 4 h, examine tubes for a pink band in the agar. To read, hold the tubes against a white background by using incident room light. 6. After 7 days of incubation, remove the second tube from the incubator and proceed as described above. Note: If the 4-day test is positive, it is not necessary to hold the second tube for the 7-day test.
TCH	*M. tuberculosis* and most of the other slowly growing mycobacteria are resistant to TCH[a] at levels of 1–5 μg/ml in the media; most species are resistant even at 10 μg/ml. This is a valuable test to differentiate niacin-positive *M. bovis* from *M. tuberculosis* since *M. bovis* is susceptible to the low concentrations with the exception of INH-resistant strains which may be resistant to TCH.	1. Prepare a 7- to 10-day-old liquid culture (or prepare a barely turbid suspension of organisms from a drug-free medium). 2. Dilute with sterile saline or distilled water to 1:1000 and 1:100,000 (10^{-3} and 10^{-5}). 3. With a sterile capillary pipette, inoculate a control and each TCH medium with 3 drops of each dilution. 4. Incubate 3 weeks in a 5–10% CO_2 atmosphere. 5. Read in the same manner as drug susceptibility test plates, comparing growth on TCH with that on control medium. Results—Record the organism as susceptible if growth on TCH medium is <1% of the growth on the control.
Tellurite	The ability to reduce potassium tellurite to metallic tellurium within 3 days is a distinctive characteristic of the *M. avium* complex, *M. celatum*, and most rapid growers. This test is most useful in conjunction with other tests for identifying the *M. avium* complex.	1. Heavily inoculate Middlebrook 7H9 liquid medium. 2. Incubate at 37°C for 7 days. The broth should be very turbid. 3. Add 2 drops of sterile potassium tellurite solution to each test culture and controls. Shake tubes to mix. 4. Reincubate at 37°C for 3 days but do not shake. 5. On the third day, examine sedimented cells in each culture tube, taking care not to shake. Results Positive—Black precipitate of metallic tellurium Negative—No black precipitate. A light brown or gray precipitate should be considered ambiguous. Compare result with control and contact supervisor.
Tween 80 hydrolysis	The ability to hydrolyze Tween 80, as indicated by a color change in the medium from yellow to red, is a valuable test for the separation of potentially pathogenic slow-growing scotochromogens (group II) and nonphotochromogens (group III) which give a negative reaction (exceptions, *M. malmoense, M. shimoidii, M. genavense*) from the potentially saprophytic groups II and III. In this test, Tween 80 acts as a lipid, binding the neutral red indicator, causing the solution to be yellow colored. If the mycobacterial lipase hydrolyzes the Tween 80, the neutral red indicator is no longer bound, and it reverts to its normal red color at pH 7.	1. Add 2 drops of the Remel Tween 80 hydrolysis substrate concentrate to 1.0 ml of sterile deionized water. 2. Take a loopful of an actively growing culture from a slant and inoculate the above Tween 80 mixture. Also, simultaneously inoculate a positive and negative control in the same manner. 3. Incubate aerobically at 35–37°C and examine at 1, 5, and 10 days. 4. Observe the number of days for the appearance of a pink color to red color (positive test) and record. Do not shake tubes while reading. 5. Discard when positive or on the 11th day. Results Positive—A pink or red color in 10 days or less Negative—An amber color remaining after 10 days of incubation

Table 4.8–1 Principles and procedures of conventional tests to identify mycobacteria *(continued)*

Test	Principle	Procedure
Urease	The ability to hydrolyze urea-releasing ammonia is especially useful in differentiating between pigmented strains of the *M. avium* complex (urease negative) from *M. scrofulaceum* (urease positive).	Transfer a young, actively growing culture of the mycobacteria to be tested to the urea broth. Inoculate to a visible turbidity. Incubate at 35–37°C without carbon dioxide up to 7 days. A dark pink to red color indicates a positive test.

[a] Abbreviations: INH, isoniazid; FAC, ferric ammonium citrate; PZA, pyrazinamide; TCH, thiophene-2-carboxylic acid hydrazide.

II. SPECIMEN

Pure culture of mycobacteria on solid media

III. RADIOMETRIC BACTEC NAP TEST

The BACTEC *p*-nitro-α-acetylamino-β-hydroxypropiophenone (NAP) test, to differentiate the *M. tuberculosis* complex from other mycobacteria, has limited utility with the advent of DNA probes. Should the reader require detailed procedures for use, we recommend referring to the *Clinical Microbiology Procedures Handbook* (7).

IV. CONVENTIONAL TESTS

(For details, see references 4–6, 8.)

See Tables 4.8–1, 4.8–2, and 4.8–3.

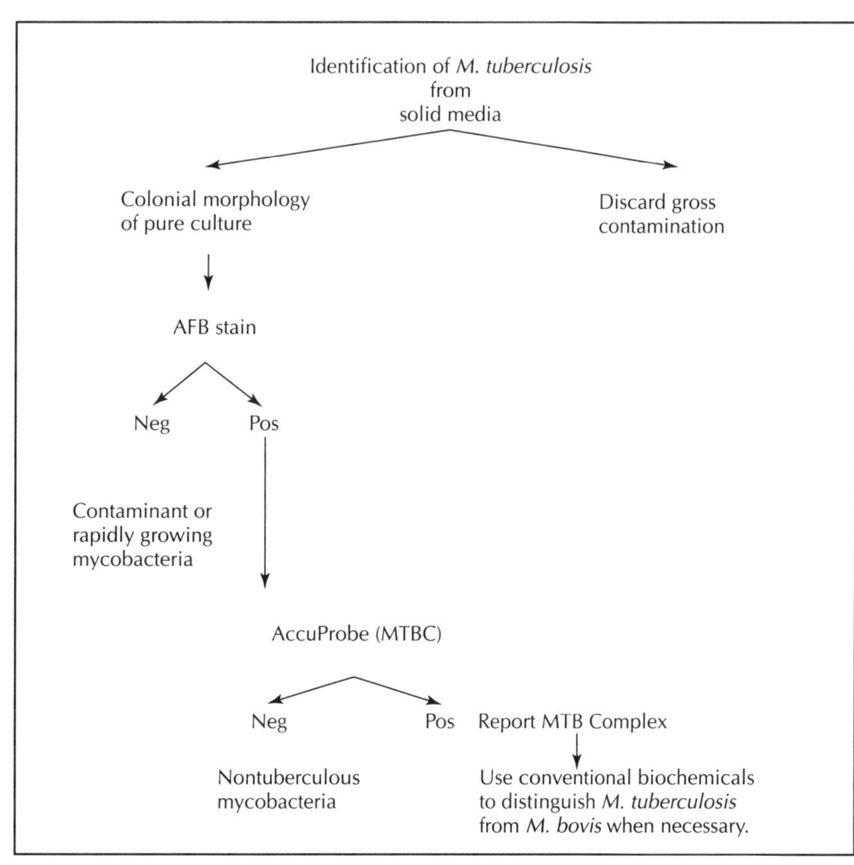

Figure 4.8–1 Identification of *M. tuberculosis* from solid media.

Table 4.8–2 Distinctive properties of cultivable mycobacteria encountered in clinical specimens[a]

Descriptive term	Species	Growth rate[b]	Optimal temp (°C)	Usual colony morph-ology[c]	Pigmenta-tion[d]	Niacin	Growth on T2H (10 μg/ml)	Nitrate reduction
TB complex	*M. tuberculosis*	S	37	R	N (100)	+ (95)	+	+ (97)
	M. africanum	S	37	R	N	−	V	−
	M. bovis	S	37	Rt	N (100)	− (4)	−	− (9)
Nonchromogens	*M. avium* complex	S	37	St/R	N (87)	− (0)	+	− (4)
	M. xenopi	S	42	S	S (21)	− (0)	+	− (7)
	M. haemophilum	S[f]	30	R	N	−	+	−
	M. malmoense	S	37	S	N (88)	− (0)	+	− (1)
	M. shimoidei	S	37	R	N	−	+	−
	M. genavense	S	37	St	N	−	+	−
	M. celatum	S	37	S/St	N (100)	−	+	− (0)
	M. ulcerans	S	30	R	N	−	+	−
	M. terrae complex	S	37	SR	N (93)	− (1)	+	± (67)
	M. triviale	M	37	R	N (100)	− (0)	+	+ (89)
	M. gastri	S	37	S/SR/R	N (100)	− (0)	+	− (0)
	M. nonchromogenicum	S	37	SR	N	−	+	−
Photochromogens	*M. kansasii*	S	37	SR/S	P (96)	− (4)	+	+ (99)
	M. marinum	M	30	S/SR	P (100)	−/+ (21)	+	− (0)
	M. simiae	S	37	S	P (90)	± (63)	+	− (28)
	M. asiaticum	S	37	S	P (86)	− (0)	+	− (5)
Scotochromogens	*M. gordonae*	S	37	S	S (99)	− (0)	+	− (1)
	M. scrofulaceum	S	37	S	S (97)	− (0)	+	− (5)
	M. szulgai	S	37	S or R	S/P (93)	− (0)	+	+ (100)
	M. flavescens	M	37	S	S (100)[h]	− (0)	+	+ (92)
Rapid growers	*M. fortuitum* group	R	28	Sf/Rf	N (100)	−/+	+	+ (100)
	M. chelonae group	R	28	S/R	N (100)	−/+	+	− (1)
	M. smegmatis	R	28	R/S	N		+	+
	M. phlei	R	28	R	S		+	+
	M. vaccae	R	28	S	S		+	+

[a] Adapted from reference 6. Plus and minus signs indicate the presence and absence, respectively, of the feature; blank spaces indicated either that the information is not currently available or that the property is unimportant. V, Variable; ±, usually present; −/+, usually absent. Percentage of CDC-tested strains positive in each test is given in parentheses, and test result is based on these percentages.

[b] S, slow; M, moderate; R, rapid.

[c] R, rough; S, smooth; SR, intermediate in roughness; t, thin or transparent; f, filamentous extensions.

[d] P, photochromogenic; S, scotochromogenic; N, nonphotochromogenic. *M. szulgai* is scotochromogenic at 37°C and photochromogenic at 24°C.

V. GLC AND HPLC

Both GLC and HPLC are useful tools for the identification of mycobacteria, by analysis of their lipid composition (9, 10). However, the high initial expense of purchasing the equipment and the technical complexity of analyzing the data are not for the routine mycobacteriology laboratory.

VI. DNA PROBES (ACCUPROBE)

A. Principle

A commercially manufactured system (AccuProbe) is available from Gen-Probe, Inc. (San Diego, Calif.) for rapid identification of *M. tuberculosis* complex (*M. tuberculosis, M. bovis, M. africanum*, and *M. microti*), *M. avium* complex (*M. avium, M. intracellulare*), *M. kansasii*, and *M. gordonae*. The DNA probes are complementary to the rRNA of each species and form species-specific DNA-RNA hybrids resulting in chemiluminescence that is measured in a luminometer. The test is rapid, sensitive, and specific and can be performed

Semi-quantitative catalase (mm of bubbles)	68°C catalase	Tween hydrolysis	Tellurite reduction	Tolerance to 5% NaCl	Iron uptake	Arysulfatase, 3 days	MAC	Urease[e]	Pyrazin-amidase, 4 days	Nucleic acid probes available
<45 (89)	− (1)	± (68)	−/+ (36)	− (0)	−	− (0)	−	± (64)	+	+
<45	−	−	−	−	−	−	−	+	−	−
<45 (69)	− (2)	− (21)		− (0)	−	− (0)	−	± (50)	−	+
<45 (98)	± (60)	− (2)	+ (81)	− (0)	−	− (1)	−/+	− (2)	+	+
>45 (85)	± (31)	− (12)	± (65)	− (0)	−	± (36)	−	− (0)	V	−
<45	−	−	−	−	−	−	−	−	+	−
<45 (99)	± (66)	+ (99)	+ (74)	− (0)	−	− (0)		− (9)	+	−
<45	−	+		−	−	−	−	−	+	−
>45	+	+			−	−		+	+	−
<45 (100)	+ (100)	− (0)	+ (100)	− (0)	− (0)	+ (100)	− (0)	− (0)	+ (100)	−
<45	+	−		−		−		V	−	−
>45 (93)	+ (92)	+ (99)	−/+ (46)	− (2)	−	− (2)	V	− (13)	V	−
>45 (100)	+ (100)	+ (100)	− (25)	+ (100)	−	± (56)	−	−/+ (33)	V	−
<45 (100)	− (11)	+ (100)	± (50)	− (0)	−	− (0)	−	−/+ (44)	−	−
>45	+	+	−	−	−	−	V	−	V	−
>45 (93)	+ (91)	+ (99)	−/+ (31)	− (0)	−	− (0)	−	−/+ (49)	−	+
<45 (98)	− (30)	+ (97)	−/+ (39)	− (0)	−	−/+ (41)[g]	−	+ (83)	+	−
>45 (93)	+ (95)	− (9)	+ (82)	− (0)	−	− (0)		± (69)	+	−
>45 (95)	+ (95)	+ (95)	− (20)	− (0)	−	− (0)		− (10)	−	−
>45 (90)	+ (96)	+ (100)	− (29)	− (0)	−	V (0)	−	V (31)	−/+	+
>45 (84)	+ (94)	− (2)	± (64)	− (0)	−	V (0)	−	V (31)	±	−
>45 (98)	+ (93)	−/+ (49)	± (53)	− (0)	−	V (0)	−	+ (72)	+	−
>45 (94)	+ (100)	+ (100)	−/+ (44)	± (62)	−	− (0)	−	+ (72)	+	−
>45 (93)	+ (90)	−/+ (43)	+ (92)	+ (85)	+	+ (97)	+	+ (70)	+	−
>45 (92)	± (53)	−/+ (39)	+ (89)	V[i]	−	+ (95)	+	+ (89)	+	−
>45	+	+	+	+	+	−	−			−
>45	+	+	+	+	+	−	−			−
>45	+	+	+	V	+	−	−			−

[e] Urease test was performed by the method of Steadham (*J. Clin. Microbiol.* **10:**134–137, 1979).
[f] Requires hemin as growth factor.
[g] Arylsulfatase reaction at 14 days is positive.
[h] Young cultures may be nonchromogenic or possess only pale pigment that may intensify with age.
[i] *M. chelonae* is negative; *M. abscessus* is positive.

from growth in liquid as well as from solid media (2). The entire assay takes 2 h.

B. Specimen

Cultures may be tested as soon as growth is visible and after up to 60 days of incubation. Morphology on solid or liquid media can be used to determine which probe will be tested first (3).

C. Materials

1. Materials provided

   Probe reagent, lysing reagent (glass beads and buffer), reagent 1 (lysis reagent), reagent 2 (hybridization buffer), reagent 3 (selection reagent), detection reagent I (0.1% H_2O_2 in 0.001 N nitric acid), detection reagent II (1 N sodium hydroxide).

2. Materials required but not provided

   Micropipettes (100 μl, 300 μl), repipettor (100 μl, 300 μl), plastic sterile inoculating loops (1 μl), vortex mixer, water bath or heating block (60 ± 1°C), water bath or heating block 2 (95 ± 5°C), sonicator and rack, and Leader Luminometer (the last three items are available from Gen-Probe).

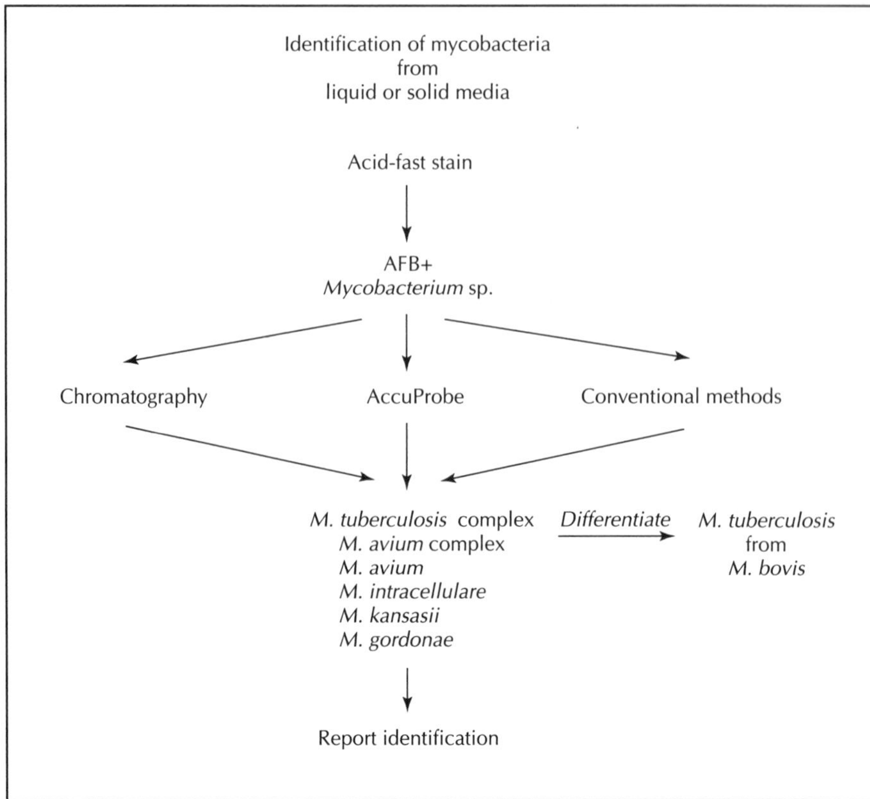

Figure 4.8–2 Identification of mycobacteria from solid or liquid media.

VI. DNA PROBES (ACCUPROBE) *(continued)*

D. QC

Run controls with each batch of cultures to be tested. Control organisms are those listed below:

Include QC information on reagent container and in QC records.

Probe	Positive	Negative
M. tuberculosis complex	*M. tuberculosis* ATCC 25177	*M. avium* ATCC 25291
M. avium complex	*M. intracellulare* ATCC 13950	*M. gordonae* ATCC 14470
M. avium	*M. avium* ATCC 25291	*M. intracellulare* ATCC 13950
M. intracellulare	*M. intracellulare* ATCC 13950	*M. avium* ATCC 25291
M. kansasii	*M. kansasii* ATCC 12478	*M. tuberculosis* ATCC 25177
M. gordonae	*M. gordonae* ATCC 14470	*M. scrofulaceum* ATCC 19073

E. Procedure

1. Fill the sonicator with hot water to 1/2 in. from the top. Sonicate for 15 min to degas the water.
2. Adjust one water bath or heating block to 60 ± 1°C, the other to 95 ± 5°C.
3. Pipette 100 μl of reagent 1 (lysing buffer) and 100 μl of reagent 2 (hybridization buffer) into labelled lysing-reagent tubes. Do not add reagent 1 to the lysing reagent tubes if broth cultures are to be tested.
4. In a biological safety cabinet, transfer growth from solid media with the 1-μl disposable plastic loop or 100 μl of a well-mixed broth culture equal to or greater than McFarland 1, into the labelled lysing-reagent tubes as described above.

Table 4.8–3 Materials and QC

Test	Media/reagents	Quality control
Arylsulfatase	Arylsulfatase broth or agar (Remel) 3-day and 14-day test media 2 N sodium carbonate (Remel)	3-Day test + Control—*Mycobacterium fortuitum* ATCC 6841 (pink or red color) − Control—*Mycobacterium intracellulare* ATCC 13950 (colorless) − Control—*Mycobacterium phlei* ATCC 11748 (solid medium) 14-Day test + Control—*Mycobacterium fortuitum* ATCC 6841 − Control—*Mycobacterium intracellulare* ATCC 13950
Catalase	LJ deep in screw-cap tubes; 30% H_2O_2 (Caution—gloves and eye protection required); 10% Tween 80; 68°C water bath; 22–25°C incubator; catalase reagent: equal amounts of 10% Tween 80 and 30% H_2O_2	68°C Test *M. tuberculosis* ATCC 25177 (bubbles: +22–25°C; −68°C) *M. fortuitum* ATCC 6841 (bubbles: +22–25°C; +68°C) Semiquantitative test High catalase—*M. fortuitum* ATCC 6841 Low catalase—*M. tuberculosis* ATCC 25177 Uninoculated medium and reagent only
Growth rate and pigment production	Dubos Tween broth or Middlebrook 7H9 broth LJ slants	(see below)
Growth on MAC plate (without crystal violet)	MAC plates (without crystal violet)	Positive—*M. fortuitum* ATCC 12478 Negative—*M. phlei* ATCC 11758
Iron uptake	LJ slant; 20% FAC (Remel); 28–30°C incubator	Positive control—*M. fortuitum* ATCC 6841 (rust brown) Negative control—*M. chelonae* ATCC 35751 (no color change) Uninoculated medium and reagent only
NaCl tolerance	LJ slant with 5% NaCl; LJ slant without 5% NaCl; 28–30°C incubator	Positive control—*M. fortuitum* ATCC 6841 Negative control—*M. gordonae* ATCC 14470
Niacin	Niacin filter paper strips (Remel) impregnated with potassium thiocyanate, chloramine T, citric acid, and sodium aminosalicylate; sterile distilled water	Positive control—*M. tuberculosis* ATCC 25177 (yellow-colored extract) Negative control—*M. intracellulare* ATCC 13950 (colorless extract) Uninoculated medium and reagent only

Growth rate and pigment production — quality control:

Organism	Growth 37°C	Growth 22–25°C	Pigment Light	Pigment Dark
M. terrae ATCC 15755	Slow	+	−	−
M. fortuitum ATCC 6841	Rapid	+	−	−
M. gordonae ATCC 14470	Slow	Slight	Orange pigment	
M. kansasii ATCC 12478	Slow	Slight	Orange (after light)	

(*continued*)

Table 4.8–3 Materials and QC *(continued)*

Test	Media/reagents	Quality control
Nitrate	Nitrate substrate broth (Remel) Reagent 1 (HCl) Concentrated hydrochloric acid (10 ml) Deionized water (10 ml) 10 ml HCl + 10 ml deionized water Reagent 2 (nitrate A for AFB) Sulfanilamide (1.0 g) Deionized water (1,000 ml) Reagent 3 (nitrate B for AFB) *n*-Napthylethylene-diamine dihydrochloride (1.0 g) Deionized water (1,000 ml) Zinc dust	Positive control—*M. tuberculosis* ATCC 25177 (pink to red) Negative control—*M. intracellulare* ATCC 13950 (no color change)
Pyrazinamidase	PZA substrate medium (Remel); 1% ferrous ammonium sulfate (prepare fresh just before use)	Positive control—*M. intracellulare* ATCC 13950 or *M. tuberculosis* (pink band in 4 days) Negative control—*M. kansasii* ATCC 12478 (no pink band) Uninoculated medium
TCH	Dubos Tween or Middlebrook 7H9 broth; TCH susceptibility media (Remel)	TCH-susceptible *M. bovis* ATCC 35734 TCH-resistant *M. tuberculosis* ATCC 25177
Tellurite reduction	7H9 broth with 0.5 ml of Tween 80 (without glycerol); potassium tellurite (0.2%)	Positive control—*M. intracellulare* ATCC 13950 Negative control—*M. tuberculosis* ATCC 25177 Uninoculated medium and reagent only
Tween 80 hydrolysis	TB Tween hydrolysis substrate concentrate (Remel)	Positive control, strong—*M. kansasii* ATCC 12478 Negative control—*M. intracellulare* ATCC 13950
Urease	Urea broth (Remel)	Positive control—*M. fortuitum* (ATCC 6841) (pink color) Negative control—*M. gordonae* ATCC 14470 (yellow color)

VI. DNA PROBES (ACCUPROBE) *(continued)*

5. Recap the lysing-reagent tubes, vortex, and place in the rack in the water bath sonicator. Caps must be above the water level, and tubes must not touch the bottom or sides of the sonicator.
6. Sonicate for 15 min.
7. Place tubes in the 95 ± 5°C heating block (or water bath) for 10 min.
8. Pipette 100 μl of the lysed specimens into corresponding probe reagent tubes, recap, and incubate for 15 min in the 60°C water or heating block.
9. Remove the tubes from the heating unit, add 300 μl of reagent 3 (selection reagent) to each tube, recap, and vortex to mix.

10. Incubate the tubes for 10 min in the 60°C water bath or block (8-min selection time for *M. kansasii* and 5 min for *M. gordonae* and MAC assays).

11. Remove tubes, keep at room temperature for 5 min, remove caps, and read in the luminometer within 1 h.

12. Select the appropriate prococol from the menu of the luminometer software, and wipe each tube before inserting into the luminometer.

F. Results

A positive signal for the Leader Luminometer is 30,000 RLU or 900 PLU for the PAL Luminometer (ACCULDR).

G. Procedural notes

1. The hybridization and selection reactions are time and temperature dependent. The water bath or heat block must be maintained at 60 ± 1°C. Hybridize at least 15 but no more than 20 min.

2. The level of the water bath should be maintained so that the entire liquid reaction volume in the probe reagent tube is submerged.

H. Limitations of the procedure

The Accuprobe *Mycobacterium tuberculosis* complex test does not differentiate between members of the *M. tuberculosis* complex.

REFERENCES

1. **Della-Latta, P.** 1996. Workflow and optimal protocols for laboratories in industrialized countries. *Clin. Lab. Med.* **16:**677–695.
2. **Ellner, P. D., T. E. Kiehn, R. Cammarata, and M. Hosmer.** 1988. Rapid detection and identification of pathogenic mycobacteria by combining radiometric and nucleic acid probe methods. *J. Clin. Microbiol.* **26:**1349–1352.
3. **Kaminski, D. A., and D. J. Hardy.** 1995. Selective utilization of DNA probes for identification of *Mycobacterium* species based on cord formation in primary BACTEC 12B cultures. *J. Clin. Microbiol.* **33:**1548–1550.
4. **Kent, P. T., and G. P. Kubica.** 1985. *Public Health Mycobacteriology. A Guide for the Level III Laboratory.* U.S. Public Health Service publication no. 86-21654. U.S. Government Printing Office, Washington, D.C.
5. **Lutz, B.** 1992. Identification tests for mycobacteria, p. 3.121–3.128. *In* H. D. Isenberg (ed.), *Clinical Microbiology Procedures Handbook,* vol. 1. American Society for Microbiology, Washington, D.C.
6. **Nolte, F. S., and B. Metchock.** 1996. *Mycobacterium*, p. 400–437. *In* P. R. Murray, E. J. Baron, M. A. Pfaller, F. C. Tenover, and R. H. Yolken (ed.), *Manual of Clinical Microbiology*, 6th ed. American Society for Microbiology, Washington, D.C.
7. **Siddiqi, S. H.** 1992. BACTEC NAP test, p. 3.13.1. *In* H. D. Isenberg (ed.), *Clinical Microbiology Procedures Handbook,* vol. 1. American Society for Microbiology, Washington, D.C.
8. **Silcox, V. A.** 1992. Identification of mycobacteria, p. 311.1–311.11. *In* H. D. Isenberg (ed.), *Clinical Microbiology Procedures Handbook,* vol. 1. American Society for Microbiology, Washington, D.C.
9. **Stockman, L.** 1992. Gas-liquid chromatography (microbial identification system) for the identification of mycobacteria, p. 3.14.1–3.14.4. *In* H. D. Isenberg (ed.), *Clinical Microbiology Procedures Handbook*, vol. 1. American Society for Microbiology, Washington, D.C.
10. **Thibert, L., and S. Lapierre.** 1993. Routine application of high-performance liquid chromatography for identification of mycobacteria. *J. Clin. Microbiol.* **31:**1759–1763.

4.9 Direct Nucleic Acid Amplification Tests from Specimens

I. PRINCIPLE

The rapid detection of *Mycobacterium tuberculosis* in respiratory specimens is of utmost importance owing to the clinical, therapeutic, and public health ramifications of active pulmonary disease. Two molecular assays with direct amplification technology to detect *Mycobacterium tuberculosis* (MTB) complex-specific nucleic acid directly from processed respiratory sediments have been U.S. Food and Drug Administration (FDA) approved, i.e., the Gen-Probe Amplified *M. tuberculosis* Direct Test (AMTD) and the Roche AMPLICOR. They have been approved for use with acid-fast, smear-positive respiratory specimens collected from untreated patients. Their sensitivity and specificity are highest with smear-positive specimens (2). These tests are recommended for use in laboratories servicing a patient population with a high prevalence of mycobacterial disease. An algorithm for incorporating this new technology into the routine workflow of the laboratories is presented in Fig. 4.9–1 (1).

II. SPECIMEN

FDA clearance for the Gen-Probe and Roche assays is limited to respiratory specimens (sputum, tracheal aspirates, and bronchoscopy samples) collected from untreated patients presenting with clinical signs consistent with active pulmonary tuberculosis. These specimens must be digested and decontaminated before the sediments are tested. The use of these direct amplification tests is expanding, i.e., inclusion of acid-fast, smear-negative specimens, other body fluids, and detection from BACTEC 12B broth cultures (5).

III. GEN-PROBE AMTD

This assay takes approximately 4 h. The technology used is transcription-mediated amplification (6).

A. Materials

1. Materials provided
   a. Specimen dilution buffer, amplification reagent, reconstitution buffer, lysing tubes, and oil reagent
   b. Enzyme reagent, enzyme dilution buffer, and termination reagent
   c. Hybridization-positive and -negative controls, hybridization buffer, selection reagent, and probe reagent
2. Materials required and available from Gen-Probe
   a. Leader Luminometer
   b. Sonicator
   c. Detection reagent kit
   d. Dry heat bath
   e. Sonicator rack
   f. Pipette tips with hydrophobic plugs
   g. Water baths (42 and 60°C)

B. Procedure (for details, *see* reference 3)

Gen-Probe AMTD

1. Add 50 μl of processed sediment to 200 μl of dilution buffer in a lysing tube; vortex.

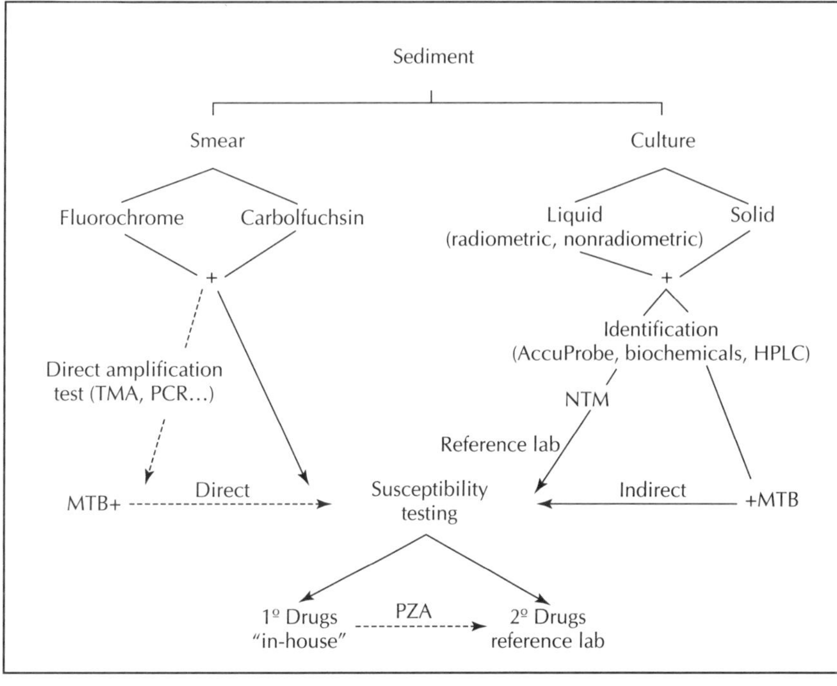

Figure 4.9–1 Algorithm for incorporation of direct nucleic acid amplification tests in today's levels I and II laboratories. Adapted from reference 1 by permission.

2. Sonicate for 15 min.
3. Transfer 50 μl to amplification tubes containing 25 μl of amplification reagent and 200 μl of oil.
4. Incubate at 95°C for 15 min.
5. Cool tubes at 42°C for 5 min.
6. Add 25 μl of enzyme mix and incubate at 42°C for 2 h.
7. Add 20 μl of termination reagent and incubate at 42°C for 10 min.
8. Add 100 μl of probe reagent, vortex, and incubate at 60°C for 15 min.
9. Add 300 μl of selection reagent, vortex, and reincubate at 60°C for 10 min.
10. Cool for 5 min.
11. Read tubes on a luminometer within 1 h.

C. QC

1. Control organisms
   - **a.** Positive—*M. tuberculosis* ATCC 21577 or ATCC 27294
   - **b.** Negative—*M. terrae* ATCC 15755 or *M. gordonae* ATCC 14470
2. Cleaning surfaces and equipment
   - **a.** Decontaminate by using a 1:1 dilution of bleach for a minimum of 15 min.
   - **b.** Rinse with water.
3. Amplification cell controls
   - **a.** Positive—RLU, ≥500,000
   - **b.** Negative—RLU, <20,000
4. Hybridization controls
   - **a.** Positive—RLU, ≥15,000
   - **b.** Negative—RLU, <5,000
5. Contamination monitoring
   Bench area, hood, and equipment

III. GEN-PROBE AMTD *(continued)*

D. Results

Patient test results	Interpretation
>500,000 RLU	Positive for MTB complex rRNA
30,000–500,000 RLU	Probable for MTB complex rRNA; repeat test to verify low-level positivity
<30,000 RLU	Negative for MTB complex rRNA

IV. ROCHE AMPLICOR

This assay takes 6 h. It uses PCR technology (6).

A. Materials

1. Materials provided

a. Specimen preparation kit
Sputum wash solution, lysis reagent, neutralization reagent

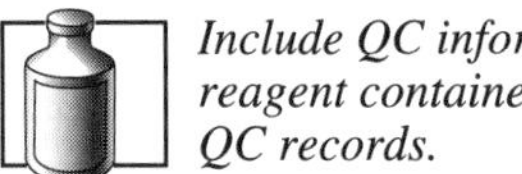

Include QC information on reagent container and in QC records.

b. Amplification kit
Mycobacterium master mix, amperase, *M. tuberculosis* (+) control, *Mycobacterium* (−) control

c. Detection kit
Denaturation solution, *Mycobacterium* hybridization buffer, avidin-HRP conjugate, substrates A and B, stop reagent, 10×-wash concentrate, and *M. tuberculosis* DNA probe coated microwell plate (MWP)

2. Materials required but not provided
MicroAmp reaction tubes, caps, base, tray, and retainer for thermocycler, microcentrifuge, 60°C dry heat block, thermal cycler (i.e., Perkin-Elmer 9600) Microwell plate washer, microwell plate reader (450 nm) and printer, pipette tips with hydrophobic plugs

B. Procedure (for details, *see* reference 4)

1. Area 1: reagent preparation

a. Add 100 μl of amperase to a tube of master mix.

b. Pipette 50 μl of working master mix to each MicroAmp tube.

2. Area 2: specimen preparation

a. Add 100 μl of processed sediment to 500 μl of wash solution.

b. Centrifuge at 12,500 × *g* for 10 min.

c. Discard supernatant and add 100 μl of lysis reagent; vortex and incubate at 60°C for 45 min.

d. Add 100 μl of neutralization reagent; vortex and pipette 50 μl of sample to a MicroAmp tube.

3. Area 3: amplification and detection

a. Place samples in a thermocycler and amplify.

b. Immediately after amplification, add 100 μl of denaturation solution; incubate at room temperature for 10 min.

c. Add 25 μl of denatured amplicons to an MWP containing 100 μl of hybridization buffer; incubate for 90 min at 37°C.

d. Wash 5 times and then add 100 μl of avidin-HRP conjugate; incubate for 15 min at 37°C.

e. Wash 5 times and add 100 μl of substrate; incubate for 10 min at room temperature in the dark.

f. Add 100 μl of stop solution; measure optical density at A_{450} nm within 1 h.

C. QC

1. Kit controls

a. Positive—Noninfectious *M. tuberculosis* DNA in a Tris-HCl EDTA solution containing nonspecific carrier DNA and 0.05% sodium azide

b. Negative—Nonspecific noninfectious DNA in a Tris-HCl EDTA solution containing 0.05% sodium azide

2. One positive control is included per run, and it should have an optical density of >3.0 A_{450}. The negative control is run in triplicate and should be <0.25 A_{450}.
3. To test the effectiveness of sample processing (recommended monthly), 10^4 *M. tuberculosis* cells should be processed and treated as a normal clinical specimen. A positive signal is above 30.0 A_{450}.

D. Results

Patient test result	Interpretation
>0.6	Positive for MTB complex DNA
0.2–0.6	Equivocal result; repeat testing in duplicate
<0.2	Presumptive negative for MTB complex DNA

V. GENERAL LIMITATIONS

A. Detection of MTB complex is dependent on several factors, including the number of organisms present in the specimen, specimen collection methods, and the patient population being tested. A negative test does not exclude the possibility of an MTB complex infection.

B. The presence of high numbers of nontuberculous mycobacteria such as *M. celatum* can affect test results (low-level false-positive AMTD), and false-negative results can occur with both AMTD and AMPLICOR in the presence of inhibitors and when high levels of nontuberculous mycobacteria mask very low numbers of MTB.

REFERENCES

1. **Della-Latta, P.** 1996. Workflow and optional protocols for laboratories in industrialized countries. *Clin. Lab. Med.* **16:**677–695.
2. **Forbes, B. A.** 1997. Critical assessment of gene amplification approaches on the diagnosis of tuberculosis. *Immunol. Invest.* **26:**105–116.
3. **Gen-Probe, Inc.** 1996. *Product Information: Amplified Mycobacterium tuberculosis Direct Test.* Gen-Probe, Inc., San Diego, Calif.
4. **Roche Molecular Systems, Inc.** 1996. *Product Information: Amplicor Mycobacterium tuberculosis Test.* Roche, Branchburg, N.J.
5. **Smith, M. B., J. S. Bergman, and G. L. Woods.** 1997. Detection of *Mycobacterium tuberculosis* in BACTEC 12B broth cultures by the Roche Amplicor PCR assay. *J. Clin. Microbiol.* **35:**900–902.
6. **Wilcott, M. J.** 1992. Advances in nucleic acid-based detection methods. *Clin. Microbiol. Rev.* **5:**370–386.

4.10 Reporting *Mycobacterium* Test Results

I. PRINCIPLE

State and local authorities vary in regulations governing reporting of *M. tuberculosis* detection. CDC guidelines advocate reporting all positive acid-fast bacteria (AFB) smears and culture results of *M. tuberculosis* complex rapidly to all appropriate authorities as well as the physician in charge of the patient (1, 3). Turnaround times expected are shown in Table 4.10–1 (2).

II. PROTOCOL

A. Communicate with the state and/or local tuberculosis control officer regarding laws governing the reporting of positive acid-fast smears and cultures.

B. Telephone diagnostic information as soon as possible to the clinician, the infection control officer, and the public health authorities. Document date, time, and person notified.

C. Issue a preliminary report each time new information is obtained and whenever isolates are referred to another laboratory for identification and susceptibility testing.

D. Verbally communicate the existence of the first positive culture (each patient, each episode) and subsequent results that may suggest dissemination, such as a sputum culture isolate followed by a blood culture isolate of suspected *Mycobacterium avium* complex.

E. Send written reports of preliminary information, final identification, and drug susceptibility results to the clinician, the infection control officer, and the medical records department.

F. Verbally communicate the first evidence of *M. tuberculosis* in culture. Call in results of additional culture-positive specimens if they were obtained during the time the patient is known or presumed to have been receiving treatment. Positive cultures at this stage will alert the public health officer to possible poor compliance with antituberculosis therapy.

G. Send a written report of the final identification and drug susceptibility results. This level of reporting will be unnecessary to laboratories that send isolates to their local public health laboratory for identification and/or antimicrobial susceptibility testing.

Table 4.10–1 Turnaround time expectations in today's typical mycobacteriology laboratory

Time to results from specimen receipt	Positive reports to physician and public health
1 Day	Acid-fast smear; direct amplification test
1–2 Weeks	Culture report
3 Weeks	MTB identification
4 Weeks	Susceptibility results

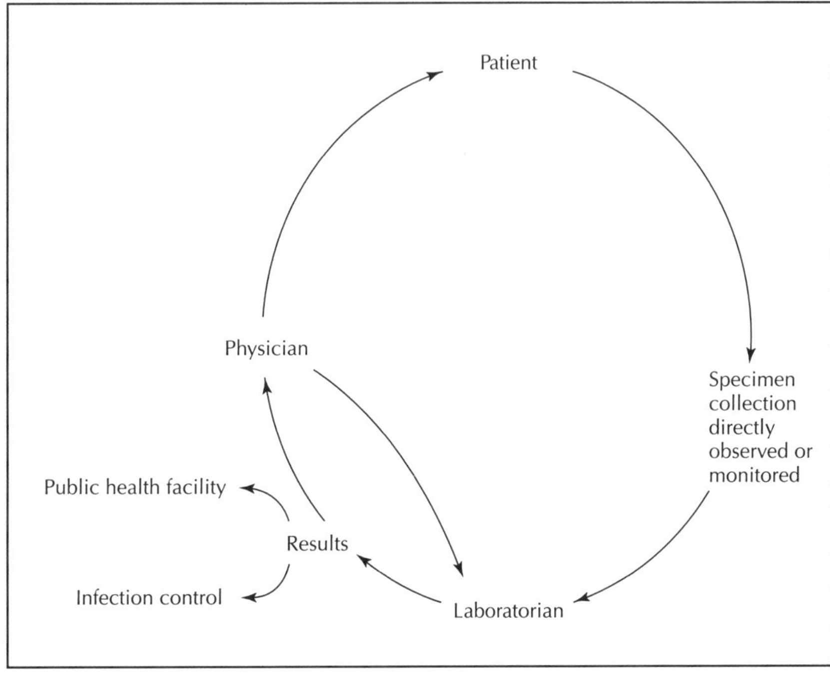

Figure 4.10–1 Effective communication network.

III. COMMUNICATION NETWORK

An example of an effective communication network, in which laboratorians, clinicians, and public health officials work as a team, is illustrated in Fig. 4.10–1 (2).

REFERENCES

1. **CDC.** 1989. A strategic plan for the elimination of tuberculosis in the United States. *Morbid. Mortal. Weekly Rep.* **38**(Suppl. S-3)**:**1–25.
2. **Della-Latta, P.** 1996. Workflow and optional protocols for laboratories in industrialized countries. *Clin. Lab. Med.* **16:**677–695.
3. **Tenover, F. C., J. T. Crawford, R. E. Huebner, L. J. Geiter, C. R. Horsburgh, Jr., and R. C. Good.** 1993. The resurgence of tuberculosis: is your laboratory ready? *J. Clin. Microbiol.* **31:**767–770.

SECTION 5

Antimicrobial Susceptibility Testing

SECTION EDITOR: *Janet Hindler*

5.1 Introduction

Many of the recommendations in the following procedures have been derived from the NCCLS disk diffusion (1) and MIC (2) standards for antimicrobial susceptibility testing. To utilize EPCM procedures, it is imperative that laboratories obtain the current NCCLS standards, which contain tables for recommending drugs for testing and reporting, interpreting results, and QC. These are updated annually and can be obtained from

NCCLS
940 West Valley Road
Suite 1400
Wayne, PA 19087
(610) 688-0100
(610) 688-0700 FAX

Additional information related to the enclosed procedures as well as as special antimicrobial susceptibility test methods can be found in CMPH section 5.

GENERAL RULES FOR PERFORMING ROUTINE ANTIMICROBIAL SUSCEPTIBILITY TESTS

1. Whenever possible, follow NCCLS standards for routine antimicrobial susceptibility testing.
2. Make certain commercial susceptibility test systems are used for their intended (and U.S. Food and Drug Administration [FDA]-approved) indications only; follow manufacturer's instructions precisely.
3. Perform susceptibility tests on clinically relevant bacteria only. Testing isolates that are unlikely to contribute to an infectious process may lead to unnecessary therapy. Protocols to define which bacteria isolated from specific body sites would be clinically relevant must be developed with the medical staff.
4. Report only those agents appropriate for the specific bacterium isolated and body site from which the organism was isolated.
5. Make certain the susceptibility results reported for a specific isolate are consistent with the identification of the isolate; verify atypical results.
6. Test QC strains in a manner identical to testing patient isolates. Investigate any out-of-control results.

REFERENCES

1. **NCCLS.** 1997. *Performance Standards for Antimicrobial Disk Susceptibility Tests*, 6th ed. Approved Standard: M2-A6 and Supplemental Tables M100-S7. NCCLS, Wayne, Pa.
2. **NCCLS.** 1997. *Methods for Dilution Antimicrobial Susceptibility Tests for Bacteria That Grow Aerobically*, 4th ed. Approved Standard: M7-A4 and Supplemental Tables M100-S7. NCCLS, Wayne, Pa.

5.2 Disk Diffusion Test

I. PRINCIPLE

A standardized inoculum of bacteria is swabbed onto the surface of a Mueller-Hinton agar (MHA) plate. Filter paper disks impregnated with antimicrobial agents are placed on the agar. After overnight incubation the diameter of the zone of inhibition is measured around each disk. By comparing these measurements with those in NCCLS disk diffusion tables (1), an organism is determined to be susceptible, intermediate, or resistant to each antimicrobial agent tested.

II. SPECIMEN

Prepare inocula from four or five isolated colonies of similar colony morphology.

A. Log-phase growth inoculum

Use colonies grown for 1 or 2 days on nonselective or selective (e.g., MAC) medium.

B. Direct colony suspension inoculum

Use colonies grown overnight on nonselective medium (e.g., BAP or CHOC).

III. MATERIALS

Include QC information on reagent container and in QC records.

A. Media and reagents

1. Agar plates (about 4 mm deep)
   Store at 2 to 8°C.
   - **a.** MHA
   - **b.** MHA with 5% sheep blood
   - **c.** Haemophilus Test Medium (HTM)
   - **d.** GC agar base with defined supplements
2. Nutrient broth (e.g., Mueller-Hinton, TSB) or 0.9% saline (3.0- to 5.0-ml aliquots)
   Store at 2 to 30°C.
3. Antimicrobial disks
   Store with desiccant at − 14 to 8°C.

B. Supplies

1. Sterile cotton-tipped swabs
2. Sterile plastic pipettes
3. McFarland 0.5 turbidity standard

C. Equipment

1. Forceps
2. Ruler, template, or sliding caliper
3. Movable light source
4. Black nonreflecting surface (e.g., sheet of black paper)
5. Vortex mixer
6. Ambient-air incubator (34 to 35°C), CO_2 incubation for some organisms (*see* Appendix 5.2–1).
7. Multidisk dispensing apparatus and container (optional)

IV. QUALITY CONTROL

A. QC strains

1. *Escherichia coli* ATCC 25922
2. *Staphylococcus aureus* ATCC 25923
3. *Pseudomonas aeruginosa* ATCC 27853
4. *Escherichia coli* ATCC 35218
5. *Enterococcus faecalis* ATCC 29212
6. (*See* appendixes to this procedure for QC of specific fastidious organisms.)

B. Monitoring accuracy

1. Test QC strains by routine procedures and record results in QC notebook. Record lot numbers and expiration dates of disks and agar.
2. Compare with expected results (*see* NCCLS QC tables [1]); note any out-of-control result and document; proceed with corrective action if necessary.

C. Frequency of QC testing

1. Perform QC daily or each time a patient isolate is tested.
2. Frequency of QC testing can be reduced from daily to weekly provided that a laboratory can document proficiency in performing this test by performing QC daily (or each day patient tests are performed) until results from 30 consecutive days of testing have been obtained.
3. Proficiency in performing QC tests is confirmed if, for *each* drug-organism combination, no more than 3 of 30 results are outside accuracy limits.
4. Document proficiency each time a new drug is added to testing protocols.
5. Perform QC testing each time a new lot or new shipment of materials (MHA, disks) is put into use.

D. Acceptable results for daily QC testing

QC testing is in control if no more than 1 out of 20 consecutive results is outside specified accuracy limits.

E. Acceptable results for weekly QC testing

1. QC testing is in control if all zone measurements are within specified accuracy limits.
2. If *any* weekly result is out of control, return to daily QC to define the problem as follows:
   a. Test the antimicrobial agent(s) with appropriate QC strain(s) for 5 consecutive test days.
   b. For each drug-organism combination, all of the five zone measurements must be within the accuracy control limits.
   c. If the problem is not resolved (if one or more zone measurements are again outside the accuracy control limits), continue daily QC. To return to weekly, document satisfactory performance for another 30 consecutive days.

V. PROCEDURE

See Appendix 5.2–1 for a summary of testing conditions for various organisms and Appendixes 5.2–2 through 5.2–4 for special procedures for testing *Haemophilus* spp., *Neisseria gonorrhoeae, Streptococcus pneumoniae,* and *Streptococcus* spp.

A. Bring agar plates and canisters of disks to room temperature before use.

B. Inoculum preparation

Using a loop or swab, transfer colonies as follows:

1. Standard method (log-phase growth)
   a. Pick four or five isolated colonies to 3.0 to 5.0 ml of broth.
   b. Incubate at 35°C for 2 to 8 h until growth reaches the turbidity at or above that of a McFarland 0.5 standard.
2. Direct colony suspension method

   Pick several colonies from a *fresh* (18 to 24 h) nonselective agar plate to broth or 0.9% saline.
3. For either method, vortex the suspension well and adjust turbidity visually with sterile broth or 0.9% saline to match a 0.5 McFarland standard (1.5 $\times$ 10^8 CFU/ml) which has been vortexed. Alternatively, standardize the suspension with a photometric device.

C. Inoculation of agar plate

1. Within 15 min of adjusting turbidity, dip a sterile cotton swab into the inoculum and rotate it against the wall of the tube above the liquid to remove excess inoculum.

V. PROCEDURE *(continued)*

2. Swab the entire surface of the agar plate three times, rotating the plate approximately 60° between streakings to ensure even distribution. Avoid hitting the sides of the petri plate and creating aerosols.
3. Allow the inoculated plate to stand for 3 to 15 min before applying disks.

D. Application of disks

1. Apply disks to the agar surface with a dispenser or manually with sterile forceps.
2. Apply gentle pressure with sterile forceps or loop to ensure complete contact of the disk with the agar (some dispensers do this automatically).
3. Do not place disks closer to each other than 24 mm from center to center.
   - **a.** No more than 12 disks on a 150-mm-diameter plate
   - **b.** No more than 5 disks on a 100-mm-diameter plate
4. *Do not* relocate a disk once it has made contact with the agar surface (discard the disk, and replace it with a new one).

E. Incubation

1. Incubate plates within 15 min of disk application.
2. Invert plates and stack them no more than five high.
3. Incubate for 16 to 18 h at 35°C in an ambient-air incubator
   - **a.** Incubate all staphylococci that are susceptible to penicillinase-resistant penicillins (oxacillin, methicillin, and nafcillin) for an additional 6 to 8 h (total of 24 h).
   - **b.** Incubate all enterococci that are susceptible to vancomycin for an additional 6 to 8 h (total of 24 h).

F. Reading plates

1. Read plates only if the lawn of growth is confluent or nearly confluent.
2. Translucent media (MHA, HTM, GC agar)
   - **a.** Rest the plate (lid down) on or hold the plate 2 to 3 inches above a black nonreflecting surface.
   - **b.** Illuminate the plate with reflected light directed from above at a 45° angle.
   - **c.** Measure the diameter of the inhibition zone from the back of the plate, to the nearest whole millimeter.
   - **d.** For staphylococci and penicillinase-resistant penicillins (oxacillin, methicillin, and nafcillin) and for enterococci and vancomycin:
     - (1) Use transmitted light, and hold the plate between your eye and light source.
     - (2) Examine closely; *any* growth indicates resistance.
3. Opaque media (blood-supplemented MHA)
   - **a.** Remove the cover of the agar plate.
   - **b.** Illuminate the plate with reflected light directed from above at a 45° angle.
   - **c.** Measure the diameter of the inhibition zone at the agar surface, to the nearest whole millimeter.
   - **d.** When testing hemolytic organisms, measure the diameter of the zone of inhibition of growth and not the zone of inhibition of hemolysis.
4. Additional considerations
   - **a.** With the exception of oxacillin and staphylococci and vancomycin and enterococci, always disregard minute colonies visible by viewing only with transmitted light or a magnifying device.
   - **b.** Disregard swarming of *Proteus* spp. and measure the edge of the obvious inhibition zone under the veil of swarming.
   - **c.** When measuring the zone for sulfonamides, trimethoprim, or trimethoprim-sulfamethoxazole, disregard light growth (read for 80% or more inhibition), and measure the edge of heavy growth.

d. Large colonies growing within the inhibition zone may represent a mixed culture or resistant variants; subculture these, reidentify, and retest for susceptibility. If repeat testing demonstrates the same results (and the culture is confirmed to be pure) or a resistant zone, report as resistant.

VI. RESULTS

A. Interpretation

Interpret sizes of zone diameters of inhibition based on criteria specified by NCCLS (1).

B. Reporting

Report the categorical result as either susceptible (S), intermediate (I), or resistant (R).

C. Important reporting rules

1. Report methicillin-, oxacillin-, or nafcillin-resistant staphylococci as resistant to all β-lactam drugs (including β-lactam–β-lactamase inhibitor combinations, all cephalosporins, all penicillins, and imipenem) regardless of the results of in vitro susceptibility testing.
2. If test organisms require a nonstandardized method or if organisms other than the *Enterobacteriaceae, Pseudomonas aeruginosa, Acinetobacter* spp., *Staphylococcus* spp., *Enterococcus* spp., *Haemophilus* spp., *Neisseria gonorrhoeae, Streptococcus pneumoniae*, and *Streptococcus* spp. are tested by these disk diffusion methods, add a notation to the report stating that susceptibility testing has been performed by a nonstandardized method (e.g., for *Corynebacterium* spp., tested by disk diffusion). For such organisms, an MIC test is recommended.
3. If CO_2 incubation is used for organisms other than *Haemophilus* spp., *Neisseria gonorrhoeae, Streptococcus pneumoniae*, or *Streptococcus* spp., add a notation to the report stating that susceptibility testing has been performed by a nonstandardized method.

VII. PROCEDURE NOTES

A. Oxacillin (methicillin)-resistant staphylococci

1. Oxacillin is the recommended penicillinase-resistant penicillin for testing, and oxacillin results can represent all penicillinase-resistant penicillins (oxacillin, methicillin, nafcillin, and also cloxacillin, dicloxacillin, and flucloxacillin).
2. Clues to detection of methicillin (oxacillin)-resistant *S. aureus*
   a. Multiple resistance to other antimicrobial agents (β-lactams, aminoglycosides, erythromycin, clindamycin, and tetracycline)
   b. Presence of a haze of growth within the zone of inhibition around an oxacillin, methicillin, nafcillin, or cephalothin disk

B. Materials

1. Disks
   a. Working supplies of disks (properly stored as described above) can be utilized for at least 1 week; verify acceptability by QC methods
   b. β-Lactams, imipenem, and clavulanic acid combinations are the most labile drugs tested.
2. Media
   a. Just before they are used, agar plates can be removed from the refrigerator and placed in a 35°C ambient-air incubator with their lids slightly ajar to evaporate excess moisture. Do not leave plates in the incubator longer than 30 min.
   b. Small MHA plates (100-mm diameter) may be used provided that the agar is approximately 4 mm deep.

VII. PROCEDURE NOTES *(continued)*

c. Thymidine can interfere with the performance of sulfonamides and trimethoprim. Monitor media by testing *Enterococcus faecalis* ATCC 29212 (a zone of ≥20 mm indicates acceptability). Blood components (other than horse blood) contain thymidine and should not be used (except sheep blood for *Streptococcus pneumoniae* and *Streptococcus* spp. as described in this procedure).

d. An increase in the cation content (Ca^{2+}, Mg^{2+}) of the medium results in a decrease in zone diameters of aminoglycosides for *P. aeruginosa* and a decrease in zone diameters of tetracycline for all organisms. A decrease in cation content has the opposite effect. Monitor cation content by testing aminoglycosides with *P. aeruginosa* ATCC 27853 and making sure results are within defined QC limits.

VIII. LIMITATIONS

A. This method is standardized only for rapidly growing aerobes, including the *Enterobacteriaceae, P. aeruginosa, Acinetobacter* spp., *Staphylococcus* spp., and *Enterococcus* spp. Modifications have been made to standardize testing of some fastidious organisms such as *Haemophilus* spp., *Neisseria gonorrhoeae, S. pneumoniae*, and *Streptococcus* spp. (*see* Appendixes).

B. Numerous factors can affect results, including inoculum size, rate of growth, media formulation and pH of media, incubation environment and length of incubation, disk content and drug diffusion rate, and measurement of endpoints. Therefore, strict adherence to protocol is required to ensure reliable results.

REFERENCE

1. **NCCLS.** 1997. *Performance Standards for Antimicrobial Disk Susceptibility Tests*, 6th ed. Approved Standard: M2-A6 and Supplemental Tables M100-S7. NCCLS, Wayne, Pa.

SUPPLEMENTAL READING

Barry, A. L., M. D. Coyle, C. Thornsberry, E. H. Gerlach, and R. W. Hawkinson. 1979. Methods of measuring zones of inhibition with Bauer-Kirby disk susceptibility test. *J. Clin. Microbiol.* **10:**885–889.

Bauer, A. W., W. M. M. Kirby, J. C. Sherrls, and M. Turck. 1966. Antibiotic susceptibility testing by a standardized single disk method. *Am. J. Clin. Pathol.* **45:**493–496.

Miller, J. M., C. Thornberry, and C. N. Baker. 1984. Disk diffusion susceptibility test troubleshooting guide. *Lab Med.* **15:**183–185.

Woods, G. L., and J. A. Washington. 1995. Antibacterial susceptibility tests: dilution and disk diffusion methods, p. 1327–1341. *In* P. R. Murray, E. J. Baron, M. A. Pfaller, P. C. Tenover, and R. H. Yolken (ed.), *Manual of Clinical Microbiology*, 6th ed. American Society for Microbiology, Washington, D.C.

APPENDIX 5.2–1

Quick reference list for performing disk diffusion tests[a]

Organism	Agar	Inoculum method	Incubation length (h)	Incubation atmosphere[b]	Notes
Enterobacteriaceae	MHA	DCS/LPG	16–18	Ambient air	
Pseudomonas aeruginosa and *Acinetobacter* spp.	MHA	DCS/LPG	16–18	Ambient air	
Other *Pseudomonas* spp.[c]	MHA	DCS/LPG	16–18	Ambient air	Presumptive
Haemophilus spp.	HTM	DCS	16–18	CO_2	*β*-Lactamase test
Neisseria gonorrhoeae	GC/CF	DCS	20–24	CO_2	*β*-Lactamase test

APPENDIX 5.2–1 *(continued)*

Organism	Agar	Inoculum method	Incubation length (h)	Incubation atmosphere[b]	Notes
Neisseria meningitidis[c]	MHA	DCS	18–24	CO_2	Presumptive sulfathiazole and/or rifampin (prophylaxis only)[d]
Staphylococcus spp.	MHA	DCS	16–18 24 (oxacillin)[e]	Ambient air	Read oxacillin by using transmitted light.
Enterococcus spp.	MHA	DCS/LPG	16–18 24 (vancomycin)	Ambient air	β-Lactamase and high-level aminoglycoside screen on blood and CSF isolates; read vancomycin by using transmitted light and perform MIC on any vancomycin "I" isolate.
Streptococcus pneumoniae	BMHA	DCS	20–24	CO_2	Oxacillin disk for penicillin resistance
Streptococcus spp.	BMHA	DCS	20–24	CO_2	Penicillin MIC on viridans from endocarditis
Moraxella catarrhalis[c]	MHA	DCS	16–24	Ambient air	Presumptive (current NCCLS recommendation is for β-lactamase testing only)
Other[c]	BMHA	DCS	16–24	Ambient air (CO_2 if necessary)	Presumptive

[a] Abbreviations: BMHA, MHA with 5% sheep blood; DCS, direct colony suspension (standardize to McFarland 0.5 turbidity standard); DCS/LPG, direct colony suspension or log-phase growth (standardize to McFarland 0.5 turbidity standard); GC/CF, GC agar, cysteine free; LPG, log-phase growth.
[b] All incubation temperatures 34 to 35°C.
[c] Obtain final results with MIC test (disk diffusion test for those not addressed by NCCLS; presumptive indicates not standardized procedure).
[d] Rifampin: S, ≥20 mm; R, ≤16 mm. Sulfathiazole: S, ≥40 mm; R, ≤36 mm (can use triple sulfa disk). Test other drugs by MIC method.
[e] And other penicillinase-resistant penicillins.

APPENDIX 5.2–2

Haemophilus spp.

A. SPECIMEN

Isolated colonies of similar colony morphology grown overnight on CHOC.

B. SUPPLEMENTAL MATERIALS

1. CHOC
   Store at 2 to 8°C.
2. HTM plates
   Store at 2 to 8°C.
3. Photometric standardizing device (e.g., bench-top photometer) for 0.5 McFarland turbidity and test tubes that will fit this device

C. QC STRAINS

1. *Haemophilus influenzae* ATCC 49247
2. *Haemophilus influenzae* ATCC 49766

D. PROCEDURE

1. Prepare inoculum by using a direct colony suspension in Mueller-Hinton broth or 0.9% saline from fresh (18–24 h) growth on CHOC.
2. Use a photometric device to adjust turbidity. This step is recommended to avoid erroneous results caused by under- or overinoculation.
3. Without delay, inoculate agar and apply disks.
4. Because *Haemophilus* spp. show large zones with some drugs (e.g., broad-spectrum cephalosporins), apply fewer disks (e.g., 9 to 10) per 150-mm agar plate.
5. Incubate in CO_2 for 16 to 18 h at 35°C.

E. READING RESULTS

1. Measure zones of inhibition from the back of the plate.
2. Refer to the NCCLS *Haemophilus* spp. interpretive table for zone diameter interpretation; refer to the NCCLS *Haemophilus* spp. QC table for QC ranges.

APPENDIX 5.2–2 *(continued)*

3. Only the susceptible interpretation is defined for several cephalosporins, aztreonam, imipenem, and several quinolones because no resistant strains have been reported. Confirm results on all isolates that appear nonsusceptible.

F. NOTES
1. β-Lactamase-positive *Haemophilus* spp. are resistant to ampicillin and amoxicillin regardless of zone sizes.
2. Occasional isolates are β-lactamase negative and ampicillin resistant. Consider these resistant to amoxicillin-clavulanic acid, ampicillin-sulbactam, cefaclor, cefetamet, cefonicid, cefprozil, cefuroxime, and loracarbef despite in vitro results.

APPENDIX 5.2–3

Neisseria gonorrhoeae

A. SPECIMEN

Isolated colonies of similar colony morphology grown overnight in 5 to 7% CO_2 on CHOC.

B. SUPPLEMENTAL MATERIALS
1. CHOC
   Store at 2 to 8°C.
2. GC agar base with defined cysteine-free GC supplement (GC/CF) plates
   Store at 2 to 8°C.

C. QC STRAIN

Neisseria gonorrhoeae ATCC 49226

D. PROCEDURE
1. Prepare inoculum by using a direct colony suspension in Mueller-Hinton broth or 0.9% saline from fresh (18–24 h) growth on CHOC.
2. Adjust turbidity visually to a McFarland 0.5 turbidity standard, or use a photometric device.
3. Without delay, inoculate agar and apply disks.
4. Since zones of inhibition tend to be large, apply no more than two to three disks per 100-mm plate, or nine disks per 150-mm plate.
5. Incubate in CO_2 for 20 to 24 h at 35°C.

E. READING RESULTS
1. Measure zones of inhibition from the back of the plate.
2. Refer to the NCCLS *N. gonorrhoeae* interpretive table for zone interpretation; refer to the NCCLS *N. gonorrhoeae* QC table for QC ranges.
3. Because resistant strains have not been reported for several cephalosporins and quinolones, only the susceptible category is defined. Confirm results on all isolates that appear nonsusceptible.

F. NOTES
1. *N. gonorrhoeae* cultures with zone diameters of $\leq$19 mm with the 10 U penicillin disk generally are β-lactamase producers. β-Lactamase tests are preferred for detecting this plasmid-mediated resistance.
2. *N. gonorrhoeae* cultures with zone diameters of $\leq$19 mm with the 30-μg tetracycline disk usually indicate plasmid-mediated resistance to tetracycline. Send any such isolate to a reference laboratory for confirmation.
3. Strains interpreted as susceptible have a >95% clinical cure rate, whereas those isolates that are intermediate have a lower cure rate of 85 to 95%.

APPENDIX 5.2–4

Streptococcus pneumoniae and *Streptococcus* spp.

A. SPECIMEN

Isolated colonies of similar colony morphology grown overnight on BAP

B. SUPPLEMENTAL MATERIALS
1. BAP
   Store at 2 to 8°C.
2. MHA supplemented with 5% sheep blood
   Store at 2 to 8°C.

C. QC STRAIN

Streptococcus pneumoniae ATCC 49619

D. PROCEDURE

1. Prepare inoculum by using a direct colony suspension in Mueller-Hinton broth or 0.9% NaCl from fresh (6 to 18 h) growth on BAP.
2. Adjust turbidity visually to a McFarland 0.5 turbidity standard or use a photometric device.
3. Inoculate agar without delay, and apply disks.
4. Incubate in CO_2 for 20 to 24 h at 35°C.

E. READING RESULTS

1. Measure zones of inhibition from the top of the plate.
2. Refer to the NCCLS *Streptococcus pneumoniae* and *Streptococcus* spp. interpretive table for zone interpretation; refer to the NCCLS *Streptococcus pneumoniae* and *Streptococcus* spp. QC table for QC ranges.
3. Penicillin (test a 1-μg oxacillin disk to determine penicillin susceptibility)
   a. Susceptible: oxacillin ≥20 mm
      Report results for *penicillin*, not *oxacillin*.
   b. (Susceptible), resistant, or intermediate: oxacillin ≤19 mm
      This test cannot separate penicillin-resistant from intermediate strains (or sometimes susceptible strains).
   c. Perform *penicillin* MIC test on isolates with oxacillin zones of ≤19 mm.

F. NOTES

1. Pneumococci may lyse rapidly in liquid (particularly water). Therefore follow inoculum preparation procedures described here precisely, and inoculate the agar plate immediately after standardization of inoculum suspension.
2. Penicillin-resistant *S. pneumoniae* cells do not produce β-lactamase; do not use a β-lactamase test to test for penicillin resistance.
3. Some isolates may have oxacillin zones below the 20-mm breakpoint and still have a penicillin-susceptible MIC of ≤0.06 μg/ml.

5.3 Broth Microdilution MIC Test

I. PRINCIPLE

The broth microdilution MIC method is used to measure (semiquantitatively) the in vitro activity of an antimicrobial agent against a bacterial isolate. A sterile plastic tray containing various concentrations of antimicrobial agents is inoculated with a standard number of test bacteria. After overnight incubation at 35°C, the MIC is determined by observing the lowest concentration of an antimicrobial agent which will inhibit visible growth of the bacterium. MIC values can be interpreted as susceptible, intermediate, or resistant based on the tables in the NCCLS MIC standard (1).

II. SPECIMEN

Prepare inocula from four or five isolated colonies of similar colony morphology.

A. Log- or stationary-phase growth inoculum

Use colonies grown for 1 or 2 days on nonselective or selective (e.g., MAC) medium.

B. Direct colony suspension inoculum

Use colonies grown overnight on nonselective medium (e.g., BAP or CHOC).

III. MATERIALS

Include QC information on reagent container and in QC records.

A. Media and reagents

1. Prepared broth microdilution trays containing an antimicrobial agent in cation-adjusted Mueller-Hinton broth (CAMHB) in volumes of 0.1 ml of solution per well
   Store at −70°C.
2. Nutrient broth (e.g., Mueller-Hinton, TSB) or 0.9% saline (3.0- to 5.0-ml aliquots)
   Store at 2 to 30°C.
3. BHI broth (0.5-ml aliquots)
   Store at 2 to 30°C.
4. Water diluent containing 0.02% Tween 80 dispensed in aliquots of 25 ml in screw-capped tubes
   Store at 25°C.
5. BAP
   Store at 2 to 8°C.

B. Supplies

1. Sterile cotton-tipped swabs
2. Sterile plastic pipettes
3. McFarland 0.5 turbidity standard (McFarland 4.0 standard for stationary-phase inoculum)
4. Sterile disposable plastic multipronged inoculator sets (include inoculum reservoir)

C. Equipment

1. Vortex mixer
2. Ambient-air incubator (34 to 35°C), CO_2 incubation for some organisms (*see* Appendix 5.3–1)
3. Adjustable micropipette and sterile pipette tips
4. Calibrated loop (0.001 ml [1 μl])
5. Viewing device to read inoculated MIC trays after incubation
6. Microdilution tray storage containers (e.g., Tupperware-type containers or plastic bags)
7. Freezer (−70°C; sometimes a −20°C, non-frost-free freezer is acceptable)

IV. QUALITY CONTROL

A. QC strains

1. *Escherichia coli* ATCC 25922
2. *Staphylococcus aureus* ATCC 29213
3. *Pseudomonas aeruginosa* ATCC 27853
4. *Escherichia coli* ATCC 35218
5. *Enterococcus faecalis* ATCC 29212
6. (*See* Appendixes to this procedure for QC of specific fasitidious organisms.)

B. Monitoring accuracy

1. Test QC strains by routine procedures and record results in a QC notebook. Record lot numbers and expiration dates of broth microdilution trays.
2. Compare with expected results (see NCCLS QC tables [1]); note any out-of-control result and document; proceed with corrective action if necessary.

C. Additional controls

1. Growth controls
   A growth control well must show 3+ to 4+ growth; a purity plate must show good growth without the presence of contaminating organisms.
2. The sterility control well should be free of any growth.
3. Inoculum controls
   An inoculum count verification plate (checked once per month) should show 30 to 70 colonies.

D. Frequency of QC testing

1. Perform QC daily or each time a patient isolate is tested.
2. Frequency of QC testing can be reduced from daily to weekly provided that a laboratory can document proficiency in performing this test by performing QC daily (or each day patient tests are performed) until results from 30 consecutive days of testing have been obtained.
3. Proficiency in performing QC tests is confirmed if, for *each* drug-organism combination, no more than 3 of 30 results are outside accuracy limits.
4. Document proficiency each time a new drug is added to testing protocols.
5. Perform QC testing each time a new lot or new shipment of materials (broth microdilution trays) is put into use.
6. Perform colony counts on QC strains at least once per month.

E. Acceptable results for daily QC testing

QC testing is in control if no more than 1 out of 20 consecutive results is outside specified accuracy limits.

F. Acceptable results for weekly QC testing

1. QC testing is in control if all MICs are within specified accuracy limits.
2. If *any* weekly result is out of control, return to daily QC to define the problem as follows:
   a. Test the antimicrobial agent(s) with appropriate QC strain(s) for 5 consecutive test days.
   b. For each drug-organism combination, all of the 5 MICs must be within the accuracy control limits.
   c. If the problem is not resolved (one or more MICs are outside the accuracy control limits), continue daily QC. To return to weekly QC, document satisfactory performance for another 30 consecutive days.

V. PROCEDURE

See Appendix 5.3–1 for a summary of testing conditions for various organisms and Appendixes 5.3–2 and –3 for special procedures for testing *Haemophilus* spp., *S. pneumoniae*, and *Streptococcus* spp.

A. Allow frozen trays to thaw at room temperature.

B. Inoculum preparation

1. Using a loop or swab, transfer colonies as follows:
   a. Standard method (log-phase growth)
      (1) Pick 4 to 5 isolated colonies to 3.0 to 5.0 ml of broth.
      (2) Incubate at 35°C for 2 to 8 h until growth reaches the turbidity at or above that of a McFarland 0.5 standard.
   b. Direct colony suspension method
      Pick several colonies from a *fresh* (18 to 24 h) nonselective agar plate to broth or 0.9% saline.

V. PROCEDURE *(continued)*

c. For either the standard or direct colony suspension method, vortex well and adjust turbidity visually with sterile broth or 0.9% saline to match a 0.5 McFarland standard (1.5×10^8 CFU/ml) which has been vortexed. Alternatively, standardize the suspension with a photometric device.

d. Stationary-phase method

(1) Transfer several isolated colonies to 0.5 ml of BHI.

(2) Incubate at 35°C to stationary phase (visual inspection must show very dense turbidity [McFarland ≥4.0 standard] corresponding to ~10^9 CFU/ml):

(a) *Enterobacteriaceae*, 2 to 8 h

(b) *Pseudomonas*, 4 to 8 h

(c) Others, 4 to 8 h

2. Calculate the volume of standardized suspension to be added to 25 ml of water–Tween 80 diluent to obtain a final organism concentration of 3 to 5 $\times 10^5$ CFU/ml in a microdilution well. For prongs that deliver 0.01 ml:

Inoculum prepn method	Equivalent organism concn (CFU/ml)	Amt (ml) of organism suspension to add to 25-ml water blank	Organism concn per ml in 0.1-ml well
Log-phase growth or direct colony suspension	1.5×10^8 (McFarland, 0.5)	1.0	5–6 $\times 10^5$
Stationary-phase growth	~10^9 (McFarland, 4 or higher)	0.15	5–6 $\times 10^5$

C. Inoculation

1. Transfer the appropriate volume of well-mixed organism suspension to the water–Tween 80 diluent with a micropipette.

2. Gently mix by inverting five to six times (try to avoid producing air bubbles).

3. Within 1 h (immediately if testing fastidious organisms), inoculate the MIC tray.

4. Inoculation with disposable plastic inoculators that deliver 0.01 ml

a. Remove the inoculator-reservoir set from its plastic packaging.

b. Dip prongs into the inoculum suspension.

(1) Orient the sterility well so that it does not get inoculated.

(2) Press down on the inoculator firmly to ensure all prongs have come in contact with the inoculum in the reservoir tray.

(3) Carefully remove prongs from inoculum.

c. Inoculate the MIC tray by dipping the filled prongs carefully into the MIC tray. Press down on the inoculator firmly to ensure that all prongs have come in contact with the antimicrobial solutions in the tray.

d. Discard the prongs.

e. Inoculate a purity plate by subculturing 0.001 ml of inoculum from the reservoir onto BAP.

5. Prepare an inoculum count verification plate monthly on each QC strain routinely tested as follows:

a. Immediately after inoculation, transfer 0.01 ml from the positive growth control well into 10 ml of sterile water or saline.

b. Vortex the mixture well, then transfer 0.1 ml to a BAP.

c. Thoroughly spread the inoculum over the surface of the plate by streaking in several directions with a loop.

D. Incubation

1. Stack inoculated MIC trays in stacks of no more than four high and place them in a plastic bag or microdilution tray storage container with the top loose.

2. Incubate MIC trays for 16 to 20 h at 35°C in an ambient-air incubator; incubate purity plates and inoculum count verification plates in CO_2.

a. Incubate all staphylococci that are susceptible to penicillinase-resistant penicillins (e.g., oxacillin) after 16 to 20 h for an additional 4 to 8 h (total of 24 h).

b. Incubate all enterococci that are susceptible to vancomycin or high concentrations of gentamicin (synergy screen) after 16 to 20 h for an additional 4 to 8 h (total of 24 h). Incubate the streptomycin synergy screen for 48 h before reporting a susceptible result. (*See* procedure 5.5.)

E. Reading MICs

1. Examine purity control plates. If mixed, repeat the corresponding MIC test; when available, examine the inoculum count verification plate, which should show 30 to 70 colonies.

2. Place each tray on the tray reading device.

3. Read the MIC tray only if the growth control well is turbid (3+ to 4+ growth).

a. ± to 1+, tiny button of very light haze in well

b. 2+, light haze in well

c. 3+ to 4+, heavy turbidity or fine granular growth throughout the well or a dense white button at the bottom of the well

Do not read the MICs if there is inadequate growth in the growth control well (often there are separate control wells if different broths are used, e.g., BHI for *Enterococcus* synergy screens).

4. Examine the uninoculated broth sterility control. Compare questionable growth to this well when necessary.

5. The MIC is the lowest concentration of antimicrobial agent showing complete inhibition of growth, or, for sulfonamides and trimethoprim, the MIC is the lowest concentration that inhibits 80% of the growth as compared with growth in the control well.

a. For trimethoprim-sulfamethoxazole, ignore very slight hazes and/or a pinpoint button (<2 mm) that persists through several dilutions.

b. When skipped wells occur, record the MIC only if the true MIC is apparent. If >1 drug skips or a skip occurs with a reportable drug, repeat the MIC test.

6. For staphylococci and penicillin, if MIC is 0.06 to 0.12 μg/ml, perform an induced β-lactamase test to determine whether the isolate is a low-level β-lactamase producer.

VI. RESULTS

A. Interpretation

Interpret MICs based on criteria specified by NCCLS (1).

B. Reporting

Report the MIC along with its categorical interpretation: susceptible (S), intermediate (I), or resistant (R).

C. Important reporting rules

1. Report methicillin-, oxacillin-, or nafcillin-resistant staphylococci as resistant to all β-lactam drugs (including β-lactam–β-lactamase inhibitor combinations, all cephalosporins, all penicillins, and imipenem) regardless of results of in vitro susceptibility testing.

2. If test organisms require a nonstandardized method or if organisms other than the *Enterobacteriaceae, Pseudomonas aeruginosa, Acinetobacter* spp., *Stenotrophomonas maltophilia, Pseudomonas* spp., *Staphylococcus* spp., *Enterococcus* spp., *Haemophilus* spp., *S. pneumoniae*, and *Streptococcus* spp. are tested by these MIC methods, add a notation to the report stating that susceptibility testing has been performed by a nonstandardized method (e.g., for *Corynebacterium* spp. tested in lysed horse blood MIC panels).

VI. RESULTS *(continued)*

3. For staphylococci and penicillin
   a. MIC ≤0.03 μg/ml, report penicillin "S."
   b. MIC ≥0.25 μg/ml, report penicillin "R."
   c. MIC 0.06 to 0.12 μg/ml and β-lactamase negative, report penicillin "S."
   d. MIC 0.06 to 0.12 μg/ml and β-lactamase positive, report penicillin "R."

VII. PROCEDURE NOTES

A. Oxacillin (methicillin)-resistant staphylococci

1. Oxacillin is the recommended penicillinase-resistant penicillin for testing, and oxacillin results can represent all penicillinase-resistant penicillins (oxacillin, methicillin, nafcillin, and also cloxacillin, dicloxacillin, and flucloxacillin). Testing of oxacillin with staphylococci should be performed in CAMHB supplemented with 2% NaCl (other drugs should not be tested in 2% NaCl-supplemented CAMHB).

2. Clues to detection of methicillin (oxacillin)-resistant *S. aureus*
   a. Multiple resistance to other antimicrobial agents (β-lactams, aminoglycosides, erythromycin, clindamycin, and tetracycline)
   b. Subtle growth in wells containing oxacillin, methicillin, or nafcillin

B. Media

1. Thymidine can interfere with the performance of sulfonamides and trimethoprim. Monitor media by testing *Enterococcus faecalis* ATCC 29212 (MIC ≤0.5/9.5 μg/ml indicates acceptability). Blood components (other than horse blood) contain thymidine and should not be used.

2. An increase in the cation content (Ca^{2+}, Mg^{2+}) of the medium results in an increase in the MICs of aminoglycosides for *P. aeruginosa* and an increase in the MICs of tetracycline for all organisms. A decrease in cation content has the opposite effect. Monitor cation content by testing aminoglycosides with *P. aeruginosa* ATCC 27853 and making sure results are within defined QC limits.

C. Materials

1. When supplements (e.g., lysed horse blood) are added to individual wells of antimicrobial broth, the diluting effect need not be considered provided that the additional volume is less than 10% of the total volume.

2. Although it is preferable to store MIC trays at −70°C, it is often acceptable to store frozen panels without imipenem and combinations with clavulanic acid at −20°C. Imipenem and clavulanic acid are more temperature labile than many other drugs and require −70°C storage. Follow the storage recommendations of the manufacturer.

D. Inoculum preparation

The stationary-phase method of inoculum preparation is not an NCCLS-recommended method; however, this method is suggested as an alternative for several commercial systems. Since the critical factor in inoculum preparation is the final concentration of viable organisms in the microdilution well ($3–5 \times 10^5$ CFU/ml), the stationary-phase method will produce satisfactory results for nonfastidious, rapidly growing organisms provided that the correct number of organisms is obtained. It is assumed that incubation to stationary phase will produce a suspension containing 10^9 CFU/ml.

E. Reading MICs

1. With bacteriostatic antimicrobial agents such as chloramphenicol, erythromycin, and clindamycin, very slight hazes may persist through several dilutions. Similarly, very slight hazes and/or pinpoint buttons (<2 mm) may persist through all dilutions of trimethoprim-sulfamethoxazole. These should be ignored.

2. When skipped wells occur, it may be necessary to repeat the MIC test. Skipped wells are indicated by growth at higher concentrations of an antimicrobial agent and no growth at one or more of the lower concentrations. This may occur as a result of:
   a. Contamination at higher dilutions
   b. Inadequate numbers of organisms inoculated into the wells
   c. Wells not inoculated properly owing to improper alignment of inoculating pins or prongs
   d. Peculiarity of the test organism (e.g., it might contain a resistant subpopulation)
   e. Improper concentrations of antimicrobial agents in the wells
   f. Each skipped well case must be evaluated individually to determine whether the test should be repeated.
3. Contamination may not always be detected with a purity plate since only a very small volume of the inoculum is sampled. The appearance of growth in the microdilution tray wells must be examined closely, and peculiar growth patterns, skipped wells, and/or atypical antibiograms should be investigated.
4. Reproducibility of broth microdilution MIC testing is generally within ±1 twofold dilution.

VIII. LIMITATIONS

A. The basic procedure described here has been standardized for testing commonly isolated bacteria that grow well after overnight incubation in cation-adjusted Mueller-Hinton broth. Various modifications have been made for testing some of the more fastidious and special-problem pathogens such as *Haemophilus* spp., *S. pneumoniae*, and *Streptococcus* spp. (*see* Appendixes to this procedure).

B. Because of the small number of organisms tested in each well ($3–5 \times 10^4$ CFU/well), resistance by a small subpopulation may not always be detected with the broth microdilution MIC method.

C. NCCLS M7-A4 establishes MIC breakpoints based primarily on blood levels after standard dosing. For optimal use of MIC values, the MIC should be correlated with the presumed antimicrobial-agent concentration at the infection site.

D. Numerous factors can affect results, including inoculum size, rate of growth, medium formulation and pH of media, incubation environment and length of incubation, drug concentration, and measurement of endpoints. Therefore, strict adherence to protocol is required to ensure reliable results.

REFERENCE

1. **NCCLS.** 1997. *Methods for Dilution Antimicrobial Susceptibility Tests for Bacteria That Grow Aerobically*, 4th ed. Approved Standard. NCCLS Publication M7-A4 and Supplement M100-S7. NCCLS, Wayne, Pa.

SUPPLEMENTAL READING

Barry, A. L., and L. E. Braun. 1981. Reader error in determining minimal inhibitory concentrations with microdilution susceptibility test panels. *J. Clin. Microbiol.* **13:**228–230.

Woods, G. L., and J. A. Washington. 1995. Antibacterial susceptibility tests: dilution and disk diffusion methods, p. 1327–1341. *In* P. R. Murray, E. J. Baron, M. A. Pfaller, F. C. Tenover, and R. H. Yolken (ed.), *Manual of Clinical Microbiology*, 6th ed. American Society for Microbiology, Washington, D.C.

APPENDIX 5.3–1

Quick reference list for performing broth microdilution MIC tests[a]

Organism	Medium	Inoculum method	Incubation length (h)	Incubation atmosphere[b]	Notes
Enterobacteriaceae	CAMHB	DCS/LPG	16–20	Ambient air	
Pseudomonas aeruginosa and *Acinetobacter* spp.	CAMHB	DCS/LPG	16–20	Ambient air	
Other *Pseudomonas* spp.	CAMHB	DCS/LPG	16–20	Ambient air	
Haemophilus spp.	HTM broth	DCS	20–24	Ambient air	β-Lactamase test
Neisseria meningitidis[c]	CAMHB with 2–5% LHB	DCS	24	CO_2	Presumptive
Staphylococcus spp.	CAMHB; CAMHB with 2% NaCl for oxacillin	DCS	16–20 24 (oxacillin)[d]	Ambient air	Final inoculum = 5×10^5 CFU/ml
Enterococcus spp.	CAMHB; BHI for GM 500 and STR 1000	DCS/LPG	16–20 24 (vancomycin and GM 500) 48 (STR 1000)	Ambient air	β-Lactamase and high-level aminoglycoside screen on blood-CSF isolates
Streptococcus pneumoniae	CAMHB with 2–5% LHB	DCS	20–24	Ambient air (CO_2 if needed)	Perform penicillin (not oxacillin) MIC when <20-mm oxacillin zones appear on the disk test.
Streptococcus spp.	CAMHB with 2–5% LHB	DCS	20–24	Ambient air (CO_2 if needed)	Perform penicillin MIC on viridans from endocarditis.
Listeria spp.[c]	CAMHB or CAMHB with 2–5% LHB	DCS	16–20	Ambient air	Presumptive
Other[c]	CAMHB or CAMHB with 2–5% LHB	DCS	16–24	Ambient air (CO_2, if needed)	Presumptive

[a] DCS, direct colony suspension (standardize to McFarland 0.5 turbidity standard); DCS/LPG, DCS or log-phase growth (standardize to McFarland 0.5 turbidity standard); HTM, *Haemophilus* test medium; LHB, lysed horse blood; LPG, log-phase growth; GM, gentamicin; STR, streptomycin.
[b] All incubation temperatures, 34–35°C.
[c] Tests for species are not addressed by NCCLS; presumptive indicates not a standardized procedure.
[d] And other penicillinase-resistant penicillins.

APPENDIX 5.3–2

Haemophilus spp.

A. SPECIMEN

Isolated colonies of similar colony morphology grown overnight on CHOC

B. SUPPLEMENTAL MATERIALS

1. CHOC
   Store at 2 to 8°C.
2. Prepared microdilution trays utilizing *Haemophilus* test medium
   Store at −70°C.
3. Photometric standardizing device (e.g., bench top photometer) for 0.5 McFarland turbidity and test tubes that will fit this device

C. QC STRAINS

1. *Haemophilus influenzae* ATCC 49247
2. *Haemophilus influenzae* ATCC 49766

D. PROCEDURE

1. Prepare inoculum by using a direct colony suspension in Mueller-Hinton broth or 0.9% saline from fresh (18- to 24-h) growth on CHOC.

APPENDIX 5.3–2 *(continued)*

2. Use a photometric device to adjust turbidity. This step is recommended to avoid erroneous results caused by under- or overinoculation.
3. Without delay, prepare intermediate dilution and inoculate the MIC tray and purity plate as previously described.
4. Incubate for 20 to 24 h at 35°C in ambient air.

E. READING RESULTS

1. Read MICs as described for nonfastidious organisms.
2. Refer to NCCLS *Haemophilus* spp. interpretive table for MIC interpretation; refer to NCCLS *Haemophilus* spp. QC table for QC ranges.
3. Only the susceptible interpretation is defined for several cephalosporins, aztreonam, imipenem, and several quinolones, because no resistant strains have been reported. Confirm results on all isolates that appear nonsusceptible.

F. NOTES

1. β-Lactamase-positive *Haemophilus* spp. are resistant to ampicillin and amoxicillin regardless of MICs.
2. Occasional isolates are β-lactamase negative and ampicillin resistant. Consider these resistant to amoxicillin-clavulanic acid, ampicillin-sulbactam, cefaclor, cefetamet, cefonicid, cefprozil, cefuroxime, and loracarbef despite in vitro results.

APPENDIX 5.3–3

***Streptococcus pneumoniae* and *Streptococcus* spp.**

A. SPECIMEN

Isolated colonies of similar colony morphology grown overnight on BAP

B. SUPPLEMENTAL MATERIALS

Prepared broth microdilution trays containing drug dilutions supplemented with 2 to 5% lysed horse blood

Store at −70°C.

C. QC STRAIN

Streptococcus pneumoniae ATCC 49619

D. PROCEDURE

1. Prepare inoculum by using a direct colony suspension in Mueller-Hinton broth or 0.9% saline from fresh (16- to 18-h) growth on BAP.
2. Adjust turbidity visually to a McFarland 0.5 turbidity standard or use a photometric device.
3. Without delay, prepare intermediate dilution and inoculate the MIC tray and purity plate as previously described.
4. Incubate for 20 to 24 h at 35°C in ambient air (CO_2 only if necessary).

E. READING RESULTS

1. Read MICs as described above for nonfastidious organisms.
2. Refer to NCCLS *Streptococcus pneumoniae* and *Streptococcus* spp. interpretive table for MIC interpretation; refer to NCCLS *Streptococcus pneumoniae* and *Streptococcus* spp. QC table for QC ranges.

F. NOTES

1. Pneumococci may lyse rapidly in liquid (particularly water). Therefore follow the inoculum preparation procedure described here precisely and inoculate the MIC tray immediately after standardization of inoculum suspension.
2. Penicillin-resistant *S. pneumoniae* do not produce β-lactamase; do not use a β-lactamase test to predict penicillin resistance.
3. When supplementing with pyridoxal (for nutritionally variant streptococci), final concentration should equal 0.001% or 1 μg/ml. (Pyridoxal stock, 1,000 μg/ml; add 0.2 ml to 9.8 ml of broth [20 μg/ml]; add 0.005 ml to each well.)
4. Wells in a panel with CAMHB can be converted to lysed horse blood wells by adding 0.005 μl of lysed horse blood (50%, vol/vol) to each well containing 0.1 ml of antimicrobial agent (final concentration is 2.5% lysed horse blood).

5.4 β-Lactamase Tests

I. PRINCIPLE

Routine β-lactamase tests are based on visual detection of the end products of β-lactamase hydrolysis, which is demonstrated with a colorimetric reaction. These tests primarily include the chromogenic cephalosporin method, the acidimetric method, and the iodometric method. The basic chromogenic cephalosporin method is described here; however, the other methods are summarized in Appendix 5.4–1. With the chromogenic cephalosporin method, most commonly, filter paper disks are impregnated with nitrocefin, a chromogenic cephalosporin. β-Lactamase, if present, hydrolyzes the amide bond in the β-lactam ring, producing a color change.

II. SPECIMEN

A. Prepare inocula from four or five isolated colonies of similar colony morphology grown overnight (18 to 24 h) on nonselective medium (e.g., BAP or CHOC).

B. Some staphylococci may require induction (exposure to a β-lactam agent) to increase production of enzyme to measurable levels.

1. Test growth from the periphery of the zone of inhibition around an oxacillin (1-μg) disk.
2. Grow the isolate overnight in broth containing a subinhibitory concentration of a β-lactam agent (e.g., 0.25 μg of cefoxitin per ml), and perform the test on this suspension.

III. MATERIALS

Include QC information on reagent container and in QC records.

A. Media and reagents

1. Chromogenic cephalosporin disks (other products are commercially available—follow manufacturer's directions)
   Store at 2 to 8°C.
2. Sterile distilled water
   Store at 25°C.

B. Supplies

1. Glass slides or empty petri plates
2. Sterile plastic pipettes
3. Sterile wooden applicator sticks or inoculating loops

IV. QUALITY CONTROL STRAINS

A. *Staphylococcus aureus* ATCC 29213
β-Lactamase positive

B. *Haemophilus influenzae* ATCC 10211
β-Lactamase negative

V. PROCEDURE

A. Dispense the required number of disks onto a clean microscope slide or an empty petri plate.

B. Moisten each disk with 1 drop of sterile distilled water.

C. With a sterile loop or applicator stick, smear several colonies onto the disk surface.

D. Observe for a color change, which usually appears within 15 s to 5 min. If no color change occurs within 5 min, the test is negative. However, positive reactions for some staphylococci may take up to 1 h.

E. Reading reactions
  1. Positive, yellow changes to red color.
  2. Negative, no change in color occurs.

VI. RESULTS

Interpret and report results as in the following example.
Haemophilus influenzae—*β*-lactamase positive (amoxicillin and ampicillin resistant)

VII. PROCEDURE NOTES

A. For some bacteria there is a direct correlation between *β*-lactamase production and resistance to specific *β*-lactam drugs that might be prescribed for treatment of infections caused by them:
  1. *Haemophilus influenzae*—amoxicillin, ampicillin
  2. *Moraxella catarrhalis*—amoxicillin, ampicillin, penicillin
  3. *Neisseria gonorrhoeae*—amoxicillin, ampicillin, penicillin
  4. *Staphylococcus* spp.—amoxicillin, ampicillin, penicillin, carbenicillin, ticarcillin, mezlocillin, piperacillin

B. For organisms other than those listed above (with the exception of some anaerobes), *β*-lactamase production cannot fully predict resistance to the various *β*-lactam drugs that may be considered for use. *β*-Lactamase testing should not be used for these (e.g., *Enterobacteriaceae, Pseudomonas* spp.) in a clinical laboratory setting.

C. *β*-Lactamase testing of staphylococci
  1. Staphylococci may require induction to demonstrate a positive *β*-lactamase reaction. Testing of induced bacteria must be performed to confirm the isolate as *β*-lactamase negative.
  2. *β*-Lactam agents other than oxacillin and cefoxitin can be used as inducing agents provided that they can be demonstrated to perform satisfactorily.
  3. Staphylococci may require up to 1 h to show a positive *β*-lactamase test.

D. Production of *β*-lactamase is not the only mechanism of *β*-lactam resistance in the species commonly tested (although it is the most common mechanism). Consequently, a positive *β*-lactamase test by definition indicates resistance to ampicillin, amoxicillin, penicillin, carbenicillin, ticarcillin, mezlocillin, piperacillin. However, a negative test does not guarantee susceptibility to these agents, and a conventional susceptibility test must be performed.

E. Microdilution MIC tests with penicillin may fail to detect penicillin-resistant (*β*-lactamase-positive) staphylococci in some isolates that produce small amounts of *β*-lactamase. An induced *β*-lactamase test is needed to confirm that an isolate is penicillin susceptible (*see* procedure 5.3).

F. The current NCCLS recommendations for routine testing of *Moraxella catarrhalis* include performance of a *β*-lactamase test only. This is because the incidence of resistance to other commonly used drugs is very low as is the incidence of ampicillin, amoxicillin, and penicillin resistance owing to a resistance mechanism other than *β*-lactamase production (1).

REFERENCE

1. **NCCLS.** 1997. *Performance Standards for Antimicrobial Disk Susceptibility Tests*, 6th ed. Approved Standard: M2-A6 and Supplemental Tables M100-S7. NCCLS, Wayne, Pa.

5 Antimicrobial Susceptibility Testing

SUPPLEMENTAL READING

Becton Dickinson Microbiology Systems. 1996. *BBL Paper Disks for the Detection of Beta-Lactamase Enzymes.* Product 88-0973-1.

Swenson, J. M., J. A. Hindler, and L. R. Peterson. 1995, Special tests for detecting antibacterial resistance, p. 1356–1367. *In* P. R. Murray, E. J. Baron, M. A. Pfaller, F. C. Tenover, and P. H. Yolken (ed.), *Manual of Clinical Microbiology*, 6th ed. American Society for Microbiology, Washington, D.C.

APPENDIX 5.4–1

Summary of β-lactamase testing methods

Test detail(s)	Testing method		
	Acidimetric	Nitrocefin (chromogenic cephalosporin)	Iodometric
Substrate	Citrate-buffered penicillin plus phenol red	Nitrocefin	Phosphate-buffered penicillin plus starch-iodine complex
Reaction	Penicilloic acid produces pH decrease.	Color change when β-lactam ring opened	Penicilloic acid reduces iodine and prevents it from combining with starch.
Results	Positive, yellow Negative, violet (red)	Positive, red Negative, no color change	Positive, colorless Negative, blue/purple

Method(s) generally satisfactory for various organisms

Organism	Testing method		
	Acidimetric	Nitrocefin	Iodometric
Haemophilus spp.	X	X	
Neisseria gonorrhoeae	X	X	X
Staphylococcus spp.	X	X	X
Moraxella catarrhalis		X	
Enterococcus faecalis		X	
Bacteroides spp.		X	

5.5 Screen Tests To Detect High-Level Aminoglycoside Resistance in *Enterococcus* spp.

I. PRINCIPLE

High-level aminoglycoside resistance (HLAR) in enterococci is most commonly detected by assessing growth at high concentrations of gentamicin (500 μg/ml) and sometimes streptomycin (1,000 μg/ml in broth and 2,000 μg/ml in agar) (1). Strains that show HLAR to gentamicin will not be synergistically killed with combinations of cell wall-active drugs (generally ampicillin, penicillin, or vancomycin) plus gentamicin. HLAR to gentamicin also means HLAR to tobramycin, netilmicin, and amikacin. HLAR to streptomycin equals resistance to combinations of cell wall-active drugs and streptomycin. The agar screen test is described in detail here, and reference is made to the disk diffusion and broth microdilution MIC test procedures for use of these methods to detect HLAR.

II. SPECIMEN

Prepare inocula from four or five isolated colonies of similar colony morphology grown overnight (18 to 24 h) on nonselective medium (e.g., BAP or CHOC).

III. MATERIALS

Include QC information on reagent container and in QC records.

A. Media and reagents

1. BHI agar screen plates prepared in quadrant plates for convenience
   - **a.** Quadrant 1, BHI agar control
   - **b.** Quadrant 2, 500 μg of gentamicin per ml
   - **c.** Quadrant 3, 2,000 μg of streptomycin per ml

   Store at 2 to 8°C.
2. Nutrient broth (e.g., Mueller-Hinton, TSB) or 0.9% saline (3.0- to 5.0-ml aliquots)

   Store at 2 to 30°C.

B. Supplies

1. Sterile cotton-tipped swabs
2. Sterile plastic pipettes
3. McFarland 0.5 turbidity standard

C. Equipment

1. Calibrated loop (10 μl [0.01 ml] or 10-μl micropipette with sterile tips
2. Vortex mixer
3. Ambient-air incubator (34 to 35°C)

IV. QUALITY CONTROL STRAINS

A. *Enterococcus faecalis* ATCC 29212

Susceptible to high concentrations of gentamicin and streptomycin

B. *Enterococcus faecalis* ATCC 51299

Resistant to high concentrations of gentamicin and streptomycin

V. PROCEDURE

A. Inoculum preparation

With a loop or swab, transfer colonies to broth or saline to obtain an organism suspension that matches a McFarland 0.5 turbidity standard (1.5×10^8 CFU/ml); vortex thoroughly.

B. Inoculation and incubation

1. Use a 10-μl micropipette with a sterile tip or calibrated loop to inoculate

V. PROCEDURE *(continued)*

the agar surface of each quadrant. The final concentration should be approximately 10^6 CFU.

2. Allow the inoculum to be absorbed into the agar.
3. Invert plates and incubate at 35°C in an ambient-air incubator for 24 h; if there is no growth at 24 h for streptomycin, reincubate streptomycin tests for an additional 24 h (total of 48 h).

C. Reading plates

1. Examine the control quadrant (no drug) for adequate growth.
2. Examine drug quadrants for the presence of growth (>1 colony, resistant).

VI. RESULTS

A. Interpretation

1. No growth (or 1 colony), no high-level resistance
2. Growth on gentamicin (500 μg/ml), high-level gentamicin resistance (also resistance to tobramycin, amikacin, netilmicin, kanamycin)
3. Growth on streptomycin (2,000 μg/ml), high-level streptomycin resistance

B. Reporting suggestion

Serious enterococcal infections require combination therapy with ampicillin, penicillin, or vancomycin plus an aminoglycoside. Bactericidal synergy occurs only when both drugs in the combination are susceptible.

VII. PROCEDURE NOTES

A. Agar method

1. Uninterpretable results may be caused by improper inoculation or the inability of the organism to grow satisfactorily on the particular agar base medium. The test should be repeated before reporting as uninterpretable.
2. Some isolates with streptomycin HLAR may not demonstrate resistance until after 48 h of incubation.
3. Most *E. faecium* isolates produce an aminoglycoside-modifying enzyme (6′-acetylyltransferase) that makes them inherently resistant to amikacin, kanamycin, netilmicin, and tobramycin; this may not be expressed as HLAR (e.g., MIC may be ≤2,000 μg/ml), but synergy will not occur with these agents.

B. Other methods

1. HLAR can also be determined by the standard disk diffusion method with special high-content disks (120 μg of gentamicin or 300 μg of streptomycin). Zones of 6 mm indicate HLAR. Zones of 7 to 9 mm are inconclusive and require that isolates be retested by another method. Zones of ≥10 mm indicate no HLAR (2).
2. HLAR can also be determined with the standard broth microdilutin MIC method by testing 500 μg of gentamicin per ml, and 1,000 μg of streptomycin per ml, respectively, diluted in BHI broth (1).
3. To detect HLAR to amikacin, use 2,000 μg of kanamycin per ml for the agar screen or broth dilution method.
4. Since gentamicin is the most widely used aminoglycoside (in combination with a cell wall-active drug) for treatment of serious enterococcal infections, an HLAR screen for gentamicin is often sufficient. However, if the isolate demonstrates high-level gentamicin resistance, screening for high-level streptomycin resistance is needed to identify those strains that might be high-level gentamicin resistant but not high-level streptomycin resistant. In this case, streptomycin could be used therapeutically.
5. Dextrose phosphate agar and broth have been shown to perform comparably to BHI in broth or agar screen tests.

VIII. LIMITATIONS

There may be occasional isolates that lack HLAR to gentamicin and streptomycin but show HLAR to amikacin and kanamycin. This resistance would be detected only with a kanamycin screen test, which is generally not available in clinical laboratories and currently not addressed by NCCLS.

REFERENCES

1. **NCCLS.** 1997. *Methods for Dilution Antimicrobial Susceptibility Tests for Bacteria That Grow Aerobically*, 4th ed. Approved Standard. NCCLS Publication M7-A4 and Supplement M100-S7. NCCLS, Wayne, Pa.
2. **NCCLS.** 1997. *Performance Standards for Antimicrobial Disk Susceptibility Tests*, 6th ed. Approved Standard: M2-A6 and Supplemental Tables M100-S7. NCCLS, Wayne, Pa.

SUPPLEMENTAL READING

Swenson, J. M., J. A. Hindler, and L. R. Peterson. 1995. Special tests for detecting antibacterial resistance, p. 1356–1367. *In* P. R. Murray, E. J. Baron, M. A. Pfaller, F. C. Tenover, and R. H. Yolken (ed.), *Manual of Clinical Microbiology*, 6th ed. American Society for Microbiology, Washington, D.C.

5.6 Agar Screen Test To Detect Vancomycin Resistance in *Enterococcus* spp.

I. PRINCIPLE

A standard number of bacteria is inoculated onto BHI containing 6 μg of vancomycin per ml. After incubation, the appearance of growth indicates that the enterococcal isolate is resistant to vancomycin (1).

II. SPECIMEN

Prepare inocula from four or five isolated colonies of similar colony morphology grown overnight (18 to 24 h) on nonselective medium (e.g., BAP or CHOC).

III. MATERIALS

Include QC information on reagent container and in QC records.

A. Media and reagents

1. BHI agar plates with 6 μg of vancomycin per ml
   Store at 2–8°C.
2. Nutrient broth (e.g., Mueller-Hinton, TSB) or 0.9% saline (3.0- to 5.0-ml aliquots)
   Store at 2–30°C.

B. Supplies

1. Sterile cotton-tipped swabs
2. Sterile plastic pipettes
3. McFarland 0.5 turbidity standard

C. Equipment

1. Calibrated loop (1.0 μl [0.001 ml] or 10 μl [0.01 ml])
2. Vortex mixer
3. Ambient-air incubator (34 to 35°C)

IV. QUALITY CONTROL STRAINS

A. *Enterococcus faecalis* ATCC 29212
Vancomycin susceptible

B. *Enterococcus faecalis* ATCC 51299
Vancomycin resistant

V. PROCEDURE

A. Inoculum preparation

With a loop or swab, transfer colonies to broth or saline to obtain an organism suspension that matches a McFarland 0.5 turbidity standard (1.5×10^8 CFU/ml); vortex thoroughly.

B. Inoculation and incubation

1. With a 0.001-ml or 0.01-ml calibrated loop, inoculate a section of the agar plate surface in a single streak (test up to six isolates per plate).
2. Allow the inoculum to be absorbed into the agar.
3. Invert plates and incubate at 35°C in an ambient-air incubator for 24 h (examine after overnight incubation, and reincubate any isolates that are susceptible).

C. Reading plates

Examine the plate for presence of growth (>1 colony, resistant).

VI. RESULTS

Interpretation

A. No growth (or 1 colony), vancomycin susceptible

B. Growth, vancomycin resistant

VII. PROCEDURE NOTES

A. There are at least three phenotypes of vancomycin-resistant enterococci:

1. VanA or high-level resistance with vancomycin MICs of ≥64 μg/ml and resistance to teicoplanin (MICs ≥16 μg/ml)
2. VanB or low-to-high-level resistance with vancomycin MICs of 16 to ≥64 μg/ml and usually without teicoplanin resistance
3. VanC or intrinsic low-level resistance with vancomycin MICs of 4 to 32 μg/ml without teicoplanin resistance and typically associated with *E. gallinarum* and *E. casseliflavus*.

B. On occasion, *E. faecalis* ATCC 29212 may show slight growth on screen plates.

C. The medium described here is to be used for testing enterococci that have been isolated in culture and not as a primary plating medium for surveillance specimens (e.g., rectal swabs).

D. Vancomycin resistance is most common in *Enterococcus faecium*, although it has been noted in *Enterococcus faecalis* and other less frequently encountered species. *E. gallinarum* and *E. casseliflavus* have inherent low-level resistance to vancomycin. The epidemiologic significance of vancomycin-resistant *E. faecium* and *E. faecalis* is great. The epidemiologic significance of *E. gallinarum* and *E. casseliflavus* has not been established, but because of the intrinsic resistance in these species, it is probably less.

E. Many organisms in addition to enterococci (e.g., gram-negative bacteria, yeasts) will grow on the screen medium; interpret with caution.

VIII. LIMITATIONS

The vancomycin screen plate does not determine the level of vancomycin resistance or the vancomycin phenotype.

REFERENCE

1. **NCCLS.** 1997. *Methods for Dilution Antimicrobial Susceptibility Tests for Bacteria That Grow Aerobically,* 4th ed. Approved Standard: M7-A4 and Supplement M100-S7. NCCLS, Wayne, Pa.

SUPPLEMENTAL READING

Swenson, J. M., J. A. Hindler, and L. R. Peterson. 1995. Special tests for detecting antibacterial resistance, p. 1356–1367. *In* P. R. Murray, E. J. Baron, M. A. Pfaller, F. C. Tenover, and R. H. Yolken (ed.), *Manual of Clinical Microbiology*, 6th ed. American Society for Microbiology, Washington, D.C.

5.7 Agar Screen Test To Detect Oxacillin (Methicillin)-Resistant *Staphylococcus* spp.

I. PRINCIPLE

A standard number of bacteria is inoculated onto Mueller-Hinton agar (MHA) containing 6 μg of oxacillin per ml and 4% NaCl. After incubation, the appearance of growth indicates that the staphylococcal isolate is resistant to oxacillin and other penicillinase-resistant penicillins (methicillin, nafcillin, cloxacillin, dicloxacillin, and flucloxacillin) (1).

II. SPECIMEN

Prepare inocula from four or five isolated colonies of similar colony morphology grown overnight (18 to 24 h) on nonselective medium (e.g., BAP or CHOC).

III. MATERIALS

Include QC information on reagent container and in QC records.

A. Media and reagents

1. MHA with 4% NaCl and oxacillin (6 μg/ml)
   Store at 2 to 8°C.
2. Nutrient broth (e.g., Mueller-Hinton, TSB) or 0.9% saline (3.0- to 5.0-ml aliquots)
   Store at 2 to 30°C.

B. Supplies

1. Sterile cotton-tipped swabs
2. Sterile plastic pipettes
3. McFarland 0.5 turbidity standard

C. Equipment

1. Calibrated loop (10 μl [0.01 ml]) or 10-μl micropipette with sterile tips (optional)
2. Vortex mixer
3. Ambient-air incubator (34 to 35°C)

IV. QUALITY CONTROL STRAINS

A. *Staphylococcus aureus* ATCC 25923
Oxacillin susceptible

B. *Staphylococcus aureus* ATCC 43300
Oxacillin resistant

V. PROCEDURE

A. Inoculum preparation

Using a loop or swab, transfer colonies to broth or saline to obtain an organism suspension that matches a McFarland 0.5 turbidity standard (1.5×10^8 CFU/ml); vortex thoroughly.

B. Inoculation and incubation

1. Method 1
   a. Prepare a 1 : 100 dilution of the standardized suspension to obtain approximately 1.5×10^6 CFU/ml (e.g., add 0.05 ml to 4.95 ml of sterile saline).

   b. Spot inoculate 0.01 ml (10 μl) (containing 10^4 CFU) onto the surface of the test plate by using a micropipette with a sterile tip.

2. Method 2

   a. Dip a fresh sterile cotton swab into the standardized organism suspension and express any excess fluid against the side of the tube.

   b. Spot inoculate or streak the surface of the plate (by drawing the swab over a 1- to 1.5-in. area).

3. Several organisms can be tested on each plate.

4. Allow the inoculum to be absorbed into the agar.

5. Invert plates and incubate at 35°C in an ambient-air incubator. Examine after overnight incubation and reincubate any isolates that are susceptible (24 h [*S. aureus*] or 48 h [coagulase-negative staphylococci]).

C. Reading plates

Examine the plate very closely for *any* indication of growth.

VI. RESULTS

A. Interpretation

1. No growth, oxacillin susceptible

2. *Any* growth, oxacillin resistant

B. For oxacillin-resistant staphylococci, report all β-lactams as resistant regardless of in vitro results. These β-lactams include all penicillins, cephalosporins, amoxicillin-clavulanic acid, ticarcillin-clavulanic acid, ampicillin-sulbactam, piperacillin-azobactam, and imipenem.

VII. PROCEDURE NOTES

A. Procedure

1. Incubation at temperatures of >35°C may adversely affect detection of oxacillin-resistant staphylococci.

2. Plates must be examined very carefully. Although some resistant isolates may demonstrate confluent growth, isolates expressing a low frequency of resistance may appear as a single colony or fine haze. Detection of growth may be enhanced by using a dissection microscope or hand lens.

B. Other

1. The typical or intrinsically oxacillin (methicillin)-resistant *S. aureus* isolates possess a *mec* gene, have high MICs (≥16.0 μg/ml), and are usually multiply resistant to other agents including clindamycin, erythromycin, and sometimes chloramphenicol, tetracycline, trimethoprim-sulfamethoxazole, and the aminoglycosides.

2. Borderline oxacillin (methicillin)-resistant *S. aureus* isolates lack the *mec* gene and are usually not multiply resistant, and oxacillin MICs for these are often at or just above the susceptible break point (>2.0 μg/ml). Occasionally, these may grow on the oxacillin screen plate. The clinical significance of borderline resistant isolates has not been established. Borderline oxacillin (methicillin)-resistant *S. aureus* cells are infrequently encountered.

3. The oxacillin screen plate method is reliable, easy, and inexpensive. It can be used as a primary method to detect oxacillin-resistant staphylococci or to arbitrate equivocal results obtained with other methods.

VIII. LIMITATIONS

The oxacillin screen plate detects intrinsically oxacillin (methicillin)-resistant staphylococci but does not usually detect those with borderline resistance.

REFERENCE

1. **NCCLS.** 1997. *Methods for Dilution Antimicrobial Susceptibility Tests for Bacteria That Grow Aerobically*, 4th ed. Approved Standard. NCCLS Publication M7-A4 and Supplement M100-S7. NCCLS, Wayne, Pa.

SUPPLEMENTAL READING

Swenson, M. J., J. A. Hindler, and L. R. Peterson. 1995. Special tests for detecting antibacterial resistance, p. 1356–1367. *In* P. R. Murray, E. J. Baron, M. A. Pfaller, F. C. Tenover, and R. H. Yolken (ed.), *Manual of Clinical Microbiology*, 6th ed. American Society for Microbiology, Washington, D.C.

5.8 Etest

I. PRINCIPLE

A standardized inoculum of bacteria is swabbed onto the surface of a Mueller-Hinton agar (MHA) plate. Etest strips containing a continuous gradient of antimicrobial-agent concentrations are placed on the agar surface. After overnight incubation, an elliptical zone of inhibition forms as the antimicrobial agent inhibits growth. The MIC is read where growth intersects the Etest strip. MIC values can be interpreted as susceptible, intermediate, or resistant based on the tables in the NCCLS MIC standard (1).

II. SPECIMEN

Prepare inocula from four or five isolated colonies of similar colony morphology.

A. Log-phase growth inoculum

Use colonies grown for 1 or 2 days on nonselective or selective (e.g., MAC) medium.

B. Direct colony suspension inoculum

Use colonies grown overnight on nonselective medium (e.g., BAP or CHOC).

III. MATERIALS

Include QC information on reagent container and in QC records.

A. Media and reagents

1. Agar plates (ca. 4 mm deep)
   Store at 2 to 8°C.
   a. MHA
   b. MHA with 5% sheep blood
   c. Haemophilus Test Medium (HTM)
   d. GC agar base with defined supplements
2. Nutrient broth (e.g., Mueller-Hinton, TSB) or 0.9% saline (3.0- to 5.0-ml aliquots)
   Store at 2 to 30°C.
3. Etest strips (AB Biodisk North America, Inc., 200 Centennial Ave., Piscataway, NJ 08854)
   Store with desiccant at −20°C based on manufacturer's instructions.

B. Supplies

1. Sterile cotton-tipped swabs
2. Sterile plastic pipettes
3. McFarland 0.5 and 1.0 turbidity standards

C. Equipment

1. Forceps
2. Sterile petri plate
3. Movable light source
4. Vortex mixer
5. Ambient-air incubator (34 to 35°C); CO_2 incubation for some organisms
6. Etest applicator (optional)

IV. QUALITY CONTROL

A. QC strains

1. *Escherichia coli* ATCC 25922
2. *Staphylococcus aureus* ATCC 29213
3. *Pseudomonas aeruginosa* ATCC 27853
4. *Escherichia coli* ATCC 35218
5. *Enterococcus faecalis* ATCC 29212

IV. QUALITY CONTROL *(continued)*

B. Monitoring accuracy

1. Test QC strains by routine procedures, and record results in QC notebook. Record the lot numbers and expiration dates of Etest strips and agar.
2. Compare with expected results (see NCCLS QC tables) (1); note any out-of-control result, and document; proceed with corrective action, if necessary.

C. Frequency of QC testing

1. Perform QC daily or each time a patient isolate is tested.
2. Frequency of QC testing can be reduced from daily to weekly provided that a laboratory can document proficiency in performing this test by performing QC daily (or each day patient tests are performed) until results from 30 consecutive days of testing have been obtained.
3. Proficiency in performing QC tests is confirmed if, for *each* drug-organism combination, no more than 3 of 30 results are outside accuracy limits.
4. Document proficiency each time a new drug is added to testing protocols.
5. Perform QC testing each time a new lot or new shipment of materials (MHA, Etest strips) is put into use.

D. Acceptable results for daily QC testing

QC testing is in control if no greater than 1 out of 20 consecutive results is outside specified accuracy limits.

E. Acceptable results for weekly QC testing

1. QC testing is in control if all MICs are within specified accuracy limits.
2. If *any* weekly result is out of control, return to daily QC to define the problem as follows:
   a. Test the antimicrobial agent(s) with appropriate QC strain(s) for 5 consecutive test days.
   b. For each drug-organism combination, all of the five MICs must be within the accuracy control limits.
   c. If the problem is not resolved (one or more MICs are outside the accuracy control limits), continue daily QC. To return to weekly, document satisfactory performance for another 30 consecutive days.

V. PROCEDURE

See Appendix 5.8–1 for a summary of testing conditions for various organisms.

A. Bring agar plates and containers of Etest strips to room temperature before use.

B. Inoculum preparation

Using a loop or swab, transfer colonies as follows:

1. Standard method (log-phase growth)
   a. Pick four to five isolated colonies to 3.0 to 5.0 ml of broth.
   b. Incubate at 35°C for 2 to 8 h until growth reaches the turbidity at or above that of a McFarland 0.5 standard.
2. Direct colony suspension method

   Pick several colonies from a *fresh* (18 to 24 h) nonselective agar plate to broth or 0.9% saline.
3. For either method, vortex the suspension well and adjust turbidity visually with sterile broth or 0.9% saline to match a 0.5 McFarland standard (1.5×10^8 CFU/ml) which has been vortexed. Alternatively, standardize suspension with a photometric device.

C. Inoculation of agar plate

1. Within 15 min of adjusting turbidity, dip a sterile cotton swab into the inoculum, and rotate against the wall of the tube above the liquid to remove excess inoculum.
2. Swab the entire surface of the agar plate three times, rotating the plate approximately 60° between streakings to ensure even distribution. Avoid hitting the sides of the petri plate to prevent the creation of aerosols.

3. Allow inoculated plate to stand for approximately 10 min. This is critical for optimum performance of the Etest.

D. Application of Etest strips
   1. Make certain that the MIC scale faces upward; do not touch the underside (antimicrobial agent) of the strip.
   2. If using forceps, grab the end of the strip labeled ''E''; use care to take one strip only.
   3. If an applicator is used, follow manufacturer's directions for this apparatus.
   4. Deposit strips with the highest concentration near the end of the petri dish. Gradually allow each strip to come in contact with the agar.
      a. For a 100-mm plate, use only one strip per plate, and position it in the center of the plate.
      b. For a 150-mm plate, place one to six strips on the plate at equal distances apart, radiating from the center of the plate. The ''E'' end of each strip should point to the edge of the petri dish.
   5. Once the strip is in place, remove large air bubbles underneath by using forceps to gently press on the strip, beginning at the lower edge of the bubble and moving up the concentration gradient of the strip toward the E. Small bubbles will not interfere.
   6. *Do not* relocate an Etest strip once it has landed on the agar surface. The antimicrobial agent is immediately released into the agar. If the strip is accidentally placed upside down, carefully pick it up, turn it over, and place it on the agar surface. If the strip touches the counter or another object, it can still be used provided that it does not contact moisture.

E. Incubation
   1. Incubate plates within 1 h of Etest application (except for anaerobes, which should be incubated immediately). See Appendix 5.8–1 for incubation recommendations and special considerations.
   2. Invert plates, and stack them no more than five high.
   3. Incubate for 16 to 18 h at 35°C in an ambient-air incubator.
      a. Incubate all staphylococci that are susceptible to penicillinase-resistant penicillins (oxacillin, methicillin, nafcillin) for an additional 6 to 8 h (total of 24 h).
      b. Incubate all enterococci that are susceptible to vancomycin for an additional 6 to 8 h (total of 24 h).

F. Reading plates
   1. Read plates only if the lawn of growth is confluent or nearly confluent.
   2. Remove the cover of the petri dish, hold the plate to a transmitted-light source, and read the MIC at the point where growth intersects the Etest strip. If using opaque media (e.g., blood MHA), use reflected light; a hand lens may help.
   3. If there is no inhibition of growth, report the MIC as greater than the highest concentration on the Etest strip.
   4. If the zone does not intersect the strip (zone is below the strip), report the MIC as less than the lowest concentration.
   5. For MICs that fall between markings, use the higher value.
   6. When testing hemolytic organisms, measure inhibition of growth and not inhibition of hemolysis.
   7. Refer to the AB Biodisk Technical Guide, which contains photographs of equivocal intersections.

VI. RESULTS

A. Interpretation

Interpret the MIC based on criteria specified by NCCLS (1).

B. Reporting

Report the MIC with categorical interpretation: susceptible (S), intermediate (I), or resistant (R).

C. If the Etest MIC falls between the twofold dilution values, round up to the next highest twofold dilution and then interpret the MIC.

D. Important reporting rules

Report methicillin-, oxacillin-, or nafcillin-resistant staphylococci as resistant to all β-lactam drugs (including β-lactam–β-lactamase inhibitor combinations, all cephalosporins, all penicillins, and imipenem) regardless of results of in vitro susceptibility testing.

VII. PROCEDURE NOTES

A. Oxacillin (methicillin)-resistant staphylococci

1. When oxacillin or methicillin and *Staphylococcus* spp. are tested, it is necessary to incorporate 2% NaCl in the medium.
2. Oxacillin is the recommended penicillinase-resistant penicillin for testing, and oxacillin results can represent all penicillinase-resistant penicillins (oxacillin, methicillin, nafcillin, and also cloxacillin, dicloxacillin, and flucloxacillin).
3. Clues to detection of methicillin (oxacillin)-resistant *S. aureus* (MRSA)
   Multiple resistance to other antimicrobial agents (β-lactams, aminoglycosides, erythromycin, clindamycin, and tetracycline)

B. Media

1. Thymidine can interfere with performance of sulfonamides and trimethoprim. Monitor media by testing trimethoprim-sulfamethoxazole and *E. faecalis* ATCC 29212 (MIC, ≤0.5/9.5 μg/ml indicates acceptability). Blood components (other than horse blood) contain thymidine and should not be used (except sheep blood for *Streptococcus pneumoniae* and *Streptococcus* spp. as described in this procedure).
2. An increase in the cation content (Ca^{2+}, Mg^{2+}) of the medium results in an increase in the MICs of aminoglycosides for *P. aeruginosa* and an increase in the MICs of tetracycline for all organisms. A decrease in cation content has the opposite effect. Monitor cation content with QC strain *P. aeruginosa* ATCC 27853.

C. Strip placement and reading results

1. Application of a strip to a wet surface often results in growth that begins at the zone-strip intersection and continues up the side of the strip. Ignore this when reading the MIC. Excessively wet swabs and incomplete swabbing of plates may result in jagged edges and an uneven intersection at the MIC.
2. The organism and antimicrobial agent tested may have an effect on the appearance of the zone edge where growth intersects the Etest strip. AB Biodisk has extensive reading guidelines (including photographic examples) which must be used.

D. Procedure

The Etest has been evaluated for other organisms in addition to those listed in Appendix 5.8–1. It is the responsibility of the laboratory to ensure there are sufficient data to warrant use of the Etest in patient testing, particularly for organism-antimicrobial agents not yet FDA cleared. Contact AB Biodisk for additional information.

VIII. LIMITATIONS

A. The Etest (an agar-based technique) correlates with the reference agar dilution procedure. However, certain discrepancies between Etest MICs and MICs from non-agar-based systems such as broth microdilution and from other automated systems based on different technical principles may occur as a consequence of the different characteristics inherent in these methods.

B. Numerous factors can affect results including inoculum size, rate of growth, media formulation and pH, incubation environment and length, drug diffusion rate, and measurement of endpoints. Therefore, strict adherence to protocol is required to ensure reliable results.

REFERENCE

1. **NCCLS.** 1997. *Methods for Dilution Antimicrobial Susceptibility Tests for Bacteria That Grow Aerobically*, 4th ed. Approved Standard. NCCLS Publication M7-A4 and Supplement M100-S7. NCCLS, Wayne, Pa.

SUPPLEMENTAL READING

Baker, C. N., S. A. Stocker, D. H. Culver, and C. Thornsberry. 1991. Comparison of the Etest to agar dilution, broth microdilution, and agar diffusion susceptibility testing techniques by using a special challenge set of bacteria. *J. Clin. Microbiol.* **29:**533–538.

NCCLS. 1997. *Performance Standards for Antimicrobial Disk Susceptibility Tests,* 6th ed. Approved Standard: M2-A6 and Supplemental Tables M100-S7. NCCLS, Wayne, Pa.

Woods, G. L., and J. A. Washington. 1995. Antibacterial susceptibility tests: dilution and disk diffusion methods, p. 1327–1341. *In* P. R. Murray, E. J. Baron, M. A. Pfaller, F. C. Tenover, and R. H. Yolken (ed.), *Manual of Clinical Microbiology*, 6th ed. American Society for Microbiology, Washington, D.C.

APPENDIX 5.8–1

Quick Reference List for Performing Etest

The table in this appendix is divided into two sections. The first part of the table lists, among other organisms, rapidly growing facultatively anaerobic organisms for which the FDA has cleared certain Etest strips for testing in the clinical laboratory. The second part of the table lists organisms for which Etest studies have been performed but for which Etest strips have not yet been granted clearance by the FDA. Testing of Etest strip-organism combinations that are not yet FDA cleared should be done for investigational purposes only.

Organism	Agar[a]	Inoculum method[a,b]	Incubation[c]	
			Length (h)	Atmosphere
Testing cleared by FDA				
Enterobacteriaceae	MHA	DCS/LPG	16–18	Ambient air
Enterococcus spp.	MHA	DCS/LPG	16–18 (24 h with vancomycin)	Ambient air
Pseudomonas aeruginosa and *Acinetobacter* spp.	MHA	DCS/LPG	16–18	Ambient air
Other *Pseudomonas* spp.	MHA	DCS/LPG	16–18	Ambient air
Haemophilus spp.[d]	HTM	DCS	18–24	CO_2
Staphylococcus spp.	MHA[e]	DCS	16–18 (24 h with oxacillin)[f]	Ambient air
Streptococcus pneumoniae[g]	BMHA	DCS	18–24	CO_2
Anaerobes	Supplemented brucella blood agar with vitamin K and hemin; Wilkins-Chalgren agar	DCS/LPG	24–72	Anaerobic environment

(continued)

APPENDIX 5.8–1 *(continued)*

Organism	Agar	Inoculum method[b]	Incubation[c]	
			Length (h)	Atmosphere
Testing not cleared by FDA as of 1 July 1997				
Corynebacterium spp.	BMHA	DCS	18–48 (if JK, 48 h)	Ambient air
Eikenella corrodens[h]	CMHA	DCS	18–24	CO_2
Listeria monocytogenes	BMHA	DCS	18–24	Ambient air
Moraxella catarrhalis[i]	MHA	DCS	18–24	Ambient air
Neisseria gonorrhoaeae	GC-CF	DCS	20–24	CO_2
Neisseria meningitidis	MHA or CMHA	DCS	18–24	CO_2
Pasteurella multocida	BMHA	DCS	18–24	Ambient air
Streptococcus spp.	BMHA	DCS	18–24	Ambient air (CO_2 if necessary)

[a] Abbreviations: BMHA, Mueller-Hinton agar with 5% sheep blood; CMHA, chocolate Mueller-Hinton agar; DCS, direct colony suspension standardized to McFarland 0.5 turbidity standard; DCS/LPG, direct colony suspension or log-phase growth standardized to McFarland 0.5 turbidity standard; GC-CF, GC agar base, cysteine free; HTM, *Haemophilus* test medium.

[b] Etest defers to NCCLS recommendations (16–18), which are listed here.

[c] All incubation temperatures are 34 to 35°C.

[d] Use a photometric device to adjust turbidity. This step is critical in avoiding erroneous results due to underinoculation or overinoculation.

[e] When testing oxacillin (or other penicillin-resistant penicillins), incorporate 2% NaCl into the medium.

[f] And other penicillinase-resistant penicillins.

[g] Pneumococci tend to lyse rapidly in liquid (particularly water). Therefore, follow the inoculum preparation procedure (in broth) described in the text, and inoculate the agar plate immediately after standardization of the inoculum preparation.

[h] For *Eikenella corrodens*, use a 1.0 McFarland standard for inoculum preparation (manufacturer's recommendation).

[i] Generally, only a β-lactamase test is warranted.

5.9 Selecting Antimicrobial Agents for Testing and Reporting

I. INTRODUCTION

Developing protocols for antimicrobial agent susceptibility testing and reporting is best done with input from the infectious diseases service, infection control, and the pharmacy and therapeutics committee. The goals are to provide clinically relevant information that will support cost-effective utilization of antimicrobial agents and to avoid reporting results that may adversely affect patient care. Because of differences in hospital formularies and laboratory functions, it is impossible to make specific recommendations, and the suggestions described herein can serve only as a guide to the decision-making processes. A listing of commonly used antibacterial agents is shown in Appendix 5.9–1.

II. BASIC STEPS IN DEVELOPING A TESTING AND REPORTING PROTOCOL

A. Assess resources available (test system[s]), numbers of drugs practically tested in each respective system, laboratory reporting capabilities [e.g., computer reporting], manpower).

B. Obtain input from the infectious diseases service, pharmacy and therapeutics committee, infection control, and others.

C. Determine which drugs will be included in routine testing "batteries." Define for example batteries for the following:

1. Gram-negative bacilli (e.g., *Enterobacteriaceae*)
2. Gram-positive cocci (e.g., *Staphylococcus* spp.)
3. *Pseudomonas* spp.
4. *Haemophilus* spp.
5. *Streptococcus pneumoniae* and *Streptococcus* spp.
6. Urine isolates
7. Very resistant isolates (include a broad-spectrum battery and/or strategy for testing agents not routinely tested; this may involve use of a reference laboratory)

D. Determine which drugs will be routinely reported for specific organism groups; determine whether selective reporting will be implemented.

E. If the clinician specifically requests that an agent be reported that is not consistent with the laboratory's routine reporting protocol, consider including a note on the laboratory report such as "Reported as requested by physician."

III. PRIMARY RESOURCES

A. NCCLS tables, "Suggested Groupings of U.S. FDA-Approved Antimicrobial Agents That Should be Considered for Routine Testing and Reporting by Clinical Microbiology Laboratories" (2, 3)

B. *The Medical Letter* provides recommendations for the drugs of first choice and alternative agents for treating infections caused by specific bacteria (1).

IV. SELECTIVE DRUG REPORTING (CASCADE REPORTING) OPTION

A. Test first-line and second-line drugs routinely.

B. Base reporting on organism identification, overall antibiogram, and site of infection.

C. Report second-line agents only if first-line agents appear inappropriate (inactive, inappropriate for specific infection or infecting species, etc.).

D. Make all results available for unusual clinical situations.

E. Examples of selective reporting based on organism identification and overall antibiogram

1. *Enterobacteriaceae*

a. Report expanded-spectrum cephalosporins only if an isolate is resistant to narrow-spectrum agents.

b. Report tobramycin or amikacin or both only if an isolate is gentamicin resistant.

c. Report fluoroquinolone only if an isolate is resistant to all other oral agents tested.

2. *Pseudomonas aeruginosa*

Report imipenem only if an isolate is amikacin and ceftazidime resistant.

3. Staphylococci

Report vancomycin only if an isolate is methicillin resistant.

F. Examples of selective reporting by body site

1. Report results of drugs (e.g., cefotaxime, ceftriaxone) that cross the blood-brain barrier on CSF isolates, if appropriate for the organism group.

2. Report results of agents (e.g., nitrofurantoin, norfloxacin) that are used only for treating urinary tract infections on urine isolates only.

3. Some antimicrobial agents are inappropriate for treatment of infections caused by certain bacteria, even if the in vitro results indicate that an isolate is susceptible. These include:

a. Narrow- or expanded-spectrum cephalosporins and aminoglycosides for *Salmonella* spp. and *Shigella* spp.

b. β-Lactams (penicillins, cephalosporins, β-lactam–β-lactamase inhibitor combinations, imipenem) for methicillin-resistant staphylococci

c. Cephalosporins, aminoglycosides (except high-level screens), clindamycin, and trimethoprim-sulfamethoxazole for enterococci

d. Cephalosporins for *Listeria* spp.

G. Selective reporting may not be practiced by institutions where there are other mechanisms in place to prevent physicians from prescribing antimicrobial agents inappropriately.

REFERENCES

1. **Abramowicz, M. (ed.).** 1996. The choice of antibacterial drugs. *The Medical Letter* **38:** 25–34.

2. **NCCLS.** 1997. *Performance Standards for Antimicrobial Disk Susceptibility Tests*, 6th ed. Approved Standard: M2-A6 and Supplemental Tables M100-S7. NCCLS, Wayne, Pa.

3. **NCCLS.** *Methods for Dilution Antimicrobial Susceptibility Tests for Bacteria That Grow Aerobically*, 4th ed. Approved Standard. NCCLS Publication M7-A4 and Supplemental Tables M100-S7. NCCLS, Wayne, Pa.

APPENDIX 5.9–1

Reference chart for commonly used antibacterial agents

Drug family/generic name (route[s] of administration)[a]	Product name(s)[b]	Primary spectrum of activity: Gram positive	Primary spectrum of activity: Gram negative	Mode of action	Clinical use; comments
Penicillins					
Natural penicillins					
Penicillin G (i.m., i.v.)	Various product names	X		Inhibition of cell wall synthesis	Community-acquired meningitis in normal hosts, caused by pneumococci, meningococci, or *Listeria* spp.; serious streptococcal infection (e.g., endocarditis, bacteremia, and osteomyelitis); serious enterococcal infections (given in combination with gentamicin); anaerobic infections other than those caused by *Bacteroides fragilis*
Penicillin V (p.o.)	Various product names	X		Inhibition of cell wall synthesis	Streptococcal pharyngitis/otitis; skin and soft tissue infections
Penicillinase-resistant penicillins		X		Inhibition of cell wall synthesis	Staphylococcal infections caused by strains susceptible to penicillinase-resistant penicillins; will cover most streptococci (but not enterococci); oral agents (cloxacillin, dicloxacillin) used for less serious infections.
Methicillin (i.m., i.v.)	Staphcillin, Celbenin				
Nafcillin (p.o., i.m., i.v.)	Nafcil, Unipen				
Isoxazolyl penicillins					
Oxacillin (p.o., i.m., i.v.)	Prostaphlin, Bactocill				
Cloxacillin (p.o.)	Tegopen				
Dicloxacillin (p.o.)	Dynapen, Pathocil, Dycill, Veracillin				
Extended-spectrum penicillins					
Aminopenicillins		X	X	Inhibition of cell wall synthesis	Same as penicillin G plus infections caused by *E. coli, Salmonella* spp., *Haemophilus influenzae* (β-lactamase negative), and *Proteus mirabilis*; amoxicillin is used for less serious infections.
Ampicillin (p.o., i.m., i.v.)	Various product names				
Amoxicillin (p.o.)	Various product names				
Carboxypenicillins			X	Inhibition of cell wall synthesis	Used in combination with an aminoglycoside for serious infections caused by susceptible *Enterobacteriaceae* (*Klebsiella* are resistant) and *Pseudomonas aeruginosa*; oral carbenicillin is used exclusively for urinary tract infections.
Carbenicillin (p.o., i.m., i.v.)	Geopen, Geocillin, Pyopen				
Ticarcillin (i.m., i.v.)	Ticar				
Ureidopenicillins			X	Inhibition of cell wall synthesis	Same as carboxypenicillins, with expanded use because of enhanced activity against *Enterobacteriaceae* (including *Klebsiella* spp.), *P. aeruginosa*, and *Bacteroides* spp.
Azlocillin (i.v.)	Azlin				
Mezlocillin (i.m., i.v.)	Mezlin				
Piperacillin (i.m., i.v.)	Pipracil				
Co-drugs (β-lactam plus β-lactamase inhibitor)		X	X	Inhibition of cell wall synthesis	Moderate to severe mixed infections due to aerobic and anaerobic bacteria (e.g. intraabdominal and pelvic infections, aspiration pneumonia, etc.); amoxicillin/clavulanic acid is used for upper and lower respiratory tract infections due to *H. influenzae, M. catarrhalis, Staphylococcus aureus*, and anaerobes; the β-lactamase-inhibiting component facilitates activity against β-lactamase-producing *Staphylococcus* spp. (methicillin-susceptible), *H. influenzae, M. catarrhalis*, and *Bacteroides* spp.
Amoxicillin/clavulanic acid (p.o.)	Augmentin				
Ticarcillin/clavulanic acid (i.v.)	Timentin				
Ampicillin/sulbactam (i.v.)	Unasyn				
Piperacillin/tazobactam (i.v.)	Zosyn				
Cephems (including cephalosporins)[c]					
Narrow spectrum		X	X	Inhibition of cell wall synthesis	Surgical prophylaxis; community-acquired pulmonary, skin, soft-tissue, and urinary tract infections; staphylococcal osteomyelitis and soft-tissue infection. Narrow-spectrum cephems are active against many gram-positive cocci; active against some *Enterobacteriaceae* including many strains of *E. coli, P. mirabilis, Klebsiella* spp., and most anaerobes (other than *B. fragilis*).
Cephalothin (i.m., i.v.)	Keflin				
Cefazolin (i.m., i.v.)	Ancef, Kefzol				
Cephapirin (i.m., i.v.)	Cefadyl				
Cephalexin (p.o.)	Keflex				
Cefadroxil (p.o.)	Duricef, Ultracef				
Cephradine (p.o., i.m., i.v.)	Velosef				

(continued)

APPENDIX 5.9–1 *(continued)*

Drug family/generic name (route[s] of administration)[a]	Product name(s)[b]	Primary spectrum of activity		Mode of action	Clinical use; comments
		Gram positive	Gram negative		
Extended spectrum		X	X	Inhibition of cell wall synthesis	The cephamycins are used for mild to moderate mixed infections (e.g., intraabdominal and pelvic infections, aspiration pneumonia, etc.) and surgical prophylaxis in gastrointestinal, obstetric, and gynecological procedures; others for community-acquired pulmonary, skin, and soft-tissue infections, and exacerbations of chronic obstructive pulmonary disease; cefaclor, loracarbef, cefuroxime, and cefprozil are often active against *Haemophilus* spp. and *Moraxella catarrhalis* and are used for respiratory tract infections and otitis media; extended-spectrum cephems are slightly more active than narrow-spectrum agents against *Enterobacteriaceae* (increased β-lactamase stability) and less active against staphylococci. The cephamycins have good activity against *B. fragilis*.
Cefamandole (i.m., i.v.)	Mandol				
Cefuroxime (p.o., i.m., i.v.)	Zinacef, Ceftin				
Cefonicid (i.m., i.v.)	Monocid				
Ceforanide (i.m., i.v.)	Precef				
Cefaclor (p.o.)	Ceclor				
Loracarbef (p.o.)	Lorabid				
Cefprozil (p.o.)	Cefzil				
Cephamycins					
Cefotetan (i.v.)	Cefotan				
Cefoxitin (i.m., i.v.)	Mefoxin				
Cefmetazole (i.v.)	Zefazone				
Broad spectrum		X	X	Inhibition of cell wall synthesis	Used in combination with an aminoglycoside for serious infections caused by susceptible *Enterobacteriaceae* and *Pseudomonas aeruginosa*; childhood meningitis due to *H. influenzae*; gram-negative bacillary meningitis; ceftriaxone is used for gonorrhea and Lyme disease; broad-spectrum cephems have greatly increased activity against gram-negative bacilli because of increased β-lactamase stability compared to other cephems, most have activity against *P. aeruginosa* (particularly cefepime, ceftazidime, and cefoperazone) and CSF penetration, however they are not as active against staphylococci. Cefepime and cefpirome have enhanced activity against some strains that are resistant to other broad-spectrum cephems. For cefepime, activity against gram positives is similar to cefotaxime and that against gram negatives is similar to ceftazidime to include good antipseudomonal activity.
Cefdinir (p.o.)					
Cefepime (i.v.)	Maxipeme				
Cefixime (p.o.)	Suprax				
Cefoperazone (i.m., i.v.)	Cefobid				
Cefotaxime (i.m., i.v.)	Claforan				
Cefpirome (i.v.)					
Cefpodoxime (p.o.)	Vantin				
Ceftizoxime (i.m., i.v.)	Cefizox				
Ceftriaxone (i.m., i.v.)	Rocephin				
Ceftazidime (i.m., i.v.)	Fortaz, Tazicef, Tazidime				
Ceftibuten (p.o.)	Cedar				
Carbapenems					
Imipenem (i.v.)	Primaxin	X	X	Inhibition of cell wall synthesis	Severe mixed infections (e.g., intra-abdominal and pelvic infections, and aspiration pneumonia); infections caused by multiply resistant gram-negative bacilli including *P. aeruginosa*; highly active against most gram-positive organisms (but not against methicillin-resistant staphylococci, *Enterococcus faecium*, and diphtheroids); and the most active β-lactams against *Enterobacteriaceae* and *P. aeruginosa*; like the cephems, carbapenems have poor activity against *Stenotrophomonas maltophilia* and *Burkholderia cepacia*; active against *B. fragilis*.
Meropenem (i.v.)	Merrem				
Monobactam					
Aztreonam (i.v.)	Azactam		X	Inhibition of cell wall synthesis	Infections caused by aerobic gram-negative bacilli in patients allergic to other β-lactams; aztreonam has activity comparable to broad-spectrum cephems against *P. aeruginosa* (except that ceftazidime, cefoperazone, and cefepime are more active) and *Enterobacteriaceae*; inactive against gram-positive organisms and anaerobes.

APPENDIX 5.9–1 *(continued)*

Drug family/generic name (route[s] of administration)[a]	Product name(s)[b]	Primary spectrum of activity		Mode of action	Clinical use; comments
		Gram positive	Gram negative		
Aminoglycosides		X	X	Interference with protein synthesis at the 30S ribosomal subunit	
Gentamicin (i.m., i.v.)	Various product names				Moderate to severe infections caused by aerobic gram-negative bacilli, often used in combination with a β-lactam agent; used in combination with ampicillin, penicillin, or sometimes vancomycin for synergistic killing of enterococci, streptococci, and staphylococci causing endocarditis
Tobramycin (i.m., i.v.)	Nebcin				Gram-negative infections, similar indications as for gentamicin; compared with gentamicin, slightly increased activity against *P. aeruginosa* and sometimes decreased activity against *Serratia marcescens*
Amikacin (i.m., i.v.)	Amikin				Gram-negative infections, similar indications as for gentamicin; increased activity against gentamicin- and tobramycin-resistant isolates
Netilmicin (i.m., i.v.)	Netromycin				Gram-negative infections, similar indications as for gentamicin; slightly less active than other aminoglycosides against *P. aeruginosa* and *Serratia marcescens*
Neomycin (p.o., topical)	Mycifradin, Neobiotic				Oral and topical use only for gastrointestinal tract decontamination and superficial infections, respectively
Kanamycin (p.o., i.m., i.v.)	Kantrex, Klebcil				Oral and topical use, similar to neomycin; parenteral forms infrequently used
Streptomycin (i.m., i.v.)	Streptomycin				Used in combination with ampicillin, penicillin, or sometimes vancomycin for synergistic killing of enterococci or streptococci causing endocarditis
Spectinomycin (i.m.)	Trobicin	*Neisseria gonorrhoeae*			Gonorrhea only; however, largely replaced by ceftriaxone or a quinolone
Tetracyclines		X	X	Interference with protein synthesis at the 30S ribosomal subunit	Not often used as directed therapy for common bacterial infections; however, used for brucellosis, chlamydial, mycoplasmal, and rickettsial infections
Tetracycline (p.o., i.m., i.v.)	Various product names				
Doxycycline (p.o., i.v.)	Various product names				
Minocycline (p.o.)	Minocin				
Sulfonamides (and trimethoprim)			X	Competitive inhibition of folic acid synthesis	Uncomplicated lower urinary tract infections and nocardiosis
Sulfamethoxazole (p.o.)	Various product names				
Sulfisoxazole (p.o.)	Various product names				
Trimethoprim-sulfamethoxazole (p.o., i.v.)	Bactrim, Septra				Acute and chronic urinary tract infections; bacterial diarrhea; infections caused by *Enterobacteriaceae*; acute and chronic upper respiratory tract infections; also used to treat *Pneumocystis carinii* infections
Trimethoprim (p.o.)	Various product names				Uncomplicated lower urinary tract infections and prophylaxis for recurrent urinary tract infections
Macrolides		X		Binds to 50S ribosomal subunit, blocking the initiation of peptide chains	Upper and lower respiratory tract infections caused by gram-positive organisms (pneumococci or group A streptococci) in penicillin-allergic patients, pertussis, *Campylobacter, Mycoplasma pneumoniae, Chlamydia trachomatis, Legionella* spp. and *Haemophilus influenzae* (azithromycin and clarithromycin); less serious staphylococcal infections; clarithromycin used in combination with other agents for mycobacterial infections
Azithromycin (p.o.)	Zithromax				
Clarithromycin (p.o.)	Biaxin				
Erythromycin (p.o., i.v.)	Various product names				
Dirithromycin (p.o.)[d]	Dynabac				

(continued)

APPENDIX 5.9–1 *(continued)*

Drug family/generic name (route[s] of administration)[a]	Product name(s)[b]	Primary spectrum of activity		Mode of action	Clinical use; comments
		Gram positive	Gram negative		
Glycopeptides Vancomycin (p.o., i.v.) Teicoplanin (i.m., i.v.)[d]	 Vancocin Targocid	X		Inhibition of cell wall synthesis	Methicillin-resistant staphylococcal and for gram-positive infections in penicillin-allergic patients; oral form for *C. difficile* toxin-associated diarrhea and colitis; inactive against *Lactobacillus* spp., *Pediococcus* spp., *Leuconostoc* spp., and *Erysipelothrix* spp.
Chloramphenicol (p.o., i.v.)	Various product names	X	X	Prevents mRNA from attaching to ribosomes	Childhood meningitis (e.g., that caused by *S. pneumoniae, H. influenzae, N. meningitidis*); typhoid fever; anaerobic infections (especially brain abscesses); salmonellosis
Clindamycin (p.o., i.m., i.v.)	Cleocin	X		Binds to 50S ribosomal subunit, blocking the initiation of peptide chains	Anaerobic infections (including those caused by *Bacteroides fragilis*); gram-positive bone and joint infections in children
Nitrofurantoin (p.o.)	Various product names	X	X	Inhibition of a variety of bacterial enzyme systems	Acute and chronic lower urinary tract infections; inactive against *Serratia marcescens, Proteus* spp., and *P. aeruginosa*
Polymyxins Polymyxin B (topical, p.o., i.m., i.v.) Colistin (topical, p.o., i.v.)	 Statrol, Aerosporin Coly-Mycin S		X	Disruption of cell membrane	Primarily topical use for infected wounds; rarely used for infections caused by multiply resistant *P. aeruginosa*
Quinolones (narrow spectrum) Nalidixic acid (p.o.) Cinoxacin (p.o.) Norfloxacin (p.o.)	 NegGram Cinobac Noroxin	X	X	Inhibition of DNA gyrase activity	Lower urinary tract infection only; norfloxacin for bacterial diarrhea and gastrointestinal tract decontamination in granulocytopenic patients; inactive against *P. aeruginosa* (norfloxacin has some activity) and limited activity against gram-positive bacteria
Quinolones (broader spectrum) Ciprofloxacin (p.o., i.v.) Enoxacin (p.o.) Fleroxacin (p.o., i.v.)[d] Grepafloxacin (p.o.)[d] Levofloxacin (p.o., i.v.) Lomefloxacin (p.o.) Ofloxacin (p.o., i.v.) Pefloxacin (p.o., i.v.)[d] Sparfloxacin (p.o.) Trovafloxacin (p.o., i.v.)[d]	 Cipro Penetrex Megalone Levaquin Maxaquin Floxin Peflacine Zagam	X	X	Inhibition of DNA gyrase activity	Upper and lower urinary tract infections, gonorrhea, chlamydial infections, bacterial diarrhea (including traveler's diarrhea), moderate to severe infections caused by gram-negative bacilli (especially pulmonary and bone and joint infections); exacerbations of chronic bronchitis; respiratory infections caused by *Legionella* spp. or *Mycoplasma pneumoniae*; some (e.g., levofloxacin, trovafloxacin, sparfloxacin) have enhanced activity against gram positives to include *S. pneumoniae*; ciprofloxacin is the most active against *P. aeruginosa*; trovafloxacin most active against anaerobes.
Metronidazole (p.o., i.v.)	Flagyl, Metryl, Protostat, Satric				Anaerobic infection; vaginitis caused by *Gardnerella vaginalis* and *Trichomonas vaginalis; C. difficile* toxin-associated diarrhea and colitis

APPENDIX 5.9–1 *(continued)*

Drug family/generic name (route[s] of administration)[a]	Product name(s)[b]	Primary spectrum of activity		Mode of action	Clinical use; comments
		Gram positive	Gram negative		
Rifamycins		X		Inhibition of DNA transcription	Used in combination with antistaphylococcal agents to enhance antibacterial activity, especially in osteomyelitis and endocarditis; first-line agent for tuberculosis in combination with isoniazid; prophylaxis for meningococcal disease; rifabutin for mycobacterial infections only
Rifampin (p.o., i.v.)	Rifadin, Rimactane				
Rifabuten (p.o., i.v.)	Ansamycin				

[a] Abbreviations: i.m., intramuscularly; p.o., orally; i.v., intravenously.
[b] Product names are from *Facts and Comparisons* or the manufacturer.
[c] Cephems have no activity against methicillin-resistant staphylococci, enterococci, and *Listeria* spp.
[d] Investigational agent; not U.S. Food and Drug Administration approved as of May 1997.

5.10 Antibiograms as a Supplemental Quality Control Measure for Antimicrobial Susceptibility Tests

I. PRINCIPLE

QC is performed to ensure optimal performance of antimicrobial susceptibility tests and to provide accurate, reproducible, and timely results to clinicians. The primary QC procedure involves testing reference strains that have defined susceptibility characteristics to the antimicrobial agents tested. Although this method controls many parameters of the antimicrobial susceptibility test, testing reference strains alone does not always ensure reliable results when testing patient isolates. Evaluation of antibiograms, inclusion of supplemental control strains, verification of technologist proficiency, and review of cumulative susceptibility statistics are some of the additional measures that can be taken to further control the quality of antimicrobial susceptibility tests.

II. ANTIBIOGRAMS

A. An antibiogram is the overall antimicrobial susceptibility profile of a bacterial isolate to a battery of antimicrobial agents. Some bacteria have typical profiles which can be used to

1. Verify antimicrobial results
2. Verify organism identification

B. In developing an antibiogram checking program to aid technologists in consistently identifying atypical (and potentially erroneous) antibiograms, provide the following:

1. Descriptions of typical antibiograms for given species (e.g., the typical *S. aureus* isolate is resistant to penicillin and susceptible to clindamycin, erythromycin, oxacillin, and vancomycin); general antibiograms for commonly encountered bacteria are shown in Appendixes 1A through 1D.
2. Descriptions of the relatedness of drugs tested (e.g., activity hierarchy). For example, the activity hierarchy of the cephalosporins against the *Enterobacteriaceae* is broad spectrum > extended spectrum > narrow spectrum.
3. Information on exceptions to ''hierarchy'' rules:
   a. Occasional gram-negative bacilli may possess extended-spectrum β-lactamases (ESBLs) that selectively inactivate cephalosporins.
      (1) **Example:** occasional *E. coli* isolates may appear susceptible in vitro to cefazolin (narrow-spectrum cephalosporin) but resistant to ceftazidime (broad-spectrum cephalosporin).
      (2) When this occurs, report all cephalosporin results so clinicians do not extrapolate that the isolate is susceptible to all cephalosporins because it is susceptible to cefazolin.
   b. Occasional gram-negative bacilli (e.g., *Acinetobacter baumanii*) may possess aminoglycoside-inactivating enzymes that inactivate gentamicin, tobramycin, and amikacin, but not netilmicin. When this occurs, report all aminoglycoside results.

4. Informal updates to advise staff of the prevalence of a particular ''atypical'' antibiogram at a given time (e.g., increased incidence of nosocomial infections caused by gentamicin-resistant *Providencia rettgeri*)
5. Descriptions of the types of antimicrobial resistance (or susceptibility) that when reported (or missed) are likely to impact on patient care

   Examples: *P. aeruginosa* resistant to amikacin, gentamicin, and tobramycin
   Oxacillin-resistant *S. aureus*
   Penicillin-resistant *S. pneumoniae* (particularly from CSF)

C. To verify atypical antibiograms, proceed as follows:

1. Check for transcription errors.
2. Re-examine the disk diffusion plate, MIC tray, purity plate, etc., to make certain the test was not misread and appears satisfactory. Subtle problems may not always be detected upon initial examination of the test.
3. Gross examination of test panels (including panels used with automated systems) may reveal blocked or empty wells, etc.
4. Check previous reports on the patient to determine whether an atypical antibiogram represents a repeat occurrence.
5. Repeat susceptibility tests, identification tests, or both (in some situations it may be helpful to use an alternative test method to verify unusual results). Obtain the assistance of a referral laboratory that specializes in antimicrobial testing to verify virtually ''unheard of'' atypical results (e.g., vancomycin-resistant *S. aureus*, cefotaxime-resistant *H. influenzae*). Some situations may warrant assistance from the local Public Health Department.

D. Examples of antibiograms that warrant verification are listed in Appendix 5.10–2.

SUPPLEMENTAL READING

Hindler, J. A. 1990. Nontraditional approaches to quality control of antimicrobial susceptibility tests. *Clin. Microbiol. Newsl.* **12:**65–69.

NCCLS. 1997. *Methods for Dilution Antimicrobial Susceptibility Tests for Bacteria That Grow Aerobically*, 4th ed. Approved Standard. NCCLS Publication M7-A4 and Supplement M100-S7. NCCLS, Wayne, Pa.

NCCLS. 1997. *Performance Standards for Antimicrobial Disk Susceptibility Tests*, 6th ed. Approved Standard: M2-A6 and Supplemental Tables M100-S7. NCCLS, Wayne, Pa.

von Graevenitz, A. 1991. Use of antimicrobial agents as tools in epidemiology, identification, and selection of microorganisms, p. 723–738. *In* V. Lorian (ed.), *Antibiotics in Laboratory Medicine*, 3rd ed. Williams & Wilkins, Baltimore.

APPENDIX 5.10–1a

Enterobacteriaceae antibiograms[a]

Organism	AMP	AMP/ SULB	CARB GRP	MEZLO GRP	TICAR/ CLAV	PIP/ TAZO	1ST CEFS	CFMDL GRP	CEFOX	CEFOT	CFTAX GRP	CARBAP GRP	AMINO GRP	CHLOR	TETRA	TMP/ SMZ	QUIN GRP	NALID GRP	NITRO-FURANT	POLY B
Aeromonas hydrophila[b]	R*	R	R*	S-R	S-R	S	R	S-R	S-R	S-R	S	S	S	S	S	S	S	S	S	S
Citrobacter koserii	R*	S-R	R	S	S	S	S	S	S	S	S	S	S	S	S	S	S	S	S	S
Citrobacter freundii	R*	S-R	S	S	S	S	R*	S-R	R	S-R	S	S	S	S	S	S	S	S	S	S
Enterobacter aerogenes *Enterobacter cloacae*	R*	S-R	S	S	S	S	R*	S-R	R	S-R	S	S	S	S	S	S	S	S	S-R	S
Pantoea agglomerans	S-R	S-R	S	S	S	S	S-R	S-R	S-R	S-R	S	S	S	S	S	S	S	S	S-R	S
E. coli 1	S	S	S	S	S	S	S	S	S	S	S	S	S	S	S	S	S	S	S	S
E. coli 2[c]	R	S-R	S-R	S-R	S-R	S	S-R	S	S	S	S	S	S	S-R	S-R	S	S	S	S	S
Klebsiella spp.[c]	R*	S-R	R*	S-R	S-R	S	S	S	S	S	S	S	S	S	S	S	S	S-R	S-R	S
Morganella morganii	R*	S-R	S	S	S	S	R	S-R	S-R	S	S	S	S	S	S-R	S	S	S	R*	R*
Proteus mirabilis	S-R	S-R	S	S	S	S	S	S	S	S	S	S	S	S	R*	S	S	S	R*	R*
Proteus vulgaris	R*	S-R	S	S	S	S	R	R	S-R	S	S	S	S	S	R*	S	S	S	R*	R*
Providencia spp.	R*	S-R	S	S	S	S	R*	S-R	S-R	S	S	S	S	S	R*	S	S	R	R*	R*
Salmonella spp.	S	S	S	S	S	S	S	S	S	S	S	S	S	S	S	S	S	S	S	S
Serratia spp.	R*	R	S	S	S	S	R	R	S-R	S-R	S	S	S	S	R	S	S	S-R	R*	R
Shigella spp.	S	S	S	S	S	S	S	S	S	S	S	S	S	S	S	S	S	S	S	S

[a] These antibiograms should serve as guidelines only; exceptions will occur. Abbreviations: AMP, ampicillin; AMP/SULB, ampicillin-sulbactam; CHLOR, chloramphenicol; TETRA, tetracycline; TMP/SMZ, trimethoprim-sulfamethoxazole; NITROFURANT, nitrofurantoin; POLY B, polymyxin B; 1ST CEFS, cefazolin, cephalothin; AMINO GRP, amikacin, gentamicin, netilmicin, tobramycin; CARB GRP, carbenicillin, ticarcillin; CARBAP GRP, imipenem, meropenem; CEFOT, cefotetan; CEFOX, cefoxitin; CFMDL GRP, cefamandole, cefonicid, cefuroxime; CFTAX GRP, aztreonam, cefepime, cefoperazone, cefotaxime, ceftazidime, ceftizoxime, ceftriaxone; MEZLO GRP, mezlocillin, piperacillin; NALID GRP, nalidixic acid, cinoxacin; PIP/TAZO, piperacillin-tazobactam; QUIN GRP, ciprofloxacin, ofloxacin, norfloxacin; TICAR/CLAV, ticarcillin-clavulanic acid. *, It is very unusual to encounter an isolate susceptible to this drug.

[b] Member of *Vibrionaceae*.

[c] Can produce extended-spectrum β-lactamases that confer resistance to all penicillins, cephems (including cephalosporins), and aztreonam.

APPENDIX 5.10–1b

Pseudomonas-Acinetobacter-Burkholderia-Stenotrophomonas antibiograms[a]

Organism	AMP	AMP/ SULB	CARB GRP	MEZLO GRP	TICAR/ CLAV	PIP/ TAZO	1ST, 2ND CEFS	CFTAX GRP	CFTAZ GRP	AZTRE	CARBAP GRP	AMINO GRP	CHLOR	TETRA	TMP/ SMZ	QUIN GRP	NALID GRP	NRFLX	NITRO-FURANT	POLY B
Acinetobacter baumanii	R	S-R	S	S	S	S	R	S-R	S	R	S	S	R	S-R	S	S	S-R	S-R	R	S
Acinetobacter lwoffi	R	S-R	S	S	S	S	R	S-R	S	R	S	S	S-R	S-R	S	S	S-R	S-R	R	S
Pseudomonas aeruginosa	R*	R*	S-R	S-R	S-R	S-R	R*	R	S	S-R	S	S	R	R	R	S-R	R	S	R	S
Burkholderia cepacia	R*	R	R	S-R	S-R	S-R	R*	S-R	S	S-R	R	R	S	R	S	R	R	R	R	R
Pseudomonas fluor-putida	R*	R*	R	S-R	S-R	S-R	R*	R	S	S-R	S	S	R	S-R	R	S-R	R	S-R	R	S
Stenotropho-monas maltophilia	R*	R*	R	S-R	S-R	S-R	R*	R	S-R	R	R	R	S	R	S	R	R	R	R	S
Pseudomonas stutzeri	S-R	S-R	S-R	S-R	S-R	S-R	R	S-R	S	S-R	S	S	S-R	S	S	S-R	R	S-R	R	S

[a] These antibiograms should serve as guidelines only; exceptions will occur. Abbreviations: 1st, 2nd CEFS, narrow- and extended-spectrum cephalosporins; AZTRE, aztreonam; NRFLX, norfloxacin. For other abbreviations, see footnote *a* of Table 5.10–1a. *, It is very unusual to encounter an isolate susceptible to this drug.

APPENDIX 5.10–1c

Gram-positive bacteria antibiograms[a]

Organism	AMP	AMP/ SULB	PEN	OXA	CEFS	CHLOR	CLIND	ERY	GENT	TETRA	TMP/ SMZ	VAN
Staphylococcus aureus 1	S	S	S	S	S	S	S	S	S	S	S	S
Staphylococcus aureus 2	R	S	R	S	S	S	S	S	S	S-R	S	S
Staphylococcus aureus 3	R	S	R	S	S	S	R	R	S	S-R	S	S
Staphylococcus aureus 4 (intrinsic OX-R)	R	R	R	R	(R)	S-R	R	R	S-R	S-R	S-R	S
Staphylococcus aureus 5 (acquired OX-R)	R	S-R	R	R	(R)	S	S	S	S	S	S	S
Coagulase-negative *Staphylococcus*	Variable results											S
Enterococcus faecalis	S	S	S	R*	R*	S	R	S-R	R*	S-R	(R)	S
Enterococcus faecium	R	R	R	R*	R*	S	R	S-R	R*	S-R	(R)	S-R
Group D non-enterococcus	S	S	S	S	S	S	S	S	R*	S	—	S
β-Hemolytic *Streptococcus* spp.	S	S	S	S	S	S	S	S	R*	S	S	S
Viridans group *Streptococcus*	S-R	S-R	S-R	S-R	S-R	S	S-R	S-R	R*	S	S	S
Streptococcus pneumoniae 1	S	S	S	S	S	S	S	S	R*	S	S	S
Streptococcus pneumoniae 2	S-R	S-R	S-R	S-R	S-R	S-R	S-R	S-R	R*	S-R	S-R	S
Corynebacterium spp.	Variable results except for *Corynebacterium* JK and D2[b]											S
Listeria monocytogenes	S	S	S	S	(R)	S	S	S	S	S	S	S

[a] These antibiograms should serve as guidelines only; exceptions will occur. Abbreviations: PEN, penicillin; OXA, oxacillin; CEFS, cephems (including cephalosporins); CLIND, clindamycin; ERY, erythromycin; GENT, gentamicin; VAN, vancomycin. For other abbreviations, see footnote *a* of Table 5.10–1a. Symbols: (R), Result often susceptible in vitro, but drug clinically ineffective against noted species; it is very unusual to encounter an isolate susceptible to this drug.

[b] JK and D2 are usually resistant to these (some ERY-S and/or TET-S).

APPENDIX 5.10–1d

Miscellaneous gram-negative bacteria antibiograms[a]

Organism	β-LAC	AMP	AMOX/CLAV	CEFS	ERY	PEN	QUINS	TETRA	TMP/SMZ
Moraxella catarrhalis 1[b]	+	R	S	S	S	R	S	S	S
Moraxella catarrhalis 2[b]	–	S	S	S	S	S	S	S	S

Organism	β-LAC	AMP	AMOX/CLAV	CEFS	CHLOR	QUINS	TMP/SMZ
Haemophilus influenzae 1	+	R	S	S	S	S	S
Haemophilus influenzae 2	–	S	S	S	S	S	S

Organism	AMOX/CLAV AMP/SULB	1ST CEFS	CHLOR	ERY	PEN	TETRA
Pasteurella multocida[c]	S	S-R	S	S-R	S	S

Organism	AMOX/CLAV and AMP/SULB	AMP	1ST CEFS	CHLOR	CLIND	ERY	PEN	TETRA
Eikenella corrodens[d]	S	S	S-R	S	R	S	S	S

[a] These antibiograms should serve as guidelines only; exceptions will occur. Abbreviations: CEFS, narrow-, expanded-, and broad-spectrum cephems (including cephalosporins); β-LAC, β-lactamase; AMOX/CLAV, amoxicillin-clavulanic acid; for other abbreviations, see Table 5.10–1a, footnote *a*. Symbols: +, positive; –, negative.
[b] Also resistant to oxacillin and clindamycin and susceptible to aminoglycosides.
[c] Also resistant to clindamycin.
[d] Also resistant to oxacillin and aminoglycosides.

APPENDIX 5.10–2

Suggested Conditions Requiring Verification of Antimicrobial Susceptibility Results

A. REQUIRES VERIFICATION BY REPEAT TESTING UNLESS PATIENT HAD SAME ISOLATE FROM ANOTHER RECENT CULTURE
1. Oxacillin-resistant *S. aureus*
2. Penicillin-resistant *S. pneumoniae*
3. Broad-spectrum cephalosporin (e.g., cefotaxime, ceftriaxone)-resistant *S. pneumoniae*
4. Penicillin-resistant or -intermediate viridans *Streptococcus* spp. from sterile body sites
5. Vancomycin-resistant or intermediate gram-positive organisms (except *Lactobacillus* spp., *Leuconostoc* spp., *Pediococcus* spp., or *Erysipelothrix rhusiopathiae*)
6. *Enterococcus* spp. with high-level resistance to gentamicin from sterile body site
7. *Klebsiella* spp. or *E. coli* with potential extended-spectrum β-lactamase (e.g., ceftazidime resistant)
8. Gentamicin-tobramycin-amikacin-resistant nonfastidious gram-negative bacilli
9. Trimethoprim/sulfamethoxazole-resistant *S. maltophilia*
10. Ampicillin-resistant (β-lactamase-negative) *Haemophilus influenzae*

APPENDIX 5.10–2 *(continued)*

11. Isolate demonstrating other than ''S'' results for those drug-organism combinations for which only ''S'' criteria are defined; isolate with resistance to a drug that has not been previously reported
12. Isolate for which the antibiogram is ''atypical'' for the species
13. Isolate resistant to all relevant drugs (obtain guidance for testing additional agents)
14. Isolate for which the results of related drugs do not correlate

B. REQUIRES VERIFICATION BY REPEAT TESTING OR RE-EXAMINATION OF TEST OR BOTH

1. Penicillin-intermediate *S. pneumoniae*
2. Aminoglycoside-resistant gram-negative bacilli (except *B. cepacia, S. maltophilia*; resistance to gentamicin or tobramycin or amikacin; if resistant to all or if results do not follow activity ''hierarchy,'' repeat test)
3. Ampicillin- and/or cefazolin (or cephalothin)-susceptible *Enterobacter* spp., *Citrobacter freundii, Serratia marcescens*, or *P. aeruginosa*
4. Ampicillin-susceptible *Klebsiella* spp.
5. Imipenem-resistant or -intermediate gram-negative bacilli (except *S. maltophilia; Proteus/Providencia* spp. are less susceptible than other *Enterobacteriaceae*)
6. Ciprofloxacin (or ofloxacin)-resistant gram-negative bacilli (except *S. maltophilia* or *B. cepacia*)

SECTION 6

Mycology and Aerobic Actinomycetes

SECTION EDITOR AND AUTHOR: *Kevin C. Hazen*

(continued)

6.1 Introduction and General Considerations

Fungi are significant, sometimes overlooked, human pathogens. Infections range in severity from merely cosmetic to life threatening. Fortunately, awareness of the role of fungi in disease is growing. Clinical laboratories must be prepared to identify fungi and realize that the range of new fungal species that cause disease is growing.

New antifungal agents less toxic than amphotericin B and variously administered have increased the requirement to identify potential pathogens to species level. (It is no longer acceptable to refer to yeast that do not produce germ tubes as "yeast, not *C. albicans*.") Although amphotericin B is effective against many invasive fungal pathogens, newer drugs with a narrower spectrum of efficacy are available. Resistance to these drugs is sometimes species characteristic and is also isolate dependent for other species.

Figure 6.1–1 provides a general overview of the approach that is taken for diagnostic mycology. The greater detail provided in Fig. 6.1–2 shows points in the diagnostic process at which individual laboratories may decide to modify steps.

More discussion between clinical mycologists and anatomic pathologists, cytopathologists, or neuropathologists about diagnostic results often facilitates interpretation of culture significance. In some fungal diseases (e.g., histoplasmosis and coccidioidomycoses), the histopathology services may provide the first diagnostic evidence of the disease, whereas the clinical laboratory provides confirmation of the etiologic entity. Table 6.1–1 provides a general overview of the typical clinical presentations associated with specific fungal diseases. Exceptions can occur. The appearance of a fungal organism in tissue material, along with an analysis of patient history and symptomatology, provides the best guidance to the likely etiology, but this should be confirmed whenever possible by mycological analysis.

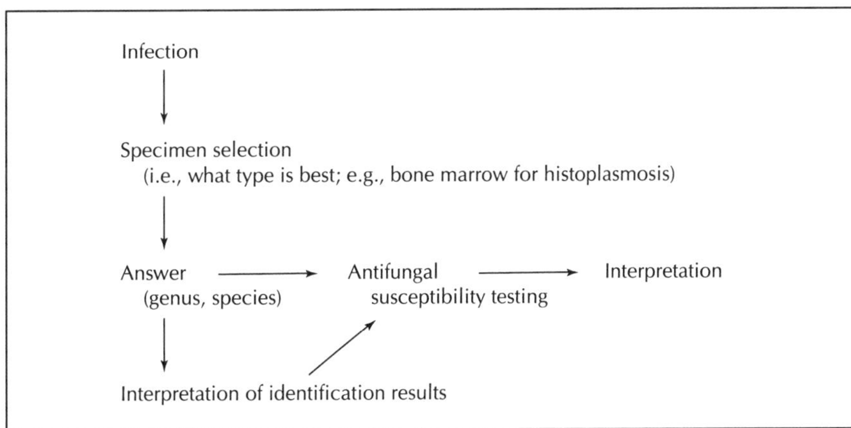

Figure 6.1–1 General approach to identifying a fungal etiologic agent from a specimen.

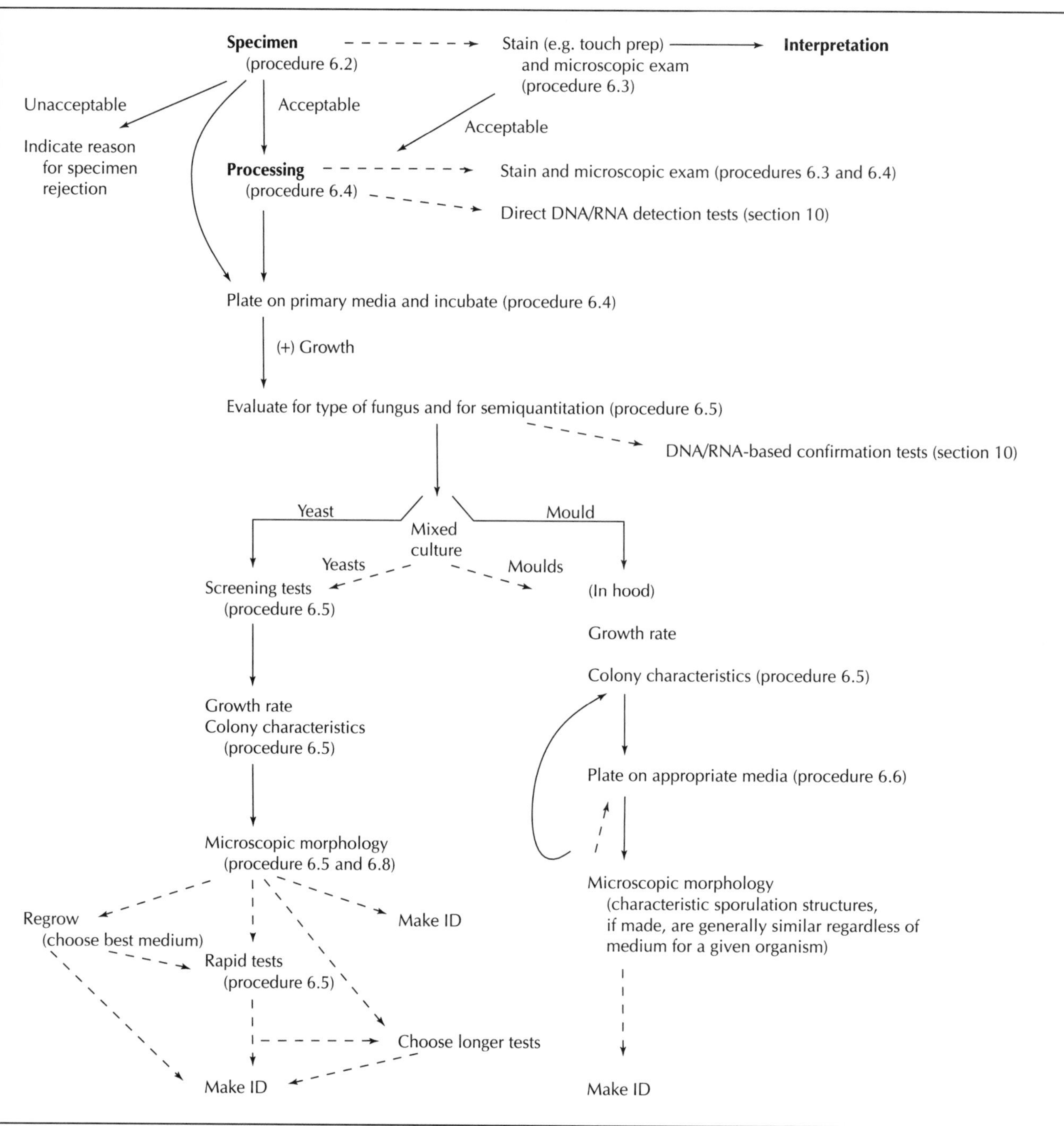

Figure 6.1–2 Greater detail for an approach to ascertaining a fungal etiologic agent from a specimen. ID, identification.

Table 6.1–1 Clinical presentations of subcutaneous and systemic mycotes[a]

Disease	Purulence	Granuloma	Fibrosis	Necrosis	Other
Aspergillosis	+	+	−	+	
Blastomycosis	+	+	+	−	
Candidiasis	+	+	−	+	
Chromoblastomycosis	−	+	+	−	Hyperplasia, acanthosis
Coccidioidomycosis	+	+	+	−	
Cryptococcosis	−	+	−	−	
Entomophthoromycosis	−	+	+	−	Eosinophilic growth
Histoplasmosis	−	+	+	−	Calcification, coin lesions
Mucormycosis	+	−	−	+	
Mycetoma	+	+	+	+	Sinus tracts, grains
Paracoccidioidomycosis	+	+	+	+	
Phaeohyphomycosis	−	+	−	−	
Sporotrichosis	+	+	+	+	Asteroid bodies

[a] Adapted from reference 1.

I. CLASSIFICATION

Fungus taxonomy is an evolving process. Nomenclatural changes also continue to have significant implications for understanding fungi in the biological community, diagnosing fungal disease, and prescribing treatment for such diseases. Table 6.1–2 is a list of recently published species names that have replaced incorrect designations.

Table 6.1–2 Revised nomenclature of selected medically important fungi[a]

Fungus	Common synonyms and obsolete names
Absidia corymbifera	*Absidia ramosa* *Mucor corymbifer*
Acremonium species	*Cephalosporium* species
Acremonium falciforme	*Cephalosporium falciforme*
Acremonium kiliense	*Cephalosporium granulomatis* *Cephalosporium infestans* *Cephalosporium madurae*
Acremonium recifei	*Cephalosporium recifei*
Ajellomyces capsulatus	*Emmonsiella capsulata*
Alternaria alternata	*Alternaria tenuis*
Alternaria infectoria	*Alternaria* state of *Pleospora infectoria*
Aphanoascus fulvescens	*Aphanoascus cinnabarinus* *Aphanoascus fulvescens* *Anixiopsis stercoraria*
Arthroderma species	*Nannizzia* species
Arthroderma borellii	*Nannizzia borellii*
Arthroderma cajetani	*Nannizzia cajetani*
Arthroderma cookiellum	*Nannizzia cookiella*
Arthroderma corniculatum	*Nannizzia corniculata*
Arthroderma fulvum	*Nannizzia fulva* *Nannizzia gypsea* var. *fulva*
Arthroderma grubyi	*Nannizzia grubyi*
Arthroderma gypseum	*Gymnoascus gypseus* *Nannizzia gypsea*
Arthroderma incurvatum	*Nannizzia incurvata*
Arthroderma obtusum	*Nannizzia obtusa*
Arthroderma otae	*Nannizzia otae*
Arthroderma persicolor	*Nannizzia quinckeani* *Nannizzia persicolor*
Arthroderma racemosum	*Nannizzia racemosa*
Arthrographis species[b]	*Geotrichum* species (misidentification)
Arthrographis kalrai	*Oidiodendron kalrai*
Aspergillus fischerianus	*Aspergillus fischeri*
Aspergillus fumigatus	*Aspergillus phialiseptus*
Aspergillus glaucus group	*Aspergillus glaucus*
Aureobasidium pullulans	*Dematium pullulans*
Basidiobolus ranarum	*Basidiobolus haptosporus* *Basidiobolus heterosporus* *Basidiobolus meristosporus*
Bipolaris australiensis	*Drechslera australiensis* *Helminthosporium australiense*
Bipolaris hawaiiensis	*Drechslera hawaiiensis* *Helminthosporium hawaiiensis*
Bipolaris spicifera	*Curvularia spicifera* *Drechslera spicifera* *Helminthosporium spiciferum*
Blastoschizomyces capitatus	*Blastoschizomyces pseudotrichosporon* *Geotrichum capitatum* *Trichosporon capitatum*

(continued)

Table 6.1–2 Revised nomenclature of selected medically important fungi[a] *(continued)*

Fungus	Common synonyms and obsolete names
Botryadiplodia theobromae	*Lasiodiplodia theobromae*
Candida albicans	*Candida claussenii* *Candida dubliniensis* *Candida langeronii* *Candida stellatoidea* (in part) *Monilia albicans*
Candida famata	*Torulopsis candida*
Candida glabrata	*Torulopsis glabrata*
Candida haemulonii	*Torulopsis haemulonii*
Candida kefyr	*Candida pseudotropicalis*
Candida tropicalis	*Candida paratropicalis*
Cerinosterus cyanescens	*Sporothrix cyanescens*
Cladophialophora bantiana	*Cladosporium bantianum* *Cladosporium trichoides* *Cladosporium trichoides* var. *chlamydosporum* *Xylohypha bantiana*
Cladophialophora boppii	*Taeniolella boppii*
Cladophialophora carrionii	*Cladophialophora ajelloi* *Cladosporium carrionii*
Cladophialophora devriesii	*Cladosporium devriesii*
Cladosporium species	*Hormodendrum* species
Canidiobolus coronatus	*Delacroixia coronata* *Entomophthora coronata*
Cryptococcus neoformans var. *gattii*	*Cryptococcus bacillisporus*
Cunninghamella bertholletiae[c]	*Cunninghamella elegans* (misidentification)
Cylindrocarpon cyanescens	*Phialophora cyanescens*
Cylindrocarpon lichenicola[d]	*Cylindrocarpon tonkinense* (misidentification)
Emmonsia parva var. *parva*	*Chrysosporium parvum*
Emmonsia parva var. *crescens*	*Chrysosporium crescens* *Chrysosporium parvum* var. *crescens* *Emmonsia crescens*
Epidermophyton floccosum	*Trichophyton cruris*
Exophiala castellanii	*Exophiala mansonii*
Exophiala jeanselmei	*Phialophora gougerotii* (in part) *Phialophora jeanselmei*
Exophiala spinifera	*Phialophora spinifera* *Rhinocladiella spinifera*
Exserohilum rostratum	*Bipolaris halodes* *Bipolaris rostratum* *Drechslera halodes* *Drechslera longirostrata* *Drechslera rostrata* *Helminthosporium rostratum*
Filobasidiella neoformans var. *bacillispora*	*Filobasidiella bacillispora*
Fonsecaea compacta	*Fonsecaea compactum* *Phialophora compactum* *Rhinocladiella compactum*
Fonsecaea pedrosoi	*Phialophora pedrosoi* *Rhinocladiella pedrosoi*
Fusarium coeruleum	*Fusarium solani* var. *coeruleum*
Fusarium dimerum (in part)	*Fusarium episphaeria*
Fusarium moniliforme[e]	*Fusarium verticillioides*
Fusarium pallidoroseum[f]	*Fusarium semitectum*
Galactomyces geotrichum	*Dipodascus geotrichum* *Endomyces geotrichum*
Geomyces pannorum	*Chrysosporium pannorum*
Geotrichum penicillatum	*Trichosporon penicillatum*
Histoplasma capsulatum var. *capsulatum*	*Histoplasma capsulatum*
Histoplasma capsulatum var. *duboisii*	*Histoplasma duboisii*
Histoplasma capsulatum var. *farciminosum*	*Histoplasma farciminosum*
Hormonema dematioides[g]	*Aureobasidium pullulans* (misidentification)
Kluyveromyces marxianus	*Kluyveromyces fragilis*
Lecythophora hoffmannii	*Phialophora hoffmannii*
Lecythophora mutabilis	*Phialophora mutabilis*
Lewia infectoria	*Pleospora infectoria*
Loboa loboi	*Blastomyces loboi* *Paracoccidioides loboi*
Malassezia furfur[h]	*Cladosporium mansonii* *Pityrosporum orbiculare* *Pityrosporum ovale*
Malassezia pachydermatis	*Pityrosporum canis* *Pityrosporum pachydermatis*
Microsporum canis var. *distortum*	*Microsporum distortum*
Microsporum persicolor	*Trichophyton persicolor*
Mucor indicus	*Mucor rouxii* (in part)
Nattrassia mangiferae	*Hendersonula toruloidea*
Neotestudina rosatii	*Pseudophaeotrichum sudanense* *Zopfia rosatii*
Ophiostoma stenoceras	*Ceratocystis stenoceras*
Paecilomyces lilacinus	*Penicillium lilacinum*
Phaeoacremonium parasiticum	*Phialophora parasitica*
Phaeoannellomyces werneckii	*Cladosporium werneckii* *Exophiala werneckii*
Phialophora verrucosa[i]	*Phialophora americana*
Pseudallescheria boydii	*Allescheria boydii* *Petriellidium boydii*
Pseudochaetosphaeronema larense	*Chaetosphaeronema larense*
Pythium insidiosum[j]	*Hyphomyces destruens* *Pythium destruens* *Pythium gracile*
Ramichloridium obovoideum	*Ramichloridium mackenziei*
Rhinocladiella aquaspersa	*Acrotheca aquaspersa* *Ramichloridium cerophilum*
Rhizomucor pusillus	*Mucor pusillus*
Rhizopus arrhizus	*Rhizopus oryzae*
Rhizopus microsporus var. *oligosporus*	*Rhizopus oligosporus*
Rhizopus microsporus var. *rhizopodiformis*	*Rhizopus rhizopodiformis*
Rhizopus stolonifer	*Rhizopus nigricans*
Sarcinosporon inkin	*Fissuricella filamenta* *Prototheca filamenta* *Sarcinomyces inkin* *Trichosporon asteroides* *Trichosporon inkin*

Table 6.1–2 Revised nomenclature of selected medically important fungi[a] *(continued)*

Fungus	Common synonyms and obsolete names	Fungus	Common synonyms and obsolete names
Scedosporium apiospermum	*Monosporium apiospermum*	*Trichophyton mentagrophytes*	*Trichophyton erinacei* *Trichophyton granulare* *Trichophyton granulosum* *Trichophyton gypseum* *Trichophyton interdigitale* *Trichophyton mentagrophytes* var. *nodulare* *Trichophyton quinckeanum*
Scedosporium prolificans	*Lomentospora prolificans* *Scedosporium inflatum*		
Scolecobasidium constrictum[k]	*Dactylaria constricta* (in part) *Ochroconis constricta*		
Scolecobasidium humicola	*Ochroconis humicola*		
Scolecobasidium tshawytschae	*Ochroconis tshawytschae*		
Scytalidium dimidiatum	*Scytalidium hyalinum* *Scytalidium* synanamorph of *Hendersonula toruloidea*	*Trichophyton persicolor*	*Microsporum persicolor*
		Trichophyton tonsurans	*Trichophyton tonsurans* var. *sulphureum*
Sporothrix schenckii[l]	*Sporotrichum gougerotii* (in part) *Sporotrichum schenckii*	*Trichosporon beigelii*[m]	*Trichosporon cutaneum*
Stenella araguta	*Cladosporium castellanii*	*Wangiella dermatitidis*	*Exophiala dermatitidis* *Fonsecaea dermatitidis* *Hormiscium dermatitidis* *Hormodendrum dermatitidis* *Phialophora dermatitidis*
Trichophyton species	*Achorion* species *Microides* species		

[a] This list of medically important fungi is neither complete nor exhaustive. Because of the nature of taxonomy and classification approaches, the treatment of the various names included in this list will not necessarily reflect the opinion of each authority on mycology. The names selected for inclusion in this list represent fungi that are important to clinicians and microbiologists in clinical microbiology laboratories. Reprinted from reference 2.

[b] A number of isolates identified as either *Geotrichum candidum* or *Geotrichum* species in the medical literature are misidentified isolates of *Arthrographis* species.

[c] *Cunninghamella bertholletiae* was incorrectly identified as *Cunninghamella elegans* in several of the early case reports.

[d] *Cylindrocarpon lichenicola* was incorrectly reported as *Cylindrocarpon tonkinense* in some case reports and textbooks.

[e] It is extremely controversial as to which name should be used. Both names apply to the same fungus.

[f] *Fusarium incarnatum* may be an earlier name for both *Fusarium pallidoroseum* and *Fusarium semitectum.*

[g] *Hormonema dematioides* was incorrectly identified as *Aureobasidium pullulans* in some case reports.

[h] It has been recently proposed that the genus *Malassezia* should contain the new species *M. globosa, M. obtusa, M. restricta,* and *M. slooffiae*. It appears premature to accept this treatment of isolates previously considered to be *M. furfur, M. pachydermatis,* and *M. sympodialis;* however, molecular taxonomic interpretations clearly merit careful attention.

[i] *Phialophora verrucosa* and *Phialophora americana* are closely related. Some investigators believe that they are separate species.

[j] *Pythium insidiosum* is an oomycete that should be classified in the kingdom Chromista. If the concept of five kingdoms is used, *P. insidiosum* would be classified in the kingdom Protoctista.

[k] The taxonomy surrounding the genera *Dactylaria, Ochroconis,* and *Scolecobasidium* is controversial.

[l] At the beginning of the 20th century, *Sporothrix schenckii* was misidentified as *Sporotrichum gougerotii.* Later, *Exophiala jeanselmei* was misidentified as *Sporotrichum gougerotii.*

[m] It has been proposed that *Trichosporon beigelii* should be discarded in favor of *Trichosporon asahii, Trichosporon cutaneum, Trichosporon mucoides, Trichosporon ovoides,* and *Sarcinosporon inkin* (synonyms *Trichosporon asteroides* and *Trichosporon inkin*) as the etiologic agents of human infections. It appears premature to accept this treatment of *T. beigelii;* however, the molecular approach to the fungal taxonomy of these yeasts is important and deserves detailed consideration.

II. SAFETY

It is imperative that these cultures be handled in a biosafety hood.

In a clinical mycology laboratory, always follow established safety measures, as described elsewhere in this book and in CMPH 6.1. Consider every mould a potential pathogen and manipulate it only in a biological safety cabinet. If a mould is seen on a bacteriology plate, tape the plate closed or seal the edge with parafilm to prevent accidental release of infectious particles. Transport the plate to the mycology area. Manipulation of fungal organisms requires that certain unique or less routine safety measures be followed (*see* CMPH 6.1).

REFERENCES

1. **De Hoog, G. S., and J. Guarro.** 1995. *Atlas of Clinical Fungi.* Centraalbureau voor Schimmelcultures, Delft, The Netherlands.
2. **McGinnis, M. R., and M. G. Rinaldi.** 1997. Selected medically important fungi and some common synonyms and obsolete names. *Clin. Infect. Dis.* **25:15**–17.

(See EPCM section 1 for tables concerning specimen selection, collection, and transport. Specific considerations for mycology are below.)

6.2.1 Specimen Selection

PRINCIPLE

Like all infectious diseases, the best specimen for determining the etiologic agent of a mycosis is material from the active infection site. In pneumonia caused by *Cryptococcus neoformans*, specimens that provide the greatest likelihood of success for detection of the fungus may be not only from the primary infection site, but from other sites, such as blood. Thus, when a fungal disease is suspected, direct specimens are the best choice, but peripheral specimens may also be useful. Some fungi are normal, occasional contaminants of specimens or are members of the normal microbiota. Because some fungi can be contaminants in specimens, the value of some specimens, especially peripheral specimens, for diagnosis of a mycosis is decreased by such contamination. For example, sputum specimens for aspergillosis in a nontransplant patient or sputum cultures for blood stream infections caused by *Candida albicans* are diagnostic challenges.

6.2.2 Specimen Collection

PRINCIPLE

Specimen collection for detection of an etiologic agent of a mycosis is very similar to specimen collection for bacteria. The amount of material for the detection of fungi may exceed that for bacteria in certain instances. Also, moulds grow as linked filaments which may or may not fragment into individual cells like bacteria. But the amount of specimen material is not controlled by the clinical laboratory. More material for the inoculum in primary culture or by concentrating a liquid or liquefied specimen may increase the number of organisms recovered from a specimen.

◪ **NOTES:** Swabs are not optimal specimen collection devices for fungal organisms.

Fingernails or toenails suspected of dermatophytosis should be cleaned extensively with 70% ethanol, clipped, and sent to the laboratory in a dry container.

6.2.3 Specimen Transport

PRINCIPLE

Fungi are hardy organisms. Their cell walls provide an effective barrier against various toxic agents. However, the viability of some fungi is affected by desiccation, elevated temperature (>37°C), low temperature (<10°C), and starvation. Bacterial overgrowth is another threat. When a specimen is obtained from a normally sterile site (e.g., blood or CSF), the specimen may be kept at 30 to 37°C. Only the infectious agent should grow. Transportation times should be kept to a minimum (<2 h) because some fungal agents are more sensitive than others to environmental stresses. Specimens from sites with possible contaminating bacteria or leukocytes should be transported rapidly and processed immediately because bacterial overgrowth and leukocyte products could cause decreased fungal viability. Specimens that cannot be transported to the clinical laboratory within 2 h should be stored at 4°C, except ''sterile'' body fluids (30 to 37°C) and dermatological specimens (15 to 30°C). The latter should not be refrigerated because some dermatophytes are sensitive to cold. Stools should not be kept at or above room temperature for more than 1 h.

Transport containers should be sterile, humidified, and leakproof. Keep the specimen moist by adding a minimum volume of sterile saline (0.85% NaCl). Avoid anaerobic transport media because they can reduce the viability of some fungi.

6.3.1 Introduction

I. PRINCIPLE

Specimens should be evaluated macroscopically and microscopically. Macroscopic evaluation can provide clues about the most probable region of a specimen to contain fungal organisms (e.g., necrotic tissue or sulfur granules). Microscopic examination reveals vital information about the relative concentration and type of fungus present in a specimen (Fig. 6.3.1–1).

Sensitivity of microscopic examination is improved when fungus-enhancing stains are used (e.g., India ink or Calcofluor white).

Liquid or liquefied specimens should be centrifuged to concentrate fungal organisms in the pellet, unless the volume of specimen is insufficient to allow both smear and culture. Liquid stools should not be concentrated. Viscid sputum specimens may be liquified (*N*-acetyl cysteine) and concentrated.

If the specimen volume is too low for smear and culture, then the specimen should only be cultured. Insufficient volumes should, in general, be rejected for microbiologic evaluation. Of course, insufficient specimen material should be accepted if procuring another specimen poses significant harm to the patient (such as by brain biopsy, endoscopic bronchial lavage, etc.).

Smears of most specimens are prepared for microscopic examination in the usual manner. Tissues such as hair, skin, nails, and intact biopsied material, appropriately minced, must be treated with KOH to digest the keratinized and other tissue structures to allow the fungi to be recognized (improved by the addition of Quink-Ink). Fungi in a specimen may be better discerned with stains, such as Calcofluor white. Other stains that are useful for tissue are periodic-acid Schiff (PAS), Gomori methenamine silver nitrate, and the Giemsa stain (see below). PAS and the silver stain are best performed by a trained histology technologist. The Giemsa stain is useful for detection of *Histoplasma capsulatum* in bone marrow and blood smears. If CSF is examined for possible meningitis caused by a fungus, an India ink stain should be performed. However, this method is insensitive. Antigen detection is preferred.

Figure 6.3.1–1 provides some of the organisms that are associated with particular macroscopic morphologies in specimens.

II. SMEARS AND STAINING OF CLINICAL SPECIMENS

A. Choice of stains

Once a smear or impression slide of a specimen is made, several stains can be used to detect fungi. While the Gram stain may allow detection of yeasts (most yeasts are stained by Gram stain) and some moulds or suggest fungi by negative staining, it is not a useful tool for mycology. Instead, Calcofluor white (*see* procedure 6.3.2) helps to distinguish fungi from host material and increases the sensitivity of staining. However, the procedure requires an epifluorescence scope. KOH digestion will allow recognition of fungal elements, but it lacks the sensitivity of Calcofluor white. Lactophenol cotton blue or its substitute, trypan blue, may aid in visualizing fungal agents.

To detect the presence of a capsule on yeast cells, a negative stain, such as the India ink stain (*see* procedure 6.3.5), may be used. The typical specimen for this stain is CSF. A caveat to this method for CSF is that tests for cryptococcal antigens must be performed on all suspected cases.

Some of the tissues (e.g., hair, nails, and cartilage) must be dissolved to help reveal fungal elements. In this case, the specimen is treated with KOH (*see* Table 6.3.1–1) before a stain is applied. Only a limited number of stains are

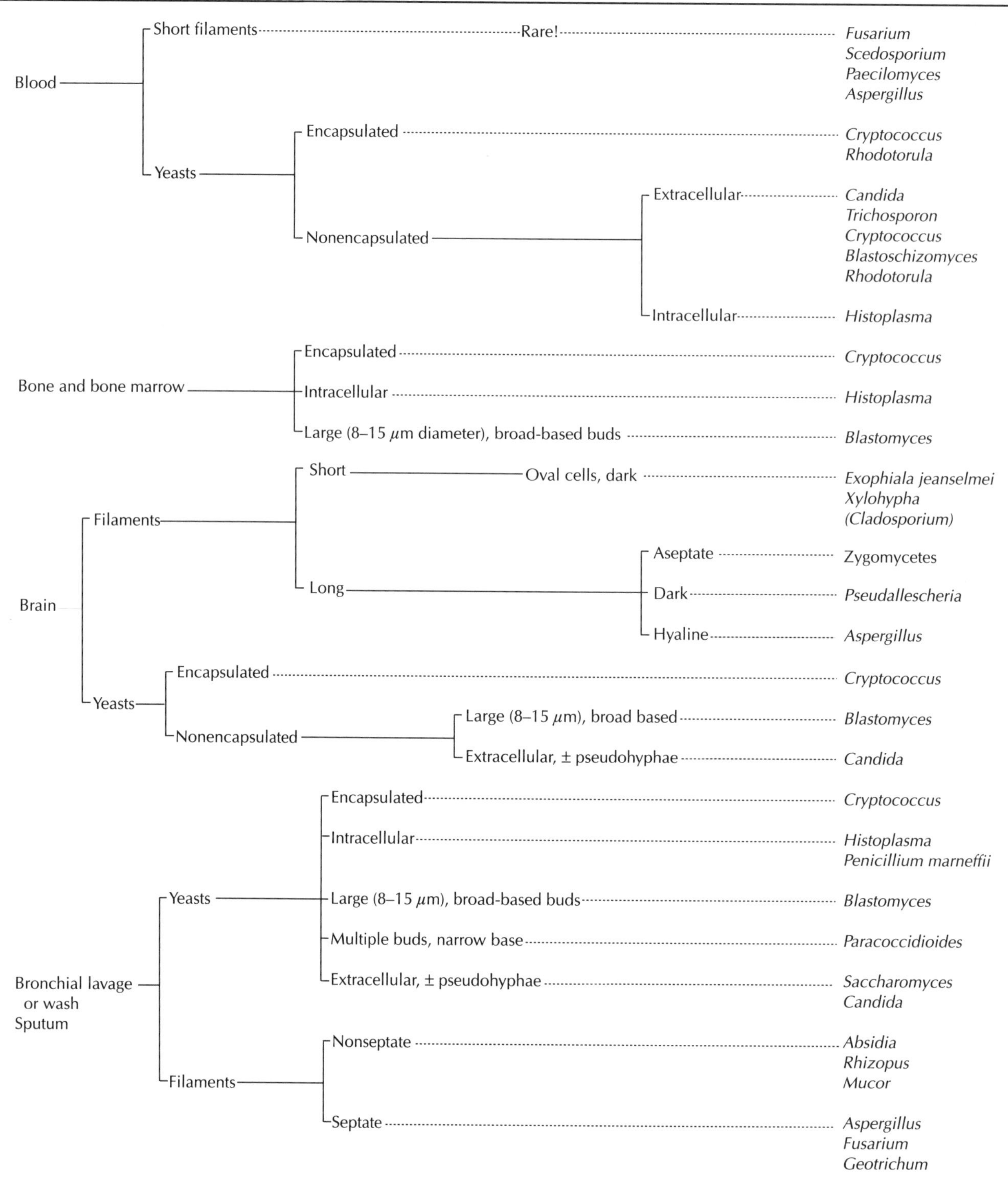

Figure 6.3–1 Most common possible fungal etiologies and morphology seen in specimens. The figure assumes that appropriate stain is used to visualize organisms (e.g., Calcofluor white) or that the specimen is unstained to see pigment.

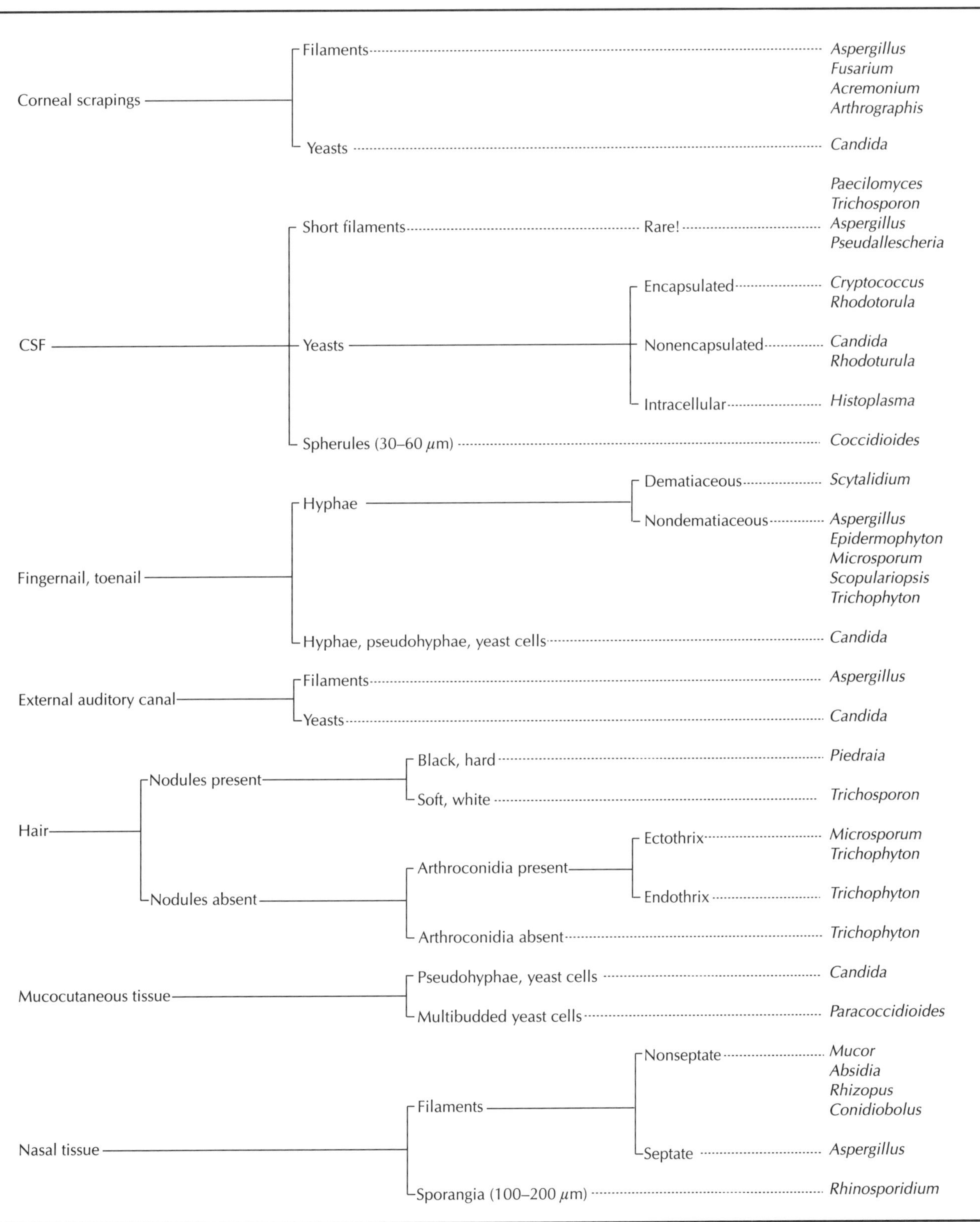

Figure 6.3–1 *(continued)*

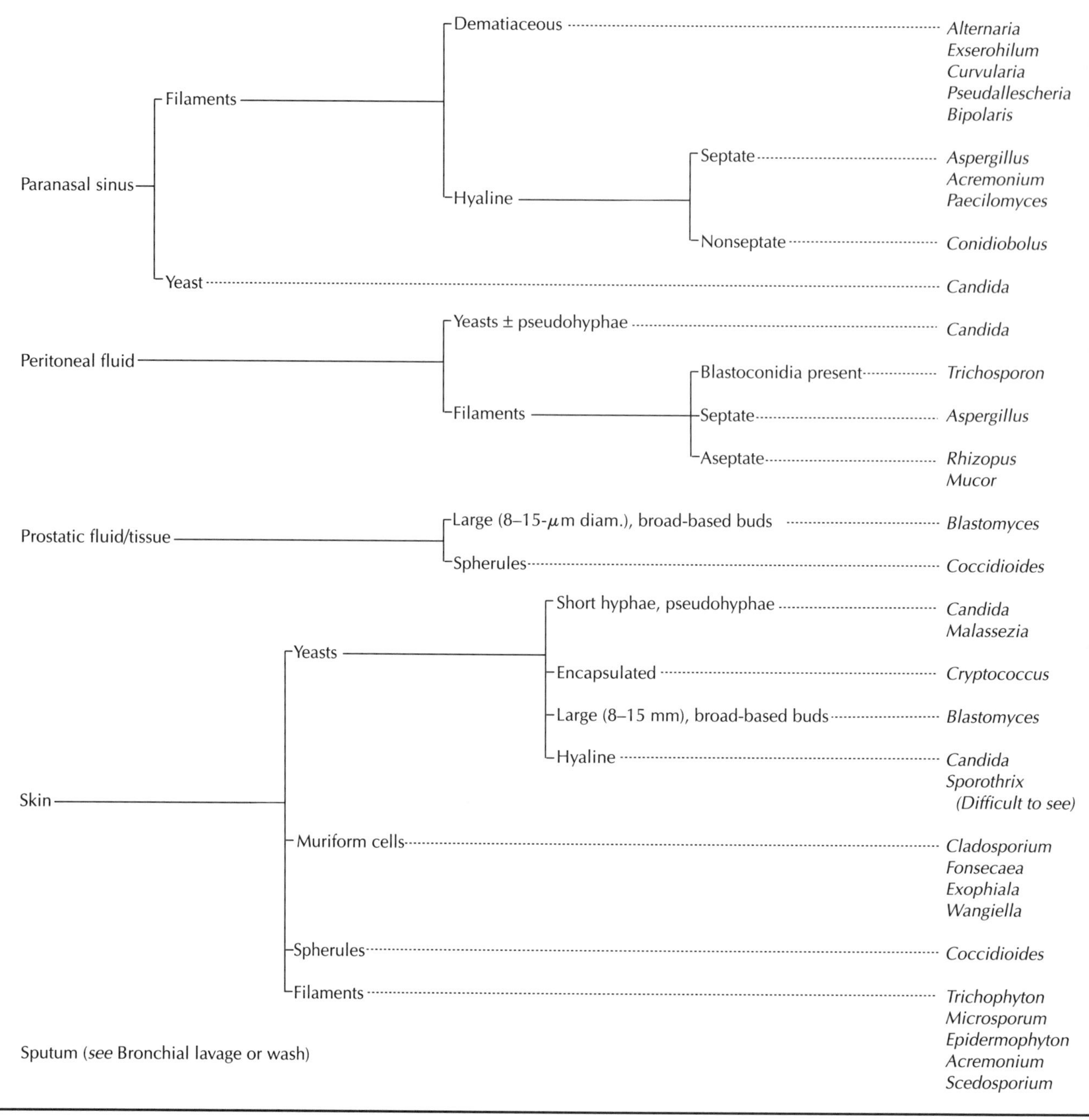

Figure 6.3–1 *(continued)*

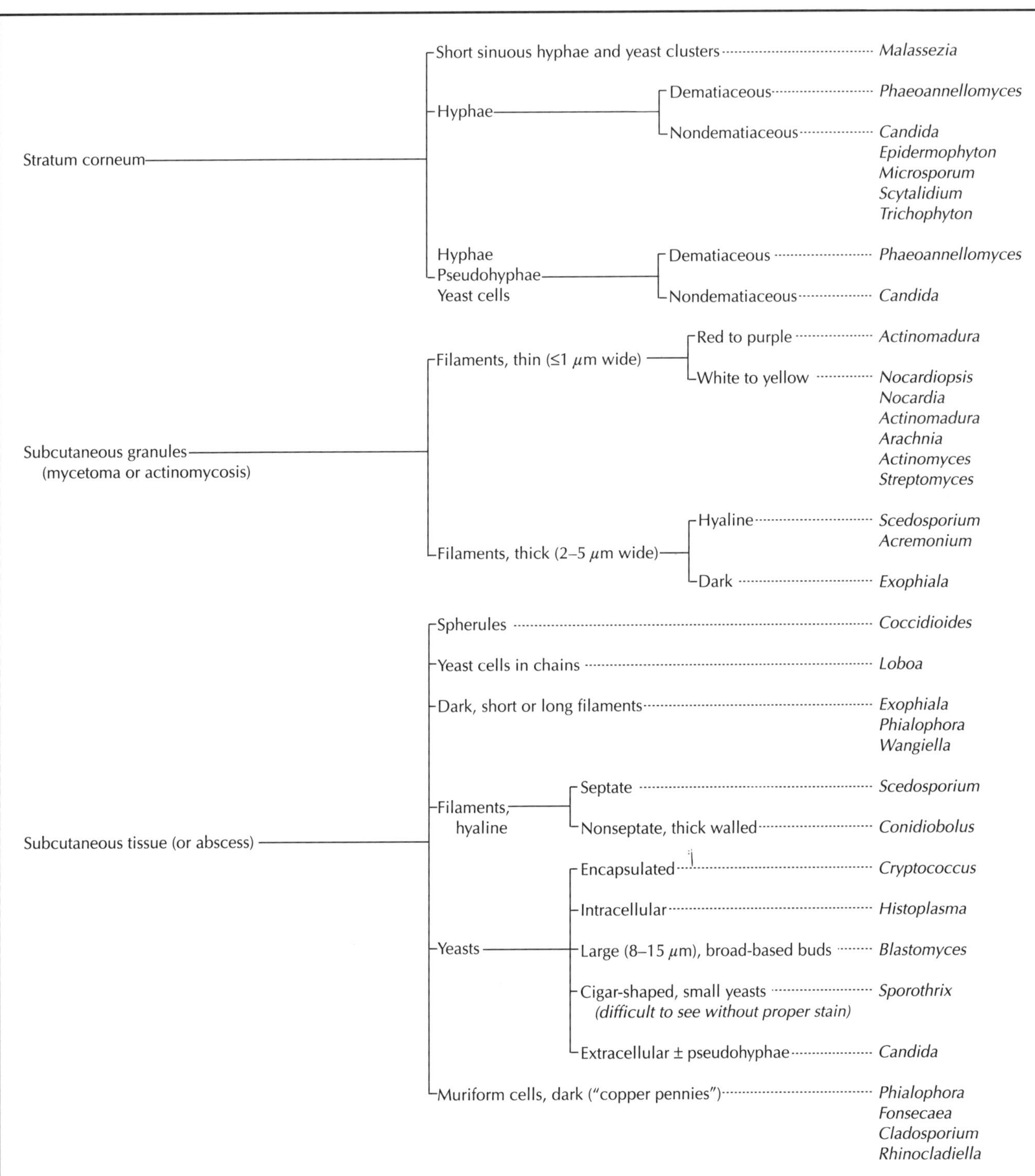

Figure 6.3–1 *(continued)*

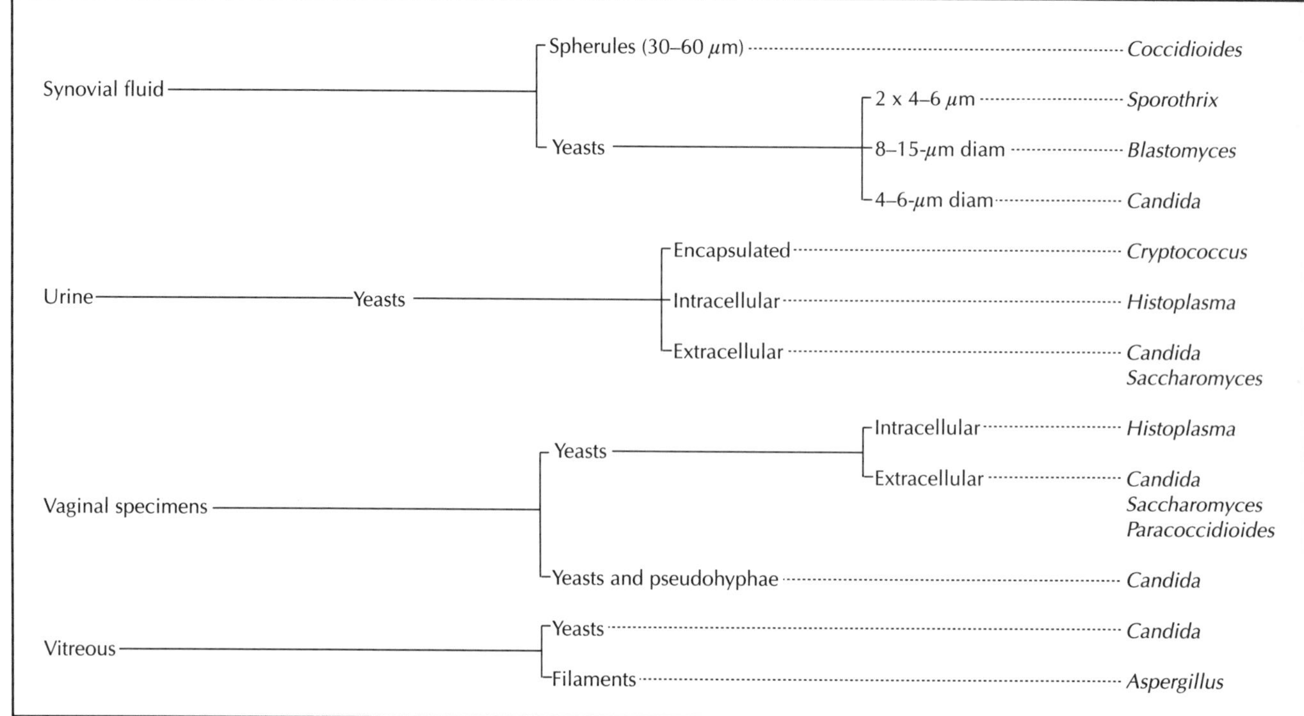

Figure 6.3–1 *(continued)*

II. SMEARS AND STAINING OF CLINICAL SPECIMENS *(continued)*

compatible with KOH. Fortunately, one stain is Calcofluor white (*see* Table 6.3.1–1).

Once organisms are seen, report the appearance of the fungal elements from the following list of distinguishing characteristics:

1. Hyphae (or pseudohyphae)
2. Hyphae and arthroconidia
3. Hyphae, arthroconidia, and yeast cells
4. Hyphae (pseudohyphae) and yeast cells
5. Yeast cells only
6. Yeast cells with capsules (possible *Cryptococcus* spp.)
7. Spherules
8. Muriform (sclerotic) cells
9. Sulfur granules with bacterial (or fungal) hyphae

Table 6.3.1–1 Stains for use on primary specimens

Stain[a]	Procedure
Gram stain	2.1
Giemsa	6.3.7
Calcofluor white	6.3.4
KOH	6.3.2
KOH with quick ink	6.3.3
KOH/DMSO	6.3.3
India ink	6.3.5
Hanks' modified acid-fast stain (for *Nocardia* spp.)	6.3.6

[a] DMSO, dimethyl sulfoxide.

II. SMEARS AND STAINING OF CLINICAL SPECIMENS
(continued)

B. Staining

Gram stain

The Gram stain is performed as it is for bacteria (*see* EPCM procedure 2.1). All fungi are gram positive (weak reactions are often seen), but some fungi (e.g., zygomycetes) can appear gram negative. If filamentous structures are suspected to be fungal, confidence is improved by the detection of cross walls (septa) or branches. The cell walls on each side of the filament should be parallel, but not invariably so, throughout the length of the filament.

6 Mycology and Aerobic Actinomycetes

6.3.2 KOH Mount

I. PRINCIPLE

KOH treatment allows rapid observation of fungal elements present in tissues and fluids, such as skin, nails, biopsy material, and sputum. Its use with brain biopsies is limited because artifactual material resembling yeast cells can result. Brain biopsies should be evaluated with a Gram stain.

II. MATERIALS

A. KOH reagent droppers from BBL may be used.
   1. Outdate established by manufacturer
   2. Stored at room temperature

B. Microscope slide

C. Pipettes

D. Coverslip

III. QUALITY CONTROL

A. A drop of KOH is placed on a slide with a coverslip.

B. The preparation is screened for contamination by bacteria and fungal spores.

IV. PROCEDURE

A. Place a drop of KOH in the center of a clear glass slide.

B. Place a fragment of tissue, purulent material, or scraping in the KOH. Tease the material well enough with the corner of a coverslip to give a thin preparation or break up the material with a sterile biological probe.

C. Mount with a coverslip.
   1. Allow preparation to digest for approximately 10 min at room temperature.
   2. Gently warm slide (do not overheat).
   3. Gently press on slide to help disperse tissue material.

V. READING AND REPORTING

A. Screen under low power.

B. Use high (40×)-power magnification to verify presence of fungal elements. Look for branching of hyphae and septa. Yeast cells with budding should be seen. This test is improved by the addition of Calcofluor white stain (*see* procedure 6.3.4, below).

6.3.3 Variations to the KOH Method

PART 1

KOH-DMSO

The addition of dimethyl sulfoxide (DMSO; Sigma Chemical Co., St. Louis, Mo.) to KOH eliminates the need to heat the specimen or to let the specimen stand at room temperature when exposed to KOH alone.

PROCEDURE

A. To 100 ml of a 60% (vol/vol) solution of DMSO in water, add 20 g of KOH. Dissolve completely.

B. Store in an airtight container in the dark. The use of amber dropper bottles is preferred.

C. Add KOH-DMSO to sample as described in procedure 6.3.1 *except do not heat* the specimen once it has been exposed to the KOH-DMSO solution.

PART 2

KOH-DMSO-Ink

A variation of the KOH-DMSO method is to add Parker's Quink Ink (blue-black). The ink enhances contrast and is particularly useful for detecting *Malassezia furfur* in skin scrapings.

To prepare the solution, mix equal volumes of Quink Ink and KOH-DMSO. Store as for KOH-DMSO.

6.3.4 Screening of Specimens for Fungal Elements by the KOH-Calcofluor Fluorescent-Stain Method

I. PRINCIPLE

Calcofluor white is a nonspecific fluorochrome stain for the rapid screening of clinical specimens for fungal elements. It binds to chitin and other large polysaccharides. Artifactual staining occurs with collagen. It can also be used to better visualize the microscopic morphology of culture isolates.

II. MATERIALS

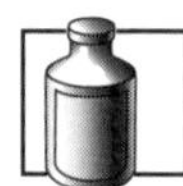

Include QC information on reagent container and in QC records.

A. Fungi-Fluor Reagent A, Remel no. 40014
B. Fungi-Fluor Reagent B, Remel no. 40014
C. Slides
D. Coverslips
E. Distilled water
F. Fluorescent microscope with appropriate filter set (fluorescein isothiocyanate filter set is satisfactory)

III. QUALITY CONTROL

A. Follow procedure below, except use a colony of either *Candida* species or *Saccharomyces cerevisiae*, place drops of solutions A and B near drops of colonial suspension, mix with coverslip, and cover.
B. Good bright white fluorescence of yeast cells should be seen.

IV. PROCEDURE

A. For patient specimens, a slide with a small amount of the specimen should be prepared.
B. Add 1 drop of Reagent A (KOH) to the patient specimen.
C. Add 1 drop of Reagent B (Calcofluor white) to the patient specimen.
D. Cover with slip (press gently on coverslip).
E. Examine under fluorescent excitation light of 300 to 412 nm after 5 to 10 min of incubation.
   1. With a 500-nm barrier (emission filter), fungi will appear yellow-green.
   2. Without an emission filter, fungi will fluoresce bright white.

V. EXPECTED RESULTS

Hyphae, yeast cells, and other fungal elements will fluoresce. Look for characteristic fungal structures, such as branching hyphae, septa, budding, thick cell walls, and bud or birth scars on yeast cells. These are distinctive features that can help eliminate possible false positives.

VI. LIMITATIONS

Vaginal secretions are difficult to interpret owing to nonspecific staining.

6.3.5 India Ink Preparation

I. PRINCIPLE

An India ink preparation can be used for the rapid detection of encapsulated yeasts based on negative staining. The capsule repels the carbon particles of the India ink, giving a clear, well-demarcated halo around each encapsulated cell.

II. MATERIAL

Include QC information on reagent container and in QC records.

Use India ink (KOH-I-NOOR, 3080-F Universal) or India ink droppers (Becton Dickinson, M7-74811).

III. QUALITY CONTROL

A. Performed daily

B. A drop of India ink is placed on a slide with a drop of distilled water.

C. The preparation is screened for contamination by bacteria or fungus spores.

IV. PREFERRED PROCEDURE

A. Place a drop of sediment (body fluids, urine) on a clean glass slide.

B. Place a coverslip on the sediment drop.

C. Place a drop of India ink on the glass slide next to the coverslip.

D. Allow the India ink to diffuse under the slide and create an ink gradient.

E. Examine under low (10×)- and high (40×)-power magnification. Read the region where the ink particle density is neither too light nor too heavy.

V. ALTERNATE PROCEDURE

A. Place a drop of sediment (body fluids, urine) on a clean glass slide.

B. Add a small drop of India Ink and mix with sediment by using the corner of a coverslip.

1. The preparation should be of a brownish color, not black.
2. If the preparation is too black, add 1 to 2 drops of sterile distilled water, mix, and mount with a coverslip.

C. Examine under low (10×)- and high (40×)-power magnification.

VI. EXPECTED RESULTS

A. Positive test

1. The presence of budding yeast cells with clear halos around them indicates capsular material. It is important that the yeast cell is in focus when interpreting the presence of a capsule.

VI. EXPECTED RESULTS *(continued)*

2. Encapsulated yeasts seen in spinal-fluid sediment are suggestive of *C. neoformans*, but cultures or antigen detection tests are needed to confirm the identity of such yeasts.

B. Negative test

1. Absence of halos
2. WBCs may also repel carbon particles, but the halo has a fuzzy, irregular appearance at the periphery. Do not confuse white cells with yeasts. Look for clear halos and budding cells. Some internal contents should be visible.

6.3.6 Hanks' Acid-Fast Stain (Modified) Technique

I. PRINCIPLE

The Hanks' modified acid-fast stain technique is used for the detection of partially acid-fast organisms such as *Nocardia* spp.

II. MATERIALS

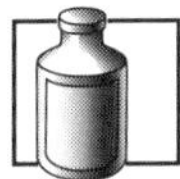

Include QC information on reagent container and in QC records.

A. Carbolfuchsin reagent (use Kinyoun reagent)

B. Sulfuric acid (1%; prepared by slow addition of 1 ml of concentrated sulfuric acid to 99 ml of distilled water)

C. Methylene blue (2.5%) in 95% ethanol

III. PROCEDURE

A. Make thin smears of the material and heat fix.

B. Use a *Nocardia* culture control slide with each procedure.

C. Flood each slide with carbolfuchsin solution for 5 min.

D. Pour off excess.

E. Flood each slide with 50% ethanol, and immediately wash with water.

F. Destain with 1% H_2SO_4, and wash with tap water.

G. Counterstain with methylene blue for 1 min.

H. Rinse with tap water and dry.

I. Examine under oil immersion.

IV. EXPECTED RESULTS

A. Acid-fast organisms will retain the carbolfuchsin and appear red on a blue background.

B. Partially acid-fast organisms require a definitive biochemical identification.

SUPPLEMENTAL READING

Larone, D. L. 1995. *Medically Important Fungi. A Guide to Identification.* ASM Press, Washington, D.C.

6.3.7 Giemsa Stain for *Histoplasma capsulatum*

I. PRINCIPLE

Giemsa stain is used for examining intracellular structures and is applied to primary specimens of bone marrow tissue and WBCs in which *Histoplasma capsulatum* is suspected.

II. SPECIMEN

Thin smears of the bone marrow specimen are prepared as follows:

A. Place a small drop of the specimen toward one end of a clean glass slide. Place the edge of a second slide on the first one near the drop, wait a moment for the bone marrow to spread along the edge, and push the second slide toward the far end. This is the thin film.

B. Allow to air dry.

III. MATERIALS

Giemsa stain solution (available from Sigma Chemical Co, St. Louis, Mo.)

IV. QUALITY CONTROL

A. This is a nonstandardized procedure. Materials are not available for QC slides of the organism.

B. However, QC of the stain itself can be performed by the appropriate staining reaction with WBCs in the smear. Normal cells will have a light blue to violet cytoplasm.

V. PROCEDURE

A. Fix smear in absolute methyl alcohol for 1 min.

B. Immediately drain off alcohol and allow the smear to dry.

C. Flood the slide with Giemsa stain that has been diluted 1 : 10 with distilled water (prepared fresh) for 5 min.

D. Wash the slide with water and air dry. *Do not blot dry.*

VI. EXPECTED RESULTS

A. Necrotic cells in the specimen will have pink cytoplasm, in contrast to light blue to violet-lavender cytoplasm for normal cells.

B. Phagocytized yeast cells will stain light to dark blue, and each will have a clear halo around it. Look for purple pseudoencapsulated yeast forms of *Histoplasma capsulatum* inside PMN cells and monocytes.

6.4 Specimen Processing

I. INTRODUCTION

Refer to previous sections of this book for processing specimens. In general, the procedures used for bacterial cultures are also used for fungal cultures.

Sampling distribution may affect the likelihood of fungal isolation. Also, the number of fungal organisms per volume (or gram) of specimen is less than that of bacteria. To compensate for these problems, increase the volume or mass of the specimen, if possible, over those needed for bacterial isolation. Also the number of fungal cells in blood may be low, and the organisms may be found inside leukocytes. Lysis of the WBCs and concentration of free-floating and released fungal cells help to increase the likelihood of success for culturing blood (e.g., lysis-centrifugation method with the Wampole Isolator system). Alternatively, broth cultures of blood can be incubated in automated blood culture instruments for extended periods (≤21 days) to recover fungal growth. Direct comparisons of broth culture systems (e.g., Organon-Teknika's BacT/Alert; Becton-Dickinson's NR660 instrument; Difco's ESP) with lysis methods indicate that the lysis-centrifugation method provides better recovery (higher positivity rate, faster detection times) of *Histoplasma capsulatum* and *Cryptococcus neoformans* from blood. Blood culture systems that include broth medium with paddles of solid medium (Septi-Chek) were found to be less useful than automated blood culture instruments but better than direct plating of blood onto solid medium.

II. PRINCIPLES

A. As noted earlier, liquid specimens may have low concentrations of fungal organisms and will require centrifugation to increase the fungal cell concentration (in the pellet). Hard specimens in which the organisms grow internally should be minced or pulverized before inoculation. Hard specimens (e.g., bone) which cannot be fragmented should be immersed and agitated in broth medium and incubated for sufficient time to allow growth of moulds (14 days).

B. Media inoculated for cultivation of fungi should be incubated at 28 to 30°C, including dimorphic systemic fungi that will not produce yeast forms on primary isolation. Only *Malassezia furfur* requires ≥35°C for adequate growth. When this organism is suspected, media should be supplemented with sterile olive oil.

C. Fungal growth can be overwhelmed by bacteria in specimens from nonsterile sites. To select for fungal pathogens, specimens from nonsterile sites should be inoculated onto media containing antibacterial agents as well as antimicrobial-agent-free media. Cycloheximide should not be used as one of the antibacterial agents because some pathogenic yeasts and moulds are inhibited by it.

D. Acidic conditions are used in some media to reduce bacterial growth, but acidic conditions also lower fungal growth rates. Primary media with pH ranges of 6.5 to 7.2 are preferred for fungal growth. Bacterial overgrowth is best reduced by the addition of β-lactam and aminoglycoside antimicrobial agents to the media.

Vaginal cultures may be incubated for only 7 days, because the agents of concern, yeasts, grow relatively quickly. Other cultures should be incubated for 4 weeks at 30°C.

Review cultures each day during the first week after specimen inoculation

II. PRINCIPLES (continued)

onto media. Review twice a week for the next 2 weeks and once in the fourth week (day 28). Most fungi will produce colonies by the end of the third week. Some systemic pathogens, *Histoplasma capsulatum* and *Blastomyces dermatitidis*, may require 4 to 8 weeks before growth is seen. Fortunately, diagnosis of these diseases is usually obtained by nonmicrobiological methods (serologic tests, histopathology). Cultures then provide unequivocal confirmation of the disease etiology.

III. MEDIA SELECTION

There is a large selection of mycologic media that could be used for primary specimen processing. However, it is cost ineffective to utilize several fungus-specific media routinely. BHI agar (BHIA) with antibacterial agents (not cycloheximide) and inhibitory mould agar (IMA) are useful media for recovering fungi from contaminated specimens. These may be used alone or together for such specimens. Additional media, such as potato dextrose agar for sterile-site specimens, may be included. IMA is a useful medium but can cause some inhibition of fungal growth and may not provide much benefit in respiratory cultures. BHIA with antimicrobial agents is adequate in most cases. Dermatophytes produce characteristic conidia on nutritionally poor media. Sabouraud glucose agar with penicillin and gentamicin is a good choice for specimens suspected of containing dermatophytic fungi.

Clinical laboratories that serve as reference centers can obtain excellent recovery of fungi from specimens by using one or more of the media listed below. Media for bacteria, such as BAP, CHOC, and buffered-charcoal yeast extract agar (BCYEA), will also support good growth of fungi. These media can serve for fungal cultivation from sterile body sites, but must be incubated for longer periods to yield optimal results. BCYEA is a useful medium for the isolation of *Histoplasma capsulatum* from contaminated respiratory specimens.

Choice of medium

Select one medium from each column for specimens containing normal bacterial flora. Only one medium from column A is needed for sterile body sites.

A (sterile body site)[a,b]	+	B (contaminated body site)
BHI[c] with antimicrobial agent[d]		SGA (Emmon's modified) + penicillin-gentamicin or chloramphenicol-gentamicin[d]
Potato dextrose agar		IMA
Mould agar[e]		Niger seed agar or equivalent if India ink reveals encapsulated yeasts

[a] Add BGYEA or YEP for recovery of *Histoplasma capsulatum*.
[b] Add CHROMagar if blood culture contains yeast cells (to detect mixed infection).
[c] Abbreviations: SGA, Sabouraud glucose agar; YEP, yeast extract-phosphate.
[d] Preferred medium. If *Malassezia* species are possible, add 2 to 3 drops of preservative-free sterile olive oil.
[e] Mould agar is IMA without antimicrobial agents.

IV. METHODS

A. Direct inoculation (for details, *see* CMPH 6.6)

Media for many specimens can be directly inoculated. Three to 5 drops of fluid should be inoculated into each tube of medium. Inoculate medium in plates with 2 or 3 drops of specimen, and then streak the specimen over the medium surface. Use up to 0.5 ml of certain sterile fluids, such as CSF, to inoculate tube medium.

1. Abscess aspirates
2. Bone marrow aspirates
3. CSF (<2.0 ml)
4. Vaginal swabs

IV. METHODS
(continued)

Vortex swab in 0.5 ml of sterile distilled water and culture the water, or directly swab the medium surface.

5. Body fluids (<2.0 ml)
   If the fluid contains a clot or membranous material, mince such inclusions with scalpels and then inoculate medium. Use multiple tubes to culture the entire specimen.
6. Hairs and skin scrapings
   Press firmly onto medium surface.
7. Nail
   Pulverize or render fingernails or toenails into shavings by using scalpels. Press fragments firmly onto medium surface.
8. Bronchial brushings
   Vortex brush in sterile water. Inoculate medium with fluid. Place brush on IMA.
9. Bronchial washings (<1.0 ml), fluid

B. Concentration (for details, *see* CMPH 6.6)

Large volumes of fluids should be concentrated by centrifugation (1,500 to 2,000 × *g* for 10 min). Use the resulting pellet for culture and KOH exam.

1. Body fluids (>2.0 ml)
   If clot or membranous material is present, mince it with scalpels and combine with concentrated fluid. Fluids that are too dense for effective contrifugation may be thinned with sterile distilled water before centrifugation.
2. Urine (>2.0 ml)
   Do not process more than 50 ml.
3. CSF (>2.0 ml)
   a. Centrifuge the CSF specimen for 10 min.
   b. Without disturbing the sediment, aseptically remove the supernatant with a sterile capillary pipette. This may be used for antigen testing.
   c. Suspend the sediment.
   d. Use the sediment to inoculate media.
4. Sputum
   Digestion of sputum with *N*-acetyl-L-cysteine is of unproven value and is not recommended.

C. Mincing or homogenization (for details, *see* CMPH 6.6)

Biopsy samples, tissues, and nails must be processed to increase the recovery of microorganisms. With sterile scalpels, mince specimens thoroughly into small pieces in a petri dish with a few drops of sterile distilled water. Inoculate the isolation medium with numerous small pieces of the tissue. If *H. capsulatum* is suspected, it may be necessary to homogenize the specimen with a tissue grinder after mincing. Inoculate appropriate media with the homogenate and pieces of the minced tissue. Place inoculated media in sealed bags or wrap in shrink seal to diminish desiccation. The incubator should be humidified.

D. Incubation

Routine cultures must be incubated at 30°C for 4 weeks. Primary cultures should be retained for the entire 4 weeks even when a fungus is isolated. This will ensure that slow-growing fungi are not overlooked. Use shrink seals or tape to reduce evaporation of moisture from agar.

Vaginal cultures

Screen routine vaginal cultures for yeasts, and incubate them at 30°C for 7 days only.

6.5.1 Examination of Fungal Growth on Primary Media

(*Manipulate cultures in biosafety cabinet.*)

It is imperative that these cultures be handled in a biosafety hood.

Read primary plates daily for the first week, every other day for the second week, and twice weekly for the remaining two weeks. When growth appears, differentiate between yeast and filamentous forms (moulds) that may require microscopic examination. Use wet mounts or stain with lactophenol cotton blue (LPCB) (*see* procedure 6.5.2). Proceed as described in procedure 6.5.3 below. If the isolate suggests an actinomycete, examine with a Gram stain and with a modified acid-fast stain (*see* procedure 6.3.6). The steps for identification of actinomycetes are summarized in procedure 6.10.

Several key characteristics of the fungus can be obtained at the time. For yeast, these include the presence of capsule, budding characteristics (single or multiple), size, morphology, and colony color. For mould, characteristics include sporulation, presence of septa, and color (dematiaceous).

For the moulds, see procedure 6.8. If CHROMagar is used in the primary setup, a presumptive identification of a yeast may be possible (*see* procedure 6.5.4).

6.5.2 Preparation of Culture Material for Microscopic Exam

Once colony formation has occurred, the organism should be viewed microscopically. For some fungi (e.g., zygomycetes), intact colonies should be viewed initially with a dissecting microscope because it provides a good indication of growth characteristics. The list below indicates the general methods for colony preparation and staining.

I. YEASTS

A. Wet preparation (with or without Tween 20)

B. Stains

1. LPCB
2. India ink

II. MOULDS

A. Wet preparation (with Tween 20)

B. Scotch Tape (or pinworm paddle)

C. Tease preparation

D. Slide culture

E. Stains

1. LPCB
2. Ascospore

6.5.3 Subsequent Testing

Based on the type of organism and special characteristics, several tests may be performed to obtain an identification of a primary culture (Table 6.5.3–1). If the fungal colonies are not well separated from bacterial colonies or there is insufficient material for testing, subculture the organism (*see* EPCM procedure 6.6, Subcultures).

PART 1 Yeast Colony Examination with Sterile Distilled Water with 0.05% Tween 80 Wetting Agent

I. PRINCIPLE

The addition of a wetting agent to any suspected pathogen before making wet preparations will help prevent infectious conidia from escaping.

II. MATERIALS

Include QC information on reagent container and in QC records.

A. Flask (125 ml)
B. Distilled water (100 ml)
C. Tween 80 (0.05 ml)
   1. Add 0.05 ml of Tween 80 to 100 ml of distilled water.
   2. Autoclave at 121°C for 15 min.
D. Store at room temperature for three months.

Table 6.5.3–1 Summary of subsequent tests

Primary growth type	Key characteristic	Further tests
Yeast	Capsulated	Caffeic acid disk, urease
	Nonencapsulated	Urease, germ tubes, *Candida albicans* screen
	CHROMagar pigmentation	Color chart Determine mixed infection May allow presumptive identification of *C. krusei, C. glabrata, C. albicans*
Mould	Nonseptate hyphae	Likely zygomycete, subculture to PDA
	Septate hyphae	Evaluate reproductive and hyphal characteristics Subculture to one or more media (*see* subculture procedure) Accu-Probe if suspect systemic pathogen (*Histoplasma, Blastomyces, Coccidioides* spp.)

III. PROCEDURE

A. Place a drop of the 0.05% Tween 80 onto a slide.

B. With a loop, touch the surface of the suspicious colony, being sure that some fungal material adheres to the loop. Gently rub the material in the drop of Tween.

C. Coverslip and examine with a microscope.

SUPPLEMENTAL READING

Haley, L. D., and C. S. Callaway. 1978. *Laboratory Methods in Medical Mycology*, p. 111–112. CDC publication #78-8361. U.S. Government Printing Office, Washington, D.C.

PART 2 LPCB Solution

I. PRINCIPLE

LPCB stain is excellent for examination of fungal material. The phenol kills the fungi, and the lactic acid increases preservation. China (cotton) blue is a stain for chitin and cellulose.

II. MATERIALS

A. Available from several sources (for example, Remel no. 40-028, Lactophenol Aniline Blue)

B. Store at room temperature for up to 6 months.

III. QUALITY CONTROL

A. Each lot of reagent should be tested. Make a tease preparation of a *Penicillium* mould in reagent. The cell walls of the organism should appear blue owing to staining by cotton blue.

B. Check stain for presence of extraneous material (fungal elements, etc.).

SUPPLEMENTAL READING

Haley, L. D., and C. S. Callaway. 1978. *Laboratory Methods in Medical Mycology*, p. 30. CDC publication #78-8361. U.S. Government Printing Office, Washington, D.C.

6.5.4 Tests Useful for Yeast Identification from Primary Cultures

Only a few tests are available to identify a yeast when first isolated. Colony color (pigment) and microscopic morphology (Table 6.5.4–1) are helpful. Tests described below provide only presumptive identification (an exception is the AccuProbe for *Cryptococcus neoformans* [Gen-Probe], which is described elsewhere in this handbook). Subculture of the yeast colony onto appropriate media will lead to yeast identification (*see* procedure 6.6, Subcultures). Isolates from medically significant specimens must be identified to species level. Stopping at the presumptive identification step could lead to inappropriate therapy.

Based on pigment, microscopic morphology, and presumptive test results summarized in Table 6.5.4–2, identification schemes are presented in Fig. 6.5.4–1 and 6.5.4–2. The germ tube test provides one of the most rapid approaches to the *presumptive* identification of *C. albicans* (Fig. 6.5.4–3).

PART 1 CHROMagar

I. PRINCIPLE

CHROMagar contains enzymatic substrates that are linked to chromogenic compounds. When specific enzymes cleave the substrates, the chromogenic substrates produce color. The action of different enzymes produced by yeast species results in color variations useful for the presumptive identification of some yeasts. Colonial variation is seen for some species. The medium is best suited for *Candida krusei, C. albicans, C. tropicalis*, and *Trichosporon* spp. The medium should not be used singly to make a definitive identification of a yeast.

It contains antimicrobial agents to inhibit bacterial contamination in primary specimens.

II. USE

CHROMagar is useful for the detection of mixed infections (especially in wounds, blood, and urine) and as an additional test to resolve difficult identifications.

III. CAUTIONS

The test provides only presumptive identification of the four yeasts listed above. Individual clinical laboratories should test at least five isolates of each of those four species to become aware of the color variation that may occur. Strict adherence to the manufacturer's incubation conditions must be maintained. Enzymatic reactivity is sensitive to temperature. Review the manufacturer's description before performing this test.

IV. SOURCE

Hardy Diagnostics (cat. no. G34)

V. QUALITY CONTROL ORGANISMS

C. albicans, C. krusei, Escherichia coli

VI. PROCEDURE

A. Plate primary specimens by procedures typical for the specimen.

B. Subcultures should be streaked for isolation.

C. Incubate at 35°C, humidified chamber, dark, 48 to 72 h (not <48 h).

Table 6.5.4–1 General list of yeast species or genera based on colony color[a]

White, cream or tan	Brown or black	Salmon, pink, or red
Blastoschizomyces	*Aureobasidium*	*Rhodotorula*
Candida	*Phaeoannellomyces elegans*	*Sporobolomyces*
Cryptococcus	*Phaeoannellomyces werneckii*[b]	
Geotrichum[c]	*Phaeococcomyces exophiala*[d]	
Kloeckera	*Ustilago*	
Malassezia		
Prototheca[c]		
Saccharomyces		
Trichosporon		
Ustilago		
Teleomorphs[e]		
Debaryomyces		
Hansenula		
Kluyveromyces		
Pichia		

[a] Based on growth on Sabouraud glucose agar (Emmon's modification), potato dextrose agar, malt extract agar, or BHI agar.

[b] Previously *Exophiala werneckii.*

[c] Not a yeast but produces yeastlike colonies on mycological media.

[d] Synanamorph of *Wangiella dermatitidis.*

[e] Sexual genera associated with some species of the anamorphic genera.

Table 6.5.4–2 Presumptive identification tests for yeasts on primary culture

Test	Procedure	Organism
CHROMagar	6.5.4 (part 1)	*C. krusei, C. albicans, C. tropicalis, Trichosporon* spp.
Germ tube	6.5.4 (part 2)	*C. albicans*
C. albicans screen	6.5.4 (part 3)	*C. albicans*
Rapid urease	6.5.4 (part 4)	*Cryptococcus* spp., *Rhodotorula* spp., *Trichosporon* spp. (variable), *C. krusei* (variable), *Malassezia pachydermatis, C. lipolytica* (i.e., basidiomycetous yeasts)
Rapid nitrate reductase	6.5.4 (part 5)	*Cryptococcus albidus* (+), *C. neoformans* (−), *C. terreus* (+)
Caffeic-acid disk	6.5.4 (part 6)	*Cryptococcus neoformans*
India ink	6.3.5	*Cryptococcus* species

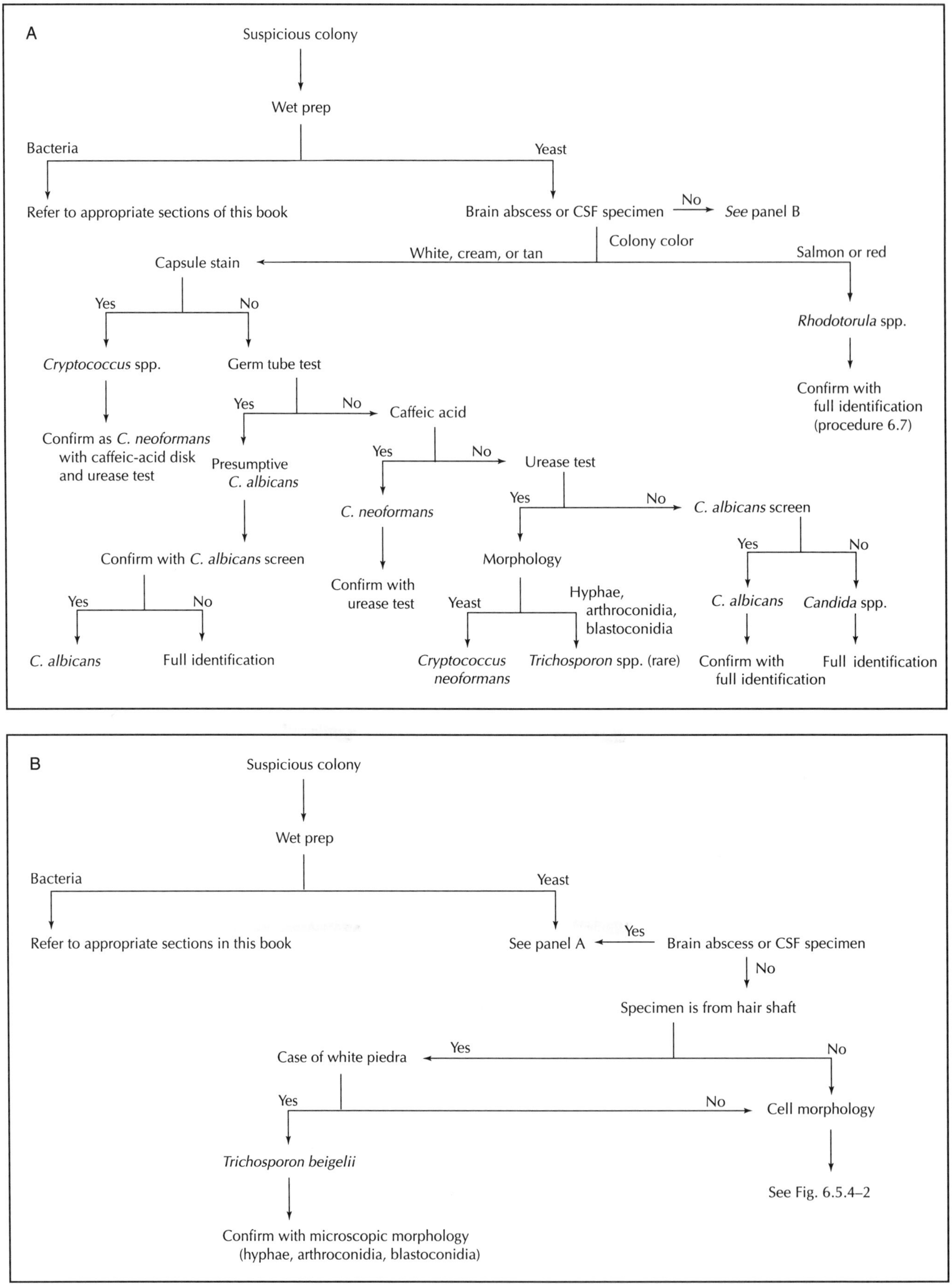

A
Suspicious colony
Wet prep
Bacteria
Yeast
Refer to appropriate sections of this book
Brain abscess or CSF specimen
No
See panel B
Colony color
White, cream, or tan
Salmon or red
Capsule stain
Rhodotorula spp.
Yes
No
Cryptococcus spp.
Germ tube test
Confirm with full identification (procedure 6.7)
Confirm as C. neoformans with caffeic-acid disk and urease test
Yes
No
Caffeic acid
Presumptive C. albicans
Yes
No
Urease test
C. neoformans
Yes
No
C. albicans screen
Confirm with C. albicans screen
Morphology
Yes
No
Confirm with urease test
Yeast
Hyphae, arthroconidia, blastoconidia
C. albicans
Candida spp.
Yes
No
C. albicans
Full identification
Cryptococcus neoformans
Trichosporon spp. (rare)
Confirm with full identification
Full identification
B
Suspicious colony
Wet prep
Bacteria
Yeast
Refer to appropriate sections in this book
See panel A
Yes
Brain abscess or CSF specimen
No
Specimen is from hair shaft
Case of white piedra
Yes
No
Yes
No
Cell morphology
Trichosporon beigelii
See Fig. 6.5.4–2
Confirm with microscopic morphology (hyphae, arthroconidia, blastoconidia)

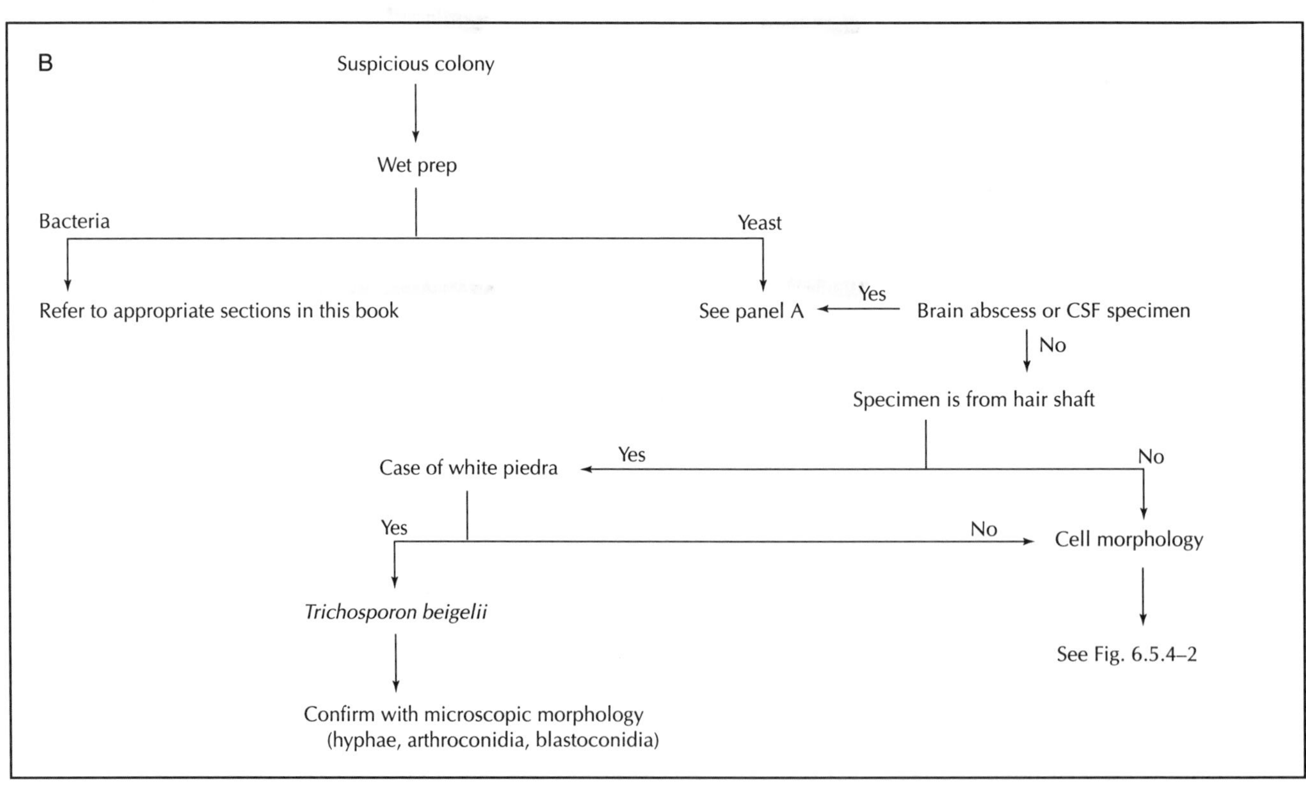

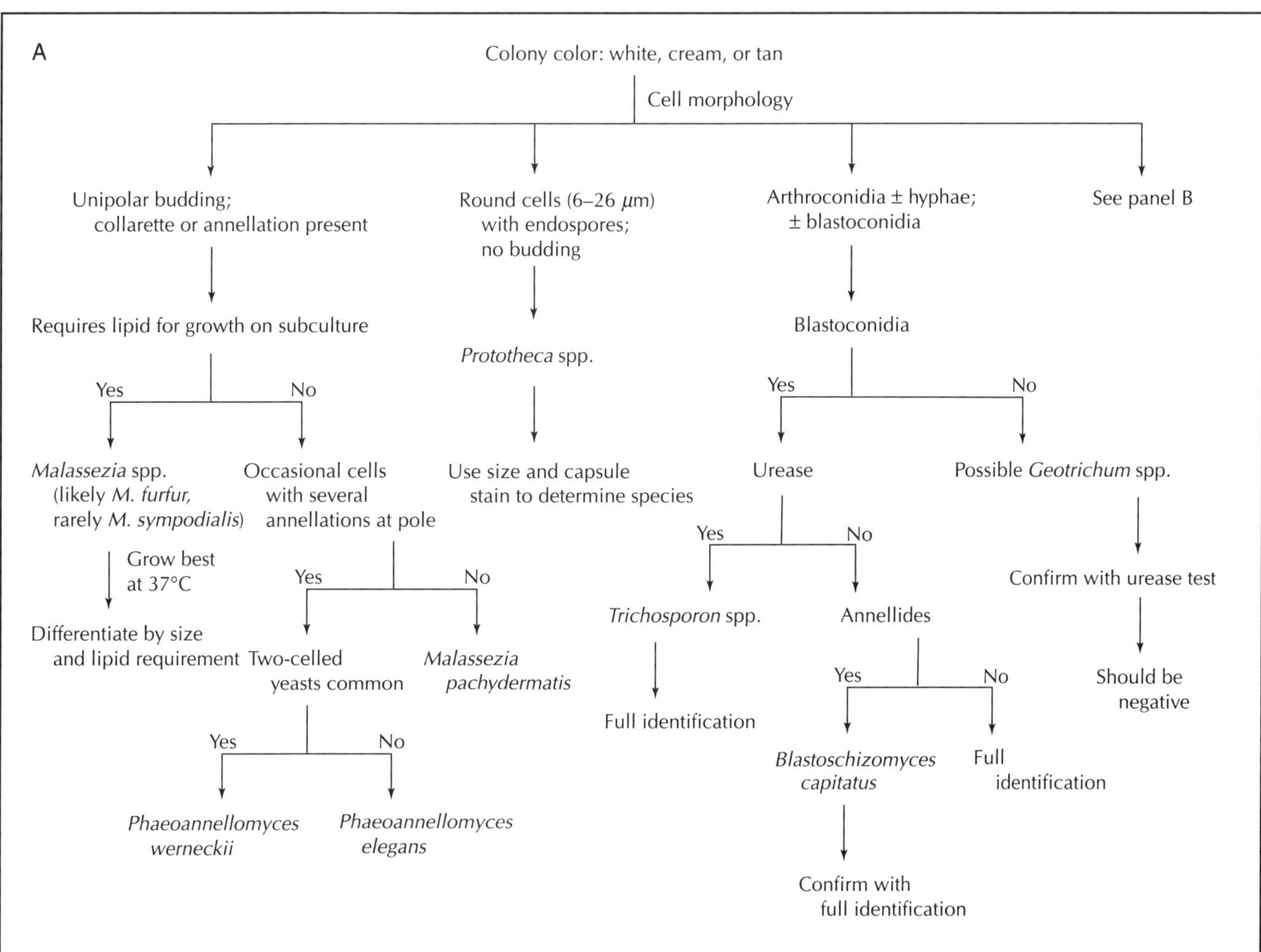

Figure 6.5.4–2 Simple tests to designate genera of yeast cells seen on primary culture. (A, B) Identification procedure when colony color is white, cream, or tan. (C) Identification procedure when colony color is salmon, pink, or red. (D) Identification procedure when colony color is brown or black.

(continued)

6 Mycology and Aerobic Actinomycetes

Figure 6.5.4–1 Flow chart for evaluating a possible yeast species on primary culture. (A) Identification procedure when yeast cells are found in brain abscess or CSF specimen. See Fig. 6.3.1–1 for common yeast agents isolated from different body sites. (B) Identification procedure for yeasts when the specimen is not brain abscess or CSF.

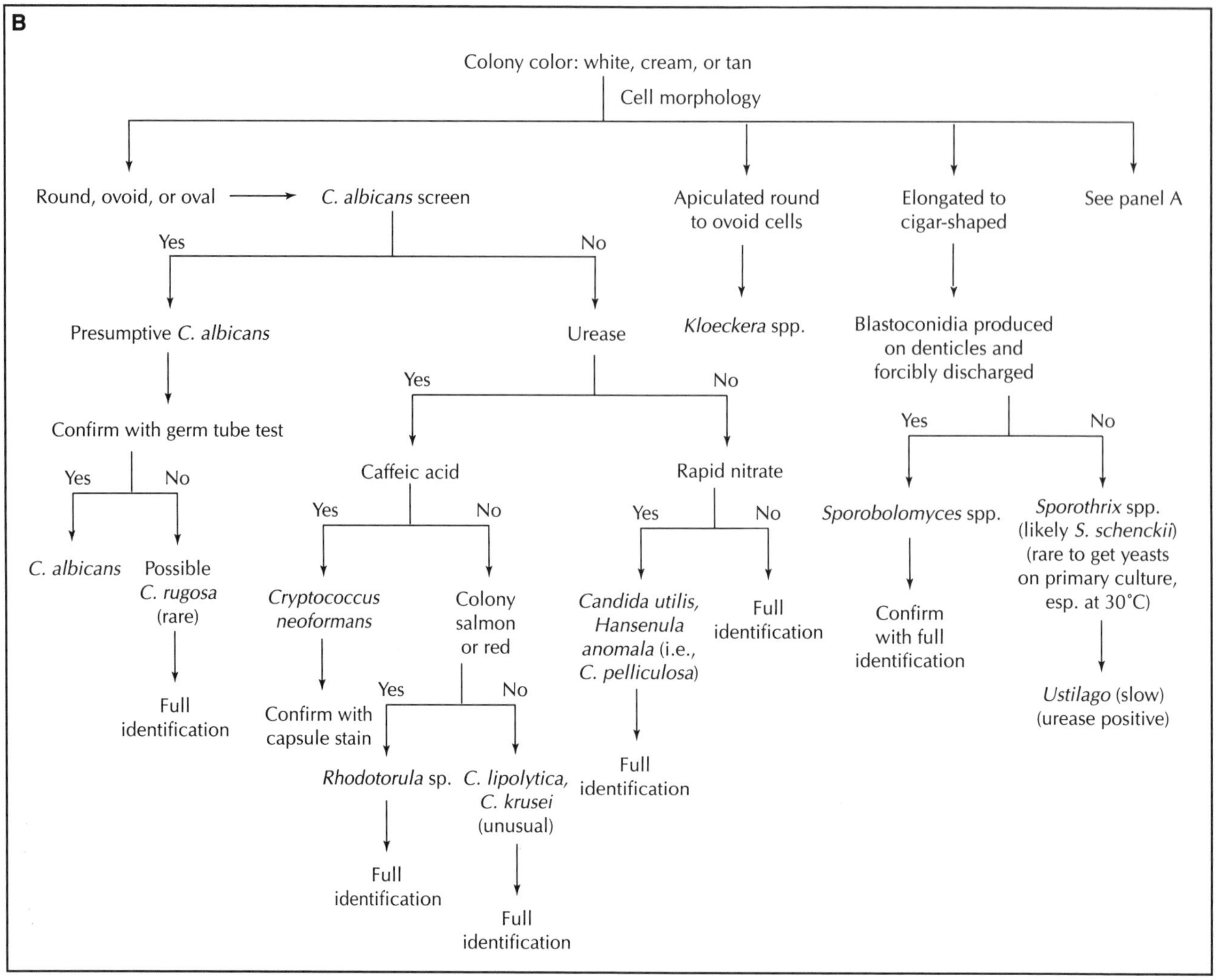

B

Colony color: white, cream, or tan — Cell morphology

- Round, ovoid, or oval → *C. albicans* screen
  - Yes → Presumptive *C. albicans* → Confirm with germ tube test
    - Yes → *C. albicans*
    - No → Possible *C. rugosa* (rare) → Full identification
  - No → Urease
    - Yes → Caffeic acid
      - Yes → *Cryptococcus neoformans* → Confirm with capsule stain
      - No → Colony salmon or red
        - Yes → *Rhodotorula* sp. → Full identification
        - No → *C. lipolytica*, *C. krusei* (unusual) → Full identification
    - No → Rapid nitrate
      - Yes → *Candida utilis*, *Hansenula anomala* (i.e., *C. pelliculosa*) → Full identification
      - No → Full identification
- Apiculated round to ovoid cells → *Kloeckera* spp.
- Elongated to cigar-shaped → Blastoconidia produced on denticles and forcibly discharged
  - Yes → *Sporobolomyces* spp. → Confirm with full identification
  - No → *Sporothrix* spp. (likely *S. schenckii*) (rare to get yeasts on primary culture, esp. at 30°C) → *Ustilago* (slow) (urease positive)
- See panel A

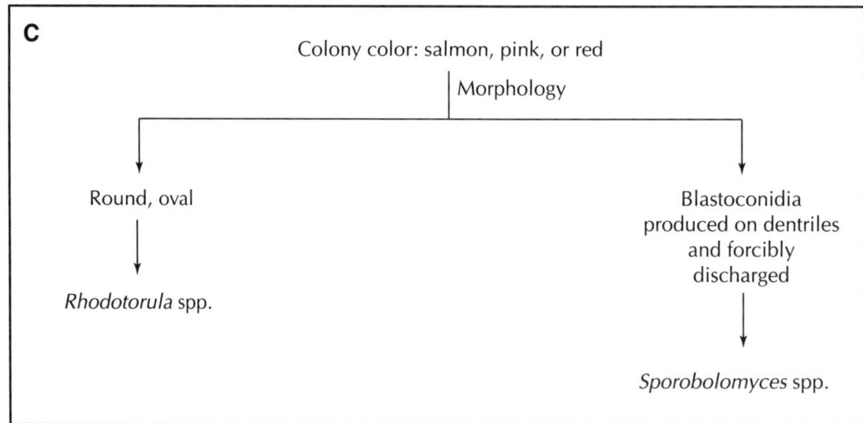

Figure 6.5.4–2 *(continued)*

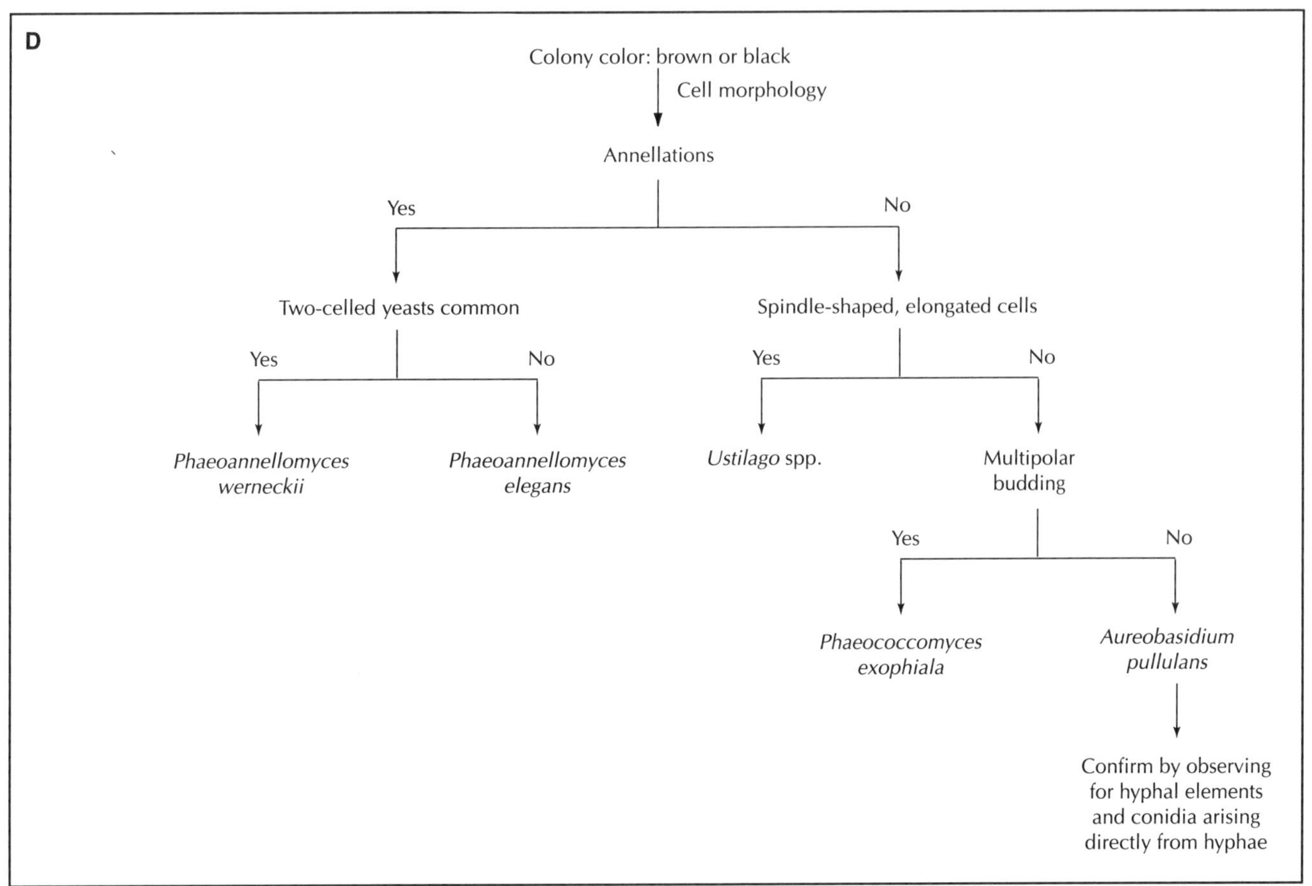

Figure 6.5.4–2 *(continued)*

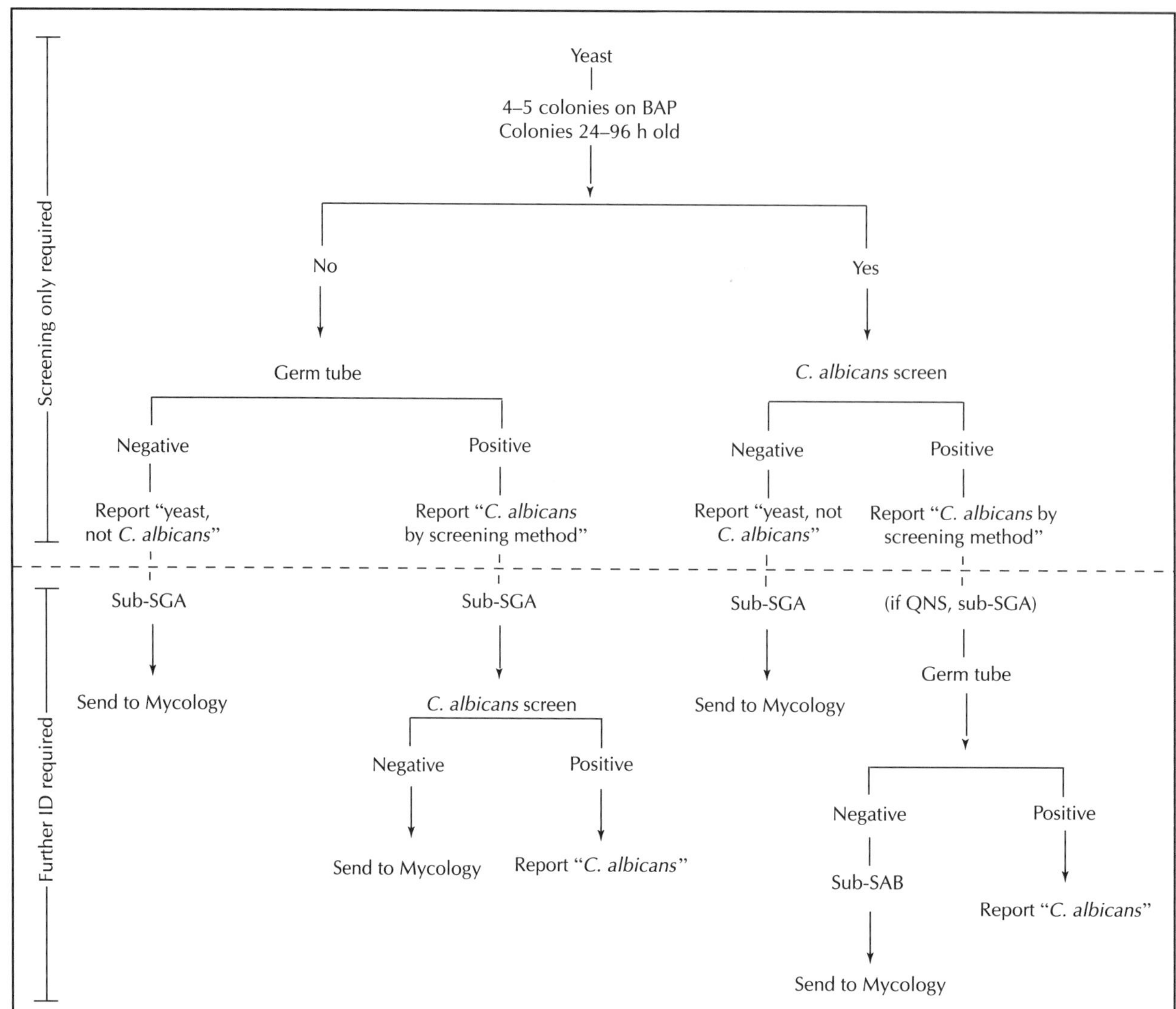

Figure 6.5.4–3 Algorithm for identification of *C. albicans* on primary culture. Follow specimen guidelines for significance of specimens submitted for routine bacteriology cultures. When a protocol requires you to do screening tests or identification for yeast, proceed as outlined in this diagram. If the yeast is from a sterile body fluid or blood, you must have two tests positive to call the organism *C. albicans* (e.g., if the *C. albicans* screen was positive, perform a germ tube test). If the organism is found in a blood culture, subculture bottles to CHROMagar to check for purity. Proceed with workflow for day 1 and check subculture plates on day 2. SGA, Sabourand glucose agar.

VI. PROCEDURE *(continued)*

D. Use only isolated colonies to determine the organism's characteristics for purposes of presumptive identification.

E. Do not use the isolated colonies to perform subsequent screening tests for identification or for biochemical identification systems. Subculture onto a nonselective medium before performing additional identification tests on the isolate.

VII. INTERPRETATION

A. ***C. albicans***

A medium-sized, green, smooth, matte colony with a very slight green halo in the surrounding medium.

B. ***C. tropicalis***

A medium-sized, smooth, matte colony which is blue to blue-grey with a paler pink edge: the colony may have a dark brown to purple halo which diffuses into the agar.

C. ***C. krusei***

A large, spreading, rough, pink colony with a pale-pink to white edge

Other yeasts produce colonial characteristics that are useful to obtain a suggestive identification. According to Odds and Bernaerts (1), *Trichosporon* species (e.g., *T. beigelii*) produce a distinctive colony, particularly with prolonged incubation. The manufacturer of CHROMagar does not include this isolate in its list of organisms that can be presumptively identified.

If the test is used to determine mixed infection in primary specimens, additional testing must be performed. If the test is used to help resolve a difficult identification, further tests may not be required.

REFERENCE

1. **Odds, F. C., and R. Bernaerts.** 1994. CHROMagar Candida, a new differential isolation medium for presumptive identification of clinically important *Candida* species. *J. Clin. Microbiol.* **32:**1923–1929.

PART 2

Germ Tube Test

I. PRINCIPLE

One of the simplest tests for the rapid presumptive identification of *Candida albicans* is the production of germ tubes in calf serum.

II. MATERIALS

Include QC information on reagent container and in QC records.

A. Sterile calf serum

1. Purchased from BioWhittaker, catalog no. 14-401E
2. Aliquots (0.5 ml) are stored at −10 to −30°C.
3. The outdate is 6 months from the date the serum is aliquoted.
4. The racks of aliquots in use are stored at 2 to 8°C.

B. Control organisms

1. *C. albicans* ATCC 60193
2. *C. tropicalis* ATCC 66029
3. Maintained on Sabouraud dextrose agar at 30°C

C. Sterile applicator sticks

D. Heat block at 37 ± 1°C

E. Microscope slides

F. Sterile Pasteur pipettes

G. Coverslips

III. QUALITY CONTROL

A. Performed each day of use

B. Controls

1. Positive, *C. albicans* ATCC 60193
2. Negative, *C. tropicalis* ATCC 66029

IV. PROCEDURE

A. Lightly touch a yeast colony with a wooden applicator stick.
B. Suspend the yeast cells in an appropriately labeled tube of calf serum.
C. Incubate at 37 ± 1°C for 2.5 to 3 h.
D. Place a drop of the suspension on a microscope slide.
E. Place a coverslip over the suspension.
F. Examine under high power for the presence or absence of germ tubes. A germ tube appears as a short lateral extension from the yeast cell and does not have a constriction (septum) where it meets the yeast cell. A constriction where the lateral extension meets the yeast cell is produced by pseudohyphae or budding cell. Observe a minimum of five germ tubes before calling the isolate positive.

V. LIMITATIONS

Sabouraud glucose agar is the best medium to isolate yeast for germ tube production; BAP is an acceptable substitute. It should be noted that the some *C. tropicalis* isolates produce pseudohyphae and some also produce apparent germ tubes that require careful observation to discriminate from *C. albicans.*

PART 3

C. albicans Screen

I. PRINCIPLE

C. albicans produces two exoenzymes, L-proline aminopeptidase and *β*-galactosaminidase, when grown on appropriate media. The detection of these enzymes with a 30-min chromogen-coupled, substrate-impregnated disk affords *presumptive* identification of *C. albicans.*

II. USE

This test serves for presumptive identification of *C. albicans* on primary culture and for confirmation of germ tube-positive yeast cells.

III. SOURCE

A. Carr-Scarborough, *C. albicans* screen (1)
B. Murex CA-50

IV. QUALITY CONTROL

A. Organisms
C. albicans ATCC 60193 and *Cryptococcus laurentii* ATCC 18803
B. Perform testing for each lot of kits received.
1. Randomly select two tubes from each lot of kits received to be tested.
2. Perform the test as described below.

V. PROCEDURE

A. Remove a disk from the vial and place it in one of the tubes provided. Add 1 drop of demineralized water. With an applicator stick, pick up a heavy visible "paste" of the yeast and rub it onto the disk. Incubate at 35°C for 30 min. After incubation, add 1 drop of 0.3% NaOH and examine against a white background for a distinct yellow color. No color is considered negative.
B. Add 1 drop of cinnamaldehyde reagent. Development of a pink to red color within 1 min is a positive proline aminopeptidase test. No color or a slight yellow color is a negative test.

VI. RESULTS

The development of a distinct yellow color upon the addition of 0.3% NaOH is a positive test for β-galactosaminidase. No yellow color indicates the absence of the enzyme. The development of a red color, within 1 min after adding cinnamaldehyde reagent, is a positive test for L-proline aminopeptidase. No color change or development of a slight yellow color is a negative test.

VII. INTERPRETATION

A positive test, i.e., both enzymes present, provides presumptive identification of *C. albicans*. Rare isolates of *C. rugosa* and *C. tropicalis* may be positive. Rule out these possibilities by performing a germ tube test. If an isolate is germ tube positive and *C. albicans* screen positive, then its identification is confirmed.

VIII. LIMITATIONS

A. A viable, fresh (18- to 24-h) culture on Sabouraud dextrose agar of each control organism is required for testing.

B. The test requires several colonies to produce the needed inoculum paste. Mixed culture could cause a false-negative result. A heavy paste of the organism on the disk is necessary. Insufficient inoculum may result in a negative test.

C. Use only pure cultures. The inoculum should come from blood agar, Sabouraud glucose agar, Mycosel (Mycophil, mycobiotic agar), or potato dextrose agar. The effect of other agars is not well known.

REFERENCE

1. **Carr-Scarborough Microbiologicals.** 1994. Package insert no. 1423-1290. Carr-Scarborough Microbiologicals, Inc., Decatur, Ga.

PART 4

Urease Disk Test for the Differentiation of Yeasts

I. PRINCIPLE

The Urease Disk Test is a rapid detection test to distinguish basidiomycetous yeasts (e.g., *Cryptococcus* spp.) from ascomycetous yeasts. Urease splits urea to ammonia and carbon dioxide, which raises the pH and causes a color shift in the phenol red indicator from amber to pinkish red.

II. MATERIALS

Include QC information on reagent container and in QC records.

A. Urea disks
   1. Source, Difco, catalog no. DF1625-33
   2. Expiration date established by manufacturer

B. Sterile distilled water stored at 2 to 8°C

C. Sterile tubes, 16 by 125 mm

D. Inoculation loop

III. QUALITY CONTROL

A. Performed on each new lot number

B. Controls
   1. Positive control, *Cryptococcus neoformans* ATCC 66031
   2. Negative control, *Candida albicans* ATCC 60193
   3. Reagent control, uninoculated

IV. PROCEDURE

A. Emulsify a colony of the yeast in 0.5 ml of sterile distilled water in a 16-by-125-mm tube.
B. Add one urea disk.
C. Incubate at 28°C for up to 72 h.
D. Check at 4 h and daily up to 72 h.

V. EXPECTED RESULTS

Record results on yeast worksheet.
A. Positive: turns from pink to red in color
B. Negative: no color change
C. Urease activity by species:

Candida albicans	−
C. tropicalis	−
C. parapsilosis	−
C. krusei	+[a]
C. guilliermondii	−
C. kefyr	−
C. stellatoidea	−
All *Cryptococcus* species	+ (strong)
Torulopsis glabrata	−
Saccharomyces spp.	−
Rhodotorula spp.	+
Trichosporon spp.	+[a]
Geotrichum spp.	−

[a] Variable depending on strain

PART 5

Nitrate Swab Rapid Test for Yeast Nitrate Reduction

I. PRINCIPLE

Nitrate reduction can be used to differentiate species of *Cryptococcus*. *C. neoformans* does not reduce inorganic nitrate. *C. albidus* and *C. terreus* are nitrate positive. Organisms that reduce nitrate do not necessarily assimilate nitrate; however, the opposite is true (if they do assimilate nitrate, they must reduce it).

Cotton swabs impregnated with KNO_3–Zephiran are used to test yeasts for the ability to form nitrite from nitrate (presence of nitrate reductase). The formation of nitrite is detected by the addition of a color reagent. The reaction (color change) occurs on the cotton swab.

II. MATERIALS

Include QC information on reagent container and in QC records.

A. Rapid nitrate swab (0.2% nitrate)
1. Purchased from Remel, catalog no. 20-355
2. Stored at 2 to 8°C
3. Expiration date established by manufacturer

B. Nitrate reagent A (0.08% sulfanilic acid in acetic acid)
1. Purchased from Remel, catalog no. 21-239
2. Stored at 2 to 8°C
3. Expiration date established by manufacturer

C. Nitrate reagent B (0.06% *N*-dimethyl-1-naphthylamine)
1. Purchased from Remel, catalog no. 21-242
2. Stored at 2 to 8°C
3. Expiration date established by manufacturer

D. Sterile screw-capped 16-by-120-mm black-top tubes

E. Control stains
1. *C. neoformans*, lab strain, negative control
2. *C. albidus*, lab strain, positive control

III. QUALITY CONTROL

QC is performed upon receipt of the medium.

IV. PROCEDURE

A. A positive and negative control are run each time the test is done.
B. Sweep the swab across several (four to five) colonies. Swirl the swab inside a sterile empty tube to ensure good contact of the organisms and the substrate.
C. Incubate the swab in the tube at 45°C for 10 min or at 35°C for 20 to 30 min.
D. Remove the tubes from the incubator, and place the swab in a tube containing 2 drops each of nitrate reagents A and B.
E. Observe the swab for the development of a pink to red color, indicating a positive test. A positive test should be visible within 3 to 5 min.

V. INTERPRETATION

A. Positive test
The development of a pink to red color on the swab indicates that nitrate has been reduced to nitrite.
B. Negative test
No color development on the swab indicates a negative test.

VI. LIMITATIONS

A. If pink to red yeasts are tested with this method, the upper part of the swab should be observed for color change (to distinguish from the yeast inoculum in the lower part of the swab).
B. Inoculate the swab with a heavy inoculum to insure a clear-cut reaction.

PART 6

Caffeic Acid Disk for *Cryptococcus neoformans* Screen

I. PRINCIPLE

The caffeic acid and ferric citrate are the substrates for the phenol oxidase enzyme, produced only by *Cryptococcus neoformans*. This enzymatic reaction produces melanin, which is absorbed by the yeast cell wall, resulting in a brown pigmentation.

II. MATERIALS

A. Remel Caffeic Acid Disk, catalog no. 21-128
Stored at 2 to 8°C. Use manufacturer's expiration date. Do not use if the color has changed. The disks should be white.

B. Remel Cornmeal Agar Plates, catalog no. 01-330
Stored at 2 to 8°C. Use manufacturer's out date

C. Bacteriological loops or sterile swabs

III. QUALITY CONTROL

A. Performed with each new lot
B. Controls
1. Positive, *C. neoformans*—brown pigment
2. Negative, *C. albicans*—no pigment

IV. PROCEDURE

A. Place the disk on the cornmeal agar plate for rehydration. (Do not use a medium containing glucose; this will inhibit the phenoloxidase activity.)
B. Inoculate the disk with several yeast colonies.
Inoculate the plate for yeast morphology.
C. Replace the lid, and incubate the plate at 35 to 37°C in ambient air.
D. Examine at 30-min intervals for up to 4 h for the production of a brown pigment indicating a positive test.

V. EXPECTED RESULTS

A. The production of a brown pigment after incubating at 35°C indicates phenoloxidase activity in the yeast cell wall.

1. If a brown pigment is observed, report ''presumptive for *Cryptococcus neoformans*'' and proceed with yeast identification workup including API confirmation (procedure 6.7.1).

2. If there is no brown pigment, but *C. neoformans* is strongly suspected, proceed with yeast identification workup. Include a birdseed agar plate.

B. Read the plate for yeast morphology and record results.

SUPPLEMENTAL READING

Hopfer, R. L., and D. Gröschel. 1989. Six-hour pigmentation test for the identification of *Cryptococcus neoformans*. *J. Clin Microbiol.* **2:**96–98.

Remel. *Technical information*, TI no. 21128. Remel, Lenexa, Kan.

6.5.5 Identification of Moulds on Primary Culture

(See procedure 6.8 for a complete guide to mould identification.)

Initial growth of moulds may require up to 4 weeks of incubation at 30°C. Some unusual dermatophytes may require up to 8 weeks. While growth may be manifest, structures required for genus or species identification may not be formed until completion of several rounds of subculture onto specific media. Both macro- and microscopic examination must be applied to mould cultures. For microscopic examination, use either a transparent tape preparation or a tease preparation (*see* procedure 6.8). When working with mould cultures, always keep them in a class II biological safety cabinet and use mounting medium containing a disinfectant (e.g., phenol).

Figure 6.3.1–1 provides a list of the more common moulds that are considered possible pathogens for a given specimen. It is not a complete list because other moulds may be involved in disease formation.

CONSIDERATIONS FOR MOULD OBTAINED ON PRIMARY CULTURE

A. The number of colonies is irrelevant when dealing with infections caused by filamentous fungi.

B. Report single-colony cultures unless the organism is clearly a contaminant. In an equivocal situation use terms such as ''probable'' and ''possible'' contaminant. When such terms are used, it is best to obtain another specimen if possible.

C. In general, the morphology of asexual reproductive structures, if produced, is similar for a given organism regardless of the growth medium. Media selection can determine whether and what kinds of identifiable structures will be produced. However, species descriptions of some genera are based on macroscopic and microscopic characteristics of colonies on a particular medium.

D. The growth rate of moulds on primary culture is not reliable. Cell growth rate is a characteristic useful for organism identification, but is based on observation with standardized media (typically Sabouraud glucose agar or malt extract agar). Growth rate should be based on monitoring colony size (diameter) once initial colony formation is observed and the colony is isolated.

E. Subcultures (*see* procedure 6.6 for an extensive discussion of this step)

If subcultures are needed, chose media that are best suited for the genus you suspect (*see* procedure 6.6). Subculture onto at least two media. Media that are formulated with plant products are best suited for nonfastidious moulds. Selection of the two media should be based on different plant groups (e.g., potato dextrose agar and V-8 Juice agar—both available commercially). Fastidious organisms, such as *Histoplasma capsulatum*, should also be subcultured on animal tissue-based media (e.g., BHI agar with 10% BAP).

F. Media that are clear are best suited for observing characteristics of the colony reverse.

G. The list provided in Table 6.6–1 indicates that subculture media could include a plate or slant of medium containing cycloheximide (CHX), accompanied by the same medium without CHX. A typical media pair is Sabouraud glucose agar with and without CHX (Mycobiotic and Mycosel agar are commercial preparations of Sabouraud glucose agar that contain CHX and chloramphenicol.)

H. Growth characteristics without reproductive-cell formation on primary media can provide helpful clues about an organism's identity. For example, width, morphology, ornamentation, cell wall thickness, and color of hyphae are useful characteristics, in addition to chlamydospores (chlamydoconidia), septa, and colony characteristics.

6.6.1 Inoculation

If an organism is not identifiable on primary culture, subcultures should be performed. Yeasts are subcultured onto specific media based on the test that is planned (e.g., Dalmau method for morphology). Generally, yeasts are subcultured onto media for isolation and purity assessment by standard streaking methods. Moulds are subcultured by several methods, including block inoculation, scarification, or slide culture. Variations of these methods are possible and are sometimes necessary for some cultures.

I. METHODS FOR SUBCULTURING MOULDS

It is imperative that these cultures be handled in a biosafety hood.

Perform all subculture methods in a class II biological safety cabinet.

A. Block inoculation

1. Cut with a sterile scalpel an approximate 0.5-by-0.5-cm block (including agar) of the fungal colony.
2. Place the block upside down onto fresh medium in the center of the plate.
3. Cover the petri dish, and place the dish right side up at 28 ± 2°C.

B. Scarification

1. With a biological probe or a bent, blunt needle, scrape the surface of the mould colony where conidia may be present. A noticeable clump of colony should be attached to the probe.
2. Scratch the surface (i.e., ~1 to 2 mm deep) of fresh medium in the center of the plate with the colony material. The material should be inserted into the scratch but also be exposed at the surface of the medium.
3. Incubate the plate right side up at 28 ± 2°C. This method is less preferable than the block method.

C. Slide culture

Never set up a slide culture on a white, filamentous colony that could be a potential pathogen.

1. Cut three 1-by-1-cm agar blocks from an agar plate and lift them onto the surface of the agar. Arrange these blocks in a triangular pattern on the plate.
2. Inoculate each side of each block with a small amount of hyphae or material from the mould colony.
3. Place a coverslip (sterilized by rapid passage through a flame) over the inoculated agar block.
4. Replace the lid of the petri plate and incubate right side up at 28 ± 2°C.
5. Examine by removing coverslips and placing them on a small drop of lactophenol cotton blue on a glass slide.

   If, after examining the coverslip, it is determined that the slide culture is

I. METHODS FOR SUBCULTURING MOULDS *(continued)*

too immature for an identification, add clean flamed coverslips to the agar blocks and reincubate.

D. Examination

1. The fungal spores will develop in the air space between the agar surface and the coverslip.
2. Monitor the development of mature structures by removing the petri dish cover and examining the edge of the agar block with the low-power objective. If this is a potential pathogen, do not remove the petri dish cover; go to step 3, below.
3. When sporulation appears to be well developed, carefully lift the coverslip off the agar plug and lay it on a drop of lactophenol cotton blue on a second slide.
   Do not drag the coverslip across the agar plug—it will disturb the identifying structural features.
4. Determine how the spores are attached: either directly from the hypha or from specialized structures.
5. Note whether the spore-bearing structures are branched or unbranched.
6. For details about identification, see reference 2 and the flow charts in procedure 6.9.

II. SELECTION OF MEDIA

Selection of media for subculture is the most crucial step for successful and timely identification of moulds when molecular or biochemical methods are unavailable. No single medium supports formation of species-identifiable structures of all moulds. It is incumbent upon the laboratory to select several media to encourage sporulation or morphologic conversion.

Yeasts must also be subcultured, typically onto Sabouraud glucose agar for subsequent biochemical testing and onto a morphology agar (e.g., corn meal-Tween 80 agar). Yeasts may also be subcultured onto differential media (e.g., CHROMagar, Niger seed agar [caffeic acid agar]), and media with and without CHX (*see* Appendix 6–1).

When mould colonies are subcultured, several considerations should be made.

A. A relatively rich medium (e.g., V-8 Juice agar) and a nutritionally poor medium (e.g., potato dextrose agar, oatmeal agar, or potato flake agar) should be used. BHI may be needed to support growth of *Nocardia* spp. (*see* procedure 6.10). Media with animal tissue components (e.g., BHI agar) are not optimal for inducing sporulation but are useful for conversion of systemic fungi.

B. A third medium may be needed to encourage fruiting-body formation (e.g., cottonseed agar, hay infusion agar, or water agar).

C. Media lacking antimicrobial agents should be used for subcultures. If a dermatophyte is subcultured from a primary medium lacking antimicrobial agents, it is useful to include a medium with antibacterial agents to prevent passenger bacteria from overgrowing the culture.

D. Slide cultures may be made with any medium, although clear or moderately translucent media are preferred.

E. Dimorphic fungi convert to yeast or spherule phases more quickly and abundantly on specific media. When conversion is attempted, two plates (or slants) of the same medium should be inoculated; one is placed at 28°C, and the other at 37°C.

F. CHX resistance of moulds may also be tested at this time, if desired. When this test is performed (*see* Appendix 6.11–1), plates of medium with and without CHX should be inoculated and incubated under identical conditions.

Table 6.6.1–1 suggests which media should be used and provides a brief

Table 6.6.1–1 Principles of selected media for fungal subculture

Medium	Source	Indicated use	Brief details of inoculation	Incubation	Comments
V-8 Juice agar	Smith-River (cat. no. 15-V8)	Conidiation of dematiaceous moulds	Standard	28–32°C	Not for pycnidia-forming fungi
Czapek agar	BBL (cat. no. 97794)	*Aspergillus, Penicillium*, and *Paecilomyces* spp.	Standard	28–32°C	Used to produce characteristic features
Cornmeal-Tween 80	Remel (cat. no. 01-328)	Yeast morphology and chlamydospore production	Two parallel lines, followed by "S" across the lines. Coverslip, but be sure some of the streak is outside the core slip	Room temperature, 72 h	Scan reverse with low-power magnification. Scan from outside of coverslip to middle to find characteristic features.
Birdseed agar (also called Niger seed agar)	BBL (cat. no. 97096)	*Cryptococcus neoformans* pigment production	Streak as desired	28–32°C, 72 h	Of the hyaline yeasts, only *C. neoformans* produces a brown pigment on this medium.

II. SELECTION OF MEDIA *(continued)*

description of methods for subculturing certain organisms. Note that all moulds should be subcultured to potato dextrose agar (PDA) (or potato flake agar [PFA] or Sabouraud glucose agar [SGA]) and to V-8 Juice agar. In selection of media for subculture, the following rules can be followed. All moulds should be subcultured to SGA, PDA, or PFA and should be cultured to V-8 Juice agar. In addition, dermatophytes, black yeasts, coelomycetes, and dematiaceous fungi should be subcultured to oatmeal agar. Cornmeal agar is useful for red yeast. SGA with olive oil is useful with *Malassezia* species. Morphology agar (ascospore agar) should be used with white yeast. Czapek agar is useful for *Aspergillus*, *Penicillium*, and *Paecilomyces* species. SGA at 45°C is useful for *Aspergillus fumigatus*. Hay infusion agar is useful for coelomycetes BHI at 28 and 37°C is useful for systemic fungi. Niger seed (caffeic acid) agar should be used for *Cryptococcus* species.

REFERENCES

1. **De Hoog, G. S., and J. Guarro.** 1995. *Atlas of Clinical Fungi.* Centraalbureau voor Schimmelcultures, Delft, The Netherlands.

2. **Larone, D. L.** 1995. *Medically Important Fungi. A Guide to Identification.* ASM Press, Washington, D.C.

6.6.2 Growth Test for *Aspergillus fumigatus*

I. PRINCIPLE

The test for *Aspergillus fumigatus* growth determines whether a particular *Aspergillus* isolate grows at 45°C. *Aspergillus fumigatus* is the only species of *Aspergillus* that is thermotolerant and will grow *rapidly* at 45 to 50°C. A forked inoculating needle is used so as not to carry over nutritional agar.

II. MATERIALS

Include QC information on reagent container and in QC records.

A. Incubator (45 to 50°C)

B. Sabouraud agar slants

1. Purchased from BBL
2. Stored in refrigerator

C. A forked inoculating needle

III. QUALITY CONTROL

Performed at the same time and in the same manner as the test being performed on the unknown.

A. Positive control, *Aspergillus fumigatus*

B. Negative control, *Aspergillus flavus*, which will not grow at 45°C

IV. PROCEDURE

A. Inoculate two Sabouraud slants with the spores of each organism to be tested, by using the forked inoculating needle.

B. Incubate one Sabouraud slant at 45 to 50°C.

C. Incubate the second Sabouraud slant at room temperature for comparative growth.

D. In 48 to 72 h, check the slants for growth.

V. RESULTS AND REPORTING

A. In 48 to 72 h the room temperature slant should show good growth of *Aspergillus* spp.

B. If growth is present at 45 to 50°C, this confirms the identification of *Aspergillus fumigatus*.

SUPPLEMENTAL READING

Haley, L. D., and C. S. Calloway. 1978. *Laboratory Methods in Medical Mycology*, p. 153–154. U.S. Government Printing Office, Washington, D.C.

McGinnis, M. R. 1980. *Laboratory Handbook of Medical Mycology*, p. 82. Academic Press, Inc., New York.

6.7.1 Full Identification of Yeasts

I. GENERAL CONSIDERATIONS

The methods provided below are designed to provide identification information for most of the commonly encountered pathogenic yeasts and include the methods used to obtain the results summarized in Tables 6.7.1–1 through 6.7.1–3.

Several points concerning yeast identification

A. Figure 6.5.4–2 provides an approach to yeast identification based on colonial and microscopic characteristics. The initial screening tests are very useful and may suffice for the extent of identification for a particular specimen and quantity of organism.

B. A common practice for some specimens is to report a yeast isolate as either *Candida albicans* or ''not *Candida albicans*'' based on a germ tube test. While the germ tube test is an excellent screening test, the identification is presumptive because *C. dubliniensis* (also considered *C. albicans*) can form germ tubes, and *C. tropicalis* can produce pseudohyphae that closely resemble germ tubes except for the location of the septum, which can be difficult to see without high magnification.

C. A variety of relatively inexpensive, commercial systems provide reliable identification of commonly encountered yeasts. These include API 20C, IDS RapID Yeast Plus, and Vitek Yeast Biochemical Card.

Procedures for supplemental tests that may be required and the organisms for which the supplemental tests are most useful are summarized in item II of this procedure, below, or in procedure 6.5.

With all commercial systems, strict adherence to manufacturer's directions should be followed. Errors such as overfilling cupules (API 20C) and variations in inoculum concentration from recommendations will cause spurious results.

In general, the following tests should be performed as part of the initial routine identification of a yeast: commercial system, urease, morphology agar, and dextrose assimilation (check for pellicle).

Tables 6.7.1–2 and 6.7.1–3 provide some of the characteristics of various pathogenic yeasts.

II. TESTS FOR YEAST IDENTIFICATION FROM PRIMARY OR SECONDARY CULTURES

Supplemental tests can sometimes provide the single result that identifies an organism. Colony color is one indication (Table 6.7.1–4) and, in combination with microscopic appearance, can be nearly sufficient for an identification (Fig. 6.5.4–2A). It is recommended that morphology agar or corn meal agar–2% Tween 80 be set up simultaneously with any commercial identification system. Yeast morphology on standard primary media and on morphology agar may, at times, be all that is

Table 6.7.1–1 General considerations of two commercial, yeast identification systems

System	Incubation period	Principle	QC organisms	Comments
API 20C	72 h	Assimilation	*Cryptococcus laurentii* (ATCC 18803) *Blastoschizomyces capitatus* (ATCC 10663)	New formulation requires strict adherence to inoculum preparation.
RapID Yeast Plus	4 h	Preformed enzyme detection	*Candida kefyr* (ATCC 2512) *Candida glabrata* (ATCC 2001)	Incubation temperature must be strictly followed.

II. TESTS FOR YEAST IDENTIFICATION FROM PRIMARY OR SECONDARY CULTURES *(continued)*

needed to obtain an identification. CHX sensitivity testing (Appendix 6–1) and ascospore formation are reserved for problematic identifications. See reference 1 for ascospore or ascus characteristics of ascomycetous yeasts (e.g., *Saccharomyces cerevisiae*).

The supplemental tests take advantage of characteristics that are restricted to a limited set of organisms (Table 6.7.1–5). When these characteristics are present, in the absence of other indicative information, consider with some caution that the test organism is part of the limited set. Other tests to consider are the screening tests described in Table 6.5–2.

REFERENCES

1. **Hazen, K. C.** 1995. New and emerging yeast pathogens. *Clin. Microbiol. Rev.* **8:**462–478.
2. **Larone, D. L.** 1995. *Medically Important Fungi. A Guide to Identification.* ASM Press, Washington, D.C.

Table 6.7.1–2 Characteristics of *Cryptococcus* spp.[a]

Organism	Color on birdseed agar[b]	Growth at 37°C on SGA	Growth with CHX at 25°C	Pseudo-hyphae (short)	Urease (25°C, 4 days)	Assimilation of:													Fermentation
						Glucose	Maltose	Sucrose	Lactose	Galactose	Melibiose	Cellobiose	Inositol	Xylose	Raffinose	Trehalose	Dulcitol	KNO_3	
C. neoformans	Brown	+	0	0^{R}	+	+	+	+	0	+	0	$+^{V}$	+	+	$+^{V}$	+	+	0	
C. uniguttulatus	White	0	0	0	+	+	+	+	0	0^{V}	0	0^{V}	+	+	$+^{V}$	$+^{V}$	0	0	
C. albidus var. *albidus*	White	0^{V}	0	$+^{V}$	+	+	+	+	$+^{V}$	0^{V}	0^{V}	+	+	+	$+^{W}$	$+^{V}$	$+^{V}$	+	
C. albidus var. *diffluens*	White	$+^{V}$	0	0^{V}	+	+	+	+	0	0^{V}	$+^{V}$	+	+	+	$+^{W}$	+	$+^{V}$	+	All species of *Cryptococcus* lack fermentative ability
C. laurentii	White or greenish	+	0^{V}	0	+	+	+	+	+	+	$+^{V}$	+	+	+	$+^{V}$	$+^{V}$	+	0	
C. luteolus	White	0	0	0	+	+	+	+	0^{V}	+	+	+	+	+	+	+	$+^{V}$	0	
C. terreus	White	+	0	0^{V}	+	+	$+^{V}$	0^{V}	0^{V}	$+^{V}$	0	+	+	+	0	$+^{V}$	$+^{V}$	+	
C. gastricus	White	0	0	0	+	+	+	0^{V}	0^{V}	+	0	+	+	+	0	+	0	0	

[a] Reprinted from reference 2. Abbreviations: SGA, Sabouraud glucose agar; +, positive; 0, negative; V, strain variation; R, occur rarely; W, reaction may be weak.
[b] The caffeic acid disk test is a rapid and sensitive alternative to birdseed agar.

Table 6.7.1–3 Characteristics of yeasts and yeast-like organisms other than *Candida* spp. and *Cryptococcus* spp.

Organism	Microscopic morphology on cornmeal-Tween 80 agar at 25°C	Growth on SGA at 37°C	Asco-spores	Urease (25°C, 4 days)	Assimilation of:													Fermentation						
					Glucose	Maltose	Sucrose	Lactose	Galactose	Melibiose	Cellobiose	Inositol	Xylose	Raffinose	Trehalose	Dulcitol	KNO_3	Glucose	Maltose	Sucrose	Lactose	Galactose	Trehalose	Cellobiose
Saccharomyces cerevisiae	Occasional short pseudohyphae	+	+	0	+	$+^V$	+	0	$+^V$	0	0	0	0	+	$+^V$	0	0	+	+	+	0	$+^V$	$+^V$	0
Hansenula anomala	May form pseudohyphae	V	+	0	+	+	+	0	$+^V$	0	+	0	$+^V$	$+^V$	+		+	+	$+^V$	+	0	$+^V$	$+^V$	V
Geotrichum candidum	True hyphae, arthroconidia, no blastoconidia	0	0	0	+	0	0	0	+	0	0	0	+	0	0	0	0	0	0	0	0	0	0	0
Blastoschizomyces capitatus[b]	Psudohyphae and true hyphae, annelloconidia, few arthroconidia	+	0	0	+	0	0	0	+	0	0	0	0	0	0	0	0	0	0	0	0	0	0	0
Trichosporon beigelii	Pseudohyphae and true hyphae, arthroconidia, blastoconidia	$+^V$	0	+	+	$+^V$	$+^V$	+	$+^V$	$+^V$	$+^V$	$+^V$	+	$+^V$	$+^V$	$+^V$	0	0	0	0	0	0	0	0
Trichosporon pullulans		0	0	+	+	+	+	+	+	+	+	$+^V$	$+^V$	+	+	0	+	0	0	0	0	0	0	0
Torulopsis glabrata	No pseudohyphae, cells small, terminal budding	+	0	0	+	0	0	0	0	0	0	0	0	0	+	0	0	+	0	0	0	0	$+^V$	0
Torulopsis sp.		0^V	0	0	+	0^V	$+^V$	0^V	0^V	0^V	0^V	0	$+^V$	$+^V$	0^V	0^V	0^V	$+^V$	0	0^V	0	0	0^V	0
Rhodotorula rubra	Usually no pseudohyphae	$+^V$	0	+	+	+	+	0	$+^V$	0	$+^V$	0	+	+	+	0	0	0	0	0	0	0	0	0
Rhodotorula glutinis		0^V	0	+	+	+	+	0	$+^V$	0	$+^V$	0	$+^V$	$+^V$	+	0	+	0	0	0	0	0	0	0
Sporobolomyces salmonicolor	Ballistoconidia, various amounts of true and pseudohyphae	0^V	0	+	+	0^V	+	0	V	0	0^V	0	+	V	+		+	0	0	0	0	0	0	0
Malassezia pachydermatis	Usually no hyphae	+	0	+	+	0	0	0	0	0	0	0	0	0	0	0	0	0	0	0	0	0	0	0
Prototheca wickerhamii	Sporangia, no hyphae	+	0	0	+	0	0	0	$+^V$	0	0	0	0	0	+	0	0	Not done						
Prototheca zopfii		+	0	0	+	0	0	0	0^V	0	0	0	0	0	0	0	0							
Prototheca stagnora		0	0	0	+	0	V	0	+	0	0	0	0	0	0	0								

[a] Reprinted from reference 2. Abbreviations: SGA, Sabouraud glucose agar; +, positive; 0, negative; V, strain variation.

[b] Annelloconidia forming clusters at the ends of hyphae of *B. capitatus* may resemble the elongated blastoconidia of *Candida krusei.*

Table 6.7.1–4 Characteristics of *Candida* spp. most commonly encountered in the clinical laboratory[a]

Organism	Microscopic morphology on cornmeal-Tween 80 agar at 25°C	Growth			Germ tubes
		In Sabouraud broth	With CHX at 25°C	On SGA at 37°C	
C. albicans *C. stellatoidea*	Pseudohyphae with terminal chlamydospores; clusters of blastoconidia at septa	NSG	+ 0	+	+
C. tropicalis (includes *C. paratropicalis*)[c]	Blastoconidia anywhere along pseudohyphae	Narrow surface film with bubbles	0^V	+	0
C. parapsilosis	Blastoconidia along curved pseudohyphae; giant mycelial cells	NSG	0	+	0
C. lusitaniae	Short chains of elongate blastoconidia along curved pseudohyphae	NSG	0	+	0
C. guilliermondii	Fairly short, fine pseudohyphae; clusters of blastoconidia at septa	NSG	+	+	0
C. kefyr (*C. pseudotropicalis*)	Elongated blastoconidia resembling "logs in a stream" along pseudohyphae	NSG	+	+	0
C. zeylanoides	Pseudohyphae give feather-like appearance at low power	Pellicle (delayed)	0	0^V	0
C. krusei	Pseudohyphae with cross-matchsticks or treelike blastoconidia	Wide surface film up sides of tube	0	+	0
C. lipolytica	Elongated blastoconidia in short chains along pseudohyphae	Pellicle (delayed)		$+^V$	0

[a] Reprinted from reference 2. Abbreviations: SGA, Sabouraud glucose agar; +, positive; 0, negative; W, reaction may be weak; V, strain variation; NSG, no surface growth.

[b] Fermentation is demonstrated by the production of gas (acid does not indicate fermentation).

[c] *C. paratropicalis* differs from *C. tropicalis* by not fermenting sucrose and melezitose, not assimilating arabinose, and having variable ability to assimilate methyl-D-glucoside, sucrose, and melezitose.

Urease (25°C)	Assimilation of:													Fermentation of:[b]						
	Glucose	Maltose	Sucrose	Lactose	Galactose	Melibiose	Cellobiose	Inositol	Xylose	Raffinose	Trehalose	Dulcitol	KNO_3	Glucose	Maltose	Sucrose	Lactose	Galactose	Trehalose	Cellobiose
0	+	+	+ 0	0	+	0	0	0	+	0	+	0	0	+	+	0	0	$+^{W}$ 0	$+^{V}$ 0	0
0	+	+	$+^{V}$	0	+	0	$+^{V}$	0	+	0	+	0	0	+	+	$+^{V}$	0	$+^{V}$	$+^{V}$	0
0	+	+	+	0	+	0	0	0	+	0	+	0	0	+	0	0	0	V	0	0
0	+	+	+	0	+	0	+	0	+	0	+	0	0	+	0	V	0	+	V	+
0	+	+	+	0	+	+	+	0	+	+	+	+	0	+	0	$+^{W}$	0	$+^{W}$	$+^{W}$	0
0	+	0	+	+	+	0	+	0	$+^{V}$	+	0	0	0	+	0	+	+	+	0	0
0	+	0	0	0	0^{V}	0	0^{V}	0	0	0	+	0	0	0^{W}	0	0	0	0	0^{V}	0
$+^{V}$	+	0	0	0	0	0	0	0	0	0	0	0	0	+	0	0	0	0	0	0
+	+	0	0	0	0	0	0	0	0	0	0	0	0	0	0	0	0	0	0	0

Table 6.7.1–5 Examples of useful supplemental tests for yeast[a]

Procedure	Yeast
Inositol assimilation	*Cryptococcus* spp., *Trichosporon* spp.
Nitrate assimilation	*C. albidus, C. terreus, Hansenula anomala, Trichosporon pullulans, R. glutinis, Sporobolomyces salmonicolor*
CHX resistance	*See* Appendix 6–1.
Ascospore agar	*Saccharomyces, Schizosaccharomyces*, and *Hansenula* spp.
Rhamnose assimilation	*C. lusitaniae*
Melibiose assimilation	*C. guilliermondii, Trichosporon* spp., ''*Torulopsis*'' spp. [non-*Candida* (*Torulopsis*) *glabrata*]
Lactose assimilation	*Candida kefyr, Trichosporon* spp., some non-*C. neoformans* spp.
Morphology agar	Various species (part of subculture routine)
Glucose fermentation[b]	Most *Candida* species (+); *Cryptococcus* species (−); ascomycetous yeasts (generally +); basidiomycetous yeasts (generally −)
Cellobiose fermentation[b]	*Candida guilliermondii* (+); *Hansenula anomala* (variable)
Maltose fermentation	*Candida albicans* (+); *Candida tropicalis* (+); other *Candida* species (−)

[a] Note that some results are components of commercial systems. They are listed because they are key tests.

[b] Fermentation tests require that the yeast first be grown on a sugar-free medium such as malt extract agar or yeast morphology agar.

6.7.2 Nitrate Assimilation Test for Identification of Yeasts to the Species Level

I. PRINCIPLE

Yeast carbon agar with the addition of various nitrogen sources tests the ability of yeasts to assimilate nitrogen. Potassium nitrate and peptone are used as the nitrogen test sources. The peptone serves as a positive growth control. The nitrate reductase test *does not* substitute for the nitrate assimilation test.

II. MATERIALS

Include QC information on reagent container and in QC records.

A. Yeast carbon agar
   1. Six tubes
   2. Purchased commercially from Remel, catalog no. 09-984
   3. Stored at 2 to 8°C
   4. Expiration date established by manufacturer

B. Three large petri dishes, 150 by 15 mm (two for controls)

C. Disks impregnated with saturated KNO_3 or peptone solutions

D. Wickerham card

III. QUALITY CONTROL

Do the procedure while using the following controls at the same time and in the same manner as the test being performed on the patient isolate.

A. *Cryptococcus albidus,* which is positive for nitrate utilization and for peptone growth

B. *Candida albicans,* which is negative for nitrate utilization and positive for peptone growth

IV. PROCEDURE

A. Melt two tubes of yeast carbon base medium per isolate or control.

B. In a separate tube of sterile distilled water, make a suspension of the yeast isolate to be identified in which the turbidity is not greater than a 1+ by Wickerham card analysis.

C. When the tubes of yeast carbon base medium have melted and then cooled to 50°C, so as not to kill the yeast, add 0.1 ml of the yeast suspension to each tube of medium.

D. Mix thoroughly but gently, taking care to avoid air bubbles, and pour the mixture into a sterile petri dish.

E. When agar has hardened, place a KNO_3 disk and a peptone disk on opposite sides of the plate.

F. Incubate the plate, inoculated surface up, at 30°C for 48 to 96 h.

V. EXPECTED RESULTS

A. Positive growth control (peptone-inoculated portion)

1. All yeasts utilize peptone as a source of nitrogen. If this portion of the plate has *no* growth after 48 to 72 h, the test is invalid.
2. If growth is present, proceed to examine the KNO_3 area.

B. Positive test

1. Growth control test positive
2. Growth also present on KNO_3 portion of plate

C. Negative test

1. Growth control test positive
2. No growth on KNO_3 portion of plate

6.7.3 Sporulation (Ascospore Production)

I. PRINCIPLE

Two of the characteristics examined to identify most fungi are the method of sporulation and the arrangement and morphology of the spores produced. In some fungi, sexual reproduction produces spores in an ascocarp. This contains small sacs, called asci, which contain ascospores. Selection of the medium to demonstrate the production of ascospores in yeast depends on which species of yeast is suspected, i.e., *Saccharomyces* spp., *Schizosaccharomyces* spp., and *Candida norvegensis* produce ascospores on acetate (ascospore) agar, but other yeasts produce ascospores on V-8 Juice agar.

II. MATERIALS

Include QC information on reagent container and in QC records.

A. Ascospore agar deeps

1. Purchased from Remel, catalog no. 09-046
2. Stored at 2 to 8°C
3. Expiration date is established by the manufacturer.

B. V-8 Juice agar deeps

1. Purchased from Smith River Biologicals, catalog no. 24V8
2. Stored at 2 to 8°C
3. Expiration date is established by the manufacturer.

C. TB Kinyoun stain kit

1. Contents of kit
   - **a.** TB Kinyoun stain carbolfuchsin
   - **b.** TB Kinyoun stain decolorizer
   - **c.** TB Kinyoun stain, brilliant green
2. Purchased from BBL, catalog no. 12318
3. Expiration date is established by the manufacturer.
4. Stored at room temperature

D. *Saccharomyces cerevisiae* ATCC 24903

Stored at 30°C

E. Microscope slides

F. Bacteriological loop or sterile applicator sticks

III. PROCEDURE

A. For those yeasts that produce ascospores on ascospore agar

1. Melt the ascospore agar deep in a boiling-water bath and cool to 45 to 50°C.
2. Mix and pour into a sterile 15-by-100-mm petri dish. Allow the agar to solidify.
3. Inoculate the medium with the yeast to be identified.
4. Incubate at 25 to 30°C for 3 to 10 days.
5. After 3 days of incubation, perform a wet preparation to check for ascospores.
6. If ascospores are seen, prepare a suspension of the yeast in a drop of water on a microscope slide.
7. Let the smear dry, then heat fix and stain by the Kinyoun's acid-fast staining method (*see* EPCM section 4).
8. Examine microscopically for acid-fast (red) ascospores.

III. PROCEDURE
(continued)

9. If no ascospores are seen, reincubate and reexamine daily for 7 more days. If the control is positive and the test is negative at the end of 7 days, the test is negative.

B. For those yeasts that produce ascospores on V-8 Juice agar

1. Melt the V-8 Juice agar deep in a boiling-water bath and cool to 45 to 50°C.
2. Mix and pour into a sterile 15-by-100-mm petri dish. Allow the agar to solidify.
3. Follow the procedure for yeasts that produce ascospores on ascospore agar.

IV. EXPECTED RESULTS

A. Positive result

Acid-fast (red) ascospores are seen among the counterstained yeast cells.

B. Negative result

No ascospores are seen.

SUPPLEMENTAL READING

Hazen, K. C. 1995. New and emerging yeast pathogens. *Clin. Micro. Rev.* **8**:462–478.

Remel, Inc. 1993. *Technical Information, Ascospore Agar*, TI no. 9046-A, 5/1/93. Remel, Lenexa, Kan.

6.7.4 Carbohydrate Fermentation Test for Yeast Identification

I. PRINCIPLE

A positive fermentation reaction is indicated by the presence of gas (bubbles) in the inverted Durham tube, not a change in bromthymol blue indicator.

II. MATERIALS

Include QC information on reagent container and in QC records.

A. Yeast fermentation broth purchased commercially from Remel with control (06-5402), dextrose (06-5314), cellobiose (06-5310), maltose (06-5330), or sucrose (06-5462)

An alternative source is PML Microbiologicals (control [T-79-00], glucose [T-79-02], maltose [T-79-08], cellobiose [not available], sucrose [T-79-10]).

1. Stored at 2 to 8°C
2. Expiration date established by manufacturer

B. Distilled water

1. Stored at 2 to 8°C
2. Expiration date, 3 months from preparation

C. No. 1 McFarland standard

1. Purchased from Remel (no. 20-351)
2. Stored at room temperature
3. Outdate established by manufacturer

D. Sterile Pasteur pipettes

E. Vaspar

1. Outdated 1 year after preparation
2. Stored at room temperature

III. QUALITY CONTROL

QC is performed at each receipt of the medium.

IV. PROCEDURE

A. Prepare a diluted suspension of a pure, 24- to 48-h-old yeast culture in distilled water equal to a no. 1 McFarland standard. Select control organisms that demonstrate a positive and negative result for each carbohydrate tested.

B. Add a single drop of this suspension to each tube of yeast fermentation broth.

C. Add a layer of molten Vaspar directly onto the top of the broth.

D. Incubate at 25 to 30°C and shake daily. Higher temperatures lead to a breakdown of disaccharides.

E. Read the tubes every 2 to 3 days for the first week, and then weekly for up to 24 days.

V. INTERPRETATION

Positive fermentation is indicated by turbidity and accumulation of gas (CO_2) in the Durham tube or underneath the Vaspar seal. The bromthymol blue indicator will turn yellow due to carbohydrate assimilation. (All fermented carbohydrates will be assimilated, but not all assimilated carbohydrates will be fermented.) Table 6.7.4–1 shows fermentation reactions for some *Candida* species.

Table 6.7.4–1 Fermentation reactions for some *Candida* species[a]

Candida species	Carbohydrate			
	Glucose	Maltose	Cellobiose	Sucrose
C. albicans	+	+	–	–
C. guilliermondii[b]	+	–	–	+ (w)
C. krusei	+	–	o	–
C. lusitaniae	+	–	+	V
C. parapsilosis[b]	+	–	–	–
C. tropicalis	+	+	–	+ (v)

[a] Key: – = negative; + = positive; o = not tested routinely; v = rare negatives; V = strain variation.

[b] Note that *C. guilliermondii* and *C. parapsilosis* produce distinctively different morphologies on morphology media.

VI. LIMITATIONS

A false-positive reaction may be caused by endogenous carbohydrates present in the yeast cells, if a heavy inoculum is used.

6.8.1 Identifying Moulds

INTRODUCTION

There are several excellent texts that illustrate the distinctive microscopic and macroscopic morphologies of the commonly encountered, clinically relevant moulds (*see* references 1 to 7). Although books and reference articles are useful tools, they are no substitute for experience. If time permits, the personnel within a clinical microbiology laboratory should take advantage of the opportunity to observe and identify plate contaminants, review recent isolates, and identify clinically important isolates to the species level.

It is imperative that these cultures be handled in a biosafety hood.

A. Mould identification rules

1. Work with all moulds in a biological safety hood (class II).
2. Evaluate the macroscopic characteristics of the mould colony: color of the top (obverse) and bottom (reverse), texture, size, and topography (e.g., wrinkled or smooth). The traits are somewhat variable within a given species.
3. If the matte surface of the obverse is one color (e.g., white or tan) and there are regions (usually central or concentric circles) which are another color or there are colored speckles, obtain material for microscopic exam from the colored regions.
4. Transparent-tape preparations (procedure 6.8.2) provide good first microscopic evaluations of a fungus. A tease preparation when performed properly can provide more information. A slide culture is also an excellent preparation method. These methods have been described earlier.
5. Start the microscopic exam from the lowest power until you find regions of interest, then increase magnification.
6. If there are conidia or spores in the microscopic preparation, look for spore-producing structures. Moulds that produce deep surface beds of conidia (e.g., *Aspergillus fumigatus*) may require a tease preparation or slide culture because cellophane tape may not show conidiogenous cells.
7. Look for features of conidia, spores, fruiting structures, hyphae, etc., that are shared by many of the same structures. Unusual or rare features can be misleading (e.g., one-celled conidia versus four- or five-celled conidia).
8. Determine whether the organism is resistant to CHX (*see* Appendix 6.1).

B. Questions when reviewing microscopic mould preparation

1. Do the hyphae display distinctive features, for example, septations at regular intervals, color, or wall thickness? If hyphae are packed together, the tease preparation was not teased sufficiently.
2. Are spores and/or conidia visible? If not, the preparation did not include spore-forming regions or is a sample from a sterile fungus. The fungus could

INTRODUCTION *(continued)*

also be making spores and/or conidia within fruiting structures but they have been released from the structure.

3. Are the hyphae large (broad)? If so, do the hyphae have septa (cross-walls) at regular intervals?

General useful characteristics	Organism
Aseptate hyphae	Zygomycetes
Lack of sexual fruiting structures	Possible deuteromycetes
Asci	Ascomycetes
Clamp connections on at least some hyphae	Basidiomycetes
Large sacs with conidia (pycnidia)	Coelomycetes

The varied nature of moulds makes describing all of their features difficult. Consult a reference dedicated to fungal identification (*see* Table 6.8.1–1).

The following parts of this procedure provide guidance for the identification of dermatophytes, dimorphic fungi (the systemic pathogens), and some of the common moulds. Identification manuals should be consulted when identifying any mould. Some of the terms are defined in Appendix 6–2.

Table 6.8.1–1 Techniques for identifying moulds[a]

Suspected mould	Tests	
	Required	Recommended
Aspergillus fumigatus		Temp. tolerance, 48°C
Blastomyces dermatitidis	Exoantigen	Probe (Accu-Probe)
Basidiobolus sp.		Forcibly ejected sporangiola
Cladophialophora bantiana (*Xylohypha bantiana)*		Temp tolerance, 42°C
Cladosporium carrionii	Temp tolerance, 35°C	
Coccidioides immitis	Exoantigen	Probe (Accu-Probe)
Conidiobolus sp.		Forcibly ejected sporangiola
Histoplasma capsulatum	Exoantigen	Probe (Accu-Probe)
Microsporum audouinii	Rice grain	
Microsporum canis	Rice grain	
Mycelia Sterilia	Temp tolerance, 35–37°C; cycloheximide sensitivity	
Paracoccidioides brasiliensis	Mould-to-yeast conversion	
Rhizomucor pusillus		Temp tolerance, 55°C
Sepedonium sp.	Morphology; negative exoantigen for *Histoplasma* spp.	
Sporobolomyces sp.		Forcibly ejected conidia
Sporothrix schenckii	Mould-to-yeast conversion	
Trichophyton concentricum	*Trichophyton* agars 1, 2, 3, and 4	
Trichophyton equinum	*Trichophyton* agars 1 and 5	
Trichophyton megninii	*Trichophyton* agars 6 and 7	
Trichophyton mentagrophytes		In vitro hair perforation; urea hydrolysis
Trichophyton rubrum		In vitro hair perforation; urea hydrolysis
Trichophyton tonsurans	*Trichophyton* agars 1, 2, 3, and 4	
Trichophyton verrucosum	*Trichophyton* agars 1, 2, 3, and 4; growth enhancement at 37°C	
Trichophyton violaceum	*Trichophyton* agars 1, 2, 3, and 4	
Trichophyton gallinae	*Trichophyton* agars 6 and 7	
Wangiella dermatitidis		Temp tolerance, 40°C

[a] Reprinted from reference 5. Abbreviation: Temp, temperature.

REFERENCES

1. **Barron, G. L.** 1968. *The Genera of Hyphomycetes from Soil.* Robert E. Krieger Publishing, Huntington, N.Y.
2. **de Hoog, G. S., and J. Guarro.** 1995. *Atlas of Clinical Fungi.* Centraalbureau voor Schimmelcultures, Delft, The Netherlands.
3. **Larone, D. L.** 1995. *Medically Important Fungi. A Guide to Identification.* ASM Press, Washington, D.C.
4. **McGinnis, M. R.** 1980. *Laboratory Handbook of Medical Mycology.* Academic Press, New York.
5. **McGinnis, M. R., and L. Pasarell.** 1992. Yeast identification using morphology, p. 6.9-1–6.9-2. *In* H. D. Isenberg (ed.), *Clinical Microbiology Procedures Handbook*, vol. 1. American Society for Microbiology, Washington, D.C.
6. **St.-Germain, G., and R. Summerbell.** 1996. *Identifying Filamentous Fungi. A Clinical Laboratory Handbook.* Star Publishing Co., Belmont, Calif.
7. **Wang, C. J. K., and R. A. Zabel.** 1990. *Identification Manual for Fungi from Utility Poles in the Eastern United States.* ATCC, Rockville, Md.

6.8.2 Cellophane Tape Mount

This procedure is from reference 1.

I. PRINCIPLES

This method usually maintains the original position of the characteristic fungal structures. However, the organism must be grown on plated medium in order to be examined by this method.

II. MATERIALS

A. Clear cellophane tape
B. Lactophenol cotton blue stain
C. Glass slides
D. Fungal culture grown on solid medium

III. PROCEDURE

A. Loop back on itself a 1.5-in. (~4-cm) strip of clear tape, sticky side out.
B. Hold the tip of the loop securely with forceps and press the lower, sticky side very firmly to the surface of the fungal colony.
C. Pull the tape gently away. Aerial hyphae will adhere to the tape.
D. Open up the tape strip and place it on a small drop of LPCB on a glass slide so that the entire sticky side adheres to the slide.
E. Examine under the microscope.

REFERENCE

1. **Larone, D. L.** 1995. *Medically Important Fungi. A Guide to Identification.* ASM Press, Washington, D.C.

6.8.3 Tease Preparation

This procedure is from reference 1.

I. PRINCIPLES

A tease preparation is the quickest and most common technique for mounting fungi for microscopic examination. Since the mould's growth is teased apart with dissecting needles, conidia or spores may be dislodged from the conidiogenous or sporogenous cells. It may be necessary to use the slide culture technique if the identification cannot be made from the tease preparation.

II. MATERIALS

A. Long-handled inoculating needle
B. Two dissecting needles
C. Lactophenol
D. Clean microscope slide and coverslip
E. Clear fingernail polish

III. PROCEDURE

A. Place a drop of lactophenol just off center on a clean microscope slide.
B. With the inoculating needle, gently remove a small portion of growth midway between the center of the colony and the edge. Place the material in the lactophenol.
C. With the two dissecting needles, gently tease the fungus apart so that it is thinly spread out in the lactophenol.
D. Place the coverslip at the edge of the lactophenol, and slowly lower it with a sharp pointed object.
E. Avoid trapping air bubbles under the coverslip. Remove excess lactophenol from the edges of the coverslip by blotting with a paper towel.
F. Seal the edges of the coverslip with fingernail polish to preserve the mount.

REFERENCE

1. **McGinnis, M. R., and L. Pasarell.** 1992. Yeast identification using morphology, p. 6.9-1–6.9-2. *In* H. Isenberg (ed.), *Clinical Microbiology Procedures Handbook*, vol. 1. American Society for Microbiology, Washington, D.C.

6.8.4 Mould Identification Based on Spore or Conidium Production

I. INTRODUCTION

Moulds are typically differentiated by their microscopic morphology in the clinical mycology laboratory. Some fungi have specialized cells called conidiogenous cells (e.g., a phialide; *see* Fig. 6.8.4–1), which give rise to conidia. A conidium is an asexual reproductive propagule that is nonmotile, usually separates readily from its parent cell, and arises in any manner other than one involving a cleavage process. When a fungus produces two conidia of distinctly different sizes, the larger is called a macroconidium, and the smaller is called a microconidium.

In contrast to conidia, spores are reproductive propagules that arise after either mitosis or meiosis. Meiotic spores are borne in or on specialized structures (e.g., asci, basidia, or zygosporangia). When mitosis is involved for zygomycetes, the propagule arises in a sporangium. A sporangium is a large saclike cell in which the entire content is cleaved into sporangiospores (Fig. 6.8.4–2). A merosporangium is a special type of sporangium in which all of the sporangiospores are aligned in a single row. A sporangium is borne on a specialized filament called a sporangiophore. The sporangium may surround a sterile, domelike, swollen area called a columella, which is found at the apex of some sporangiophores. A number of zygomycetes have rootlike structures called rhizoids. The location of the rhizoids relative to the sporangiophore is a useful clue to differentiate *Rhizopus, Rhizomucor*, and *Absidia* spp.

A conidiophore is a specialized hypha that bears either conidia or conidiogenous cells. During adverse conditions, fungi may form chlamydoconidia or chlamydospores. These are thick-walled resting structures which serve as a survival unit.

Although identification of many medically important moulds is best performed based on analysis of conidiogenous characteristics (that is, how the conidia are formed), conidial characteristics (shape, size, arrangement, etc.) provide a very useful and effective approach to mould identification. Below are some of the types of conidia that may be seen. The terminology used below is not considered contemporary but it is useful for making identifications.

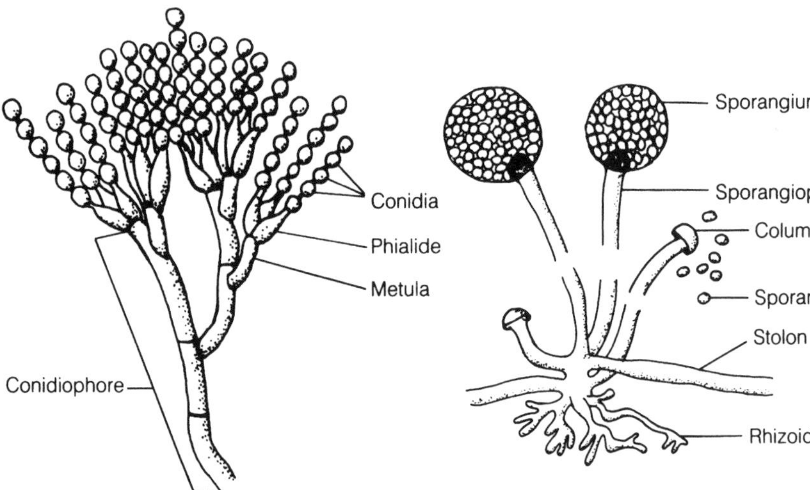

Figure 6.8.4–1 Illustration of phialides associated with *Penicillium* species (reprinted from reference 1).

Figure 6.8.4–2 Illustration of sporangial structures and rhizoids associated with zygomycetes within the mucorales (reprinted from reference 1).

II. FORMS OF CONIDIA

A. Aleurioconidia

Aleurioconidia are usually borne singly, either directly (sessile) on the hyphae, on short hyphal projections (pedicels), or on longer, usually undifferentiated conidiophores.

Major pathogenic fungi included in this group include *B. dermatitidis, H. capsulatum,* and the dermatophytes (*Microsporum, Epidermophyton*, and *Trichophyton* species).

II. FORMS OF CONIDIA
(continued)

B. Annelloconidia

Annelloconidia arise from the apex of annellides (flask-shaped or cylindrical conidiogenous cells) and are produced, with the youngest conidium at the base (basipetal), in balls or chains. The annellide increases in length at the apex as successive conidia are produced, leaving an elongated tip with successive rings.

The following fungi, which may be pathogenic, are included in this group: *Exophiala werneckii, Exophiala spinifera*, and *Scopulariopsis* species.

C. Arthroconidia

Arthroconidia are produced by septations and the disjunction or rounding off of simple or branched conidiogenous hyphae, or by basipetal elongation to form a chain.

Major pathogenic fungi include *Coccidioides immitis* and *Trichosporon* species.

D. Blastoconidia

Blastoconidia are blown out from another cell, from a hypha, or from a conidiophore. Blastoconidia may produce either other blastoconidia by simple budding or germ tubes that develop into hyphae which may produce other blastoconidia.

Pathogenic fungi in this group include *Fonsecaea pedrosoi, Cladosporium* species, and *Candida* species.

E. Phialoconidia

Phialoconidia are produced within a conidiogenous cell called a phialide. A phialide is a rounded, or elongated, vaselike structure, which is not altered as conidia are produced, unlike annellides. In a chain of phialoconidia, the youngest will be at the base (basipetal).

Many fungi, both pathogens and nonpathogens, commonly encountered in a medical laboratory are included in this group: *Aspergillus* species, *Acremonium* species, *Phialophora* species, *Penicillium* species, *Paecilomyces* species, and *Wangiella* (*Exophiala*) *dermatitidis*.

F. Poroconidia

Poroconidia are produced through minute pores in the outer wall of the conidiophore or previously formed conidia. The pores are difficult to see without excellent optics. The conidia may be septate, pigmented, and solitary or in chains with the youngest at the tip (acropetal).

Examples of some moulds producing poroconidia are *Alternaria* species, *Curvularia* species, and *Bipolaris* species.

G. Sympoduloconidia

Sympoduloconidia are conidia that are successively blown out from a conidiophore that enlarges as each new conidium is produced. The primary conidium is produced at the tip of the conidiophore. Successive conidia are produced to the side and above each preceding conidium. As these conidia break away, scars are seen at the growing points on the conidium.

Examples of major pathogens in this group are *Fonsecaea pedrosoi* and *Sporothrix schenckii*, and *Fusarium* species. Also included are *Beauvaria* species and *Rhinocladiella* species.

REFERENCE

1. **Larone, D. L.** 1995. *Medically Important Fungi. A Guide to Identification.* ASM Press, Washington, D.C.

6.8.5 Species Confirmation of Dimorphic Moulds

I. INTRODUCTION

The dimorphic moulds (also called the dimorphs or the systemic pathogens) cause a range of disease from mild, asymptomatic to life threatening. To confirm that a suspicious mould is a dimorphic pathogen, additional definitive testing is needed. Do not assume that an isolate is a dimorphic pathogen because its mould morphology is consistent with that of a dimorphic pathogen. Species of morphologically similar moulds may be isolated as contaminants (Table 6.8.5–1).

The typical tests available for confirming dimorphic-mould species include traditional mould-to-yeast conversion methods (1 to 4 weeks), exoantigen production (several days), and DNA-RNA amplification methods (1 day). The first two tests are described below (procedures 6.8.5 and 6.8.7). It is noteworthy that the three methods have specificities of nearly 100% but the most rapid method, genetic probe technology, is the most definitive.

II. GENERAL PROCEDURE FOR DIMORPHS

The mould form of the organism will be obtained at 30°C, but conversion is induced at elevated temperature on appropriate media. For exoantigen and probe testing, subculturing the mould onto Sabouraud dextrose agar at 25 to 30°C is suitable. These latter two tests are preferred in the clinical laboratory because of their rapidity and specificity.

Table 6.8.5–1 Mould imitators of systemic dimorphic pathogens

Systemic dimorph	Mould imitator(s)	Alternate morphology of systemic dimorph
Blastomyces dermatitidis[a]	*Chrysosporium* spp.; *Scedosporium* spp.	Yeast
Coccidioides immitis	*Malbranchea* spp.	Spherule
Histoplasma capsulatum[a]	*Sepedonium* spp.; *Chrysosporium* spp.	Yeast
Paracoccidioides brasiliensis	Sterile hyphae	Yeast
Sporothrix schenckii	*Ophiostoma stenoceras*	Yeast

[a] The notable resemblance to *Chrysosporium* spp. by *B. dermatitidis* and *H. capsulatum* is possibly a reflection that both species belong to the same teleomorphic genus, *Ajellomyces*.

6.8.6 Conversion of Dimorphic, Systemic Moulds to Their Parasitic Morphologies

I. INTRODUCTION

It is imperative that these cultures be handled in a biosafety hood.

All white, tan, buff, or brown moulds should be considered potential dimorphic pathogens. Do not work with these organisms except in a class II biosafety cabinet. Never open the lid or cap of a culture vessel outside of a cabinet. Only trained personnel should handle these cultures.

II. PRINCIPLE

Infections caused by dimorphic fungi may be life-threatening. A hallmark of the pathogenesis caused by these organisms is the morphologic conversion from the form seen in the environment to the parasitic form. Morphologic conversion can be induced in vitro assuming appropriate incubation temperature (37°C for all except *C. immitis*, which is done at 40°C) and suitable growth medium.

The vegetative forms induced at the high temperature (37°C, 40°C) are unstable and are not easily maintained without strict adherence to appropriate growth conditions. When working with these forms, maintain the cultures at the elevated temperature. Do not let a culture become exposed to lower temperature for more than 15 min. Otherwise, the yeast form may revert to the mould form.

III. MATERIALS

It is imperative that these cultures be handled in a biosafety hood.

A. **Conversion medium (Table 6.8.6–1) on slants**
B. **Inoculating needle**
C. **Bacti-Cinerator or flame**
D. **37°C incubator (40°C for *C. immitis*)**
E. **Class II biosafety cabinet** (Do not work with the organisms outside a biosafety cabinet!)

IV. PROCEDURE

A. Inoculate conversion medium with a 4- to 5-mm^2 fragment of the suspected colony. Pick a region of the colony near the growing edge but having dense growth. Make sure the surface of the conversion medium is moist.
B. Incubate at suitable temperature aerobically or in the presence of 5% CO_2. Do not use 35°C when 37°C is indicated. Conversion may not occur or will proceed slowly.
C. Check for radiating growth from the inoculum fragment after 3 to 7 days. If growth is seen, make a slide preparation with lactophenol cotton blue and examine for yeasts or spherules, which should conform to the characteristics provided in Table 6.8.5–1. If growth is absent, subculture onto fresh conversion medium. Again, choose a piece of the colony from the periphery.

Table 6.8.6–1 Morphologies of dimorphs causing systemic disease[a]

Dimorph	Mould form	Parasitic form	Conversion media[b]	Temp (°C)
Blastomyces dermatitidis	Hyphae, aleurioconidia	Round yeasts (~15 μm), thick-walled buds with broad base of attachment	BHI blood (other media work), cotton seed agar	37
Coccidioides immitis	Hyphae, arthroconidia	Spherules (10–80 μm), thick-walled with endospores (2–5 μm)	Converse medium, 20% CO_2	40
Histoplasma capsulatum	Hyphae, micro- and macroconidia	Round to oval yeasts (2–4 μm), intracellular	BHI blood, preferably with cysteine, blood-glucose-cysteine agar, 7H9 broth, CYE agar	37
Paracoccidioides brasiliensis	Hyphae, interalary and terminal chlamydospores	Round yeasts (30–60 μm) with multiple buds (2–10 μm)	BHI blood, McVeigh-Morton broth modified	37
Penicillium marneffei	Hyphae, phialoconidia	Fission yeasts, intracellular	Sabouraud dextrose agar, Emmons	37
Sporothrix schenckii	Hyphae, sigmoidal conidia arrayed in a flower pattern	Round to fusiform (1–3 × 3–10 μm) yeasts, polar budding	BHI blood	37

[a] Abbreviation: CYE, charcoal-yeast extract agar.

[b] A variety of conversion media are available, and no one medium works best for all. The media listed in the table are commercially available.

V. INTERPRETATION

A. The dimorphic fungi will produce a yeast phase at 37°C or, in the case of *C. immitis*, a spherule phase at 40°C.

B. This differentiates the dimorphic fungi from other fungi that resemble them at room temperature.

6.8.7 Exoantigen Test for Identification of Dimorphic Moulds

This procedure is reprinted from reference 1. Materials are available from various sources, including Meridian Diagnositics, Carr-Scarborough, and Immuno-Mycologics, Inc.

I. INTRODUCTION

The exoantigen test is a specific immunodiffusion (ID) test developed to provide rapid information on the immunoidentity of an unknown isolate. This is done by reacting a concentrate of the organism's soluble antigens against paired positive-control reagents for each of the systemic fungi. The test consists of a ≥6-day-old slant of the unknown organism, which is extracted overnight in a Merthiolate solution and then concentrated 25 to 50 times. The concentrate is placed in two of the peripheral wells of a hexagonal ID matrix, a concentrate derived from one of the deep fungi (control antigen) is placed at the top and bottom of the ID pattern, and both unknown and control are reacted against an antiserum directed toward the control antigen. The test is read at 24 h, and any line of identity with the positive control is considered a positive identification.

The advantages of this procedure are several. Final identification can be made within 24 h after the fungus on the slant has developed a characteristic morphology (i.e., 6 to 10 days) without having to wait an additional 2 to 6 weeks for conversion to the tissue phase. Sterile Mycelia, i.e., those organisms that do not form recognizable conidial structures, may be extracted, and their relationships to the deep fungi can be determined. This is particularly important since primary isolates of *Histoplasma, Blastomyces*, and *Coccidioides* spp. sometimes do not exhibit their characteristic microscopic morphologies. Finally, any laboratory that has a biological safety cabinet and access to the control reagents can do the test, thus eliminating delay of the final report while a reference laboratory confirms the identification.

II. PROCEDURE

A. Preparation of ID agar (1% phenolized-agar medium, pH 6.3 to 6.4)

sodium chloride	0.9 g
sodium citrate ($Na_3C_6H_5O_7 \cdot 2H_2O$)	0.4 g
phenol (aqueous 90% [wt/wt])	0.25 ml
glycine	7.5 g
purified agar (Difco)	1.0 g
distilled water	to 100 ml

1. Add 50 ml of distilled water to a 125-ml screw-cap Erlenmeyer flask.
2. Add 0.9 g of sodium chloride and 0.4 g of sodium citrate to the flask. Completely dissolve both salts.
3. Add 0.25 ml of liquid phenol (90% [wt/wt]) and 7.5 g of glycine to the flask. Mix thoroughly.
4. Add 1 g of purified agar (Difco), and make up to 100 ml with distilled water.

II. PROCEDURE *(continued)*

5. Autoclave for 10 min at 120°C. (Loosen cap on flask before autoclaving.) The final pH of the agar must be 6.3 to 6.4. Adjust pH if necessary.

B. Preparation of ID plates

1. Prepare the microimmunodiffusion agar plate by pipetting 6.5 ml of melted glycine-phenol agar (60 to 65°C) into a plastic petri dish (100 by 15 mm). This agar forms the base layer.
2. Let the base layer gel for at least 30 min.
3. Pipette 3.5 ml of hot glycine-phenol agar onto one side of the solidified base layer of the agar. Immediately lower and press a microimmunodiffusion template into the hot agar in a manner that avoids trapping air bubbles. This secures the template to the basal agar layer.
4. Let the plate sit for at least 30 min before using it.
5. After agar has gelled, clear the extra agar off the template with a 1-mm-wide spatula. Putting the plates in a refrigerator for an additional 5 to 10 min before cleaning them facilitates the cleaning of agar from the wells.

   NOTE: These plates can be stored for up to 4 weeks if kept in a moist chamber at 4°C. Write the expiration date on the label and in the work record.

C. Extraction of exoantigens

1. Find a mature culture with at least 15-by-30-mm of growth on a Sabourand glucose agar slant (in a tube [20 by 150 mm] with 10 ml of agar). Use a Pasteur pipette to cover this growth with 8 to 10 ml of a 1:5,000 aqueous solution of thimerosal.
2. Allow the solution to extract with mycelial growth for at least 24 h at room temperature.
3. Use a Pasteur pipette to transfer 5 ml of each cellular extract to an Amicon Minicon macrosolute B-15 concentrator. Place 1 drop of remaining extract on any medium for sterility test.
4. Prepare a wet mount from the killed cultures.
5. Concentrate extracts derived from cultures demonstrating arthroconidia and suspected of being *C. immitis* 25-fold.
6. Concentrate extracts from cultures demonstrating tuberculated macroconidia and suspected of being *H. capsulatum,* as well as those exhibiting pyriform microconidia consistent with the mould phase of *B. dermatitidis,* 50-fold.
7. Test cultures without characteristic conidia for antigens against all three pathogens.

D. Performance of the ID test for the identification of fungal cultures

1. Put control antiserum in the center well of the pattern to be used. Before proceeding to the next step, allow the serum to diffuse for 1 h at room temperature.
2. Place the control antigens in the upper and lower wells, and place duplicate unknown supernatant antigens in the lateral wells on the same slide.

   NOTE: Set up extracts of the same culture on the same side of the control antigen, since an extract containing a strong antigen may produce such an intense reaction that a negative extract in an adjacent may well appear weakly positive.
3. Place the charged ID plates in a moist chamber at room temperature for 24 h.
4. After 24 h, carefully remove the ID templates. Wash the agar surface, and cover it with distilled water. Then read the plates. If the reactions are not clear, wash the agar surface free of excess agar, cover it with distilled water, and reexamine it.

E. Interpretation

1. A fungus is identified as *H. capsulatum* when its concentrated extract contains antigens identical (lines of identity) to the precipitinogens.

II. PROCEDURE *(continued)*

2. An isolate is identified as *B. dermatitidis* when lines of identity occur between the *Blastomyces* control reagents and the unknown.
3. An isolate is identified as *C. immitis* when its concentrate produces lines of identity with positive reagents.

REFERENCE

1. **McGinnis, M. R., and L. Pasarell.** 1992. Mould identification, p. 6.11.1–6.11.17. *In* H. D. Isenberg (ed.), *Clinical Microbiology Procedures Handbook*, vol. 1. American Society for Microbiology, Washington, D.C.

6.8.8 Dermatophytes

I. INTRODUCTION

Dermatophytes are keratinophilic and infect three cutaneous areas: skin, nails, and hair. Recent reports indicate dermatophytes can cause mycetoma and deep-seated infections. Such manifestations are rare, however.

Dermatophytes can be identified based on colonial morphology, biochemical characteristics, and microscopic morphology. The distinctive microscopic features are macroconidia, microconidia, hyphal structures (e.g., pectinate hyphae, spirals, favic chandeliers) and chlamydoconidia. Biochemical characteristics involve urease production, alkalinization of media owing to breakdown of proteins and amino acids, and vitamin requirements. Colonial morphology concerns colony color, topography, and growth rate.

The value of identifying dermatophytes to species level is debatable. In regions where *Trichophyton favus* occurs, species identification of an agent causing tinea capitis can provide information to guide therapeutic treatment and help in implementation of appropriate infection control measures. Species identification of an agent causing tinea corporis in an AIDS patient can help indicate whether the patient may develop a more severe form of dermatophyte infection. Also, with the development of new antifungal agents and their respective differences in efficacy against the various dermatophyte species, identification to species level becomes more important.

As a laboratorian becomes familiar with organism characteristics, identification based simply on colony characteristics becomes tempting. However, caution should be heeded because species can occasionally mimic other species.

II. MEDIA

Include QC information on reagent container and in QC records.

Media that support conidial formation by dermatophytes include Sabouraud glucose agar, potato dextrose agar, V-8 Juice agar, and oatmeal agar (available from Smith-River Biologicals). Once formed, microscopic features are the foremost method for identifying dermatophytes. Initial evaluations are based on the micro- and macroconidia and hyphal structures.

III. INTERPRETATION

A. *Trichophyton* species

Microconidia are dominant. There are few, if any, macroconidia. When present, macroconidia are narrow and have smooth, thin walls.

B. *Microsporum* species

Microconidia typically absent. The surface of macroconidia are rough. The macroconidia are wide and thick-walled.

C. *Epidermophyton* species

Clavate, multicelled (two to six cells), thin- to thick-walled, smooth-surfaced macroconidia. No microconidia are produced.

IV. OTHER PRINCIPLES FOR IDENTIFICATION OF DERMATOPHYTES

Dermatophyte colonies need to be transferred weekly to avoid pleomorphism. Pleomorphism is the result of spontaneous mutation in the colony, causing it to lose the ability to form conidia. Pleomorphism begins in one part of the culture as a sterile white growth. It will eventually consume the entire colony and eliminate the characteristic pigmentation, sporulation, and normal physiological characteristics. It is irreversible.

V. SPECIMENS

The actual specimen may be submitted to the laboratory for culture. Specimens of skin, hair, and nails can be expected to contain a dermatophyte species. Dermatophyte species to expect from each specimen type include the following:

Hair	Skin	Nails
—	*Epidermophyton floccosum*	*E. floccosum*
Microsporum spp.	*Microsporum* spp.	—
Trichophyton	*Trichophyton* spp.	*Trichophyton* spp.

Microsporum species generally present an abundance of macroconidia, but few microconidia. The size and shape of the macroconidia are used to differentiate the species. *Trichophyton* species contain an abundance of microconidia, but few macroconidia. The macroconidia are thin walled and pencil shaped. The size, shape, and arrangement of microconidia are used to determine the species. *Epidermophyton* species contain beaver tail-shaped macroconidia in clusters, whereas microconidia are absent.

VI. TESTS TO IDENTIFY DERMATOPHYTES

These tests are based on the features listed in Tables 6.8.8–1 and 6.8.8–2.

REFERENCES

1. **De Houg, G. S., and J. Guarro.** 1995. *Atlas of Clinical Fungi.* Centraalbureau voor Schimmelcultures, Dielft, The Netherlands.
2. **St. Germain, G., and R. Summerbell.** 1996. *Identifying Filamentous Fungi. A Clinical Laboratory Handbook.* Star Publishing Co., Belmont, Calif.

Table 6.8.8–1 A more detailed table to differentiate dermatophytes[a]

Genus and species	Growth rate	Conidia: Macro	Conidia: Micro	Hair perforation	BCPMSG	Urease (7 days)	Growth factor requirements
Microsporum							
audouinii	Mod. rapid	Fusoid, deformed, very rare	Absent or numerous	Neg.	Neg.	Neg.	
canis	Rapid	Fusoid, apex recurved, numerous	Mod. numerous	Pos.	Neg.	Variable	
cookei	Mod. rapid	Fusoid, numerous	Numerous	Pos.		Neg.	
ferrugineum	Slow	Absent	Absent	Neg.			
gallinae	Mod. rapid	Club-shaped, rare or numerous	Rare or numerous	Neg.			
gypseum	Rapid	Fusoid, symmetrical, numerous	Mod. numerous	Pos.	Neg.	Variable	
nanum	Mod. rapid	Ovoid, 2-celled, numerous	Mod. numerous	Pos.		+	
persicolor	Rapid (poor at 37°C)	Fusoid, club-shaped, sometimes present	Numerous	Pos.	Neg.	Variable	
Trichophyton							
ajelloi	Mod. rapid	Fusiform, thick walled, numerous	Pyriform, absent or rare	Pos.		Pos.	None
concentricum	Very slow	Absent	Absent	Neg.		??	None or thiamine
equinum	Mod. rapid	Club-shaped, absent or rare	Pyriform, numerous	Neg. (pos.)	Alk	Pos.	Nicotinic acid
megninii	Mod. rapid	Cylindrical, absent or rare	Pyriform, club-shaped, numerous	Neg.	Alk	Pos.	Histidine
mentagrophytes	Mod. rapid	Club-shaped, absent or ± numerous	Pyriform, round, numerous or rare	Pos.	Alk	Pos.	None
rubrum	Slow to mod. rapid	Cylindrical, absent or ± numerous	Club-shaped, pyriform, ± numerous	Neg.	Neg.	Neg.	None
schoenleinii	Very slow	Absent	Absent	Neg.	Alk	??	None
soudanense	Slow	Absent	Pyriform, ovoid, rare or absent	Neg.	Alk with clearing	Neg.	None (?)
terrestre	Mod. rapid	Cylindrical, ± numerous	Club-shaped, numerous	Pos.	Alk	Pos.	None
tonsurans	Slow	Club-shaped, sinuous, absent or rare	Club-shaped, balloon, numerous	Neg. (pos.)	Alk (var.)	Pos.	Thiamine
verrucosum	Very slow	"Rat tail," absent or rare	Club-shaped, absent or rare	Neg.	Neg.	Neg.	Thiamine, inositol ±
violaceum	Very slow	Absent	Absent or rare	Neg.		Neg.	Thiamine
yaoundei	Very slow	Absent	Absent	Neg.		??	None

[a] Adapted from references 1 and 2. For details about identification tests, see the following EPCM items; Urease, EPCM 6.8.9; BCPMSG, EPCM 6.8.10; growth factor (*Trichophyton* agars), EPCM 6.8.11; hair perforation, EPCM 6.8.12; cornmeal agar dextrose, EPCM 6.8.13; rice grain, EPCM 6.8.14. Mod., moderately; neg., negative; pos., positive.

Table 6.8.8–2 Differentiation of similar conidia-producing *Trichophyton* spp. (predominantly microconidia)[a]

Organism	Agar no. 1 (casein base)	Agar no. 4 (casein + thiamine)	Urease (4 days)	Urease (7–14 days)	In vitro hair perforation	Red pigment on cornmeal with 1% dextrose	Growth at 37°C	BCPMSG (7 days)
T. mentagrophytes	4+	4+	+		+	0	+	+ (alkaline)
T. rubrum	4+	4+		0 or W	0	+	+	− (no pH change)
T. tonsurans	± or +	4+		+	0^v	0	+	
T. terrestre	4+	4+		+	+	V	0	

[a] Symbols: + = positive; 0 = negative; W = weak; ± = trace; 4+ = maximum growth; V = variable.

6.8.9 Hydrolysis of Urea

I. PRINCIPLES

Urea hydrolysis in Christensen's medium causes a rise in pH after the formation of ammonia. This characteristic is sometimes variable and cannot be used as a conclusive criterion in the identification of dermatophytes. It is used primarily to distinguish *Trichophyton rubrum* and *T. mentagrophytes*.

II. MATERIALS

Include QC information on reagent container and in QC records.

A. Christensen's urea agar slants
B. Fungal colony on solid medium

III. QUALITY CONTROL

A. ***T. mentagrophytes*, positive reaction (red)**
B. ***T. rubrum*, negative reaction (yellow)**

IV. PROCEDURE

A. Inoculate a slant of Christensen's urea agar with a fragment of the colony.
B. Incubate at 25°C for 7 days.

V. EXPECTED RESULTS

A positive reaction is seen in a change of the original yellowish color of the medium to red.

6.8.10 Bromocresol Purple-Milk Solids-Glucose Medium for *Trichophyton* Species Differentiation

I. PRINCIPLE

Bromocresol purple-milk solids-glucose medium (BCPMSG) can be used as an aid in the rapid determination of *Trichophyton rubrum* and *T. mentagrophytes*. Typical pure cultures of *T. mentagrophytes* produce spreading growth and a pronounced alkaline pH change at 7 days. *T. rubrum* typically shows a restricted pattern of growth with no pH changes within the same period.

II. MATERIALS

A. BCPMSG
1. Purchased from Smith River Biologicals
2. Stored at 2 to 8°C
3. Expiration date established by manufacturer

B. Stock culture of the following organisms maintained at room temperature:
1. *T. mentagrophytes* (lab strain)
2. *T. rubrum* (lab strain)

C. Mycological probe

III. QUALITY CONTROL

QC is performed with each receipt of the media and each day of use.

IV. PROCEDURE

A. Inoculate isolate and controls onto slants of BCPMSG medium by using a mycology probe (heavy-gauge probe such as a biological probe).

B. Incubate for 7 days at 25°C.

V. INTERPRETATION

A. A change in the pH indicator (bromocresol purple) indicates alkalinity (medium changes from pale blue to violet purple on tube reverse of growth area). *T. mentagrophytes* will produce an alkaline reaction as well as profuse growth.

B. No change in the pH indicator and restricted growth are typical for *T. rubrum*.

C. *Trichophyton* spp. responses on BCPMSG after a 7-day incubation:

Organism	(+)Alkaline reaction	(−)Alkaline reaction	Type of growth
T. mentagrophytes	√		Profuse
T. rubrum		√	Restricted

D. The microscopic appearance of the isolate must be correct for the organism in question.

VI. LIMITATIONS

A. Some *T. rubrum* isolates (melanoid variants) may secrete a weak pinkish brown pigment on BCPMSG, but this should not be mistaken for a positive alkaline reaction.

B. Typical *T. rubrum* isolates may produce the typical red colony reverse pigment of the species within 4 to 6 days on BCPMSG. This pigmentation can be seen through the colony from above or from the side, but the underside of the colony is obscured by the opaque milk solids of the medium.

C. The majority of *T. mentagrophytes* will produce profuse growth and an alkaline reaction on BCPMSG medium. Slow-growing nodular isolates of *T. mentagrophytes* may be detectable but only slightly positive on the BCPMSG medium after 7 days. They become fully positive within 10 days.

D. *Microsporum persicolor* shows profuse growth but does not produce an alkaline reaction on BCPMSG medium.

SUPPLEMENTAL READING

Summerbell, R. C., S. A. Rosenthal, and J. Kane. 1988. Rapid method for differentiation of *Trichophyton rubrum, Trichophyton mentagrophytes*, and related dermatophyte species. *J. Clin. Microbiol.* **26:**2279–2282.

6.8.11 Nutritional Tests for the Differentiation of Some *Trichophyton* Species

I. PRINCIPLE

Occasionally, *Trichophyton* species are difficult to differentiate from one another solely on microscopic and morphologic characteristics. Members of this genus have specific nutritional requirements that are essential for definitive identification. Nutritional requirements are determined by inoculating a control medium and a medium enriched with a specific vitamin or amino acid with *Trichophyton* spp. isolates that have been presumptively identified by gross colony characteristics and microscopic morphology. Moderate to heavy growth in the vitamin- or amino acid-enriched medium compared with little or no growth in the basal medium indicates that the isolate requires that nutrient.

II. MEDIA

Include QC information on reagent container and in QC records.

Four types of nutritional media can be used to differentiate *Trichophyton* species.

A. *Trichophyton* agar no. 1 is a vitamin-free basal medium consisting of acid hydrolysate of casein to provide the basic nutrients necessary to support growth. Dextrose provides an energy source, and a phosphate buffer maintains the pH of the medium.

B. *Trichophyton* agar no. 2 consists of the basal medium supplemented with inositol.

C. *Trichophyton* agar no. 3 consists of the basal medium supplemented with inositol and thiamine.

D. *Trichophyton* agar no. 4 consists of the basal medium supplemented with thiamine.

III. MATERIALS

Include QC information on reagent container and in QC records.

A. ***Trichophyton* agar no. 1**
   1. Purchased commercially from BBL, catalog no. 97613
   2. Stored at 2 to 8°C
   3. Expiration date established by manufacturer

B. ***Trichophyton* agar no. 2**
   1. Purchased commercially from BBL, catalog no. 97614
   2. Stored at 2 to 8°C
   3. Expiration date established by manufacturer

C. ***Trichophyton* agar no. 3**
   1. Purchased commercially from BBL, catalog no. 97615
   2. Stored at 2 to 8°C
   3. Expiration date established by manufacturer

D. ***Trichophyton* agar no. 4**
   1. Purchased commercially from BBL, catalog no. 97616
   2. Stored at 2 to 8°C
   3. Expiration date established by manufacturer

E. **A forked inoculating needle**

F. **Mycological inoculation probe**

G. **Positive and negative controls**
   1. *T. mentagrophytes* lab strain, positive control
   2. *T. tonsurans* lab strain, negative control

IV. QUALITY CONTROL

QC is performed at each receipt of the medium.

V. PROCEDURE

The test is performed in parallel with tests on negative and positive control strains.

A. For each organism to be tested, use one set of four agar plates (*Trichophyton* agars 1, 2, 3, and 4). Based on the isolate and the microscopic morphology, *Trichophyton* agars 1 and 4 may be adequate to differentiate *T. tonsurans*.
   1. The BBL *Trichophyton* agar is stored as tube deeps. Place the tubes in a boiling-water bath to melt the agar and then pour the agar into a petri dish.
   2. Allow the plates to cool and solidify.

B. Use a probe or a forked inoculator to inoculate the center of each plate in a set.
   1. Use an inoculum no larger than a pin head.
   2. This method allows the smallest fragment of a mold culture to be consistently inoculated onto the medium.
   3. Avoid transferring agar with the inoculum, since the agar is not likely vitamin free.

C. Incubate the plates at room temperature for 7 to 10 days. If growth is inadequate at 10 days, it may be necessary to incubate the plates for 14 days.

D. Other tests should be performed at the same time. See the urease test.

VI. INTERPRETATION

A. Examine the slants for growth within 7 to 10 days.

B. Interpretation of nutritional test
   1. Use the following list to check the control strains and to determine the identity of the unknown isolate.
   2. Grade the amount of growth in the following manner:
      a. 4+, maximum growth for the series of tests when comparing them with colony growth in the other tests
      b. +/−, trace of submerged growth around the inoculum

	No. 1 (basal medium only)	No. 2 (Inositol)	No. 3 (Thiamin + inositol)	No. 4 (Thiamin)
T. mentagrophytes	4+	4+	4+	4+
T. rubrum	4+	4+	4+	4+
T. tonsurans	+ to 1+	− to +	3+	4+

VII. LIMITATIONS

A. For identification, organisms must be in pure culture.

B. The control strains must perform properly for the test to be valid.

C. Additional morphological and biochemical tests should be performed for final identification, such as the urease test and BCPMSG, especially to differentiate *T. rubrum* (urease negative or weak) and *T. mentagrophytes* (urease positive).

6.8.12 Hair Perforation Test

I. PURPOSE

The hair perforation test is used for the differentiation of *T. mentagrophytes* from *T. rubrum*.

II. PRINCIPLE

A. Hairs exposed to *T. mentagrophytes* are penetrated perpendicularly by hyphae, resulting in wedge-shaped perforations.

B. Hairs exposed to *T. rubrum* are not perforated.

C. This technique can be used to differentiate *T. mentagrophytes* and *T. rubrum* only.

D. Other dermatophytes and keratinophilic fungi will perpendicularly penetrate hairs in vitro.

III. MATERIALS

A. Sterile, healthy, clean prepubertal human hair

1. Hair bleached, permanented, or treated with hair spray is unsuitable.
2. Sterilize hair by autoclaving (in a test tube) at 15 psi and 121°C for 15 min.

B. Sterile 10% yeast extract

1. 10 g of yeast extract in 100 ml of dH_2O
2. Filter sterilize.
3. Refrigerate at 4°C.

IV. PROCEDURE

A. Run test in parallel with stock *T. mentagrophytes* as a positive control.

B. Into a petri dish, place several fragments of sterile hair.

1. Add 20 to 25 ml of sterile distilled water.
2. Add 0.1 ml of 10% sterile yeast extract.
3. Place several fragments of fungus culture into the suspension.

C. Incubate the culture at room temperature.

D. Examine at weekly intervals for a period of 4 weeks for hair perforation.

V. EXPECTED RESULTS

A. Place hair from the culture into a drop of lactophenol cotton blue and mount with a coverslip.

B. In a positive test, hair will be penetrated perpendicularly in wedge-shaped perforations.

SUPPLEMENTAL READING

Haley, L. D. 1978. *Laboratory Methods in Medical Mycology.* ODC publication no. 78-8361. U.S. Government Printing Office, Washington, D.C.

6.8.13 Cornmeal Agar for *Trichophyton* Species Differentiation

I. PRINCIPLE

Cornmeal agar with dextrose is used in differentiating some species of *Trichophyton* on the basis of pigmentation. A deep ruby red color will be evident in this agar in 3 weeks.

Dextrose is added to cornmeal agar to differentiate some species of *Trichophyton* on the basis of pigmentation.

II. MATERIALS

A. Cornmeal agar with dextrose
1. Purchased commercially from Remel, catalog no. 09-286
2. Stored at 2 to 8°C
3. Expiration date established by manufacturer

B. Stock cultures
1. Include
   a. *T. rubrum* (ATCC 28188)
   b. *T. mentagrophytes* (lab strain)
2. Maintained at room temperature

C. Mycological inoculation probe

III. QUALITY CONTROL

QC is performed at each receipt of the medium.

IV. PROCEDURE

A. The media is received as an agar deep. Heat the agar deep in a boiling-water bath to melt the agar. Slant the tube and cool to solidify.
B. Inoculate the middle of the slant with the mold to be tested. For comparison, inoculate two other slants with *T. rubrum* ATCC 28188 and a *T. mentagrophytes* lab strain.
C. Incubate at 28 ± 2°C for up to 4 weeks.
D. Examine at regular intervals for growth and pigmentation.
E. *T. rubrum* is characterized by development of a deep red pigment on the reverse of the colony.
F. Other common dermatophytes do not produce this color on cornmeal agar. (The nonpathogenic *T. terrestre* may also produce a deep red pigment.)

V. INTERPRETATION

A. Positive, colony develops a deep red pigment.
B. Negative, colony remains colorless, pale yellow to brown.

VI. LIMITATIONS

A. It can take up to 4 weeks for color development.
B. *T. terrestre* may also produce a deep red pigment.

6.8.14 Rice Grain Test

I. PRINCIPLE

The sterile-rice-grain test is useful in distinguishing atypical isolates of *Microsporum canis* from isolates of *M. audouinii*.

II. MATERIALS

A. Flask containing sterile rice grains
B. Long-handled inoculating needle

III. PROCEDURE

A. With the inoculating needle, transfer a small portion of the isolate to a flask containing sterile rice grains.
B. Incubate the flask at 30°C, and examine for growth for 8 to 10 days.

IV. QUALITY CONTROL

Known isolates of *M. canis* and *M. audouinii* must be run concurrently with each rice grain test.

V. INTERPRETATION

A. Positive
Rapid growth on rice grains, typically with many conidia and a bright, yellow pigment: *M. canis*

B. Negative
Absence of growth with or without a brown discoloration of the rice grains at the site of inoculation: *M. audouinii*

REFERENCES

1. **De Hoog, G. S., and J. Guarro.** 1995. *Atlas of Clinical Fungi.* Centraalbureau voor Schimmelcultures, Delft, The Netherlands.
2. **St-Germain, G., and R. Summerbell.** 1996. *Identifying Filamentous Fungi. A Clinical Laboratory Handbook.* Star Publishing Co., Belmont, Calif.

6.9 Antifungal Susceptibility Testing

Susceptibility testing for fungi has not been well developed and lacks the historical richness of antibacterial-agent-susceptibility testing. A standard protocol (NCCLS M27-A) for broth dilution antifungal-agent-susceptibility testing (AFST) of yeasts has been published; however, there is no available standard applicable to moulds.

The choice to pursue AFST should be made cautiously. Although it is a standard method, the procedure requires maintenance of competency. Unless a laboratory routinely receives at least five requests per week for susceptibility testing of yeasts, the AFST should be performed by a reference laboratory. The more common agents of disease have relatively well-established antibiograms (Table 6.9–1). Breakpoints have been proposed for fluconazole, itraconazole, and 5-fluorocytosine (Table 6.9–2). This fact allows empiric selection of therapeutic treatment without the need for testing unless clear indications suggest otherwise (e.g., refractory infection, contraindications).

Diseases for which susceptibility testing should be performed include aspergillosis, fusariosis, and zygomycosis. These diseases are caused by several species within each genus of fungal pathogen, and significant variation within species has been observed. Each of these diseases is caused by a mould. Unfortunately, as noted above, a standard AFST for moulds has not been established. However, an NCCLS working committee has begun to publish suggestions for an AFST for moulds (1).

Table 6.9–1 General susceptibility patterns of the more common yeast species (based on 48-h MIC_{50} values)[a]

Drug	Susceptibility pattern
Fluconazole	*C. albicans* < *C. parapsilosis* = *C. lusitaniae* < *C. guilliermondii* = *C. reugosa* = *Cryptococcus neoformans* < *C. tropicalis* < *C. krusei*[b]
Amphotericin B	*C. parapsilosis*, tolerant *C. lusitaniae*, resistant *C. krusei* and *C. glabrata*, likely resistant

[a] Adapted from reference 2.

[b] In order from susceptible to resistant.

Table 6.9–2 Interpretive guidelines for susceptibility testing in vitro of *Candida* species[a]

Antifungal agent	Susceptible (S)	Susceptible-dose dependent (S-DD)[b]	Intermediate (I)[c]	Resistant (R)
Fluconazole[d]	≤8	16–32		≥64
Itraconazole	≤0.125	0.25–0.5		≥1
Flucytosine	≤4		8–16	≥32

[a] Adapted from NCCLS M27-A. If MICs are measured by a scale that yields results falling between categories, the next higher category is implied. For example, a fluconazole MIC of 12.5 μg/ml is placed in the S-DD category.

[b] S-DD, Susceptible but dose dependent.

[c] The susceptibility of these isolates is not certain, and the available data do not permit them to be clearly categorized as either susceptible or resistant.

[d] Isolates of *C. krusei* are assumed to be intrinsically resistant to fluconazole and their MICs should not be interpreted using this scale.

REFERENCES

1. **Espinel-Ingroff, A., M. Bartlett, R. Bowden, N. X. Chin, C. Cooper, Jr., A. Fothergill, M. R. McGinnis, P. Menezes, A. Messer, P. W. Nelson, F. C. Odds, L. Pasarell, J. Peter, M. A. Pfaller, J. H. Rex, M. G. Rinaldi, G. S. Shankland, T. J. Walsh, and I. Weitzman.** 1997. Multicenter evaluation of proposed standardized procedure for antifugal susceptibility testing of filamentous fungi. *J. Clin. Microbiol.* **35:**139–143.

2. **Pfaller, M. A., and A. L. Barry.** 1994. Evaluation of novel colorimetric broth microdilution method for antifungal susceptibility testing of yeast isolates. *J. Clin. Microbiol.* **32:** 1992–1996.

6.10.1 Introduction

I. PRINCIPLE

(Adapted from reference 2.)

The genera of aerobic actinomycetes addressed below are those most frequently encountered in the clinical microbiology laboratory. These genera, members of the order *Actinomycetales*, include *Nocardia, Streptomyces, Actinomadura, Arachnia, Nocardiopsis, Rhodococcus*, and *Dermatophilus*. These organisms range from aerobic, gram-positive diphtheroids to branching, filamentous bacteria capable of growth at 25 to 27°C. Some genera are partially acid fast, thermophilic, and capable of forming well-developed aerial and vegetative growth composed of branched filaments approximately 1 μm in diameter.

They are closely related to *Mycobacterium* species and may have the raised, dry, wrinkled surface similar to that of many *Mycobacterium* species.

Nocardia spp. are partially acid fast. They are widely distributed in soil and are normally saprophytic. *N. brasiliensis* may produce disease in healthy people, whereas *N. asteroides* usually infects only debilitated persons and is considered opportunistic.

Streptomyces and *Actinomadura* species usually exhibit extensive hyphal formation with no fragmentation into bacillary or coccoid forms. There is more abundant aerial mycelium, and these mycelial elements are not acid fast. The spores are acid fast, oval to round, and produced in chains from the ends of specialized hyphal elements. The principal disease with which *Actinomadura* and *Streptomyces* spp. are associated is actinomycotic mycetoma, a chronic, pus-forming, and granulomatous mycotic disease of the subcutaneous tissue and bones.

Identification of these organisms may range from quick and easy (Tables 6.10.1–1 and 6.10.1–2) to extremely complicated reference-level procedures such as defining chemotaxonomic markers. The following schema comprises direct microscopic techniques and a minimum of biochemical reactions that will permit the identification of most aerobic actinomycetes encountered in the clinical laboratory.

II. SPECIMEN PROCESSING AND CULTURING

(Adapted from reference 2.)

A. Specimens for the isolation of the aerobic actinomycetes are usually submitted for bacterial and/or fungal culture, except for urine and genital specimens. The pathogenic actonomycetes cause acute, necrotizing pyogenic or chronic exudative inflammatory infections. In either case, the usual specimen is pus or necrotic tissue. Lesions may be deep, soft abscesses or burrowing sinus tracts. If infection is respiratory, cutaneous, or mucocutaneous, exudates may be contaminated with normal microbiota.

B. Processing

1. Specimens suspected of containing aerobic actinomycetes are processed as usual for bacteria.
2. Do not decontaminate respiratory specimens in the manner of mycobacteria. If an actinomycete is suspected, rely on antimicrobial-agent-containing medium that will inhibit normal microbiota and allow the selective recovery of actinomycetes.
3. If an anaerobic actinomycete is considered a possibility, collect specimens in anaerobic transport systems or containers containing appropriate media and environments (*see* section 2 of this book).
4. Transport specimens to the laboratory immediately.

Table 6.10.1–1 Identification of the aerobic actinomycetes[a]

Organism	Acid fast	Biochemical reaction						
		LR	C	X	T	HX	U	N
Nocardia asteroides	+	R	−	−	−	−	+	+
Nocardia brasiliensis	+	R	+	−	+	+	+	+
Nocardia otitidiscaviarum	+	R	−	+	−	+	+	+
Actinomadura madurae	−	S	+	−	+	+	−	+
Actinomadura pelletieri	−	S	+	−	+	+	−	+
Nocardiopsis dassonvillei	−	S	+	+	+	+	+	+
Streptomyces spp.	−	S	+	+	+	+	V	−
Streptomyces somaliensis	−	S	+	−	+	−	−	V
Streptomyces albus	−	S	+	+	+	+	+	−
Streptomyces griseus	−	S	+	+	+	+	V	−
Rhodococcus spp.	+/−	R/S	−	−	+/−	−	V	V
Mycobacterium group IV	+	R	−	−	−	−	V	V
Dermatophilus congoliensis	−	S	+	−	−	−	+	−

[a] Adapted from reference 2. Decomposition reactions: C, casein; X, xanthine; T, tyrosine; HX, hypoxanthine. Other biochemical reactions: U, urea; N, nitrate reduction. V, ≤15%; LR, lysozyme resistance.

III. DIRECT MICROSCOPIC EXAMINATION OF SPECIMENS

(Adapted from reference 2.)

A. For minced tissue, body fluids, pus, and exudates, make smears in the usual manner.

1. Gram stain reveals gram-positive organisms.
2. *Nocardia* spp., the rapidly growing mycobacteria, and some *Rhodococcus* spp. are positive by modified Kinyoun acid-fast stain (*see* procedure 6.10.2).
3. Morphology
   - **a.** *Streptomyces, Actinomyces,* and *Nocardiopsis* spp., <1-μm-diameter filamentous bacteria
   - **b.** Rapidly growing mycobacteria and *Nocardia* spp., <1-μm-diameter branching filaments fragmenting into coccobacillary forms
   - **c.** Granules

     Branching filaments intertwined to form granules that vary in texture and color depending on the etiologic agent

B. Presumptive identifications based on staining and morphology in smears and tissue sections

1. *Nocardia* spp. and *Mycobacterium* spp.
   - **a.** Gram-positive, branched diphtheroids to filamentous bacteria
   - **b.** Partially or completely acid fast

Table 6.10.1–2 Hydrolysis and selected biochemical profiles of pathogenic species of *Nocardia*[a]

Organism	Decomposition of:						Growth at 45°C	Arylsulfatase (14 day)	Acid from:	
	Casein	Hypoxanthine	Tyrosine	Xanthine	Gelatin	Starch			Glucose	Rhamnose
N. asteroides	−	−	−	−	−	−	+/−	−	+	−/+
N. farcinica	−	−	−	−	−	−	+	−	+	+/−
N. nova	−	−	−	−	−	−	−	+	+	−
N. brasiliensis	+	+/−	+	−	+	−	−	−	+	−
N. otitidiscaviarum	−	+	−	+	−	−	+/−	−	+	−
N. transvalensis	−/+	+/−	−/+	+/−	−	+	−	−	+	−

[a] Adapted from reference 1. Symbols: +, predominantly positive; +/−, mostly positive with some species or strains negative; −/+, mostly negative with some species or strains positive; −, predominantly negative.

III. DIRECT MICROSCOPIC EXAMINATION OF SPECIMENS *(continued)*

c. Granules

Small (25- to 150-μm), soft, white or yellow, and lobulated; rarely occur with nocardioforms.

2. *Rhodococcus* spp.

a. Gram-positive, coccobacillary to rudimentary diphtheroids

b. Some species partially acid fast

c. Granules

Small, white or yellow, and soft; resemble bacterial colonies

3. All other aerobic actinomycetes

a. Gram-positive, intertwined branching filaments; rarely, fragmenting forms to diphtheroids

b. Non-acid fast

c. Granules

The granules of each species are unique in hyphal appearance, texture, color, and size.

4. Anaerobic actinomycetes

a. Gram-positive, intertwined branching filaments, fragmenting forms to diphtheroids

b. Non-acid fast

c. Granule

Commonly called sulfur granules

REFERENCES

1. **Beaman, B. L., M. A. Saubolle, and R. J. Wallace.** 1995. *Nocardia, Rhodococcus, Streptomyces, Oerskovia*, and other aerobic actinomycetes of medical importance, p. 379–399. *In* P. R. Murray, E. J. Baron, M. A. Pfaller, F. C. Tenover, and R. H. Yolken (ed.), *Manual of Clinical Microbiology*, 6th ed. ASM Press, Washington, D.C.

2. **Land, G. A.** 1992. Identification of the aerobic actinomycetes, p. 4.1.1–4.1.9. *In* H. D. Isenberg (ed.), *Clinical Microbiology Procedures Handbook*, vol. 1. American Society for Microbiology, Washington, D.C.

6.10.2 Modified Kinyoun Acid-Fast Stain

I. MATERIALS

Include QC information on reagent container and in QC records.

A. Kinyoun carbolfuchsin (Remel 40–104)

B. Decolorization reagent (1.0% aqueous sulfuric acid)

concentrated sulfuric acid ....1.0 ml
distilled water ....99.0 ml

Slowly add the acid to the water (exothermic reaction)

C. Methylene blue counterstain (Remel 40–110)

◪ **NOTE:** The Kinyoun stain kit from BBL may be substituted for these reagents. The decolorizer, sulfuric acid, listed above must be used.

II. QUALITY CONTROL

A. Positive control

Smears containing *Nocardia asteroides* or rapidly growing mycobacteria

B. Negative control

Smears containing bacteria other than mycobacteria or nocardia

III. PROCEDURE

A. Prepare a smear, and fix it over a flame or on a slide warmer set to 56°C.
B. Flood the slide with excess Kinyoun carbolfuchsin for 5 min.
C. Pour off excess stain.
D. Flood the slide with 50% alcohol, and immediately wash it with water.
E. Decolorize the smear with 1% aqueous sulfuric acid.
F. Wash the slide in water.
G. Counterstain the smear with methylene blue for 1 min.
H. Rinse with water, dry, and examine at 1,000× magnification.

IV. INTERPRETATION

The filaments of *Nocardia* spp. and some *Rhodococcus* spp. typically appear partially acid fast.

6.10.3 Media, Reagents, and Stains

This item is adapted from reference 1.

I. INTRODUCTION

Unless specified, the primary and biochemical media listed below are commercially available from numerous sources (*see* Appendix A for media suppliers) as prepared plates or in deep-butt tubes, which may be melted and poured as plates immediately before testing. The former have a shelf life of 20 to 30 days, and the latter last for up to 6 months before pouring.

II. SAFETY CONSIDERATIONS

It is imperative that these cultures be handled in a biosafety hood.

Carry out all subculturing and inoculations within a certified biological safety cabinet.

III. PRIMARY CULTURE MEDIA

Include QC information on reagent container and in QC records.

A. Characteristics of plating battery

1. Media inhibiting normal microbiota but not aerobic actinomycetes
   a. BHI agar with chloramphenicol and cycloheximide
   b. Columbia agar with colistin and nalidixic acid
2. Anaerobic media (*see* EPCM section 3)
3. Routine media
   a. Sabouraud glucose agar
   b. BAP
   c. BHI agar
   d. TSA

B. Other media

1. Lowenstein-Jensen, Middlebrook 7H10 or 7H11
2. Selective Middlebrook 7H10 or 7H11
3. *Nocardia* spp. only: chemically defined medium containing paraffin and/or gelatin as the sole carbon source

C. Incubation

1. Aerobic media
   Incubate all primary and biochemical media discussed in this section at 30°C for 14 to 21 days. Examine the plates every 3 or 4 days.
2. Anaerobic media (*see* section 3)
   Incubate media at 37°C for 10 to 14 days. Examine the plates at 48 h postinoculation and every 3 or 4 days thereafter.

REFERENCE

1. **Land, G. A.** 1992. Identification of the aerobic actinomycetes, p. 4.1.1–4.1.9. *In* H. D. Isenberg (ed.), *Clinical Microbiology Procedures Handbook*, vol. 1. American Society for Microbiology, Washington, D.C.

6.10.4 Biochemical Identification

Use Tables 6.10.4–1 and 6.10.1–2 for identification to species level. The following are some biochemical identification tests.

PART 1

Casein, Xanthine, and Tyrosine Decomposition for Aerobic Actinomycetes

I. PRINCIPLE

Biochemical tests demonstrating the hydrolysis of casein, tyrosine, or xanthine may be used to differentiate *Nocardia* species and to differentiate *Nocardia* from *Streptomyces* species, which appear morphologically similar to *Nocardia* spp. in clinical materials and in culture.

Casein agar consists of agar with skim milk as a source of casein. Tyrosine agar and xanthine agar consist of nutrient agar supplemented with tyrosine or xanthine. *Nocardia* spp. may be differentiated from *Streptomyces* spp. based on patterns of enzymatic hydrolysis of casein, tyrosine, and xanthine. Clear zones under colony growth and in the surrounding medium indicate hydrolysis of the substrate.

II. MATERIALS

Include QC information on reagent container and in QC records.

A. Casein agar
   1. Tubed media purchased from BBL catalog no. 96188
   2. Stored at 2 to 8°C

B. Tyrosine and xanthine agar
   1. Tubed media purchased from BBL catalog
      a. Catalog no. 97222, tyrosine
      b. Catalog no. 97224, xanthine
   2. Stored at 2 to 8°C

C. Control organisms
   1. *Nocardia asteroides* lab strain
   2. *Streptomyces* sp. lab strain

D. Sterile small petri dishes

E. Inoculating probe

III. QUALITY CONTROL

QC is performed with each receipt of the medium.

IV. PROCEDURE

A. Melt casein, tyrosine, and xanthine agar in a hot-water bath and pour into sterile petri dishes. Allow to solidify.

B. Divide plates into quadrants. Label each quadrant.
   1. Laboratory number of test organism
   2. Control strains

C. Inoculate the middle of each quadrant with a good portion of the organisms.
   1. Use a wedge-shaped slice of culture if the colony adheres to the medium.
   2. If the inoculum is insufficient, the organism will fail to grow.

IV. PROCEDURE *(continued)*

D. Incubate at 30°C.
E. Plates should be checked every other day to determine initially that the organisms are growing.
F. The plates should be checked weekly for decomposition of the above substrates.
G. Consult table for organism identification.

PART 2

Lysozyme Broth Test

I. PRINCIPLE

A. *Nocardia* species are resistant to lysozyme. Therefore, they will grow in this medium, while *Streptomyces* species do not.
B. This test is useful in separating *Nocardia* species from other actinomycetes.

II. MATERIALS

Include QC information on reagent container and in QC records.

A. Lysozyme broth
1. Purchased commercially from BBL, catalog no. 927231
2. Stored at 2 to 8°C
3. Expiration date established by manufacturer

B. Basal broth
1. Purchased commercially from BBL, catalog no. 927232
2. Stored at 2 to 8°C
3. Expiration date established by manufacturer

C. Mycological inoculation probe

D. Stock cultures
1. *Nocardia asteroides* ATCC 19247
2. *Streptomyces griseus* ATCC 10971

NOTE: Maintain at room temperature.

III. QUALITY CONTROL

QC is performed at each receipt of the medium.

IV. PROCEDURE

A. Perform the test in parallel with tests of negative and positive control strains.
B. Inoculate several fragments of the culture into both broth tubes.
C. Inoculate a control set of broths (tubes 1 to 4).
1. Positive control
   N. asteroides should show growth in both tubes.
2. Negative control
   Streptomyces species should have growth in the control tube but not in the lysozyme tube.
3. Include an uninoculated set of broths for sterility.

D. Incubate at 30°C.
1. Check weekly for growth for 4 weeks.
2. Record results weekly on worksheet.
   Tubes 1 and 3 contain glycerol broth with lysozyme. Tubes 2 and 4 contain glycerol broth without lysozyme. Tubes 1 and 2 have been inoculated with *Nocardia asteroides* which is resistant to lysozyme. Tubes 3 and 4 have been inoculated with *Streptomyces griseus*, a lysozyme-sensitive actinomycete.

SUPPLEMENTAL READING

Haley, L. 1978. *Laboratory Methods in Medical Mycology.* CDC publication #78-8361. U.S. Government Printing Office, Washington, D.C.

PART 3

Urea Decomposition for Aerobic Actinomycetes

I. PRINCIPLE

The organism's ability to decompose urea is tested.

II. MATERIALS

Include QC information on reagent container and in QC records.

A. Urea agar slants
1. Purchased from various commercial sources (*see* Appendix A for list of media suppliers).
2. Stored at 2 to 8°C
3. Expiration date established by manufacturer

B. Control strains of *Nocardia asteroides* and *Streptomyces* species

III. QUALITY CONTROL

Sterility and performance tests are performed with each new lot number.

IV. PROCEDURE

A. There will be one slant used for each organism plus one uninoculated slant.
B. Label slants with laboratory numbers and control strains.
C. Inoculate slants.
1. Inoculate a wedge-shaped slide of the culture to the middle of the slant.
2. Inoculate control strains to the slant.

D. Incubate at 30°C.
E. Examine weekly for the development of an alkaline reaction. A pink to red color will develop.
F. Record the reactions weekly on the worksheet.
G. Consult Table 6.10.1–1 for organism identification.

SUPPLEMENTAL READING

Moore, G. S., and D. M. Jaciow. 1979. *Mycology for the Clinical Laboratory.* Reston Publishing Company, Reston, Va.

PART 4

Gelatin Growth Test for Identification of Aerobic Actinomycetes

I. PRINCIPLE

Nocardiosis is a disease of humans and animals caused by *Nocardia* species. The disease may resemble tuberculosis when the organism is inhaled or may produce granulomatous abscesses when the organism is introduced into tissues at the time of an injury.

Biochemical tests demonstrating growth in 0.4% gelatin may be used to differentiate *Nocardia* species. The gelatin test is used in the presumptive differentiation of *Nocardia brasiliensis* from *N. asteroides*. *N. brasiliensis* is the only *Nocardia* species capable of decomposing gelatin into free amino acids.

II. MATERIALS

A. Gelatin broth (0.4%)
1. Purchased commercially from Becton Dickinson, catalog no. 97130
2. Stored at 2 to 8°C
3. Expiration date established by manufacturer

B. Stock culture maintained at 28 ± 2°C
1. *N. brasiliensis* (ATCC 19296) (+ reaction)
2. *N. asteroides* lab strain (− reaction)

C. Mycological probe

6 Mycology and Aerobic Actinomycetes

III. QUALITY CONTROL

QC is performed with each receipt of the media and each day of use.

IV. PROCEDURE

A. Inoculate isolate and controls into tubes of 0.4% gelatin just below the surface of the medium, using a mycology probe.

B. Incubate the media at 35°C in air.

C. Growth on 0.4% gelatin medium should be visible within 10 days of incubation. Incubate for 2 weeks, reading weekly and recording reactions.

V. INTERPRETATION

A. *N. brasiliensis* grows well in the 0.4% gelatin medium, producing compact, round colonies.

B. *N. asteroides* fails to grow or grows poorly and produces thin, flaky growth.

C. Refer to identification chart for reactions (Tables 6.10–1 and 6.10–2).

VI. LIMITATIONS

A. *N. asteroides* may produce a thin flaky growth in the medium.

B. This test should not be the sole basis for identification. Check the other biochemical tests performed.

6.10.5 Disk Diffusion Susceptibility Testing of Rapidly Growing *Mycobacterium* and *Nocardia* spp.

I. PURPOSE

A method for susceptibility testing of rapid growers of *Mycobacterium* and *Nocardia* species.

II. MATERIALS

(See Appendix A for media suppliers.)

A. TSB
Stored at 2 to 8°C

B. Large Mueller-Hinton plates
Stored at 2 to 8°C

C. Antimicrobial disks
Stored at 2 to 8°C

III. PROCEDURE

A. Inoculate TSB with the patient's isolate and with control strains (*M. chelonae, M. fortuitum*, or *Nocardia asteroides* ATCC 23824).

B. Vortex daily (*it is very important to do this step*).

C. On day 5 vortex again; then allow the broth to settle for 15 min. Remove the middle suspension to a sterile screw-cap tube.

D. Dilute this suspension to a 0.5 McFarland standard with saline.

E. Streak this diluted 0.5 standard suspension onto a large Mueller-Hinton plate (three-way streak).

F. Apply selected antibiotic disks.

G. Incubate plates at 37°C in a hot-air incubator.

IV. INTERPRETATION

A. Read plates at 48 and 72 h.

B. Record zone sizes.

C. Report to the requesting physician that this test was performed by nonstandardized methods: ''Nonstandardized Disk Diffusion Susceptibility Test Suggests Resistance to/Susceptibility to...''

SUPPLEMENTAL READING

Wallace, R. J., Jr., D. B. Jones, and K. Wiss. 1981. Sulfonamide activity against *Mycobacterium fortuitum* and *Mycobacterium chelonei*. *Rev. Infect. Dis.* **3:**898–904.

Mycology and Aerobic Actinomycetes

APPENDIX 6–1 Cycloheximide Resistance and Maximum Growth Temperature

Cycloheximide resistance

Some fungi are able to grow in the presence of cycloheximide (CHX). Resistance to CHX is a species-specific characteristic and, therefore, helps to distinguish species within a genus. Although this characteristic is helpful, there is not much information about which species are resistant (Tables A6–1–1 and A6–1–2). Confounding this problem is the lack of a standardized protocol for testing CHX resistance. When correlating CHX resistance with available CHX resistance information, one should be aware of how CHX resistance was determined.

Below are two methods for testing CHX resistance. These methods differ in the concentration of CHX that is used and in the format of the test, solid versus broth media. We prefer the solid-medium method (method 1) because it is easy to perform, can be or may be included as part of the specimen setup, and is familiar to many clinical mycology laboratories.

Method 1

1. Inoculate Sabouraud glucose agar (SGA) and a commercial medium containing CHX (0.05%), e.g., mycobiotic medium (Remel, cat. no. 01-630), with specimen or a fragment of test colony, and incubate both plates at 30°C.
2. Inspect daily until good growth is obtained on the SGA. Resistant organisms will produce nearly as much growth on the CHX plate as that obtained on the SDA. Sensitive organisms will show no growth or only initial growth on the CHX plate.

Method 2

1. This is tested at CHX concentrations of 0.01 and 0.1% in tubes of liquid nitrogen base (+0.5% glucose) broth. Inoculate tubes with specimen or a fragment of test colony and incubate at 30°C.
2. Inspect tubes daily for up to 7 days before recording negative results. (This test is useful for a number of basidiomycetous yeasts.)

Table A6–1–1 CHX sensitivity and maximum growth temperature of various fungal genera or genus-species

Organism	CHX sensitivity[a,b]	Max growth temp (°C)
Dermatophytes		
Microsporum nanum	R	
M. persicolor		37 (works better at 25)
Trichophyton terrestre		30
T. verrucosum		37 (works better at 25)
Zygomycetes		
Absidia corymbifera	S	48–52
Apophysomyces spp.	R	42
Cunninghamella bertholettiae	S	40
C. elegans		<40
Mucor spp.		Most <37
Rhizomucor spp.	S	54–58
Rhizopus arrhizus (*R. oryzae*)		45 (weak)
Saksenaea vasiformis		44
Dematiaceous fungi		
Bipolaris spicifera		40
Cladophialophora bantiana		42–43
Cladosporium sp.		Most <37
Dactylaria constricta	R	
D. gallopava	S	40
Exophiala jeanselmei		≤37
E. mcginnisii		40
Fonsecaea pedrosoi		37
Madurella grisea		≤37
Scytalidium dimidiatum	S	
Wangiella dermatitidis		40
Other		
Aspergillus sp.	S	
Aspergillus fumigatus	S	48
Arthrinium sp.	S	
Arthrographis kalreae		45
Blastomyces dermatitidis (hyphae)	R	
Chrysosporium sp.	R	
Coccidioides immitis	R	42
Histoplasma capsulatum (hyphae)	R	
Penicillium marneffei	S	
Pseudallescheria boydii	S	
Scedosporium apiospermum	R	<40
S. prolificans	S	45
Scolecobasidium constrictum	R	30
Sporotrichum sp.	S mostly	
Sporothrix sp.	R mostly	

[a] Based on growth on Mycosel agar (500 mg of CHX/ml).

[b] R, resistant; S, susceptible.

Table A6–1–2 CHX sensitivity and maximum growth temperature of various yeast genera or species

Organism	CHX sensitivity[a,b]	Max growth temp (°C)
Candida sp.		
C. albicans	R	
C. stellatoidea	S	
C. famata	V	
C. glabrata	S	
C. guilliermondii	V	
C. kefyr	V	
C. krusei	S	
C. lusitaniae	S	
C. haemulonii	S	
C. lusitaniae	V	
C. norvegensis	S	
C. parapsilosis	S	
C. rugosa	S	
C. tropicalis	V	
C. zeylanoides	S	<37
Other		
Blastoschizomyces spp.	R	45
Blastomyces dermatitidis (yeast)	S	
Cryptococcus laurentii	S (V)	
Cryptococcus neoformans	S	
Geotrichum spp.	S	<37
Hansenula spp.	S	
Malassezia spp.	R (V)	
Prototheca spp.	S	
Prototheca stagnora	S	<37
Rhodotorula spp.	V	
Saccharomyces spp.	S	
Trichosporon spp.	V	
Trichosporon pullulans	V	<37

[a] Based on growth on Mycosel agar (500 mg of CHX/ml).

[b] R, resistant; S, susceptible; V, variable reports (likely caused by methodology problems).

APPENDIX 6–2

Useful Definitions (adapted from references 1 and 2)

Anamorph An asexual reproductive or somatic structure of an organism.

Annellide A cell that produces and extrudes conidia; the tip tapers, lengthens, and acquires a ring of cell wall material as each conidium is released; oil immersion magnification may be required to see the rings.

Apiculated Having a short projection at one end or both ends of the cell.

Ascus (pl. asci) A round or elongate saclike structure usually containing two to eight ascospores. The asci are often formed within a fruiting body, such as a cleistothecium or perithecium.

Collarette A small collar of cell wall material at the tip of a phialide which is formed by rupture of the phialide's tip during release of the first conidium.

Columella The enlarged, dome-shaped tip of a sporangiophore that extends into the sporangium. Often the sporangium bursts, leaving the columella bare and readily visible upon microscopic examination.

Dematiaceous The characteristic of having structures that are brown to black, caused by a melanotic pigment in the cell walls.

Denticle Short, narrow, peglike structure bearing a conidium.

Nodal rhizoid Rhizoid that emanates from a point directly below the base of a sporangiophore.

-phore (suffix) A filament extending from a hypha that directly gives rise to a spore-producing structure (e.g., sporangiophore, conidiophore).

Rhizoid Rootlike, branched hypha extending into the medium.

Synanamorph Term usually applied to a specific anamorph that produces another distinct anamorph.

Teleomorph The sexual form of an organism.

Whorl A group of cells radiating from a common point.

References

1. **Kwon-Chung, K. J., and J. E. Bennett.** 1992. *Medical Mycology.* Lea and Febiger, Philadelphia.
2. **Larone, D. L.** 1995. *Medically Important Fungi. A Guide to Identification.* ASM Press, Washington, D.C.

APPENDIX 6–3 Illustrations of Selected Features of Moulds (reprinted from reference 1)

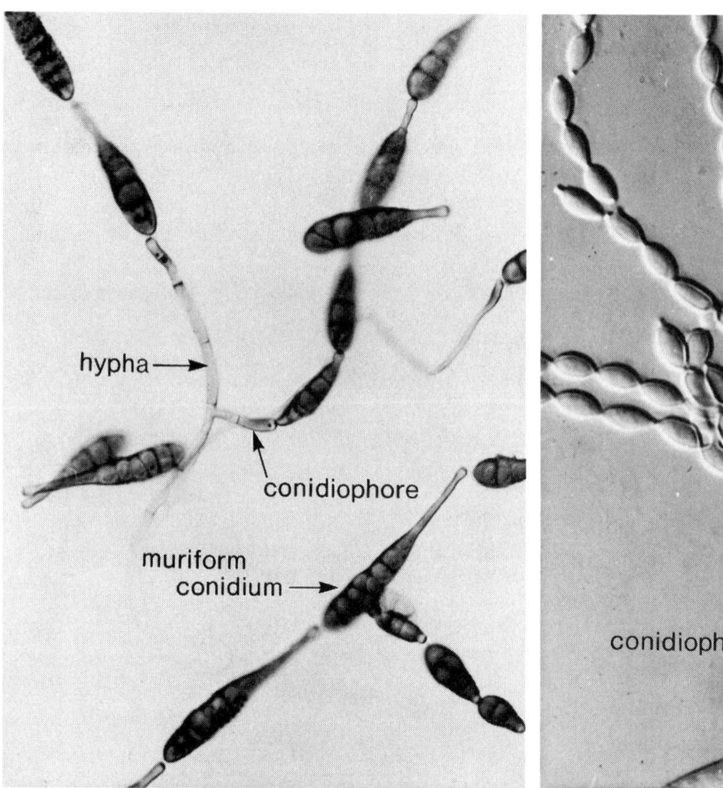

Figure A6–3–1 *Alternaria* sp.

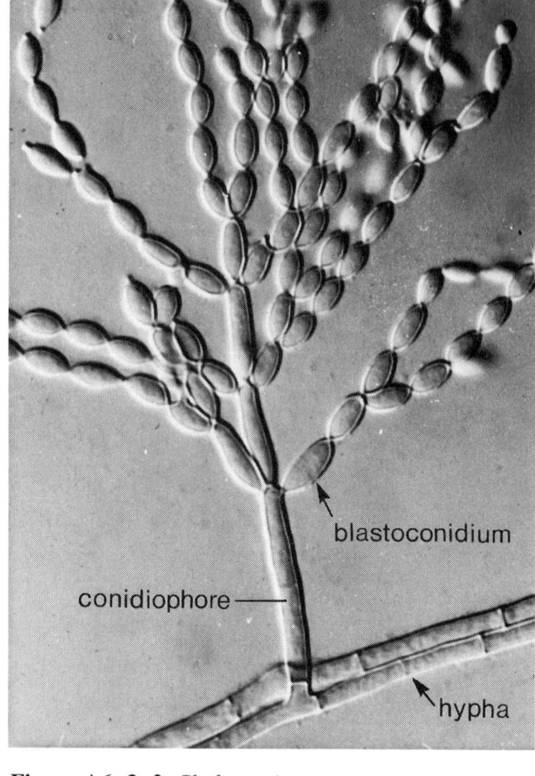

Figure A6–3–2 *Cladosporium carrionii*

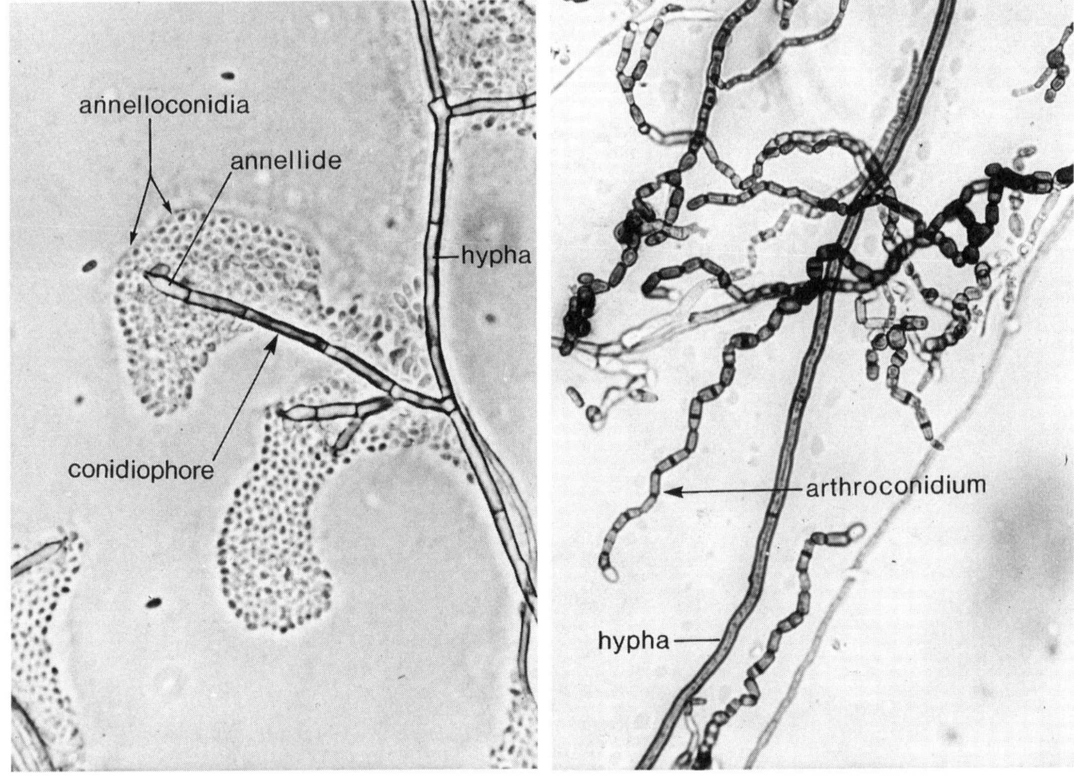

Figure A6–3–3 *Exophiala spinifera*

Figure A6–3–4 *Scytalidium dimidiatum*

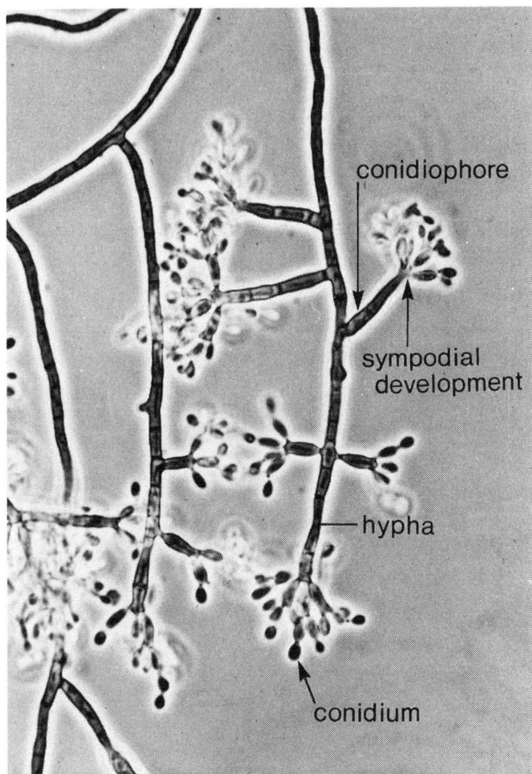

Figure A6–3–5 *Fonsecaea pedrosi*

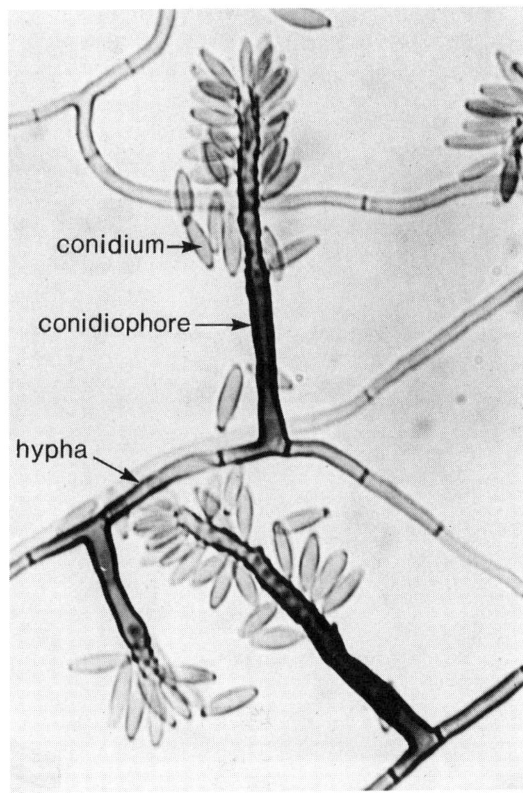

Figure A6–3–6 *Rhinocladiella aquaspersa*

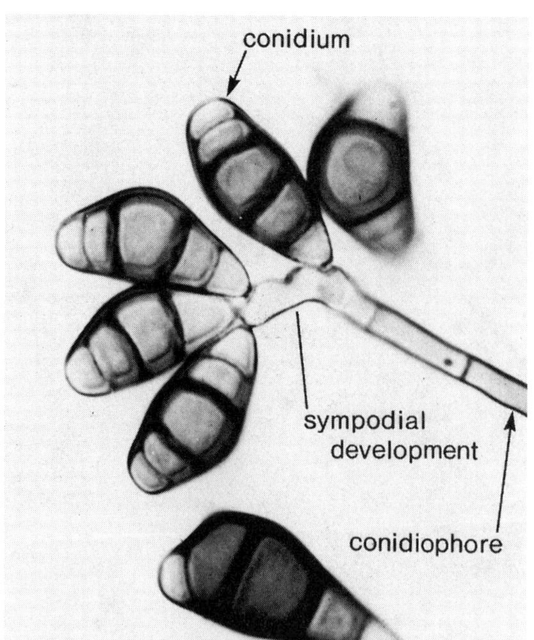

Figure A6–3–7 *Curvularia* sp.

Reference

1. **McGinnis, M. R., and L. Pasarell.** 1992. Mould identification, p. 6.11.1–6.11.17. *In* H. D. Isenberg (ed.), *Clinical Microbiology Procedures Handbook*, vol. 1. American Society for Microbiology, Washington, D.C.

SECTION 7

Parasitology

SECTION EDITOR: *Lynne S. Garcia*

Numerous other procedures can be found in CMPH, and readers are referred to this publication for procedures in diagnostic parasitology that are less frequently used in the routine clinical laboratory.

(continued)

7.1 Introduction

PART 1 Equipment

The majority of equipment used in diagnostic parasitology can be found in any clinical laboratory. However, information contained in this section may help the reader understand the proper use of such equipment within the context of procedures designed to confirm parasitic infections. Also, the most important piece of equipment used in diagnostic parasitology is the regular light microscope, and the necessity for good optics cannot be overemphasized.

I. MICROSCOPE

Good microscopes and light sources are mandatory for the examination of specimens for parasites. Identification of the majority of organisms depends on morphologic differences, most of which must be seen using stereoscopic or regular microscopes. A procedure for calibration of the microscope can also be found later in this section.

A. Stereoscopic microscope

A stereoscopic microscope should be available for larger specimens (arthropods, tapeworm proglottids, various artifacts). The total magnification usually ranges from ×10 to ×45. Some microscopes have a zoom capacity from ×10 to ×45, and others have fixed objectives (.66×, 1.3×, 3×) that can be used with 5× or 10× oculars. Direct the light source either from under the specimen or onto the top of the specimen.

B. Regular microscope

The light microscope should be equipped with the following:

1. Head

A monocular head can be used, but a binocular type is recommended to reduce fatigue during lengthy examinations. The binocular head should contain a diopter adjustment to compensate for variation in focus between the eyes. Tilt heads are also available and highly recommended if different personnel use the microscope on a continuous basis.

2. Oculars

10× are required; 5× can be helpful, but are not essential.

3. Objectives

10× (low power), 40× to 45× (high power), 97× to 100× (oil immersion). Also, a 60× (oil immersion) objective is very helpful for screening stained smears.

4. Stage

A mechanical stage for X and Y movement is necessary. Graduated stages can be helpful and are recommended for recording the exact location of an organism in a permanently stained slide. This capability is essential for consultation and teaching responsibilities.

I. MICROSCOPE *(continued)*

5. Condenser
   a. A brightfield condenser equipped with an iris diaphragm is required. The condenser Numerical Aperture (NA) should be equal to or greater than the highest objective NA.
   b. An adjustable condenser is not required with the newer microscopes.
6. Filters, both clear blue-glass and white ground-glass, are recommended.
7. Light source
   a. The light source, along with an adjustable voltage regulator, should be contained in the microscope base. Align this light source as directed by the manufacturer.
   b. If the light source is external, the microscope must be equipped with an adjustable mirror and a condensing system containing an iris diaphragm which can be released or lowered.
   c. The light source should be a 75- to 100-W bulb.

C. Comments on routine use

1. Remove dust from all optical surfaces with a camel hair brush. Remove oil and finger marks immediately from the lenses with a clean soft cloth or several thicknesses of lens tissue. Single-thickness lens tissue may permit corrosive acids from the fingers to damage the lens. *Do not use any type of tissue other than lens tissue; otherwise you may scratch the lens.* Use very little pressure to clean the lenses to avoid removing the coatings on their external surfaces.
2. Use water-based cleaning fluid for normal cleaning. Use organic solvents sparingly and only if absolutely necessary to remove oil from the lens. Since microscope manufacturers do not agree on a common solvent, the manufacturer of each microscope should be consulted. One solvent that is recommended is 1,1,1-trichloroethane; it is a good solvent for immersion oil and mounting media, but it will not soften the lens sealers and cements. Do not use xylene, any alcohols, acetone, or any other ketone.
3. Clean the lamp (cool, in *off* position) with lens tissue moistened in 70% isopropyl or ethyl alcohol (to remove oil from fingers), after the lamp has been installed into the lamp holder.
4. Clean the stage with a small amount of disinfectant (70% isopropyl or ethyl alcohol) if it becomes contaminated with fecal material.
5. Cover the microscope when not in use. In extremely humid climates (a relative humidity of more than 50%), good ventilation is necessary to prevent fungal growth on the optical elements.
6. Schedule a complete general cleaning and readjustment at least annually to be performed by a factory-trained and authorized individual. Record all data related to preventive maintenance and repair.

II. CENTRIFUGE

Either a table or floor model centrifuge, which should be able to accommodate 15-ml centrifuge tubes (for concentration procedures), is acceptable. Some laboratories prefer to use 50-ml tubes; both sizes are available commercially. Regardless of the model, a free-swinging or horizontal head type is recommended. With this type of centrifuge, the sediment will deposit evenly on the bottom of the tube. The flat upper surface of the sediment facilitates the removal of the supernatant fluid from a loosely packed pellet.

III. FUME HOOD

Although a fume hood is not required, many laboratories prefer to keep their staining reagents in a fume hood. Even with the substitution of dehydrating reagents other than xylene, fume hoods may be preferred to eliminate odors. In keeping with good laboratory practice, the placement of reagents and equipment into a hood should not interfere with the proper operation of the hood.

IV. BIOLOGICAL SAFETY CABINET

A biological safety cabinet (BSC) is not required for processing routine specimens in a diagnostic parasitology laboratory. However, some laboratories use class I (open-face) or class II (laminar-flow) biological safety cabinets to process all unpreserved specimens. Disinfect the work area after each use and do not depend on ultraviolet (UV) irradiation to decontaminate the work surface. UV radiation has very limited penetrating powers (12). Have a class II biological safety cabinet certified to meet Standard Number 49 of the National Sanitation Foundation (Ann Arbor, Mich.), when it is installed. Recertify the cabinet to meet Standard 49 when the cabinet is moved out after filters are replaced, the exhaust motor is repaired or replaced, or any gaskets are removed or replaced. Recertify at least annually, and record the date of service recertification along with the name of the individual or company performing the service.

V. REFRIGERATOR/FREEZER

A general purpose laboratory (non-explosion-proof) or household-type refrigerator/freezer is adequate for use in the parasitology laboratory. The temperature should be approximately 5°C. Solvents with flash points below refrigeration temperature should not be stored, even in modified (explosion-proof) refrigerators.

PART 2 Safety

The standard safety considerations of any clinical laboratory (3) should be practiced in a parasitology laboratory. Specific attention should be given to personal practices, specimen and reagent handling, and equipment.

I. PERSONAL PRACTICES

Observe universal (standard) precautions.

Since many parasitic infections are acquired via the oral route, it is imperative that the parasitologist practice good personal hygiene. Hand washing with soap and water and adherence to laboratory technique standards are two of the most important ways to guard against infection in a clinical laboratory. Gloves, although not mandatory, are recommended and Universal Precautions must be followed when handling blood specimens.

II. IMMUNIZATIONS

In addition to the routine childhood immunizations against polio and rubella, an immunization against hepatitis B is recommended for those individuals processing fresh unpreserved fecal specimens. At the present time, there are no effective immunizations for parasitic diseases.

III. SPECIMENS

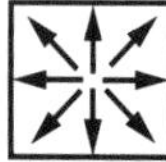

Observe universal (standard) precautions.

Fecal specimens should be collected and transported to the laboratory so that anyone handling the container does not come in direct contact with the specimen. The test request slip should be bagged separately from the specimen. Usually the specimen is placed in a waxed cardboard container with a tightly fitting lid. The unpreserved stool specimen should be considered potentially infectious; gloves are recommended. Protozoan cysts; *Cryptosporidium* oocysts; eggs of *Taenia solium, Enterobius vermicularis,* and *Hymenolepis nana;* and larvae of *Strongyloides stercoralis* may be infective. If stool collection vials (containing fixatives) are included in the collection kit, the manufacturers' directions should be followed. Blood, other body fluids, and tissue specimens should be collected in a manner consistent with Universal Precautions (7). In processing the unpreserved specimen, avoid direct contact between the specimen and the parasitologist and/or the equipment being used. Use of a fume hood is often recommended.

IV. REAGENTS

Include QC information on reagent container and in QC records.

Know what chemical hazards are in the workplace (1). One way to make the hazards known is to make available the Material Safety Data Sheets on the chemical(s) in use. The availability and location of these sheets should be well known by all laboratory personnel. Usually they are kept filed in a notebook rather than being posted. The financial impact of various state and federal regulations governing the use and disposal of hazardous laboratory reagents serves as the impetus to search for less hazardous reagents. Some of the reagents for which there are substitutes are the following:

A. Ethyl acetate (10) with a higher flash point (4°C) can replace ether (flash point of −45°C) in the Formalin-Ether concentration technique. Hemo-De (Medical Industries, Los Angeles, Calif., or Fisher Scientific, Los Angeles, Calif.) can also replace ethyl acetate in the concentration procedure (11). This reagent has a flash point of 57.8°C and is generally regarded as safe by the U.S. Food and Drug Administration.

B. Xylene, as used in the trichrome or iron hematoxylin staining of polyvinyl alcohol (PVA)-fixed fecal smears, possesses potential toxic and fire hazards. Again, Hemo-De has successfully replaced xylene in both the carbol-xylene and xylene steps of the trichrome procedure (9, 10). There are other substitutes available as well (e.g., Hemo-Sol, Fisher Scientific). Check with your local pathology departments or reagent suppliers for other alternatives.

C. Mercuric chloride, used in PVA fixative and in Schaudinn's fixative, presents both a toxic hazard and a disposal problem. Because of regulatory requirements, very few laboratories are using mercuric chloride-based PVA; incineration and land fill options are generally no longer available. Currently there is an ongoing review regarding the use of mercury compounds in the health care setting; whether these discussions will lead to a change in regulations regarding disposal options is unclear. Since state regulations vary, please contact your State Department of Public Health for additional information. Copper sulfate has been suggested as a substitute for mercuric chloride (2, 4); however, protozoan morphology will not be as clear and precise with this formula. Zinc sulfate is now being used as the base in several of the newer single-vial collection kit systems.

V. EQUIPMENT

If the stage of the microscope becomes contaminated with fecal material, it should be wiped clean with a disinfectant (70% isopropyl or ethyl alcohol). The general safety precautions governing the use of electrical equipment should be observed; i.e., don't use near water, check for bare wires, turn off power before servicing, etc. Plastic tubes should be used to reduce the hazards of breaking during centrifugation. If glass centrifuge tubes are used, they should be inspected, and only those free of defects should be used. The centrifuge should always be balanced before operation, and the centrifuge lid must be closed when the centrifuge is in operation. Should noise or vibration develop, the centrifuge should be stopped and checked for symmetrical loading of the head. If a fume hood is used to store flammable solvents such as ethyl acetate, ether, or xylene, the hood must be operating to prevent the build-up of dangerous or toxic levels of these reagents. An adequate face velocity of 60 to 100 ft/min is recommended. If the hood is used for other purposes, the placement of these reagents should not interfere with the proper operation of the hood.

PART 3 Quality Control

I. INTERNAL QUALITY CONTROL

This component of a QC program generally consists of documenting the proper functioning of reagents and equipment at prescribed intervals and evaluating performance on split samples (day-to-day, batch-to-batch) and within batch reproducibility studies.

System QC

A. The permanent staining procedure (trichrome or iron hematoxylin portion of the ova and parasite examination) should be checked with known positive control specimens when the stain is prepared (or a new lot number is purchased) and at least monthly. Quality control smears must be used every time staining is performed for particular organisms (e.g., modified trichrome for microsporidia and modified acid-fast stains for *Cryptosporidium parvum* or *Cyclospora cayetanensis*). All QC results must be recorded.

B. If positive specimens are not available, smears of feces containing epithelial or pus cells can be used.

C. Stains for blood parasites should be checked for staining quality at the time of preparation, when put into use, and quarterly. If positive blood smears are not available, a negative smear can be used.

D. Previously identified specimens can be introduced as blind specimens to evaluate the overall performance of the parasitology laboratory. The use of these samples is of particular value in those laboratories that process few specimens and in those that receive a low number of positive specimens or lack a variety of specimens.

II. EXTERNAL QUALITY CONTROL (PROFICIENCY TESTING)

Every parasitology laboratory should subscribe to a proficiency testing program to provide an unbiased evaluation of its performance.

III. REFERENCE MATERIALS

A. Reference materials should be available for comparison with unknown organisms, refresher training, and the training of additional personnel.

B. Ideal reference materials include Formalin-preserved specimens of helminth eggs, larvae, and protozoan cysts; stained fecal smears of protozoan oocysts, cysts, and trophozoites; and positive blood smears.

C. Color slides and atlases are recommended although the level of microscopic focus cannot be changed.

D. Reference books and manuals from a number of publishers are available, and selected ones should be part of the parasitology library.

PART 4

Quality Assurance

I. QUALITY ASSURANCE

QA is the sum of all the activities necessary to produce consistently accurate test results. These include the preanalytical, analytical, and postanalytical activities.

A. Preanalytical activities

Preanalytical activities are the sum of all activities, up to but not including the actual laboratory manipulation of the specimen, that influence the quality of laboratory results. During personnel training, the person advising the patient or medical staff must be adequately trained in all facets of specimen collection (preparation of the patient, specimen collection times, sample quality and volume, condition of specimen container, use of preservatives, and proper labeling). The person performing the parasitologic examination must be familiar with appropriate technical procedures to be used for each type of specimen and with morphologic recognition and differentiation of parasites. Preparation of the patient involves the use of proper laxatives and/or enemas and the proper method for handling the patient before collecting a pinworm specimen, i.e., before bowel movement and bathing. Consideration must be given to the sequence of collecting multiple specimens, i.e., the time intervals between administering antibiotics, consumption of X-ray dyes (barium) or nonabsorbable antidiarrheal preparations, collection of the specimen, and subsequent collections. Blood collections for bloodborne parasites should be optimized for the suspected parasites (8).

B. Analytical activities

Analytical activities are technical or laboratory procedures necessary to produce accurate test results (9) and include detailed descriptions of the procedures necessary to process the specimens and to identify the various parasites and their diagnostic stages. This information must be part of a written procedure manual (6) which, when reviewed and approved by the laboratory director, becomes the legal protocol for the laboratory by which each laboratorian is held responsible for performing tests *exactly* as described. The procedure manual should be reviewed by the laboratory director at least annually and before procedural changes are adopted.

C. Postanalytical activities

Postanalytical activities include transfer of information, either verbally, in writing, or by electronic means, from the laboratory to the clinician, providing the clinician with a meaningful laboratory report necessary to optimally manage the patient. This includes the extent of laboratory procedures performed, i.e., complete workup or limited to direct smear, presence of abnormalities seen in the specimen, excessive amounts of blood, yeast, etc., in addition to the identity of the parasite(s) found. Any qualifying statements regarding the quality of the specimen, such as "inadequately preserved when received in the laboratory" or "contaminated with water or urine," should be included.

II. QUALITY ASSURANCE DOCUMENTATION

All quality assurance monitors must be properly documented. Use of the 10-step protocol developed by the Joint Commission on the Accreditation of Healthcare Organizations (JCAHO) is one option (5).

REFERENCES

1. **Code of Federal Regulations.** 1989. Title 29, part 1910.1200. U.S. Government Printing Office, Washington, D.C.
2. **Garcia, L. S., R. Y. Shimizu, T. C. Brewer, and D. A. Bruckner.** 1983. Evaluation of intestinal parasite morphology in polyvinyl alcohol preservative: comparison of copper sulfate and mercuric chloride bases for use in Schaudinn's fixative. *J. Clin. Microbiol.* **17:** 1092–1095.
3. **Gröschel, D. H. M.** 1986. Safety in clinical microbiology laboratories, p. 33. *In* B. M. Miller, D. H. M. Gröschel, J. H. Richardson, D. Vesley, J. R. Songer, R. D. Housewright, and W. E. Barkley (ed.), *Laboratory Safety: Principles and Practices.* American Society for Microbiology, Washington, D.C.
4. **Horen, W. P.** 1981. Modification of Schaudinn's fixative. *J. Clin. Microbiol.* **13:** 204–205.
5. **JCAHO.** 1987. *Monitoring and Evaluation of Pathology and Medical Laboratory Services.* JCAHO, Chicago, Ill.
6. **NCCLS.** 1984. *Clinical Laboratory Procedure Manuals.* Approved Guideline GP-2A. NCCLS, Wayne, Pa.
7. **NCCLS.** 1989. *Protection of Laboratory Workers from Infectious Disease Transmitted by Blood, Body Fluids, and Tissue.* Proposed Guideline M29-T. NCCLS, Wayne, Pa.
8. **NCCLS.** 1990. *Use of Blood Film Examination for Parasites.* Proposed Guideline M15-P. NCCLS, Wayne, Pa.
9. **NCCLS.** 1993. *Procedures for the Recovery and Identification of Parasites from the Intestinal Tract.* Proposed Guideline M28-P. NCCLS, Wayne, Pa.
10. **Neimeister, R. J., A. L. Logan, and J. H. Egleton.** 1985. Modified trichrome staining technique with a xylene substitute. *J. Clin. Microbiol.* **22:**306–307.
11. **Neimeister, R., A. L. Logan, B. Gerber, J. H. Egleton, and B. Kleger.** 1987. Hemo-De as a substitute for ethyl acetate in Formalin-ethyl acetate concentration technique. *J. Clin. Microbiol.* **25:**425–426.
12. **Vesley, D., and J. Lauer.** 1986. Decontamination, sterilization, disinfection, and antisepsis in the microbiology laboratory, p. 194. *In* B. M. Miller, D. H. M. Gröschel, J. H. Richardson, D. Vesley, J. R. Songer, R. D. Housewright, and W. E. Barkley (ed.), *Laboratory Safety: Principles and practices.* American Society for Microbiology, Washington, D.C.

SUPPLEMENTAL READING

Bloom, H. M. 1986. Designs to simplify laboratory construction and maintenance, improve safety, and conserve energy, p. 142. *In* B. M. Miller, D. H. M. Gröschel, J. H. Richardson, D. Vesley, J. R. Songer, R. D. Housewright, and W. E. Barkley (ed.), *Laboratory Safety: Principles and Practices.* American Society for Microbiology, Washington, D.C.

Code of Federal Regulations. 1989. Title 29, part 1910.106. U.S. Government Printing Office, Washington, D.C.

Melvin, D. M., and M. V. Brooke. 1982. *Laboratory Procedures for the Diagnosis of Intestinal Parasites.* Department of Health and Human Services publication no. (CDC) 82-8282, p. 59. U.S. Government Printing Office, Washington, D.C.

National Institutes of Health. 1979. *NIH Guidelines for Recombinant DNA Research Supplement: Laboratory Safety Monograph.* National Institutes of Health, Bethesda, Md.

National Sanitation Foundation. 1976. *NSF Standard No. 49 for Class II (Laminar Flow) Biohazard Cabinetry.* National Sanitation Foundation, Ann Arbor, Mich.

National Sanitation Foundation. 1987. *NSF Standard No. 49 for Class II (Laminar Flow) Biohazard Cabinetry.* National Sanitation Foundation, Ann Arbor, Mich.

NCCLS. 1986. *Clinical Laboratory Hazardous Waste.* Proposed Guideline GP5-P. NCCLS, Wayne, Pa.

Young, K. H., S. L. Bullock, D. M. Melvin, and C. L. Sprull. 1979. Ethyl acetate as a substitute for diethyl ether in the Formalin-ether sedimentation technique. *J. Clin. Microbiol.* **10:**852–853.

7.2.1 Fresh-Specimen Collection

I. PRINCIPLE

One of the most important steps in the diagnosis of intestinal parasites is the proper collection of specimens (3). Improperly collected specimens can result in inaccurate results. Fresh specimens are mandatory for the recovery of motile trophozoites. However, unless strict collection and delivery times are adhered to, the specimen may have little value for diagnostic testing. Fresh specimens held too long before preservation or examination will not be acceptable, even for the recovery of protozoan cysts, coccidian oocysts, microsporidian spores, helminth eggs and larvae, and whole or partial adult worms.

II. SPECIMEN

A. Collect all fecal specimens before the administration of antimicrobial agents or antidiarrheal agents. The use of mineral oil, bismuth, and barium should also be avoided before fecal collection, since any of these substances may interfere with the detection or identification of intestinal parasites (1). The examination of purged specimens is less frequently performed and will not be included in this section. However, the same time limits for fixation and/or examination of diarrheic stools can be used for purged specimens submitted for examination.

B. Fecal specimens should be collected in a clean, wide-mouthed container or on newspaper and transferred to a container with a tight-fitting lid. Some use layered paper towels in the toilet bowl; however, if possible, the specimen should not come in contact with the toilet bowl water.

C. Contamination with urine or water from the toilet should be avoided.

D. The specimen should be transported to the laboratory as soon as possible or kept refrigerated until transport is possible. Obviously, dried-out specimens (diarrheic, semi-formed, formed) are not acceptable for fecal examination. Commercial transport options are available and products and suppliers are listed in Appendix 7–4.

III. QUALITY CONTROL

A. There is no maximum limit on the amount of stool collected.

B. Several grams (or teaspoon amounts) should be collected as a minimum amount. Smaller amounts can be examined, but the specimen is likely to dry out before examination (unacceptable for testing). Commercial transport options are recommended for specimen preservation (*see* procedure 7.2.2), and products and suppliers are listed in Appendix 7–4.

IV. PROCEDURE

A. Areas of the feces that appear bloody, purulent, or watery should be sampled for examination as direct wet smears and permanent stained smears.

B. For adequate sampling of a formed fecal specimen, material from the sides, ends, and middle should be collected and examined using direct wet smears, fecal concentration procedures, and permanent stained smears. It is recommended that a complete examination be performed (direct wet smears, concentration, permanent stained smears) on all types of fresh fecal specimens. Some laboratories may also perform some of the newer immunological assays for organism detection and/or confirmation (enzyme immunoassays, fluorescent procedures).

V. PROCEDURE NOTES

A. To ensure the recovery of parasitic organisms that are passed intermittently and in fluctuating numbers, the examination of a minimum of three specimens, collected over a 7- to 10-day period, is recommended (1–3). However, there are also other options available. When specific collection protocols are selected, it is very important to understand the pros and cons of each approach.

B. Infections with *Entamoeba histolytica/E. dispar* or *Giardia lamblia* may require the examination of up to six specimens before the organism is detected.

C. Liquid specimens should be received, examined, and/or preserved by the lab within 30 min of passage, soft or semiformed specimens within 1 h, and formed specimens on the same day of passage.

VI. LIMITATIONS OF PROCEDURE

A. Protozoan trophozoites will not survive if the stool specimen begins to dry out. Cysts will not form once the specimen has been passed. Fresh specimens held too long before preservation or examination are not acceptable, even for the recovery of protozoan cysts, coccidian oocysts, microsporidian spores, helminth eggs and larvae, or whole or partial adult worms.

B. Unless guidelines for delivery times are followed and inappropriate specimens are rejected, the laboratory results may be incorrect. The laboratory report should include a statement indicating that such results may be incorrect.

REFERENCES

1. **Garcia, L. S., and D. A. Bruckner.** 1997. *Diagnostic Medical Parasitology,* 3rd ed., p. 593–607. ASM Press, Washington, D.C.
2. **Melvin, D. M., and M. M. Brooke.** 1982. *Laboratory Procedures for the Diagnosis of Intestinal Parasites,* 3rd ed. U.S. Dept. of Health, Education, and Welfare publication no. (CDC) 82-8282. U.S. Government Printing Office, Washington D.C.
3. **Parasitology Subcommittee/Microbiology Section of Scientific Assembly, ASMT.** 1978. Recommended procedures for the diagnosis of parasitic infections. *Am. J. Med. Technol.* **44:** 1101–1105.

7.2.2 Preservation of Specimens

I. PRINCIPLE

Fecal specimens that cannot be processed and examined in the recommended time period should be placed in an appropriate preservative or combination of preservatives for later examination (1). Preservatives will prevent the deterioration of any parasites that may be present. A number of fixatives are available for preserving protozoan and helminth organisms. Each preservative has specific limitations, and no single solution enables all techniques to be performed with optimal results. The choice of preservative should give the laboratory the capability to perform a concentration technique and prepare a permanent stained smear on every specimen submitted for fecal examination (1).

II. SPECIMEN

Collect the fecal specimen as described previously (*see* procedure 7.2.1). Place a portion of the specimen in the preservative immediately after passage. Excellent directions are available with stool collection vials and/or kits, and this approach eliminates many of the problems encountered with fresh stool specimen collection.

III. PRESERVATIVES

Include QC information on reagent container and in QC records.

Various preservatives are available commercially (*see* Appendix 7–4); the most commonly used mercury substitutes are sodium acetate-acetic acid-formalin (SAF) and zinc sulfate-based modified polyvinyl alcohol (PVA). The routine use of stool fixatives for specimen collection also provides more consistent specimen quality and, thus, more accurate results. Every fixative has advantages and disadvantages which are indicated below:

A. Schaudinn's fixative

This preservative is used with fresh stool specimens or samples from the intestinal mucosal surface. Many laboratories that receive specimens from in-house patients (no problem with delivery times) may select this approach. Permanent stained smears are then prepared from fixed material.

1. Advantages
   - **a.** Designed to be used for the fixation of slides prepared from fresh fecal specimens or samples from the intestinal mucosal surfaces
   - **b.** Prepared slides can be stored in the fixative for up to a week without distortion of protozoan organisms
   - **c.** Easily prepared in the laboratory
   - **d.** Available from a number of commercial suppliers
2. Disadvantages
   - **a.** Not recommended for use in concentration techniques
   - **b.** Has poor adhesive properties with liquid or mucoid specimens
   - **c.** Contains mercury compounds (mercuric chloride) which may cause disposal problems (*see* section 7.1, part 2, topic IV, Reagents)

B. PVA (mercury base) (1)

PVA is a plastic resin that is normally incorporated into Schaudinn's fixative. The PVA powder serves as an adhesive for the stool material; i.e., when the stool/PVA mixture is spread onto the glass slide, it adheres because of the PVA component. Fixation is still accomplished by the Schaudinn's fluid itself. Perhaps the greatest advantage in the use of PVA is the fact that a permanent stained smear can be prepared. PVA fixative solution is highly recommended as a means of preserving cysts and trophozoites for examination at a later time. The use of PVA also permits specimens to be shipped (by regular mail service) from any location in the world to a laboratory for subsequent examination. PVA is particularly useful for liquid specimens and should be used in the ratio of 3 parts PVA to 1 part fecal specimen.

1. Advantages
   a. Ability to prepare permanent stained smears and perform concentration techniques
   b. Good preservation of protozoan trophozoites and cyst stages
   c. Long shelf life (months to years) in tightly sealed containers at room temperature
   d. Commercially available from a number of sources
   e. Allows for shipment of specimens
2. Disadvantages
   a. Some organisms are not concentrated as well from PVA as from formalin-based fixatives (*Trichuris trichiura* eggs, *Giardia lamblia* cysts, *Isospora belli* oocysts), and morphology of some ova and larvae may be distorted.
   b. Contains mercury compounds (Schaudinn's fixative) which may cause disposal problems
   c. May turn white and gelatinous when aliquoted into small amounts (begins to dehydrate) or if refrigerated
   d. Difficult to prepare in the laboratory

C. Modified PVA (copper base) (1, 2, 4, 5)

Although there has been a great deal of interest in developing preservatives without the use of mercury compounds, substitute compounds usually do not provide the same quality of preservation necessary for good protozoan morphology on the permanent stained smear. Copper sulfate has been tried, but does not provide results equal to those seen with mercuric chloride (2, 4).

1. Advantages
   a. Can be used for concentration techniques and stained smears
   b. Contains no mercury compounds (usually prepared with copper sulfate)
   c. Commercially available from a number of suppliers. These products (commercial formulas) apparently contain other "fixation agents" which may produce better overall fixation than original formula.
2. Disadvantages
   a. Does not provide the quality of preservation for good protozoan morphology on stained slides (2)
   b. Staining characteristics of protozoan organisms are variable; identification may be difficult, particularly when compared with staining characteristics seen from mercuric chloride-based fixatives.

D. Modified PVA (zinc base) (3)

A number of these fixatives are now available commercially, although they do not provide the same overall quality of fixation seen with mercury-based fixatives and the specific formulas are proprietary. Morphologic differences based on organism fixation can be placed in perspective as follows:

III. PRESERVATIVES *(continued)*

Mercury-based fixatives: How beautiful is the organism?
Zinc-based fixatives: Can you identify the organism?
(In terms of clinical relevance, the last is the critical question.)

Although some of the organisms do not present with "textbook-quality" morphologic features, the majority of the time most organisms can be identified. In general, trophozoites tend to fix well, while cyst forms do not. The most difficult organisms to identify in zinc-preserved specimens are *Endolimax nana* cysts.

1. Advantages
   a. Can be used for concentration techniques and stained smears
   b. Contains no mercury compounds (usually prepared with zinc sulfate)
   c. Commercially available from a number of suppliers. These products (commercial formulas) apparently contain other "fixation agents" which are proprietary.
2. Disadvantages
   a. Does not provide the same quality of preservation for good protozoan morphology on stained slides as seen in mercury-fixed solutions (2).
   b. Staining characteristics of protozoan organisms are variable; identification may be difficult, particularly when compared with staining characteristics seen from mercuric chloride-based fixatives.

E. Sodium acetate-acetic acid-formalin (SAF) (6)

SAF lends itself to both the concentration technique and the permanent stained smear and has the advantage of not containing mercuric chloride, as found in Schaudinn's fluid and PVA. It is a liquid fixative, much like 10% formalin. The sediment is used to prepare the permanent smear, and it is recommended that the stool material be placed on an albumin-coated slide to improve adherence to the glass. SAF is considered to be a "softer" fixative than mercuric chloride. The organism morphology will not be quite as sharp after staining as will that of organisms originally fixed in solutions containing mercuric chloride. The pairing of SAF-fixed material with iron-hematoxylin staining appears to provide better organism morphology than staining SAF-fixed material with trichrome.

1. Advantages
   a. Can be used for concentration techniques and stained smears
   b. Contains no mercury compounds
   c. Long shelf life
   d. Easily prepared or is commercially available from a number of suppliers
2. Disadvantages
   a. Has poor adhesive properties; albumin-coated slides recommended for stained smears
   b. Protozoan morphology with trichrome stain not as clear as with PVA smears. Hematoxylin staining generally gives better results.
   c. More difficult for inexperienced workers to use

F. Merthiolate-iodine-formalin (MIF) (1, 5)

Merthiolate (thimerosal)-iodine-formalin (MIF) is a good stain preservative for most kinds and stages of parasites found in feces and is useful for field surveys. It is used with all common types of stools and aspirates; protozoa, eggs, and larvae can be diagnosed without further staining in temporary wet mounts. Many laboratories using this fixative examine the material only as a wet preparation (direct smear and/or concentration sediment). The MIF preservative is prepared in two stock solutions, stored separately and mixed immediately before use.

1. Advantages
   a. A combination of preservative and stain (Merthiolate), especially useful in field surveys
   b. Protozoan cysts and helminth eggs and larvae can be diagnosed from temporary wet mount preparations
2. Disadvantages
   a. Difficult to prepare permanent stained smears
   b. Iodine component unstable, needs to be added immediately before use
   c. Concentration techniques may give unsatisfactory results
   d. Morphology of organisms becomes distorted after prolonged storage

G. 5% or 10% formalin (1, 5)

Formalin is an all-purpose fixative that is appropriate for helminth eggs and larvae and protozoan cysts. Two concentrations are commonly use, 5%, which is recommended for preservation of protozoan cysts, and 10%, which is recommended for helminth eggs and larvae. Most commercial manufacturers provide a 10% concentration, which is most likely to kill all helminth eggs. To help maintain organism morphology, the formalin can be buffered with sodium phosphate buffers, e.g., neutral formalin.

1. 10% Formalin
   formaldehyde (USP) ....100 or 50 ml (for 5%)
   formaldehyde is normally purchased as a 37 to 40% HCHO solution; however, for dilution it should be considered to be 100%.
   0.85% NaCl ....900 or 950 ml (for 5%)
   Dilute 100 ml of formaldehyde with 900 ml of 0.85% NaCl solution. (Distilled water may be used instead of NaCl solution.)
   If you want to use buffered formalin, the following approach is recommended.
   Na_2HPO_4 ....6.10 g
   NaH_2PO_4 ....0.15 g
   Mix the two thoroughly and store the dry mixture in a tightly closed bottle. Prepare 1 liter of either 10 or 5% formalin and add 0.8 g of the buffer salt mixture.
2. Advantages
   a. Good routine preservative for protozoan cysts and helminth eggs and larvae; materials can be preserved for several years.
   b. Can be used for concentration techniques (sedimentation techniques)
   c. Long shelf life and commercially available
   d. Neutral formalin (buffered with sodium phosphate) helps maintain organism morphology with prolonged storage.
3. Disadvantage
   Not able to prepare a permanent stained smear from formalin-preserved fecal specimens.

IV. QUALITY CONTROL

A. Obtain a fresh, anticoagulated blood specimen and prepare a buffy coat sample.
B. Mix approximately 2 g of soft, fresh fecal specimen (normal stool, containing no parasites) with several drops of the buffy coat cells.
C. Prepare several fecal smears and fix immediately in the Schaudinn's fixative to be quality controlled.
D. Mix the remaining feces-buffy coat mixture in 10 ml of PVA, modified PVA, or SAF preservative to be quality controlled.
E. Allow 30 min for fixation, then prepare several fecal smears. Allow to dry thoroughly (30 to 60 min) at room temperature.
F. Stain slides by using normal staining procedure.

IV. QUALITY CONTROL *(continued)*

G. After staining, if the white blood cells appear well fixed and display typical morphology, assume that any intestinal protozoa placed in the same lot number of preservative would also be well fixed, provided the fecal sample was fresh and fixed within the recommended time limits.

V. PROCEDURE

A. Add a portion of fecal material to the preservative vial to give a 3:1 or 5:1 ratio of preservative to fecal material (a grape-sized formed specimen or about 5 ml of liquid specimen).
B. Mix well to give a homogeneous solution.
C. Allow to stand for 30 min at room temperature to allow adequate fixation.
D. If using commercial collection systems, follow the manufacturer's directions concerning shaking the vials, etc.

VI. PROCEDURE NOTES

A. Most of the commercially available kits have a "fill to" line on the vial label to indicate how much fecal material to add to assure adequate preservation of the fecal material.
B. Although the two-vial system (one vial of 10% buffered formalin concentration and one vial of PVA [permanent stained smear]) has always been the "gold standard," laboratories are beginning to use other options. Changes in the selection of fixatives are based on the following.
  1. Problems with disposal of mercury-based fixatives
  2. The cost of a two-vial system compared with that of a single collection vial
  3. Selection of specific stains (trichrome, iron hematoxylin) to use with specific fixatives
  4. Immunoassay procedures (EIA, fluorescent antibody) cannot be performed on specimens preserved in certain fixatives; check with the manufacturer.

7 Parasitology

VII. LIMITATIONS OF PROCEDURE

A. Adequate fixation still depends on the following.
  1. Meeting recommended time limits for lag time between passage of the specimen and fixation
  2. Use of the correct ratio of specimen to fixative
  3. Thorough mixing of the preservative and specimen
B. Unless the appropriate stain is used with each fixative, the final permanent stained smear may be difficult to examine (organisms hard to see and/or identify).

REFERENCES

1. **Garcia, L. S., and D. A. Bruckner.** 1997. *Diagnostic Medical Parasitology,* 3rd ed. ASM Press, Washington, D.C.
2. **Garcia, L. S., R. Y. Shimizu, T. C. Brewer, and D. A. Bruckner.** 1983. Evaluation of intestinal parasite morphology in polyvinyl alcohol preservative: comparison of copper sulfate and mercuric chloride base for use in Schaudinn's fixative. *J. Clin. Microbiol.* **17:** 1092–1095.
3. **Garcia, L. S., R. Y. Shimizu, A. Shum, and D. A. Bruckner.** 1993. Evaluation of intestinal protozoan morphology in polyvinyl alcohol preservative: comparison of zinc sulfate- and mercuric chloride-based compounds for use in Schaudinn's fixative. *J. Clin. Microbiol.* **31:** 307–310.
4. **Horen, W. P.** 1981. Modification of Schaudinn's fixative. *J. Clin. Microbiol.* **13:** 204–205.
5. **NCCLS.** 1993. *Procedures for the Recovery and Identification of Parasites from the Intestinal Tract.* Proposed Guideline M28-P. NCCLS, Wayne, Pa.
6. **Scholten, T. H., and J. Yang.** 1974. Evaluation of unpreserved and preserved stools for the detection and identification of intestinal parasites. *Am. J. Clin. Pathol.* **62:**563–567.

APPENDIX 7.2–1

Common preservatives used in diagnostic parasitology (stool specimens)

Preservative	Concn	Permanent stained smear
5 or 10% formalin	×	Not appropriate for permanent stain
5 or 10% buffered formalin	×	Not appropriate for permanent stain
MIF	×	× (limited use, requires different stain)
SAF	×	× (hematoxylin probably best)
PVA (mercury-based)		× (trichrome or hematoxylin)
PVA (copper-based)		× (trichrome or hematoxylin)
PVA (zinc-based)	×	× (trichrome or hematoxylin)
Single-vial systems	×	× (trichrome or hematoxylin)[a]
Schaudinn's (without PVA)		× (trichrome or hematoxylin)

[a] Single-vial systems may be used with specific stains developed by the manufacturer and recommended for use with specific fixatives.

Considerations when selecting a stool fixative

A. Overall fixation efficacy for trophozoites, cysts, oocysts, microsporidia spores, eggs, and larvae
B. Ability to perform both the concentration and permanent stained smear from the preserved specimen
C. Selection of a one- or two-vial collection system
D. Preparation of reagents in-house or commercial purchase
E. Decision to perform immunoassay procedures (EIA, fluorescent antibody) from preserved specimen
F. Need to perform special stains from the preserved specimen (modified acid-fast stains for *Cryptosporidium, Cyclospora, Isospora;* and modified trichrome stains for microsporidia)

7.3 Calibration of the Microscope

I. PRINCIPLE

The identification of protozoa and other parasites depends on several factors, one of which is size. Any laboratory doing diagnostic work in parasitology should have a calibrated microscope available for precise measurements. Measurements are made by using a micrometer disk that is placed in one of the oculars of a binocular microscope; the disk is usually calibrated as a line divided into 50 units. Depending on the objective magnification used, the divisions in the disk represent different measurements. The ocular disk division must be compared with a known calibrated scale, usually a stage micrometer with a scale of 0.1- and 0.01-mm divisions (1).

II. MATERIALS

A. Supplies

1. Ocular micrometer disk (line divided into 50 U) (any laboratory supply distributor: Fisher, Baxter, Scientific Products, VWR, etc.)
2. Stage micrometer with a scale of 0.1- and 0.01-mm divisions (Fisher, Baxter, Scientific Products, VWR, etc.)
3. Immersion oil
4. Lens paper

B. Equipment

1. Binocular microscope with 10×, 40×, and 100× objectives. Other objective magnifications may also be used (50× oil or 60× oil immersion lenses).
2. Oculars should be 10×. Some may prefer 5×; however, smaller magnification may make final identifications more difficult.
3. Single 10× ocular to be used to calibrate all laboratory microscopes (to be used when any organism is being measured)

III. QUALITY CONTROL

A. The microscope should be recalibrated once each year, particularly if the scope receives heavy use or is moved throughout the laboratory.

B. Often the measurement of RBCs (approximately 7.5 μm) is used to check the calibrations of the three magnifications (×100, ×400, ×1,000).

C. Latex or polystyrene beads of a standardized diameter can be used to check the calculations and measurements (Sigma, J. T. Baker, etc.). Beads of 10 and 90 μm are recommended.

D. Record all measurements in QC records.

IV. PROCEDURE

A. Unscrew the eye lens of a 10× ocular and place the micrometer disk (engraved side down) within the ocular. Use lens paper to handle the disk; keep all surfaces free of dust or lint.

B. Place the calibrated micrometer on the stage and focus on the scale. You should be able to distinguish the difference between the 0.1- and 0.01-mm divisions. Make sure you understand the divisions on the scale before proceeding.

C. Adjust the stage micrometer so that the ''0'' line on the ocular micrometer is exactly lined up on top of the 0 line on the stage micrometer.

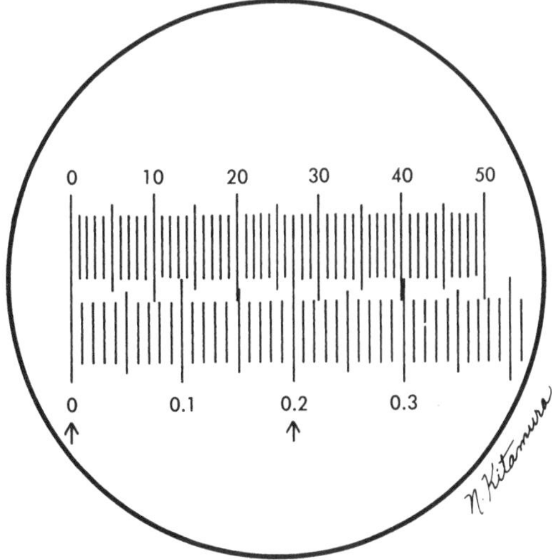

Figure 7.3–1 Ocular micrometer, top scale; stage micrometer, bottom scale. (From L. S. Garcia and D. A. Bruckner, *Diagnostic Medical Parasitology,* 3rd ed., ASM Press, Washington, D.C., 1997.)

D. After these two 0 lines are lined up, do not move the stage micrometer any farther. Look to the right of the 0 lines for another set of lines superimposed on each other. The second set of lines should be as far to the right of the 0 lines as possible; however, the distance varies with the objectives being used (Fig. 7.3–1).

E. Count the number of ocular divisions between the 0 lines and the point where the second set of lines is superimposed. Then, on the stage micrometer, count the number of 0.1-mm divisions between the 0 lines and the second set of superimposed lines.

F. Calculate the portion of a millimeter that is measured by a single small ocular unit.

G. When the high dry and oil immersion objectives are used, the 0 line of the stage micrometer will increase in size, whereas the ocular 0 line will remain the same size. The thin ocular 0 line should be lined up in the center or at one edge of the broad stage micrometer 0 line. Thus, when the second set of superimposed lines is found, the thin ocular line should be lined up in the center or at the corresponding edge of the broad stage micrometer line.

Examples:

1. $\frac{\text{stage reading (mm)}}{\text{ocular reading}} \times \frac{1{,}000\ \mu\text{m}}{1\ \text{mm}} = \text{ocular units } (\mu\text{m})$
2. Low power (10×): $\frac{0.8\ \text{mm}}{100\ \text{units}} \times \frac{1{,}000\ \mu\text{m}}{1\ \text{mm}} = 8.0\ \mu\text{m (factor)}$
3. High dry power (40×): $\frac{0.1\ \text{mm}}{50\ \text{units}} \times \frac{1{,}000\ \mu\text{m}}{1\ \text{mm}} = 2.0\ \mu\text{m (factor)}$
4. Oil immersion (100×): $\frac{0.05\ \text{mm}}{62\ \text{units}} \times \frac{1{,}000\ \mu\text{m}}{1\ \text{mm}} = 0.8\ \mu\text{m (factor)}$

Examples: If a helminth egg measures 15 ocular units by 7 ocular units (high dry objective) multiply the measurement by the factor of 2.0 μm. The egg measures 30 × 14 μm and is probably *Clonorchis sinensis.*

If a protozoan cyst measures 27 ocular units (oil immersion objective), multiply the measurement by the factor 0.8 μm. The cyst then measures 21.6 μm.

V. RESULTS

A. For each objective magnification, there will be a factor generated (1 ocular unit = certain number of micrometers).

B. If standardized latex or polystyrene beads or an RBC is measured by using various objectives, the size for the object measured should be the same (or very close), regardless of the objective magnification.

VI. REPORTING RESULTS

A. The multiplication factor for each objective should be posted (either on the base of the microscope or on a nearby wall or bulletin board) for easy reference.

B. Once the number of ocular lines per width and length of the organism is measured, then, depending on the objective magnification, the factor (1 ocular unit = certain number of micrometers) can be applied to the number of lines to obtain the width and length of the organism.

C. Comparison of these measurements with reference measurements in various books and manuals should confirm the organism identification.

VII. PROCEDURE NOTES

A. The final multiplication factors will be only as good as your visual comparison of the ocular 0 and stage micrometer 0 lines.

B. As a rule of thumb, the high dry objective (40×) factor should be approximately 2.5 times more than the factor obtained from the oil immersion objective (100×). The low-power objective (10×) factor should be approximately 10 times that seen by using the oil immersion objective (100×).

VIII. LIMITATIONS OF THE PROCEDURE

A. After each objective has been calibrated, the oculars containing the disk and/or these objectives cannot be interchanged with corresponding objectives or oculars on another microscope.

B. Each microscope used to measure organisms must be calibrated as a unit. The original oculars and objectives that were used to calibrate the microscope must also be used when an organism is measured.

C. The objective containing the ocular micrometer can be stored until needed. This single ocular can be inserted when measurements are taken. However, this particular ocular containing the ocular micrometer disk must have also been used as the ocular during microscope calibration.

REFERENCE

1. **Garcia, L. S., and D. A. Bruckner.** 1997. *Diagnostic Medical Parasitology,* 3rd ed. ASM Press, Washington, D.C.

7.4 Macroscopic Examination of Fecal Specimens: Age and Physical Description

I. PRINCIPLE

The age (elapsed time from collection to arrival in the laboratory) of fresh fecal specimens is an important factor in the diagnosis of parasitic infections (1). The date and time of passage must be recorded and provided for each specimen submitted to the laboratory. The physical characteristics of a fresh fecal specimen can aid in determining what types of organisms might be present (1, 2). Fecal specimens are described as formed, semiformed, soft, loose, or watery. Loose or watery specimens may contain trophozoites, whereas formed or semiformed specimens are more likely to contain cyst stages. Helminth eggs or larvae can be found in any type of specimen, but are more difficult to find in liquid specimens because of the dilution factor. One can also see whether blood and/or mucus is present, although if present, neither necessarily indicates a parasitic infection. When the fresh specimen is examined visually in the collection container, adult pinworms (*Enterobius vermicularis*) and tapeworm proglottids might also be seen.

II. SPECIMEN

(See EPCM procedure 7.2.)

A. Fresh fecal specimen in waterproof container

B. Preserve or examine liquid or soft stools containing blood or mucus within 30 min of passage.

C. Preserve or examine semiformed specimens within 1 h of passage.

D. Preserve or examine formed specimens within the same day.

E. Fecal specimens that cannot be processed and examined in the recommended time period should be placed in an appropriate preservative or combination of preservatives for examination at a later time (1). Preservatives will prevent the deterioration of any parasites are present. A number of fixatives are available commercially (*see* Appendix 7–1) for preserving protozoan and helminth organisms. The choice of preservative should give the laboratory the capability to perform a concentration technique and prepare a permanent stained smear on every specimen submitted for fecal examination.

III. PROCEDURE

A. Examine the specimen macroscopically to determine consistency.

B. Examine the surface of the fecal specimen for the presence of blood or mucus.

C. Areas of blood or mucus should be sampled for examination for trophic amebae.

D. Examine the surface of the specimen and the area underneath the specimen for possible organisms (adult pinworms or tapeworm proglottids).

E. Check information for date and time of passage.

IV. RESULTS

A. Adult helminths or portions of helminths may be recovered and seen with the naked eye. Examples would include *Enterobius vermicularis* adult worms, *Ascaris lumbricoides* adult worms, and tapeworm proglottids.

B. Occasionally, other helminths may be recovered (hookworm, *Strongyloides stercoralis*) but identification would require the use of the microscope.

V. REPORTING RESULTS

A. Report the presence of adult helminths or portions of helminths. Morphology and size are usually adequate for identification of pinworm and *Ascaris* adults and tapeworm proglottids.

Examples: *Ascaris lumbricoides* adult worm identified.
Enterobius vermicularis adult worm identified.
Taenia saginata gravid proglottid identified.

B. Report the presence of blood on or in (occult blood positive) the fecal specimen.

Example: Fresh blood seen on stool specimen.

VI. PROCEDURE NOTES

A. Trophic amebae or flagellates are found most frequently in liquid or soft specimens and tend to disintegrate rapidly at room temperature.

B. Trophozoites and cyst stages can be found in semiformed specimens.

C. Cyst stages are found most frequently in formed specimens and will not lose characteristic morphology at room temperature for approximately 1 day.

D. Eggs and larvae in fresh fecal specimens do not lose characteristic morphology at room temperature as rapidly as trophozoites or cysts. Some eggs (hookworm) may hatch if the specimen is kept unpreserved at room temperature for more than a day. Coccidian oocysts and microsporidial spores tend to maintain their morphology.

E. Refrigeration of the fresh fecal specimen will delay the deterioration of the parasitic organisms.

F. Freezing of the fecal specimen is not recommended as characteristic morphology of the parasitic organisms will be altered.

G. Fecal specimens should never be incubated.

H. Laboratories that receive the fecal specimen in preservative vials must rely on information that is submitted with the specimen as to the consistency of the specimen.

I. Many commercially available kits contain vials with labeling which allows the patient to indicate the original consistency of the specimen.

J. A clean vial (containing no preservative) can be provided to patients for submitting a portion of the fresh specimen for determining consistency.

K. Dark or tar-colored fecal specimens usually indicate bleeding in the upper gastrointestinal tract.

L. Bright-colored blood indicates bleeding at a lower level or around the rectum.

M. Barium causes feces to be light tan to white in color. These specimens should be rejected.

N. Ingested iron and some antidiarrheal compounds can cause the specimen to be dark to black in color.

O. Yellowish specimens may be noted in cases of fat malabsorption, which is seen commonly in infection with *Giardia lamblia.*

P. Vegetable material is frequently seen in fecal specimens and needs to be differentiated from helminths. Size and gross morphology are used to differentiate vegetable material from helminth parasites.

VII. LIMITATIONS OF PROCEDURE

Although there are some benefits (organism motility) associated with the examination of fresh fecal specimens, many laboratories have switched to stool preservative collection kits. Many intestinal parasites tend to disintegrate soon after collection, particularly if there is a lag time between specimen collection and preservation. In order to eliminate lag time problems and ensure adequate organism morphology, stool collection kits are recommended. Also, remember that organisms other than parasites can cause diarrhea; you might want to check whether bacterial cultures have been ordered.

REFERENCES

1. **Garcia, L. S., and D. A. Bruckner.** 1997. *Diagnostic Medical Parasitology,* 3rd ed. ASM Press, Washington, D.C.
2. **NCCLS.** 1993. *Procedures for the Recovery and Identification of Parasites from the Intestinal Tract.* Proposed Guideline M28-P. NCCLS, Wayne, Pa.

SUPPLEMENTAL READING

Ash, L. R., and T. C. Orihel. 1987. *Parasites: A Guide to Laboratory Procedures and Identification.* ASCP Press, Chicago.

Melvin, D. M., and M. M. Brooke. 1982. *Laboratory Procedures for the Diagnosis of Intestinal Parasites.* U.S. Department of Health and Human Services publication no. (CDC) 82-9393. U.S. Government Printing Office, Washington, D.C.

7.5.1 Direct Smears

I. PRINCIPLE

The microscopic examination of a direct smear has several purposes: to assess the worm burden of a patient, to provide a quick diagnosis of a heavily infected specimen, to check organism motility, and to diagnose parasites that may be lost in concentration techniques (1–3).

II. SPECIMEN

Any fresh stool specimen that has not been refrigerated is acceptable. Since trophozoites within preserved specimens would exhibit no motility on a direct smear, it is not necessary to perform this procedure on specimens submitted in preservatives (5% or 10% formalin, sodium acetate-acetic acid-formalin [SAF], various types of PVA).

III. REAGENTS

Include QC information on reagent container and in QC records.

The reagents indicated below are available commercially.

A. 0.85% NaCl

B. D'Antoni's iodine or Lugol's iodine

Aliquot some of the iodine into a brown dropper bottle. The working solution should resemble a strong tea color and should be discarded when it lightens in color (usually within 10 to 14 days).

IV. QUALITY CONTROL

A. Check the working iodine solution each time it is used or periodically (once a week).

1. The iodine should be free of any signs of bacterial and/or fungal contamination.
2. The color should be that of strong tea (discard if too light).
3. Protozoan cysts should contain yellow-gold cytoplasm, brown glycogen material, and paler refractile nuclei. Human WBCs mixed with negative stool can be used as a QC specimen. The human cells will stain with the same color as that seen in the protozoa.

B. The microscope should have been calibrated within the last 12 months (if moved frequently or in constant use), and the original optics used for the calibration should be in place on the microscope. The calibration factors for all objectives should be posted on the microscope for easy access (*see* procedure 7.3).

C. All QC results should be appropriately recorded.

V. PROCEDURE

A. Place 1 drop of 0.85% NaCl on the left side of the slide and 1 drop of iodine (working solution) on the right side of the slide.

B. Take a very small amount of fecal specimen (about the amount picked up on the end of an applicator stick when introduced into the specimen) and thoroughly emulsify the stool in the saline and iodine preparations (use separate sticks for each).

C. Place a coverslip (22 by 22 mm) on each suspension.

D. Systematically scan both suspensions using the 10× objective. Examine the entire coverslip area.

E. If you see something suspicious, use the 40× objective for more detailed study. At least one-third of the coverslip should be examined with the 40× objective, even if nothing suspicious has been seen.

VI. RESULTS

Protozoan trophozoites and/or cysts and helminth eggs and larvae may be seen and identified.

VII. REPORTING RESULTS

A. Motile trophozoites may or may not be identified to the species level (depending on the clarity of the morphology).
Example (positive report): *Giardia lamblia* trophozoites present.

B. Protozoan cysts may or may not be identified to the species level (depending on the clarity of the morphology).
Example (positive report): *Entamoeba coli* cysts present.

C. Helminth eggs and/or larvae may be identified.
Examples: *Ascaris lumbricoides* eggs present.
Strongyloides stercoralis larvae present.

D. Artifacts and/or other structures may also be seen and reported as follows.
Examples: Moderate Charcot-Leyden crystals present.
Few RBCs present.

VIII. PROCEDURE NOTES

A. In preserved specimens, formalin replaces the saline and can be used directly; however, you will not be able to see any organism motility (organisms are killed by 5% or 10% formalin).

B. Some workers prefer to make the saline and iodine mounts on separate slides. Often there is less chance of getting fluids on the microscope stage if separate slides are used (less total fluid on the slide and under the coverslip).

C. The microscope light should be reduced for low-power observations, since most organisms will be overlooked with bright light. Illumination should be regulated so that some of the cellular elements in the feces show refraction. Most protozoan cysts will refract under these light conditions.

IX. LIMITATIONS OF THE PROCEDURE

A. Results obtained with wet smears should usually be confirmed by permanent stained smears. Some protozoa are very small and difficult to identify to the species level by direct wet smears alone.

B. Confirmation is particularly important in the case of *Entamoeba histolytica* versus *Escherichia coli.*

REFERENCES

1. **Garcia, L. S., and D. A. Bruckner.** 1997. *Diagnostic Medical Parasitology,* 3rd ed. ASM Press, Washington, D.C.
2. **Markell, E. K., and M. Voge.** 1981. *Medical Parasitology,* 5th ed. W. B. Saunders Co., Philadelphia.
3. **NCCLS.** 1993. *Procedures for the Recovery and Identification of Parasites from the Intestinal Tract.* Proposed Guideline M28-P. NCCLS, Wayne, Pa.

7.5.2 Concentration Methods

Fecal concentration has become a routine procedure as a part of the complete ova and parasite examination for parasites and allows the detection of small numbers of organisms that may be missed by using only a direct wet smear. There are two types of concentration procedures, flotation and sedimentation, both of which are designed to separate protozoan organisms and helminth eggs and larvae from fecal debris by centrifugation and/or differences in specific gravity.

PART 1 Sedimentation: Formalin-Ethyl Acetate

I. PRINCIPLE

By centrifugation, this concentration procedure leads to the recovery of all protozoa, eggs, and larvae present; however, the preparation contains more debris than the flotation procedure. Ethyl acetate is used as an extractor of debris and fat from the feces and leaves the parasites at the bottom of the suspension. The formalin-ethyl acetate sedimentation concentration is recommended, because it is the easiest to perform, allows recovery of the broadest range of organisms, and is least subject to technical error (1–3).

II. SPECIMEN

A. The specimen must be fresh or formalinized stool (5 or 10% buffered or non-buffered formalin or SAF). (*See* procedure 7.2.)

B. PVA-preserved specimens can be used (*see* Procedure Notes below).

III. REAGENTS

Include QC information on reagent container and in QC records.

A. Ethyl acetate

B. Formalin (5 or 10% buffered or non-buffered or SAF)

C. 0.85% NaCl

D. D'Antoni's or Lugol's iodine

V. QUALITY CONTROL

A. Examine the reagents each time they are used. The formalin and saline should appear clear, without any visible contamination.

B. The microscope should have been calibrated within the last 12 months (if moved frequently or in constant use), and the objectives and oculars used for the calibration procedure should be used for all measurements on the microscope. The calibration factors for all objectives should be posted on the microscope for easy access (multiplication factors can be pasted right on the body of the microscope). (*See* procedure 7.3.)

C. Known positive specimens should be concentrated and organism recovery verified at least quarterly, and particularly after the centrifuge has been recalibrated. Record results on appropriate sheet.

V. PROCEDURE

A. Transfer 1/2 teaspoon (about 4 g) of fresh stool into 10 ml of 10% formalin in a shell vial, unwaxed paper cup, or round-bottom tube (container may be modified to suit individual laboratory preferences). Mix the stool and formalin thoroughly. Let the mixture stand a minimum of 30 min for fixation. If the specimen is already in 5 or 10% formalin (or SAF), restir the stool-formalin mixture.

B. Depending on the amount and viscosity of the specimen, strain a sufficient quantity through wet gauze into a conical 15-ml centrifuge tube to give the desired amount of sediment (1/2 to 1 ml) in step C below. Usually 8 ml of the stool and formalin mixture prepared in step A will be sufficient. If the specimen is received in vials of preservative (5% or 10% formalin or SAF), then approximately 3 to 4 ml will be sufficient unless the specimen has very little stool in the vial.

C. Add 0.85% NaCl (*see* Procedure Notes below) almost to the top of the tube and centrifuge for 10 min at 500 × *g*. The amount of sediment obtained should be approximately 1/2 to 1 ml.

D. Decant the supernatant fluid and suspend the sediment in saline. Add saline almost to the top of the tube and centrifuge again for 10 min at 500 × *g*. This second wash may be eliminated if the supernatant fluid after the first wash is light tan or clear (use of more than one wash step varies among laboratories).

E. Decant the supernatant fluid and suspend (vortex is acceptable) the sediment on the bottom of the tube in 10% formalin. Fill the tube half full only. If the amount of sediment left in the bottom of the tube is very small or the original specimen contained a lot of mucus, *do not add ethyl acetate* in step F; merely add the formalin, spin, decant, and examine the remaining sediment.

F. Add 4 to 5 ml of ethyl acetate. Stopper the tube and shake vigorously for at least 30 s. Hold the tube so the stopper is directed away from your face.

G. After a 15- to 30-s wait, carefully remove the stopper.

H. Centrifuge for 10 min at 500 × *g*. Four layers should result: a small amount of sediment (containing the parasites) in the bottom of the tube; a layer of formalin; a plug of fecal debris on top of the formalin layer; and a layer of ethyl acetate at the top.

I. Free the plug of debris by ringing the plug with an applicator stick; decant all of the supernatant fluid. After proper decanting, a drop or two of fluid remaining on the side of the tube may run down into the sediment. Mix this fluid with the sediment.

J. If the sediment is still somewhat solid, add a drop or two of saline to the sediment, mix, add a small amount of material to a slide, add a coverslip, and examine.

K. Systematically scan using the 10× objective. The entire coverslip area should be examined.

L. If something suspicious is seen, the 40× objective can be used for more detailed study. At least one-third of the coverslip should be examined with the 40× objective, even if nothing suspicious has been seen. As in the direct wet smear, iodine can be added to enhance morphological detail.

VI. RESULTS

Protozoan trophozoites and/or cysts and helminth eggs and larvae may be seen and identified. Protozoan trophozoites are less likely to be seen.

VII. REPORTING RESULTS

A. Protozoan cysts may or may not be identified to the species level (depending on the clarity of the morphology).
Example (positive report): *Giardia lamblia* cysts present.

B. Helminth eggs and/or larvae may be identified.
Example (positive report): *Trichuris trichiura* eggs present.

C. Artifacts and/or other structures may also be seen and reported as follows.
Examples: Few Charcot-Leyden crystals present.
Moderate PMNs present.

VIII. PROCEDURE NOTES

A. Tap water may be substituted for 0.85% NaCl throughout this procedure, although the addition of water to fresh stool will cause *Blastocystis hominis* cyst forms to rupture. In addition to the original 5 or 10% formalin fixation, some workers prefer to use 5 or 10% formalin for all the rinses throughout the procedure.

B. Ethyl acetate is widely recommended as a substitute for ether. It can be used the same way in the procedure and is much safer. Hemo-De can also be used (3).

1. After the plug of debris is rimmed and excess fluid decanted, while the tube is still upside down, the sides of the tube can be swabbed with a cotton-tipped applicator stick to remove excess ethyl acetate. This is particularly important if you are working with plastic centrifuge tubes. If the sediment is too dry after swabbing the tube, then add several drops of saline before preparing your wet smear for examination.
2. If you have excess ethyl acetate in the smear of the sediment prepared for examination, there will be bubbles present which will obscure the material you are trying to see.

C. If specimens are received in SAF, begin at topic V of this procedure, step B.

D. If specimens are received in PVA, the first two steps of the procedure (V.A and B above) should be modified as follows:

1. Immediately after stirring the stool-PVA mixture with applicator sticks, pour approximately one-half of the mixture into a tube (container optional) and add 0.85% NaCl to approximately the top of the tube.
2. Filter the stool-PVA-saline mixture through wet gauze into a 15-ml centrifuge tube. Standard steps are followed from this point to completion, beginning with step V.C of this procedure.

E. Too much or too little sediment will result in an ineffective concentration.

F. The centrifuge should reach the recommended speed before the centrifugation time is monitored. If the centrifugation time at the proper speed is reduced, some of the organisms (*Cryptosporidium* spp. oocysts) might not be recovered in the sediment.

IX. LIMITATIONS OF THE PROCEDURE

A. Results obtained with wet smears should usually be confirmed by permanent stained smears. Some protozoa are very small and difficult to identify to species level by using just the direct wet smears.

B. Confirmation is particularly important in the case of *Entamoeba histolytica/E. dispar* versus *Entamoeba coli.*

C. Certain organisms, *Giardia lamblia,* hookworm eggs, and occasionally *Trichuris* eggs may not concentrate as well from PVA-preserved specimens as they do from those preserved in formalin. However, if there are enough *Giardia* present to concentrate from formalin, then PVA should contain enough for detection on the permanent stained smear. In clinically important infections, the number of helminth eggs present would ensure detection regardless of the type of preservative used. Also, the morphology of *Strongyloides stercoralis* larvae is not as clear from PVA as from specimens fixed in formalin.

D. For unknown reasons, *Isospora belli* oocysts are routinely missed in the concentrate sediment when concentrated from PVA-preserved specimens.

E. At the recommended centrifugation speed and time in this procedure, there is anecdotal evidence to strongly indicate *Cryptosporidium* oocysts and microsporidia spores should be recovered. The current recommendation is centrifugation at 500 × *g* for a minimum of 10 min for the recovery of the coccidia and microsporidia.

REFERENCES

1. **Garcia, L. S., and D. A. Bruckner.** 1997. *Diagnostic Medical Parasitology,* 3rd ed. ASM Press, Washington, D.C.
2. **Melvin, D. M., and M. M. Brooke.** 1982. *Laboratory Procedures for the Diagnosis of Intestinal Parasites.* U.S. Department of Health and Human Services publication no. (CDC) 82-8282. U.S. Government Printing Office, Washington, D.C.
3. **Neimeister, R., A. L. Logan, B. Gerber, J. H. Egleton, and B. Kleger.** 1987. Hemo-De as substitute for ethyl acetate in formalin-ethyl acetate concentration technique. *J. Clin. Microbiol.* **25:**425–426.

PART 2

Flotation: Zinc Sulfate

I. PRINCIPLE

The flotation procedure permits the separation of protozoan cysts and certain helminth eggs from excess debris through the use of a liquid with a high specific gravity. The parasitic elements are recovered in the surface film, and the debris remains in the bottom of the tube. This technique yields a cleaner preparation than the sedimentation procedure; however, some helminth eggs (operculated eggs and/or very dense eggs such as unfertilized *Ascaris* eggs) do not concentrate well with the flotation method (1–3). The specific gravity of the zinc sulfate can be increased, although this usually causes more distortion in the organisms present and is not recommended for routine clinical use. To ensure detection of all possible organisms, both the surface film and the sediment must be examined. For most laboratories, this is not a practical approach.

II. SPECIMEN

The specimen must be fresh or formalinized stool (5 or 10% buffered or nonbuffered formalin or sodium acetate-acetic acid-formalin [SAF]).

III. REAGENTS

Include QC information on reagent container and in QC records.

A. Formalin (5 or 10% buffered or non-buffered) or SAF

B. 0.85% NaCl

C. Zinc sulfate (33% aqueous solution)

IV. EQUIPMENT

Hydrometer (with a range that includes 1.18 to 1.20), required for checking the specific gravity of the zinc sulfate solution.

V. QUALITY CONTROL

A. Examine the reagents each time they are used. The formalin, saline, and zinc sulfate should appear clear, without any visible contamination.

B. The microscope should be calibrated within the last 12 months (if moved frequently or in constant use), and the objectives and oculars used for the calibration procedure should be used for all measurements on the microscope. The calibration factors for all objectives should be posted on the microscope for easy access (multiplication factors can be placed right on the body of the microscope) (*see* procedure 7.3).

V. QUALITY CONTROL *(continued)*

C. Known positive specimens should be concentrated and organism recovery verified at least quarterly, particularly after the centrifuge has been recalibrated. Record results on appropriate sheets.

VI. PROCEDURE

A. Transfer 1/2 teaspoon (about 4 g) of fresh stool into 10 ml of 10% formalin in a shell vial, unwaxed paper cup, or round-bottom tube (container may be modified to suit individual laboratory preferences). Mix the stool and formalin thoroughly. Let the mixture stand a minimum of 30 min for fixation. If the specimen is already in 5 or 10% formalin (or SAF), restir the stool-formalin mixture.

B. Depending on the size and density of the specimen, strain a sufficient quantity through wet gauze into a conical 15-ml centrifuge tube to give the desired amount of sediment (0.5 to 1 ml) in step C below. Usually 8 ml of the stool-formalin mixture prepared in step A will be sufficient. If the specimen is received in vials of preservative (10% formalin or SAF), then approximately 3 to 4 ml will be sufficient unless the specimen has very little stool in the vial.

C. Add 0.85% NaCl (*see* Procedure Notes below) almost to the top of the tube and centrifuge for 10 min at 500 × *g*. The amount of sediment obtained should be approximately 0.5 to 1 ml. Too much or too little sediment will result in an ineffective concentration.

D. Decant the supernatant fluid and suspend (vortex is acceptable) the sediment in 0.85% NaCl almost to the top of the tube and centrifuge for 10 min at 500 × *g*. This second wash may be eliminated if the supernatant fluid after the first wash is light tan or clear.

E. Decant the supernatant fluid and suspend the sediment on the bottom of the tube in 1 to 2 ml of zinc sulfate. Fill the tube within 2 to 3 mm of the rim with additional zinc sulfate.

F. Centrifuge for 1 min at 500 × *g*. Allow the centrifuge to come to a stop without interference or vibration. Two layers should result: a small amount of sediment in the bottom of the tube and a layer of zinc sulfate. The protozoan cysts and some helminth eggs will be in the surface film; some operculated and/or heavy eggs will be in the sediment.

G. Without removing the tube from the centrifuge, remove 1 or 2 drops of the surface film with a Pasteur pipette or a freshly flamed (and then cooled) wire loop and place them on a slide. Do not use the loop as a "dipper"; simply touch the surface (bend the loop portion of the wire 90° so the loop is parallel with the surface of the fluid). Make sure the pipette tip or wire loop is not below the surface film.

H. Add a coverslip to the preparation. Iodine can be added to the preparation.

I. Systematically scan using the 10× objective. The entire coverslip area should be examined.

J. If something suspicious is seen, the 40× objective can be used for more detailed study. At least one-third of the coverslip should be examined with the 40× objective, even if nothing suspicious has been seen. As in the direct wet smear, iodine can be added to enhance morphological detail.

VII. RESULTS

Protozoan trophozoites and/or cysts and helminth eggs and larvae may be seen and identified.

VIII. REPORTING RESULTS

A. Protozoan cysts may or may not be identifiable to the species level (depending on the clarity of the morphology).
Example (positive report): *Giardia lamblia* cysts present.

B. Helminth eggs and/or larvae may be identified.
Example (positive report): *Trichuris trichiura* eggs present.

C. Artifacts and/or other structures may also be seen and reported as follows.
Examples: Few Charcot-Leyden crystals present.
Moderate PMNs present.

IX. PROCEDURE NOTES

A. Tap water (*see* note D below) may be substituted for 0.85% NaCl throughout this procedure; some workers prefer to use 5 or 10% formalin for all the rinses throughout the procedure.

B. If fresh stool is used (nonformalin preservatives), then the zinc sulfate should be prepared with a specific gravity of 1.18.

C. If specimens are received in SAF, then begin at topic VI, step B.

D. If fresh specimens are received, the standardized procedure requires the stool to be rinsed in distilled water before the addition of zinc sulfate in step F. However, the addition of fresh stool to distilled water will destroy any *Blastocystis hominis* present and is not a recommended approach.

E. Some workers prefer to remove the tubes from the centrifuge before sampling the surface film. This is acceptable; however, there is more chance the surface film will be disturbed before sampling.

F. Some workers prefer to add a small amount of zinc sulfate to the tube, so the fluid forms a slightly convex meniscus. A coverslip is then placed on top of the tube so that the under surface touches the meniscus, and this is left undisturbed for 5 min. The coverslip is then carefully removed and placed on a slide for examination.

G. When using the hydrometer (solution at room temperature), mix the solution well. Float the hydrometer in the solution, giving it a slight twist to see that it is completely free from the sides of the container. Read the bottom meniscus and correct the figure for temperature, if necessary. Most hydrometers are calibrated at 20°C. A difference of 3°C between the solution temperature (room temperature) and the hydrometer calibration temperature requires a correction of 0.001 to the specific gravity, to be added if above and subtracted if below 20°C.

X. LIMITATIONS OF THE PROCEDURE

A. Results obtained with wet smears should usually be confirmed by permanent stained smears. Some protozoa are very small and difficult to identify to species level by using just the direct wet smears.

B. Confirmation is particularly important in the case of *Entamoeba histolytica/E. dispar* vs. *Entamoeba coli.*

C. Protozoan cysts and thin-shelled helminth eggs are subject to collapse and distortion when left for more than a few minutes in contact with the high-specific-gravity zinc sulfate. The surface film should be removed for examination within 5 min of the time the centrifuge comes to a stop. The longer the organisms are in contact with the zinc sulfate, the more distortion you will see on microscopic examination of the surface film.

D. If zinc sulfate is the only concentration method used, both the surface film and the sediment should be examined to ensure detection of all possible organisms.

REFERENCES

1. **Garcia, L. S., and D. A. Bruckner.** 1997. *Diagnostic Medical Parasitology,* 3rd ed. ASM Press, Washington, D.C.
2. **Melvin, D. M., and M. M. Brooke.** 1982. *Laboratory Procedures for the Diagnosis of Intestinal Parasites.* U.S. Department of Health and Human Services publication no. (CDC) 82-8282. U.S. Government Printing Office, Washington, D.C.
3. **NCCLS.** 1993. *Procedures for the Recovery and Identification of Parasites from the Intestinal Tract.* Proposed Guideline M28-P. NCCLS, Wayne, Pa.

7.5.3 Permanent Stained Smears

PART 1 Trichrome

It is generally recognized that stained fecal films are the single most productive means of stool examination for intestinal protozoa. The permanent stained smear facilitates detection and identification of cysts and trophozoites and affords a permanent record of the protozoa encountered. Small protozoa, missed by direct smear and concentration techniques, are often seen on the stained smear. It also provides laboratories the ability to refer the slide to a specialist for help when they have encountered an organism with an unusual morphology or have difficulty with the identification. For the above reasons, the permanent stained smear is recommended for use with every stool specimen submitted for a routine parasite examination (1,3). The use of specimen collection vial systems supports good specimen fixation and, thus, good organism morphology after permanent staining.

I. PRINCIPLE

The Trichrome technique of Wheatley (5) for fecal specimens is a modification of Gomori's original staining procedure for tissue (2). It is a rapid, simple procedure which produces uniformly well-stained smears of the intestinal protozoa, human cells, yeast, and artifact material.

II. SPECIMEN

A. The specimen usually consists of fresh stool smeared on a microscope slide which is immediately fixed in Shaudinn's fixative or polyvinyl alcohol (PVA)-preserved stool smeared on a slide and allowed to air dry (*see* topic V below).

B. Stool preserved in sodium acetate-acetic acid-formalin (SAF) or any of the new single-vial fixatives for parasitology can also be used.

III. REAGENTS

Include QC information on reagent container and in QC records.

A. Trichrome stain

B. 70% ethanol

C. 70% ethanol plus iodine

Prepare a stock solution by adding iodine crystals to 70% alcohol until you obtain a dark solution (1 to 2 g/100 ml). To use, dilute the stock solution with 70% alcohol until a dark reddish brown (port wine) or strong tea color is obtained. As long as the color is acceptable, new working solution does not have to be replaced. Replacement time will depend on the number of smears stained and the size of the container (1 week to several weeks).

D. 90% ethanol, acidified

90% ethanol............................99.5 ml
acetic acid (glacial) ....0.5 ml

Prepare by combining.

E. 100% ethanol

F. Xylene or xylene substitute

IV. QUALITY CONTROL

A. For QC of Schaudinn's, SAF, or PVA fixatives, see procedure 7.2.2.

B. Stool samples used for QC can be fixed stool specimens known to contain protozoa or PVA-preserved negative stools to which buffy coat cells have been added. A QC smear prepared from a positive PVA or PVA containing buffy coat cells should be used when new stain is prepared or at least once each month. Cultured protozoa can also be used.

C. A QC slide can be included with each run of stained slides (optional).

D. If the xylene becomes cloudy or has an accumulation of water in the bottom of the staining dish, use fresh 100% ethanol and xylene.

E. All staining dishes should be covered to prevent evaporation of reagents.

F. Depending on the volume of slides stained, staining solutions will have to be changed on an as-needed basis.

G. When the smear is thoroughly fixed and the stain is performed correctly, the cytoplasm of protozoan trophozoites will have a blue green color sometimes with a tinge of purple. Cysts tend to be slightly more purple. Nuclei and inclusions (chromatoid bodies, RBCs, bacteria, and Charcot-Leyden crystals) have a red color sometimes tinged with purple. The background material usually stains green, providing a nice color contrast with the protozoa. This contrast is more distinct than that obtained with the hematoxylin stain.

H. The microscope should have been calibrated within the last 12 months (if moved frequently or in constant use), and the objectives and oculars used for the calibration procedure should be used for all measurements on the microscope. The calibration factors for all objectives should be posted on the microscope for easy access (multiplication factors can be pasted right on the body of the microscope). See procedure 7.3.

I. Known positive microscope slides, Kodachrome 2 × 2 projection slides and photographs (reference books) should be available at the work station.

J. Record all QC results.

V. PROCEDURE

A. Slide preparation

1. Fresh fecal specimens
   - **a.** When the specimen arrives, prepare two slides with applicator sticks and *immediately* (without drying) place them in Schaudinn's fixative. Allow the specimen to fix for a minimum of 30 min; overnight fixation is acceptable. The stool smeared on the slide should be thin enough that newsprint can be read through the smear. Proceed with the trichrome staining method at step V.B below (if using mercuric chloride-based Schaudinn's fluid).
   - **b.** If the fresh specimen is liquid, place 3 or 4 drops of PVA on the slide, mix several drops of fecal material with the PVA, spread the mixture, and allow it to dry for several hours in a 37°C incubator or overnight at room temperature (25°C). Proceed with the trichrome staining method at step V.C below (if using mercuric chloride-based PVA). *This approach is recommended for liquid specimens only.*
2. PVA-preserved fecal specimens (mercuric chloride base)
   - **a.** Allow the stool specimens that are preserved in PVA to fix for at least 30 min. Thoroughly mix the contents of the PVA vial with two applicator sticks.
   - **b.** Pour some of the PVA-stool mixture onto a paper towel, and allow it to stand for 3 min to absorb the PVA. *Do not eliminate this step.*
   - **c.** With an applicator stick, apply some of the stool material from the paper towel to two slides, and allow them to dry for several hours in a 37°C incubator or overnight at room temperature (25°C).
   - **d.** Begin the trichrome staining process at step V.C, below.

3. Modified PVA-preserved fecal specimens (copper or zinc base)
   a. Allow the stool specimens that are preserved in PVA to fix for at least 30 min. Thoroughly mix the contents of the PVA vial with two applicator sticks.
   b. Pour some of the PVA-stool mixture onto a paper towel, and allow it to stand for 3 min to absorb the PVA. *Do not eliminate this step.*
   c. With an applicator stick, apply some of the stool material from the paper towel to two slides, and allow them to dry for several hours in a 37°C incubator or overnight at room temperature (25°C).
   d. Begin the trichrome staining process at step V.E or V.F, below.

4. SAF-preserved fecal specimens
   a. Thoroughly mix the SAF-stool mixture, and strain through gauze into a 15-ml centrifuge tube.
   b. After centrifugation (10 min at 500 $\times$ *g*), decant the supernatant fluid. The final sediment should be about 0.5 to 1.0 ml. If necessary, adjust by repeating this step or by suspending the sediment in saline (0.85% NaCl) and removing part of the suspension.
   c. Prepare a smear from the sediment for later staining by placing 1 drop of Mayer's albumin on the slide and adding 1 drop of SAF-preserved specimen. Allow the smear to air dry at room temperature (25°C) for 30 min before staining. Begin the trichrome staining procedure at step V.E, below.

B. Remove slide from Schaudinn's fixative and place slide in 70% ethanol for 5 min.

C. Place slide in 70% ethanol plus iodine for 1 min for fresh specimens or 5 to 10 min for PVA air-dried smears. *All slides exposed to mercuric chloride-based fixatives must be placed in the iodine dish to remove the mercury. The subsequent rinses in ethanol remove the iodine. At the point the slide is placed into trichrome stain, both the mercury and iodine have been removed from the fecal smear.*

D. Place slide in 70% ethanol for 5 min.

NOTE: Slides can be held up to 24 h in 70 or 100% ethanol or in xylene (*see* steps V.E and V.I through K) without harming the quality of the smear or stainability of organisms.

E. Place in second 70% ethanol for 3 min. Fecal smears prepared from SAF-preserved stool material do not require the iodine step (V.C, above) and can be placed in this alcohol dish before trichrome staining.

F. Place in Trichrome stain for 10 min. Fecal smears prepared from modified PVA-fixed material (copper or zinc base) do not require the iodine step (V.C, above) or subsequent alcohol rinses (V.D and E, above) but can be placed directly into the trichrome stain (this step). One alcohol rinse may be used (*see* V.E) before this trichrome step; some labs prefer this approach.

G. Place in 90% ethanol plus acetic acid for 1 to 3 s. Immediately drain the rack (*see* topic VIII below [Procedure Notes]) and proceed to the next step. Do not allow slides to remain in this solution.

H. Dip several times in 100% ethanol. Use this step as a rinse.

I. Place in two changes of 100% ethanol for 3 min each.

J. Place in xylene for 5 to 10 min.

K. Place in second xylene for 5 to 10 minutes.

L. Mount with coverslip (no. 1 thickness), using mounting medium (e.g., Permount).

M. Allow the smear to dry overnight or after 1 h at 37°C.

N. Examine the smear microscopically with the 100$\times$ oil immersion objective. Examine at least 200 to 300 oil immersion fields.

VI. RESULTS

A. Protozoan trophozoites and cysts will be readily seen if present.
B. Helminth eggs and larvae might not be easily identified; therefore, examine wet mounts of concentrates.
C. Yeast and human cells can be identified. Human cells include macrophages, PMNs, and RBCs. Yeasts include single and budding cells and pseudohyphae.

VII. REPORTING RESULTS

A. Report the organism and stage (do not use abbreviations).
   Example: *Entamoeba histolytica/E. dispar* trophozoites
B. Quantitate the number of *Blastocystis hominis* seen (rare, few, moderate, many). Do not quantitate other protozoa.
C. Note and quantitate the presence of human cells.
   Examples: Moderate WBCs, many RBCs, few macrophages, rare Charcot-Leyden crystals
D. Report and quantitate yeast cells.
   Examples: Moderate budding yeast cells and few pseudohyphae
E. Save positive slides for future reference. Label before storage (name, patient number, organisms present).
F. Quantitation of parasites, cells, yeast cells, and artifacts
   Few = $\leq$2 per 10 oil immersion fields ($\times$1,000)
   Moderate = 3 to 9 per 10 oil immersion fields ($\times$1,000)
   Many = $\geq$10 per 10 oil immersion fields ($\times$1,000)

VIII. PROCEDURE NOTES

A. Fixation of specimens is important. Improperly fixed specimens will result in protozoan forms that are nonstaining or predominantly red.
B. Slides should always be drained between solutions. Touch the end of the slide to a paper towel for 2 s, to remove excess fluid, before proceeding to the next step.
C. Incomplete removal of mercuric chloride (Schaudinn's fixative and PVA) may cause the smear to contain highly refractive granules which may prevent finding or identifying any organisms present (1, 3). Since the 70% ethanol-iodine solution removes the mercury complex, it should be changed at least weekly to maintain the port wine or strong-tea color.
D. To restore weakened trichrome stain, remove cap and allow the ethanol to evaporate (carried over on staining rack from previous dish). After a few hours, fresh stain can be added to restore lost volume. Older, more concentrated stain produces more intense colors and might require slightly longer destaining times (an extra dip).
E. Smears that are predominantly green may be caused by the inadequate removal of iodine by the 70% ethanol (steps V.D and E above). Lengthening the time of these steps or changing the 70% ethanol more frequently will help.
F. In the final stages of dehydration (steps V.I through K above) the 100% ethanol and the xylenes should be kept as free from water as possible. Coplin jars must have tightly fitting caps to prevent both evaporation of reagents and absorption of moisture. If the xylene becomes cloudy after addition of slides from the 100% ethanol, return the slides to fresh 100% ethanol, and replace the xylene.
G. If the smears peel or flake off, the specimen might have been inadequately dried on the slide (in the case of PVA-fixed specimens) or the slides might have been greasy. Slides do not have to be cleaned with alcohol before use.

H. If the stain appears unsatisfactory and it is not possible to obtain another slide to stain, it can be restained. Place the slide in xylene to remove the coverslip, and reverse the dehydration steps, adding 50% ethanol as the last step. Destain the slide in 10% acetic acid for several hours, and then wash it thoroughly first in water (major) and then in 50% and 70% ethanol. Place the slide in the trichrome stain for 8 min and complete the staining procedure (3).

IX. LIMITATIONS OF THE PROCEDURE

A. The permanent stained smear is not recommended for staining helminth eggs or larvae. However, occasionally they can be recognized and identified.

B. The smear should be examined by using the oil immersion lens (100×) for the identification of protozoa, human cells, Charcot-Leyden crystals, yeast cells, and artifact material.

C. This high-magnification examination is recommended for protozoa.

D. Screening the smear with low magnification (10×) might reveal eggs or larvae; however, this is not recommended as a routine approach.

E. Helminth eggs and larvae and *Isospora belli* oocysts are best seen in wet preparations.

F. *Cryptosporidium parvum* and *Cyclospora cayetanensis* are generally not seen on a trichrome-stained smear (modified acid-fast stains are recommended).

G. Microsporidia spores will not be seen on a trichrome-stained smear (modified trichrome stains are recommended).

REFERENCES

1. **Garcia, L. S., and D. A. Bruckner.** 1997. *Diagnostic Medical Parasitology,* 3rd ed., p. 623–631. ASM Press, Washington, D.C.
2. **Gomori, G.** 1950. A rapid one-step trichrome stain. *Am. J. Clin. Pathol.* **20:**661.
3. **NCCLS.** 1993. *Procedures for the Recovery and Identification of Parasites from the Intestinal Tract.* Proposed Guideline M28-P. NCCLS, Wayne, Pa.
4. **Smith, J. W., and M. S. Bartlett.** 1991. Diagnostic parasitology: introduction and methods, p. 701–716. *In* A. Balows, W. J. Hausler, Jr., K. L. Herrmann, H. D. Isenberg, and H. J. Shadomy (ed.), *Manual of Clinical Microbiology,* 5th ed. American Society for Microbiology, Washington, D.C.
5. **Wheatley, W. B.** 1951. A rapid staining procedure for intestinal amoebae and flagellates. *Am. J. Clin. Pathol.* **21:**990–991.

APPENDIX 7.5.3 (part 1)–1

As an alternative to using mounting fluid on every slide, the following method can be used. This approach saves time (drying the slides after being mounted) and eliminates the need for routine use of mounting fluids.

A. Remove the stained slides from the last dehydrating dish (step V.K above).

B. Allow the slide to air dry (minimum of 30 min).

C. Place a drop of immersion oil directly onto the dry stool smear.

D. Allow the oil to "sink in" for a minimum of 15 min.

E. Place a number 1 coverslip onto the oil-covered stool smear.

F. Add 1 drop of immersion oil onto the coverslip and proceed to examine the smear using the 100× oil immersion objective. A 50× or 60× oil immersion objective can be used for screening.

G. *Do not use this approach unless you add the coverslip before examination of the smear. The dry stool material is quite hard; the objective lens could accidentally be scratched if the stool smear is not covered before reading.*

PART 2

Iron Hematoxylin Stain (Modified Spencer-Monroe Method)

I. PRINCIPLE

The iron hematoxylin stain is one of a number of stains used to make a permanent stained slide for detecting and quantitating parasitic organisms. Iron hematoxylin was the stain used for most of the original morphological descriptions of intestinal protozoa found in humans (1). On oil immersion power (×1,000), one can examine the diagnostic features used to identify the protozoan parasite.

II. SPECIMEN

A. The specimen usually consists of fresh stool smeared on a microscope slide and immediately fixed in Shaudinn's fixative or preserved stool (polyvinyl alcohol [PVA], PVA-modified) smeared on a slide and allowed to air dry (*see* procedure 7.5.3, part 1).

B. Stool preserved in sodium acetate-acetic acid-formalin (SAF) can also be used (*see* procedure 7.5.3, part 1).

III. REAGENTS

Include QC information on reagent container and in QC records.

A. Iron hematoxylin stain
B. Ethanol: 70, 95, or 100%
C. D'Antoni's iodine solution
D. Xylene or xylene substitute

IV. QUALITY CONTROL

A. For QC of Schaudinn's SAF or PVA fixatives, see procedure 7.2.2.

B. Stool samples used for QC can be fixed stool specimens known to contain protozoa or PVA-preserved negative stools to which buffy coat cells have been added. A QC smear prepared from a positive PVA or PVA containing buffy coat cells should be used when new stain is prepared or at least once per month. Cultured protozoa can also be used.

C. A QC slide can be included with each run of stained slides (optional).

D. If the xylene becomes cloudy or has an accumulation of water in the bottom of the staining dish, use fresh 100% ethanol and xylene.

E. All staining dishes should be covered to prevent evaporation of reagents.

F. Depending on the volume of slides stained, change staining solutions on an as-needed basis.

G. Background material will stain a blue-gray color. Cells and organisms will stain various intensities of blue-gray. Inclusions, chromatoidal bodies, and nuclear structures will stain darker than the surrounding cytoplasm.

H. The microscope should have been calibrated within the last 12 months (if moved frequently or in constant use), and the objectives and oculars used for the calibration procedure should be used for all measurements on the microscope. The calibration factors for all objectives should be posted on the microscope for easy access (multiplication factors can be pasted right on the body of the microscope) (*see* procedure 7.3).

I. Known positive microscope slides, Kodachrome 2-inch-by-2-inch projection slides, and photographs (reference books) should be available at the work station.

J. Record all QC results.

V. PROCEDURE

A. Prepare slide to be stained as previously described (*see* procedure 7.5.3, part 1, item V.A).
B. Place slide in 70% ethanol for 5 min.
C. Place slide in an iodine-70% ethanol (70% alcohol to which is added enough D'Antoni's iodine to obtain a strong tea color) solution for 2 to 5 min (fecal smears preserved with mercuric chloride-based fixatives).
D. Place in 70% ethanol for 5 min. Begin procedure for SAF- or modified-PVA (copper or zinc sulfate-based)-fixed slides at this point.

NOTE: Slides may be held up to 24 h in these solutions, in 70, 95, or 100% ethanol, or in xylene without harming the quality of the smear or stainability of organisms.

E. Wash slide in running tap water (constant stream of water into the container) for 10 min.
F. Place slide in iron hematoxylin working solution for 4 to 5 min.
G. Wash slide in running tap water (constant stream of water into the container) for 10 min.
H. Place slide in 70% ethanol for 5 min.
I. Place slide in 95% ethanol for 5 min.
J. Place slide in two changes of 100% ethanol for 5 min each.
K. Place slide in two changes of xylene for 5 min each.
L. Add Permount to the stained area of the slide and cover with a coverslip.
M. Examine the smear microscopically with the 100× objective. Examine at least 200 to 300 oil immersion fields.

VI. RESULTS

A. Protozoan trophozoites and cysts will be readily seen if present.
B. Helminth eggs and larvae might not be easily identified; therefore, examine wet mount concentrates.
C. Yeast and human cells can be identified. Human cells include macrophages, PMNs, and RBCs. Yeast cells include single and budding cells and pseudohyphae.

VII. REPORTING RESULTS

A. Report the complete scientific name (genus and species) of the organism and the stage seen.

Example: *Entamoeba histolytica/E. dispar* trophozoites

B. Quantitate the number of *Blastocystis hominis* seen (rare, few, moderate, many). Do not quantitate other protozoa.
C. Note and quantitate the presence of human cells.

Examples: Moderate WBCs, many RBCs, few macrophages, rare Charcot-Leyden crystals

D. Report and quantitate yeast cells.

Example: Moderate budding yeast cells and few pseudohyphae

E. Save positive slides for future reference. Label before storage (name, patient number, organisms present).
F. Quantitation of parasites, cells, yeast cells, and artifacts

Few = ≤2 per 10 oil immersion fields (×1,000)
Moderate = 3 to 9 per 10 oil immersion fields (×1,000)
Many = ≥10 per 10 oil immersion fields (×1,000)

VIII. PROCEDURE NOTES

A. Once the staining process has begun, the slides should not be allowed to dry until they have been placed in xylene.

B. Slides should always be drained between solutions. Touch the end of the slide to a paper towel for 2 s, to remove excess fluid, before proceeding to the next step.

C. Incomplete removal of mercuric chloride (Schaudinn's fixative and PVA) can cause the smear to contain highly refractive granules which might prevent finding or identifying any organisms present (2, 3). Since the 70% ethanol-iodine solution removes the mercury complex, it should be changed at least weekly to maintain the port wine or strong tea color.

D. When large numbers of slides are stained, the working hematoxylin solution might become diluted and affect the quality of the stain. If dilution occurs, discard the working solution and prepare a fresh working solution.

E. The shelf life of the stock hematoxylin solutions can be extended by keeping the solutions in the refrigerator at 4°C. Because of crystal formation in the working solutions, it might be necessary to filter them before preparing a new working solution.

F. In the final stages of dehydration (steps V.I through K above) the 100% ethanol and the xylenes should be kept as free from water as possible. Coplin jars must have tightly fitting caps to prevent both evaporation of reagents and absorption of moisture. If the xylene becomes cloudy after addition of slides from the 100% ethanol, return the slides to fresh 100% ethanol and replace the xylene.

G. If the smears peel or flake off, the specimen might have been inadequately dried on the slide (for PVA-fixed specimens), or the slides might have been greasy. Slides do not have to be cleaned with alcohol before use.

IX. LIMITATIONS OF THE PROCEDURE

A. The permanent stained smear is not recommended for staining helminth eggs or larvae. However, occasionally they might be recognized and identified.

B. The smear should be examined with the oil immersion lens (100×) for the identification of protozoa, human cells, Charcot-Leyden crystals, yeast cells, and artifact material.

C. This high-magnification examination is recommended for protozoa.

D. Screening the smear with low magnification (10×) might reveal eggs or larvae; however, this is not recommended as a routine approach.

E. Helminth eggs and larvae and *Isospora belli* oocysts are best seen in wet preparations.

F. *Cryptosporidium parvum* will not be seen on an iron hematoxylin-stained smear (modified acid-fast stains are recommended).

G. Microsporidia spores will not be seen on an iron hematoxylin-stained smear (modified trichrome stains are recommended).

REFERENCES

1. **Garcia, L. S., and D. A. Bruckner.** 1997. *Diagnostic Medical Parasitology,* 3rd ed. ASM Press, Washington, D.C.
2. **NCCLS.** 1993. *Procedures for the Recovery and Identification of Parasites from the Intestinal Tract.* Proposed Guideline M28-P. NCCLS, Wayne, Pa.
3. **Spencer, L. S., and L. S. Monroe.** 1961. *The Color Atlas of Intestinal Parasites.* Charles C Thomas, Publisher, Springfield, Ill.

APPENDIX 7.5.3 (part 2)–1

Modified Iron Hematoxylin Stain (Incorporating Carbol Fuchsin Step)

The following combination staining method for SAF-preserved fecal specimens was developed to allow the microscopist to screen for acid-fast organisms in addition to other intestinal parasites. For those laboratories where iron hematoxylin stains are used in combination with SAF-fixed material and modified acid-fast stains for *C. parvum, Cyclospora cayetanensis,* and *I. belli,* this modification represents an improved approach to current staining methods. This combination stain provides a saving in both time and personnel use.

Any fecal specimen submitted in SAF fixative can be used. Fresh fecal specimens after fixation in SAF for 30 min can also be used. This combination stain approach is not recommended for specimens preserved in Schaudinn's fixative or PVA.

I. REAGENTS

A. Mayer's albumin

Add an equal quantity of glycerin to a fresh egg white. Mix gently and thoroughly. Store at 4°C and indicate an expiration date of 3 months. Mayer's albumin from commercial suppliers can normally be stored at 25°C for 1 year (e.g., product 756, E. M. Diagnostic Systems Inc., Gibbstown, N.J.).

B. Stock solution of hematoxylin stain

hematoxylin powder .... 10 g
ethanol (95 or 100%) .... 1,000 ml

1. Mix well until dissolved.
2. Store in a clear glass bottle in a light area. Allow to ripen for 14 days before use.
3. Store at room temperature, with an expiration date of 1 year.

C. Mordant

ferrous ammonium sulfate [$Fe(NH_4)_2(SO_4)_2 \cdot 6H_2O$] .... 10 g
ferric ammonium sulfate [$FeNH_4(SO_4)_2 \cdot 12H_2O$] .... 10 g
hydrochloric acid (concentrated) .... 10 ml
distilled water to .... 1,000 ml

D. Working solution of hematoxylin stain

1. Mix equal quantities of stock solution of stain and mordant.
2. Allow mixture to cool thoroughly before use (prepare at least 2 h before use). The working solution should be made fresh every week.

E. Picric acid

Mix equal quantities of distilled water and an aqueous saturated solution of picric acid to make a 50% saturated solution.

F. Acid-alcohol decolorizer

hydrochloric acid (concentrated) .... 30 ml
alcohol to .... 1,000 ml

G. 70% alcohol and ammonia

70% alcohol .... 50 ml
ammonia .... 0.5 to 1.0 ml

Add enough ammonia to bring the pH to approximately 8.0.

H. Carbol fuchsin

1. To make basic fuchsin (solution A), dissolve 0.3 g of basic fuchsin in 10 ml of 95% ethanol.
2. To make phenol (solution B), dissolve 5 g of phenol crystals in 100 ml of distilled water. (Gentle heat might be needed.)
3. Mix solution A with solution B.
4. Store at room temperature. Solution is stable for 1 year.

II. PROCEDURE

A. Prepare slide.

1. Place 1 drop of Mayer's albumin on a labeled slide.
2. Mix the sediment from the SAF concentration well with an applicator stick.
3. Add approximately 1 drop of the fecal concentrate to the albumin and spread the mixture over the slide.

B. Allow slide to air dry at room temperature (smear will appear opaque when dry).

C. Place slide in 70% alcohol for 5 min.

D. Wash in container of tap water (not running water) for 2 min.

APPENDIX 7.5.3 (part 2)–1
(continued)

E. Place slide in Kinyoun's stain for 5 min.
F. Wash slide in running tap water (constant stream of water into container) for 1 min.
G. Place slide in acid-alcohol decolorizer for 4 min.

NOTE: This step can also be performed as follows.

1. Place slide in acid-alcohol decolorizer for 2 min.
2. Wash slide in running tap water (constant stream of water into container) for 1 min.
3. Place slide in acid-alcohol decolorizer for 2 min.
4. Wash slide in running tap water (constant stream of water into container) for 1 min.
5. Continue staining sequence with step II.I below (iron hematoxylin working solution).

H. Wash slide in running tap water (constant stream of water into container) for 1 min.
I. Place slide in iron hematoxylin working solution for 8 min.
J. Wash slide in distilled water (in container) for 1 min.
K. Place slide in picric acid solution for 3 to 5 min.
L. Wash slide in running tap water (constant stream of water into container) for 10 min.
M. Place slide in 70% alcohol plus ammonia for 3 min.
N. Place slide in 95% alcohol for 5 min.
O. Place slide in 100% alcohol for 5 min.
P. Place slide in two changes of xylene for 5 min.

III. PROCEDURE NOTES

A. The first 70% alcohol step acts with the Mayer's albumin to "glue" the specimen to the glass slide. The specimen may wash off if insufficient albumin is used or if the slides are not completely dry before staining.
B. The working hematoxylin stain should be checked each day of use by adding a drop of stain to alkaline tap water. If a blue color does not develop, prepare fresh working stain solution.
C. The picric acid differentiates the hematoxylin stain by removing more stain from fecal debris than from the protozoa and removing more stain from the organism cytoplasm than from the nucleus. When properly stained, the background should be various shades of gray-blue and protozoa should be easily seen with medium blue cytoplasm and dark blue-black nuclei.

Supplemental Reading

Palmer, J. 1991. Modified iron hematoxylin/kinyoun stain. *Clin. Microbiol. Newsl.* **13:**39–40.

7.6.1 Modified Kinyoun's Acid-Fast Stain (Cold) (Coccidia)

I. PRINCIPLE

Cryptosporidium parvum, Cyclospora cayetanensis, and *Isospora belli* have been recognized as the cause of severe diarrhea in immunocompromised hosts and can also cause diarrhea in immunocompetent hosts. Oocysts in clinical specimens can be difficult to detect without special staining. Modified acid-fast stains are recommended to demonstrate these organisms. Unlike the Ziehl-Neelsen modified acid-fast stain, this stain does not require the heating of reagents for staining (1–3).

II. SPECIMEN

Concentrated sediment of fresh or formalin-preserved stool can be used. PVA-preserved fecal specimens are not recommended. Other types of clinical specimens such as duodenal fluid, bile, or those from pulmonary sources (induced sputum, bronchial wash, biopsies) can also be stained after centrifugation.

III. REAGENTS

Include QC information on reagent container and in QC records.

A. Absolute methanol

B. 50% ethanol

C. Kinyoun carbol fuchsin

1. Dissolve 4 g of basic fuchsin in 20 ml of 95% ethanol (solution A).
2. Dissolve 8 g of phenol crystals in 100 ml of distilled water (solution B).
3. Mix solutions A and B together.
4. Store at room temperature. Stable for 1 year.

   NOTE: Note expiration date on label.

D. 1% sulfuric acid

1. Add 1 ml of concentrated sulfuric acid to 99 ml of distilled water.
2. Store at room temperature. Stable for 1 year.

   NOTE: Note expiration date on label.

E. Methylene blue

1. Dissolve 0.3 g methylene blue in 100 ml of 95% ethanol.
2. Store at room temperature. Stable for 1 year.

   NOTE: Note expiration date on label.

IV. QUALITY CONTROL

A. A control slide of *Cryptosporidium parvum* from a 10% formalin-preserved specimen is included with each staining batch run. If the *Cryptosporidium* cells stain well, any *Isospora* or *Cyclospora* spp. present will also take up the stain.

B. *Cryptosporidium* spp. stain pink-red. Oocysts are 4 to 6 μm in diameter and four sporozoites might be present internally. The background should stain uniformly blue.

C. The specimen is also checked (macroscopically) for adherence to the slide.

D. Record all QC results.

V. PROCEDURE

A. Smear 1 to 2 drops of specimen on the slide and allow it to air dry. Do not make the smears too thick (you should be able to see through the wet material before it dries). Prepare two smears.

B. Fix with absolute methanol for 1 min.

C. Flood slide with Kinyoun's carbol fuchsin and stain for 5 min.

D. Rinse slide briefly (3 to 5 s) with 50% ethanol.

E. Rinse thoroughly with water.

F. Decolorize by using 1% sulfuric acid for 2 min or until no more color runs from the slide.

G. Rinse slide with water. Drain.

H. Counterstain with methylene blue for 1 min.

I. Rinse slide with water. Air dry.

J. Examine with low or high dry objectives. To see internal morphology, use the oil immersion objective (100×).

VI. RESULTS

A. With this cold Kinyoun acid-fast method, the oocysts of *Cryptosporidium parvum, Cyclospora cayetanensis,* and *Isospora belli* will stain pink to red to deep purple. Some of the four sporozoites might be visible in the *Cryptosporidium* oocysts. *Cyclospora* oocysts will exhibit a great deal of stain variability, some staining almost clear and some deep purple; there is more variation with this organism than with *Cryptosporidium* spp. Some of the *Isospora* immature oocysts (entire oocyst) will stain, while those that are mature will usually appear with the two sporocysts within the oocyst wall stained pink to purple and a clear area between the stained sporocysts and the oocyst wall. The background will stain blue.

B. There will usually be a range of color intensity in the organisms present; not every oocyst will appear deep pink to purple.

VII. REPORTING RESULTS

A. Report the organism and stage (oocyst). Do not use abbreviations.
Example: *Cryptosporidium parvum* oocysts or *Isospora belli* oocysts or *Cyclospora cayetanensis* oocysts

B. Call the physician when these organisms are identified.

VIII. PROCEDURE NOTES

A. Routine stool examination stains are not recommended; however, the sedimentation concentration is recommended (500 × *g* for 10 min) for the recovery and identification of *Cryptosporidium* and *Cyclospora* species. The routine concentration (formalin-ethyl acetate) should be used to recover *Isospora* sp. oocysts, but routine permanent stains are not reliable for this purpose.

B. PVA-preserved specimens are not acceptable for staining with the modified acid-fast stain.

C. Other organisms stain positive, such as acid-fast bacteria and *Nocardia* spp.

D. It is very important that smears are not too thick. Thicker smears might not adequately destain.

E. Concentration of the specimen is essential to demonstrate organisms. The number of organisms seen in the specimen can vary from numerous to very few.

F. Some specimens require treatment with 10% KOH owing to their mucoid consistency. Add 10 drops of 10% KOH to the sediment, and vortex until homogeneous. Rinse with 10% formalin and centrifuge (500 × *g* for 10 min). Without decanting the supernatant, take one drop of the sediment, and smear it thinly on a slide.

G. Commercial concentrators and reagents are available.

H. Concentrations of sulfuric acid (1.0 to 3.0%) are normally used. Stronger concentrations (acid-alcohol) will remove too much stain.

I. There is some debate about whether organisms lose their ability to take up the acid-fast stain after long-term storage in 10% formalin. Use of the hot modified acid-fast method might eliminate this problem (1).

J. Specimens should be centrifuged in capped tubes, and gloves should be worn during all phases of specimen processing.

IX. LIMITATIONS OF PROCEDURE

A. Light infections might be missed (low number of oocysts). The monoclonal antibody methods are more sensitive.

B. Multiple specimens must be examined since the numbers of oocysts present in the stool will vary from day to day. A series of three specimens submitted on alternate days is recommended.

REFERENCES

1. **Garcia, L. S., and D. A. Bruckner.** 1997. *Diagnostic Medical Parasitology,* 3rd ed. ASM Press, Washington, D.C.
2. **Ma., P., and R. Soave.** 1983. Three step stool examination for cryptosporidiosis in 10 homosexual men with protracted diarrhea. *J. Infect. Dis.* **147:**824–828.
3. **Miller, J. M.** 1991. Quality control of media, reagents, and stains, p. 1203–1225. *In* A. Balows, W. J. Hausler, Jr., K. L. Herrmann, H. D. Isenberg, and H. J. Shadomy (ed.), *Manual of Clinical Microbiology,* 5th ed. American Society for Microbiology, Washington, D.C.

7.6.2 Modified Trichrome Stain (Microsporidia)

I. PRINCIPLE

Modified trichrome stain for the microsporidia (Weber-Green)

The diagnosis of intestinal microsporidiosis (*Enterocytozoon bieneusi, Encephalitozoon intestinalis, Encephalitozoon* spp., *Nosema* spp., *Vittaforma* spp., *Pleistophora* spp., *Trachipleistophora* sp., and *Microsporidium* spp.) has depended on the use of invasive procedures and subsequent examination of biopsy specimens, often by using electron microscopy methods. However, the need for a practical method for the routine clinical laboratory has stimulated some work in the development of additional methods. Slides prepared from fresh or formalin-fixed stool specimens can be stained by a new chromotrope-based technique and can be examined with light microscopy. This staining method is based on the fact that stain penetration of the microsporidial spore is very difficult; thus, the dye content in the chromotrope 2R is higher than that routinely used to prepare Wheatley's modification of Gomori's trichrome method, and the staining time is much longer (90 min) (1–3). Several of these stains are available commercially from a number of suppliers.

II. SPECIMEN

The specimen can be fresh stool or stool that has been preserved in 5 or 10% formalin, SAF, or some of the newer single-vial system fixatives. Actually, any specimen other than tissue thought to contain microsporidia could be stained by these methods. PVA-preserved fecal material is not recommended.

III. REAGENTS (Weber-Green)

Include QC information on reagent container and in QC records.

A. Modified trichrome (Weber-Green formulation)

B. Acid alcohol

90% ethyl alcohol ....995.5 ml
acetic acid (glacial).........4.5 ml

Prepare by combining the two solutions.

IV. QUALITY CONTROL (Weber-Green)

A. Unfortunately, the only way to perform acceptable QC procedures for this method is to use actual microsporidial spores as the control organisms. Obtaining these positive controls can be somewhat difficult. It is particularly important to use the actual organisms because the spores are difficult to stain and the size is very small (1 to 1.5 μm).

B. A QC slide should be included with each run of stained slides, particularly if the staining setup is used infrequently.

C. All staining dishes should be covered to prevent evaporation of reagents (screw-cap Coplin jars or glass lids).

D. Depending on the volume of slides stained, staining solutions will have to be changed on an as-needed basis.

E. When the smear is thoroughly fixed and the stain is performed correctly, the spores will be ovoid and refractile, with the spore wall being bright pinkish red. Occasionally, the polar tubule can be seen either as a stripe or as a diagonal line across the spore. The majority of the bacteria and other debris will tend to stain green or blue (see Ryan-Blue method below). However, there will still be some bacteria and debris that will stain red.

F. The specimen is also checked for adherence to the slide (macroscopically). Can you see stained material (color) on the glass slide?

G. The microscope should have been calibrated within the last 12 months, and the objectives and oculars used for the calibration procedure should be used for all measurements on the microscope. The calibration factors for all objectives should be posted on the microscope for easy access (multiplication factors can be pasted on the body of the microscope). Although recalibration every 12 months may not be necessary, this will vary from laboratory to laboratory, depending on equipment care and use.

H. Known positive microscope slides, Kodachrome 2-inch-by-2-inch projection slides, and photographs (reference books) should be available at the work station.

I. Record all QC results; the laboratory should also have an action plan for "out of control" results.

V. PROCEDURE (Weber-Green)

A. Using a 10-μl aliquot of concentrated (10 min at 500 $\times$ *g*), preserved liquid stool (5 or 10% formalin, SAF, or one of the single vial systems [zinc-based]), prepare the smear by spreading the material over an area of 45 by 25 mm.

B. Allow the smear to air dry.

C. Place the smear in absolute methanol for 5 min.

D. Allow the smear to air dry.

E. Place in trichrome stain for 90 min.

F. Rinse in acid-alcohol for no more than 10 s.

G. Dip slides several times in 95% alcohol. Use this step as a rinse.

H. Place in 95% alcohol for 5 min.

I. Place in 100% alcohol for 10 min.

J. Place in xylene substitute for 10 min.

K. Mount with coverslip (no. 1 thickness), using mounting medium.

L. Examine smears under oil immersion (1,000$\times$) and read at least 100 fields; the examination time will probably be at least 10 min per slide.

VI. RESULTS

A. Microsporidia spores might be seen. The spore wall should stain pinkish to red, with the interior of the spore being clear or perhaps showing a horizontal or diagonal stripe which represents the polar tubule. The background will appear green (Weber stain) or blue (Ryan stain [see below]), depending on the method.

B. Other bacteria, some yeast cells, and some debris will also stain pink to red; the shapes and sizes of the various components can be helpful in differentiating the spores from other structures.

C. The results from this staining procedure should be reported only if the positive control smears are acceptable.

VII. REPORTING RESULTS

A. Report the organism and stage (do not use abbreviations).

Examples (stool specimens): Microsporidia spores present.

Enterocytozoon bieneusi or *Encephalitozoon intestinalis* present (if from fecal specimen); the two organisms cannot be differentiated on the basis of size or morphology.

VII. REPORTING RESULTS (continued)

Example from urine: Microsporidia spores present.
Encephalitozoon intestinalis present (identification to species highly likely); generally this organism is involved in disseminated cases from GI tract to kidneys and will be found in urine.

B. Quantitate the number of spores seen (rare, few, moderate, many).

C. Note and quantitate the presence of human cells.
Examples: Moderate WBCs, many RBCs, few macrophages, rare Charcot-Leyden crystals

D. Save positive slides for future reference. Label before storage (name, patient number, and organisms present).

E. Quantitation of parasites, cells, yeast cells, and artifacts
Few = ≤2 per 10 oil immersion fields (×1,000)
Moderate = 3 to 9 per 10 oil immersion fields (×1,000)
Many = ≥10 per 10 oil immersion fields (×1,000)

VIII. PROCEDURE NOTES

A. It is mandatory that positive control smears be stained and examined each time patient specimens are stained and examined.

B. Because of the difficulty in getting stain penetration through the spore wall, prepare thin smears and do not reduce the staining time in trichrome (unless using the hot method [see below]). Also make sure the slides are not left too long in the decolorizing agent (acid-alcohol). If the control organisms are too light, leave them in the trichrome longer and shorten the time to two dips in the acid-alcohol solution. Also remember that the 95% alcohol rinse after the acid-alcohol should be performed quickly to prevent additional destaining from the acid-alcohol reagent.

C. When you purchase the chromotrope 2R, obtain the highest dye content available. Two sources are Harleco (Gibbstown, N.J.) and Sigma Chemical Co. (St. Louis, Mo.) (dye content among the highest [85%]). Fast green and aniline blue can be obtained from Allied Chemical and Dye (New York, N.Y.).

D. In the final stages of dehydration, the 100% ethanol and the xylenes (or xylene substitutes) should be kept as free from water as possible. Coplin jars must have tight-fitting caps to prevent both evaporation of reagents and absorption of moisture. If the xylene becomes cloudy after addition of slides from 100% alcohol, return the slides to 100% alcohol and replace the xylene with fresh stock.

IX. LIMITATIONS OF THE PROCEDURE

A. Although this staining method will stain the microsporidia, the range of stain intensity and the small size of the spores will cause some difficulty in identifying these organisms. Since this procedure will result in many other organisms or objects staining in stool specimens, differentiation of the microsporidia from surrounding material will still be very difficult. There also tends to be some slight size variation among the spores.

B. If the patient has severe watery diarrhea, there will be less artifact material in the stool to confuse with the microsporidial spores; however, if the stool is semiformed or formed, the amount of artifact material will be much greater; thus, the spores will be much harder to detect and identify. Also, remember that the number of spores will vary based on the stool consistency (the more diarrhetic, the more spores that will be present).

C. Those who developed some of these procedures feel that concentration procedures result in an actual loss of microsporidial spores; thus there is a strong recommendation to use unconcentrated, formalinized stool. However, there are no data indicating what centrifugation speeds, etc., were used in the study.

D. In the UCLA Clinical Microbiology Laboratory, we have generated data (unpublished) to indicate that centrifugation at 500 × *g* for 10 min increases dramatically the number of microsporidial spores available for staining (from the concentrate sediment). This is the same protocol we use for centrifugation of all stool specimens, regardless of the suspected organism.

E. Avoid the use of wet gauze filtration (an old, standardized method of filtering stool before centrifugation), which uses too many layers of gauze which can trap organisms and not allow them to flow into the fluid to be concentrated. No more than two layers of gauze should be used. Another option is the commercially available concentration systems that use metal or plastic screens for filtration.

F. Because the microsporidia spores are so small (1 to 2 μm), it might be very difficult to differentiate among bacteria, small yeasts, and actual spores.

G. The smear should be examined by using the oil immersion lens (100×); use of lower magnification will prevent identification of spores.

H. *Cryptosporidium, Cyclospora,* and *Isospora* spp. will not be visible with this stain; modified acid-fast stains are recommended.

REFERENCES

1. **Garcia, L. S., and D. A. Bruckner.** 1997. *Diagnostic Medical Parasitology,* 3rd ed. ASM Press, Washington, D.C.
2. **NCCLS.** 1993. *Procedures for the Recovery and Identification of Parasites from the Intestinal Tract.* Proposed Guideline M28-P. NCCLS, Wayne, Pa.
3. **Weber, R., R. T. Bryan, R. L. Owen, C. M. Wilcox, L. Gorelkin, G. S. Visvesvara, and The Enteric Opportunistic Infections Working Group.** 1992. Improved light detection of microsporidia spores in stool and duodenal aspirates. *N. Engl. J. Med.* **326:**161–166.

APPENDIX 7.6.2–1

Two additional modified-trichrome staining methods are in use; the order in which they are listed in this procedure does not indicate preference, but the order in which they were developed for staining microsporidial spores. Each laboratory must select the method that works best for them. Some have found that there is be more contrast between the spores and the background with the Ryan-Blue stain.

A. Modified-trichrome stain for the microsporidia (Ryan-Blue) (*see* appendix reference 2 below)

A number of variations to the modified-trichrome staining method (Weber-Green) were tried to improve the contrast between the color of the spores and background staining. Optimal staining was achieved by modifying the composition of the trichrome solution. This stain is also available commercially from a number of suppliers.

The specimen can be fresh stool or stool that has been preserved in 5 or 10% formalin, SAF, or some of the newer single-vial system fixatives. Actually, any specimen other than tissue thought to contain microsporidia could be stained by this method.

1. Trichrome stain (modified for microsporidia; Ryan-Blue [different formula from Weber-Green])
   Acid-alcohol (See Weber-Green)
2. QC for the modified-trichrome staining method (Ryan-Blue)
   Review comments under the Weber-Green staining method above.
3. Procedure for modified-trichrome staining method (Ryan-Blue)
   a. Using a 10-μl aliquot of unconcentrated, preserved liquid stool (5 or 10% formalin or SAF), prepare the smear by spreading the material over an area of 45 by 25 mm.
   b. Allow the smear to air dry.
   c. Place the smear in absolute methanol for 5 or 10 min.
   d. Allow the smear to air dry.
   e. Place in trichrome stain for 90 min.
   f. Rinse in acid-alcohol for ≤10 s.
   g. Dip slides several times in 95% alcohol. Use this step as a rinse (no more than 10 s).

APPENDIX 7.6.2–1 *(continued)*

h. Place in 95% alcohol for 5 min.
i. Place in 100% alcohol for 10 min.
j. Place in xylene substitute for 10 min.
k. Mount with coverslip (no. 1 thickness), using mounting medium.
l. Examine smears under oil immersion (1,000×) and read at least 100 fields; the examination time will probably be at least 10 min per slide.

B. Modified-trichrome stain for the microsporidia (Kokoskin-Hot Method) (*see* appendix reference 1 below)

Changes in temperature from room temperature to 50°C and in the staining time from 90 to 10 min have been recommended as improvements for the modified-trichrome staining methods. The procedure is as follows.

1. With a 10-μl aliquot of unconcentrated, preserved liquid stool (5 or 10% formalin or SAF), prepare the smear by spreading the material over an area of 45 by 25 mm.
2. Allow the smear to air dry.
3. Place the smear in absolute methanol for 5 min.
4. Allow the smear to air dry.
5. Place in trichrome stain (Weber-Green or Ryan-Blue) for 10 min at 50°C.
6. Rinse in acid-alcohol for no more than 10 s.
7. Dip slides several times in 95% alcohol. Use this step as a rinse (no more than 10 s).
8. Place in 95% alcohol for 5 min.
9. Place in 100% alcohol for 10 min.
10. Place in xylene substitute for 10 min.
11. Examine smears under oil immersion (1,000×) and read at least 100 fields; the examination time will probably be at least 10 min per slide.

References

1. **Kokoskin, E., T. W. Gyorkos, A. Camus, L. Cedilotte, T. Purtill, and B. Ward.** 1994. Modified technique for efficient detection of microsporidia. *J. Clin. Microbiol.* **32:** 1074–1075.

2. **Ryan, N. J., G. Sutherland, K. Coughlan, M. Globan, J. Doultree, J. Marshall, R. W. Baird, J. Pedersen, and B. Dwyer.** 1993. A new trichrome-blue stain for detection of microsporidial species in urine, stool, and nasopharyngeal specimens. *J. Clin. Microbiol.* **31:** 3264–3269.

Supplemental Reading

Moura, H., J. L. Nunes Da Silva, F. C. Sodre, P. Brasil, K. Wallmo, S. Wahlquist, S. Wallace, G. P. Croppo, and G. S. Visvesvara. 1996. Gram-chromotrope: a new technique that enhances detection of microsporidial spores in clinical samples. *J. Eukaryot. Microbiol.* **43:**94S–95S.

7.7.1 Introduction

During some stages in their life cycle, species of the genera *Plasmodium* (malaria), *Babesia, Trypanosoma, Leishmania donovani,* and filariae are detectable in human blood. *Plasmodium* and *Babesia* species are found within RBCs. Trypanosomes and microfilariae, the larval stage of filariae, are found outside RBCs. *Leishmania* amastigotes are occasionally found within monocytes. Trypanosomes and microfilariae, which frequently are present in low numbers, exhibit motility in freshly collected blood films, and this can aid in their detection. However, species identifications of all blood parasites are usually made from either or both of two types of stained blood films: a thin film and a thick film. These films can be made from whole or anticoagulated blood or from the sediment of a variety of procedures designed to concentrate trypanosomes and microfilariae in the blood. Although the films are best when stained with Giemsa stain, many infections are detected and diagnosed by using Wright's stain. Delafield's hematoxylin is used to enhance the morphologic features of microfilariae.

Microscopic examination of stained blood films is best accomplished by beginning with a thorough search of both the thin and thick films with low-power magnification for microfilariae. If larvae are found, magnification at a higher power will reveal the finer morphological details necessary to make a definitive identification. Other blood parasites require examination with oil-immersion magnification of both the thin and thick films. Trypanosomes, even those detected in thick films, are more frequently identified in the thicker portion of the thin film. *Plasmodium* and *Babesia* spp., being intracellular parasites, are detected in the thick film but are more frequently identified in the thinner portion of the thin film. Depending on the amount of experience the microscopist has, satisfactory examination usually requires 5 to 10 min for the thick film (about 100 fields) and 15 to 20 min for the thin film (about 200 to 300 fields) at 1,000× (oil immersion) magnification. All species of parasites found in a blood specimen should be reported to the attending physician as soon as possible. Notification of appropriate governmental authorities should be made expeditiously where required by state law.

THIN AND THICK BLOOD FILMS

A. Purpose

To date, stained blood films are the most reliable and efficient means for definitive diagnosis of nearly all blood parasites. They provide a permanent record and can be sent to a reference laboratory for consultation or verification of diagnosis. Ordinarily, when a laboratorian tests for blood parasites, two types of blood films are prepared: a thin film and a thick film. These can be made on the same microscope slide, that is, the thin film on one end of the slide and the thick film on the other, or they can be made on separate slides. When malaria is suspected, the recommended procedure is to prepare a thin film on one slide, a thick film on another, and a combination thin and thick on a third slide. The thin film can be stained within a few minutes and will afford a quick diagnosis of malaria if the patient has a high parasitemia; the thick film can be stained in a few hours and will afford a diagnosis of lighter infection; and the combination film is stained several hours later when the blood has dried for a longer period of time, thus resulting in a better differential stain. The combination film is then used to verify the quick diagnosis and is kept as the permanent record. This three-slide procedure can be used for detecting all blood parasites.

B. Thin blood film

The thin film is identical to a differential blood cell count film. It provides a good area for examining the morphology of parasites and RBCs and is used

THIN AND THICK BLOOD FILMS *(continued)*

to confirm the identity of parasites when one cannot identify them in thick films. Most parasitologists concur that a thin film must be used to differentiate *Plasmodium ovale* from *Plasmodium vivax* and *Babesia* species from the ring forms of *Plasmodium falciparum*. The thin film is also better for identifying *Trypanosoma cruzi,* because these organisms become distorted in thick films. The thin film, however, is less sensitive than the thick film in light parasitemias.

C. Thick blood film

The thick film essentially condenses into an area suitable for examination about 20 times more blood than is examined in the thin film. In this respect, the thick film is a concentration procedure. Here, the RBCs are lysed during the staining process so that only parasites, platelets, and WBCs remain visible. The thick film, then, has two advantages over the thin film. It saves time in examining the blood and increases the chance of detecting light infections. Therefore, the thick film is recommended for the routine *detection* of all parasites where the diagnostic stages occur in blood, and it can be used, in most instances, for identifying to the species level all microfilariae, the African trypanosomes, and all *Plasmodium* spp. except *Plasmodium ovale.*

D. Blood specimen

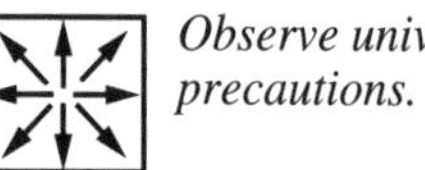

Observe universal (standard) precautions.

Blood for detecting parasites is obtained either by finger puncture or venipuncture. These procedures are described in detail elsewhere (1, 2). Universal precautions must be used in the collecting, handling, and disposing of these blood samples. If blood is obtained by finger puncture, care must be taken not to squeeze tissue juice into the sample and risk diluting a possible light parasitemia to below the level of detection.

E. Use of anticoagulants

Blood samples for malaria are preferably collected without anticoagulants, but if anticoagulants must be used for other testing, films for reliable staining of malaria parasites should be made immediately or within 1 h of collection. Trypanosomes and microfilariae are usually not affected by use of anticoagulants. Although several anticoagulants have been used, EDTA (0.020 g per 10 ml of blood) is recommended by most parasitologists for use with malaria parasites. Heparin (0.002 g per 10 ml of blood) and sodium citrate (0.050 g per 10 ml of blood) are often used for concentration procedures for trypanosomes and microfilariae.

F. Collection guidelines

1. The blood sample must be labeled as to the date and time of collection so that findings can be correlated with symptoms. Some parasites appear more frequently in the blood during certain periods than others. To accommodate this, blood for detecting of parasites is usually collected as follows.
   - **a.** On admission or when first suspected of parasite infection
   - **b.** If no parasites are found in the first sample, blood is collected every 6 to 12 h until a diagnosis is made or infection is no longer suspected (usually 3 to 5 days). *Infection with a blood parasite cannot be ruled out on a single blood sample.*
   - **c.** If either trypanosomes or microfilariae are suspected, each sample should be concentrated by an appropriate technique in an effort to detect a low parasitemia.
   - **d.** *The examination of blood for malaria should be considered a STAT request. Preparation and examination of smears and reporting of results should be performed on a STAT basis.*
2. Blood films for parasites should be made on clean standard 1-by-3-inch glass microscope slides. All slides, even new "precleaned" ones, should be dipped in alcohol and polished with a lint-free towel to remove any grease or dirt before the preparations are made.

3. Instrument methods, either for preparing thin films or staining films, have not been reliable for blood parasites. At this time, manual techniques yield better results.

G. Staining of blood films

1. All blood films, regardless of the stain to be used, should be stained as soon as possible after they are thoroughly dry. Prolonged storage can cause erratic staining.
2. After staining and thorough drying, films can be examined by using oil directly on the film or a number 1-thickness coverslip can be applied to the film with a mounting medium at pH 7.0.

REFERENCES

1. **NCCLS.** 1986. *Procedures for the Collection of Diagnostic Blood Specimens by Skin Puncture,* vol. 6, no. 7. NCCLS, Wayne Pa.
2. **NCCLS.** 1984. *Procedures for the Collection of Diagnostic Blood Specimens by Venipuncture,* vol. 4, no. 5. NCCLS, Wayne, Pa.

7.7.2 Preparation of Thin Blood Films

I. PRINCIPLE

The thin film is prepared like that for a differential blood cell count and provides an area for examination in which the RBCs are neither overlapping nor distorted. Here the morphologies of parasites and infected RBCs are most typical (1,2).

II. SPECIMEN

The specimen usually consists of fresh whole blood collected by finger puncture or whole blood containing EDTA (0.020 g per 10 ml of blood) that is less than 1 h post collection by venipuncture (1). Occasionally a buffy coat (for leishmaniasis) or the sediment from a special concentration procedure (triple centrifugation for trypanosomes) is prepared as a thin film.

III. REAGENTS

Include QC information on reagent container and in QC records.

Absolute methanol (for Giemsa stain)

IV. QUALITY CONTROL

A. Visually, the thin film should be rounded, feathered, and progressively thinner toward the middle of the slide.

B. There should not be any clear areas or smudges in the film itself (indicating that grease or fingerprints were on the glass).

V. PROCEDURE

A. Wear gloves when performing this procedure.

B. For specimen collection by finger puncture, after wiping off the first drop of blood, touch a clean 1-by-3-inch glass microscope slide, about 1/2 inch from the end, to a small drop of blood (10 to 15 μl) standing on the finger, remove the slide from the finger, turn it blood side up, and place it on a horizontal surface. For specimens collected by venipuncture, place a clean 1-by-3-inch glass microscope slide on a horizontal surface. Place a small drop (10 to 15 μl) of specimen onto the center of the side about 1/2 inch from the end.

C. Holding a second clean glass slide at a 40° angle, touch the angled end to the midlength area of the specimen slide.

D. Pull the angled slide back into the blood, and allow the blood to almost fill the end area of the angled slide.

E. Continue contact with the blood under the lower edge, quickly and steadily moving the angled slide toward the opposite end of the specimen slide until the blood is used up.

F. The result will be a thin film that is rounded, feathered, and progressively thinner towards the center of the slide.

G. Label the slide appropriately and allow it to air dry while protected from dust for a least 10 min.

H. If the film will be stained with Giemsa, after the film is completely dry, fix it by dipping the slide into absolute methanol and allow the film to air dry in a vertical position. If the film will be stained with Wright's, it does not need to be fixed because Wright's contains the fixative and stain in one solution.

VI. PROCEDURE NOTES

A. A diamond marking pen is recommended.

B. An indelible ink pen can be used.

C. Pencil can be used if the information is actually written in the thick part of the smear (where the original drop of blood was placed).

D. Do not use wax pencils; the material may fall off during the staining procedure.

E. Make sure the films are protected from dust (while drying).

F. The last few drops of blood remaining in the needle after a venipuncture can also be used to prepare thin blood films. However, remember if you are preparing thick films, the blood has not been in contact with the anticoagulant so you will need to follow directions seen in the protocol for thick blood films (from finger puncture).

VII. LIMITATIONS OF THE PROCEDURE

A. A light infection might be missed in a thin film, whereas the increased volume of blood present on a thick film can allow the detection of the infection, even with a low parasitemia.

B. If the smears are prepared from anticoagulated blood which is over 1 h old, morphology of both parasites and infected RBCs might not be typical.

C. *Plasmodium vivax* and *P. ovale* should be identifiable, even with the absence of Schüffner's dots (stippling).

REFERENCES

1. **Garcia, L. S., and D. A. Bruckner.** 1997. *Diagnostic Medical Parasitology,* 3rd ed. ASM Press, Washington, D.C.
2. **NCCLS.** 1992. *Use of Blood Film Examination for Parasites.* Proposed Guideline M15-T. NCCLS, Wayne, Pa.

7.7.3 Preparation of Thick Blood Films

I. PRINCIPLE

The thick-film method samples more blood than the thin-film method and therefore is more likely to demonstrate a low parasitemia. The RBCs are lysed during staining, making the preparation more or less transparent and leaving only parasites, platelets, and WBCs for examination (1–3).

II. SPECIMEN

The specimen usually consists of fresh whole blood collected by finger puncture or whole blood containing EDTA (0.02 g per 10 ml of blood) that is less than 1 h post collection by venipuncture. Heparin (2 mg per 10 ml of blood) or sodium citrate (0.05 g per 10 ml of blood) can be used as an anticoagulant if trypanosomes or microfilariae are suspected. The sediment from a concentration procedure for trypanosomes or microfilariae is frequently prepared as a thick film that is stained, examined, and kept as a permanent record.

III. QUALITY CONTROL

A. The thick smear should appear round to oval, approximately 2.0 cm across.

B. One should be able to barely read newsprint through the wet or dry film.

C. The film itself should have no clear areas or smudges (indicating that grease or fingerprints were on the glass).

IV. PROCEDURE

A. Wear gloves when performing this procedure.

B. For specimen collection by *finger puncture*, after wiping off the first drop of blood, touch a clean 1-by-3-inch glass microscope slide to a large drop of blood standing on the finger, and rotate the slide on the finger until the circle of blood is nearly the size of a dime or nickel (1.8 to 2.0 cm). Without breaking contact with the blood, rotate the slide back to the center of the circle. Remove the slide from the finger, quickly turn it blood side up, and place it on a horizontal surface. The blood should spread out evenly over the surface of the circle and be of the thickness that fine print (''newsprint'' size) can be just barely read through it. If not, take the corner of a second clean slide or an applicator stick and expand the size of the circle until the print is just readable. The final thickness of the film is important. If too thick, it might flake off while drying or wash off while staining. If too thin, the amount of blood available for examination is insufficient to detect a low parasitemia.

For specimen collection by *venipuncture*, place a clean 1-by-3-inch glass microscope slide on a horizontal surface. Place a drop (30 to 40 μl) of blood onto the center of the slide about 1/2 inch from the end. Using either the corner of another clean glass slide or an applicator stick, spread the blood into a circle about the size of a dime or nickel (1.8 to 2.0 cm). Immediately place the thick film over some small print, and be sure that the print can be just barely read through it. If not, expand the size of the film until the print can be read.

Three or four small drops of blood can be used in place of the larger drop, and the small ones can be pooled together into a thick film by using the corner of a clean slide or an applicator stick. Be sure that small print can be read through it.

C. Allow the film to air dry in a horizontal position and protected from dust for several hours (6 to 8 h) or overnight. Do not attempt to speed the drying process by applying any type of heat, because heat will fix the RBCs and they subsequently will not lyse in the staining process.

D. *Do not fix the thick film.* If thin and thick films are made on the same slide, do not allow the methanol or its vapors to contact the thick film when fixing the thin film.

E. Label the slide appropriately.

F. If staining with Giemsa will be delayed for more than 3 days or the film will be stained with Wright's, lyse the RBCs in the thick film by placing it in buffered water, pH 7.0 to 7.2, for 10 min, then removing it from the water and placing it in a vertical position (thick film down) to air dry.

V. PROCEDURE NOTES

A. A diamond marking pen is recommended.

B. An indelible ink pen can be used.

C. Do not use wax pencils; the material might fall off during the staining procedure.

D. Make sure the films are protected from dust (while drying).

E. The last few drops of blood remaining in the needle after a venipuncture can also be used to prepare thick blood films. However, remember when you are preparing thick films, the blood has not been in contact with the anticoagulant so you will need to follow directions given in the protocol for thick blood films (from finger puncture).

VI. LIMITATIONS OF THE PROCEDURE

A. If the smears are prepared from anticoagulated blood that is over 1 h old, morphology of the parasites might not be typical, and the film might wash off the slide during the staining procedure.

B. Identification to species, particularly between *P. ovale* and *P. vivax* and between the ring forms of *P. falciparum* and *Babesia* spp., might be impossible without examining the stained thin blood film. Also, *T. cruzi* trypomastigotes are frequently distorted in thick films.

C. Excess stain deposition on the film can be confusing and make the detection of organisms difficult.

REFERENCES

1. **Garcia, L. S., and D. A. Bruckner.** 1997. *Diagnostic Medical Parasitology,* 3rd ed. ASM Press, Washington, D.C.
2. **NCCLS.** 1992. *Use of Blood Film Examination for Parasites.* Proposed Guideline M15-T. NCCLS, Wayne, Pa.
3. **Wilcox, A.** 1960. *Manual for the Microscopical Diagnosis of Malaria in Man.* U.S. Public Health Service publication no. 796. U.S. Government Printing Office, Washington, D.C. (Out of print.)

SUPPLEMENTAL READING

Ash, L. R., and T. C. Orihel. 1987. *Parasites: a Guide to Laboratory Procedures and Identification.* American Society of Clinical Pathologists, Chicago.

7.7.4 Giemsa Stain

I. PRINCIPLE

Giemsa stain is used to differentiate nuclear and/or cytoplasmic morphology of platelets RBCs, WBCs, and parasites (1–5). The most dependable stain for blood parasites, particularly in thick films, is obtained with Giemsa stain containing azure B. Liquid stock is available commercially or can be made from dry stain powder. Either must be diluted for use with water buffered to pH 6.8 or 7.0 to 7.2, depending on the specific technique used. Either should be tested for proper staining reaction before use. The stock is stable for years, but must be protected from moisture because the staining reaction is oxidative. Therefore, the oxygen in water will initiate the reaction and ruin the stock stain. The aqueous working dilution of stain is good only for one day.

Although not essential, the addition of Triton X-100, a nonionic surface-active agent, to the buffered water used to dilute the stain enhances the staining properties of Giemsa (2) and helps to eliminate possible transfer of parasites from one slide to another (1). For routine staining of thin films and combination thin and thick films, a 0.01% (vol/vol) final concentration of Triton is best. For staining thick films for microfilariae, use a 0.1% (vol/vol) concentration.

II. SPECIMEN

The specimen can consist of a thin blood film that has been fixed in absolute methanol and allowed to dry, a thick blood film that has been allowed to dry thoroughly and is not fixed, or a combination of a fixed thin film and an adequately dried thick film (not fixed) on the same slide.

III. REAGENTS

Include QC information on reagent container and in QC records.

A. Stock Giemsa stain

B. Stock solution of Triton X-100 (10% aqueous solution)

C. Buffered water (for diluting stain and washing films), **pH 7.0 to 7.2**

D. Buffered water, pH 6.8 (called for by some commercial stains for diluting stain and washing films)

E. Triton-buffered water solutions (optional)

1. For thin blood films or combination thin and thick blood films, *after* determining the pH of the buffered water, add 1 ml of the stock 10% aqueous dilution of Triton X-100 to 1 liter of buffered water (pH 7.0 to 7.2, 0.01% final concentration). Label appropriately and store in tightly stoppered bottle. The solution can be used as long as the pH is within limits listed for the procedure.
2. For thick blood films, *after* determining the pH of the buffered water, add 10 ml of the stock 10% aqueous dilution of Triton X-100 to 1 liter of buffered water (pH 7.0 to 7.2, 0.1% final concentration). Label appropriately and store in a tightly stoppered bottle. The solution can be used as long as the pH is within limits listed for the procedure.

F. Methyl alcohol, absolute

IV. QUALITY CONTROL

A. The stock buffer solutions and buffered water should appear clear, without any visible contamination.

B. Check the Giemsa stain reagents, including the pH of the buffered water, before each use. If Triton X-100 has been added to the buffered water, do not use a colorimetric method to determine the pH because Triton X-100 interferes with the color indicators. Use a pH meter to test buffered water that contains Triton

X-100. The buffered water is usable as long as the pH is within the limits listed for the procedure. (*See* CMPH 12.16 for pH meter.)

C. Prepare and stain films from ''normal'' blood, and microscopically evaluate the staining reactions of the RBCs, platelets, and WBCs.

1. Macroscopically, blood films appear purplish. If blue, the buffered water was too alkaline; if pink to red, the buffered water was too acid.
2. Microscopically, RBCs appear pinkish gray, platelets appear deep pink, and WBCs have purple-blue nuclei and lighter cytoplasm. Eosinophilic granules are bright purple-red, and neutrophilic granules are purple. Basophilic stippling within uninfected RBCs is blue.
3. Slight variations may appear in the colors described above, depending on the batch of stain used and the character of the blood itself, but if the various morphological structures are distinct, the stain is satisfactory.

D. The microscope should have been calibrated within the last 12 months (if moved frequently or in constant use), and the objectives and oculars used for the calibration procedure should be used for all measurements on the microscope. The calibration factors for all objectives should be posted on the microscope for easy access (multiplication factors can be pasted right on the body of the microscope).

E. Record results on appropriate sheets.

V. PROCEDURE

A. Thin blood films (only)

1. Fix air-dried film in absolute methyl alcohol by dipping the film briefly (two dips) in a Coplin jar containing methyl alcohol.
2. Remove and let air dry.
3. Stain with diluted Giemsa stain (1:20 [vol/vol]) for 20 min. (Add 2 ml of stock Giemsa to 40 ml of buffered water containing 0.01% Triton X-100 in a Coplin jar.)
4. Wash by briefly dipping the slide in and out of a Coplin jar of buffered water (one to two dips).

   ◪ **NOTE:** Excessive washing will decolorize the film.
5. Let air dry in a vertical position.

B. Thick blood films (only)

1. Allow film to air dry thoroughly for several hours or overnight. Do not dry films in an incubator or by heat because this will fix the blood and interfere with the lysing of the RBCs.

   ◪ **NOTE:** If a rapid diagnosis of malaria is needed, thick films can be made slightly thinner than usual, allowed to dry for 1 h, and then stained.
2. *Do not fix.*
3. Stain with diluted Giemsa stain (1:50 [vol/vol]) for 50 min. Add 1 ml of stock Giemsa to 50 ml of buffered water containing 0.01% Triton X-100 (if staining microfilariae, use 0.1% Triton X-100) in a Coplin jar.
4. Wash by placing film in buffered water for 3 to 5 min.
5. Let air dry in a vertical position.

C. Combination thin and thick blood films

1. Allow the thick film to air dry thoroughly.
2. Fix the thin film by placing only the thin film in methyl alcohol. Be sure not to get the alcohol or its fumes on the thick film (remember, only two dips).
3. Let air dry in a vertical position with the thick film up. Be sure slide is thoroughly dry before staining. Introducing even a minute amount of methyl alcohol into the stain dilution will interfere with the lysing of RBCs in the thick films.

V. PROCEDURE *(continued)*

4. Stain with diluted Giemsa stain (1:50 [vol/vol]) for 50 min. Place the slide in the stain, *thick film down,* to prevent the debris caused by dehemoglobinization from falling onto the thin film. (Add 1 ml of stock Giemsa to 50 ml of buffered water containing 0.01% Triton X-100 in a Coplin jar.)
5. Rinse the thin film by briefly dipping the film in and out of a Coplin jar of buffered water (one to two dips). Wash the thick film for 3 to 5 min. Be sure that the thick film is immersed but *do not allow the water to cover any part of the thin film.*
6. Let dry in a vertical position with the *thick film down.*

VI. RESULTS

A. If *Plasmodium* spp. are present, the cytoplasm of the organisms stains blue, and the nuclear material stains red to purple-red.

B. Schüffner's stippling and other inclusions in RBCs infected by *Plasmodium* spp. stain red.

C. Nuclear and cytoplasmic colors that are seen in the malarial parasites will also apply to the trypanosomes and any intracellular leishmaniae that might be present.

D. The sheath of microfilariae might or might not stain with Giemsa, but the body will usually appear blue to purple.

VII. REPORTING RESULTS

A. Any parasite, including the stage(s) seen, should be reported (do not use abbreviations).

Examples: *Plasmodium falciparum* rings and gametocytes, or rings only
Plasmodium vivax rings, trophozoites, schizonts, and gametocytes
Wuchereria bancrofti microfilariae
Trypanosoma brucei gambiense/rhodesiense trypomastigotes
Trypanosoma cruzi trypomastigotes
Leishmania donovani amastigotes

B. Any laboratory providing malaria diagnoses should be able to identify *Plasmodium vivax* and *P. ovale,* even in the absence of Schüffner's stippling.

VIII. PROCEDURE NOTES

A. It is important that if blood films must be prepared from venipuncture blood (use of anticoagulant), they be prepared within 1 h of collection. Otherwise, certain morphologic characteristics of both parasites and infected RBCs might be atypical. Also, thick blood films might wash off the slide during the staining procedure.

B. The correct pH of all buffered water and staining solutions is also important. Solutions with an incorrect pH will prevent certain morphological characteristics from being visible (stippling) and will not give typical nuclear and cytoplasmic colors on the stained film.

C. A QC slide should be stained each time patient blood films are stained. If several patient specimens are stained on the same day (with the same reagents), only one control slide needs to be stained and examined.

IX. LIMITATIONS OF THE PROCEDURE

A. *Finding no parasites in one set of blood films does not rule out a parasitic infection with* Plasmodium *spp.*

B. A minimum of 300 oil immersion (×1,000) fields should be examined before reporting "no parasites found."

C. The entire smear should be examined under low power (100×) for the presence of microfilariae. Remember that the sheath might not be visible (*W. bancrofti*).

REFERENCES

1. **Brooke, M. M., and A. W. Donaldson.** 1950. Use of a surface-active agent to prevent transfer of malarial parasites between blood films during mass staining procedures. *J. Parasitol.* **36:** 84.
2. **Garcia, L. S., and D. A. Bruckner.** 1997. *Diagnostic Medical Parasitology,* 3rd ed. ASM Press, Washington, D.C.
3. **Melvin, D. M., and M. M. Brooke.** 1955. Triton X-100 in Giemsa staining of blood parasites. *Stain Technol.* **30:**269–275.
4. **NCCLS.** 1992. *Use of Blood Film Examination for Parasites.* Proposed Guideline M15-T. NCCLS, Wayne, Pa.
5. **Wilcox, A.** 1960. *Manual for the Microscopical Diagnosis of Malaria in Man.* U.S. Public Health Service publication no. 796. U.S. Government Printing Office, Washington, D.C. (Out of print.)

SUPPLEMENTAL READING

Ash, L. R., and T. C. Orihel. 1987. *Parasites: A Guide to Laboratory Procedures and Identification.* American Society of Clinical Pathologists, Chicago.

Beaver, P. C., R. C. Jung, and E. W. Cupp. 1984. *Clinical Parasitology,* 9th ed. Lea and Febiger, Philadelphia.

7.7.5 Wright's Stain

I. PRINCIPLE

Wright's stain can be used to stain thin blood films for detecting blood parasites, but it is inferior to Giemsa for staining thick films. The liquid, ready-to-use stain is available commercially, or the stain can be made from dry stain powder and is ready for use in about a week. The staining reaction is somewhat similar to that of Giemsa and is achieved by using buffered water with a pH of 6.8. The stain contains the fixative and stain in one solution (1, 2).

II. SPECIMEN

A. The specimen usually consists of a dry, unfixed thin blood film.

B. Thick blood films may be stained with Wright's stain if the RBCs are lysed before staining.

III. MATERIALS

Include QC information on reagent container and in QC records.

A. Reagents

1. Wright's stain

Wright's stain .... 0.9 g
methyl alcohol, absolute
(acetone free) .... 500.0 ml

a. Grind 0.9 g of stain powder with a portion (10 to 15 ml) of methyl alcohol (anhydrous, acetone free) in a clean mortar. Gradually add methyl alcohol while grinding. As the dye is dissolved in the alcohol, pour that solution off and add more alcohol to the mortar. Repeat the process until the alcohol is used up.

b. Store the stain solution in a tightly stoppered glass bottle (1 liter) at room temperature.

c. Shake the bottle several times daily for at least 5 days.

d. Before use, filter through Whatman no. 1 paper into a brown bottle.

e. Label appropriately and store, protected from light, at room temperature. The shelf life is 36 months, provided that QC criteria are met.

2. Stock buffers (for preparing buffered water)

a. Alkaline buffer—disodium hydrogen phosphate (dibasic), 0.067 M solution

Na_2HPO_4 (anhydrous) .. 9.5 g
distilled water, to
make .... 1.0 liter

In a 1-liter volumetric flask, dissolve Na_2HPO_4 in about three-fourths of the water. Add water to make 1 liter of solution. Store in tightly stoppered bottle and label appropriately. The shelf life is 24 months.

b. Acid buffer, sodium dihydrogen phosphate (monobasic) 0.067 M solution

$NaH_2PO_4 \cdot H_2O$ .... 9.2 g
distilled water, to
make .... 1.0 liter

In a 1-liter volumetric flask, dissolve $NaH_2PO_4 \cdot H_2O$ in about three-fourths of the water. Add water to make 1 liter of solution. Store in tightly stoppered bottle and label appropriately. The shelf life is 24 months.

3. Buffered water, pH 6.8 (called for by some commercial stains for diluting stain and washing films)
   alkaline buffer
   (Na_2HPO_4).................... 50 ml
   acid buffer
   ($NaH_2PO_4 \cdot H_2O$)......... 50 ml
   distilled water................. 900 ml
   a. Combine above liquids in a 1-liter bottle.
   b. Mix thoroughly and test pH of solution. If pH is not 6.8 ± 0.1, discard and remake. The solution can be used as long as the pH is within limits listed for the procedure.

B. Supplies

1. Coplin jars, 50 ml, for dehemoglobinizing thick films (2)
2. Mortar and pestle
3. Flask, volumetric, 1 liter
4. Bottle, brown, 500 ml
5. Bottle, tight stopper, 1 liter
6. Staining rack
7. Three or four pipettes, disposable
8. Three or four pipette bulbs
9. Gauze for wiping slide (per slide)
10. Filter paper, Whatman no. 1
11. Funnel (glass) to hold filter paper

C. Equipment

1. Microscope, binocular with mechanical stage; low (10×), high dry (40×), and oil immersion (100×) objectives; 10× oculars; a calibrated ocular micrometer; a light source equivalent to a 20-W halogen or 100-W tungsten bulb; and blue and white ground-glass diffuser filters
2. Timer, 1 h or more, in 1-min increments
3. pH meter

IV. QUALITY CONTROL

A. The stock buffer solutions and buffered water should appear clear, without any visible contamination.

B. Check the Wright's stain reagents, including the pH of the buffered water, before each use for diagnosis of blood parasites. The buffered water is usable as long as the pH is within the limits listed for the procedure. (*See* CMPH 12.16 for pH meter.)

C. Prepare and stain films from "normal" blood, and microscopically evaluate the staining reactions of the RBCs, platelets, and WBCs.

1. Macroscopically, blood films appear pinkish purple. If blue, the buffered water was too alkaline; if pink to red, the buffered water was too acid.
2. Microscopically, the RBCs appear tan to pinkish-red, the platelets appear deep pinkish red, and the WBCs have bright blue nuclei and lighter cytoplasm.
3. Slight variations might appear in the colors described above, depending on the batch of stain used and the character of the blood itself, but if the various morphological structures are distinct, the stain is satisfactory.

D. The microscope should have been calibrated within the last 12 months (if moved frequently or in constant use), and the objectives and oculars used for the calibration procedure should be used for all measurements on the microscope. The calibration factors for all objectives should be posted on the microscope for easy access (multiplication factors can be pasted right on the body of the microscope).

E. Record results on appropriate sheets.

V. PROCEDURE

A. Thin blood films (only) (1)

1. Place air-dried films on a level staining rack.
2. Use a pipette to cover the surface of the slide with stain, adding stain drop by drop. Count the number of drops needed to cover the surface. Let stand 1 to 3 min (optimal staining time [color range and intensity] will vary with each batch of stain).
3. Add the same number of drops of buffered water, as were used for step 2 above, to the slide, and mix the stain and water by blowing on the surface of the fluid.

V. PROCEDURE *(continued)*

4. After 4 to 8 min, flood the stain from the slide with buffered water. Do not pour the stain off before flooding, or a precipitate will be deposited on the slide.
5. Wipe the underside of the slide to remove excess stain.
6. Let air dry in a vertical position.

B. Combination thin and thick blood films

1. Lyse the RBCs in the thick film by immersing it for 10 min in buffered water. *Be sure that the water does not touch the unfixed thin film.*
2. Remove the slide, and rinse the thick film by dipping in additional buffered water (two to three dips).
3. Let film air dry thoroughly.
4. Stain both thin and thick with Wright's stain as directed for thin films.

VI. RESULTS

A. If *Plasmodium* spp. are present, the cytoplasm of the organisms stains pale blue, and the nuclear material stains red.

B. Schüffner's stippling in the RBCs infected by malaria species usually does not stain or stains very pale red with Wright's.

C. Nuclear and cytoplasmic colors that are seen in the malarial parasites will also apply to the trypanosomes and any intracellular leishmaniae that might be present.

D. The sheath of microfilariae might or might not stain with Wright's, while the body will usually appear pale to dark blue.

VII. REPORTING RESULTS

A. Any parasite, including the stages found, should be reported (do not use abbreviations).

Examples: *Plasmodium falciparum* rings and gametocytes, or rings only
Plasmodium vivax rings, trophozoites, schizonts, and gametocytes
Wuchereria bancrofti microfilariae
Trypanosoma brucei gambiense/rhodesiense trypomastigotes
Trypanosoma cruzi trypomastigotes
Leishmania donovani amastigotes

B. Any laboratory providing malaria diagnoses should be able to identify *Plasmodium vivax* and *P. ovale,* even in the absence of Schüffner's stippling.

VIII. PROCEDURE NOTES

A. It is important that if blood films must be prepared from venipuncture blood (use of anticoagulant), they be prepared within 1 h of collection. Otherwise, certain morphologic characteristics of both parasites and infected RBCs can be atypical.

B. The correct pH of all buffered water and staining solutions is also important. Solutions with the incorrect pH will prevent certain morphological characteristics from being visible (stippling) and will not give typical nuclear and cytoplasmic colors on the stained film.

C. A QC slide should be stained each time patient blood films are stained. If several patient specimens are stained on the same day (with the same reagents), only one control slide needs to be stained and examined.

IX. LIMITATIONS OF THE PROCEDURE

A. *Finding no parasites in one set of blood films does not rule out a parasitic infection with* Plasmodium *spp.*

B. A minimum of 300 oil immersion (1,000×) fields should be examined before reporting "no parasites found."

C. The entire smear should be examined under low power (100×) for the presence of microfilariae. Remember that the sheath may not be visible (*W. bancrofti*).

REFERENCES

1. **Garcia, L. S., and D. A. Bruckner.** 1997. *Diagnostic Medical Parasitology,* 3rd. ed. ASM Press, Washington, D.C.
2. **Smith, J. W., and M. S. Bartlett.** 1991. Diagnostic parasitology: introduction and methods, p. 701–716. *In* A. Balows, W. J. Hausler, Jr., K. L. Herrmann, H. D. Isenberg, and H. J. Shadomy (ed.), *Manual of Clinical Microbiology,* 5th ed. American Society for Microbiology, Washington, D.C.

SUPPLEMENTAL READING

Beaver, P. C., R. C. Jung, and E. W. Cupp. 1984. *Clinical Parasitology,* 9th ed. Lea and Febiger, Philadelphia.

Parasitology

APPENDIX 7–1 Flow Charts

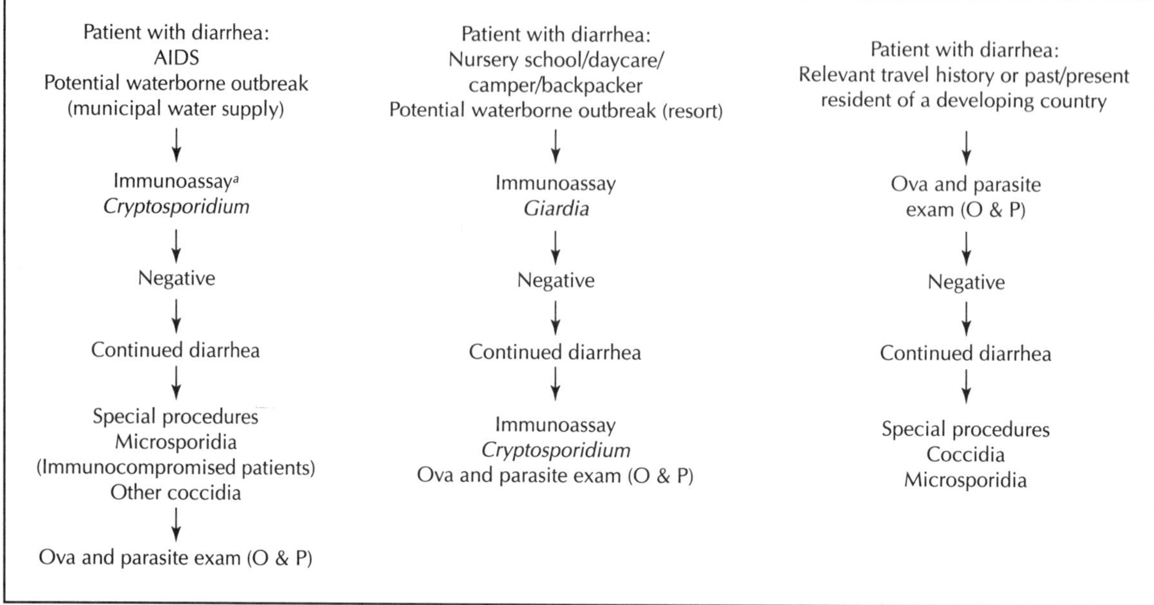

Figure A7–1–1 Ordering algorithm for the laboratory examination for intestinal parasites. [a]EIA, FA.

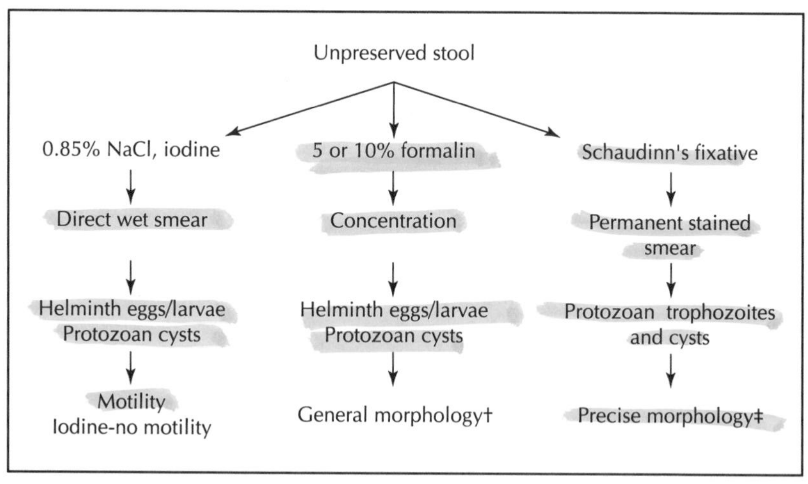

Figure A7–1–2 Procedure for processing fresh stool for the ova and parasite examination. Special stains are necessary for *Cryptosporidium* and *Cyclospora* spp. (modified acid-fast stain) and the microsporidia (modified-trichrome stain; Calcofluor). Immunoassay kits are now available for some of these organisms. If the permanent staining method (iron hematoxylin) contains a carbol fuchsin step, the coccidia will stain pink. Symbols: †, some protozoa might not be identified by using the wet examination only; ‡, protozoa (primarily trophozoites) can be identified and cysts confirmed.

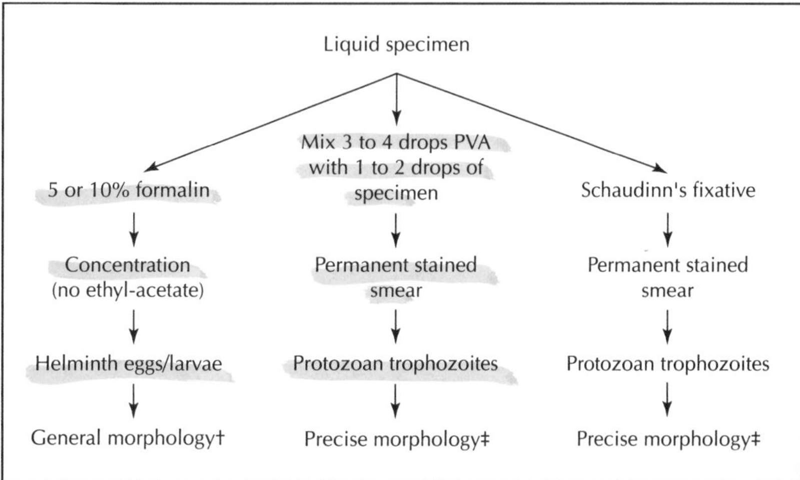

Figure A7–1–3 Procedure for processing liquid specimens for the ova and parasite examination. PVA and specimen will be mixed together on the slide, allowed to air dry, and then stained (fixation sufficient for liquid specimen, but not formed stool). Symbols: †, some protozoa may not be identified from the concentration procedure; ‡, protozoa (trophozoites) can be identified.

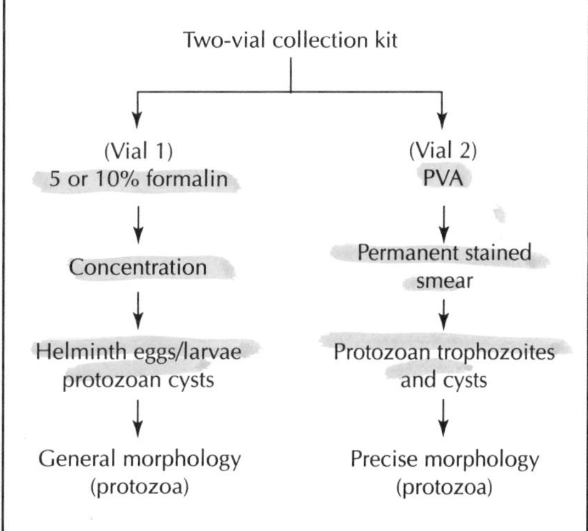

Figure A7–1–4 Procedure for processing preserved stool for the ova and parasite examination. Bases and effects: mercuric chloride, best PVA (trichrome, iron hematoxylin); zinc, current best substitute (trichrome, iron hematoxylin probably okay); copper sulfate, fair substitute (trichrome, iron hematoxylin—both fair to poor); SAF, good substitute for PVA fixative (iron hematoxylin is best, and trichrome is okay).

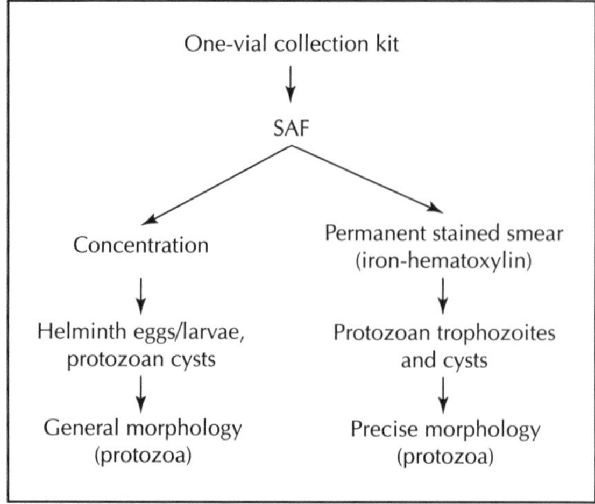

Figure A7–1–5 Use of SAF-preserved stool for the ova and parasite examination. SAF can also be used with EIA and fluorescent-antibody immunoassay kits and the modified trichrome stain for microsporidia.

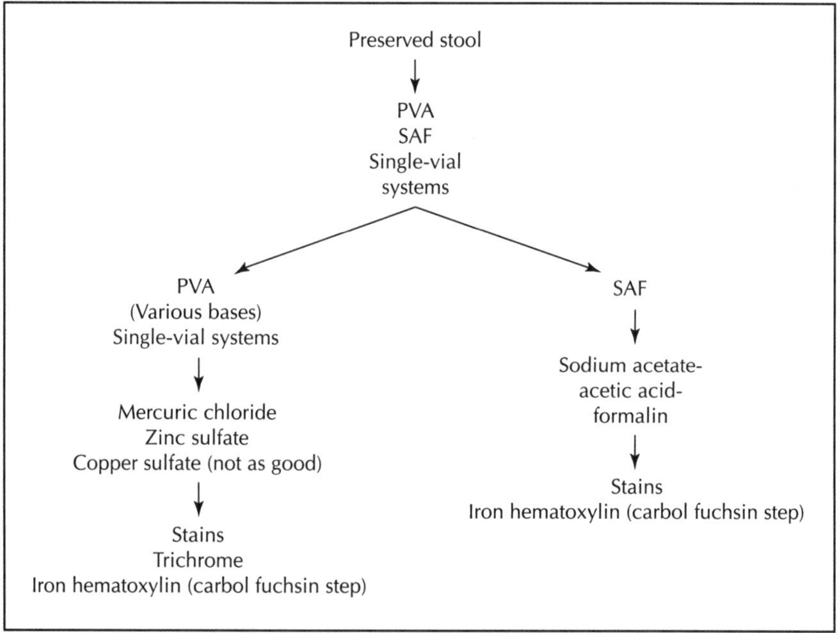

Figure A7–1–6 Use of various fixatives and their recommended stains. Formalin (5 or 10%), SAF, and *some* of the single-vial systems can also be used with EIA and fluorescent-antibody monoclonal kits and the modified trichrome stains for microsporidia.

APPENDIX 7–2 Tables

Table A7–2–1 O & P examination options: pros and cons[a] (1)

Testing option[b]	Pros	Cons
Stools from patients who have been in-house (inpatients) for at least 3 days should not be submitted for routine O & P examinations without consulting the laboratory.	Normally there are very few parasitic infections that can be acquired while an inpatient. Although rare, one of these infections is cryptosporidiosis, which has been documented in nosocomial transmission.	Etiologic agents other than parasites, as well as other causes such as therapy, etc. (noninfectious agents or reasons), are more likely to cause diarrhea in patients who have been hospitalized for 3 days or more. There are always exceptions, but this is a reasonable guideline to follow.
Three separate stools collected within 10 days; concentration and permanent stained smear on all specimens If specimen is fresh (unpreserved), direct wet smear can be performed (soft, liquid specimens) for the detection of organism motility.	Maximize recovery of all organisms, including those in rare numbers; concentration for recovery of protozoan cysts, helminth eggs and larvae; permanent stain for confirmation of protozoa and finding protozoa missed on concentration wet examination. Special stains or immunoassay reagents may be used for the recovery of coccidia and microsporidia.	Although this approach is the most labor intensive, the results are the most complete.[c]
Three separate stools collected within 10 days; concentration on pooled specimen, permanent stained smear on all specimens (performed as three separate examinations)	By pooling three specimens and performing one concentration and examination, time is saved. Since intestinal protozoa tend to be the most common parasites found in many areas of the United States, it is important to perform permanent stained smears on the three individual specimens.	If helminth eggs or larvae are present in rare numbers, they might be missed. Protozoa that could be missed in low numbers will probably be identified on the individual permanent stained smears.
One stool; concentration and permanent stained smear; additional specimens may be examined after results of the first specimen are reported.	Saves time.	Organisms in low numbers might be missed; the collection of specimens number 2 or 3 might be outside the 10-day recommended time frame. The series of three specimens now becomes three series of a single specimen (depending on collection time frames).
One stool; *Giardia*/*Cryptosporidium* screen	This can be ordered as a screen after complete discussion with physicians so they recognize what the results (negative) mean in terms of limitations; if negative with a patient who continues to be symptomatic, O & P examinations can be ordered. In areas where *Giardia* spp. are commonly found or in a suspect waterborne outbreak (*Cryptosporidium* or other organisms), this is an acceptable option.	If complete information on the relevance of a negative report is not given to the physician, this approach can be used incorrectly. It is very important that immunoassay procedures be thoroughly discussed regarding the clinical relevance of results and the specificity and sensitivity of the methods.
One stool; concentration performed; permanent stained smear performed *only* if something suspicious is seen on the concentration examination.	This is thought to be a time saver; however, this approach is not acceptable, neither technically nor in terms of quality patient care.	Numerous studies have confirmed the importance of the permanent stain, not only in confirming the identification of intestinal protozoa, *but in finding those organisms that were missed in the concentration examination.* This technical approach can lead to reported results that are very misleading and should be discouraged. If the laboratory is using this approach, the parasitology testing should be sent to another laboratory.

(continued)

Table A7–2–1 O & P examination options: pros and cons[a] *(continued)*

Testing option[b]	Pros	Cons
Examination approach: *Direct wet smear:* Unpreserved, soft or liquid stools; total 22-by-22-mm coverslip with 10× objective (low power); 1/4 to 1/3 area covered by using 40× objective (high dry power). *Concentration:* Total 22-by-22-mm coverslip with 10× objective (low power); 1/4 to 1/3 area covered by using 40× objective (high dry power). *Permanent stained smear:* minimum of 300 fields with 100× objective (oil immersion). A 50 or 60× oil immersion lens can also be used in the screening process; however, some small organisms will be missed without the use of the 100× oil immersion lens.	This represents the most complete O & P approach; however, the important point is to understand the pros and cons of any modifications to this approach. The most common change is the transition from fresh, unpreserved specimens being submitted to the laboratory to fecal specimens that are transported in fixatives. Once received by the laboratory, the direct wet smear is eliminated from the standard O & P procedure (no motility visible in preserved specimens) and only the concentration and permanent stained smear will be performed.	Any modification(s) from the recommended procedures *must* be discussed thoroughly with your physician clients. It is critical that they understand the advantages and disadvantages of each approach. Information can be transmitted through various communication channels: telephone, newsletters, information updates, in-service presentations, verbal and written reminders, information via computer system, etc.

[a] NOTE: It is mandatory that physicians know exactly what procedures are performed for the diagnosis of parasitic infections in your specific laboratory and how the results should be interpreted in terms of clinical relevance. It is mandatory that they be given information regarding the pros and cons (limitations) of any diagnostic approach. The complete O & P examination includes direct wet smear, concentration, and permanent stained smear.

[b] A direct wet smear should be examined on fresh specimens (soft or liquid) for motile trophozoites. If the specimen is formed or is received in preservative, eliminate the direct wet smear examination and proceed directly to the concentration and permanent stained smear.

[c] In a previous work, the sensitivity of one stool examination was compared with that of three examinations in a group of symptomatic patients. The additional examinations increased the percent positives as follows: 22.7% increase, *Entamoeba histolytica;* 11.3% increase, *Giardia lamblia;* and 31.1% increase, *Dientamoeba fragilis.* Even in symptomatic patients, these data indicate that a single stool specimen examination will miss a large number of infections with pathogenic protozoa (2).

References

1. **Garcia, L. C., and D. A. Bruckner.** 1997. *Diagnostic Medical Parasitology,* 3rd ed. ASM Press, Washington, D.C.
2. **Hiatt, R. A., E. K. Markell, and E. Ng.** 1995. How many stool specimens are necessary to detect pathogenic intestinal protozoa? *Am. J. Trop. Med. Hyg.* **53:**36–39.

Table A7–2–2 Diagnostic laboratory report information that should be relayed to the physician (from reference 1 above)

Organisms/other	Genus/species	Life cycle stages	Quantitation[b]	Pathogenic/non-pathogenic listed	Viable/nonviable
Protozoa					
Intestinal	Yes	Yes (C, T)	No	Both	NA
Blastocystis hominis		NA	Yes	NA	NA
Tissue	Yes	NA	No	NA	NA
Blood	Yes	Yes (G)	Yes	NA	NA
GU	Yes	NA	No	NA	NA
Trematodes					
Intestinal	Yes	NA	Yes	NA	NA
Liver	Yes	NA	Yes	NA	NA
Lung	Yes	NA	Yes	NA	NA
Blood	Yes	NA	Yes	NA	Egg viability
Cestodes					
Intestinal	Yes	NA	NA	NA	NA
Tissue	Yes	NA	NA	NA	NA
Nematodes					
Intestinal	Yes	NA	Rarely[c]	NA	NA
Filarial	Yes	NA	Yes	NA	NA

Table A7–2–2 Diagnostic laboratory report information that should be relayed to the physician[a] *(continued)*

Organisms/other	Genus/species	Life cycle stages	Quantitation[b]	Pathogenic/non-pathogenic listed	Viable/nonviable
Nonparasitic					
Yeast cells	NA	Yes[d]	Yes	NA	NA
Human cells (WBCs, RBCs)	NA	NA	Yes	NA	NA
Crystals (C-L, barium)	NA	NA	Yes	NA	NA

[a] Abbreviations: C, cyst; T, trophozoite; G, gametocyte; GU, genitourinary; C-L, Charcot-Leyden crystals.

[b] Could require numbers of infected cells, eggs, etc. (either as specific count or rare, few, moderate, many).

[c] *Trichuris trichiura:* if eggs are rare or few, the infection might not be treated.

[d] Specimen must be fresh and examined immediately or freshly preserved in a stool collection kit *by the patient* (no lag time between stool passage and preservation); if these requirements are not met, the report (presence of budding yeast and/or pseudohyphae/hyphae) could be misleading and should not be reported to the physician. Report budding yeast cells and/or the presence of pseudohyphae/hyphae.

APPENDIX 7–3 Figures

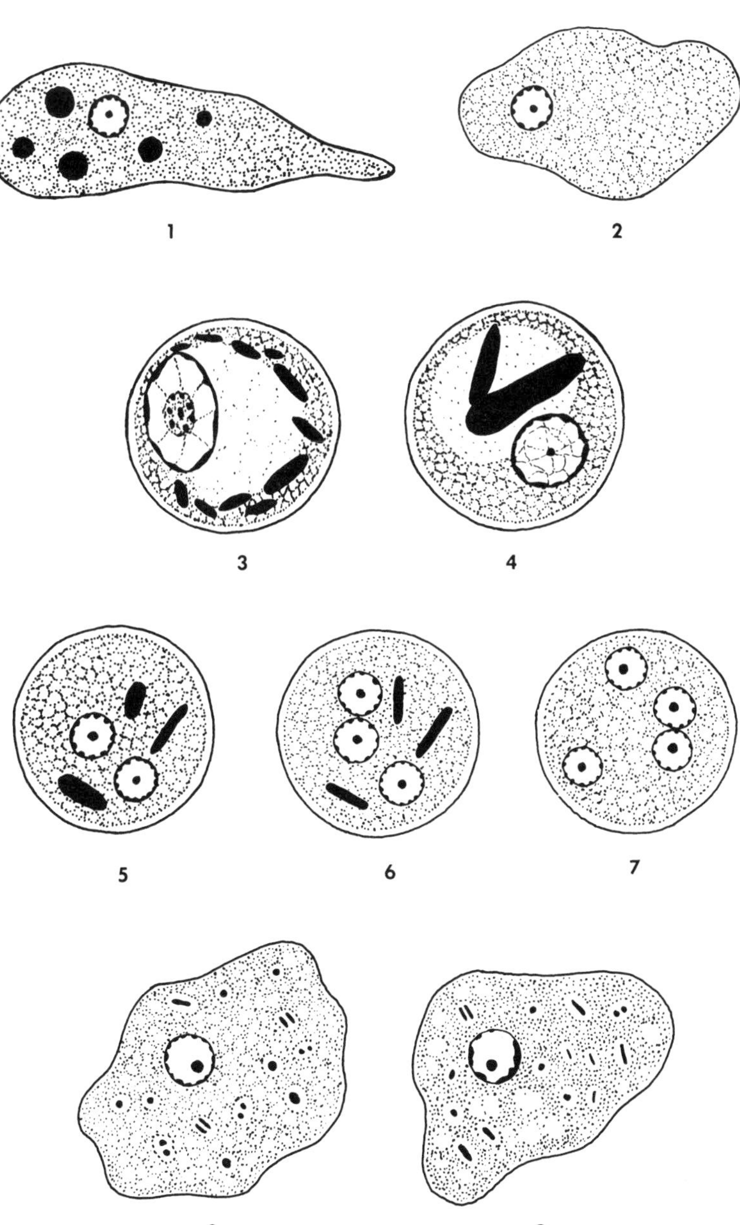

Figure A7–3–1 *Entamoeba* spp. (1) Trophozoite of *Entamoeba histolytica;* (2) trophozoite of *Entamoeba histolytica/E. dispar;* (3 and 4) early cysts of *E. histolytica;* (5 through 7) cysts of *E. histolytica/E. dispar;* (8 and 9) trophozoites of *Entamoeba coli;* (10 and 11) early cysts of *E. coli;* (12 through 14) cysts of *E. coli;* (15 and 16) trophozoites of *Entamoeba hartmanni;* (17 and 18) cysts of *E. hartmanni.* (Reprinted from L. S. Garcia and D. A. Bruckner, *Diagnostic Medical Parasitology,* 3rd ed., 1997, ASM Press, Washington, D.C.) *(continued)*

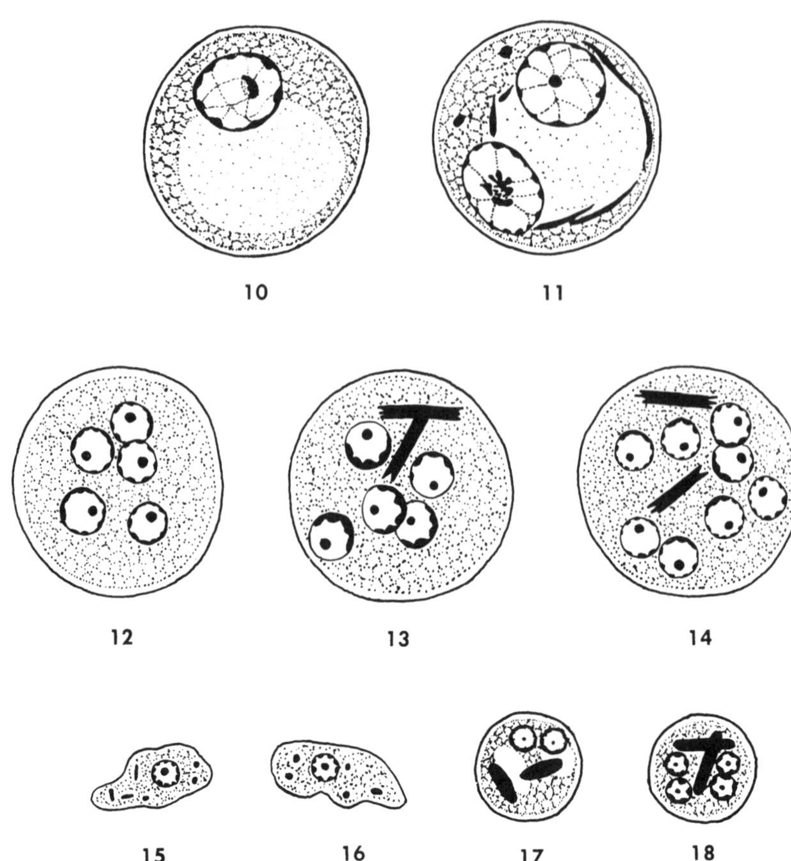

Figure A7–3–1 *(continued)*

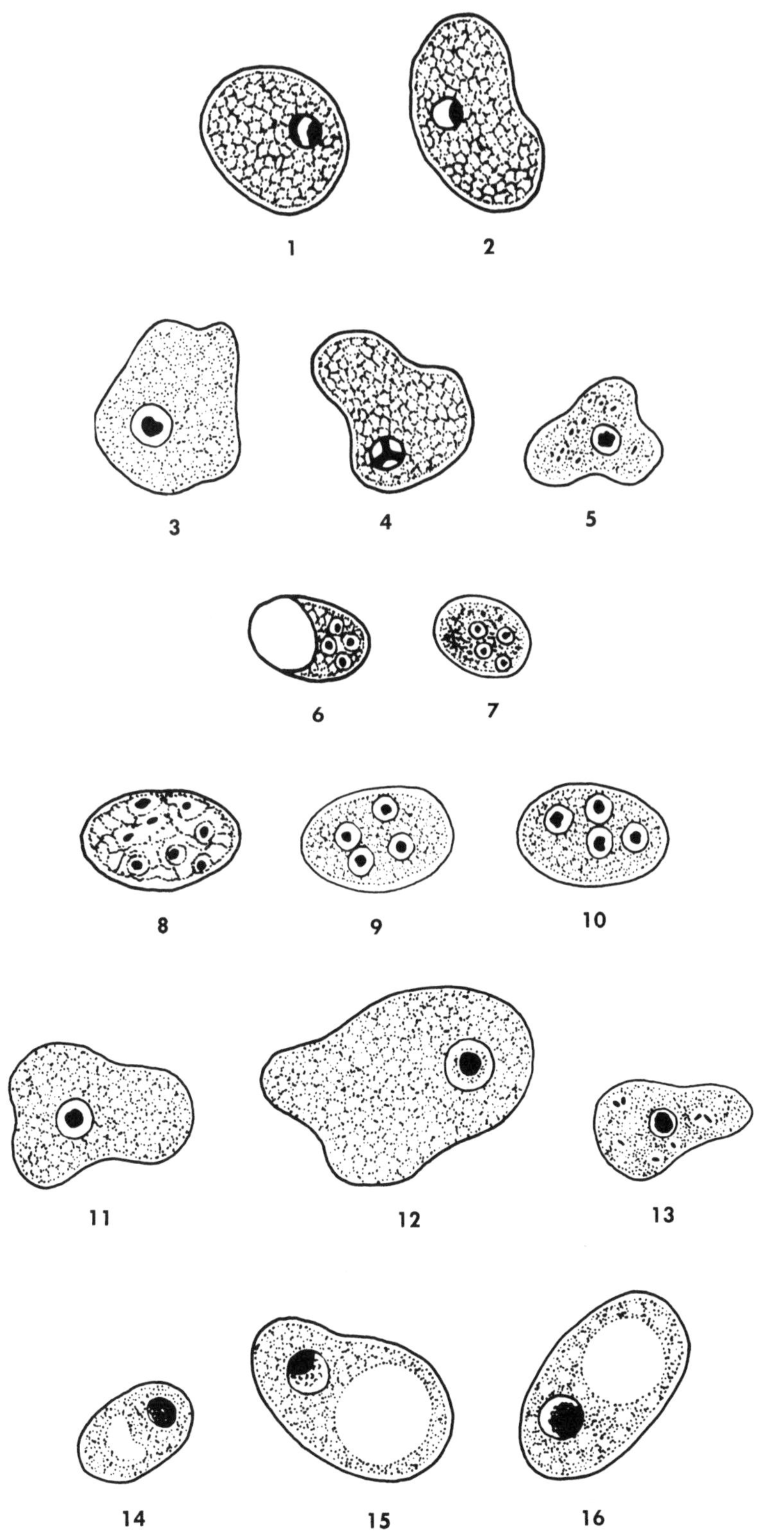

Figure A7–3–2 Other intestinal amebae. (1 through 5) Trophozoites of *Endolimax nana;* (6 through 10) cysts of *E. nana;* (11 through 13) trophozoites of *Iodamoeba bütschlii;* (14 through 16) cysts of *I. bütschlii.* (Reprinted from L. S. Garcia and D. A. Bruckner, *Diagnostic Medical Parasitology,* 3rd ed., 1997, ASM Press, Washington, D.C.)

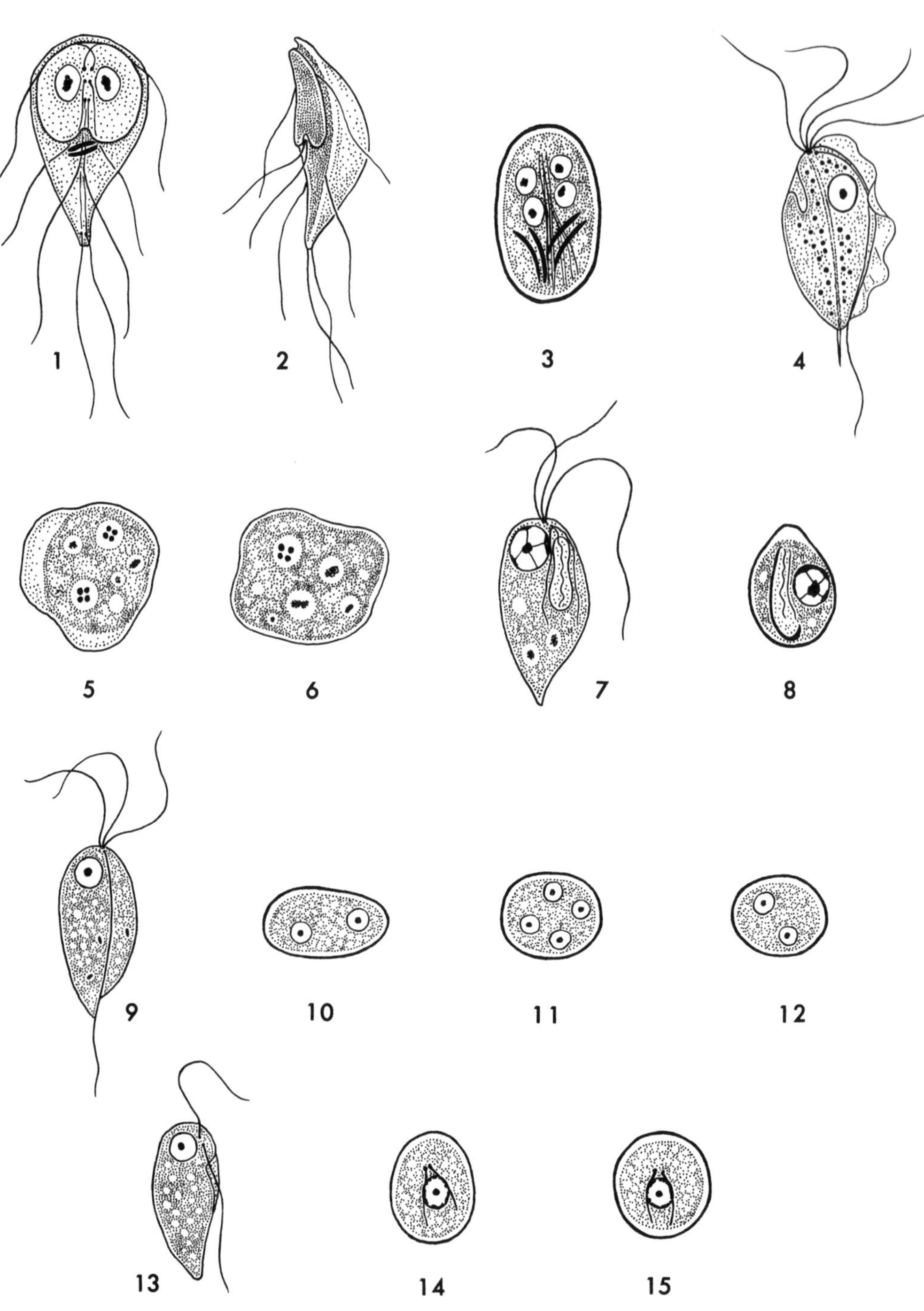

Figure A7–3–3 Intestinal flagellates. (1) *Giardia lamblia* trophozoite (front view); (2) *G. lamblia* trophozoite (side view); (3) *G. lamblia* cyst; (4) *Trichomonas hominis* trophozoite (no cyst form known); (5 and 6) *Dientamoeba fragilis* trophozoites (no cyst form known); (7) *Chilomastix mesnili* trophozoite; (8) *C. mesnili* cyst; (9) *Enteromonas hominis* trophozoite; (10 through 12) *E. hominis* cysts; (13) *Retortamonas intestinalis* trophozoite; (14 and 15) *R. intestinalis* cysts. (Illustration by Sharon Belkin.)

Cilia
Micronucleus
Macronucleus
Micronucleus
Macronucleus

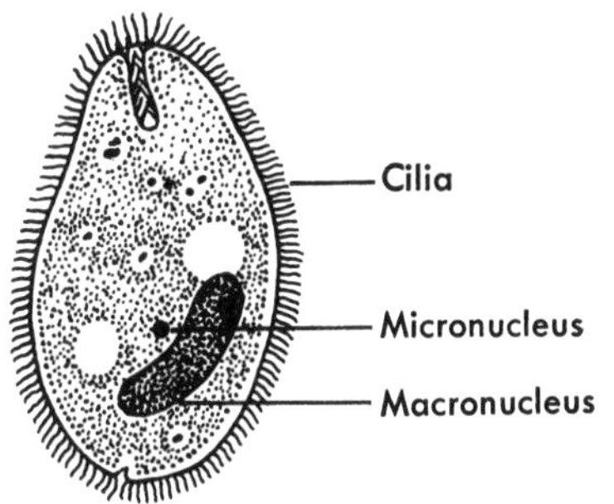

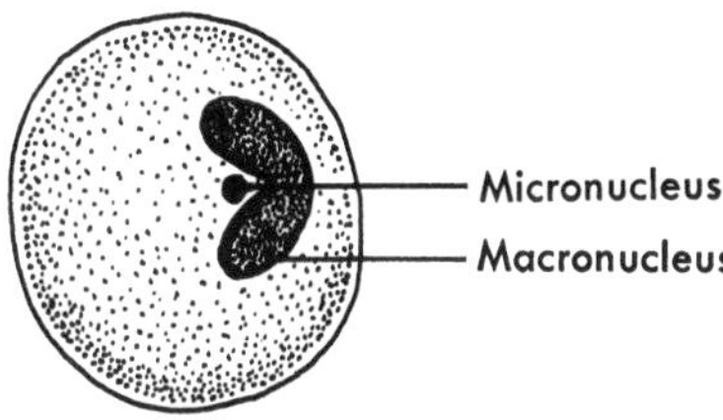

igure A7–3–4 *Balantidium coli.* (1) Tro-hozoite; (2) cyst. (Illustration by Sharon elkin.) (Reprinted from L. S. Garcia and D. .. Bruckner, *Diagnostic Medical Parasitol-gy,* 3rd ed., 1997, ASM Press, Washington, .C.)

Clonorchis sinensis
27-35 µm long
12-19 µm wide
Taenia spp.
30-47 µm diameter
Hymenolepis nana
31-43 µm diameter
Trichuris trichiura
50-54 µm long
20-23 µm wide
Enterobius vermicularis
70-85 µm long
60-80 µm wide
Ascaris lumbricoides (fertile egg)
45-75 µm long
35-50 µm wide
Hookworm
56-75 µm long
36-40 µm wide
Diphyllobothrium latum
58-75 µm long
40-50 µm wide
Hymenolepis diminuta
70-85 µm long
60-80 µm wide

Figure A7–3–5 Relative sizes of helminth eggs. (Reprinted from L. S. Garcia and D. A. Bruckner, *Diagnostic Medical Parasitology,* 3rd ed., 1997, ASM Press, Washington, D.C.)

(continued)

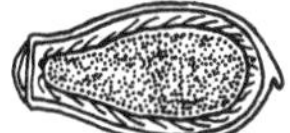

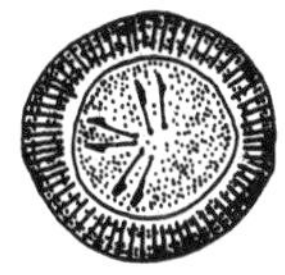

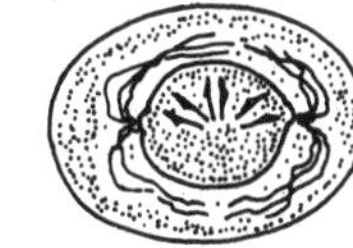

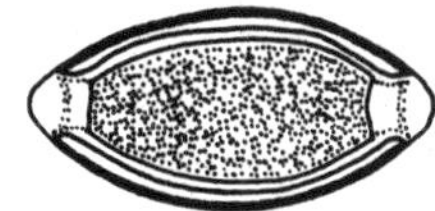

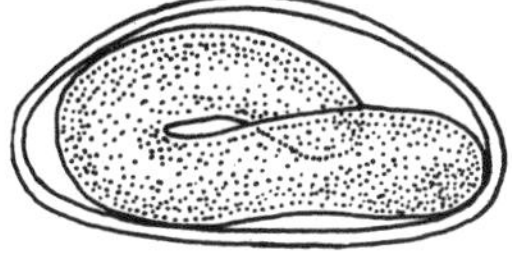

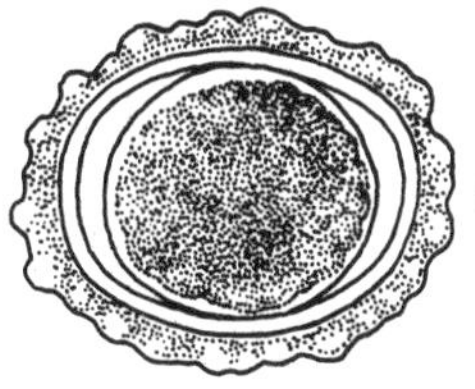

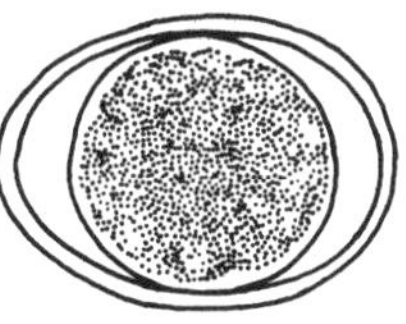

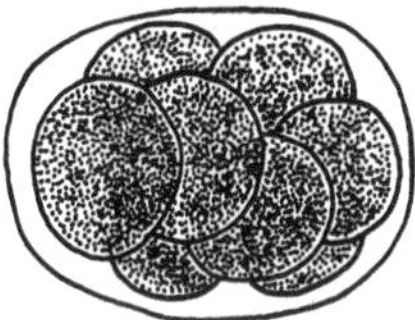

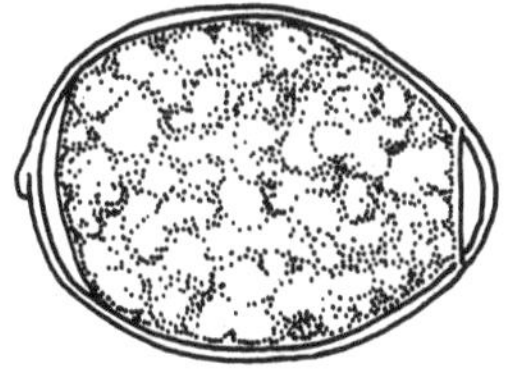

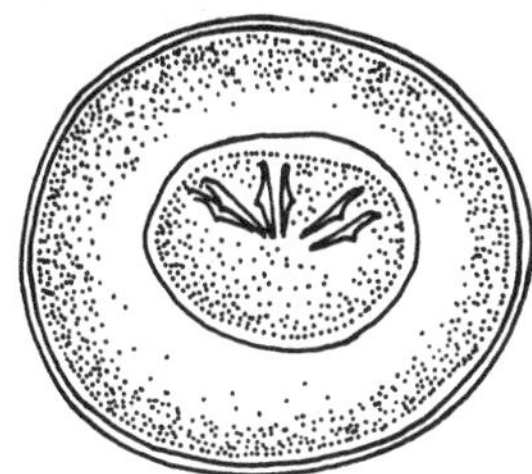

Paragonimus westermani

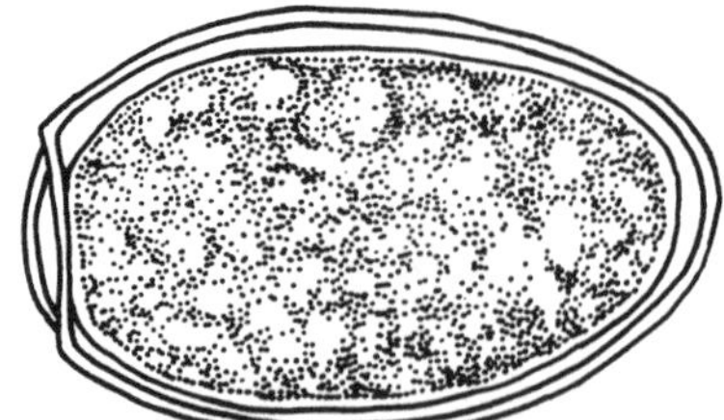

80-120 μm long
48-60 μm wide

Schistosoma haematobium

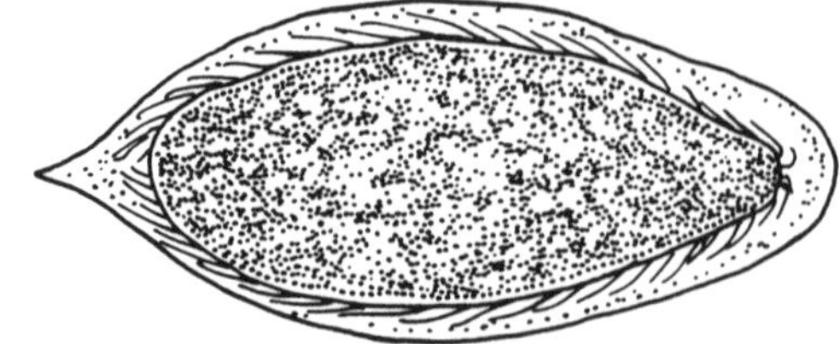

112-170 μm long
40-70 μm wide

Trichostrongylus

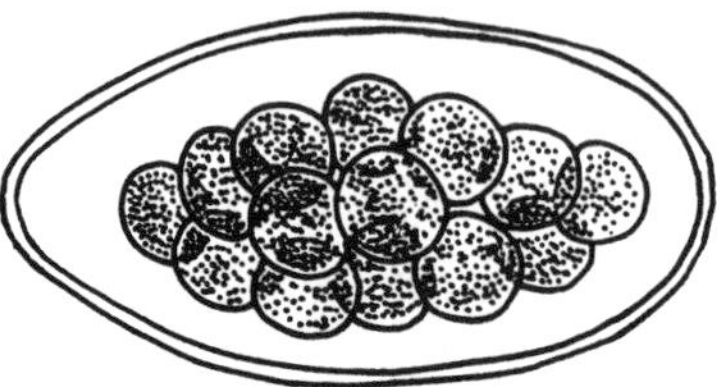

73-95 μm long
40-50 μm wide

Schistosoma mansoni

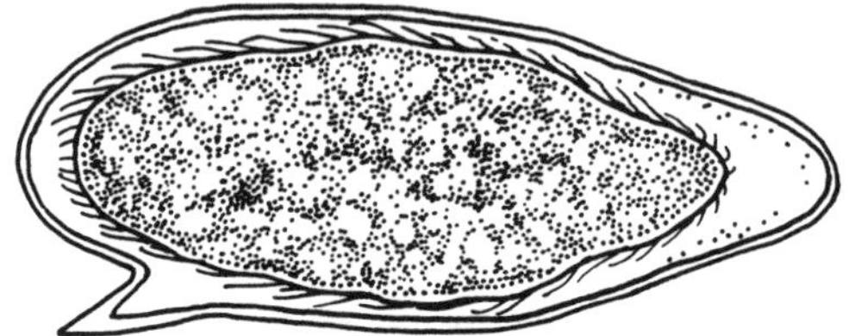

114-180 μm long
45-70 μm wide

Ascaris lumbricoides
(Unfertilized egg)

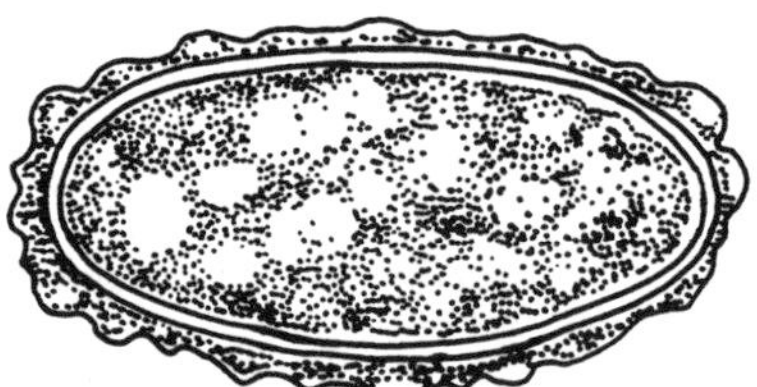

85-95 μm long
43-47 μm wide

Fasciola hepatica or
Fasciolopsis buski

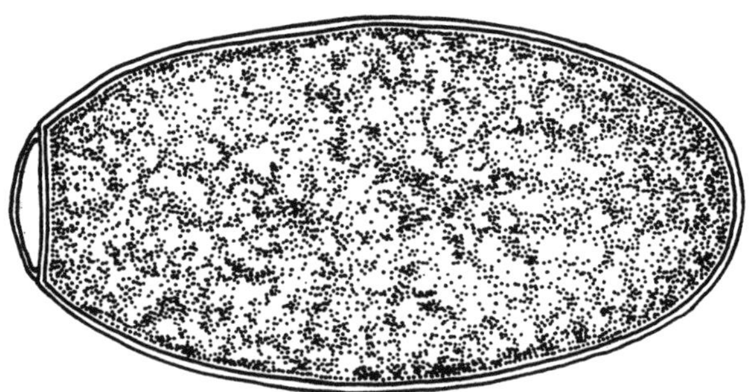

130-140 μm long
80-85 μm wide

Schistosoma japonicum

70-100 μm long
55-65 μm wide

Figure A7–3–5 *(continued)*

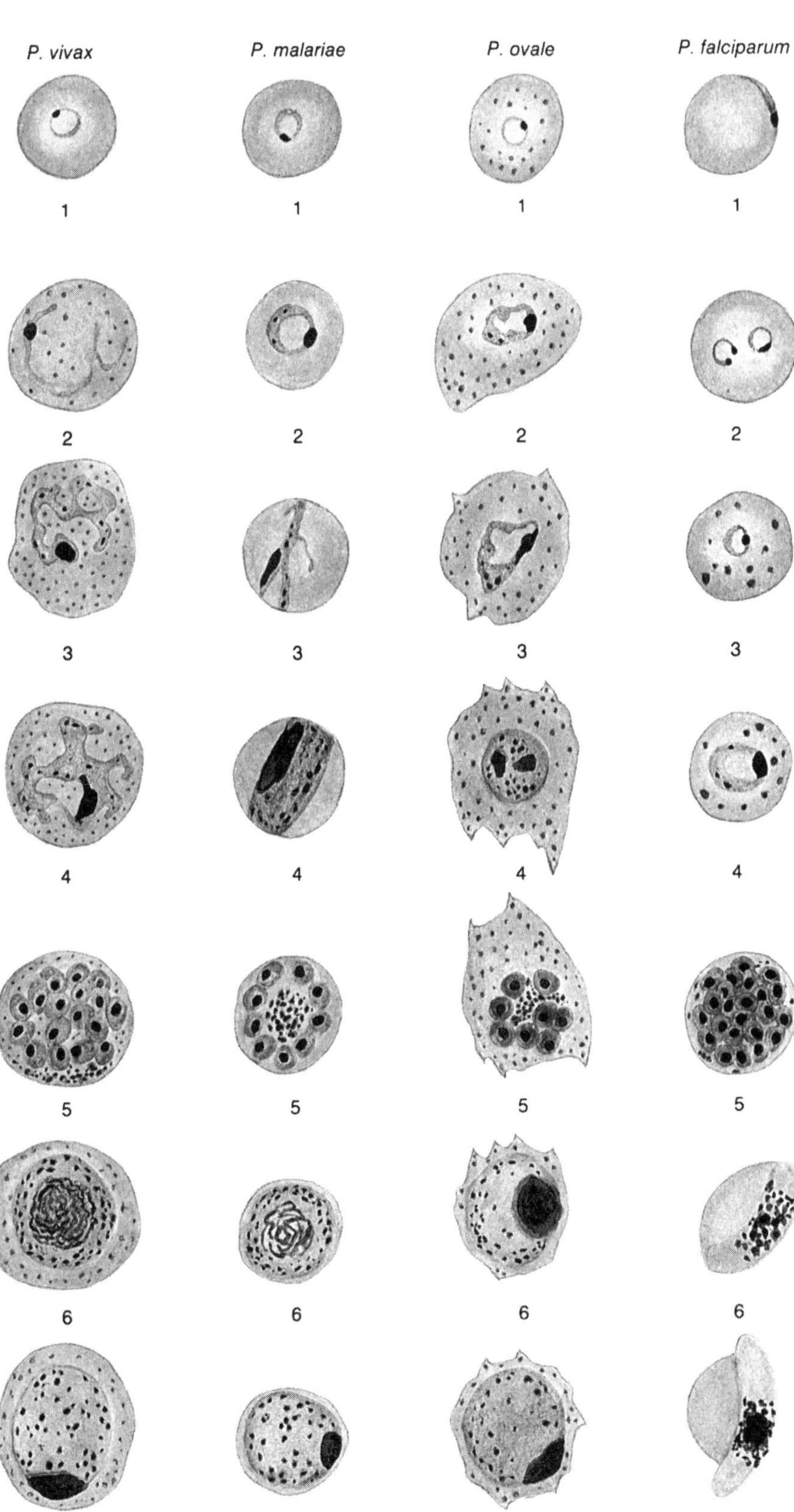

Figure A7–3–6 Morphology of malaria parasites. *Plasmodium vivax:* (1) early trophozoite (ring form); (2) late trophozoite with Schüffner's dots (note enlarged RBC); (3) late trophozoite with ameboid cytoplasm (very typical of *P. vivax*); (4) late trophozoite with ameboid cytoplasm; (5) mature schizont with 18 merozoites and clumped pigment; (6) microgametocyte with dispersed chromatin; (7) macrogametocyte with compact chromatin. *Plasmodium malariae:* (1) early trophozoite (ring form); (2) early trophozoite with thick cytoplasm; (3) early trophozoite (band form); (4) late trophozoite (band form) with heavy pigment; (5) mature schizont with nine merozoites arranged in a rosette; (6) microgametocyte with dispersed chromatin; (7) macrogametocyte with compact chromatin. *Plasmodium ovale:* (1) early trophozoite (ring form) with Schüffner's dots; (2) early trophozoite (note enlarged RBC); (3) late trophozoite in RBC with fimbriated edges; (4) developing schizont with irregular-shaped RBC; (5) mature schizont with eight merozoites arranged irregularly; (6) microgametocyte with dispersed chromatin; (7) macrogametocyte with compact chromatin. *Plasmodium falciparum:* (1) early trophozoite (accolé or appliqué form); (2) early trophozoite (one ring is in the headphone configuration with double chromatin dots); (3) early trophozoite with Maurer's dots; (4) late trophozoite with larger ring and Maurer's dots; (5) mature schizont with 24 merozoites; (6) microgametocyte with dispersed chromatin; (7) macrogametocyte with compact chromatin. *Note:* Without the appliqué form, Schüffner's dots, multiple rings per cell, and other developing stages, differentiation among the species can be very difficult. It is obvious that the early rings of all four species can mimic one another very easily. *Remember: One set of negative blood films cannot rule out a malaria infection.* (Illustration by Sharon Belkin.) (Reprinted from L. S. Garcia and D. A. Bruckner, *Diagnostic Medical Parasitology,* 3rd ed., 1997, ASM Press, Washington, D.C.)

Common Problems in Parasite Identification

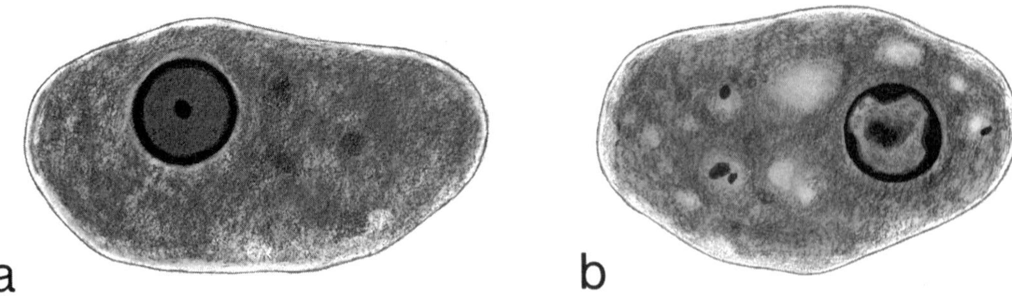

Figure A7–3–7 (a) *Entamoeba histolytica/E. dispar* trophozoite. Note the evenly arranged nuclear chromatin, central compact karyosome, and relatively clean cytoplasm. (b) *Entamoeba coli* trophozoite. Note the unevenly arranged nuclear chromatin, eccentric karyosome, and messy cytoplasm. These characteristics are very representative of the two organisms. (Illustration by Sharon Belkin. Reprinted from CMPH 7.10.3.)

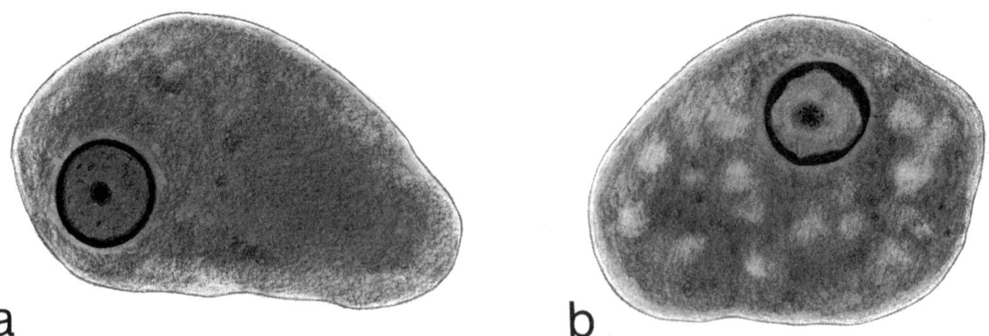

Figure A7–3–8 (a) *Entamoeba histolytica/E. dispar* trophozoite. Note the evenly arranged nuclear chromatin, central compact karyosome, and clean cytoplasm. (b) *Entamoeba coli* trophozoite. Note that the nuclear chromatin appears to be evenly arranged, the karyosome is central (but more diffuse), and the cytoplasm is messy, with numerous vacuoles and ingested debris. The nuclei of these two organisms tend to resemble one another (a very common finding in routine clinical specimens). (Illustration by Sharon Belkin. Reprinted from CMPH 7.10.3.)

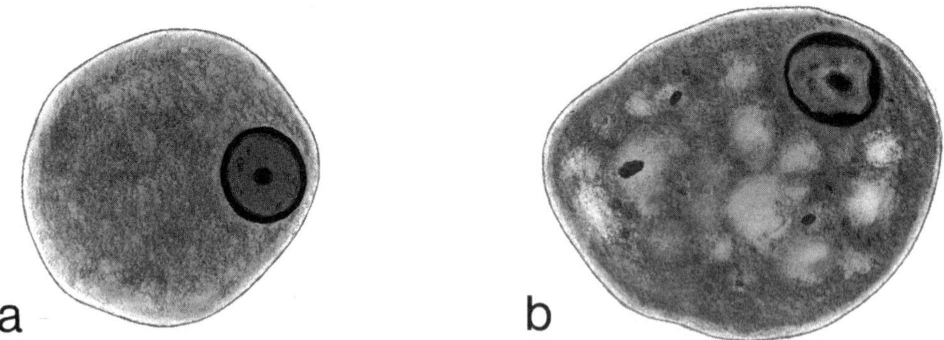

Figure A7–3–9 (a) *Entamoeba histolytica/E. dispar* trophozoite. Again, note the typical morphology (evenly arranged nuclear chromatin, central compact karyosome, and relatively clean cytoplasm). (b) *Entamoeba coli* trophozoite. Although the nuclear chromatin is eccentric, note that the karyosome seems to be compact and central. However, note the various vacuoles containing ingested debris. These organisms show some characteristics that are very similar (very common in clinical specimens). (Illustration by Sharon Belkin. Reprinted from CMPH 7.10.3.)

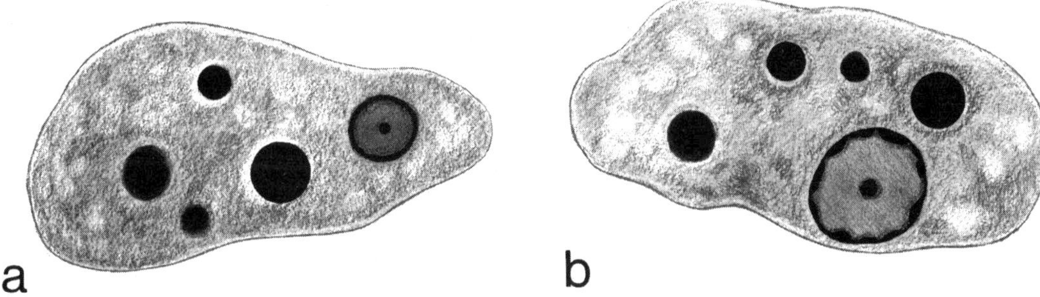

Figure A7–3–10 (a) *Entamoeba histolytica* trophozoite. Note the evenly arranged nuclear chromatin, central compact karyosome, and red blood cells (RBCs) in the cytoplasm. (b) Human macrophage. The key difference between the macrophage nucleus and that of *E. histolytica* is the size. Usually the ratio of nucleus to cytoplasm in a macrophage is approximately 1 : 6 or 1 : 8, while the true organism has a nucleus-to-cytoplasm ratio of approximately 1 : 10 or 1 : 12. The macrophage also contains ingested RBCs. In patients with diarrhea or dysentery, trophozoites of *E. histolytica* and macrophages are often confused, occasionally leading to a false-positive diagnosis of amebiasis when no parasites are present. Both the actual trophozoite and the macrophage may also be seen without ingested RBCs and can mimic one another. (Illustration by Sharon Belkin. Reprinted from CMPH 7.10.3.)

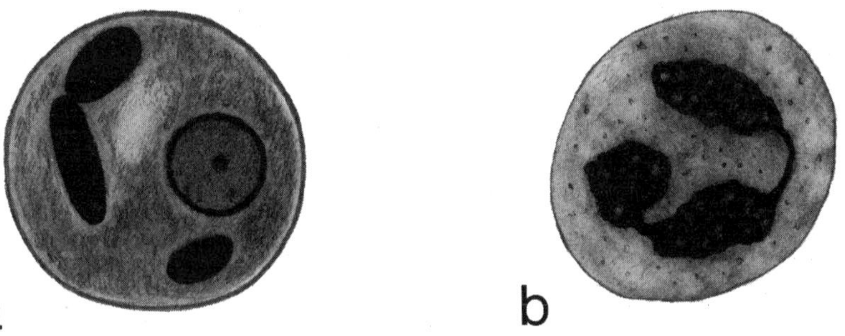

Figure A7–3–11 (a) *Entamoeba histolytica/E. dispar* precyst. Note the enlarged nucleus (prior to division) with evenly arranged nuclear chromatin and central compact karyosome. Chromatoidal bars (rounded ends with smooth edges) are also present in the cytoplasm. (b) Polymorphonuclear leukocyte (PMN). The nucleus is somewhat lobed (normal morphology) and represents a PMN that has not been in the gut very long. Occasionally, the positioning of the chromatoidal bars and the lobed nucleus of the PMN will mimic one another. The chromatoidal bars will stain more intensely, but the shapes can overlap, as seen here. (Illustration by Sharon Belkin. Reprinted from CMPH 7.10.3.)

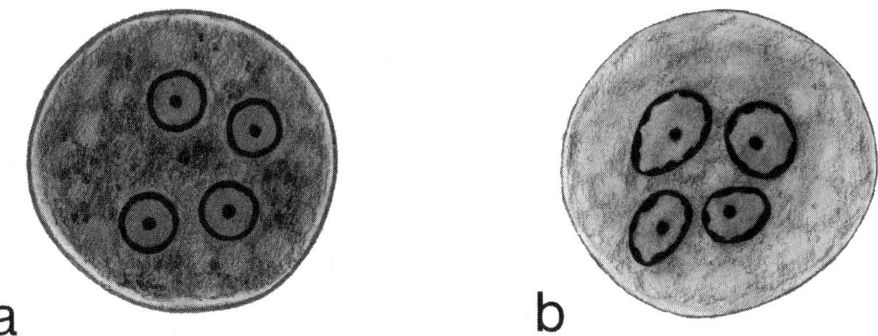

Figure A7–3–12 (a) *Entamoeba histolytica/E. dispar* cyst. Note that the four nuclei are very consistent in size and shape. (b) PMN. Note that the normal lobed nucleus has now broken into four fragments, which mimic four nuclei with peripheral chromatin and central karyosomes. When PMNs have been in the gut for some time and have begun to disintegrate, the nuclear morphology can mimic that seen in an *E. histolytica/E. dispar* cyst. However, human cells are often seen in the stool in patients with diarrhea; with rapid passage of the gastrointestinal tract contents, there will not be time for *E. histolytica/E. dispar* cysts to form. Therefore, for patients with diarrhea and/or dysentery, if ''organisms'' that resemble the cell in panel b are seen, think first of PMNs, not *E. histolytica/E. dispar* cysts. (Illustration by Sharon Belkin. Reprinted from CMPH 7.10.3.)

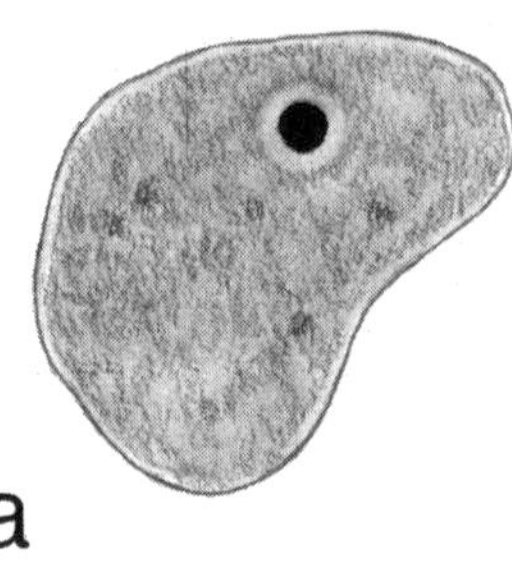

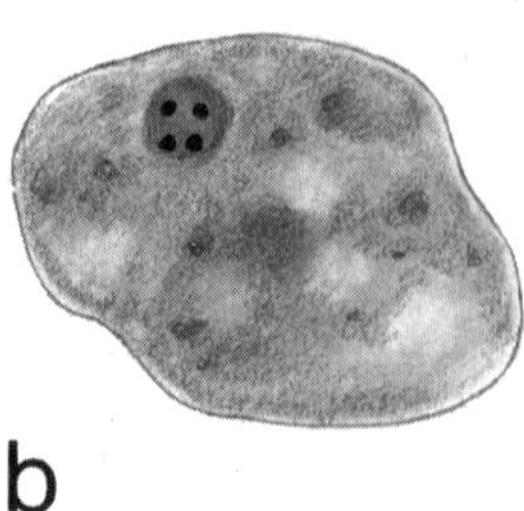

Figure A7–3–13 (a) *Endolimax nana* trophozoite. This organism is characterized by a large karyosome with no peripheral chromatin, although many nuclear variations are normally seen in any positive specimen. (b) *Dientamoeba fragilis* trophozoite. Normally, the nuclear chromatin is fragmented into several dots (often a "tetrad" arrangement). The cytoplasm is normally more "junky" than that seen in *E. nana.* If the morphology is typical, as in these two illustrations, differentiating between these two organisms is not very difficult. However, the morphologies of the two will often be very similar. (Illustration by Sharon Belkin. Reprinted from CMPH 7.10.3.)

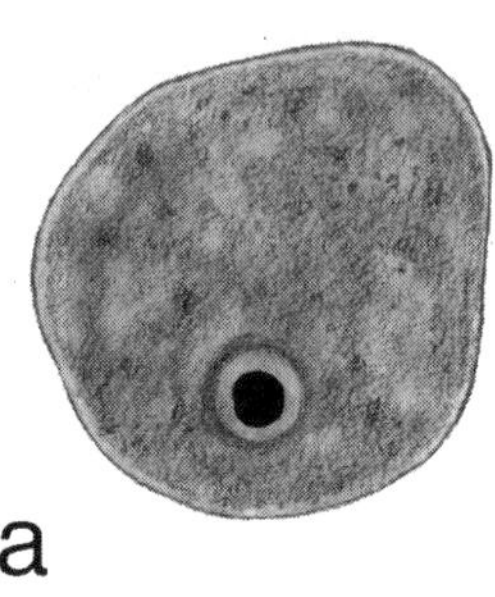

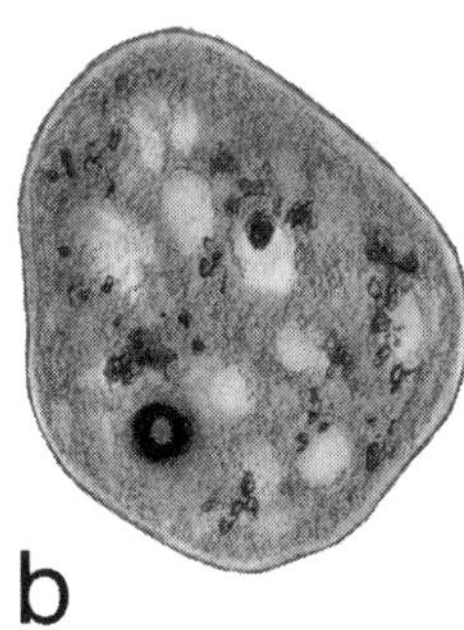

Figure A7–3–14 (a) *Endolimax nana* trophozoite. Notice that the karyosome is large and surrounded by a "halo," with very little, if any, chromatin on the nuclear membrane. (b) *Dientamoeba fragilis* trophozoite. In this organism, the karyosome is beginning to fragment and there is a slight clearing in the center of the nuclear chromatin. If the nuclear chromatin has not become fragmented, *D. fragilis* trophozoites can very easily mimic *E. nana* trophozoites. This could lead to a report indicating that no pathogens were present when, in fact, *D. fragilis* is now considered a definite cause of symptoms. (Illustration by Sharon Belkin. Reprinted from CMPH 7.10.3.)

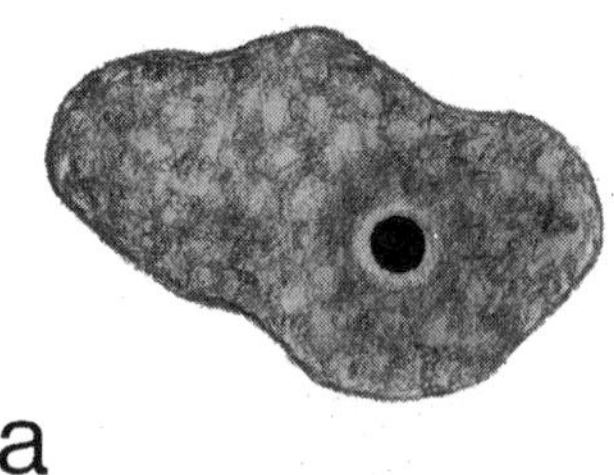

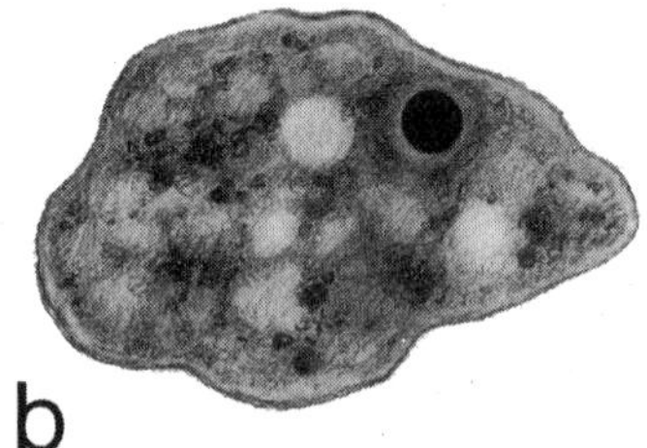

Figure A7–3–15 (a) *Endolimax nana* trophozoite. Note the large karyosome surrounded by a clear space. The cytoplasm is relatively clean. (b) *Iodamoeba bütschlii.* Although the karyosome is similar to that of *E. nana,* note that the cytoplasm in *I. bütschlii* is much more heavily vacuolated and contains ingested debris. Often, these two trophozoites cannot be differentiated. However, the differences in the cytoplasm are often helpful. There will be a definite size overlap between the two genera. (Illustration by Sharon Belkin. Reprinted from CMPH 7.10.3.)

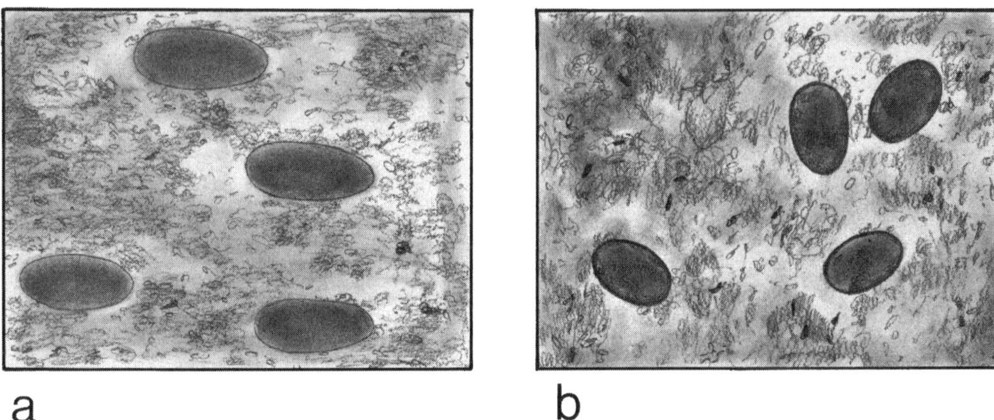

Figure A7–3–16 (a) RBCs on a stained fecal smear. Note that the cells are very pleomorphic but tend to be positioned in the direction the stool was spread onto the slide. (b) Yeast cells on a stained fecal smear. These cells tend to remain oval and are not aligned in any particular way on the smear. These differences are important when the differential identification is between *Entamoeba histolytica* containing RBCs and *Entamoeba coli* containing ingested yeast cells. If RBCs or yeast cells are identified in the cytoplasm of an organism, they must also be visible in the background of the stained fecal smear. (Illustration by Sharon Belkin. Reprinted from CMPH 7.10.3.)

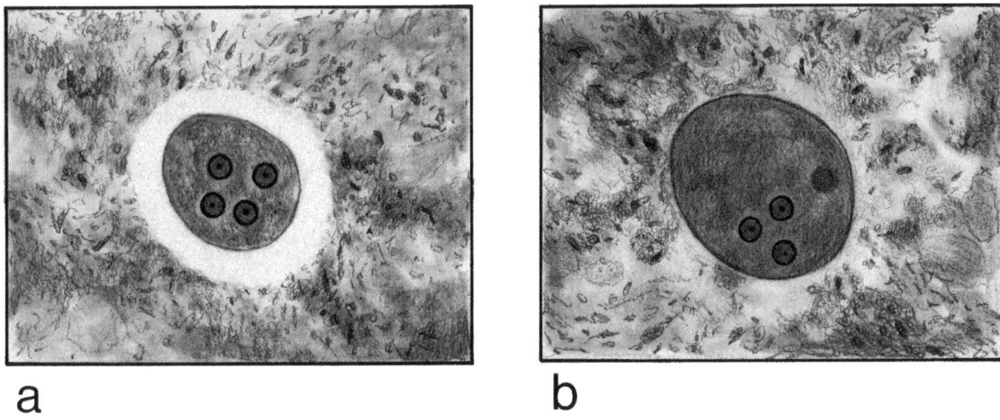

Figure A7–3–17 (a) *Entamoeba histolytica/E. dispar* cyst. Note the shrinkage due to dehydrating agents in the staining process. (b) *E. histolytica/E. dispar* cyst. In this case, the cyst exhibits no shrinkage. Only three of the four nuclei are in focus. Normally, this type of shrinkage is seen with protozoan cysts and is particularly important when a species is measured and identified as either *E. histolytica/E. dispar* or *Entamoeba hartmanni.* The whole area, including the halo, must be measured prior to species identification. If just the cyst is measured, the organism would be identified as *E. hartmanni* (nonpathogenic) rather than *E. histolytica/E. dispar* (may be pathogenic). (Illustration by Sharon Belkin. Reprinted from CMPH 7.10.3.)

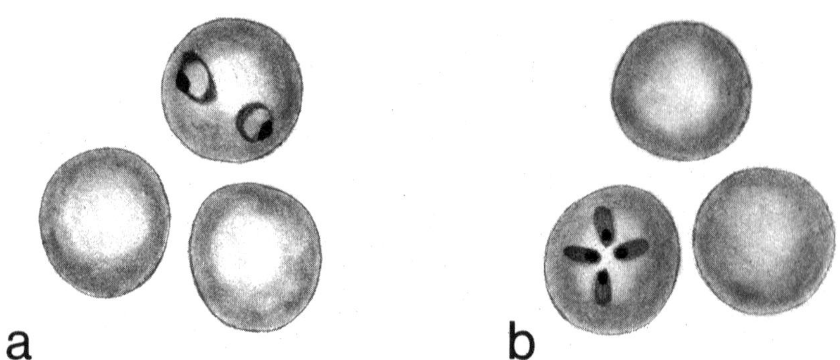

Figure A7–3–18 (a) *Plasmodium falciparum* rings. Note the two rings in the RBC. Multiple rings per cell are more typical of *P. falciparum* than of the other species causing human malaria. (b) *Babesia* rings. One of the RBCs contains four small *Babesia* rings. This particular arrangement is called the Maltese cross and is diagnostic for *Babesia* spp. *Babesia* infections can be confused with cases of *P. falciparum* malaria, primarily because multiple rings can be seen in the RBCs. Another difference involves ring morphology. *Babesia* rings are often of various sizes and tend to be very pleomorphic, while those of *P. falciparum* tend to be more consistent in size and shape. (Illustration by Sharon Belkin. Reprinted from CMPH 7.10.3.)

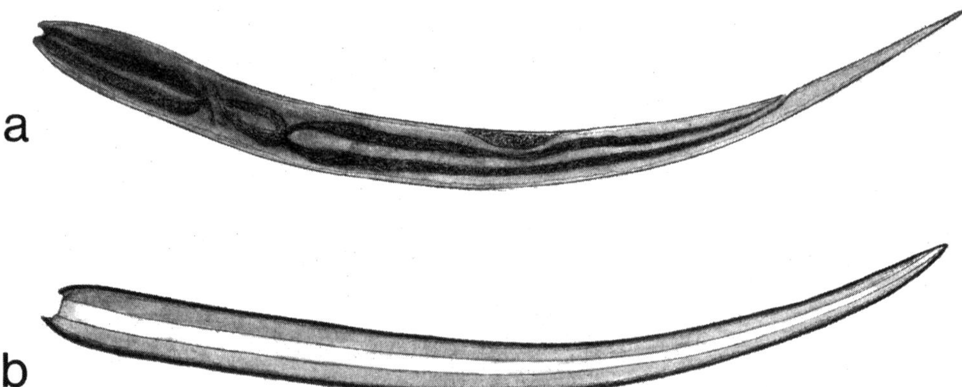

Figure A7–3–19 (a) *Strongyloides stercoralis* rhabditiform larva. Note the short buccal capsule (mouth opening) and the internal structure, including the genital primordial packet of cells. (b) Root hair (plant material). Note that there is no specific internal structure and the end is ragged (where it was broken off from the main plant. Plant material often mimics some of the human parasites. This comparison is one of the best examples. These artifacts are occasionally submitted as proficiency testing specimens. (Illustration by Sharon Belkin. Reprinted from CMPH 7.10.3.)

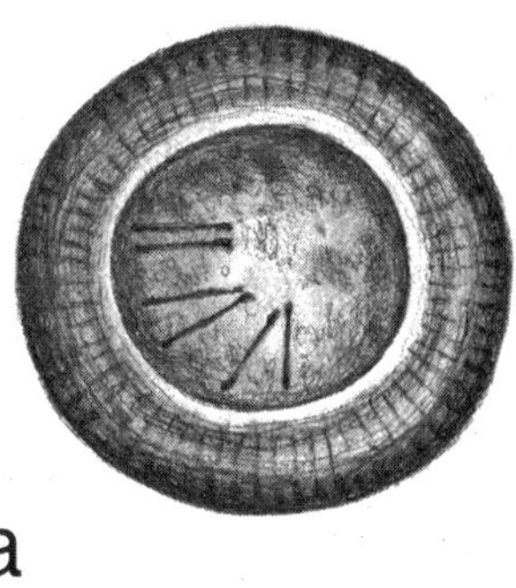

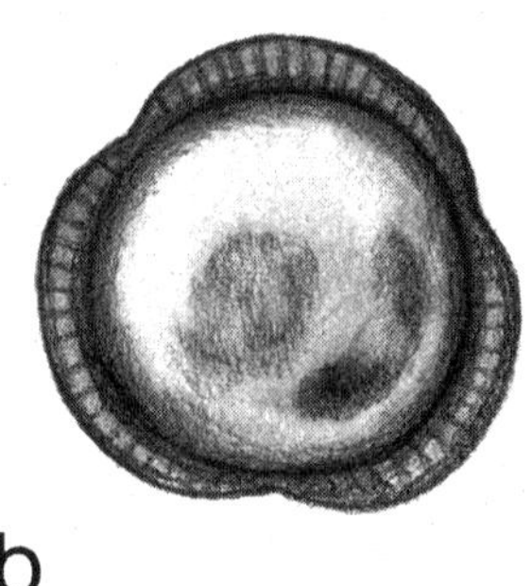

Figure A7–3–20 (a) *Taenia* egg. This egg has been described as having a thick, radially striated shell containing a six-hooked embryo (oncosphere). (b) Pollen grain. Note that this trilobed pollen grain has a similar type of "shell" and, if turned the right way, could resemble a *Taenia* egg. This similarity represents another source of confusion between a helminth egg and a plant material artifact. When examining fecal specimens in a wet preparation, tap on the coverslip to get objects to move around. As they move, you can see more morphological detail. (Illustration by Sharon Belkin. Reprinted from CMPH 7.10.3.)

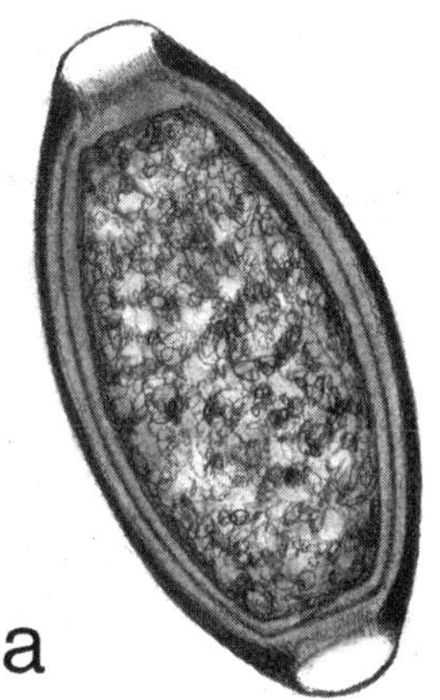

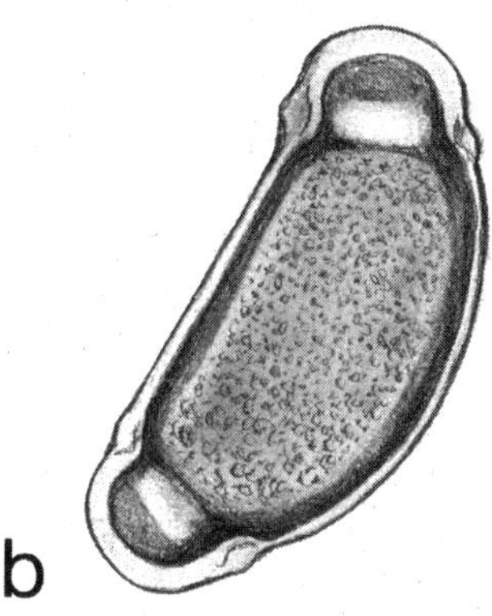

Figure A7–3–21 (a) *Trichuris trichiura* egg. This egg is typical and is characterized by the barrel shape with thick shell and two polar plugs. (b) Bee pollen. This artifact certainly mimics the *T. trichiura* egg. However, note that the shape is somewhat distorted. This is an excellent example of a parasite look-alike that could be confusing. (Illustration by Sharon Belkin. Reprinted from CMPH 7.10.3.)

APPENDIX 7-4 Sources of Commercial Reagents and Supplies

Table A7–4–1 Sources of commercial reagents and supplies[a]

Reagent or supply	AJP Scientific	ALPHA-TEC	J. T. Baker	A. B. Becker	Bio-Spec	City Chemical Corp.	Evergreen Scientific	Fisher Scientific
Specimen collection kit(s)								
Formalin, PVA		X			X		X	
PVA, modified (copper base)		X						
PVA, modified (zinc base)		X					X	
MIF		X						
SAF		X						X
Pinworm paddles								
Preservatives (bulk)								
Schaudinn's fixative solution		X+/−			X+/−			
PVA fixative solution	X	X			X			
PVA powder[b]			X		X			X
MIF solution	X	X						
SAF solution		X						
Concentration systems								
Formalin-ethyl acetate		X					X	
Zinc sulfate (sp gr, 1.18 or 1.20)		X	X		X			
Stains								
Trichrome, solution	X (W)	X (W)			X (W)			
Trichrome, modified[c]								
Trichrome, dye powders								
Chromotrope 2R			X					
Fast green FCF			X					X
Light green SF yellowish			X					X
Hematoxylin, solution					X			
Hematoxylin, powder			X					X
Chlorazol black E, powder						X		
Giemsa, solution	X		X		X			X
Giemsa, powder			X					X
Carbol fuchsin, Kinyoun's	X	X			X			X
Modified acid fast with DMSO		X						
Auramine-rhodamine								X
Acridine orange								
Toluidine blue O								
Miscellaneous								
Lugol's iodine, dilute 1:5	X	X			X			
Dobell & O'Connor's iodine		X						X
Diatex				X				
Triton X-100		X	X					
Eosin-saline solution (1%)					X			X
Mayer's albumin		X						
Control slides or suspensions		X						

[a] Adapted from CMPH 7.10.6. PVA, polyvinyl alcohol; MIF, merthiolate-iodine-formalin; SAF, sodium acetate-acetic acid-formalin; W, Wheatley; G, Gomori; DMSO, dimethyl sulfoxide; +, with acetic acid; −, without acetic acid.

[b] Use grade with high hydrolysis and low viscosity for parasite studies.

[c] Used for the identification of microsporidian spores in stool or other specimens.

Hardy Diagnostics	Harleco	Medical Chemical Corp.	Meridian Diagnostic	MML Diagnostic	PML Microbio-logicals	Poly Science	Remel	Rohm & Haas	Rowley Biochemical	Scientific Device Laboratory	Trend Scientific	Volu-Sol
X		X	X	X	X		X					
X		X	X	X	X		X					X
X		X	X		X		X					
X		X	X		X		X					
X		X	X	X	X		X					
X										X		
X+		X+	X+				X−		X−			X−
X		X	X		X		X			X		X
X		X	X									
					X							
X		X	X									
X		X								X		
X (W)		X (W)	X (W)		X		X (W)		X (W, G)	X (W)	X (W)	X (W, G)
X		X	X				X			X	X	X
	X								X			
	X								X			
	X								X		X	
X		X			X				X			X
	X								X		X	
						X			X			
X	X	X							X		X	X
	X								X			
X	X	X			X		X		X		X	X
					X						X	X
		X			X		X				X	X
					X		X				X	X
		X									X	
X	X	X	X		X		X		X		X	X
X							X			X		
X		X	X		X		X	X		X		
		X									X	
X							X					
X			X		X		X			X	X	

Table A7–4–2 Addresses of suppliers listed in Table A7–4–1

AJP Scientific, Inc.
P.O. Box 1589
Clifton, NJ 07015
(800) 922-0223
(201) 472-7200
Products also available through various distributors

ALPHA-TEC Systems, Inc.
P.O. Box 5435
Vancouver, WA 98668
(800) 221-6058
(360) 260-2779
Fax: (360) 260-3277
Products also available through various distributors

J. T. Baker
222 Red School Lane
Phillipsburg, NJ 08865
(908) 859-2151
Fax: (908) 859-9385
Products also available through American Scientific Products or VWR

A. B. Becker
21124 Malmö, Sweden
Products available through American Scientific Products

Bio-Spec, Inc.
179 Mason Circle, Suite A
Concord, CA 94520-1213
(415) 689-0771
Products also available through VWR Scientific

City Chemical Corp.
32 West 22nd Street
New York, NY 10011
(212) 929-2723
Products also available through Amfak Drug or Bergen-Brunswick

Evergreen Scientific
2300 East 49th Street
P.O. Box 58248
Los Angeles, CA 90058-0248
(800) 421-6261
(800) 372-7300 in California
(213) 583-1331
Fax: (213) 581-2503
Products also available through various distributors

Fisher Scientific
2761 Walnut Avenue
Tustin, CA 92681
(800) 766-7000 (national no.)
Fax: (800) 926-1166 (national no.)
Products available through regional sales offices

Hardy Diagnostics
1430 West McCoy Lane
Santa Maria, CA 93455
(800) 266-2222
Fax: (805) 346-2760
Products also available through various distributors

Harleco (New owner: E. Merck Co.)
Darmstadt, West Germany
Products available through American Scientific Products; product numbers remain unchanged

Medical Chemical Corp.
1909 Centinela Avenue
Santa Monica, CA 90404
(310) 829-4304
Fax: (310) 453-1212
Products also available through various distributors

Meridian Diagnostics, Inc.
3741 River Hills Drive
Cincinnati, OH 45244
(800) 543-1980
(513) 271-3700
Fax: (513) 271-0124
Products available only through Baxter MicroScan and Scientific Products Division, CMA, and various other regional distributors

MML Diagnostic Packaging
P.O. Box 458
Troutdale, OR 97060
(503) 666-8398
(800) 826-7186
Products available through various distributors

PML Microbiologicals
P.O. Box 459
Tualatin, OR 97062
(800) 547-0659
(503) 639-1500 in Oregon
Fax: (800) 765-4415-
Products also available through other distributors

Poly Science, Inc.
400 Valley Road
Warrington, PA 18976
(800) 523-2575
(215) 343-6484
Products available only through company

Remel
12076 Santa Fe Drive
Lenexa, KS 66215
(800) 255-6730
Fax: (800) 477-5781
Products available through Rupp & Bowman and certain other distributors

Rohm & Haas
Philadelphia, PA 19105
Products also available through various distributors

Rowley Biochemical Institute
U.S. Route 1
Rowley, MA 01969
(508) 948-2067
Fax: (508) 948-2206
Products available only through company

Table A7–4–2 Addresses of suppliers listed in Table A7–4–1 *(continued)*

Scientific Device Laboratory, Inc.
411 E. Jarvis Avenue
Des Plaines, IL 60018
(847) 803-9495
Fax: (847) 803-8251
Products also available through various distributors

Trend Scientific, Inc.
368 W. County Road
P.O. Box 120266
St. Paul, MN 55112-0266
(800) 328-3949
(612) 633-0925
Fax: (612) 633-6073
Products also available through various distributors

Volu-Sol, Inc. (a division of Biomune, Inc.)
5095 West 2100 South
Salt Lake City, UT 84120
(800) 821-2495
(801) 974-9474
Fax: (800) 860-4317
Products also available through various distributors

Table A7–4–3 Commercial suppliers of diagnostic parasitology products[a]

Alexon-Trend
1190 Borregas Avenue
Sunnyvale, CA 94089-1302
(800) 366-0096
Fax: (408) 747-7011
Stains, slides, *Giardia lamblia* (EIA, FA), *Cryptosporidium* spp. (EIA, FA)

ALPHA-TEC Systems, Inc.
P.O. Box 5435
Vancouver, WA 98668
(800) 221-6058
(360) 260-2779
Fax: (360) 260-3277
Control slides, reagents, stains (including polychrome IV), collection system, concentration system, QC suspensions

American Micro Scan[b]
1584 Enterprise Blvd.
West Sacramento, CA 95691
(916) 372-1900
Stains, reagents, collection system, concentration system

Antibodies Inc.[b]
P.O. Box 1560
Davis, CA 95617
(800) 824-8540
(916) 758-4400 in California
Giardia lamblia (ELISA and FA)

BIOMED Diagnostics
1430 Koll Circle, Suite 101
San Jose, CA 95112
(800) 964-6466
InPouch (visual identification and culture system for *Trichomonas vaginalis*)

Diagnostic Technology, Inc.[b]
240 Vanderbilt Motor Parkway
Hauppauge, NY 11788
(516) 582-4949
Fax: (516) 582-4694
Toxoplasma gondii (EIA, IgG, IgM)

Diamedix Corp.
2140 N. Miami Avenue
Miami, FL 33127
(800) 327-4565
Fax: (305) 324-2395
Entamoeba histolytica (EIA), *Toxoplasma gondii* (EIA, IgG, IgM)

Eastman Kodak Co.
Rochester, NY 14650
(800) 225-5352
(716) 458-4014
PVA powder, Giemsa powder (products available only through American Scientific Products, Fisher Scientific, VWR, and through the company)

Empyrean Diagnostics, Inc.
2761 Marine Way
Mountain View, CA 94043
(415) 960-0516
Fax: (415) 960-0515
Visual identification and culture system for *Trichomonas vaginalis*

Evergreen Scientific
2300 East 49th Street
P.O. Box 58248
Los Angeles, CA 90058
(800) 421-6261
(213) 583-1331
Fax: (213) 581-2503
Collection system, concentration systems

General Biometrics, Inc.[b]
15222 Avenue of Science
Bountiful, UT 84117
(800) 288-4368
Fax: (619) 592-9400
Toxoplasma gondii (EIA, IgG; IFA, IgG, IgM)

Gull Laboratories, Inc.
1011 East Murray Holladay Road
Salt Lake City, UT 84117
(801) 263-3524
Fax: (801) 265-9268
Toxoplasma gondii (EIA, IgG, IgM; IFA, IgG, IgM), *Trypanosoma cruzi* (EIA, IgG; IFA, IgG), formalin suspensions (parasites), parasite slides (stained and unstained)

(continued)

Table A7–4–3 Commercial suppliers of diagnostic parasitology products[a] *(continued)*

Hardy Diagnostics
1430 West McCoy Lane
Santa Maria, CA 93455
(800) 266-2222
Fax: (805) 346-2760
Stains, reagents, collection system, concentration systems, control slides

HDC Corp.
2109 O'Toole Avenue, Suite M
San Jose, CA 95131
(408) 954-1909
Fax: (408) 954-0340
Entero-Test capsules (adult and pediatric) (method of sampling upper gastrointestinal tract)

INCStar Corp.[b]
1990 Industrial Blvd.
P.O. Box 285
Stillwater, MN 55082
Toxoplasma gondii (EIA, IgG, IgM; IFA, IgG, IgM)

Interfacial Dynamics Corp.
17300 SW Upper Boones Ferry Road, Suite 120
Portland, OR 97224
(503) 256-0076
(800) 323-4810
Fax: (503) 255-0989
Uniform-sized polystyrene microspheres (can be used to check microscope calibrations)

Medical Chemical Corp.
1909 Centinela Avenue
Santa Monica, CA 90404
(310) 829-4304
Fax: (310) 453-1212
Reagents, modified trichrome stain for microsporidia, other stains, collection system, concentration system

Meridian Diagnostics, Inc.
3741 River Hills Drive
Cincinnati, OH 45244
(800) 543-1980
(513) 271-3700
Fax: (513) 271-0124
Stains, reagents, collection systems, concentration systems, *Cryptosporidium* spp. (EIA, FA), *Giardia lamblia* (EIA, FA)

MicroBioLogics Inc.
217 Osseo Avenue North
Saint Cloud, MN 56303
(800) 599-2847
Fax: (320) 253-6250
Control slides (*Cryptosporidium* spp., *Plasmodium* spp., *Pneumocystis* spp., intestinal parasites, microsporidia)

MML Diagnostic Packaging
P.O. Box 458
Troutdale, OR 97060
(503) 666-8398
(800) 826-7186
Collection system

Pharmacia Diagnostics, Inc.[b]
800 Centennial Avenue
P.O. Box 2803
Piscataway, NJ 08854
(800) 346-4364
Fax: (201) 227-5432
Entamoeba histolytica (EIA), *Toxoplasma gondii* (EIA, IgG, IgM)

PML Microbiologicals
P.O. Box 459
Tualatin, OR 97062
(800) 547-0659
(503) 639-1500 in Oregon
Fax: (800) 765-4415
Stains, reagents, collection system, concentration system, control slides

Remel (Regional Media Laboratories)
12076 Santa Fe Drive
Lenexa, KS 66215
(800) 255-6730
Fax: (800) 477-5781
Stains, reagents, collection system, concentration system, control slides

Sanofi Pasteur[b]
1000 Lake Hazeltine Dr.
Chaska, MN 55318
(800) 666-5111
Fax: (612) 368-1110
Toxoplasma gondii (EIA, IgG, IgM)

Scientific Device Laboratory, Inc.
411 E. Jarvis Avenue
Des Plaines, IL 60018
(847) 803-9495
Fax: (847) 803-8251
Stains, stain kit, modified trichrome stain for microsporidia, fixative (formalin-free), formalin-fixed protozoa and helminth eggs and larvae, control slides (*Cryptosporidium* spp., *Isospora* spp., microsporidia, *Pneumocystis carinii*), stained slides

TechLab
VPI Research Park
1861 Pratt Drive
Blacksburg, VA 24060-6364
(540) 231-3943
Fax: (540) 231-3942
Giardia lamblia (EIA, FA), *Cryptosporidium* spp. (EIA, FA), *Entamoeba histolytica/E. dispar* (EIA), *Entamoeba histolytica* (EIA)

Volu-Sol, Inc. (a division of Biomune, Inc.)
5095 West 2100 South
Salt Lake City, UT 84120
(800) 821-2495
(801) 974-9474
Fax: (800) 860-4317
Parasitology starter kits, stains, reagents, control slides

Table A7–4–3 Commercial suppliers of diagnostic parasitology products[a] *(continued)*

VWR Scientific, Inc.[b] P.O. Box 7900 San Francisco, CA 94120 (415) 468-7150 in northern California (213) 921-0821 in southern California Rest of United States, use either number Stains, reagents, collection system	Wampole Laboratories[b] Half Acre Road P.O. Box 1001 Cranbury, NJ 08512 (609) 655-6000 Fax: (609) 655-6660 *Entamoeba histolytica* (hemagglutination inhibition)

[a] Modified from CMPH 7.10.6. Abbreviations: EIA, enzyme immunoassay; ELISA, enzyme-linked immunosorbent assay; FA, fluorescent-antibody assay; IgG, immunoglobulin G; IgM, immunoglobulin M; QC, quality control.
[b] No updated information received for this publication.

Table A7–4–4 Sources of parasitologic specimens (catalogs of available materials and price lists available from the companies and person listed)[a]

Ann Arbor Biological Center 6780 Jackson Road Ann Arbor, MI 48103	Trend Scientific, Inc. 368 W. County Road P.O. Box 120266 St. Paul, MN 55112-0266 (800) 328-3949 (612) 633-0925 Fax: (612) 633-6073
Dr. David A. Bruckner Clinical Microbiology A2-179 UCLA Medical Center (171315) Los Angeles, CA 90095-1713 (310) 794-2751 Fax: (310) 794-2765	Triarch, Inc. N8028 Union St. P.O. Box 98 Ripon, WI 54971 (414) 748-5125 (800) 848-0810 Fax: (414) 748-3034
Carolina Biological Supply Co. 2700 York Road Burlington, NC 27215 (800) 334-5551 (919) 584-0381 Fax: (919) 584-3399	Tropical Biologicals P.O. Box 139 Guaynabo, PR 00657
Gull Laboratories, Inc. 1011 East Murray Holladay Road Salt Lake City, UT 84117 (801) 263-3524 Fax: (801) 265-9268	TURTOX P.O. Box 92912 Rochester, NY 14692 (800) 826-6164
Scientific Device Laboratory, Inc. 411 E. Jarvis Avenue Des Plaines, IL 60018 (708) 803-9495 Fax: (708) 803-8251	Ward's Natural Science Establishment, Inc. P.O. Box 92912 Rochester, NY 14692 (800) 962-2660

[a] Adapted from CMPH 7.10.6.

Table A7–4–5 Sources of Kodachrome slides (35 mm, 2 × 2)[a]

Armed Forces Institute of Pathology Dept. ID/Parasitic Disease Pathology 6825 16th Street, NW Bldg. 54, Room 4015 Washington, DC 20306-6000
American Society of Clinical Pathologists ASCP Press 2100 West Harrison Street Chicago, IL 60612-3798 (312) 738-4890 Human Parasitology, teaching slide set (Ash and Orihel, 1990), Supplement to Human Parasitology, teaching slide set (in preparation), Parasites in Human Tissues, teaching slide set (Orihel and Ash, 1996)
Clinical Diagnostic Parasitology Visual Teaching Aids (set of 100 2 × 2 slides) Diagnostic Protozoa, Helminths, and Blood Parasites (Lynne S. Garcia) 512 12th St. Santa Monica, CA 90402
W. H. Curtin and Co. P.O. Box 1546 Houston, TX 77001
TURTOX P.O. Box 92912 Rochester, NY 14692
Dr. Herman Zaiman (A Pictorial Presentation of Parasites) P.O. Box 543 Valley City, ND 58072

[a] Adapted from CMPH 7.10.6.

SECTION 8

Viruses, Rickettsiae, Chlamydiae, and Mycoplasmas

SECTION EDITOR AND AUTHOR: *Lorraine Clarke*

8.1 Laboratory Diagnosis of Viral Infections: Introduction

The virology laboratory uses several diagnostic modalities, including culture, antigen and nucleic-acid assays, cytohistopathology, and serology, to aid the physician in the diagnosis of viral infections. Many variables influence the choice of methods used by a particular laboratory, including the type of patient population (e.g., adult, pediatric, or immunocompromised), the test application (e.g., detecting infection, identifying the etiology of disease, or assessing the response to therapy), the setting (e.g., outpatient clinic, emergency room, or transplant unit), specific viral characteristics, and the resources available. Thus, no single approach can satisfy all needs, and the laboratory scientist must carefully assess these variables and develop testing algorithms suitable for the particular laboratory.

This section provides useful information not only for a full-service virology laboratory but also for those laboratories performing only limited cultures or nonculture assays, as well as for those facilities dispatching specimens to off-site laboratories for viral testing. To ensure successful and proper use of viral-testing facilities, the laboratory must provide written guidelines to personnel at collection sites (e.g., nursing stations, clinics, and physician offices). These guidelines should include tests available; laboratory hours of operation and contact person(s); instructions for specimen collection, labeling, storage, and transport; sources and storage conditions for transport media and containers; information required for adequate testing; turnaround time; reporting procedures and values; and assay limitations.

Knowledge of the natural history and pathogenesis of viral infections is essential for the proper use of laboratory assays and accurate interpretation of results. Several sources, including those cited below, can be consulted regarding the epidemiology, pathogenesis, clinical manifestations, and diagnosis of viral infections. Human viral infections and their laboratory diagnostic approaches are summarized in Table 8.1–1; zoonotic viral infections and their laboratory diagnosis are summarized in Table 8.1–2.

Table 8.1–1 Clinical manifestations of human viral diseases[a]

Viral agent and diagnostic methods[b]	Respiratory	CNS and neuromuscular	Urinary tract	Mucocutaneous	Gastrointestinal	Ocular	Other
Adenoviruses (>42 types) A: 1, 2, 7	URI (common cold, pharyngitis, tonsillitis), pertussis syndrome, acute respiratory disease, bronchitis, pneumonia	Rare: meningitis, encephalomyelitis	Acute hemorrhagic cystitis	Exanthems	Gastroenteritis (types 40 and 41)	Conjunctivitis, pharyngoconjunctival fever, follicular conjunctivitis, epidemic keratoconjunctivitis, acute hemorrhagic conjunctivitis	Immunocompromised patients (6): pneumonia, gastroenteritis, meningoencephalitis, disseminated disease
Coronaviruses (strains 229E and OC43) NA	Common cold						
Enteroviruses							
Coxsackie group A (types 1–22, 24) A: 1 (not all types), 5, 7 (not all types)	Common cold, pharyngitis, pneumonitis in infants	Aseptic meningitis, acute rhabdomyelosis, paralysis (uncommon)		Herpangina; lymphonodular pharyngitis; gingivostomatitis; hand, foot, and mouth disease; exanthems (vesicular, petechial, maculopapular)	Infantile diarrhea, hepatitis	Acute hemorrhagic conjunctivitis	Undifferentiated febrile illness; severe disease in neonates and in individuals with agammaglobulinemia
Coxsackie group B (6 types) A: 1, 5, 7	URI, pneumonia	Meningoencephalitis (infants), paralysis (uncommon)		Exanthems (maculopapular, vesicular, and petechial)	Hepatitis		Undifferentiated febrile illness, myocarditis, pericarditis, pleurodynia; severe disease in neonates and in individuals with agammaglobulinemia
Echoviruses (types 1–9, 11–27, and 29–34) A: 1, 5, 7 (not all types)	URI	Aseptic meningitis, encephalitis, paralysis, acute rhabdomyelosis		Exanthems (maculopapular, vesicular, petechial, hemangioma-like)			Undifferentiated febrile illness, diabetes/pancreatitis, orchitis; severe disease in neonates and in individuals with agammaglobulinemia
Enterovirus type 68 A: 1, 5	Pneumonia, bronchiolitis						
Enterovirus type 70 A: 1, 5		Paralytic myelitis				Acute hemorrhagic conjunctivitis	
Enterovirus type 71 A: 1, 5		Aseptic meningitis, meningoencephalitis, paralysis		Hand, foot, and mouth disease			
Polioviruses (3 types) A: 1, 5, 7		Aseptic meningitis, poliomyelitis, paralysis					
Gastroenteritis viruses[c] (Norwalk agent and other caliciviruses and astroviruses)					Common-source outbreaks of gastroenteritis		Associated with contaminated drinking water and shellfish

Hepatitis viruses							
Types A, E, and Delta A: 7					Hepatitis		HBV and HCV associated with hepatocellular carcinoma
Type B A: 2, 3, 4, 7							
Type C A: 5, 7							
Herpesviruses							
CMV A: 1, 2, 3, 6, 7 S:4[d]	Neonatal pneumonitis						Heterophile-negative mononucleosis; congenital CMV syndrome; immunocompromised individuals: pneumonia, hepatitis, retinitis, encephalitis, myelitis, polyradiculopathy, gastrointestinal disease (esophagitis, gastritis, colitis, papillary stenosis, cholangitis, hepatitis, proctitis), adrenalitis, bone marrow graft suppression
Epstein-Barr virus A: 7 S: 4[d]					Hepatitis		Infectious mononucleosis (complications include splenic rupture, chronic mononucleosis syndrome); associated with Burkitt's lymphoma, nasopharyngeal carcinoma; AIDS-associated CNS lymphoma
Herpes simplex (2 types) A: 1, 2, 6, 7 S: 4[d]	Pharyngitis	Meningoencephalitis, encephalitis, Mollaret's meningitis (recurrent aseptic meningitis)	Anogenital vesicular/ ulcerative mucocutaneous lesions, cervicitis	Vesicular and ulcerative lesions, herpetic whitlow, gingivostomatitis, erythema multiforme	Proctitis	Conjunctivitis, keratitis	Infection may be severe in neonates; CNS and disseminated disease can occur in the absence of mucocutaneous lesions; immunocompromised patients: severe/persistent mucocutaneous lesions, proctitis, esophagitis, encephalitis, hepatitis, pneumonitis
HHV-6 A: 7 S: 4[d]		Encephalitis in young children		Exanthem subitum (roseola infantum, sixth disease)			Febrile illness
HHV-7 NA				May be associated with exanthem subitum (see HHV-6)			Disease association remains to be determined
HHV-8 NA							Identified in association with AIDS-associated KS; role in KS remains to be elucidated

(continued)

Table 8.1–1 Clinical manifestations of human viral diseases[a] *(continued)*

Viral agent and diagnostic methods[b]	Respiratory	CNS and neuromuscular	Urinary tract	Mucocutaneous	Gastrointestinal	Ocular	Other
Varicella-zoster virus A: 1, 2, 6, 7 S:4[d]	Pneumonia	Rare: cerebellar ataxia		Chickenpox, shingles	Hepatitis		Neonatal disease may be severe; congenital disease is rare; immunocompromised patients: zoster, visceral disease, retinitis, encephalitis, myelitis
Human papillomaviruses (>75 types) A: 3, 6	Laryngeal papillomatosis		Condylomata, anogenital intraepithelial lesions	Common warts, plantar warts, epidermodysplasia verruciformis			Immunocompromised patients: disease may be more severe or progressive, e.g., AIDS patients; associated with anogenital carcinomas
Influenza virus types A, B, and C A: 1, 2, 7	Influenza, pneumonia (particularly in elderly and debilitated), otitis media, sinusitis						Severely immunocompromised adults may be at risk for pneumonia
Paramyxoviruses							
Mumps A: 1, 2, 7		Meningoencephalitis, aseptic meningitis, encephalomyelitis					Orchitis, oophoritis, pancreatitis
Parainfluenza viruses (4 types) A: 1, 2, 7	Laryngotracheobronchitis (croup), bronchiolitis, tracheobronchitis, pneumonia	Rare: meningitis					Severely immunocompromised adults may be at risk for pneumonia
Respiratory syncytial virus A: 1, 2, 7	URI, bronchiolitis, pneumonia, croup						Severe disease may occur in elderly or immunocompromised patients (e.g., bone marrow transplant recipients)
Measles virus (rubeola) A: 1, 2, 7		Encephalitis, SSPE		Measles			Immunocompromised patients: giant cell pneumonia
Parvovirus B19 A: 7 S: 4		Arthralgia		Erythema infectiosum (fifth disease)			Aplastic crisis in sickle cell patients; immunocompromised patients: chronic anemia; congenital disease (fetal hydrops)
Polyomaviruses (BK and JC viruses) S: 1, 4		JC virus is associated with progressive multifocal leukoencephalopathy (PML) in immunocompromised patients	May be associated with hemorrhagic cystitis in renal transplant recipients				Asymptomatic infection appears common in immunocompromised individuals

Poxviruses							
Molluscum contagiosum A: 6				Pearly nodular exanthem with caseous exudate			Most common cutaneous manifestation of advanced HIV infection, occurring most commonly in the head and neck area (11)
Vaccinia virus A: 1, 7				Mucocutaneous vesicles			Potential for exposure confined to laboratories working with this agent
Reoviruses A: 1	URI, pharyngitis			Maculopapular rashes	Gastroenteritis		Association with disease is not well documented; rarely isolated from clinical samples
Retroviruses							
HIV-1 A: 2, 4, 5, 7 S: 1		Encephalomyopathy and dementia					AIDS
HIV-2 A: 7							AIDS
HTLV-1 A: 7 S: 4		HTLV-associated myopathy, tropical spastic paraparesis					Associated with adult T-cell leukemia
Rhinoviruses A: 1	Common cold; lower respiratory tract disease may occur with predisposing conditions						
Rotaviruses A: 2					Gastroenteritis, primarily among infants		Infection may be chronic in immunocompromised patients
Rubella virus A: 1, 7		Arthralgia, polyarthritis (these occur primarily in females)		German measles			Congenital rubella syndrome

NOTE: Coxsackievirus A23 shown to be echovirus type 9; echovirus type 10 reclassified as a reovirus; echovirus type 28 reclassified as rhinovirus type 1; echovirus type 34 is related to coxsackievirus A24 as a prime strain; disease association has not been demonstrated for enterovirus type 69.

[a] For additional information, see references 2, 5, 8, and 9. Abbreviations: CNS, central nervous system; URI, upper respiratory infection; HBV, hepatitis B virus; HCV, hepatitis C virus; CMV, cytomegalovirus; HHV, human herpesvirus; KS, Kaposi's sarcoma; HIV, human immunodeficiency virus; HTLV, human T-cell leukemia virus; SSPE, subacute sclerosing panencephalitis.

[b] Symbols: A, Available through clinical virology laboratories or reference laboratories; S, performed by a limited number of laboratories, including specialized reference laboratories; NA, not available in the diagnostic setting; 1, culture; 2, antigen assays; 3, unamplified DNA assays; 4, amplified DNA assays; 5, amplified RNA assays; 6, cytohistopathology; 7, serology (antibody assays).

[c] Electron microscopy and polymerase chain reaction are available through the CDC or public health laboratories.

[d] Availability of polymerase chain reaction for the detection of herpesviruses in CSF is increasing.

Table 8.1–2 Zoonotic viral infections associated with human disease (2, 5, 8, 9, 10)

Virus	Respiratory	Central nervous system and neuromuscular[a]	Mucocutaneous	Gastrointestinal	Other	Lab diagnosis	Comments
Arboviruses (3)							
Arthropod-borne viruses; include viruses belonging to several major families including *Togaviridae, Flaviviridae, Bunyaviridae, Reoviridae, Rhabdoviridae, Othomyxoviridae,* and *Poxviridae*		Encephalitis (e.g., EEE, WEE, VEE, SLE, Everglades, California, LaCrosse, and Powassan viruses); myalgia, arthralgia (e.g., Colorado tick fever, dengue)	Maculopapular rashes (several, including dengue)	Hepatitis (e.g., yellow fever), hemorrhagic fevers	Undifferentiated febrile illness (several, including Colorado tick fever); hemorrhagic fever (e.g., Rift Valley fever, dengue, yellow fever)	Serology: immunoglobulin M testing (MACEIA) on serum, plasma, and CSF; follow-up with convalescent-phase serum samples to demonstrate diagnostic titer increases and/or confirmation with other assays (e.g., neutralization) Isolation, antigen, etc.: blood (viremia usually brief); CSF; autopsy tissues (brain, spleen, lung, liver)	Man is an incidental host with transmission by hematophagous arthropod vectors (primarily mosquitoes but also ticks or flies); some (e.g., Powassan, a tick-borne infection) may be acquired by ingestion of milk from viremic cattle, goats, and sheep. The World Health Organization centers (Vector-borne Virus Diseases Division, CDC, Fort Collins, Colo.; Arbovirus Research Unit, Yale University, New Haven, Conn.) perform special diagnostic tests and offer consultation. BSL 2, BSL 3, or BSL 4 levels required, with agents causing encephalitis or hemorrhagic fever usually requiring the higher biosafety levels.
Arenaviruses							
Lassa fever virus and other viruses causing hemorrhagic fevers					Lassa fever, hemorrhagic fever	Serology: LCMV testing more widely available than other agents. For information and testing of suspected hemorrhagic fever viruses, contact CDC, Special Pathogens	Maintained in nature in rodent hosts. Transmission is usually via an aerosol route from human to human or following exposure to infected rodent excreta.

				Branch, Atlanta, Ga. Isolation, antigen, etc.: blood, nasopharyngeal washing, CSF (LCMV), autopsy tissues	Except for LCMV, these viruses are geographically restricted to tropical Africa and South America and require BSL 4 containment practices. LCMV is widely distributed, with the house mouse (*Mus musculus*) being its natural host. Infected mice and hamsters are important sources to laboratory workers and pet owners.
LCMV (1)	Flulike illness	Aseptic meningitis, meningoencephalitis	Undifferentiated febrile illness	See above	See above
Filoviruses Ebola, Marburg virus				Same as for arenaviruses	Same as for arenaviruses
Hantaviruses *Bunyaviridae*	HPS: sin nombre virus (7)		Hemorrhagic fever with renal syndrome (e.g., Korean hemorrhagic fever virus)	Serology (serum, plasma): IgM-capture enzyme, IgG assay Isolation: viral culture has not been successful in recovering virus from cases of HPS Polymerase chain reaction and immunocytochemistry are available through public health laboratories or the CDC.	Mice (e.g., deer mouse [*Permyscus maniculatus*]) are the natural hosts. Transmission is by the aerosol route after exposure to infected rodent excreta; person-to-person transmission does not appear to occur.
Newcastle disease virus (*Paramyxoviridae*)	Influenzalike illness		Conjunctivitis	Isolation in cell culture	Transmission is by the aerosol route from infected fowl to humans in the poultry industry.

(continued)

Table 8.1–2 Zoonotic viral infections associated with human disease (2, 5, 8, 9, 10) (*continued*)

Virus	Respiratory	Central nervous system and neuromuscular[a]	Mucocutaneous	Gastrointestinal	Other	Lab diagnosis	Comments
Zoonotic poxviruses (para- and orthopoxviruses)			Milker's nodule (pseudocowpox), orf (pustular dermatitis; sheep and goats), monkeypox			Serology: acute and convalescent serum samples Cytohistopathology: detection of typical eosinophilic cytoplasmic inclusion bodies (Guarnieri bodies) Isolation, antigen, etc: lesion scrapings	Occurs in individuals exposed to infected animals; limited testing available through public health laboratories or the CDC.
Rabies		Encephalitis				Antigen by immunofluorescence assay: animals; brain; humans; skin biopsy with several (~10) hair follicles from the posterior region of the neck just above the hairline; saliva; brain at autopsy Cytohistopathology: Negri bodies Serology (serum, CSF); antibody not present until >1 week after onset	Man is an accidental host after exposure to infected animals. In North America rabies occurs in bats, cats, dogs, cattle, deer, foxes, raccoons, skunks, wolves, coyotes, opossums, weasels, and other animals; bat bites may be imperceptible; contact the local public health laboratory or the CDC for information regarding handling and testing of potentially infected animals and for vaccination recommendations for individuals at high risk of exposure or following a potential exposure; BSL 2 practices are recommended for handling potentially infectious materials. Individuals preparing animals for shipment to a testing laboratory should wear heavy rubber gloves and protective clothing, including a face shield.

Simian B virus (*Herpesviridae*)	Encephalitis	Vesicular lesions	Isolation: lesion specimen (vesicular fluid, lesion swab or scraping). Although isolation is possible in cells routinely available in the diagnostic laboratory (e.g., PMK), testing should be referred to a specialized laboratory. Serology: EIA, Western blot	Simian B virus (herpesvirus simiae) is indigenous to Old World monkeys such as rhesus (*Macaca mulatta*), cynomolgus (*Macaca fascicularis*), and other Asiatic monkeys of the genus *Macaca*. Infection in primates is asymptomatic or manifested by dermal, oral, ocular, or genital lesions. Transmission to humans occurs primarily by bites or by direct contact with saliva or infected tissues, although human-to-human transmission has been reported (4). Assistance and information can be obtained from CDC, Division of Viral and Rickettsial Diseases, Atlanta, Ga., (404) 639-1338; Southwest Foundation for Biomedical Research, San Antonio, Tex., (512) 674-1410; Virus Reference Laboratory Inc., San Antonio, Tex., (512) 696-5510. BSL 2 practices are recommended for all activities involving the use or manipulation of tissues, body fluids, and primary tissue culture material from macaques, with BSL 3 practices recommended for activities involving the use or manipulation of any materials known to contain this virus.

[a] Abbreviations: EEE, eastern equine encephalitis; WEE, western equine encephalitis; VEE, Venezuelan equine encephalitis; SLE, St. Louis encephalitis; LCMV, lymphocytic choriomeningitis virus; HPS, hantavirus pulmonary syndrome; MACEIA, IgM antibody capture EIA.

REFERENCES

1. **Barton, L. L.** 1996. Lymphocytic choriomeningitis virus: a neglected central nervous system pathogen. *Clin. Infect. Dis.* **22:**197.
2. **Belshe, R. B. (ed.).** 1991. *Textbook of Human Virology*, 2nd ed. The C. V. Mosby Co., St. Louis.
3. **Calisher, C. H.** 1994. Medically important arboviruses of the United States and Canada. *Clin. Microbiol. Rev.* **7:**89–116.
4. **Centers for Disease Control.** 1987. B virus infection in humans—Pensacola, Florida. *Morbid. Mortal. Weekly Rep.* **36:**289–290, 295–296.
5. **Chonmaitree, T., C. D. Baldwin, and H. L. Lucia.** 1989. Role of the virology laboratory in diagnosis and management of patients with central nervous system disease. *Clin. Microbiol. Rev.* **2:**1–14.
6. **Hierholzer, J. C.** 1992. Adenoviruses in the immunocompromised host. *Clin. Microbiol. Rev.* **5:**262–274.
7. **Khan, A. S., T. G. Ksiazek, and C. J. Peters.** 1996. Hantavirus pulmonary syndrome. *Lancet* **347:**739–741.
8. **Lennette, E. H., D. A. Lennette, and E. T. Lennette (ed.).** 1995. *Diagnostic Procedures for Viral, Rickettsial, and Chlamydial Infections*, 7th ed. American Public Health Association, Washington, D.C.
9. **Murray, P. M., E. J. Baron, M. A. Pfaller, F. C. Tenover, and R. H. Yolken (ed.).** 1995. *Manual of Clinical Microbiology*, 6th ed. American Society for Microbiology, Washington, D.C.
10. **Public Health Service, CDC, and NIH.** 1993. *Biosafety in Microbiological and Biomedical Laboratories.* Publication no. (NIH) 93-8395, 3rd ed. U.S. Department of Health and Human Services, U.S. Government Printing Office, Washington, D.C.
11. **Tappero, J. W., B. A. Perkins, J. D. Wenger, and T. G. Berger.** 1995. Cutaneous manifestations of opportunistic infections in patients infected with human immunodeficiency virus. *Clin. Microbiol. Rev.* **8:** 440–450.

SUPPLEMENTAL READING

Evans, A. S., and R. A. Kaslow (ed.). 1997. *Viral Infections of Humans: Epidemiology and Control*, 4th ed. Plenum Publishing Corp., New York.

Gorbach, S. L., J. G. Bartlett, and N. R. Blacklow (ed.). 1992. *Infectious Diseases.* The W. B. Saunders Co., Philadelphia.

Hsiung, G. D., C. K. Y. Fong, and M. L. Landry. 1994. *Hsiung's Diagnostic Virology*, 4th ed. Yale University Press, New Haven, Conn.

Mandell, G. L., J. E. Bennett, and R. Dolin (ed.). 1995. *Principles and Practice of Infectious Diseases*, 4th ed. Churchill Livingstone, New York.

Remington, J. S., and J. O. Klein (ed.). 1994. *Infectious Diseases of the Fetus and Newborn Infant,* 4th ed. The W. B. Saunders Co., Philadelphia.

8.2 Viral Culture: Selection, Assessment, Quality Control, and Maintenance of Uninoculated Cell Cultures

I. PRINCIPLE

Viruses are obligate intracellular parasites requiring metabolically active cells to support their replication. These organisms vary in their ability to be cultivated, from those that can be conveniently cultured in the diagnostic setting to those for which a laboratory culture system has not yet been identified (Table 8.2–1). All cell cultures routinely used in the diagnostic laboratory are available commercially and consist of single layers (monolayers) of metabolically active cells which are adherent to either the side of a glass tube or the surface of a cover slip contained in a flat-bottomed shell vial.

Monolayers are derived either from tissue obtained from a donor organ (primary cell cultures) or by subculture from established monolayers of diploid or heteroploid cell lines (3). Tissues or cells are treated with proteolytic enzymes and/or chelating agents to dissociate the cells, and the resultant cell suspension is used to seed the tubes or shell vials. The viable cells attach to the surface of the culture vessel within a few hours of seeding and then begin to replicate; monolayers generally reach confluence within 1 to 4 days, depending on cell density used for seeding as well as characteristics of the cells themselves. Adherent (monolayer) cell cultures most frequently utilized in the diagnostic laboratory and their characteristics are summarized in Table 8.2–2. Viruses exhibit a selective pattern of replication in cell cultures (Table 8.2–3), and the number of different cell culture types used by a particular laboratory depends on the range of anticipated isolates (Fig. 8.2–1).

This procedure provides information necessary for laboratories in which commercially supplied cell cultures are used. For on-site preparation of cell cultures, see CMPH 8.20.

II. MATERIALS

Include QC information on reagent container and in QC records.

A. Cell cultures

Cell culture monolayers (received or prepared weekly)

Most full-service viral culture laboratories utilize a combination of conventional tube cultures and shell vial cultures. Shell vials are particularly valuable for performing cytomegalovirus (CMV) (9, 20) and varicella-zoster virus (VZV) (8, 23, 30) cultures but have been applied to the rapid detection of several other viruses including herpes simplex virus (HSV) (10), respiratory viruses (5–7, 13, 16, 19, 25, 29), and measles (18). Although the shell vial technique is rapid and provides good sensitivity in many cases, some laboratories have experienced lower recovery rates of CMV from blood (14, 20) and of adenoviruses from respiratory samples (19) with shell vial cultures. See Appendix 8.2–1 for commercial cell culture sources.

B. Culture medium

Eagle's minimum essential medium in either Earle's or Hanks' balanced salt solution (EBSS or HBSS, respectively) and containing 2% heat-inactivated fetal bovine serum (FBS) is a medium suitable for the maintenance of routinely used, uninoculated and inoculated monolayer cell cultures (*see* EPCM 8.6 for details).

C. Supplies

1. Racks designed to hold culture tubes in a horizontal position, with the tube necks slightly raised
2. Sterile individually wrapped pipettes (5 and 10 ml) and safety pipetting devices
3. Protective clothing
4. Sodium hypochlorite solution (0.5%)
5. Autoclavable containers for discarding infectious supplies and cultures

D. Equipment

1. Microscope (100×)

   Shell vials can be examined by using an inverted microscope with a glass microscope slide to support the vial. Tube cultures can be examined with a standard or inverted mi-

Table 8.2–1 Viral culture systems and availability

Culture system	Culture availability	Viruses
Routinely available monolayer cell cultures[a]	Widely available	Adenoviruses, group A coxsackieviruses (not all types), group B coxsackieviruses, CMV, echoviruses, enterovirus types 68–71, HSV, influenza viruses, measles virus (rubeola), mumps virus, parainfluenza viruses, polioviruses, reoviruses, respiratory syncytial virus, rhinoviruses, rubella virus, vaccinia virus, VZV
	Limited availability	Polyomaviruses (BK)
Leukocyte cultures	Limited availability	Human herpesvirus type 6[b], human immunodeficiency virus type 1
	Not available in the diagnostic setting	Epstein-Barr virus
Specialized cell cultures (e.g., organ cultures)	Research settings	Coronaviruses, hepatitis type A virus (enterovirus type 72), para- and yatapoxviruses, rotaviruses
Animal hosts (suckling mice)	Not available in the diagnostic setting	Several types of group A coxsackieviruses
	Specialized reference labs, state laboratories, CDC	Rabies, arboviruses[c], arenaviruses[c], filoviruses[c]
Noncultivatable		Papillomaviruses, molluscum contagiosum, parvovirus B19, hepatitis viruses (types B, C, etc.), Norwalk agent and other viral gastroenteritis viruses (e.g., caliciviruses, astroviruses)

[a] Orthopoxviruses (vaccinia, cowpox, monkeypox) and simian B virus are also cultivatable in routinely used cell cultures, but these cultures are not frequently requested; BSL 3 practices are recommended for handling cultures to be submitted to reference laboratories for simian B virus culture.

[b] Has been cultivated in human lung fibroblast shell vial monolayer cultures.

[c] Replicate in some cell lines (e.g., Vero, LLC-MK_2). Several require BSL 3 or BSL 4 containment; clinical presentation and travel and/or exposure history should alert the laboratorian to the possibility of these agents and the need to use appropriate precautions. Contact a local public health laboratory or the Special Pathogens Branch, CDC, regarding information and testing.

Table 8.2–2 Monolayer cell cultures used in diagnostic virology (3)

Culture type	Source	Subpassages	Characteristics	Examples[a] and comments
Primary	Donor tissue	Few	Epithelial-like; karyotype same as tissue of origin; can be subpassaged for one or two generations, but original characteristics may be lost, including viral susceptibility	PMK, HNK, RK, AGMK; primary simian cultures are frequently contaminated with endogenous viruses that may produce cytopathic effects and/or hemadsorption
Diploid (semicontinuous)	Monolayer	30–50	≥75% of cells have same karyotype (normal) as tissue of origin	MRC-5, WI-38, HLF, HNF; viral sensitivity may diminish with increasing passage level; must be monitored periodically for mycoplasma contamination
Heteroploid (continuous, immortalized, established, transformed)	Monolayer	Unlimited	<75% of cells have the same karyotype as the normal cells of the tissue of origin	HEp-2, A-549, H292, BGMK, RD, Graham-293, LLC-MK_2, MDCK, HeLa, Vero, McCoy, ML; must be monitored periodically for mycoplasma contamination

[a] A-549, human epidermoid lung carcinoma; AGMK and Vero, African green monkey kidney; BGMK, Buffalo green monkey kidney; Graham-293, adenovirus-transformed human kidney; H292, human pulmonary epidermoid carcinoma; HeLa, human cervical carcinoma; HEp-2, human laryngeal carcinoma; HLF, MRC-5, and WI-38, diploid human lung fibroblasts; HF or HNF, human fetal or neonatal foreskin; HNK, human neonatal kidney; LLC-MK_2, rhesus monkey kidney; McCoy, mouse fibroblasts; MDCK, Madin-Darby canine kidney; ML, mink lung; RD, rhabdomyosarcoma; RK, rabbit kidney; PMK, primary monkey kidney (rhesus, cynomolgus).

Table 8.2–3 Cell cultures used for viral culture[a]

Virus	Cell cultures (references)	Comments (references)
Adenovirus types 1–39	HNK, A-549	Primary or low-passage human kidney cells are preferred to continuous cell lines; isolation in PMK is variable.
Adenovirus types 40–41 (enteric)	Graham-293 (1, 12), A-549	
BK virus	MRC-5, WI-38, HF, HNK	CPE may require several weeks to develop; can use shell vial cultures. Antibody to simian virus 40 T antigen cross-reacts with human polyomaviruses.
CMV	MRC-5, WI-38, HF	Shell vials are recommended.
Enteroviruses (poliovirus types 1–3; echovirus types 1–9, 11–27, 29–34; group A coxsackievirus types 1–22, 24; group B coxsackievirus types 1–6; enterovirus types 68–71)	Maximum recovery achieved with a panel of 4 cell types (primary human and monkey kidney cells, BGMK, RD (11, 17, 22); diploid human lung cells (e.g., MRC-5) have limited sensitivity	RD cells support the replication of a number of group A coxsackieviruses that cannot be cultivated in other cell lines (15, 24). However, this cell line monolayer overgrows rapidly and is difficult to interpret.
HSV types 1 and 2	A-549, RK, ML, HNK, MRC-5, WI-38	Isolation in less sensitive cell lines (e.g., MRC-5) may be enhanced by using shell vial cultures. A genetically engineered baby hamster kidney cell line (ELVIS), which contains a reporter transgene for HSV, has been shown to be useful for detecting this virus in shell vials (21, 26); also available cocultured with MRC-5 cells.
Influenza virus types A, B, & C	PMK	MDCK cells provide good sensitivity with incorporation of trypsin (1–2 μg/ml) in cell culture medium (7, 27, 28).
Measles	PMK, HNK, A-549 shell vials (18)	
Mumps	PMK, HNK	
Parainfluenza virus types 1–4	PMK, HNK, LLC-MK_2	
Poxviruses (orthopoxviruses [e.g., vaccinia, monkeypox, cowpox] and parapoxviruses [e.g., orf, milker's nodule)	PMK, MRC-5, WI-38	

(continued)

Table 8.2–3 Cell cultures used for viral culture[a] *(continued)*

Virus	Cell cultures (references)	Comments (references)
Respiratory syncytial virus	HEp-2, H292	PMK and MRC-5 are also useful (2), but do not develop characteristic CPE.
Reovirus types 1–3	PMK, PHK	Not commonly isolated
Rhinoviruses (>100 types)	MRC-5, WI-38, PMK	
Rubella virus	AGMK	
Vaccinia virus	PMK, MRC-5, WI-38, HNK	
VZV	MRC-5, WI-38, A-549 (4)	Shell vials recommended

NOTE: Coxsackie A23 shown to be echovirus type 9; echovirus type 10 reclassified as a reovirus; echovirus type 28 reclassified as rhinovirus type 1; echovirus type 34 is related to coxsackievirus A24 as a prime strain.

[a] For definitions of cell culture designations, see Table 8.2–2, footnote *a*. CPE, cytopathic effect.

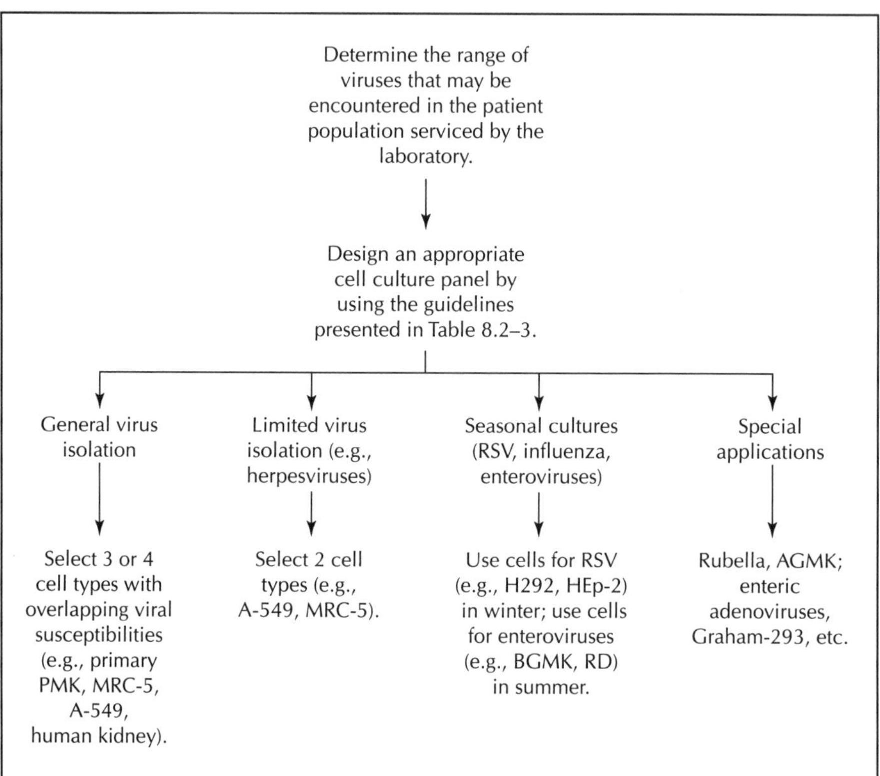

Figure 8.2–1 Selection of cell cultures for virus isolation. RSV, respiratory syncytial virus.

II. MATERIALS *(continued)*

croscope; tubes can be held in place with a holder consisting of two parallel tracks glued to the stage.

◪ **NOTE:** One end of the holder should be slightly higher so that the culture medium does not flow into the caps during observation.

2. Class II biological safety cabinet (certified semiannually) (*see* procedure 12.6)

Wipe down all interior cabinet surfaces with 70% ethanol or other decontaminating material before and after each use.

◪ **NOTE:** 10% Clorox is used by many laboratories to wipe down bench surfaces; this material will cause pitting and deterioration of the stainless-steel surfaces of the biological safety cabinet.

Do not impede the airflow within the unit by blocking airflow vents. Avoid disrupting the air curtain at the front of the hood; minimize movements into and out of the hood, and avoid outside air turbulence such as that created by air conditioning units or open doors or windows. Do not use the unit for storage; remove all supplies at the completion of work.

◪ **NOTE:** If separate units are not available for handling uninoculated and inoculated cell cultures, perform uninoculated cell culture maintenance before processing samples in the cabinet.

3. Refrigerator at 2 to 8°C

4. Incubator (35 to 37°C)

◪ **NOTE:** CO_2 atmosphere is not necessary for closed containers.

III. QUALITY CONTROL AND ASSESSMENT OF UNINOCULATED CELL CULTURES

Cell cultures may exhibit extensive lot-to-lot variation in monolayer quality, appearance, and viral susceptibility. Assess and record cell culture quality upon receipt of cell cultures and throughout the time that they are in use (Fig. 8.2–2). Troubleshooting guidelines and suggested corrective actions for cell culture monolayers are provided in Table 8.2–4.

A. Uninoculated-cell-culture incubation and maintenance

1. Incubation

After initial assessment (Fig. 8.2–2), incubate the cell cultures as described in Fig. 8.2–2.

2. Monitor the quality of the uninoculated cell cultures as described in Fig. 8.2–2 for initial assessment throughout the time that they are in use.

3. Refeeding

Replace the cell culture medium with fresh medium if the pH is low (*see* Table 8.2–4) and at the time of inoculation.

Procedure

a. Discard the medium from culture tubes and vials (decant, aspirate, or pipette).

◪ **NOTE:** Be careful not to scrape the monolayer with the pipette.

b. Add 1 to 2 ml of cell culture maintenance medium to each tube.

◪ **NOTE:** Use separate pipettes for each lot of each cell line; change pipettes frequently.

B. Procedure notes

Use aseptic techniques for all cell culture manipulations (Appendix 8.2–2). A class II biological safety cabinet is suitable for cell culture work.

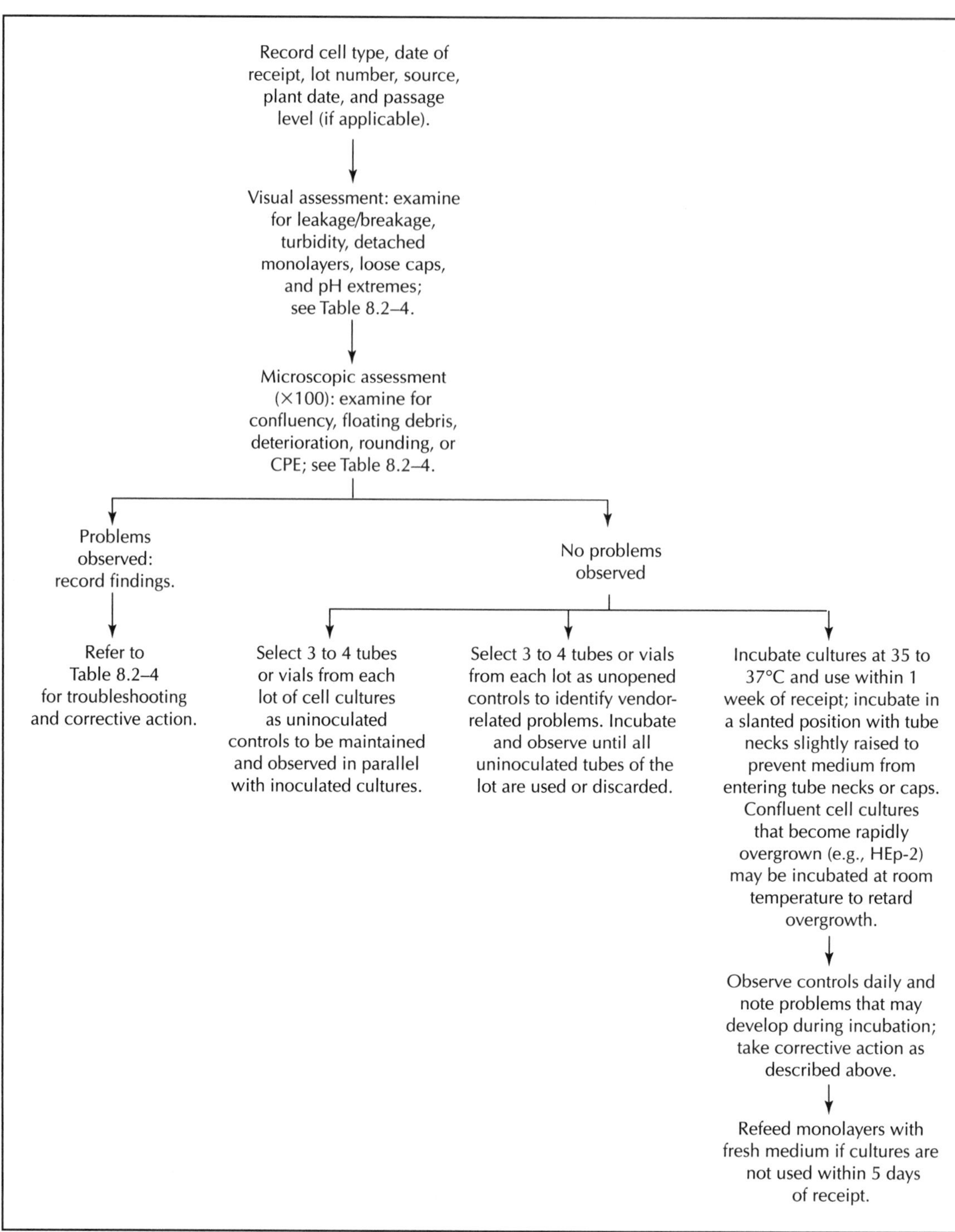

Figure 8.2–2 Assessment, incubation, and maintenance of uninoculated cell cultures.

Table 8.2–4 Assessment of monolayered cell cultures and troubleshooting

Component	Appearance	Possible cause(s)	Corrective action
Medium[a]	Clear yellow (low pH)	CO_2 accumulation	Examine microscopically for evidence of microbial contamination. If none, refeed with fresh maintenance medium and incubate. Re-evaluate after 18–24 h. Retain approximately four to six tubes without medium change for observation.
	Clear purple (high pH)	Loose caps	Tighten caps; incubate. Notify vendor.
		Shipping trauma, chilling	Incubate for several hours or overnight and re-evaluate.
		Poor cell quality, metabolically inactive cells; sparse monolayer	Notify vendor. Incubate; reject if not satisfactory within 24 h.
	Turbid or organisms visible	Bacterial/fungal contamination[b]	Notify vendor; reject. Do not open.
Monolayer[c]	Sparse (less than 60–70%)	Seeded too sparsely, cells failed to divide, medium or glassware problem	Notify vendor. Incubate and reject if not satisfactory within 24–48 h.
	Cells piled/detached	Seeded too heavily, old monolayer or poor cell quality, mycoplasma contamination[d], medium or glassware problem, temperature extremes, inverted during shipment	Notify vendor. Reject if extensive.
	Cellular granularity and/or rounding	Shipping trauma, chilling, poor cell quality, bacterial/fungal contamination[b], mycoplasma contamination[d], medium or glassware problem	Incubate for several hours or overnight, and re-evaluate. Notify vendor; reject if improvement not observed.
	CPE and/or hemadsorption (HAd)	Endogenous or contaminating viruses, mycoplasma	Primary simian cells are frequently contaminated with endogenous viruses. Perform appropriate QC to monitor extent of contamination and effect on sensitivity of cultures. May mimic human viral CPE, cause HAd, or interfere with the recovery of viruses from clinical samples. If a herpeslike CPE is observed, notify the vendor immediately and do not open the tubes because cell cultures may contain simian B virus, which is associated with a high mortality rate in humans. NOTE: Primary simian cell cultures may be obtained with medium containing antibodies to simian viruses (SV5 and SV40) which may temporarily retard the replication of these viruses.

[a] Should be clear and orange-red to cherry pink (pH 6.8 to 7.2).

[b] Perform Gram stain of sediment obtained from centrifuged culture supernatant; inoculate broth medium (e.g., THIO, TSB) or agar (e.g., blood, Sabouraud-dextrose) with sediment and culture medium, and observe for bacterial or fungal growth for 10 to 14 days.

[c] Monolayered cell cultures should consist of a single layer of adherent cells exhibiting normal morphology at or near confluency (100% monolayer). Monolayers that are slightly subconfluent are preferred for viral culture.

[d] Mycoplasmal contamination can have a profound detrimental effect on viral isolation. Contamination with these organisms frequently does not result in obvious changes of the cell monolayer but may present as a generalized deterioration of the monolayer; occasionally a cytopathic effect (CPE) mimicking viral growth may be observed. Obtain written certification from the vendor that cultures are mycoplasma-free as determined periodically (e.g., at least every 3 months) by an appropriate method such as culture, nucleic acid hybridization, or the Hoechst 33158 DNA fluorochrome stain. *See* CMPH 8.20.10 for additional information.

REFERENCES

1. **Albert, M. J.** 1986. Enteric adenoviruses. *Arch. Virol.* **88:**1–17.
2. **Arens, M. Q., E. M. Swierkosz, R. R. Schmidt, T. Armstrong, and K. A. Rivetna.** 1986. Enhanced isolation of respiratory syncytial virus in cell culture. *J. Clin. Microbiol.* **23:**800–802.
3. **Bird, B. R., and F. T. Forrester.** 1981. *Basic Laboratory Techniques in Cell Culture.* U.S. Department of Health and Human Services, U.S. Government Printing Office, Washington, D.C.
4. **Coffin, S. E., and R. L. Hodinka.** 1995. Utility of direct immunofluorescence and virus culture for detection of varicella-zoster virus in skin lesions. *J. Clin. Microbiol.* **33:** 2792–2795.
5. **Espy, M. J., C. Hierholzer, and T. F. Smith.** 1987. The effect of centrifugation on the rapid detection of adenovirus in shell vials. *J. Clin. Pathol.* **88:**358–360.
6. **Espy, M. J., T. F. Smith, M. W. Haromon, and A. P. Kendal.** 1986. Rapid detection of influenza virus by shell vial assay with monoclonal antibodies. *J. Clin. Microbiol.* **24:** 677–679.
7. **Frank, A. L., R. B. Couch, C. A. Griffis, and B. D. Baxter.** 1979. Comparison of different tissue cultures for isolation and quantitation of influenza and parainfluenza viruses. *J. Clin. Microbiol.* **10:**32–36.
8. **Gleaves, C. A., C. F. Lee, C. I. Bustamante, and J. D. Meyers.** 1988. Use of murine monoclonal antibodies for laboratory diagnosis of varicella-zoster infection. *J. Clin. Microbiol.* **26:**1623–1625.
9. **Gleaves, C. A., T. F. Smith, E. A. Shuster, and G. R. Pearson.** 1984. Rapid detection of cytomegalovirus in MRC-5 cells inoculated with urine specimens by using low-speed centrifugation and monoclonal antibody to an early antigen. *J. Clin. Microbiol.* **19:**917–919.
10. **Gleaves, C. A., D. J. Wilson, A. D. Wold, and T. F. Smith.** 1985. Detection and serotyping of herpes simplex virus in MRC-5 cells by use of centrifugation and monoclonal antibodies 16 h postinoculation. *J. Clin. Microbiol.* **21:**29–32.
11. **Grandien, M., M. Forsgren, and A. Ehrnst.** 1995. Enteroviruses, p. 279–297. *In* E. H. Lennette, D. A. Lennette, and E. T. Lennette (ed.), *Diagnostic Procedures for Viral, Rickettsial, and Chlamydial Infections,* 7th ed. American Public Health Association, Washington, D.C.
12. **Hierholzer, J. C., R. Wigand, and J. C. de Jong.** 1988. Evaluation of human adenoviruses 38, 39, 40, and 41 as new serotypes. *Intervirology* **29:**1–10.
13. **Johnston, S. L. G., and C. S. Siegel.** 1990. Evaluation of direct immunofluorescence, enzyme immunoassay, centrifugation culture, and conventional culture for the detection of respiratory syncytial virus. *J. Clin. Microbiol.* **28:**2394–2397.
14. **Landry, M. L., and D. Ferguson.** 1993. Comparison of quantitative cytomegalovirus antigenemia assay with culture methods and correlation with clinical disease. *J. Clin. Microbiol.* **31:**2851–2856.
15. **Lipson, S. M., R. Walderman, P. Costello, and K. Szabo.** 1988. Sensitivity of rhabdomyosarcoma and guinea pig embryo cell cultures to field isolates of difficult-to-cultivate group A coxsackieviruses. *J. Clin. Microbiol.* **26:**1298–1303.
16. **Matthey, S., D. Nicholson, S. Ruhs, B. Alden, M. Knock, K. Schultz, and A. Schmuecker.** 1992. Rapid detection of respiratory viruses by shell vial culture and direct staining by using pooled and individual monoclonal antibodies. *J. Clin. Microbiol.* **30:** 540–544.
17. **Menegus, M. D., and G. E. Hollick.** 1982. Increased efficiency of group B coxsackievirus isolation from clinical specimens by use of BGM cells. *J. Clin. Microbiol.* **15:** 945–948.
18. **Minnich, L. L., F. Goodenough, and C. G. Ray.** 1991. Use of immunofluorescence to identify measles virus infections. *J. Clin. Microbiol.* **29:**1148–1150.
19. **Olsen, M. A., K. M. Shuck, A. R. Sambol, S. M. Flor, J. O'Brien, and B. J. Cabrera.** 1993. Isolation of seven respiratory viruses in shell vials: a practical and highly sensitive method. *J. Clin. Microbiol.* **31:**422–425.
20. **Paya, C. V., A. D. Wold, and T. F. Smith.** 1987. Detection of cytomegalovirus infections in specimens other than urine by the shell vial assay and conventional tube cell cultures. *J. Clin. Microbiol.* **25:**755–757.
21. **Proffitt, M. R., and S. A. Schindler.** 1995. Rapid detection of HSV with an enzyme-linked virus inducible system (ELVIS™) employing a genetically modified cell line. *Clin. Diagn. Virol.* **4:**175–182.
22. **Rotbart, H. A.** 1995. Enteroviruses, p. 1004–1011. *In* P. R. Murray, E. J. Baron, M. A. Pfaller, F. C. Tenover, and R. H. Yolken (ed.), *Manual of Clinical Microbiology,* 6th ed. American Society for Microbiology, Washington, D.C.
23. **Schirm, J., J. M. Janneke, G. W. Pastoor, P. C. vanVoorst Vader, and F. P. Schroder.** 1989. Rapid detection of varicella-zoster virus in clinical specimens using monoclonal antibodies on shell vials and smears. *J. Med. Virol.* **28:**1–6.
24. **Schmidt, N. J., H. H. Ho, and E. H. Lennette.** 1975. Propagation and isolation of group A coxsackieviruses in RD cells. *J. Clin. Microbiol.* **2:**183–185.
25. **Smith, M. C., C. Creutz, and Y. T. Huang.** 1991. Detection of respiratory syncytial virus in nasopharyngeal secretions by shell vial technique. *J. Clin. Microbiol.* **29:**463–465.
26. **Stabell, E. C., S. R. O'Rourke, G. A. Storch, and P. D. Olivo.** 1993. Evaluation of a genetically engineered cell line and a histochemical β-galactosidase assay to detect herpes simplex virus in clinical specimens. *J. Clin. Microbiol.* **31:**2796–2798.
27. **Takiff, H. E., S. E. Straus, and C. F. Garon.** 1981. Propagation and in vitro studies of previously noncultivatable enteral adenoviruses in 293 cells. *Lancet* **ii:**832–834.

28. **Tobita, K., A. Sugiura, C. Enomoto, and M. Furuyama.** 1975. Plaque assay and primary isolation of influenza A viruses in an established line of canine kidney cells (MDCK) in the presence of trypsin. *Med. Microbiol. Immunol.* **162:**9–14.
29. **Waris, M., T. Aiegler, M. Kivivirta, and O. Ruuskananen.** 1990. Rapid detection of respiratory syncytial virus and influenza A virus in cell cultures by immunoperoxidase staining with monoclonal antibodies. *J. Clin. Microbiol.* **28:**1159–1162.
30. **West, P. G., B. Aldrich, R. Hartwig, and G. J. Haller.** 1988. Increased detection rate for varicella-zoster virus with combination of two techniques. *J. Clin. Microbiol.* **26:**2680–2681.

SUPPLEMENTAL READING

Hodinka, R. L. 1992. Cell culture techniques: serial propagation and maintenance of monolayer cell cultures, p. 8.20.2–8.20.14. *In* H. D. Isenberg (ed.), *Clinical Microbiology Procedures Handbook,* vol. 2. American Society for Microbiology, Washington, D.C.

APPENDIX 8.2–1

Commercial sources of monolayer cell cultures

Bartels Inc.
Diagnostic Division of Intracel Corp.
Issaquah, WA 98027
(800) BARTELS

NeoGenex
Everett, WA 98204
(800) 334-4297

Viromed, Inc.
Minnetonka, MN 55343
(800) 582-0077

Whittaker Bioproducts, Inc.
Walkersville, MD 21793
(800) 638-8174

APPENDIX 8.2–2

Summary of Aseptic Technique

A. Bacterial and fungal contamination may originate from many sources, including contaminated reagents or supplies, room air, or the technologist's hands. When preparing reagents or handling cell cultures, laboratory personnel must be aware of the need to practice aseptic techniques.
B. Perform all cell culture manipulations in a class II biological safety cabinet and follow good operating practices (*see* equipment description above).
C. Test all cell culture reagents for sterility before use and handle in a way to maintain sterility throughout their use (for details, *see* procedure 8.6).
D. Use only sterile supplies (e.g., pipettes, containers, caps, and stoppers); individually wrapped sterile pipettes are recommended.
E. Dispense sterile reagents in a manner that maintains sterility; pouring, rather than dispensing with a pipette or other dispenser, increases the risk for contamination.
F. Avoid touching the exposed necks, caps, or stoppers of opened containers.
G. Keep the work area and surfaces clean and free of dust.
H. Arrange materials and equipment within the safety cabinet to provide easy access and to minimize reaching.
I. Wear clean long-sleeved laboratory gowns or smocks when working with cell cultures and cell culture reagents.
J. Wipe reagent and medium bottles with 70% ethanol upon removal from storage; flaming bottle, tube, or flask necks is generally not performed when using a biological safety cabinet. Avoid leaving containers uncapped.

8.3 Viral Culture: Specimen Selection, Collection, Transport, Evaluation, and Storage

I. PRINCIPLE

Successful viral culture requires careful attention to the selection, collection, transport, and assessment of specimens (Fig. 8.3–1). The information contained in this section is essential not only to those laboratories performing viral cultures on site but also to those outsourcing specimens. The laboratory must provide written guidelines to collection sites (nursing station, clinic, physician's office, and emergency room) detailing cultures that are available and instructions for submitting specimens. Instructions should include laboratory hours of operation and contact person(s), instructions for specimen collection, labeling, storage, and transport; source and storage conditions for transport media and containers; information required for adequate testing; turnaround time; reporting procedures and values; and testing limitations.

II. SPECIMEN SELECTION

For best correlation between viral recovery and disease etiology, the specimen should reflect the target organ whenever possible. For example, isolation of an enterovirus from a throat or rectal swab obtained from an individual with aseptic meningitis is less meaningful than isolation from CSF. Specimens most frequently submitted for viral testing are presented in Table 8.3–1.

A. Timing

Collect specimens as soon after onset of symptoms as possible, because the likelihood of obtaining positive results is generally greatest within the first 3 days after onset of symptoms and diminishes rapidly as the course of infection proceeds in otherwise healthy and immunocompetent individuals. Viruses may be recovered from clinical samples for prolonged periods with disseminated or persistent infections.

Collect autopsy specimens as soon after death as possible.

B. Collection

Refer to Table 8.3–1 for specimen collection guidelines. Collect specimens as aseptically as possible since microbial contamination will interfere with testing. Place each specimen into a separate sterile leak-proof container labeled with the patient's name and identification number, the collection site, and the date and time of collection. Collect fluid (e.g., CSF, urine) and bulk (e.g., stool, autopsy tissues) specimens in a sterile, dry, leak-proof container.

◪ **NOTE:** Viral transport medium (VTM) can be added to prevent drying of the latter specimens if transport to the laboratory is not immediate.

Collect swab specimens by using sterile cotton-, Dacron-, or rayon-tipped swabs with plastic or aluminum shafts; do not use calcium alginate swabs (1) or wooden-shafted swabs (3). Use flexible aluminum-shafted, small-tipped swabs for sampling sites (urethra, nasopharynx) for which larger swabs with rigid shafts would be inappropriate. Place swabs, scrapings, and small pieces of tissue into tubes or vials containing VTM.

Transport media are available commercially from several sources including those listed in Appendix 8.3–1. VTM prevents specimen drying, maintains

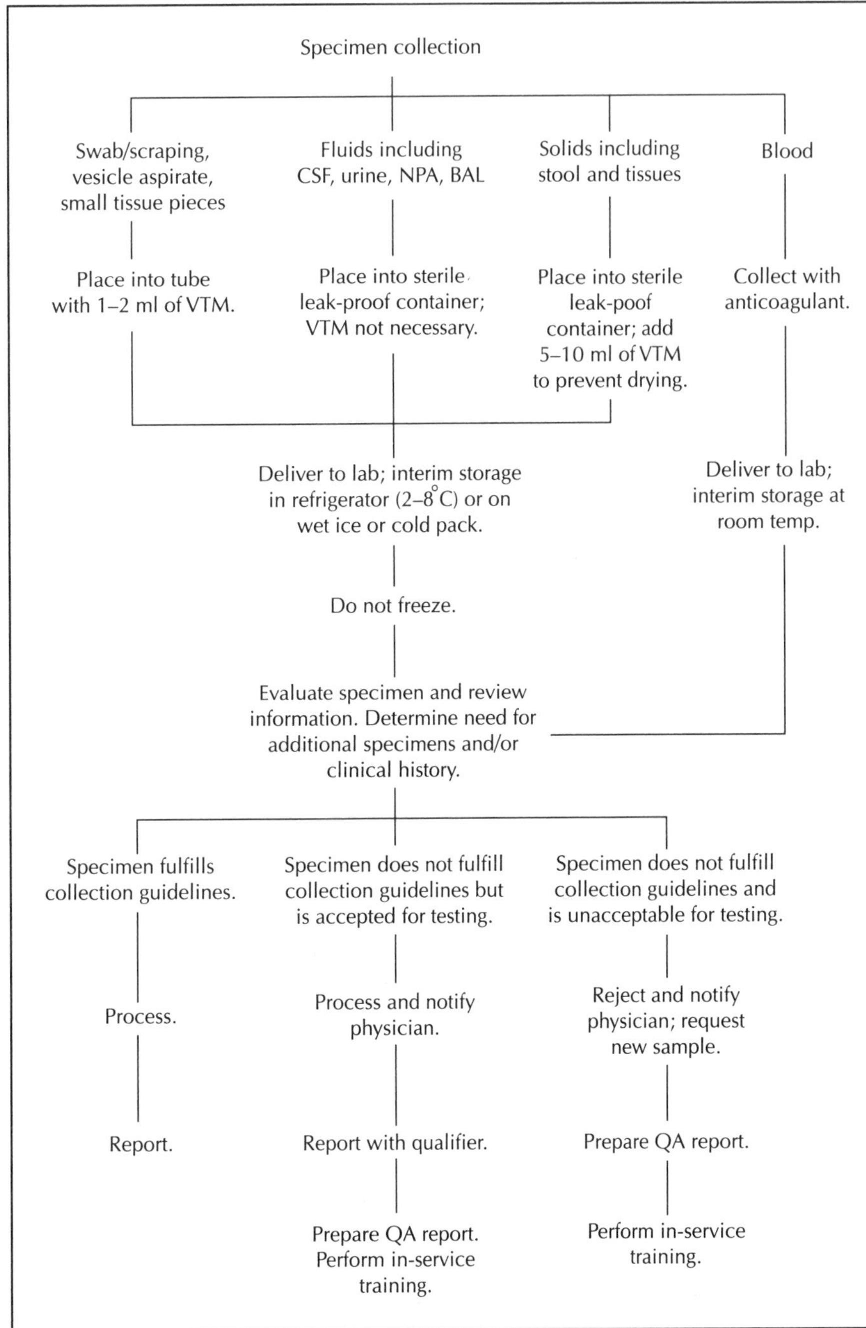

Figure 8.3–1 Summary of collection, transport, and evaluation of specimens for viral culture. NPA, nasopharyngeal aspirate; BAL, bronchoalveolar lavage fluid.

viral viability, and retards the growth of microbial contaminants. Although few comprehensive studies have been performed, no significant differences in the recovery of selected viruses by using several different transport media have been observed (reviewed in reference 3). Most laboratories use a liquid transport medium but other types are also used.

1. Liquid transport media

   VTM typically consists of a protein source such as bovine serum albumin, gelatin, or serum and antimicrobial agents in a buffered salt or sucrose solu-

Table 8.3–1 Specimen collection[a]

Specimen	Collection	Representative viruses	Comments
Amniotic fluid	Collect 5–8 ml in a sterile container.		
Blood	Collect 8–10 ml in anticoagulant (sodium citrate, EDTA, or heparin). For pediatric specimens smaller volumes are acceptable, but less than 1–2 ml may be inadequate for testing.	Arboviruses, arenaviruses, CMV, HSV, HIV-1, and other retroviruses; serum or plasma useful for enteroviruses	Useful for persistent and disseminated infections in immunocompromised patients and neonates; do not refrigerate if cell separation techniques are to be performed. Do not freeze blood.
Bone marrow	Collect at least 2 ml in anticoagulant (sodium citrate, EDTA, heparin).	CMV	
CSF	Collect 2–5 ml in a sterile container.	CMV, enteroviruses[b] other than poliovirus, HHV-6, HSV, LCMV, mumps virus	CMV: culture useful with polyradiculopathy, but not encephalitis; HSV: Culture useful with meningitis or meningoencephalitis, but not encephalitis; useful for some arboviruses.
Gastrointestinal			
Rectal swab	Insert swab 4–6 cm into rectum, and roll swab against mucosa. Place swab into VTM.	Adenovirus, enteroviruses,[b] CMV, HSV	Not equivalent to stool sample. Useful for proctitis.
Stool	Place 2–4 g (ca. 3 tsps) into a sterile container. Add 8–10 ml of VTM if transport to laboratory is not immediate.	Adenoviruses, enteroviruses,[b] reoviruses	
Genitourinary			
Cervical swab	Remove exocervical mucus with swab and discard swab. Insert fresh swab at least 1 cm into cervical canal, and rotate swab for 10 s. Place into VTM.	HSV, CMV	Asymptomatic HSV in women: shedding, collection of a vulvar swab in addition to a cervical swab may increase viral recovery (7).
Urine	Collect 10–20 ml of midstream clean-voided urine in a sterile container.	Adenoviruses, BK virus, CMV, HSV, mumps virus, rubella virus	CMV shedding may be intermittent; two or three sequential urine specimens increase chance of a positive culture; low clinical specificity for CMV disease in immunocompromised individuals
Lesion, dermal swab	Wipe vesicle with saline. Disrupt vesicle, and collect fluid with swab. With same swab, collect cells from base of the lesion. For nonvesicular lesions, collect cells from base of lesion by using a swab premoistened with saline. Place swab into VTM.	Group A coxsackieviruses, echoviruses, HSV, poxviruses, VZV	Recovery of HSV may be diminished with ulcerative and crusted lesions (4).
Vesicle aspirate	Wipe area with sterile saline. Aspirate fluid from vesicle with a 26- or 27-gauge needle attached to a tuberculin syringe. Immediately rinse the syringe in 1–2 ml of VTM.		Preferred specimen for VZV
Lesion, mucosal swab	Swab back and forth over lesion and place swab into VTM.	Oral: group A coxsackieviruses, HSV; anogenital: HSV	

Table 8.3–1 Specimen collection[a] *(continued)*

Specimen	Collection	Representative viruses	Comments
Ocular			
Conjunctival swab	Swab lower conjunctiva with flexible fine-shafted swab premoistened with sterile saline. Place swab into VTM.	Adenoviruses, group A coxsackieviruses, CMV, enterovirus type 70, HSV	
Corneal or conjunctival scraping	Place scraping into VTM. Specimen should be obtained only by an ophthalmologist or other trained individual.	Same as for conjunctival swab	
Pericardial fluid	Place at least 2 ml in a sterile container.	Group B coxsackieviruses	
Pleural fluid	Place at least 2 ml in a sterile container.	Group B coxsackieviruses	
Respiratory swabs			
Nasal	Insert flexible fine-shafted swab into nostril, rotate swab, and let it rest for several seconds to absorb secretions. Place into VTM. Use separate swabs for each nostril; both swabs may be placed into the same transport vial.	Rhinoviruses	Not recommended for other respiratory viruses; prompt inoculation into cell culture is recommended.
Nasopharyngeal	Insert flexible fine-shafted swab through nostril into nasopharynx and rotate the swab gently a few times. Place swab into VTM.	Adenoviruses, CMV, enteroviruses,[b] HSV, influenza virus, measles virus, mumps virus, parainfluenza viruses, RSV, reoviruses, rhinoviruses, rubella virus, VZV	Preferred to pharyngeal swabs for respiratory viruses; prompt inoculation into cell culture is recommended for RSV.
Pharyngeal (throat)	Vigorously swab tonsillar areas and posterior nasopharynx. Use tongue blade to depress tongue to prevent contamination of swab with saliva. Place swab into VTM.	Same as nasopharyngeal swab	
Respiratory fluids			
Lower respiratory tract: bronchoalveolar lavage, bronchial wash	Place 8–10 ml in sterile container.	Viruses associated with pneumonia in otherwise healthy (e.g., influenza, RSV, adenoviruses) or immunocompromised (e.g., CMV, VZV) patients	
Nasopharyngeal aspirate	Using mucus collection device, insert appropriate-sized catheter nasally into posterior nasopharynx. Apply suction, using intermittent suction as catheter is withdrawn. Wash aspirate through tubing with 5–8 ml of VTM or sterile saline, and transport material from trap to sterile container.	Same as nasopharyngeal swab	Preferred upper respiratory tract specimen
Saliva	Collect saliva with 1 or 2 swabs from anterior floor of mouth and near Stenson's ducts. Place swabs into VTM.	CMV, mumps, rabies virus	

(continued)

Table 8.3–1 Specimen collection[a] *(continued)*

Specimen	Collection	Representative viruses	Comments
Tissues	Place small samples (e.g., lung biopsy specimen) into VTM to prevent drying. Place larger specimens (1–2 g; e.g., autopsy tissues) into sterile container. Add 8–10 ml of VTM.	Many viruses can be recovered from tissues during disseminated or visceral disease (e.g., CMV, HSV, adenoviruses); arboviruses, rabies virus, HSV, and other viruses associated with encephalitis can be recovered from brain tissue.	An understanding of the pathogenesis of viral infections is essential for selecting appropriate specimens. Autopsy tissues: several organs are usually sampled (liver, lung, spleen) and any organs or tissues for which involvement is suggested by clinical history or pathologic findings (e.g., brain, adrenal glands)

[a] Adapted from Table 1, CMPH 8.2.3. Abbreviations: tsps, teaspoons; HHV-6, human herpesvirus type 6; HIV-1, human immunodeficiency virus type 1; LCMV, lymphocytic choriomeningitis virus; RSV, respiratory syncytial virus; VZV, varicella-zoster virus.

[b] Includes group A and group B coxsackieviruses, echoviruses, enterovirus types 68–71, and polioviruses.

II. SPECIMEN SELECTION *(continued)*

tion. A sucrose-containing medium such as 2-SP is frequently used as a chlamydial transport medium and appears to protect labile viruses and may be particularly useful in situations requiring extended refrigeration or frozen shipment (2, 3). Transport tubes usually contain 2 to 3 ml of medium and small glass beads to facilitate specimen preparation; some sources package collection tubes and swabs together as a convenient collection kit. Use larger volumes (5 to 7 ml) of VTM to transport bulk (e.g., tissue) samples; these volumes should not be used for swabs because of the greater dilution effect.

2. Transporter (Bartels, Inc.)

The transporter system from Bartels, Inc., consists of viable human diploid fibroblast cells with cell culture medium. Greater viral recovery and earlier detection of cytomegalovirus (CMV) and herpes simplex virus (HSV) have been reported by using the transporter compared with a sucrose-phosphate-glutamate transport medium (6). However, the transporter system has a limited shelf life and is not suitable for viruses that do not replicate in human fibroblasts.

3. Agar-containing transport systems (Stuart, Amies)

Stuart and Amies media are sometimes used to reduce the number of transport systems in an institution. Upon receipt in the laboratory, swabs are transferred to one of the liquid transport media described above. Stuart transport medium may be less satisfactory than liquid medium (3).

III. TEST ORDERING AND PATIENT INFORMATION

Information must be provided that uniquely identifies the patient and the ordering physician and states the viral agent(s) for which the specimen is to be tested. Laboratory utilization and cost containment issues preclude the practice of ordering "viral studies" on all specimens for which a viral laboratory diagnosis is being sought, and most requests now state the specific virus(es) being considered. However, a general "viral studies" request is usually ordered if the candidate etiologic agent(s) is not clear or symptoms are not typical, or if multiple agents are associated with the same clinical presentation. In these situations it is important to obtain information (clinical history, date of onset, history of recent exposure or vaccination, pertinent travel history, and animal or arthropod bite or exposure) that will enable the laboratory scientist to select an appropriate testing algorithm and determine the need for additional assays or specimens.

IV. TRANSPORT AND STORAGE

Place the tightly capped specimen container and the laboratory requisition form into separate compartments of a plastic specimen transport bag. Deliver all specimens to the laboratory as soon after collection as possible; a loss of infectivity occurs over time, resulting in the diminished likelihood of a positive result. Loss of viability is slower at refrigeration temperatures than at ambient temperature; samples containing labile viruses (e.g., respiratory syncytial virus [RSV] or CMV) at low titers are those most likely to show loss of infectivity with delayed transport. If immediate delivery to the laboratory is not possible, store specimens in a refrigerator (2 to 8°C) or place them on wet ice or a cold pack. Store and transport anticoagulated blood samples at ambient temperature because chilling may interfere with cell separation techniques.

Store specimens for which culture will be delayed beyond 48 h after collection at −70°C or lower; avoid freezing at higher temperatures and freeze-thaw cycles. If possible, snap-freeze specimens in a slurry of dry ice and acetone prior to storage. With few exceptions (e.g., arboviral studies), frozen blood specimens are unsuitable for testing. Transport frozen specimens on sufficient dry ice to ensure that specimens remain frozen until receipt by the laboratory. Pack, label, and ship specimens in compliance with published guidelines (for details, *see* procedure 14.12).

V. SPECIMEN ACCESSIONING AND EVALUATION

It is essential to utilize a manual or computerized accessioning system that uniquely identifies each specimen and stores information required for laboratory records, including patient and physician identification, specimen type, date and time of collection, specimen condition, and the test(s) requested. A system must also be in place for bringing unusual requests to the immediate attention of the laboratory director or supervisor.

No laboratory procedure can compensate for improper or inappropriate specimen collection and handling. Establish written criteria for specimen rejection, and determine situations in which to include a qualifier with the report. Upon receipt, determine whether the specimen is acceptable, unacceptable, or "compromised" (Appendix 8.3–2). Contact the physician of record immediately regarding rejected or compromised specimens or the need for additional information; request a replacement sample if possible and appropriate.

Maintain a record of rejected specimens, and review the information at least monthly to identify areas where in-service training or other intervention is necessary. Periodically review collection and transport guidelines with laboratory staff, and develop a system for identifying and correcting problems with specimen collection and transport. Prepare QA reports consistent with institutional policy, and include the corrective action taken. Establish a system to follow up with corrective action and monitor compliance with recommendations.

VI. SAFETY CONSIDERATIONS

While viruses encountered in the diagnostic laboratory usually encompass BSL 2 agents, a laboratory may be called on to assist in collection and referral of specimens requiring BSL 3 or BSL 4 practices (5). Fortunately, infections involving the latter category (e.g., arboviruses, arenaviruses, filoviruses, and rabies) are usually associated with specific clinical presentations, travel history, and/or animal or insect vector exposures. (For details, *see* section 14.)

VII. SUPPLIES AND REAGENTS

A. Transport medium

1. On-site preparation
   Refer to EPCM procedure 8.6 for selected formulations and instructions for preparing transport medium (Hanks' balanced salt solution [HBSS] with bovine serum albumin; 2-SP [0.2 M sucrose–0.02 M phosphate]).
2. Commercial sources
   Transport media are commercially available from several sources including those listed in Appendix 8.3–1. Several formulations have been developed, and some products are suitable for chlamydia as well as viruses. Formulations can be obtained from the manufacturer, or refer to CMPH 8.2.6 to 8.2.7 for selected formulations.

B. Swabs

Sterile, individually wrapped cotton-, Dacron-, or rayon-tipped swabs with plastic or aluminum shafts

C. Tuberculin syringes with 26- or 27-gauge needles

Vesicle fluid aspiration

D. Blood collection tubes containing anticoagulant

Heparin, EDTA, sodium citrate

E. Acetone and dry ice

For quick freezing samples, if necessary

F. Crushed ice and/or cold packs

For storage and transport

G. Dry ice

If necessary for frozen shipment

H. Equipment

1. Refrigerator (2 to 8°C) for storing transport medium before use
2. Freezer (−70°C or lower) for long-term specimen storage

RERERENCES

1. **Crane, L. R., P. A. Gutterman, T. Chapel, and A. M. Lerner.** 1980. Inoculation of swab materials with herpes simplex virus. *J. Infect. Dis.* **131:**531.
2. **Howell, C. L., and M. J. Miller.** 1983. Effect of sucrose phosphate and sorbitol on infectivity of enveloped viruses during storage. *J. Clin. Microbiol.* **18:**658–662.
3. **Johnson, F. B.** 1990. Transport of viral specimens. *Clin. Microbiol. Rev.* **3:**120–131.
4. **Moseley, R. C., C. L. Corey, D. Benjamin, C. Winter, and M. Remington.** 1981. Comparison of viral isolation, direct immunofluorescence and indirect immunoperoxidase techniques for detection of genital herpes simplex virus infection. *J. Clin. Microbiol.* **13:** 913–918.
5. **Public Health Service, CDC, and NIH.** 1993. *Biosafety in Microbiological and Biomedical Laboratories.* Publication no. (NIH) 93-8395, 3rd ed. U.S. Department of Health and Human Services, U.S. Government Printing Office, Washington, D.C.
6. **Warford, A. L., W. G. Eveland, C. A. Strong, R. A. Levy, and K. A. Rekrut.** 1984. Enhanced virus isolation by use of the transporter for a regional laboratory. *J. Clin. Microbiol.* **19:**561–562.
7. **Warford, A. L., R. A. Levy, K. A. Rekrut, and E. Steinberg.** 1986. Herpes simplex virus testing of an obstetric population with an antigen enzyme-linked immunosorbent assay. *Am. J. Obstet. Gynecol.* **154:**21–28.

SUPPLEMENTAL READING

Leonardi, G. P., and C. A. Gleaves. 1992. Selection, collection, and transport of specimens for viral and rickettsial cultures, p. 8.2.1.–8.2.10. *In* H. D. Isenberg (ed.), *Clinical Microbiology Procedures Handbook,* vol. 2. American Society for Microbiology, Washington, D.C.

APPENDIX 8.3–1

Commercial sources of VTM

Bartels, Inc.
Diagnostic Division of Intracel Corp.
Issaquah, WA
(800) BARTELS

BBL Microbiology Systems
Cockeysville, MD
(800) 638-8663

Carr-Scarborough
Stone Mountain, GA
(800) 241-0998

Micro Test, Inc.
Lilburn, GA
(800) 646-6678

Whittaker Bioproducts, Inc.
Walkersville, MD
(800) 638-8174

APPENDIX 8.3–2

Suggested Criteria for Rejecting Specimens and Qualifying Negative Results

A. Unacceptable specimen
   Reject for testing; obtain another sample if possible.
   1. Quantity not sufficient (QNS)
   2. Formaldehyde-fixed specimens
   3. Improper collection (e.g., contains material which inactivates virus or renders specimen unprocessable)
   4. Improper storage (e.g., high temperature)
   5. Dried specimen
   6. Excessive microbial contamination or toxicity
   7. Culture system not available (e.g., collected for human papillomavirus [HPV])

B. Compromised specimen
   Test and report with qualifier for negative results
   1. Not optimally collected but difficult to recollect (e.g., CSF)
   2. Delayed transport
   3. Freezing at inappropriate temperature
   4. Wrong swab (or failure to retain swab)
   5. Diluted sample (e.g., excessive volume of transport medium)

8.4 Viral Culture: Specimen Preparation

I. PRINCIPLE

Specimen preparation varies with the type of specimen (i.e., fluid, swab in viral transport medium [VTM], or tissue). With blood samples, processing is dictated by the fraction with which the suspected virus is associated. Preinoculation specimen preparation serves to maximize the number of viable viral particles in the inoculum, eliminate microbial contaminants, and minimize specimen toxicity.

II. SPECIMEN REQUIREMENTS

Collect, transport, and store specimens before testing, as described in procedure 8.3. Do not allow samples to stand at room temperature for extended periods during processing.

III. QUALITY CONTROL

Include QC information on reagent container and in QC records.

A. Do not use reagents beyond their expiration date.

B. Reagents used for specimen preparation, including VTM, must be compatible with cell cultures; monitor reagents for toxicity as described for cell culture medium (*see* procedure 8.6) if a problem is suspected or if a reagent substitution is to be made.

IV. PROCEDURES

A. Fluids and aspirates

◩ **NOTE:** Process vesicle aspirates collected in VTM as described for swab and scraping specimens.

Fluid specimens contain various amounts and types of cells, with virus both extra- and intracellular. Laboratories vary with respect to preinoculation specimen preparation procedures. For example, the addition of antimicrobial agents and thorough suspension of the sample is a simple but satisfactory preparatory method for most fluid specimens. Alternatively, fluid specimens may be processed by suspending the cells in a small amount of VTM, diluent, or supernatant, after pelleting them by low-speed centrifugation. Yet another approach is to use clarified supernatant after cell disruption by sonication or vigorous vortexing with glass beads. Although one might suspect that the greatest culture positivity, particularly with a cell-associated virus such as CMV, would be achieved by using the cellular component, it has been demonstrated that clarified supernatant or a well-suspended, unprocessed bronchoalveolar lavage fluid (BAL) sample can yield higher isolation rates than do pelleted cells suspended in VTM (1).

1. Materials

   a. Pipettes (assorted) and pipetting device

   b. Antimicrobial-agent concentrate (*see* procedure 8.6)

   c. Vortex mixer

2. For the procedure, see Fig. 8.4–1.

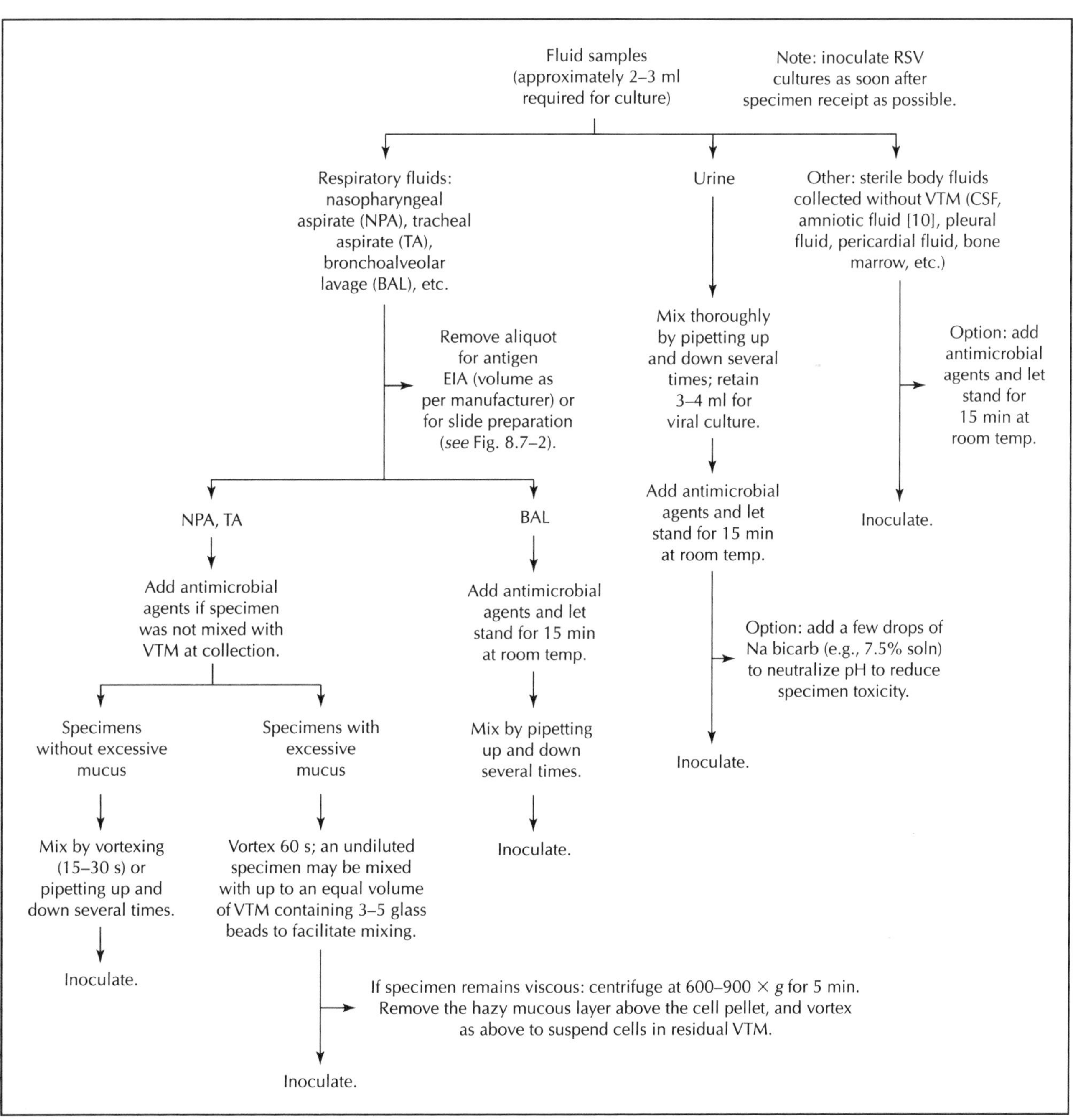

Figure 8.4–1 Preparation of fluid specimens. RSV, respiratory syncytial virus; soln, solution.

B. Swabs, scrapings, and aspirates collected in VTM
   1. Materials
      a. Forceps
      b. Vortex mixer
   2. For the procedure, see Fig. 8.4–2.

C. Solid specimens
   1. Materials
      a. Pipettes (assorted) and pipetting device

IV. PROCEDURES *(continued)*

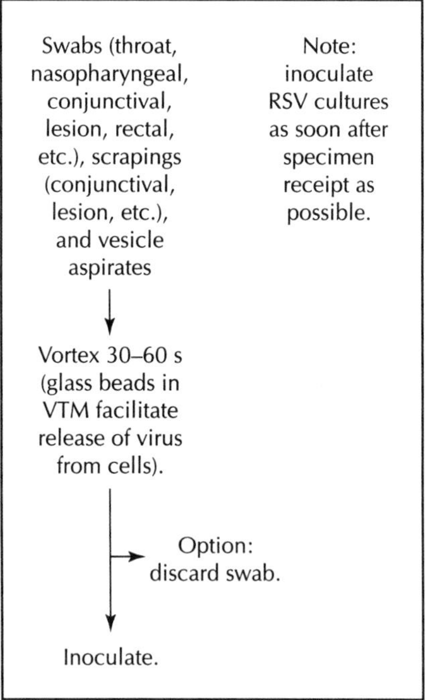

Figure 8.4–2 Preparation of swabs, scrapings, and vesicle aspirates collected in VTM. RSV, respiratory syncytial virus.

b. 15-ml sterile screw-cap centrifuge tubes
c. Tongue blades
d. Sterile petri plates
e. Scalpels (for tissue samples)
f. Sterile disposable tissue grinders (Sage Products, Inc., Crystal Lake, IL 60014)
g. Diluent: VTM or Hanks' balanced salt solution (HBSS) with antimicrobial agents (*see* procedure 8.6)
h. Vortex mixer
i. Centrifuge, refrigerated, with swinging-bucket rotor

2. Procedure
   a. For details on tissue samples, see Fig. 8.4–3.
   NOTE: A Tekmar Stomacher Blender may be used to homogenize tissue samples (9).
   b. For details on stool samples, see Fig. 8.4–4.

D. Anticoagulated blood

Viruses may be associated with various components of the blood including polymorphonuclear leukocytes (CMV), plasma (enteroviruses, arboviruses), or

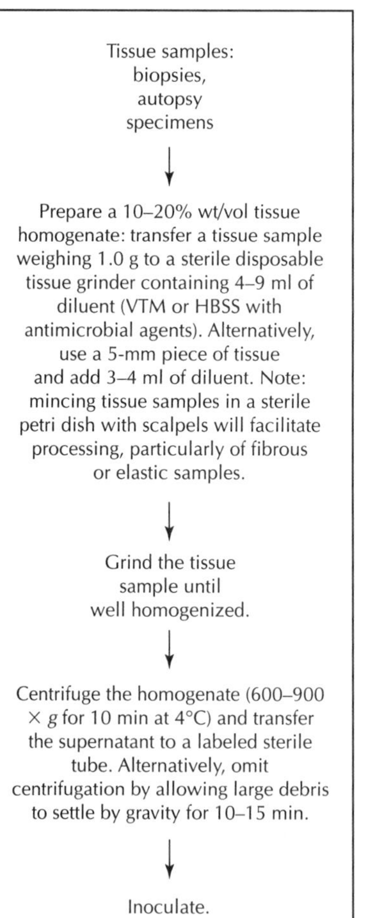

Figure 8.4–3 Preparation of tissue samples.

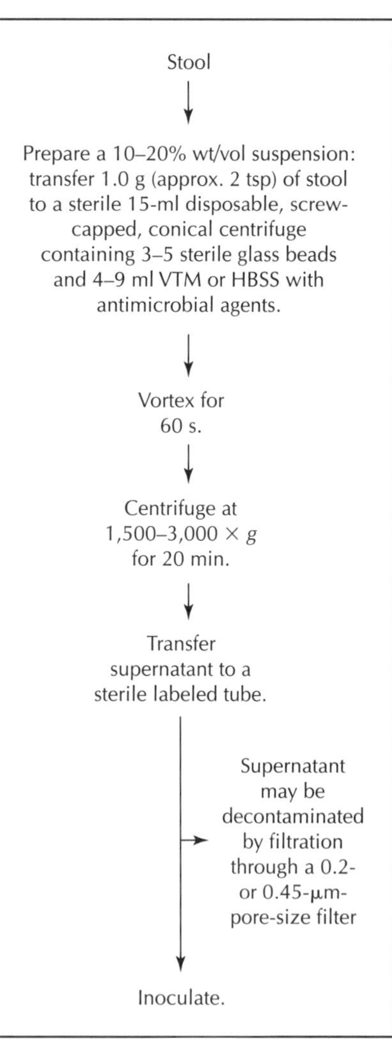

Figure 8.4–4 Preparation of stool samples. tsp, teaspoons.

Anticoagulated blood (3–7 ml); PBS may be added to bring sample to 3 ml if necessary.

↓

Mix 2 volumes of blood with 1 volume of room temp. 6% dextran in a 15-ml centrifuge tube; cap and invert several times to mix thoroughly.

↓

Allow the sample to stand in a vertical position for approximately 45 min at room temp or 30 min at 37°C (until most of the RBCs have settled and the WBCs are suspended in the plasma).

↓

Collect the WBC-rich fluid phase and transfer to a 15-ml sterile conical centrifuge tube.

↓

Cetrifuge at 600 × *g* for 10 min at room temp.

↓

Remove and discard the supernatant, tap the tube to loosen the cell pellet, and suspend cells in 10 ml of PBS.

↓

Centrifuge at 600 × *g* for 10 min at room temp.

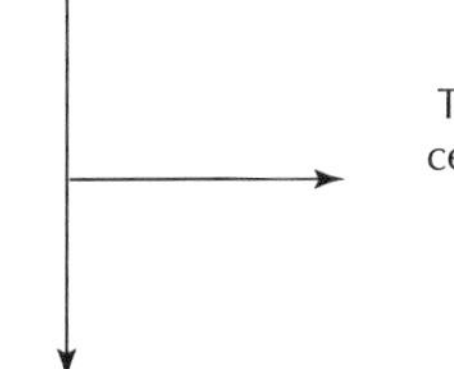

→ To remove contaminating RBCs, suspend cells in 10 ml of 0.155 M NH_4Cl in water; incubate for 5 min at room temp, and centrifuge at 600 × *g* for 10 min.

↓

Discard the supernatant and suspend the cells in 10 ml of PBS.

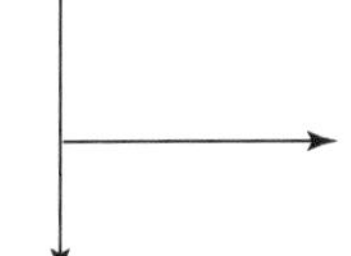

→ Remove an aliquot for cell counting (Coulter counter, hemacytometer) if the inoculum dose is to be adjusted for cell number.

↓

Centrifuge cells at 600 × *g* for 10 min at room temp.

↓

Discard supernatant.

↓ (left branch)

Suspend the cell pellet with 1–2 ml of VTM or cell culture medium.

↓

Inoculate without determining cell number.

↓ (right branch)

Suspend the cell pellet with VTM or cell culture medium to contain approximately 3 × 10^6 cells per ml.

↓

Inoculate known cell number.

Figure 8.4–5 Preparation of leukocytes by dextran sedimentation. PBS, phosphate-buffered saline; temp, temperature.

IV. PROCEDURES *(continued)*

Thoroughly mix blood collected in EDTA by gentle inversion.

↓

Transfer 2.5 ml of blood into a 15-ml conical polypropylene centrifuge tube containing 10.0 ml of ammonium chloride solution (0.155 M with 10 mM sodium bicarbonate, pH 7.2).

↓

Rock blood for 5 min at room temp.

↓

Centrifuge at 160 × *g* for 10 min at 4°C.

↓

Remove and discard supernatant fluid.

↓

Suspend the cell pellet in 1.0 ml of ammonium chloride solution and add 9.0 ml more of the ammonium chloride to the tube. Mix thoroughly.

↓

Centrifuge at 160 × *g* for 10 min at 4°C.

↓

Remove and discard the supernatant fluid.

↓

Suspend the WBC pellet in approximately 1.2 ml of PBS.

↓

Inoculate.

Figure 8.4–6 Preparation of leukocytes from anticoagulated blood by the ammonium chloride method. For details, see reference 7.

mononuclear cells (human immunodeficiency virus, enteroviruses, measles virus, Epstein-Barr virus) (6, 11, 12). Mixed leukocyte populations (PMNs, lymphocytes, monocytes) can be recovered from anticoagulated blood by allowing RBCs to settle by gravity and harvesting the leukocyte-rich plasma. Dextran, an erythrocyte-aggregating agent, accelerates RBC sedimentation, and this method is preferred to sedimentation of untreated anticoagulated blood. More recently, a modification of a rapid NH_4Cl treatment method (2) has been shown to recover WBC types in proportions similar to those found in whole blood and was useful for preparation of leukocytes for CMV culture (7).

Leukocytes separated by gradient centrifugation methods (3, 4) have been shown to yield higher rates of viral recovery compared with buffy coat preparations derived by gravity separation alone (6). Several formulations are commercially available (Appendix 8.4–1) for either simultaneously isolating mononuclear cells and PMNs (6, 8) or for isolating mononuclear cells (6, 8). With the former reagents (e.g., Mono-Poly Resolving Medium, Polymorphprep), differential migration of cells during centrifugation yields distinct mononuclear cell and PMN layers (bands), with RBCs pelleted at the tube bottom. With the latter reagents (e.g., Ficoll-Paque, lymphocyte separation medium), mononuclear cells yield a layer of cells, with PMNs and RBCs pelleted at the tube bottom; the PMNs can then be separated from the RBCs by mixing the sedimented cells with 6% dextran in a 2:1 ratio (6) as described in Fig. 8.4–5 for whole blood.

1. Materials
- **a.** Sterile pipettes (assorted) and pipetting device
- **b.** Sterile 15-ml screw-cap centrifuge tubes
- **c.** Phospate-buffered saline (PBS) or HBSS, pH 7.2–7.4
- **d.** Dextran sedimentation method: 6% dextran (*see* Appendix 8.4–1)
- **e.** NH_4Cl lysis: 0.155 M NH_4Cl, pH 7.2 (Appendix 8.4–1)
- **f.** Centrifuge, swinging bucket

2. Procedure
- **a.** Dextran sedimentation: Fig. 8.4–5
- **b.** NH_4Cl lysis: Fig. 8.4–6

E. Semen

Semen is not generally collected for diagnostic viral cultures, but is sometimes used in the research setting (e.g., CMV, human immunodeficiency virus [HIV] type 1). Toxicity is extensive with this sample and has been shown to be associated with the enzyme-rich fluid fraction rather than the pelleted cellular elements when the sample is fractionated by centrifugation (5, 9).

V. PROCEDURE NOTES

A. Use aseptic technique (Appendix 8.2–2) and sterile materials for specimen processing.

B. Follow appropriate safety practices, including the use of protective clothing and gloves, biological safety cabinets, and adequate decontamination techniques.

C. Hold specimens at 2–8°C before inoculation; inoculate specimens as soon after preparation as possible.

REFERENCES

1. **Clarke, L. M., B. J. Daidone, R. Inghida, M. Kirwin, and M. F. Sierra.** 1992. Differential recovery of cytomegalovirus from cellular and supernatant components of bronchoalveolar lavage specimens. *J. Clin. Pathol.* **97:** 313–317.
2. **Eggleton, P., R. Gargan, and D. Fisher.** 1989. Rapid method for the isolation of neutrophils in high yield without the use of dextran or density gradient polymers. *J. Immunol. Methods* **121:**105–113.
3. **English, D., and B. R. Anderson.** 1974. Single-step separation of red blood cells, granulocytes and mononuclear leukocytes on discontinuous density gradients of Ficoll-Hypaque. *J. Immunol. Methods* **5:**249–259.
4. **Ferrante, A., and Y. H. Thong.** 1980. Optimal conditions for simultaneous purification of mononuclear and polymorphonuclear leukocytes from human peripheral blood by the Hypaque-Ficoll method. *J. Immunol. Methods* **36:**109–117.
5. **Howell, C. L., M. J. Miller, and D. A. Bruckner.** 1986. Elimination of toxicity and enhanced cytomegalovirus detection in cell cultures inoculated with semen from patients with acquired immunodeficiency syndrome. *J. Clin. Microbiol.* **24:**657–660.
6. **Howell, C. L., M. J. Miller, and W. J. Martin.** 1979. Comparison of rates of virus isolation from leukocyte populations separated from blood by conventional and Ficoll-Paque/Macrodex methods. *J. Clin. Microbiol.* **10:**533–537.
7. **Menegus, M. A., C. M. Mayer, C. F. Mellen, and J. R. Zeller.** 1994. A simple and rapid method for the preparation of white blood cells (WBC) suitable for CMV culture. Abstr. Tenth Annual Clinical Virology Symposium, Clearwater, Fla.
8. **Miller, M. J.** 1992. Isolation of leukocytes from anticoagulated peripheral blood, p. 8.4.1–8.4.4. *In* H. D. Isenberg (ed.), *Clinical Microbiology Procedures Handbook,* vol. 2. American Society for Microbiology, Washington, D.C.
9. **Miller, M. J., and A. L. Warford.** 1992. Preparation of specimens for inoculation of cell cultures, p. 8.3.1–8.3.8. *In* H. D. Isenberg (ed.), *Clinical Microbiology Procedures Handbook,* vol. 2. American Society for Microbiology, Washington, D.C.
10. **Mulongo, K. N., M. E. Lamy, and M. Van Lierde.** 1995. Requirements for diagnosis of prenatal cytomegalovirus infection by amniotic fluid culture. *Clin. Diagn. Virol.* **4:** 231–238.
11. **Prater, S. L., R. Dagan, J. A. Jenista, and M. A. Menegus.** 1984. The isolation of enteroviruses from blood: a comparison of four processing methods. *J. Med. Virol.* **14:** 221–227.
12. **Schmidt, N. J., and R. W. Emmons.** 1989. *Diagnostic Procedures for Viral, Rickettsial, and Chlamydial Infections*, 6th ed. American Public Health Association, Washington, D.C.

APPENDIX 8.4–1

Reagents for Processing Anticoagulated Blood Specimens

A. Dextran (6%) with a molecular weight of 70,000
   1. Place 6 g of dextran and 0.9 g of NaCl in a 100-ml volumetric flask, add approximately 50 ml of double deionized water, and swirl to dissolve, then add a quantity sufficient (q.s.) to bring the final volume to 100 ml.
   2. Sterilize by Millipore filtration using a 0.22-μm-pore-size filter.
   3. Store at room temperature (up to 6 months). Do not use if cloudiness or a precipitate develops.

B. Ammonium chloride lysing solution (0.155 M), with 10 mM $NaHCO_3$, pH 7.2
   1. Place 4.15 g of NH_4Cl and 0.42 g of $NaHCO_3$ in a 500-ml volumetric flask, add approximately 300 ml of double deionized water, swirl to dissolve, then q.s. to 500 ml.
   2. Sterilize by filtration using a 0.22-μm-pore-size filter.
   3. Store at 2–8°C (up to 6 months).

C. Sources of reagents

Sterile 6% Dextran 70 in 0.9% NaCl (Macrodex)
Medisan Pharmaceutical Inc.
Parsippany, NJ 07054
(800) 763-3472

Mono-poly resolving medium (M-PRM) (*see* CMPH 8.4); lymphocyte separation medium (*see* CMPH 8.15), Lymphosep
ICN Biomedical Corp.
McLean, VA 22102
(800) 854-0530

Ficoll-Paque (*see* CMPH 8.15)
Pharmacia Biotech, Inc.
Piscataway, NJ 08855
(800) 526-3593

Polymorphprep (*see* CMPH 8.4)
Robbins Scientific Corp.
Sunnyvale, CA
(800) 752-8585

8.5 Viral Culture: Isolation of Viruses and *Chlamydia trachomatis* in Cell Cultures

I. PRINCIPLE

Most viral culture laboratories use a combination of tube and shell vial cultures. A number of variables can influence the sensitivity of viral cultures, including cell culture type, age, and confluence, the number of tubes or vials inoculated, the inoculation and incubation conditions, and the method and reagent used for isolate detection or identification. *Chlamydia trachomatis* cultures (Appendix 8.5–8) are performed in a manner similar to that described for viral shell vial cultures.

A. Conventional tube cultures

Adherent cell culture monolayers are inoculated and observed for evidence of viral replication, most commonly observed as cytopathic effect (CPE) or hemadsorption (HAd). Once viral CPE or HAd is observed, isolate identification is confirmed, usually by immunofluorescence (IF).

B. Shell vial cultures

The shell vial culture method uses centrifugation-enhanced inoculation of cells monolayered on a coverslip contained in a flat-bottomed shell vial. After incubation for 1 to 5 days, the monolayers are fixed and examined, most frequently by IF, for evidence of viral infection. The overnight pre-CPE detection of cytomegalovirus (CMV) in shell vial cultures, by using a monoclonal antibody directed against a viral protein expressed in the cell nucleus within several hours after infection, introduced the concept of rapid viral cultures (18, 33). Unlike most CMV shell vial cultures, CPE may be apparent for other viral cultures by the time staining is performed, depending on the length of incubation.

Despite the great utility of shell vial cultures, this methodology does not replace the need for tube cultures. Furthermore, the use of shell vial cultures does not eliminate the need to develop cell culture expertise, including the ability to identify viral CPE and differentiate it from problems that may arise in inoculated cell cultures, even during brief incubation.

II. SPECIMENS

A. Prepare specimens as described in procedure 8.4.

B. Thaw frozen specimens rapidly and completely by swirling in a 37°C water bath until they are *almost* thawed; do not allow thawed specimens to remain at 37°C.

III. MATERIALS

Include QC information on reagent container and in QC records.

A. Supplies

1. Racks designed to hold culture tubes in a slightly slanted horizontal position with the necks slightly raised
2. Sterile individually wrapped pipettes (1-, 2-, 5-, and 10-ml sizes) and safety pipetting devices
3. Cold blocks or ice
4. Personal protective gear
5. Waste containers
Discard specimens, cell cultures, and other infectious waste based on local regulations.
6. Microscope slides (1 × 3 in.) for mounting shell vial coverslips
7. Teflon-coated microscope slides with 5- to 8-mm wells and no. 1 coverslips for IF
8. 0.22-μm-pore-size filters for culture decontamination
9. Sterile screw-capped vials (2 and 4 ml) suitable for freezing isolates at −70°C
10. Sterile 15-ml polypropylene conical centrifuge tubes
11. Labels for tubes and marking pens
12. Dry ice

B. Cell cultures

Select cell lines based on the guidelines presented in procedure 8.2.

C. Reagents

1. Cell culture medium (*see* procedure 8.6).
2. Immunologic reagents
Reagents are available from several sources including those listed in Appendix 8.5–1. Most antibody reagents used for viral identification are monoclonal antibodies (MAbs); because of their high specificity and lower binding efficiencies, a mixture of MAbs directed against different viral epitopes is generally more useful than individual MAb preparations. Mixtures of MAbs (blends, pools) directed against two or more viruses are also available for detecting multiple viruses with a single reagent (23, 27, 30, 34). Recently, mixed antibody preparations labeled with different fluorescing reagents have been introduced for the simultaneous identification of common virus pairs (e.g., influenza types A and B [1, 6], herpes simplex virus [HSV], and CMV).
3. Sterile phosphate-buffered saline (PBS), pH 7.2 to 7.6 (*see* EPCM procedure 8.6)
4. Hanks' balanced salt solution (HBSS) (*see* EPCM procedure 8.6)
5. Buffered glycerol mounting medium, pH 8.0, for IF
6. Acetone or other fixative recommended by the antibody manufacturer
7. Guinea pig blood in Alsever's solution for HAd
8. Disinfectants
Hypochlorite solution (0.5%), Wescodyne, 70% ethanol

D. Equipment

1. Microscope (100×)
Shell vials can be examined by using an inverted microscope with a glass microscope slide over the objective to support the vial. Tube cultures can be examined by using a standard or inverted microscope; tubes can be held in place with a holder consisting of two parallel tracks glued to the stage.
NOTE: One end of the holder should be slightly higher so that the culture medium does not flow into the caps during observation.
2. Class II biological safety cabinet (certified semiannually)
Wipe down all interior cabinet surfaces with 70% ethanol or another appropriate reagent before and after each use.
NOTE: 10% Clorox is used by many laboratories to wipe down bench surfaces; this material will cause pitting and deterioration of the stainless-steel surfaces of the biological safety cabinet.
3. Centrifuge
Swinging-bucket rotor with carriers to accommodate 15-ml centrifuge tubes and shell vials (15 by 45 mm); size 00 rubber stoppers may be used as cushions to adapt carriers to hold shell vials.
4. Incubator
A CO_2 environment is not necessary for tube or shell vial cultures since these vessels are generally tightly capped during incubation. However, cell culture methods with plastic dishes, microtiter plates, or vented flasks require an atmosphere containing 5 to 8% CO_2 (*see* CMPH 8.20).

III. MATERIALS *(continued)*

5. Roller drum (optional) to hold tubes horizontally in incubator at 12 to 15 rph
6. Rotator (15 to 20 rpm) or rocker (for IF)
7. Fluorescence microscope equipped with filters suitable for fluorescein isothiocyanate (FITC) or fluorochrome of choice
8. Refrigerator at 2 to 8°C
9. Ultra-low freezer, −70°C or lower

IV. QUALITY CONTROL

A. Incubators

1. Temperature stability is important. If a universal incubation temperature of 35°C is selected, the acceptable range should be 34 to 36°C. Check incubators for hot spots and utilize a continuous temperature-monitoring system.
2. Temperatures above 36°C may adversely affect respiratory-virus recovery, particularly that of rhinoviruses, which prefer 33°C.

B. Cell cultures

1. Record the lot number of cell cultures used for each specimen.
2. Uninoculated (negative) controls
   a. Select several uninoculated controls from each lot. Inoculate, observe, and process in parallel with inoculated specimens.
   b. Utilize uninoculated lot-matched cultures as negative controls for assays (e.g., HAd and IF).
3. Inoculated (positive) controls
   a. Use viral QC stocks to assess cell culture sensitivity. Cell culture susceptibility can be affected by several variables (e.g., monolayer density and age, mycoplasma contamination, endogenous virus contamination), and quality can vary from lot to lot. A typical QC panel for a general viral-culture laboratory consists of those viruses most likely to be encountered in the patient population served by the laboratory. Cell culture sensitivity is best determined by low virus doses or, ideally, by titrating characterized stocks (Appendix 8.5–2). However, the time and cost of performing titrations on each lot of cells are prohibitive and laboratories generally limit use of titrations to special circumstances (e.g., reduction in isolation rate; evaluation of new cell lines or vendors; or development of a new isolation procedure). Communication among laboratories regarding suspected cell culture problems is invaluable for identifying problems with cell culture performance.
   b. Use virus-inoculated cell cultures as controls for assays (e.g., HAd or IF).

      ◩ **NOTE:** Characterized viral QC material can be obtained from several sources, including proficiency testing samples, reference laboratories, or the ATCC (Rockville, Md., [800] 638-6597). Prepare stock material by inoculating several tubes and harvesting positive cultures as described for viral isolation; dispense in single-use aliquots and store at −70°C or lower.

C. Fetal bovine serum (FBS) and other cell culture reagents

1. These reagents must be sterile and nontoxic to cell culture monolayers.
2. Cell culture medium must maintain cells in a metabolically active condition and must not be inhibitory to viral replication.

V. PROCEDURES

A. Inoculation (Fig. 8.5–1)

1. The recommended method consists of adsorption of the prepared specimen onto a drained monolayer as described in Fig. 8.5–1. Alternatively, the specimen can be inoculated directly into the tube containing 1 to 2 ml of cell culture medium; refeed tubes with fresh medium before inoculation.

Select 2 tubes/vials of each cell culture type to be inoculated; single tubes/vials may be used if two or more cell cultures with overlapping sensitivities are used. Select uninoculated controls to be incubated and observed in parallel with cultures. (Refer to Table 8.2–3 for cell line selections.)

Note: cell culture sensitivity can vary over time. Therefore, even for isolation of a single virus, using a minimum of 2 cell lines or at least one cell line obtained from 2 sources is recommended.

↓

Determine that each tube has an acceptable monolayer with no visible evidence of contamination (*see* Table 8.2–4). Fresh, slightly subconfluent to confluent monolayers are recommended.

↓

Label each tube/vial to identify the specimen and the date of inoculation. Record the cell culture lot numbers used for each specimen.

↓

Decant or aspirate and discard the medium from each monolayer. Note: handle uninoculated cell cultures in the same manner as infectious materials.

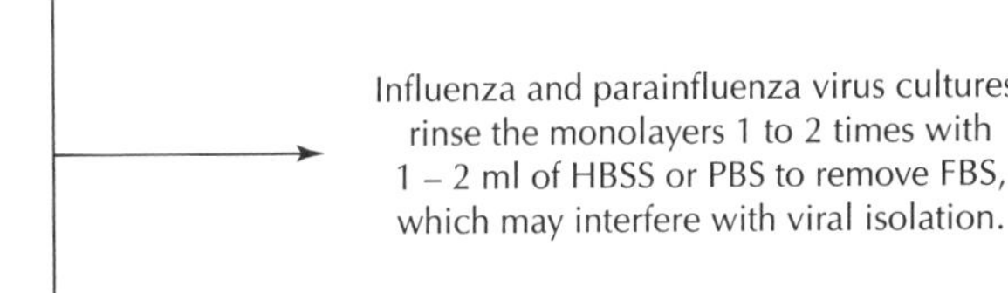

Deliver 0.2 to 0.3 ml of prepared specimen to each monolayer[a]. To avoid cross-contamination of cultures with detached cells or endogenous viruses, inoculate cultures in the following sequence; fibroblast, primary nonsimian, transformed, primary simian.

↓

Tube culture adsorption: incubate the tubes in a slanted position, with the inoculum overlaying the monolayer, for 45 to 90 min at 35°C. Shell vial cultures: centrifuge at 700 × *g* for 45 min at room temperature to 35°C; do not allow temperature to reach greater than 36°C during centrifugation.

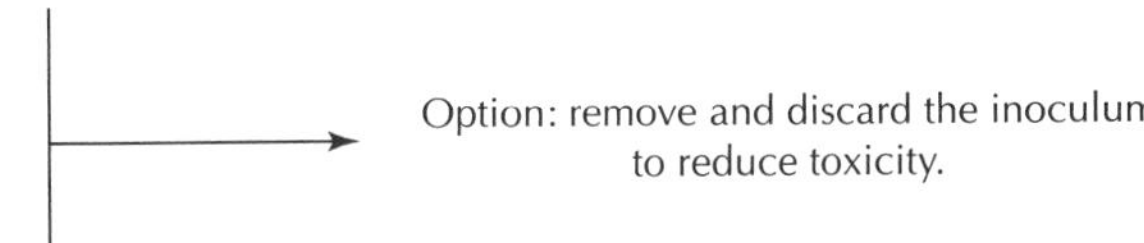

Refeed the monolayers with 1–2 ml of the appropriate cell culture medium[b] warmed to room temperature; feed monolayers in the same order as used for inoculation. Note: use caution when opening shell vials since cross-contamination may occur as a result of aerosols created when opening vials (32).

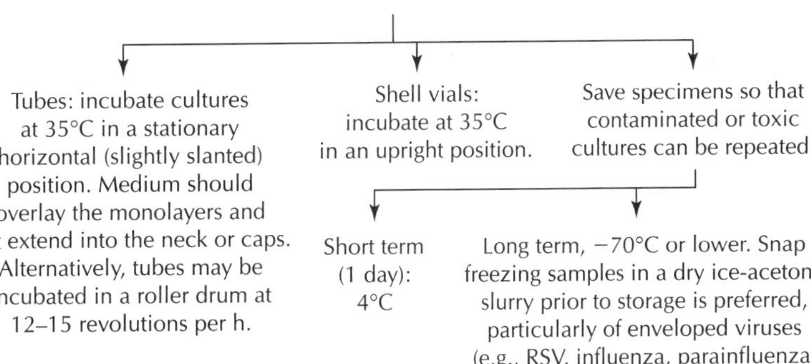

Figure 8.5–1 Inoculation of tube and shell vial cell cultures for the detection of cytopathic and hemadsorbing viruses (*see* Table 8.5–1).

[a] Blood specimens with known cell concentrations: inoculate three vials with approximately 1.5×10^6 leukocytes per vial. It has been reported that a total of at least 4×10^6 leukocytes is recommended for the sensitive detection of CMV viremia (5); concentrations $> 2 \times 10^6$ cells per vial are not generally recommended because of toxicity.

[b] Eagle minimal essential medium (EMEM) supplemented with 2% heat-inactivated FBS is a medium suitable for all or most viral cultures performed in the diagnostic setting with monolayered cell cultures; serum-free EMEM or other serum-free cell culture medium is used by many laboratories for the isolation of influenza and parainfluenza viruses. Medium containing trypsin is recommended for influenza and parainfluenza virus cultures with MDCK and LLC-MK_2 cells, respectively (*see* EPCM 8.6).

V. **PROCEDURES** *(continued)*

This latter method may be particularly useful for specimens known to be very toxic to cell monolayers (stool suspensions, blood, bone marrow, and urine).

2. A variety of centrifugation forces and times have been used for shell vial cultures (reviewed in reference 20). Centrifugation forces of 700 to 900 × *g* for 30 to 45 min are most commonly used, although a shorter centrifugation at 3,500 × *g* has provided good results without adversely affecting monolayer integrity (10).

B. Incubation

1. Many laboratories use an incubation temperature of 35°C (34 to 36°C) for all viral cultures, while others incubate respiratory virus cultures at 33 to 34°C.
2. Incubate shell vials in an upright position. Tubes may be incubated in a stationary horizontal position or in a roller drum. Low-speed rolling (12 to 15 rph) has been shown to enhance recovery of several viruses (reviewed in reference 20).

   NOTE: Tissue degeneration may accelerate for some cell lines during roller incubation.
3. Except for CMV cultures, observe inoculated tube cultures for 10 to 14 days after inoculation; observe CMV cultures for at least 3 weeks and up to 4 weeks or longer. Time points for terminating shell vial cultures also vary. For example, some laboratories perform staining for CMV at 24 h, while others stain at 48 h, and yet others stain at both time points. Incubation times for shell vial cultures for other viruses range from 24 h to 4 to 5 days, depending on the virus and the laboratory's experience with sensitivity at different time points compared with that of tube cultures.

C. Observation of inoculated monolayers

1. Observe monolayers for the development of viral CPE and/or HAd (Fig. 8.5–2) and problems and conditions sometimes encountered with cell cultures (Fig. 8.5–3); record all observations and manipulations.
2. Most of the human viruses encountered in the diagnostic-virology laboratory produce a characteristic CPE (Table 8.5–1). CPE represents a composite of morphological changes that may include swelling, rounding, clumping, and increased refractility of cells; cytoplasmic vacuolation, granulation, and nuclear condensation; and the development of multinucleated giant cells (syncytia). It is usually progressive and may lead to complete monolayer destruction. The appearance, the time required for the development of CPE, and the rate of progression are determined not only by the type and dose of virus in the inoculum but also by the monolayer type, age, and density. These characteristics also provide a basis for tentative identification of the viral isolate. Commonly encountered CPEs are illustrated in Fig. 8.5–4, parts 1 to 7.
3. Endogenous viruses can also produce CPE, including many simian viruses. Simian virus 40 (SV40) (vacuolating virus) produces cytoplasmic vacuoles, and simian foamy virus produces large vacuoles resembling soap foam. Simian adenoviruses, herpes B virus, enteroviruses, reovirus, and measles can produce CPE mimicking the CPE produced by their human counterparts and can cause human infections associated with high mortality (herpes B virus, Marburg virus). See reference 19 for a discussion of endogenous viruses and photographs.

D. HAd procedure

1. HAd with guinea pig RBCs (*see* Fig. 8.5–8) is a simple and inexpensive way to screen inoculated cell cultures to detect influenza, parainfluenza,

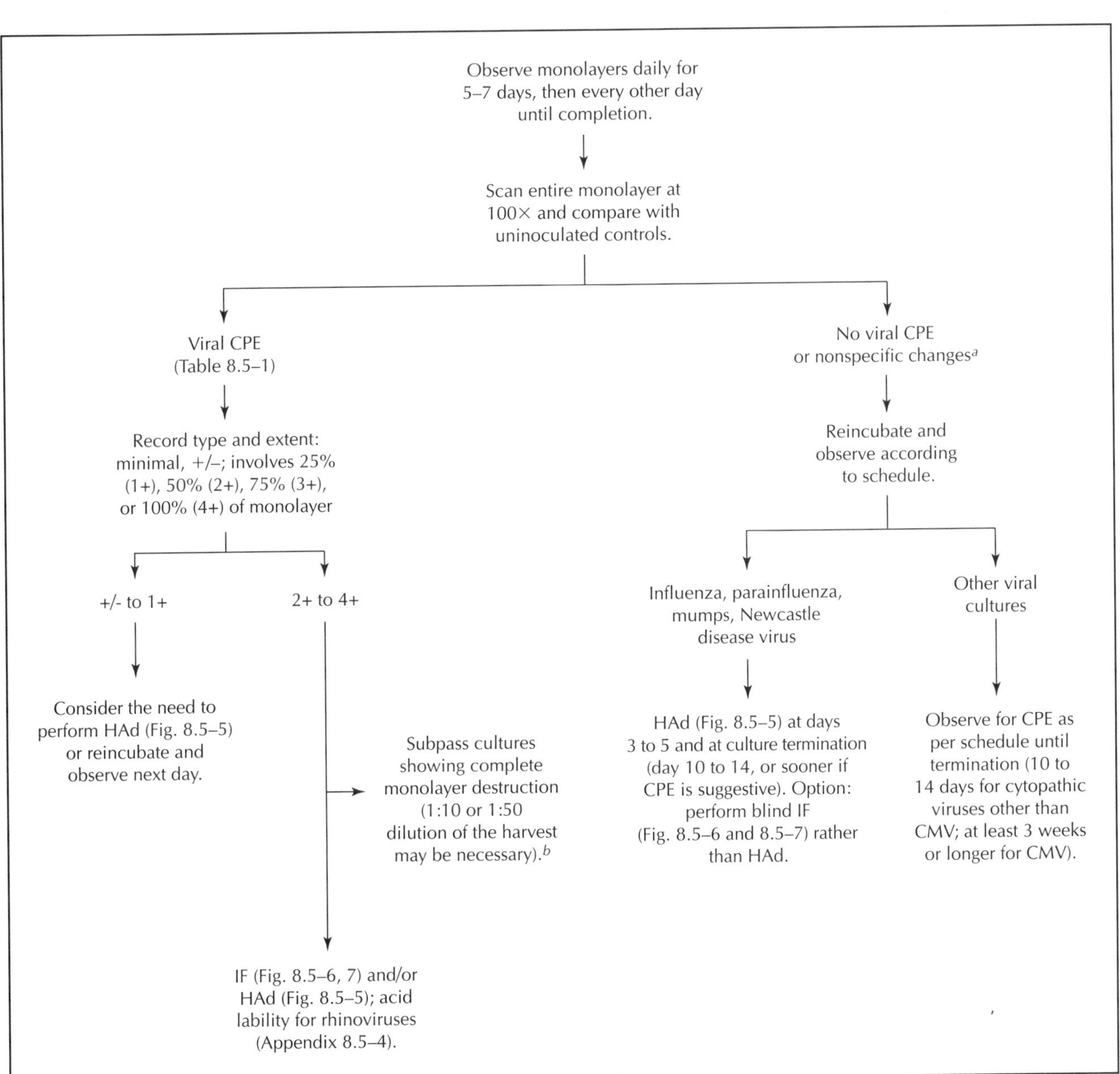

Figure 8.5–2 Observation of inoculated monolayers: CPE and HAd. NOTE: Rarely, cells exhibiting minimal CPE revert to a normal appearance. Subpassage or reinoculation may be helpful in this instance.
[a] See Fig. 8.5–3.
[b] Viral viability may decline rapidly with incubation in the absence of cells, particularly with labile viruses (e.g., RSV, CMV).

mumps, and Newcastle disease viruses. Perform this procedure as shown in Fig. 8.5–5.

2. SV5, a simian parainfluenza virus, hemadsorbs guinea pig RBCs and is an important endogenous contaminant of primary monkey kidney cells.
3. HAd, the attachment of RBCs to infected cells, is mediated by hemagglutinins, viral envelope proteins that are inserted into the plasma membrane of infected cells during viral replication.
4. Not all hemadsorbing viruses produce CPE (Table 8.5–1).

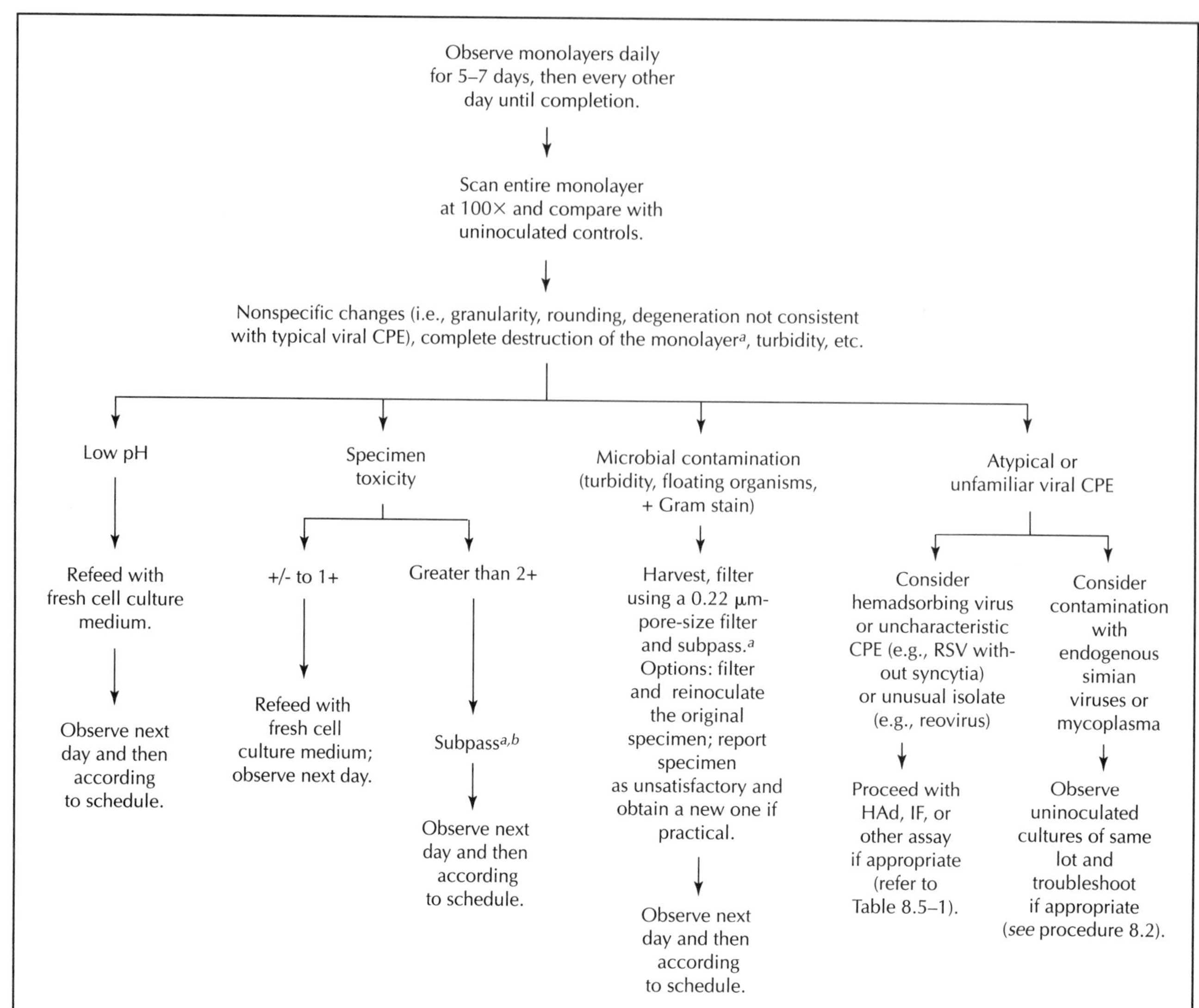

Figure 8.5–3 Observation of inoculated cell cultures: toxicity, microbial contamination, and atypical changes.
[a] Viral viability may decline rapidly with incubation in the absence of cells, particularly with labile viruses (e.g., RSV, CMV).
[b] In rare cases, specimen toxicity may be so great that two or three subpassages are required. Unlike viral CPE, toxicity may sometimes be observed at 3 to 4 h after inoculation.

V. PROCEDURES *(continued)*

E. Isolate detection and identification

1. Direct or indirect IF with MAbs is the method most frequently used by diagnostic laboratories to identify viral isolates in tube and shell vial cultures. FITC is the most commonly used fluorochrome, with specific staining appearing as bright to brilliant apple green fluorescence.
2. Prepare slides from tube cultures for isolate identification by IF or immunoperoxidase staining as described in Fig. 8.5–6. This procedure yields cell spots suitable for culture confirmation of isolates obtained in tube cultures. Other methods described in CMPH 8.10 yield better quality slides but are labor intensive and are used for special applications (e.g., preparing teaching materials).

Table 8.5–1 Viral culture characteristics

Virus	Characteristic CPE				Isolate identification	Shell vial applications
	Appearance	Development[a] (days)	Progression	Comments		
Adenoviruses	Enlarged, rounded cells in tightly associated grapelike clusters. Some isolates may produce a lattice-type arrangement of rounded cells.	4–7	Moderate	CPE is less characteristic in diploid fibroblasts.	IF (group-specific antibody), neutralization[b] useful for typing (Appendix 8.5–3)	Sensitivity may be low compared with that of tube cultures (27, 30).
CMV	Plump, rounded cells in elongated foci parallel to the long axis of the cell	7–10	Slow	CPE may take 2 to 3 weeks or longer to develop with low viral concentrations; CPE may develop rapidly with specimens containing high viral concentrations. Some isolates may not progress beyond a few patches of CPE.	IF; identification of isolates with minimal CPE may be facilitated by subpassing harvested cells and fluid into centrifugation-enhanced shell vial cultures.	Shell vial method preferred, with staining generally performed at 16 to 48 h. Low sensitivity has been reported by some labs (11, 21, 28, 29, 31) for blood cultures, including the use of ML cells (21) which have been reported to be useful for circumventing toxicity problems (15). Inoculating 3 (28) or 4 shell vials may improve sensitivity.
Enteroviruses	Rounded, highly refractile cells in loose clusters or dispersed throughout monolayer: varies with enterovirus group.	2–5	Moderate to rapid, depending on the virus		IF (panenterovirus MAb, polio blends, type-specific MAbs); isolates not confirmable by IF may be identified by neutralization[b] (Appendix 8.5–3). A panenterovirus MAb has been shown to be useful for culture confirmation of many isolates (39).	

(continued)

Table 8.5–1 Viral culture characteristics (*continued*)

Virus	Characteristic CPE				Isolate identification	Shell vial applications
	Appearance	Development[a] (days)	Progression	Comments		
					Ruling out poliovirus is important and can be accomplished by using a blend of MAbs to the three poliovirus types.	
HSV	Clusters of rounded, ballooned cells with or without syncytia. Early CPE is focal but then progresses throughout the monolayer.	1–3	Moderate to rapid	CPE may develop more slowly and be less characteristic in human fibroblasts. Simian B virus produces a similar CPE in simian cells.	IF (type-specific MAbs, type 1/type 2 blends)	Shell vials may yield a less dramatic improvement over tube cultures since most isolates can be detected from genital specimens collected from symptomatic individuals within 24 to 72 h after inoculation of tube cultures. However, shell vial cultures may improve the recovery of HSV in less sensitive cell lines and with specimens containing low viral concentrations (40).
Influenza	Variable. No CPE may be produced or may include granular and vacuolated appearance or nonspecific degeneration. Rounded refractile cells may be associated with type B.	3–5	Moderate	HAd with guinea pig RBCs; degree of HAd is independent of the presence or degree of CPE. Blind IF (e.g., at days 3 to 5 and/or at culture termination) rather than HAd may enhance isolation. Incorporate trypsin in medium when using MDCK cells (13, 26, 36).	IF (type-specific MAbs, dual-labeled mixtures [1, 6]), pools containing antibodies against several respiratory viruses. Subtyping by IF, HI (see CMPH 8.12)	Shell vials have been shown to be useful for respiratory viruses (12, 22), including the use of MAb pools (23, 27, 30, 34). Staining performed at 1–4 days

Measles	Syncytia develop by fusion of cells. Nuclei may encircle granular mass of giant cell. Extensive vacuolization may also be present.	5–10	Slow to moderate		IF	A-549 useful (urine specimens required longer [up to 5 days] than NP swabs [18–36 h] [25])
Mumps	Cell rounding and syncytia formation. May appear as nonspecific-looking granularity with progressive degeneration.	3–7	Moderate	HAd with guinea pig RBCs	IF	
Parainfluenza viruses	Variable, with increased rounding, granularity, and progressive degeneration; syncytia formation associated with types 2 and 3	3–7 days; type 4 may require incubation beyond the usual 10 to 14 days (7, 8).	Moderate	HAd with guinea pig RBCs; type 4 hemadsorbs better at room temperature or 37°C than at the usual incubation temperature of 4°C (8, 26, 35). Blind IF (e.g., at days 3 to 5 and/or at culture termination) may enhance isolation. Incorporate trypsin in medium when using LLC-MK_2 cells (13).	IF (type-specific MAbs, pools containing antibodies against several respiratory viruses including parainfluenza virus types 1, 2, and 3)	Shell vials have been shown to be useful for respiratory viruses (22, 23, 27, 30, 34). Staining performed at 1–4 days
Reoviruses	Nonspecific-looking granular appearance with progressive degeneration and detaching of the monolayer	7–10	Slow to moderate	CPE may be difficult to distinguish from nonspecific monolayer degeneration.	Neutralization[b] (Appendix 8.5–3)	

(continued)

Table 8.5–1 Viral culture characteristics (*continued*)

Virus	Characteristic CPE				Isolate identification	Shell vial applications
	Appearance	Development[a] (days)	Progression	Comments		
RSV	Syncytia develop in some cell lines, particularly subconfluent HEp-2 cells. May also appear as granular progressive degeneration.	3–5	Moderate	Blind IF (e.g., at days 3 to 5 and/or at culture termination) may enhance isolation.	IF (type-specific MAbs, pools containing antibodies against several respiratory viruses)	Shell vials have been shown to be useful for respiratory viruses (22, 23, 27, 30, 34). Staining performed at 1–4 days; longer incubation shown to detect more isolates (27).
Rhinoviruses	Enterovirus-like	5–7	Moderate	Incubation at 33°C preferred	Acid lability assay (Appendix 8.5–4)	
Rubella	CPE not produced on primary isolation				Interference assay (Appendix 8.5–5), IF	Shell vial technique appears useful (38).
Vaccinia	Syncytia and clusters of rounded, enlarged, refractile cells with cytoplasmic strands frequently bridging foci of CPE.	1–3	Moderate to rapid	HAd with chicken red blood cells	IF, neutralization[b] (Appendix 8.5–3)	
VZV	Foci of enlarged, rounded, refractile cells with or without syncytia Cytoplasmic strands and granularity may be prominent as CPE progresses.	4–7	Slow to moderate		IF; identification of isolates with minimal CPE may be facilitated by subpassing harvested cells and fluid into centrifugation-enhanced shell vial cultures.	Shell vials (17) preferred; stain 1 vial at 24 to 48 h and second vial at day 5 or sooner if CPE develops.

[a] Time generally required for majority of isolates to produce CPE or HAd.

[b] Neutralization assay involves mixing the isolate with antiserum containing neutralizing antibodies and, after a brief incubation, inoculating the cell cultures with the virus-antibody mixture(s). Neutralization by an antibody is determined by the failure to develop CPE, thereby establishing viral identity.

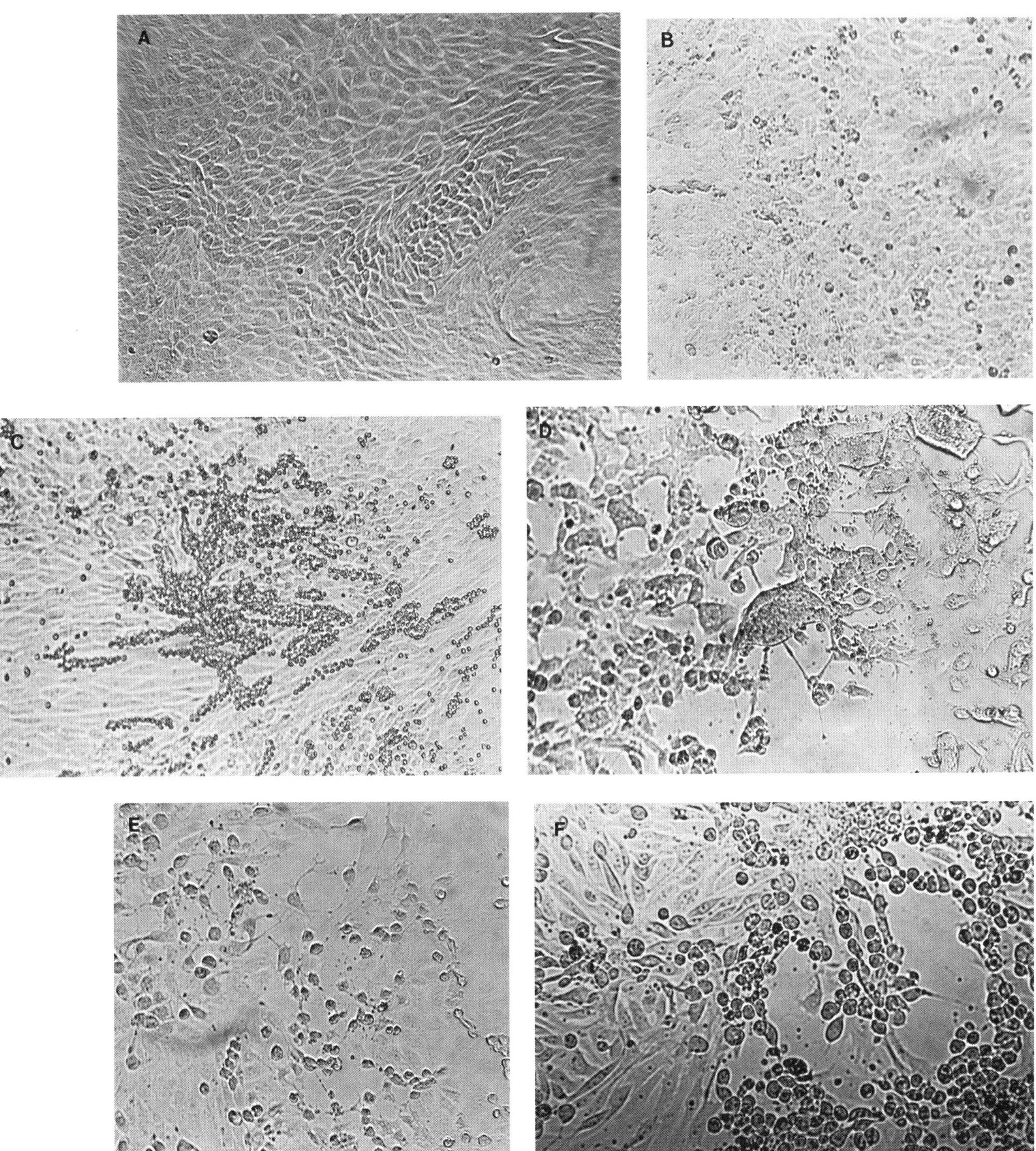

Figure 8.5–4 (part 1) Primary rhesus monkey kidney cell cultures. Original magnification, ×200. (A) Uninoculated confluent monolayer; (B) influenza virus type B-infected culture showing granular cytopathic changes produced by some influenza virus strains; (C) culture infected with influenza virus type A, showing HAd of guinea pig RBCs; (D) cellular rounding and syncytium formation produced by measles virus; (E) culture infected with coxsackie virus type B3 showing scattered rounded refractile cells frequently referred to as enterovirus-like CPE; (F) CPE produced by echovirus type 11 (courtesy of CDC). (Reprinted from CMPH 8.7.10.)

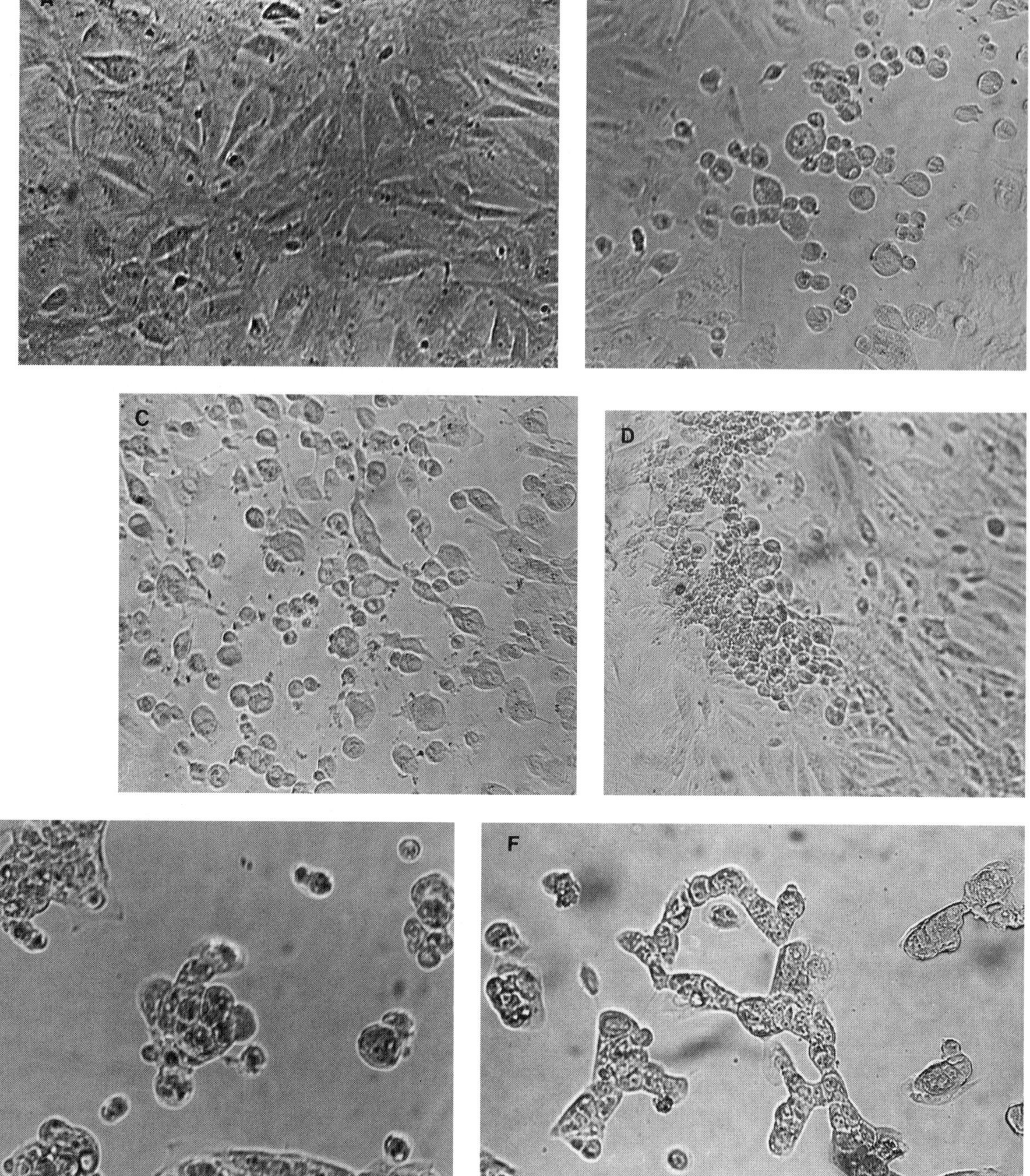

Figure 8.5–4 (part 2) Low-passage-number human neonatal kidney cell cultures. Original magnification, ×200. (A) Uninoculated confluent monolayer. (B) Early focus of enlarged rounded cells produced by herpes simplex virus type 2. (C) Extensive involvement of the monolayer by the rapidly progressive CPE of herpes simplex virus type 2. (D) Focal area of CPE produced by VZV. Progression of cytopathic changes produced by this virus is slower than with infection by herpes simplex virus. (E) Culture infected with adenovirus type 3 showing the tightly overlapped rounded cells frequently referred to as grapelike clusters. (F) Latticed arrangement of cells that may result from infection by adenoviruses.

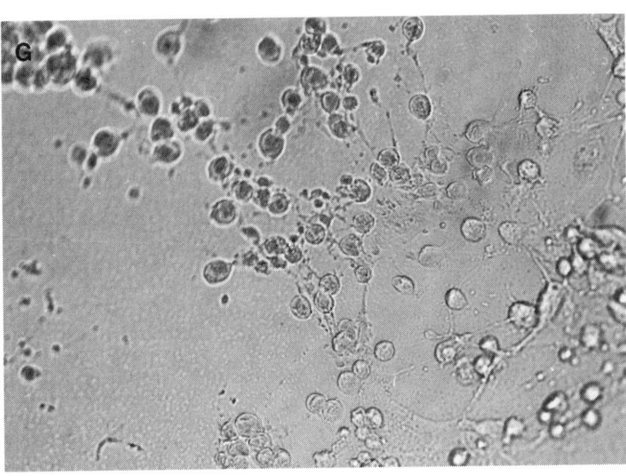

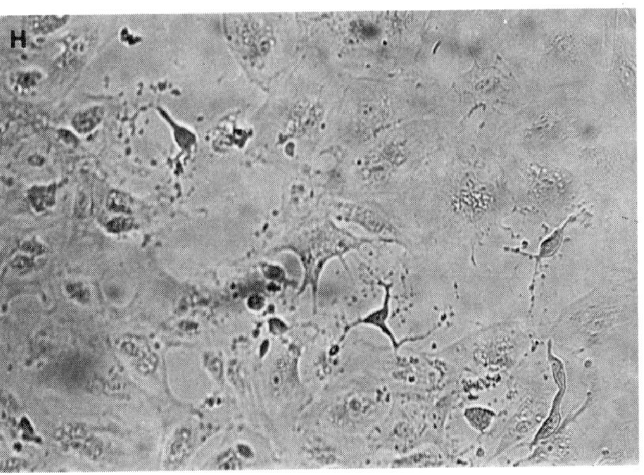

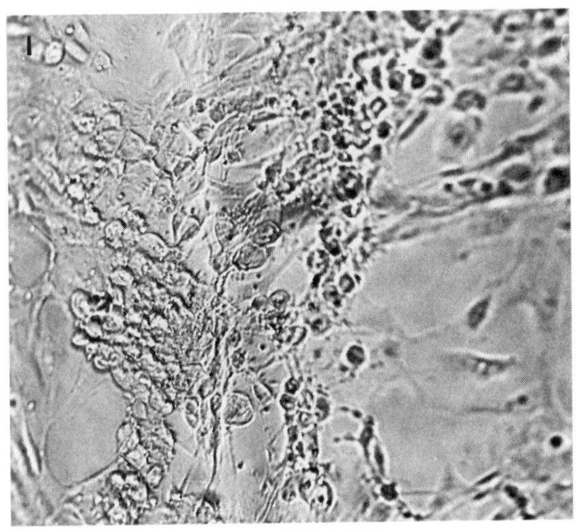

Figure 8.5–4 (part 2) *(continued)* (G) Generalized rounding and extensive monolayer destruction caused by echovirus type 6. (H) Early cytopathic changes produced by coxsackievirus type B3. (I) Nonspecific type of CPE observed with infection by reoviruses. Monolayer shows cellular rounding, degeneration, and lifting from the glass surface. (Reprinted from CMPH 8.7.11–8.7.12.)

V. PROCEDURES *(continued)*

3. Perform IF for tube culture confirmation as described in Fig. 8.5–7. IF staining patterns vary with the virus and the antibody preparation used (Fig. 8.5–9A to L).
4. For shell vial cultures, establish the time points for fixing and staining monolayers and follow the procedure described in Fig. 8.5–10.
5. Perform IF on shell vial coverslips as described in Fig. 8.5–11. IF staining patterns vary with the virus and the antibody preparation used (Fig. 8.5–9A to L).
6. Instructions for determining the optimal working dilutions of antibody reagents and methods for assessing heterologous viral cross-reactivity are provided in CMPH 8.9. However, the diagnostic laboratory generally uses commercial reagents supplied either at the appropriate working dilution or with instructions for preparing the optimal working dilution; products generally contain a counterstain such as trypan blue which stains cells to aid in visualization of cells and provide a contrasting background color for the IF staining.

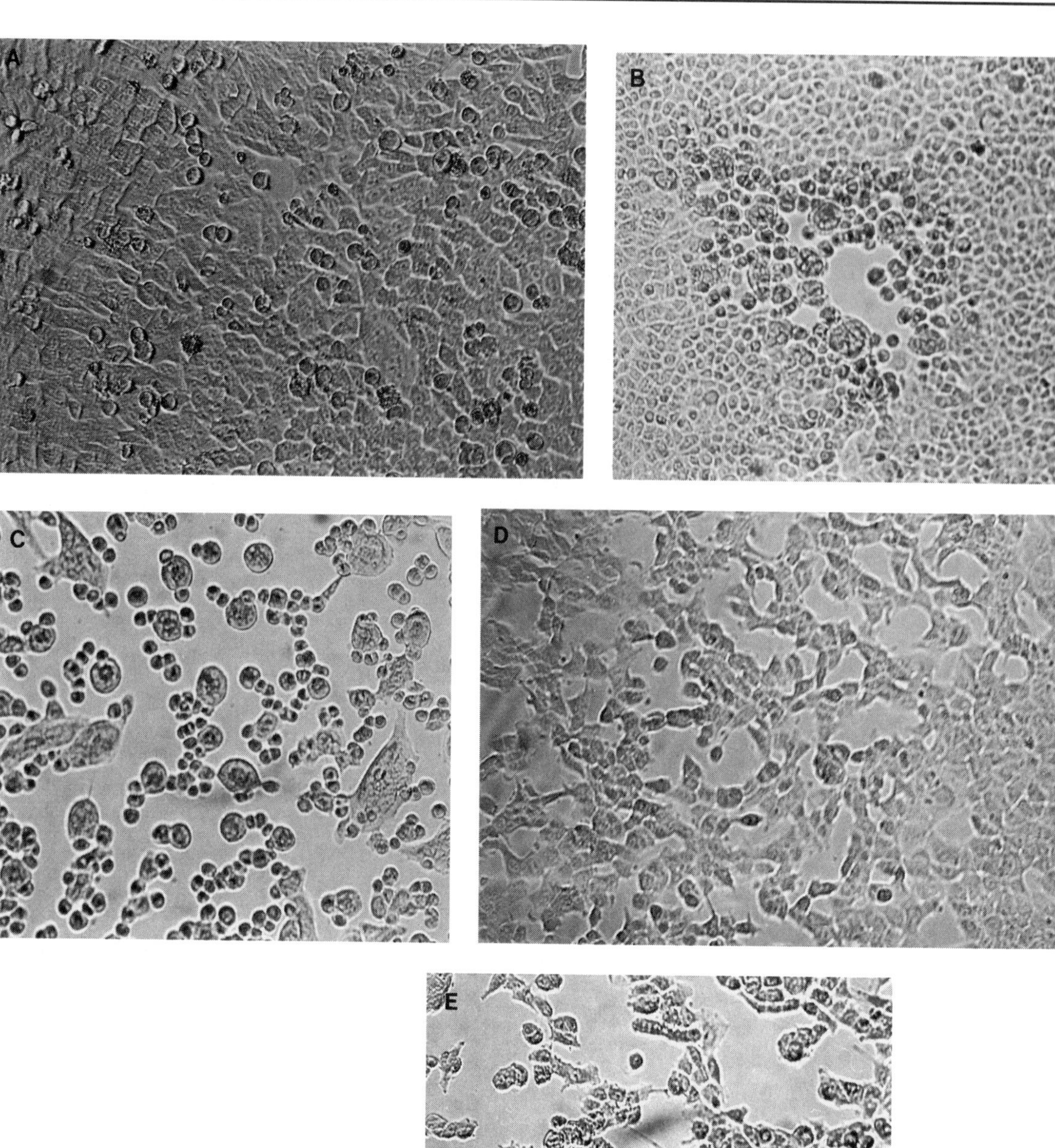

Figure 8.5–4 (part 3) A-549 cells. Original magnification, ×200. (A) Uninoculated culture showing overgrowth of the monolayer typical of rapidly growing continuous cell lines; (B) early focus of CPE produced by herpes simplex virus type 2 against a background of uninfected confluent cells; (C) advanced CPE of herpes simplex virus type 2; (D) early CPE of adenovirus type 3; (E) advanced CPE of adenovirus type 3. (Reprinted from CMPH 8.7.13.)

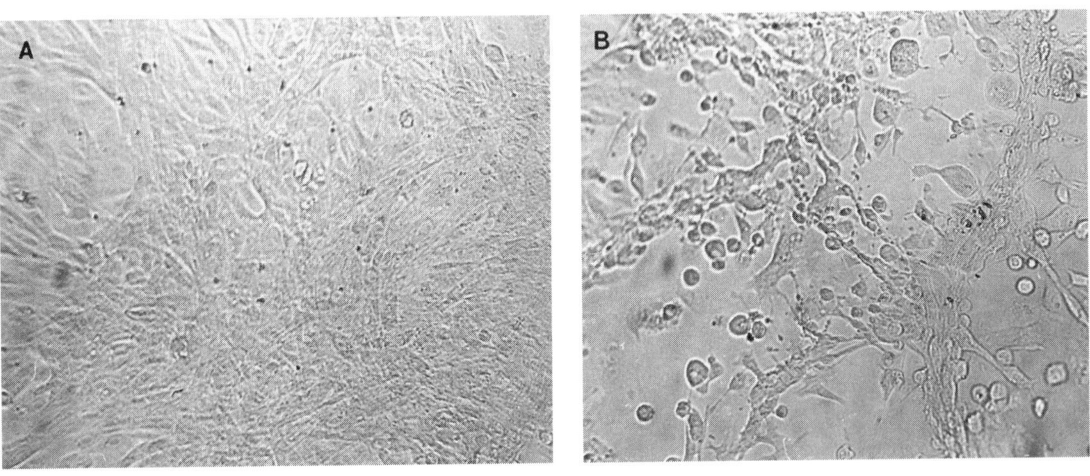

Figure 8.5–4 (part 4) Primary rabbit kidney cell cultures. Original magnification, ×200. (A) Uninoculated confluent monolayer; (B) rapidly progressive CPE produced by herpes simplex virus type 2. (Reprinted from CMPH 8.7.14.)

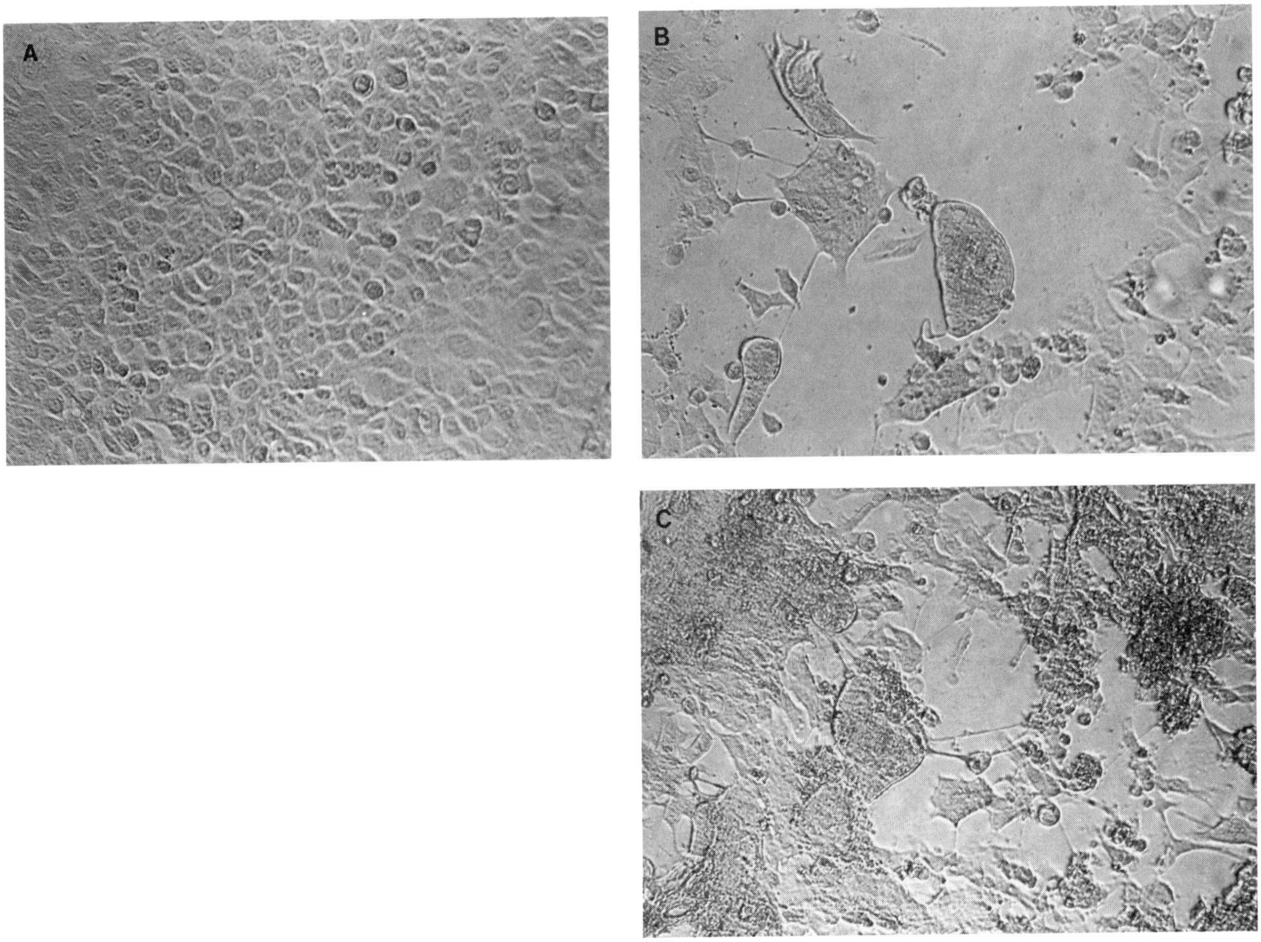

Figure 8.5–4 (part 5) HEp-2 cell cultures. Original magnification, ×200. (A) Uninoculated confluent monolayer; (B) CPE produced by respiratory syncytial virus showing numerous syncytia formed as a result of cell fusion; (C) syncytium formation resulting from infection with measles virus. (Reprinted from CMPH 8.7.14.)

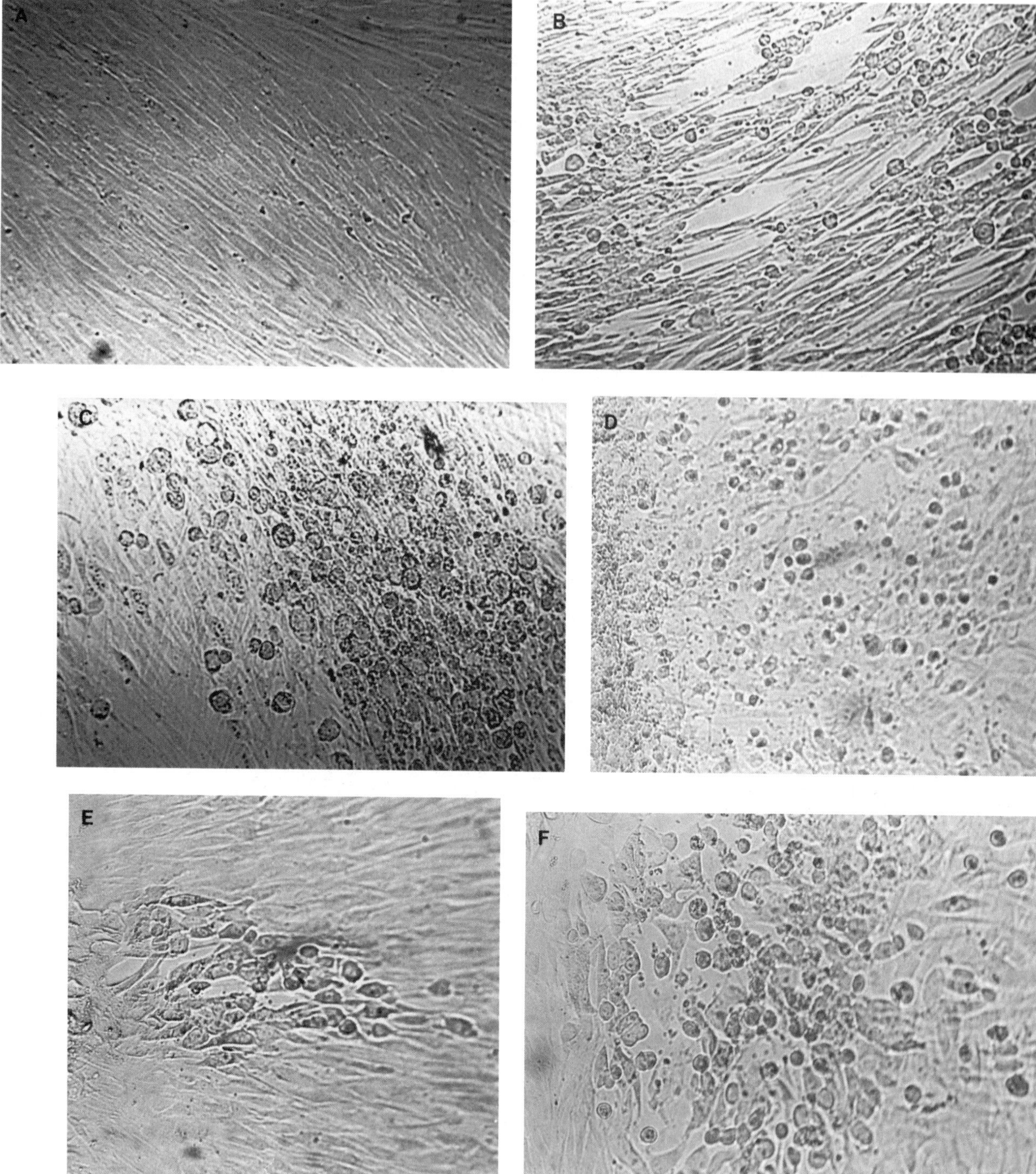

Figure 8.5–4 (part 6) Diploid human lung fibroblast cultures. Magnification, ×200. (A) Uninoculated confluent monolayer. (B) Advanced CPE produced by herpes simplex virus type 2 (courtesy of CDC). (C) Focal area of CPE produced by VZV (courtesy of CDC). (D) Advanced CPE of VZV. (E) Focal area of CPE produced by CMV. This CPE usually progresses slowly. (F) Large focal area of late CPE produced by CMV.

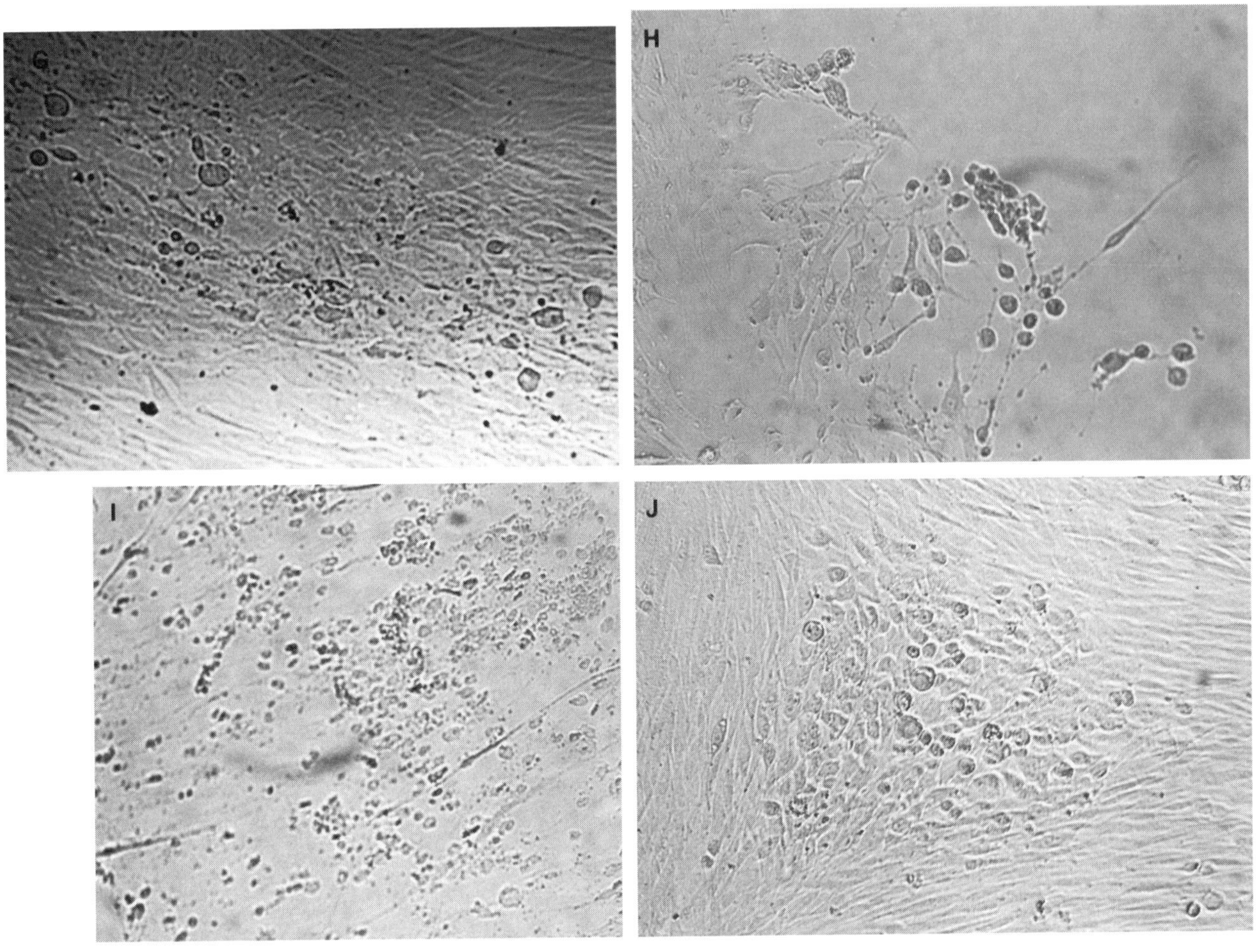

Figure 8.5–4 (part 6) *(continued)* (G) Late CPE produced by adenovirus (courtesy of CDC). (H) CPE of echovirus type 11 starting at the monolayer edge. (I) Late CPE of echovirus type 11 showing complete involvement of the monolayer. (J) Focal area of replicating A-549 cells that have inadvertently cross-contaminated an MRC-5 monolayer. (Reprinted from CMPH 8.7.15–8.7.16.)

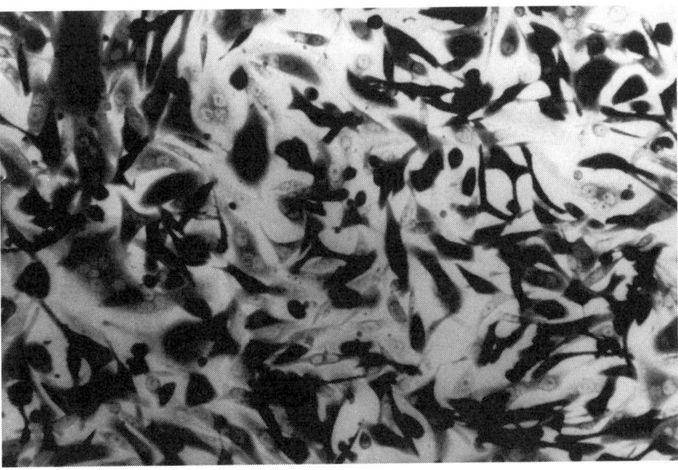

Figure 8.5–4 (part 7) HSV-1-infected ELVIS HSV cells (black cells in this photo, ×200) stained 16 h after infection. ELVIS (enzyme-linked virus-inducible system) HSV cells are genetically engineered BHK cells in which HSV proteins induce the intracellular accumulation of the reporter enzyme β-galactosidase, which is then histochemically detected. Provided by Diagnostic Hybrids, Inc.

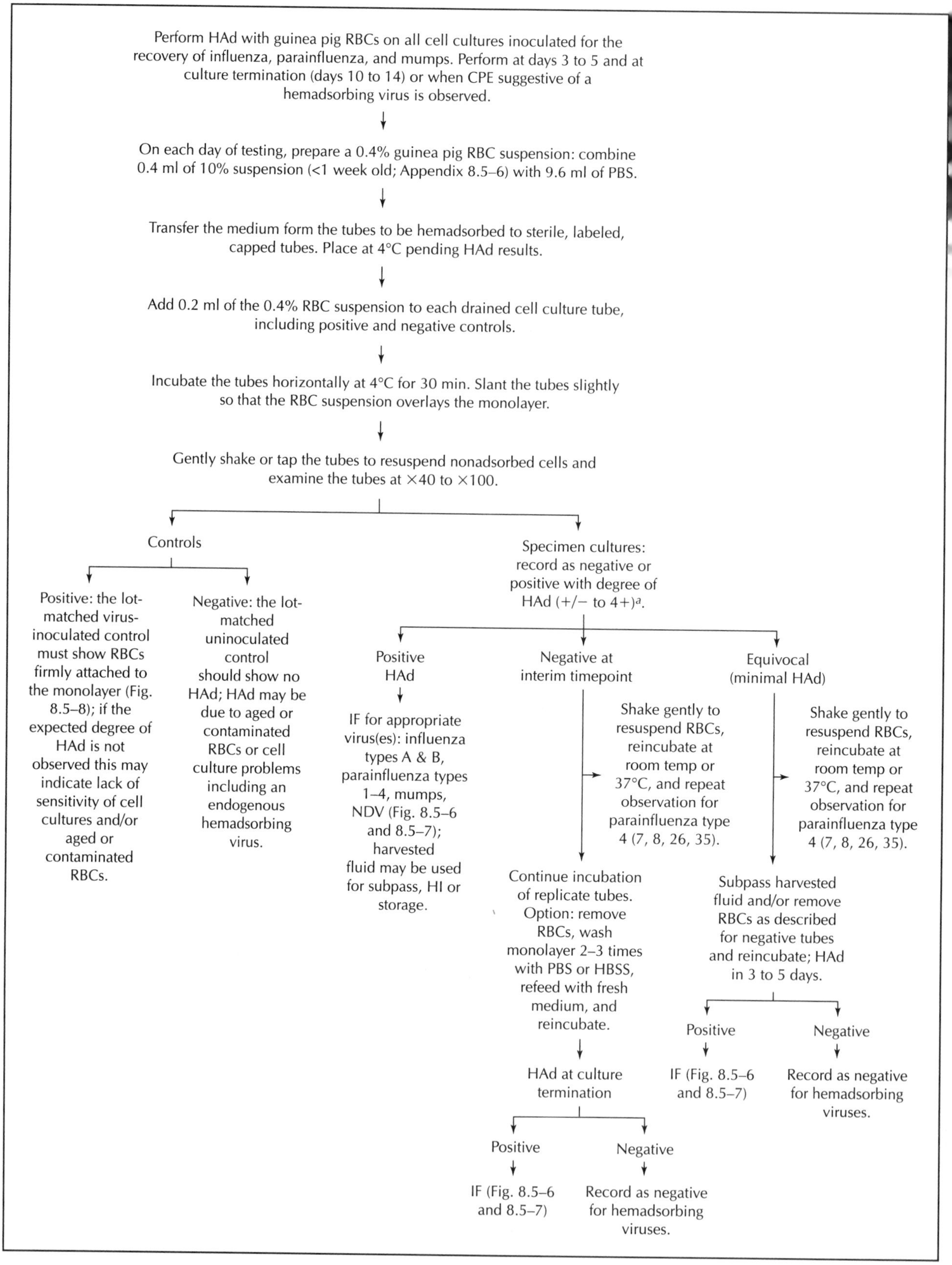

Perform HAd with guinea pig RBCs on all cell cultures inoculated for the recovery of influenza, parainfluenza, and mumps. Perform at days 3 to 5 and at culture termination (days 10 to 14) or when CPE suggestive of a hemadsorbing virus is observed.

↓

On each day of testing, prepare a 0.4% guinea pig RBC suspension: combine 0.4 ml of 10% suspension (<1 week old; Appendix 8.5–6) with 9.6 ml of PBS.

↓

Transfer the medium form the tubes to be hemadsorbed to sterile, labeled, capped tubes. Place at 4°C pending HAd results.

↓

Add 0.2 ml of the 0.4% RBC suspension to each drained cell culture tube, including positive and negative controls.

↓

Incubate the tubes horizontally at 4°C for 30 min. Slant the tubes slightly so that the RBC suspension overlays the monolayer.

↓

Gently shake or tap the tubes to resuspend nonadsorbed cells and examine the tubes at ×40 to ×100.

→ Controls

- Positive: the lot-matched virus-inoculated control must show RBCs firmly attached to the monolayer (Fig. 8.5–8); if the expected degree of HAd is not observed this may indicate lack of sensitivity of cell cultures and/or aged or contaminated RBCs.
- Negative: the lot-matched uninoculated control should show no HAd; HAd may be due to aged or contaminated RBCs or cell culture problems including an endogenous hemadsorbing virus.

→ Specimen cultures: record as negative or positive with degree of HAd (+/− to 4+)[a].

- Positive HAd → IF for appropriate virus(es): influenza types A & B, parainfluenza types 1–4, mumps, NDV (Fig. 8.5–6 and 8.5–7); harvested fluid may be used for subpass, HI or storage.
- Negative at interim timepoint → Shake gently to resuspend RBCs, reincubate at room temp or 37°C, and repeat observation for parainfluenza type 4 (7, 8, 26, 35).
  - → Continue incubation of replicate tubes. Option: remove RBCs, wash monolayer 2–3 times with PBS or HBSS, refeed with fresh medium, and reincubate.
  - ↓ HAd at culture termination
    - Positive → IF (Fig. 8.5–6 and 8.5–7)
    - Negative → Record as negative for hemadsorbing viruses.
- Equivocal (minimal HAd) → Shake gently to resuspend RBCs, reincubate at room temp or 37°C, and repeat observation for parainfluenza type 4 (7, 8, 26, 35).
  - → Subpass harvested fluid and/or remove RBCs as described for negative tubes and reincubate; HAd in 3 to 5 days.
    - Positive → IF (Fig. 8.5–6 and 8.5–7)
    - Negative → Record as negative for hemadsorbing viruses.

Figure 8.5–5 HAd procedure. NDV, Newcastle disease virus; HI, hemagglutination inhibition.

[a] Agglutinated RBCs may be observed in the fluid phase (hemagglutination).

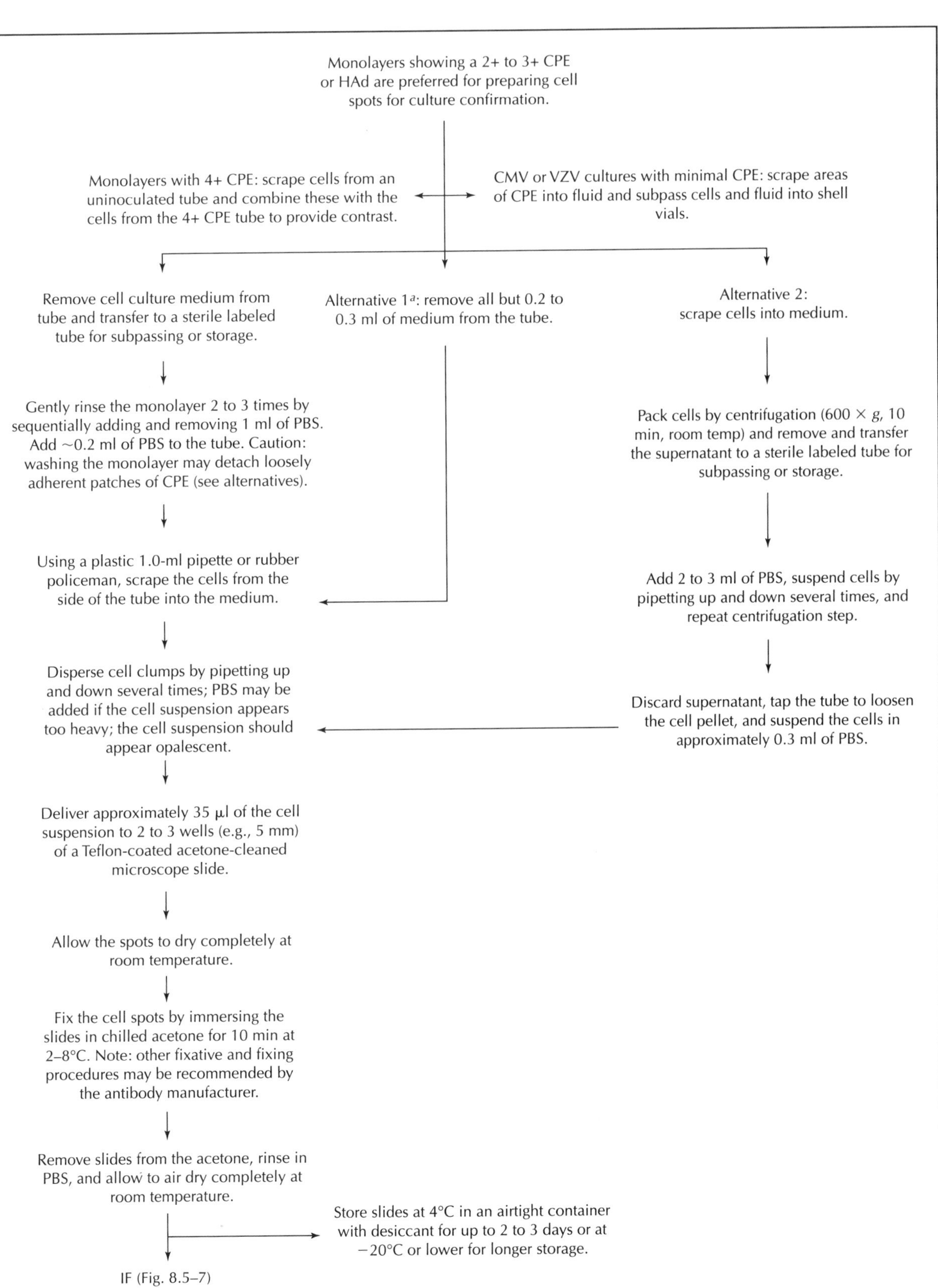

Figure 8.5–6 Isolate identification: slide preparation.
[a] This method is rapid but may yield higher levels of nonspecific staining than observed when the monolayers or packed cells are washed before spotting.

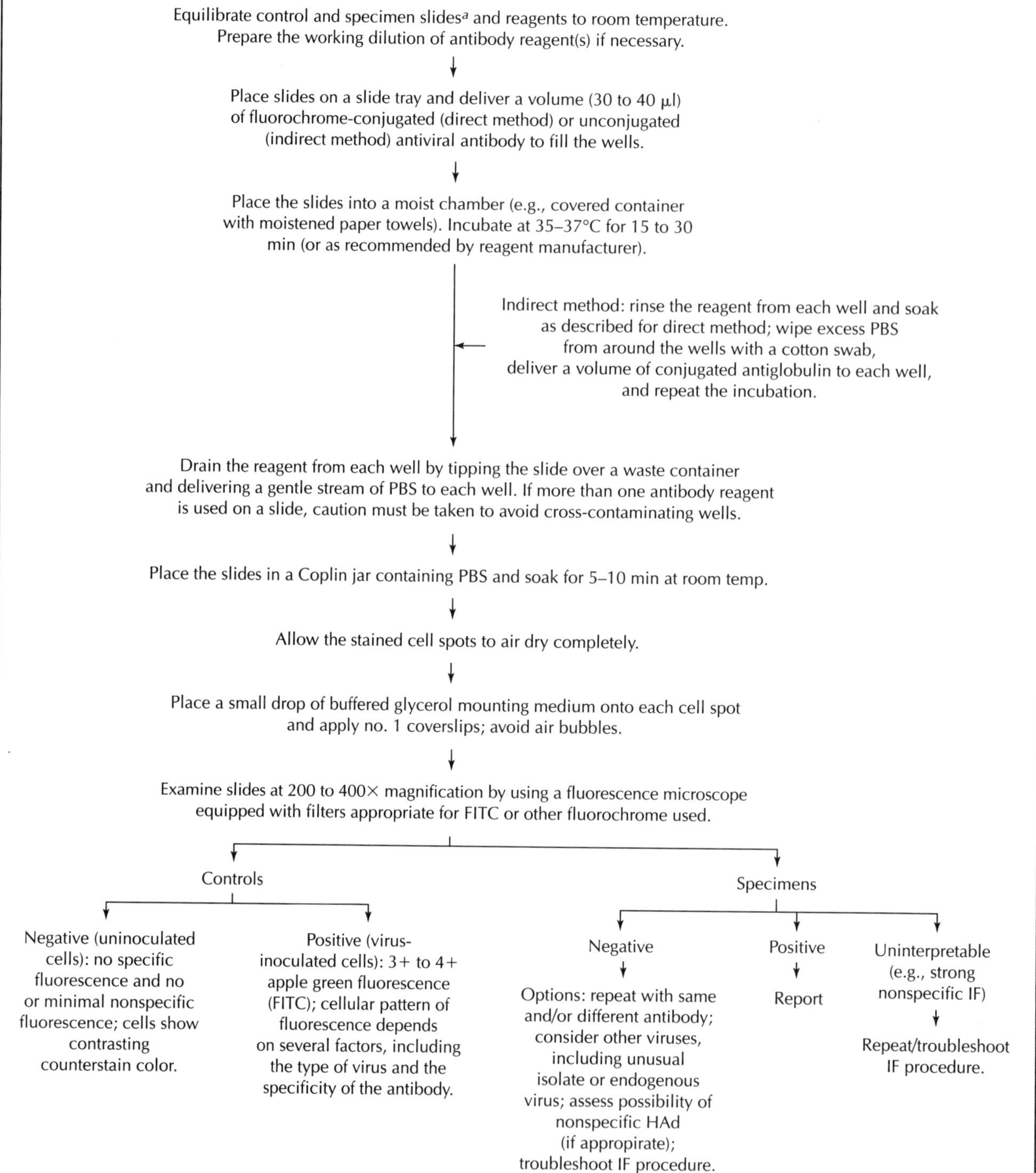

Figure 8.5–7 Isolate identification: IF.

[a] When removing slides from cold storage, allow them to equilibrate to room temperature before opening the storage container.

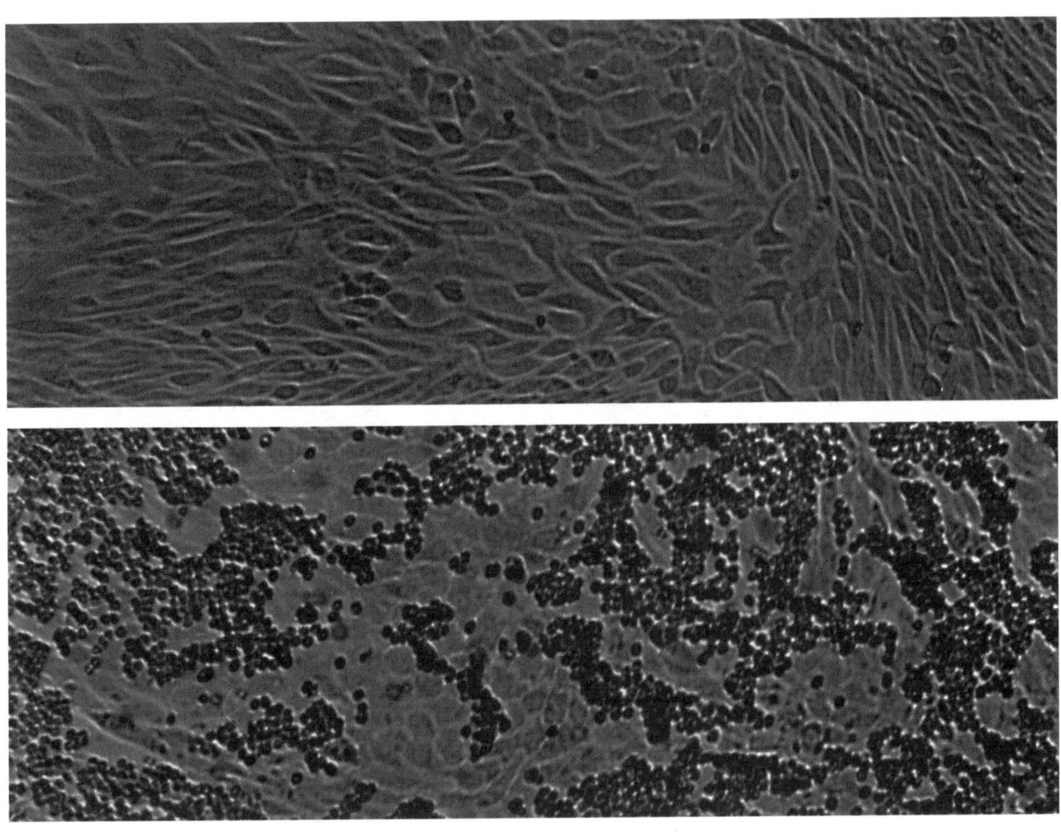

Figure 8.5–8 HAd with guinea pig RBCs. (Top) Uninoculated primary rhesus monkey kidney cells; (bottom) primary rhesus monkey kidney cells infected with parainfluenza virus type 3. Magnification, ×100. (Reprinted from CMPH 8.8.2.)

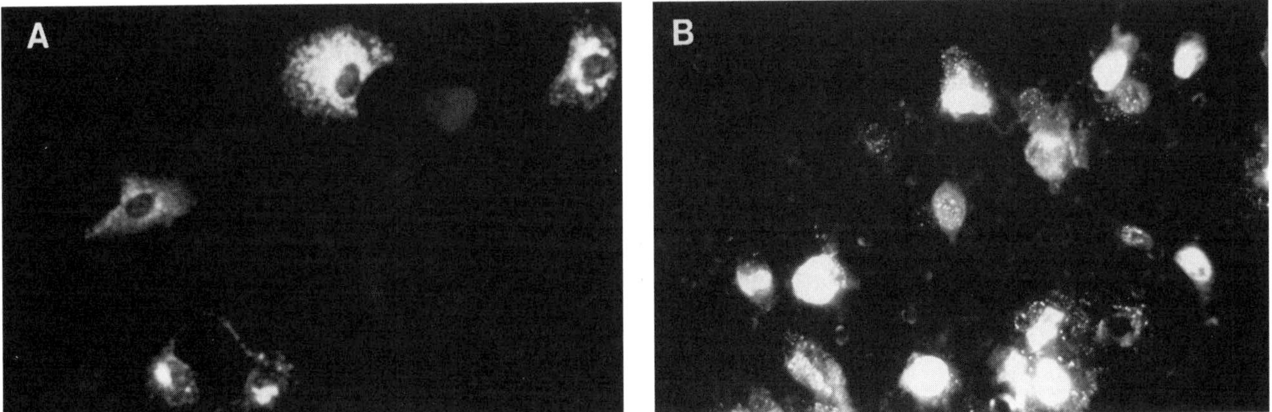

Figure 8.5–9 (A) LLC-MK_2 cells infected with parainfluenza virus (magnification, ×200). Provided by Bartels, Inc. (B) Adenovirus culture showing nuclear and cytoplasmic staining (magnification, ×400) (reprinted from CMPH 8.6.6).

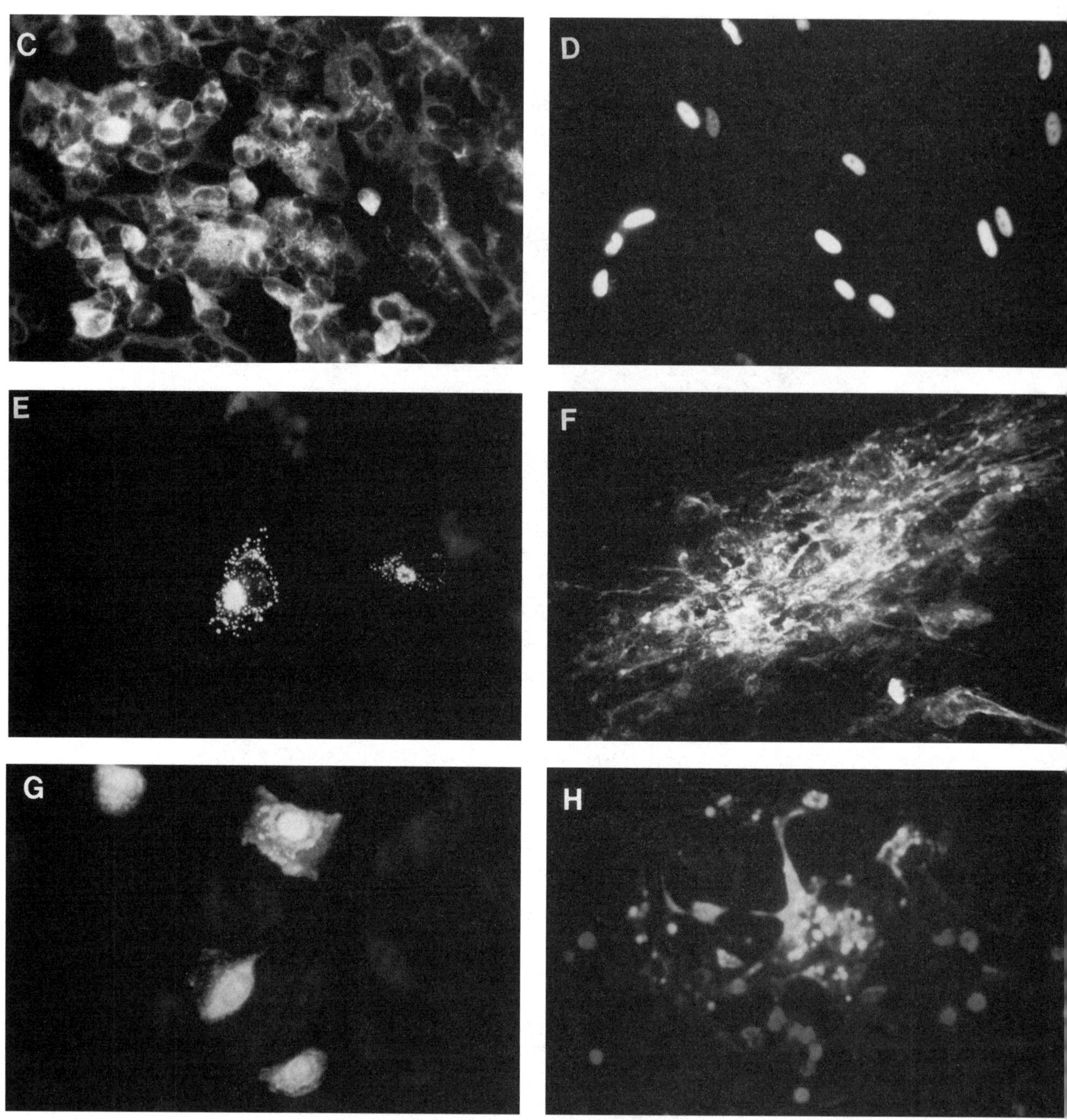

Figure 8.5–9 *(continued)* (C) Perinuclear and cytoplasmic staining characteristic of herpes simplex virus-infected cells (magnification, ×400) (reprinted from CMPH 8.6.6). (D) Brilliant oval nuclei of human fibroblast cells infected with CMV after being stained with a monoclonal antibody to CMV early nuclear protein (magnification, ×200) (reprinted from CMPH 8.6.7). (E) MRC-5 cells stained after 72 h with a mixture of monoclonal antibodies directed against intermediate early nuclear and late cytoplasmic antigens (×200). Provided by Bartels, Inc. (F) VZV in human fibroblast cell culture (magnification, ×400) (reprinted from CMPH 8.6.7). (G) LLC-MK_2 cells infected with influenza A virus (×200). Provided by Bartels, Inc. (H) HEp-2 cells infected with RSV (×200). Provided by Bartels, Inc.

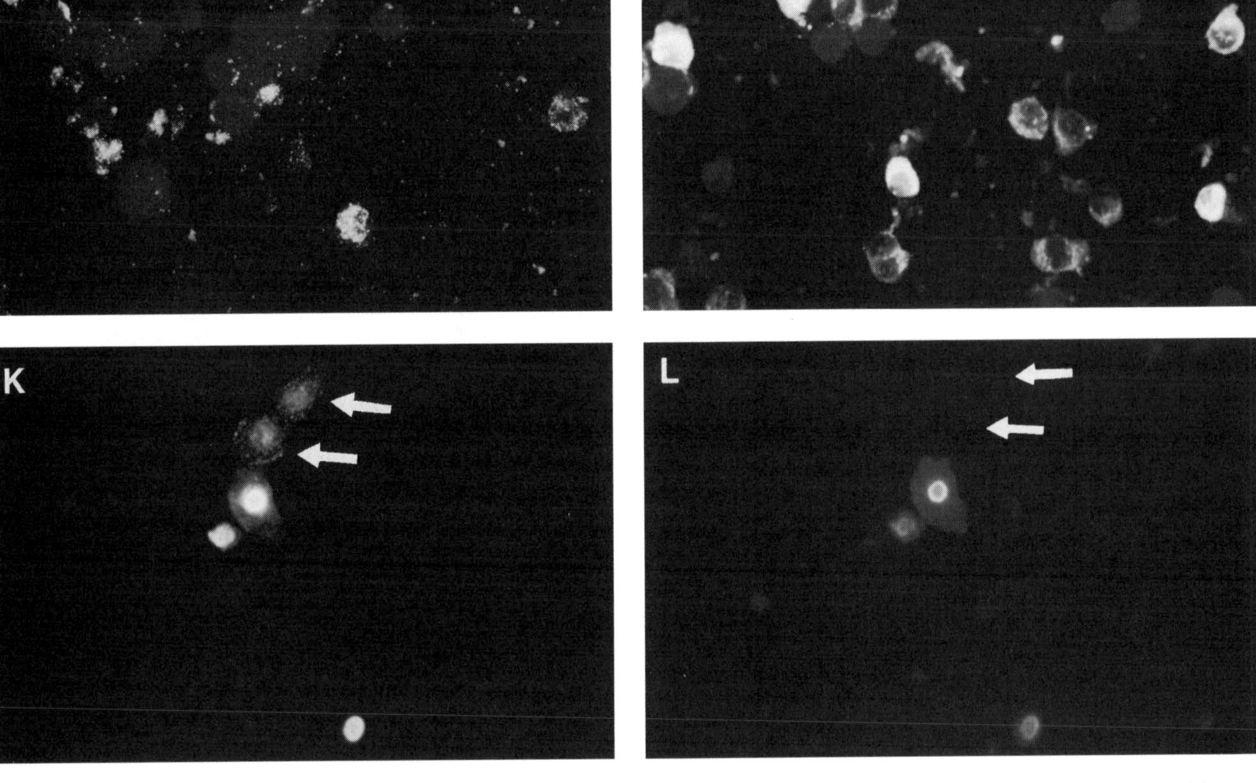

Figure 8.5–9 *(continued)* (I) Echovirus type 4-infected BGMK cells stained with pan-enterovirus blend of monoclonal antibodies (×200). Provided by Chemicon International. (J) Echovirus type 4-infected BGMK cells stained using echovirus type 4 typing reagent (×200). Provided by Chemicon International. (K) Influenza A- and influenza B-infected LLC-MK_2 cells stained with SimulFluor influenza A/influenza B reagent and observed using filters for FITC. Influenza A-infected cells (arrows) stain apple green; influenza B-infected cells are yellow-orange. Provided by Chemicon International. (L) Same field as panel K viewed with a TRITC/rhodamine filter. Influenza B-infected cells appear hot pink to red (×200); arrows indicate the location of influenza A-infected cells observed in panel K but which are not visible with this filter system. Provided by Chemicon International.

V. PROCEDURES *(continued)*

7. In addition to IF, immunoperoxidase staining (2–4, 14, 37), solid-phase EIA (2, 9, 24), or nucleic acid hybridization (16) can also be used to identify isolates.
8. Other immunologic methods are occasionally used by the diagnostic laboratory (e.g., neutralization for typing enteroviruses, adenoviruses [Appendix 8.5–3]; hemagglutination inhibition [HI] for subtyping influenza isolates [*see* CMPH 8.12]). Alternatively, laboratories may arrange to send isolates to reference or public health laboratories when typing or subtyping is important (e.g., influenza surveillance).
9. Immunologic reagents are not available for identification of rhinoviruses; these viruses are distinguished from enteroviruses with the acid lability assay (Appendix 8.5–4).

F. Rubella interference assay (*see* Appendix 8.5–5)

G. Refeeding inoculated cultures

1. Pipette or aspirate the medium from the culture tubes or vials.

   NOTE: Be careful not to scrape the monolayers.

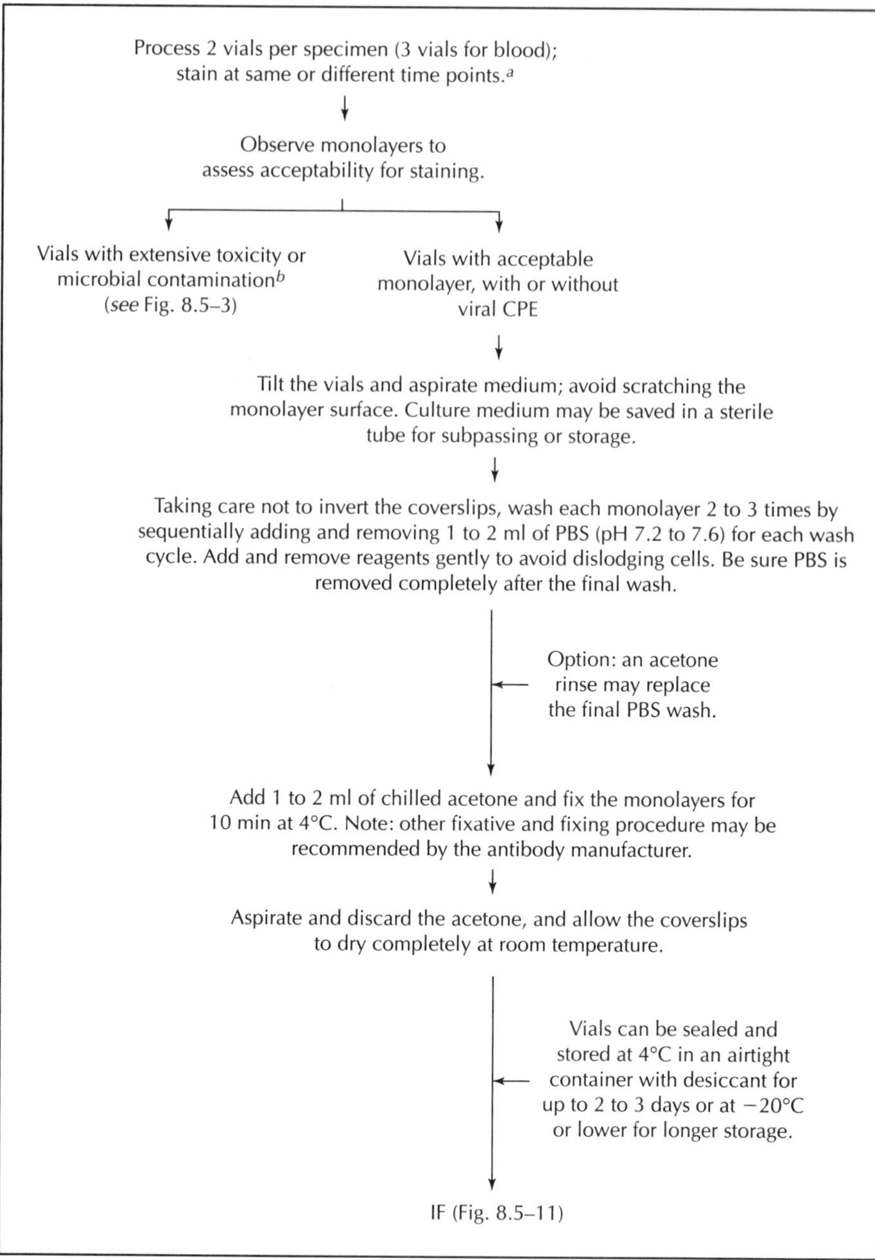

Figure 8.5–10 Shell vial processing.
[a] The sensitivity of shell vial cultures is influenced by the length of incubation, and each laboratory must determine appropriate staining timepoints.
[b] Complete monolayer destruction may also be caused by a rapid CPE.

V. PROCEDURES *(continued)*

2. Using a separate pipette for each specimen, add 1 to 2 ml of cell culture maintenance medium to each tube.

   NOTE: To reduce cross-contamination with detached cells or endogenous viruses, aspirate and refeed in the following sequence: fibroblast, primary nonsimian, transformed, and primary simian.

H. Harvesting and subpassing inoculated cultures

1. Isolates are frequently harvested and stored for many reasons, such as to provide teaching and QC material and to provide material for additional

Equilibrate control and specimen vials[a] and reagents to room temperature. Determine that coverslips are cell side up. Prepare the working dilution of antibody reagent(s) if necessary.

↓

Perform staining in the shell vials by adding 50 to 100 µl of the conjugated (direct method) or unconjugated (indirect method) antiviral antibody to each vial. Alternatively, coverslips may be removed and glued cell side up to microscope slides for staining.

↓

Cover the vials with Parafilm and incubate for 30 min at 35 to 37°C (or as specified by the reagent manufacturer). Incubation on a rotator or rocker at low speed will facilitate distribution of the reagent over the monolayer.

← Indirect method: wash the monolayer twice by sequentially adding and removing 1 to 2 ml of PBS, add 50 to 100 µl of conjugated antiglobulin, and repeat the incubation.

↓

Wash the monolayers twice by sequentially adding and removing 1 to 2 ml of PBS. Note: distilled water may be used for the final wash.

↓

Allow the stained monolayers to air dry completely.

↓

Using slightly bent fine-tipped forceps, remove the coverslips from the vials and mount them on a slide, cell side down, onto a small drop of buffered glycerol mounting medium.

↓

Examine the entire monolayer at ×200; observe questionable areas at ×400.

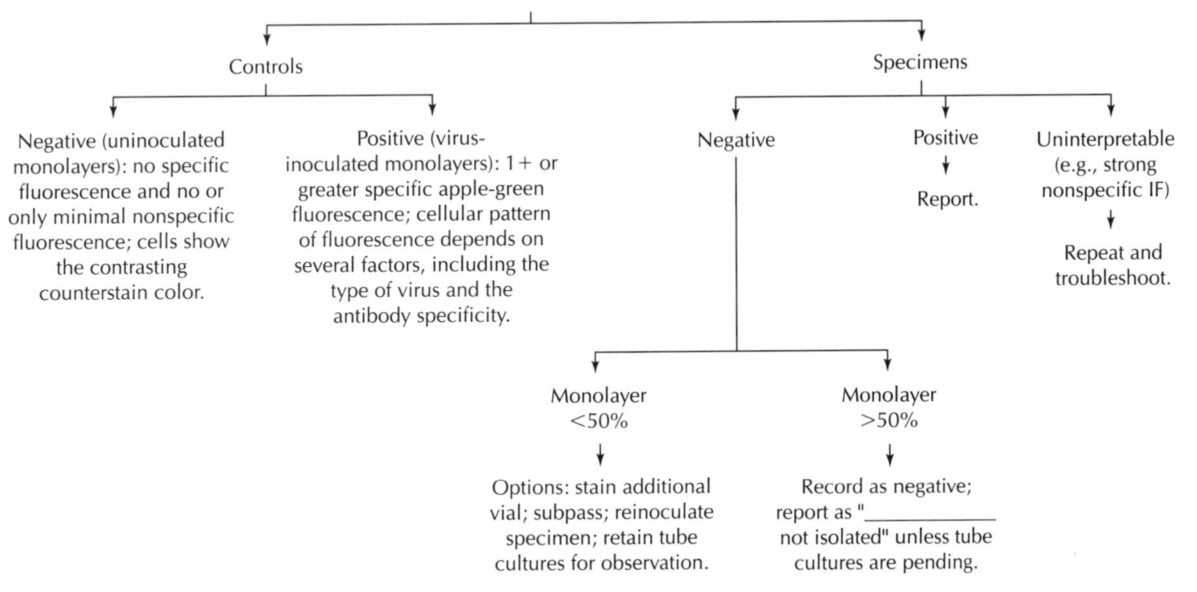

Figure 8.5–11 Shell vial cultures, IF.
[a]When removing vials from storage, allow them to equilibrate to room temperature before opening the storage container.

V. PROCEDURES *(continued)*

identification procedures (e.g., neutralization, acid lability, or HI). The extent to which viruses are shed or released into culture medium depends on the particular virus. For example, while most enteroviruses are readily released into the culture fluid, CMV and varicella-zoster virus (VZV) are cell associated, and successful harvesting and subpassing require infected cells; subpassage of CMV or VZV cultures with a CPE less than 2+ may not be efficient.

2. To harvest isolates, use a plastic 1-ml pipette or sterile rubber policeman to dislodge cells from the tube surface, and mix cells and fluid by pipetting up and down several times.

   NOTE: Successful subpassage of isolates may generally be accomplished by using the fluid only for isolates other than CMV, VZV, or adenoviruses.

3. To subpass cultures, deliver 0.2 to 0.3 ml of the harvested material to each new tube to be inoculated.

   NOTE: Isolates exhibiting very strong and rapid CPE may first be diluted (e.g., 1:10, 1:100, or greater) with HBSS.

4. To store isolates, transfer harvested material to a tube suitable for low-temperature storage (e.g., Nunc cryotube) and store at −70°C or lower.

   NOTE: Rapid freezing in a dry ice-acetone slurry is recommended; adding a cryopreservative is recommended for CMV and VZV harvests (add dimethyl sulfoxide [DMSO] to a final concentration of 10%).

I. Culture decontamination

1. Harvest cells and fluids as described above.
2. Draw the harvested material into a 3- to 5-ml disposable plastic syringe, attach a disposable 0.2-μm-pore-size filter (e.g., Millipore), and expel the material into a sterile tube.

   NOTE: Filtration of CMV and VZV cultures may markedly reduce the viral titer of recovered material, particularly with cultures exhibiting less than a 3+ CPE. Cell disruption by addition of glass beads and vortexing the material for 60 s before filtration will facilitate viral release. Viruses not adversely affected by freezing and thawing (e.g., unenveloped viruses such as adenoviruses and enteroviruses) may be subjected to three rapid freeze-thaw cycles to facilitate viral release.

VI. CULTURE RESULTS

A. A positive viral culture is one exhibiting evidence of viral replication by the development of CPE, positive HAd, and/or the presence of a viral antigen or nucleic acid.

NOTE: Identification on the basis of CPE or HAd is tentative, and reports should not be based on these findings.

B. A negative viral culture is one that fails to develop CPE and/or HAd or one in which no viral antigen or nucleic acid is detected after incubation of an acceptable monolayer for the prescribed period of time.

C. An incomplete or unsatisfactory culture is one that cannot be completed or processed in the appropriate manner for reasons that include extensive microbial contamination.

VII. PROCEDURE NOTES

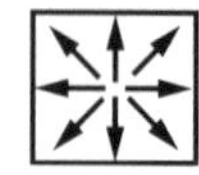

Observe universal (standard) precautions.

A. Observe universal precautions.

B. Cell culture fluids may contain large concentrations of virus capable of aerosolization. Place a pipette discard jar or bucket containing 2 to 3 in. (ca. 5 to 7.5 cm) of liquid disinfectant (e.g., Wescodyne) inside the biological safety cabinet; to discard contaminated pipettes, immerse the tip in the disinfectant and then disengage it from the pipetting device. If the discard container is to be reused, seal it or place it into a sealable container and autoclave. If a suction apparatus is used, include an aerosol trap containing disinfectant between the collection flask and the vacuum source.

C. Use aseptic techniques for all cell culture manipulations (*see* Appendix 8.2–2).

D. Use appropriate measures to minimize the potential risk of cross-contamination of cultures and infection of personnel from manipulations that generate aerosols or may contaminate gloved hands (e.g., disengaging pipettes from pipetting devices; removing shell vial caps or stoppers; pouring).

E. The sensitivity of viral culture is affected by a number of variables, including the viral concentration; specimen type and quality; cell culture type, age, density, and quality; the number of tubes or vials inoculated; incubation conditions; and the choice of detection or identification reagent.

F. Blind subpassage of negative tube cultures midway through the incubation-observation period may slightly increase the number of positive cultures. However, this practice has been largely discontinued because of the expense and time involved. Subpassage is now generally limited to specimens showing toxicity or contamination or to prepare larger volumes of isolates for additional studies (e.g., antiviral susceptibility testing).

G. Simian B virus, arboviruses, arenaviruses, and filoviruses replicate in some heteroploid cell lines that may be in use in the routine diagnostic setting. While not frequently encountered, the symptoms and travel or exposure history should alert the laboratorian to the special safety considerations required for processing these specimens and the need to contact the appropriate state health department laboratory, CDC, or specialized reference laboratory (*see* procedure 8.1).

H. Monolayers exhibiting CPE or HAd but not needed for identification or harvesting may be preserved indefinitely for teaching purposes (Appendix 8.5–7).

VIII. LIMITATIONS OF THE PROCEDURE

A negative viral culture does not rule out the possibility of a viral etiology.

REFERENCES

1. **Bankowski, M. J., K. Erickson, A. Alvarez, P. Horton, J. Rogers, S. Maxwell, and M. A. Neumann.** 1997. The SimulFluor™ (Chemicon) assay compared to Bartels IFA reagents for direct testing and viral culture confirmation of influenza types A and B. Proc 13th Annual Clinical Virology Symposium, Clearwater, Fla.

2. **Bartholoma, N. Y.** 1992. Identification of viral isolates by enzyme immunoassay, p. 8.11.1–8.11.9. *In* H. D. Isenberg (ed.), *Clinical Microbiology Procedures Handbook,* vol. 2. American Society for Microbiology, Washington, D.C.

3. **Benjamin, D. R., and C. G. Ray.** 1974. Use of immunoperoxidase for the rapid identification of human myxoviruses and paramyxoviruses in tissue culture. *Appl. Microbiol.* **28:** 47–51.

4. **Bonville, C. A., B. A. Forbes, N. Y. Bartholoma, J. A. McMillan, and L. B. Weiner.** 1987. Rapid detection of herpes simplex virus in MRC-5 cells using low-speed centrifugation enhancement and immunoperoxidase staining 16 h post-inoculation. *Diagn. Microbiol. Infect. Dis.* **8:**251–254.

5. **Buller, R. S., T. C. Bailey, N. A. Ettinger, M. Keener, T. Langlois, J. P. Miller, and G. A. Storch.** 1992. Use of a modified shell vial technique to quantitate cytomegalovirus viremia in a population of solid-organ transplant recipients. *J. Clin. Microbiol.* **30:** 2620–2624.

6. **Canas, L., T. Ottensmeier, C. Douglas, P. Asbury, and J. Treadwell.** 1997. Evaluation of the single reagent Simulfluor™ influenza A/influenza B fluorescence assay (Light Diagnostics) for the simultaneous detection and identification of influenza A and/or influenza B for use in a world wide influenza surveillance program. Thirteenth Annual Clinical Virology Symposium, Clearwater, Fla.

7. **Canchola, J., A. J. Vargosko, H. W. Kim, et al.** 1964. Antigenic variation among newly isolated strains of parainfluenza type 4 virus. *Am. J. Hyg.* **79:**357–364.

8. **Canchola, J. G., R. M. Chanock, B. C. Jeffries, E. E. Christmas, H. W. Kim, J. A. Vargosko, and R. H. Parrott.** 1965. Recovery and identification of human myxoviruses. *Bacteriol. Rev.* **29:**496–503.

Viruses, Rickettsiae, Chlamydiae, Mycoplasmas 8

REFERENCES *(continued)*

9. **Doing, K. M., and J. C. Jellison.** 1997. Comparison of an enzyme immunoassay and direct fluorescent antibody staining for the detection of herpes simplex virus in shell vial cultures using two different cell lines. Thirteenth Annual Clinical Virology Symposium, Clearwater, Fla.
10. **Engler, H. D., and S. T. Selepak.** 1994. Effect of centrifuging shell vials at 3,500 $\times$ g on detection of viruses in clinical specimens. *J. Clin. Microbiol.* **32:**1580–1582.
11. **Erice, A., M. A. Holm, P. C. Gill, S. A. Henry, C. L. Dirksen, D. L. Dunn, R. P. Hillan, and H. H. Balfour.** 1992. Cytomegalovirus (CMV) antigenemia assay is more sensitive than shell vial cultures for rapid detection of CMV in polymorphonuclear blood leukocytes. *J. Clin. Microbiol.* **30:** 2822–2825.
12. **Espy, M. J., T. F. Smith, M. W. Harmon, and A. P. Kendal.** 1986. Rapid detection of influenza virus by shell vial assay with monoclonal antibodies. *J. Clin. Microbiol.* **24:** 677–679.
13. **Frank, A. L., R. B. Couch, C. A. Griffis, and B. D. Baxter.** 1979. Comparison of different tissue cultures for isolation and quantitation of influenza and parainfluenza viruses. *J. Clin. Microbiol.* **10:**32–36.
14. **Gay, H., and J. J. Doherty.** 1986. Immunoperoxidase detection of viral antigens in cells, p. 147–158. *In* S. Specter and G. J. Lancz (ed.)., *Clinical Virology Manual.* Elsevier, New York.
15. **Gleaves, C. A., D. A. Hursh, and J. D. Myers.** 1992. Detection of human cytomegalovirus in clinical specimens by centrifugation culture with a nonhuman cell line. *J. Clin. Microbiol.* **30:**1045–1048.
16. **Gleaves, C. A., D. A. Hursh, D. H. Rice, and J. D. Meyers.** 1989. Detection of cytomegalovirus from clinical specimens in centrifugation culture by in situ DNA hybridization and monoclonal antibody staining. *J. Clin. Microbiol.* **27:**21–23.
17. **Gleaves, C. A., C. F. Lee, C. I. Bustamante, and J. D. Myers.** 1988. Use of murine monoclonal antibodies for laboratory diagnosis of varicella-zoster virus infection. *J. Clin. Microbiol.* **26:**1623–1625.
18. **Gleaves, C. A., T. F. Smith, E. A. Shuster, and G. R. Pearson.** 1984. Rapid detection of cytomegalovirus in MRC-5 cells inoculated with urine specimens by using low-speed centrifugation and monoclonal antibody to an early antigen. *J. Clin. Microbiol.* **19:**917–919.
19. **Hsiung, G. D., and V. F. Chan.** 1994. Endogenous viral and mycoplasmal contaminants in cell cultures, p. 346–362. *In* G. D. Hsiung, C. K. Y. Fong, and M. L. Landry (ed.) *Hsiung's Diagnostic Virology,* 4th ed. Yale University Press, New Haven, Conn.
20. **Hughes, J. H.** 1993. Physical and chemical methods for enhancing rapid detection of viruses and other agents. *Clin. Microbiol. Rev.* **6:**150–175.
21. **Landry, M. L., and D. Ferguson.** 1993. Comparison of quantitative cytomegalovirus antigenemia assay with culture methods and correlation with clinical disease. *J. Clin. Microbiol.* **31:**2851–2856.
22. **Lee, S. H. S., J. E. Boutilier, M. A. MacDonald, and K. R. Forward.** 1992. Enhanced detection of respiratory viruses using the shell vial technique and monoclonal antibodies. *J. Virol. Methods* **39:**39–46.
23. **Matthey, S., D. Nicholson, S. Ruhs, B. Alden, M. Knock, K. Schultz, and A. Schmuecker.** 1992. Rapid detection of respiratory viruses by shell vial culture and direct staining by using pooled and individual monoclonal antibodies. *J. Clin. Microbiol.* **30:** 540–544.
24. **Michalski, F. J., M. Shaikh, F. Sahraie, S. Desai, L. Verano, and J. Vallabhaneni.** 1986. Enzyme-linked immunosorbent assay spin amplification technique for herpes simplex virus antigen detection. *J. Clin. Microbiol.* **24:**310–311.
25. **Minnich, L. L., F. Goodenough, and C. G. Ray.** 1991. Use of immunofluorescence to identify measles virus infections. *J. Clin. Microbiol.* **29:**1148–1150.
26. **Mufson, M. A.** 1989. Parainfluenzaviruses, mumps virus, and Newcastle disease virus, p. 669–691. *In* N. J. Schmidt and R. W. Emmons (ed.), *Diagnostic Procedures for Viral, Rickettsial, and Chlamydial Infections,* 6th ed. American Public Health Association, Washington, D. C.
27. **Olsen, M. A., K. M. Shuck, A. R. Sambol, S. M. Flor, J. O'Brien, and B. J. Cabrera.** 1993. Isolation of seven respiratory viruses in shell vials: a practical and highly sensitive method. *J. Clin. Microbiol.* **31:**422–425.
28. **Paya, C. V., A. D. Wold, D. M. Ilstrup, and T. F. Smith.** 1988. Evaluation of number of shell vial cell cultures per clinical specimen for rapid diagnosis of cytomegalovirus infection. *J. Clin. Microbiol.* **26:**198–200.
29. **Paya, C. V., A. D. Wold, and T. F. Smith.** 1987. Detection of cytomegalovirus infections in specimens other than urine by the shell vial assay and conventional tube culture. *J. Clin. Microbiol.* **25:**755–757.
30. **Rabalais, G. P., G. G. Stout, K. L. Ladd, and K. M. Cost.** 1992. Rapid diagnosis of respiratory viral infections by using a shell vial assay and monoclonal antibody pool. *J. Clin. Microbiol.* **30:**1505–1508.
31. **Rabella, N., and W. L. Drew.** 1990. Comparison of conventional and shell vial cultures for detecting cytomegalovirus infection. *J. Clin. Microbiol.* **28:**806–807.
32. **Salmon, V. C., B. R. Kenyon, and J. C. Overall, Jr.** 1990. Cross contamination of viral specimens related to shell vial caps. *J. Clin. Microbiol.* **28:**2820–2822.
33. **Shuster, E. A., J. S. Beneke, G. E. Tegtmeier, G. R. Pearson, C. A. Gleaves, A. D. Wold, and T. F. Smith.** 1985. Monoclonal antibody for rapid laboratory detection of cytomegalovirus infections: characterization and diagnostic application. *Mayo Clin. Proc.* **60:**577–585.
34. **Stout, C., M. D. Murphy, S. Lawrence, and S. Julian.** 1989. Evaluation of a monoclonal antibody pool for rapid diagnosis of respiratory viral infections. *J. Clin. Microbiol.* **27:** 448–452.

35. **Swierkosz, E., T. Bonnot, and D. Regan.** 1993. Recovery of parainfluenza 4 from pediatric patients in the metropolitan St. Louis area over the past 3 respiratory virus seasons, abstr. C-45, p. 43. *Abstr. 79th Annu. Meet. Am. Soc. Microbiol. 1979.* American Society for Microbiology, Washington, D.C.

36. **Tobita, K., A. Sugiura, C. Enomoto, and M. Furuyama.** 1975. Plaque assay and primary isolation of influenza A viruses in an established line of canine kidney cells (MDCK) in the presence of trypsin. *Med. Microbiol. Immunol.* **162:**9–14.

37. **Waris, M., T. Ziegler, M. Kivivirta, and O. Ruuskanen.** 1990. Rapid detection of respiratory syncytial virus and influenza A virus in cell cultures by immunoperoxidase staining with monoclonal antibodies. *J. Clin. Microbiol.* **28:**1159–1162.

38. **Wiedbrauk, D. L., J. P. Gibson, R. Bollinger, and E. Ostler.** 1993. Centrifugation-enhanced (shell vial) method for detecting rubella virus. Ninth Annual Clinical Virology Symposium, Clearwater, Fla.

39. **Young, S., B. Strong, and R. Radloff.** 1994. Genus and type specific monoclonal antibodies for the identification of enteroviruses by immunofluorescence, abstr. C-520, p. 583. *Abstr. 94th Gen. Meet. Am. Soc. Microbiol. 1994.* American Society for Microbiology, Washington, D.C.

40. **Zhao, L., M. L. Landry, E. S. Balkovic, and G. D. Hsiung.** 1987. Impact of cell culture sensitivity and virus concentration on rapid detection of herpes simplex virus by cytopathic effects and immunoperoxidase staining. *J. Clin. Microbiol.* **25:**1401–1405.

SUPPLEMENTAL READING

Hodinka, R. L. 1992. Cell culture techniques: preparation of cell culture medium and reagents, p. 8.19.1.–8.19.15. *In* H. D. Isenberg (ed.), *Clinical Microbiology Procedures Handbook,* vol. 2. American Society for Microbiology, Washington, D.C.

Hodinka, R. L. 1992. Cell culture techniques: serial propagation and maintenance of monolayer cell cultures, p. 8.20.1–8.20.14. *In* H. D. Isenberg (ed.), *Clinical Microbiology Procedures Handbook*, vol. 2. American Society for Microbiology, Washington, D.C.

Hsiung, G. D., C. K. Y. Fong, and M. L. Landry (ed.). 1994. *Hsiung's Diagnostic Virology,* 4th ed. Yale University Press, New Haven, Conn.

Keller, E. W. 1992. Detection and identification of viruses by immunofluorescence, p. 8.9.1.–8.9.10. *In* H. D. Isenberg (ed.), *Clinical Microbiology Procedures Handbook,* vol. 2. American Society for Microbiology, Washington, D.C.

Keller, E. W. 1992. Preparation of cell spots for immunofluorescence, p. 8.10.1–8.10.9. *In* H. D. Isenberg (ed.), *Clinical Microbiology Procedures Handbook,* vol. 2. American Society for Microbiology, Washington, D.C.

Lennette, E. H., D. A. Lennette, and E. T. Lennette (ed.). 1995. *Diagnostic Procedures for Viral, Rickettsial, and Chlamydial Infections,* 7th ed. American Public Health Association, Washington, D.C.

Murray, P. K., E. J. Baron, M. A. Pfaller, F. C. Tenover, and R. H. Yolken (ed.). 1995. *Manual of Clinical Microbiology,* 6th ed. American Society for Microbiology, Washington, D.C.

Swenson, P. D. 1992. Hemagglutination inhibition test for the identification of influenza viruses, p. 8.12.1–8.12.11. *In* H. D. Isenberg (ed.), *Clinical Microbiology Procedures Handbook,* vol. 2. American Society for Microbiology, Washington, D.C.

APPENDIX 8.5–1

Sources for IF and EIA reagents for culture confirmation

Bartels Inc.
2005 NW Sammamish Road
Suite 107
Issaquah, WA 98027
(800) BARTELS

BioWhittaker
8830 Biggs Road
Walkersville, MD 21793
(800) 638-8174

Chemicon International, Inc.
28835 Single Oak Drive
Temecula, CA 92590
(800) 437-7500

DAKO Corp.
6392 Via Real
Carpinteria, CA 93013
(800) 424-0021

Meridian Diagnostics, Inc.
3471 River Hills Drive
Cincinnati, OH 45244
(800) 343-3858

Novocastra Laboratories Ltd.
U. S. distributor: Vector Laboratories, Inc.
30 Ingold Road
Burlingame, CA 94010
(800) 227-6666

Sanofi Diagnostics Pasteur, Inc.
1000 Lake Hazeltine Drive
Chaska, MN 55318
(800) 666-5111

APPENDIX 8.5–2

Viral Titration and Determination of $TCID_{50}$

Reprinted from **Lipson, S. M.** 1992. Neutralization test for the identification of viral isolates, p. 8.14.1–8.14.8. *In* H. D. Isenberg (ed.), *Clinical Microbiology Procedures Handbook*, vol. 2. American Society for Microbiology, Washington, D.C.

A. Viral quantitation

The viral isolate to be identified in the neutralization assay must be quantitated to ensure optimal performance of the assay. Viral quantitation is achieved by preparing a series of dilutions of the material and assaying this material by using either quantitative or quantal assay systems. In the quantitative assay, the number of infectious viral particles is established by using a plaque or infectious-focus assay (2). Although this type of assay is more precise, it is, like the plaque reduction neutralization assay, extremely cumbersome, costly, and labor intensive. Thus, these quantitative procedures are reserved for the research setting or are limited to special procedures.

The viral quantitation method used in the diagnostic laboratory is the endpoint dilution method. This is a quantal assay, since it measures an "all-or-none" effect and does not determine the number of infectious particles (1). Thus, the quantal assay determines only that the material contains a sufficient dose of infectious virus to produce infection in a susceptible host system. With the endpoint dilution method, the viral titer is expressed as the 50% tissue culture infective dose ($TCID_{50}$) or, in animals, as the 50% lethal dose.

B. Preparation of the viral titration (Fig. 8.5–A1)

1. Harvest the viral isolate to be titrated (*see* V.H.1–4) or thaw a frozen aliquot.
2. Determine the number of serial 10-fold dilutions to be inoculated.
   a. With slower-growing low-titered viruses (e.g., adenoviruses), prepare dilutions ranging from 10^{-1} through 10^{-5}, and inoculate tubes starting with the undiluted material.
   b. With high-titered material (e.g., many enteroviruses), prepare dilutions ranging from 10^{-1} through 10^{-7}, and inoculate tubes starting with 10^{-1} or 10^{-2}.
3. Label sterile dilution tubes for each virus to be titrated.
4. Add 1.8 ml of HBSS to each dilution tube.
5. Add 0.2 ml of each virus to be titrated to the corresponding tube labeled 10^{-1}, and mix by pipetting up and down several times.
6. With a new pipette, remove 0.2 ml from the 10^{-1} dilution, transfer it to the tube labeled 10^{-2}, and mix.
7. Continue preparing dilutions through the highest dilution, using a new pipette for each dilution.
8. Label four to six cell culture tubes for each dilution to be inoculated; select two tubes as negative (uninoculated controls).
9. Refeed each tube with 1.0 ml of maintenance medium.
10. Inoculate 0.1 ml of each virus dilution into the corresponding culture tubes.

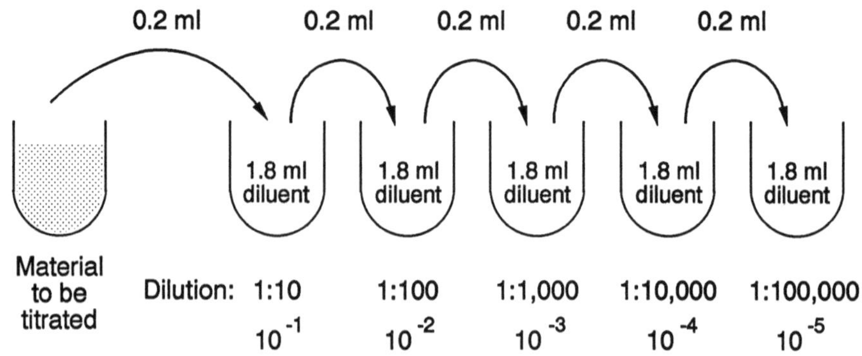

Figure 8.5–A1 Viral titration with serial 10-fold dilutions.

Table 8.5–A1 Calculation of $TCID_{50}$ by the Reed-Muench method (1)

Dilution	Observed value[a]		Cumulative value			
	(1) Pos	(2) Neg	(3) Pos[b]	(4) Neg[c]	(5) Ratio[d]	(6) % Pos[e]
10^{-5}	5	0	10	0	10/10	100.0
10^{-6}	4	1	5	1	5/6	83.3
10^{-7}	1	4	1	5	1/6	16.7
10^{-8}	0	5	0	10	0/10	0.0

[a] Pos, tube showing CPE; Neg, tube not showing CPE.
[b] Obtained by adding numbers in column 1, starting at the bottom.
[c] Obtained by adding numbers in column 2, starting at the top.
[d] Value in column 3/(value in column 3 + value in column 4).
[e] $\frac{\text{\% CPE next above 50\%} - 50}{\text{\% CPE next above 50\%} - \text{\% CPE next below 50\%}} =$ proportional distance between the two dilutions at which 50% CPE occurs

From the above example: $\frac{83.3 - 50}{83.3 - 16.7} = 0.5$

One $TCID_{50}$ in this case is $10^{-6.5}$, and therefore 100 $TCID_{50}$ equals $10^{-4.5}$.

11. Incubate the culture tubes at 35 to 37°C for 7 to 10 days. Observe the cell culture monolayers daily for the development of CPE, and record the presence or absence of CPE (*see* Fig. 8.5–2).
12. NOTE: Some viruses grow more slowly than others. For example, culture readings should be terminated after 10 to 12 days with the slower-growing adenoviruses, while a 7-day incubation period would be appropriate for the faster-growing enteroviruses.

C. Calculation of $TCID_{50}$ by the Reed-Muench method (1)

The $TCID_{50}$ is that dilution that produces CPE in one-half of the cumulative number of cell cultures (Table 8.5–A1).

1. Record the number of positive (column 1) and negative (column 2) values.
2. Calculate the cumulative numbers of positive (column 3) and negative (column 4) tubes.
3. Calculate the percentage of positive tubes (column 5) by using the cumulative values.
4. Calculate the proportionate distance between the dilution showing >50% CPE and the dilution showing <50% CPE.
5. Calculate the $TCID_{50}$ by adding the proportionate distance factor (as a negative value) to the dilution (negative logarithm) showing >50% CPE.
6. Since each tube was inoculated with 0.1 ml of each virus dilution, the $TCID_{50}$ is expressed as $TCID_{50}$/0.1 ml.
7. NOTE: The proportionate distance must be multiplied by a correction factor, which is the logarithm of the dilution factor. In this procedure, with 10-fold dilutions, that factor is equal to 1, and the step can be disregarded. However, if other dilution series are used, the factor must be considered. For example, the correction factor for a twofold dilution series would be 0.3, and that for a fivefold dilution series would be 0.7.

References

1. **Reed, L. J., and H. Muench.** 1938. A simple method of estimating fifty percent endpoints *Am. J. Hyg.* **27:**493–497.
2. **Schmidt, N. J.** 1989. Cell culture procedures for diagnostic virology, p. 51–100. *In* N. J. Schmidt and R. W. Emmons (ed.), *Diagnostic Procedures for Viral, Rickettsial, and Chlamydial Infections,* 6th ed. American Public Health Association, Washington, D.C.

APPENDIX 8.5–3

Viral Neutralization Assay

Reprinted from **Lipson, S. M.** 1992. Neutralization test for the identification of viral isolates, p. 8.14.1–8.14.8. *In* H. D. Isenberg (ed.), *Clinical Microbiology Procedures Handbook,* vol. 2. American Society for Microbiology, Washington, D.C.

I. PRINCIPLE

The neutralization test is a highly sensitive assay applicable to the identification of all viral isolates. However, it is costly and labor intensive to perform and has been largely supplanted by more rapid and convenient methods. Currently, the major applications of the neutralization test in the diagnostic virology laboratory are for the identification of enteroviruses and the typing of adenoviruses (1, 3, 4).

The neutralization test is a biphasic system consisting of (i) an antigen-antibody reaction step, in which the virus to be identified is mixed and incubated with the appropriate antibody reagent(s), and (ii) an inoculation step, in which the mixture is inoculated into the appropriate host system (e.g., cell cultures, animals). The absence of infectivity, for example, the failure of cytopathic effect (CPE) or hemadsorption to develop in cell cultures, constitutes a positive neutralization reaction and indicates the identity of the virus.

II. SPECIMEN

Cell culture harvest of isolate to be identified (*see* V.H.1–4)

A. Titrate the viral isolate and determine the 50% tissue culture infective dose ($TCID_{50}$) as described in Appendix 8.5–2.

B. Alternatively, select the test dilution empirically. For example, use a 1:10 or 1:50 dilution for isolates with a slowly progressing CPE (e.g., adenovirus) and a 1:500 or 1:1,000 dilution for viruses with a rapidly progressing CPE (e.g., enteroviruses).

III. MATERIALS

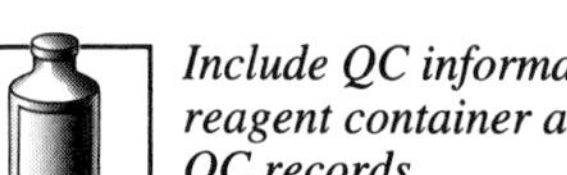

Include QC information on reagent container and in QC records.

A. Cell cultures and reagents

1. Cell culture monolayers (*see* procedure 8.2)
   a. Perform the neutralization test with the same type of cell culture used for determination of the test dose ($TCID_{50}$).
   b. Refeed the cell cultures with maintenance medium prior to performing the assay.
2. Cell culture maintenance medium (*see* procedure 8.6). For example, use Eagle minimum essential medium supplemented with 50 μg of gentamicin per ml, 1.25 μg of amphotericin B per ml, and 2% heat-inactivated fetal bovine serum (*see* procedure 8.6).
3. Antiviral neutralizing antisera
   a. Availability

      Antisera suitable for neutralization tests for many viruses are available from numerous commercial sources and reference laboratories. Pools for the identification of enterovirus isolates are available on a more limited basis. There are differences in the quality of products obtained from different sources or different lots. It is therefore the responsibility of the individual laboratory to monitor reagent quality by using homologous and heterologous viruses.
   b. Preparation and storage
      (1) Reconstitute lyophilized antiserum according to the recommendation of the manufacturer. Product data sheets should include information on reagent potency as well as storage and shelf life guidelines. Label containers with expiration dates.
      (2) Heat inactivate the serum in a 56°C water bath for 30 min to inactivate nonspecific heat-labile inhibitors of infectivity. Alternatively, heat inactivate the serum just prior to use.
      (3) Dispense antiserum into aliquot volumes consistent with the needs of the laboratory. Freeze and store the aliquots at −70°C. Avoid repeated freeze-thaw cycles. Label containers with expiration dates.
      (4) Record all pertinent information, including source, viral specificity and cross-reactivity, donor species, lot number, potency (neutralizing units),

aliquot volume and quantity, recommended test dilution and suggested diluent, lot number of matching normal (nonimmune) serum (if applicable), and dates of receipt, opening, use, heat inactivation, and expiration.

(5) Detailed step-by-step instructions for the reconstitution, storage, and use of the Lim Benyesh-Melnick (LBM) intersecting serum pools for the identification of enteroviruses are provided with the reagents. See Appendix 3 in CMPH procedure 8.14 for a description of these reagents.

(6) Handle all sera aseptically to maintain sterility.

(7) Monitor antisera for deterioration by performing neutralization assays with the matching (homologous) viruses. Perform these assays periodically (e.g., annually) on frozen stock, whenever the material has been subjected to improper storage conditions, or if deterioration is suspected based on performance.

4. Hanks' balanced salt solution (HBSS), pH 7.2 to 7.6 (*see* procedure 8.6)

B. Supplies

1. Sterile capped tubes (12 by 75 mm)
2. Assorted sterile pipettes and pipetting device
3. Autoclavable containers for discarding cultures

C. Equipment

1. Class II biological safety cabinet
2. Water bath, 56°C
3. Incubator, 35 to 37°C
4. Inverted microscope or standard microscope fitted with tracks for the observation of culture tubes; 10× to 20× objectives
5. Freezer, −70°C

IV. QUALITY CONTROL

A. Test each lot of antiserum at the working dilution prior to use to determine whether the reagent has a toxic effect on the cell cultures used for neutralization tests.

1. Inoculate 0.1 ml of each working dilution of serum into two tubes of the appropriate cell culture type.
2. Observe the cultures for 7 to 10 days for evidence of toxicity (cell rounding and degeneration of the monolayer).

B. Include a normal serum control, a positive serum control, a virus check titration, and a negative cell control with each assay.

1. Normal serum control

   Include a normal (nonimmune, negative) serum to determine whether the virus is nonspecifically inactivated by antiserum components.

   a. Include a nonimmune serum obtained from the same animal species used to produce the antiserum. Ideally, normal serum should be obtained as a preimmune serum from the animal prior to inoculation.

   b. Use the normal serum at the same dilution as the matching viral antiserum.

   c. If more than one antiserum from the same animal species is used in a neutralization test (i.e., an antiserum with a different [heterologous] viral specificity), the heterologous serum may substitute for the normal serum.

2. Positive serum control

   Include a reference virus incubated (neutralized) with its matching (homologous) serum as a positive control. Since this is not practical for each antiserum when multiple antisera are used in a neutralization test, select the antiserum to the prevalent or suspected virus.

3. Virus check titration

   Include a confirmatory (check) titration of the virus consisting of 3 or 4 serial $\log_{10}$ dilutions, starting with the test dilution of virus.

4. Negative cell control

   Select two uninoculated lot-matched cell cultures to be incubated and observed in parallel with the neutralization test.

V. PROCEDURE

A. Preparation of antiserum working dilution

1. Dilute each type-specific and normal antiserum 1:10 (or other predetermined dilution) with HBSS. The final working dilution of the type-specific antiserum

APPENDIX 8.5–3 *(continued)*

should contain 20 (or more) neutralization units per 0.1 ml of antiserum. Determine the appropriate dose on the basis of the manufacturer's specifications and in-house QC results.

2. If the serum has not been heat inactivated, place the diluted serum in a 56°C water bath for 30 min.
3. The working dilutions of antisera may be stored at 4°C for up to 5 days. Diluted antiserum may be stored at −70°C for extended periods (i.e., months to years).
4. Do not repeatedly freeze-thaw undiluted or working dilution sera. Aliquot in single-use volumes, and freeze at −70°C.

B. Preparation of virus test dilution and check titration

1. Determine the dilution of each unknown and control virus to be used.
   a. Prepare a dilution of virus in HBSS that contains 100 $TCID_{50}$/0.1-ml volume (EPCM Appendix 8.5–2). For example, if the $TCID_{50}$ of a viral harvest is 10^{-3}/0.1 ml (1:1,000 dilution), use the material at 10^{-1} (1:10 dilution) in the neutralization test by combining 1 part undiluted material with 9 parts diluent.
   b. Alternatively, prepare a dilution of virus based on culture characteristics (item II.B).
2. Check titration
   Starting with the virus test dilution, prepare three additional serial $\log_{10}$ dilutions with HBSS as diluent. For example, if the test dilution is 10^{-1}, prepare dilutions through 10^{-4} (EPCM Appendix 8.5–2).

C. Performance of the neutralization test

1. For each virus to be neutralized, add 0.3 ml of the diluted normal serum to a labeled tube.
2. For each virus to be neutralized, add 0.3 ml of the appropriate diluted immune antiserum to a labeled tube.
3. Add 0.3 ml of the viral dilution containing 100 $TCID_{50}$/0.1 ml to the corresponding serum tubes.
4. Gently agitate the virus-serum mixtures, and incubate them and the virus check titration at room temperature for 1 h or as specified by the antiserum manufacturer.
5. Inoculate 0.2 ml of each virus-serum mixture into two cell culture tubes, each containing 1 to 2 ml of maintenance medium.
6. Inoculate 0.1 ml of the test dose and each dilution of the check titration into two or three cell culture tubes, each containing 1 to 2 ml of maintenance medium.
7. Incubate the cell culture tubes at 35 to 37°C. Observe the monolayers daily for the appearance of CPE. Record the presence or absence of CPE for each tube (*see* Fig. 8.5–2).

VI. RESULTS

A. The neutralization test is completed when the anticipated endpoint of the check titration is achieved. The endpoint should be attained after an incubation period of 7 to 10 days, depending on the virus.

B. Controls

1. The degree of CPE in the normal serum control should be similar to that observed in the virus test dilution of the check titration.
2. The positive control should show no or only minimal (breakthrough) CPE compared with CPE in the virus control.
3. The virus test dose of 100 $TCID_{50}$ is confirmed by the development of CPE in the virus test dilution and in the two subsequent serial dilutions of the check titration. In most cases, a test dose of 3 to 300 $TCID_{50}$ is acceptable.
4. The negative cell control should not show any CPE, and the monolayer quality should be acceptable.

C. Specimens

Compare the development and degree of CPE observed in the neutralized tubes with that observed in the normal serum controls and in the test dilution of the check titration.

1. The absence of CPE in tubes inoculated with a particular virus-antiserum mixture indicates a positive neutralization reaction and thus identifies the viral isolate.

Table 8.5–A2 Problems associated with interpretation of the neutralization test

Problem	Definition	Possible cause(s)	Solution
No neutralization	CPE develops at same rate with all or most antisera.	Test dose of virus too strong	Redetermine endpoint titers.
		Serum deteriorated	Obtain new antisera. Check storage conditions.
Breakthrough	"Neutralized" virus produces delayed and reduced CPE.	Test dose of virus too strong	Redetermine endpoint titers.
		Cross-reactivity	Check product specifications for cross-reactive properties; obtain non-cross-reactive serum if possible.
		Antiserum deteriorated or failed to attain expected efficacy	Check against reference virus; use lower dilution. Other methods, such as kinetic neutralization assay (2), required.
		Clumps or aggregates in virus suspension	Remove by ultrafiltration.
Neutralization by more than one antiserum	CPE is inhibited by two or more antisera.	Cross-reactivity	Check product specifications for cross-reactive properties.
		Test dose of virus too weak	Redetermine endpoint titers.
		Antiserum dose too strong	Check suggested working dilution. Heat inactivate serum (56°C, 30 min). Obtain new antiserum.
Nonspecific virus inactivation	Virus is neutralized (no CPE) by normal serum.	Serum not heat inactivated or contains nonspecific viral inhibitors	
		Virus inactivated during incubation step	Check recommended temp. Check incubator-water bath temp.
Monolayer toxicity	Nonspecific degeneration of culture monolayer	Serum not heat inactivated or is toxic to cells	Heat inactivate serum (56°C, 30 min). Obtain new antiserum.

2. The simultaneous development of CPE in the unknown virus-antiserum mixture, the normal serum control, and the test dilution indicates that there is no relationship between the unknown virus and the antiserum.
3. Occasionally, the neutralization test may be difficult to interpret. In such cases, consider the factors presented in Table 8.5–A2 or the possibility of a mixed infection.

References

1. **Horstmann, D. M., and G. D. Hsiung.** 1965. Principles or diagnostic virology, p. 405–429. *In* F. L. Horsfall, Jr., and I. Tamm (ed.), *Viral and Rickettsial Infections of Man.* J. B. Lippincott Co., Philadelphia.
2. **McBride, W. D.** 1959. Antigenic analysis of poliovirus by kinetic studies of serum neutralization. *Virology* **7:**45–48.
3. **Melnick, J. L., H. A. Wenner, and C. A. Phillips.** 1979. Enteroviruses, p. 471–534. *In* E. H. Lennette and N. J. Schmidt (ed.), *Diagnostic Procedures for Viral Rickettsial, and Chlamydial Infections*, 5th ed. American Public Health Association, Washington, D.C.
4. **Schmidt, N. J.** 1989. Cell culture procedures for diagnostic virology, p. 51–100. *In* N. J. Schmidt and R. W. Emmons (ed.), *Diagnostic Procedures for Viral, Rickettsial, and Chlamydial Infections.* 6th ed. American Public Health Association, Washington, D.C.

APPENDIX 8.5–4

Acid Lability Assay

Reprinted from **Aarnaes, S.** 1992. Differentiation of rhinoviruses from enteroviruses: acid lability test, 8.13.1–8.13.3. *In* H. D. Isenberg (ed.), *Clinical Microbiology Procedures Handbook*, vol. 2. American Society for Microbiology, Washington, D.C.

I. PRINCIPLE

The differentiation of rhinoviruses from enteroviruses cannot be made on the basis of routine cultural characteristics, since both groups may be isolated from respiratory samples and produce a similar cytopathic effect (CPE). Furthermore, routine confirmation

APPENDIX 8.5–4 *(continued)*

of a suspected rhinovirus isolate is not feasible because of the very numerous rhinovirus types. While extreme lability, short-term shedding, and preference for cultivation at 33°C limit the number of rhinovirus isolates, the possibility that a rhinovirus may be isolated must be considered. The suspected rhinovirus isolate can be conveniently differentiated from an enterovirus on the basis of acid lability (1, 2). Enteroviruses have been shown to be stable at pHs 3.0 to 5.0. Rhinoviruses, by contrast, are inactivated at a low pH.

II. SPECIMEN

Cell culture supernatants from cultures showing a 3 to 4+ CPE are suitable specimens (*see* Fig. 8.5–2).

III. MATERIALS

Include QC information on reagent container and in QC records.

A. Reagents and supplies

1. Human diploid fibroblast cell cultures (*see* procedure 8.2)
2. Citrate buffer, 0.1 M, pH 4.0
   a. Stock solutions
      (1) Citric acid solution, 0.1 M
      citric acid .... 21.01 g
      deionized water .... to 1,000.00 ml
      (2) Sodium citrate solution, 0.1 M
      sodium citrate .... 29.41 g
      deionized water .... to 1,000.00 ml
      (3) Sterilize by filtration through a 0.2-μm-pore-size filter. Store at 2 to 8°C. Shelf life is 6 months or longer.
   b. Citrate buffer, pH 4.0
      (1) Ingredients
      citric acid, 0.1 M .... 33.0 ml
      sodium citrate, 0.1 M .... 17.0 ml
      (2) Aseptically combine sterile stock solutions. Prepare fresh on day of use.
3. Phosphate buffer, 0.1 M, pH 7.0
   a. Ingredients
      Na_2HPO_4 .... 8.7 g
      KH_2PO_4 .... 5.3 g
      deionized water .... to 1,000.0 ml
   b. Sterilize by filtration through a 0.2-μm-pore-size filter. Store at 2 to 8°C. Shelf life is 1 year.
4. Phosphate buffer, 0.5 M, pH 7.2
   a. Ingredients
      Na_2HPO_4 .... 51.1 g
      KH_2PO_4 .... 19.0 g
      deionized water .... to 1,000.0 ml
   b. Sterilize by filtration through a 0.2-μm-pore-size filter. Store at 2 to 8°C. Shelf life is 1 year.
5. Cell culture maintenance medium: Eagle's minimal essential medium supplemented with 2% heat-inactivated fetal bovine serum (*see* procedure 8.6)
6. Sterile disposable capped tubes, 12 by 75 mm and 13 by 100 mm
7. Sterile pipettes (assorted) and pipetting device
8. Cell culture racks

B. Equipment

1. Incubator, 33°C
2. Inverted microscope or standard microscope fitted with stage tracks to accommodate cell culture tube

IV. QUALITY CONTROL

A. Positive control

1. Determine the 50% tissue culture infective dose ($TCID_{50}$) of a reference rhinovirus (*see* Appendix 8.5–2).
2. Treat the reference virus with the buffer solutions in the same manner as the unknown virus. Use the reference virus at approximately 1,000 $TCID_{50}$/ml.

B. Negative control

Treat supernatant fluid from negative (uninoculated) cell culture controls with buffer solutions in the same manner as the unknown virus.

V. PROCEDURE

A. Add 0.5 ml of each unknown virus, the reference rhinovirus, and the negative control supernatant to a labeled tube (13 by 100 mm) containing 0.5 ml of citrate buffer (pH 4.0).

B. Add 0.5 ml of each unknown virus, the reference rhinovirus, and the negative control supernatant to a labeled tube (13 by 100 mm) containing 0.5 ml of phosphate buffer (pH 7.0).

C. Incubate the tubes containing the virus and control mixtures at 37°C for 1 h.

D. Following the incubation, add 4.0 ml of 0.5 M phosphate buffer (pH 7.2) to each mixture of virus and control. This represents a 1:10 (10^{-1}) dilution of the original material.

E. Prepare serial 10-fold dilutions (10^{-2} to 10^{-5}) of each unknown and reference virus mixture. Use cell culture maintenance medium as diluent. Serial titration of the negative control is not necessary.

1. Add 0.9 ml of diluent to four labeled tubes (12 by 75 mm) for each mixture to be titrated.
2. Transfer 0.1 ml of the 1:10 dilution of each material to the appropriate 10^{-2} dilution tube, and mix by pipetting up and down. This represents a 1:100 (10^{-2}) dilution of the original material.
3. Prepare the remaining dilutions in the same manner. Use a separate pipette for each dilution.

F. Inoculate two or three diploid fibroblast cultures containing 1.0 ml of maintenance medium with 0.2 ml of each of the unknown and reference virus dilutions.

G. Inoculate two diploid fibroblast cultures containing 1.0 ml of maintenance medium with 0.2 ml of the negative control.

H. Incubate the cell cultures at 33°C in a stationary rack, and examine them daily for CPE. Record the titers.

VI. RESULTS

A. Rhinoviruses are acid labile and will show either a total inactivation or a reduction in titer in the material treated with acid (pH 4.0) compared with activity and titer of viruses treated with the pH 7.0 buffer. High-titered isolates may not be completely inactivated, and in such cases, a ≥100-fold reduction in titer would be interpreted as acid lability. Although definitive confirmation requires a neutralization test, the combination of enterovirus-type CPE and acid lability is considered sufficient to designate an isolate a rhinovirus.

B. An enterovirus will show no reduction in titer after treatment at pH 4.0 compared with the titer after treatment at pH 7.0.

C. The rhinovirus control should show either total or a ≥100-fold reduction in titer after treatment at pH 4.0.

D. Negative controls should show no toxicity or CPE.

References

1. **Ballew, H., F. T. Forrester, H. C. Lyerla, W. M. Velleca, and B. R. Bird.** 1977. *Differentiation of Enteroviruses from Rhinoviruses*, p. 112–113. U.S. Department of Health, Education and Welfare. Center for Disease Control, Atlanta.

2. **Gwaltney, J. M., Jr., R. J. Colonno, V. V. Hamparian, and R. B. Turner.** 1989. Rhinovirus, p. 579–614. *In* N. J. Schmidt and R. W. Emmons (ed.) *Diagnostic Procedure for Viral, Rickettsial, and Chlamydial Infections,* 6th ed. American Public Health Association, Inc., Washington D.C.

APPENDIX 8.5–5

Rubella Interference Assay

Reprinted from **Aarnaes, S., and B. J. Daidone.** 1992. Observation and maintenance of inoculated cell cultures, p. 8.7.8–8.7.9. *In* H. D. Isenberg (ed.)., *Clinical Microbiology Procedures Handbook,* vol. 2. American Society for Microbiology, Washington, D.C.

A. Cells

Primary African green monkey kidney (AGMK) cells are the cells of choice for isolation of rubella virus from clinical specimens. A CPE is not produced, and a positive rubella culture is detected by interference with the growth of a superinfected challenge virus.

APPENDIX 8.5–5 *(continued)*

B. Materials (in addition to materials used for routine viral culture)
   1. Five specimen-inoculated tubes and four each of lot-matched negative and positive (rubella-inoculated) controls
   2. Challenge virus (echovirus type 11 or coxsackievirus A9) diluted with maintenance medium or Hanks' balanced salt solution to contain 100 50% tissue culture infective doses ($TCID_{50}$) per 0.1 ml.
      Determine endpoint titer and appropriate dilution as described in CMPH procedure 8.14. Do not inoculate specimens until a titered challenge virus is available.
   3. Incubator, 35 to 37°C

C. Interference assay
   1. Procedure
      a. Remove the culture medium from the tubes to be challenged, and save it. Store the harvested fluid at −70°C; discard negative harvests upon completion of the interference assay.
      b. Refeed tubes to be challenged with 1 ml of maintenance medium.
      c. Deliver 0.1 ml of the challenge virus dilution containing 100 $TCID_{50}$ to each specimen and control tube to be challenged.
      d. Refeed unchallenged tubes weekly with 1 ml of culture maintenance medium.
      e. Incubate challenged and unchallenged cultures at 35°C. Observe challenged cultures daily for 7 days for the development of enterovirus CPE.
   2. Schedule summary
      a. At 1 week
         (1) Challenge one specimen tube, one lot-matched positive control, and one lot-matched negative control.
         (2) Refeed unchallenged tubes with maintenance medium.
      b. At 2 weeks
         (1) Challenge one specimen tube, one lot-matched positive control, and one lot-matched negative control.
         (2) Subpassage harvested cells and fluid from one specimen tube into two AGMK tubes. Prepare two lot-matched positive and negative controls.
         (3) Refeed unchallenged tubes with maintenance medium.
      c. At 3 weeks
         (1) Challenge two specimen tubes (one from primary inoculation and one subpassage) and the lot-matched positive and negative controls.
         (2) Refeed unchallenged tubes with maintenance medium.
      d. At 4 weeks
         Challenge two specimen tubes (one from primary inoculation and one subpassage) and the lot-matched positive and negative controls.

D. Interpretation of interference assay
   1. Controls
      a. The challenged positive control should show no enterovirus CPE.
      b. The challenged negative control should show enterovirus CPE.
      c. The unchallenged negative control should show no CPE.
   2. Specimens
      a. Cultures that exhibit enterovirus CPE are negative for rubella virus.
      b. Cultures that do not exhibit enterovirus CPE are considered positive for rubella virus and may be confirmed by neutralization (*see* Appendix 8.5–3) or other immunoassay.
      c. Subpassage cultures exhibiting partial interference.
         ◩ **NOTE:** Infection of AGMK by simian viruses may be associated with reduced sensitivity for the recovery of rubella virus (1). Therefore, in addition to the positive and negative controls included in the interference assay, select 8 to 10 uninoculated AGMK tubes from each lot, and observe these in parallel with specimen-inoculated tubes. Observe for the development of spontaneous CPE, and use guinea pig RBCs to test for HAd in two tubes each week (*see* Fig. 8.5–5).

Reference

1. **Herrmann, K. L.** 1979. Rubella virus, p. 725–766. *In* E. H. Lennette and N. J. Schmidt (ed.), *Diagnostic Procedures for Viral, Rickettsial, and Chlamydial Infections*, 5th ed. American Public Health Association, Inc., Washington, D.C.

APPENDIX 8.5–6

Preparation of 10% Guinea Pig RBC Suspension

NOTE: Use aseptic technique.

1. Obtain guinea pig RBCs collected aseptically in Alsever's solution. Store at 4°C and use within 7 days of receipt.
2. Transfer 5 to 7 ml of suspended blood to a sterile, graduated, 15-ml conical centrifuge tube.
3. Add an equal volume of PBS and centrifuge at 700 to 900 × *g* for 5 min at room temperature.
4. Discard the supernatant, suspend the RBC pellet in 10 ml of PBS by gently pipetting up and down several times, and centrifuge the suspension at 700 to 900 × *g* for 5 min at room temperature.
5. Repeat step 4 twice.
6. Discard the supernatant, and measure the packed cell volume.
7. Add a volume of PBS equal to nine times the packed cell volume.
8. Store at 4°C and use within 7 days. Discard if evidence of hemolysis or contamination occurs.

APPENDIX 8.5–7

Preservation of Cell Culture Monolayers (1)

1. Prepare buffered formaldehyde preservative medium by combining the following ingredients. Store at room temperature.
   formaldehyde solution (37 to 40%)....100 ml
   distilled water....900 ml
   sodium phosphate, monobasic ($NaH_2PO_4 \cdot H_2O$)....4 g
   sodium phosphate, dibasic (Na_2HPO_4)....6.5 g
2. Aspirate the medium from the cell culture tubes to be preserved.
3. Add 2 to 4 ml of preservative to each tube.
4. Store tubes indefinitely at room temperature.

Reference

1. **Gurtler, J. H., C. Ballew, C. M. Preissner, and T. F. Smith.** 1982. Cell culture medium for preserving cytopathic effects in cell cultures. *Lab. Med.* **13:**244–245.

APPENDIX 8.5–8

Cultivation of *C. trachomatis*

I. INTRODUCTION

Chlamydiae are obligate intracellular bacteria that contain RNA and DNA, have a cell wall resembling those of gram-negative bacteria, and multiply by binary fission in a manner distinct from those of other bacteria (1). The 300- to 400-nm spherical elementary body (EB) is the infectious form of the organism. Following cellular infection, the EB reorganizes into a larger, metabolically active reticulate body, which divides repeatedly by binary fission for 24 to 48 h and eventually develops into the characteristic intracytoplasmic inclusion. Human infections (4) associated with the genus *Chlamydia* are summarized in Table 8.5–A3. Despite the introduction of numerous nonculture assays, including amplified assays, culture remains an important assay for the detection of *C. trachomatis* infections and, because of its specificity, is recommended for laboratory testing in cases of sexual abuse and medicolegal situations.

APPENDIX 8.5–8 *(continued)*

Table 8.5–A3 Human chlamydial infections

Organism	Disease(s)
C. trachomatis	
Serotypes A, B, Ba, and C	Trachoma
Serotypes D through K	Cervicitis, epididymitis, inclusion conjunctivitis, nongonococcal disease, pelvic inflammatory disease, neonatal pneumonitis, proctitis
Serotypes L-1, L-2, and L-3	Lymphogranuloma venereum
C. psittaci	Psittacosis, ornithosis following contact with infected birds
C. pneumoniae	Upper and lower respiratory tract disease

II. SPECIMEN

Collect, transport, and store specimens in the same manner as for viral culture, taking care to use transport medium suitable for these organisms. 2-SP, as described for viral transport, is suitable for use as a chlamydial transport medium (*see* EPCM procedure 8.6).

NOTE: Use antimicrobial concentrations suitable for chlamydiae. Antimicrobial agents frequently used include gentamicin (50 μg/ml), vancomycin (100 μg/ml), and amphotericin B (25 μg/ml); nystatin (25 U/ml) may be used in place of amphotericin B. Do not include penicillin in chlamydial transport medium.

III. PROCEDURE SUMMARY

Refer to CMPH 8.23 for a complete procedure. Perform chlamydial culture by using BGMK or McCoy cell shell vial cultures as described for viral culture (Fig. 8.5–1 and 8.5–10), using the following conditions:

A. Centrifuge inoculated shell vials at 1,500 × *g* for 45 to 60 min at 22 to 37°C.
B. Refeed cultures after centrifugation with chlamydial cell culture medium (EPCM procedure 8.6).
C. Perform IF at 48 to 72 h, using antibodies directed against the genus-specific lipopolysaccharide (LPS) antigen or the species-specific major outer membrane protein (MOMP) antigen to detect characteristic intracytoplasmic inclusion (Fig. 8.5–A2).

   NOTE: Use methanol or acetone fixative, as specified by the antibody manufacturer.

Figure 8.5–A2 Typical chlamydial inclusions in a culture stained at 48 h with a monoclonal antibody to *C. trachomatis* (reprinted from CMPH 8.23.8).

IV. RESULTS

A. Control cultures

1. The positive control must show strongly fluorescent (brilliant apple green) characteristic inclusions.
2. The negative control should have a satisfactory monolayer and show minimal to no nonspecific fluorescence.

B. Specimen cultures

1. A positive culture is one showing one or more characteristic intracytoplasmic inclusions. Report as positive for *Chlamydia* spp. or *C. trachomatis*, depending on the MAb used for detection.
2. A negative culture is one that has a satisfactory monolayer, minimal or no nonspecific staining, and no chlamydial inclusions.
3. An unsatisfactory culture is one showing strong nonspecific fluorescence that interferes with reading or one that does not contain an adequate monolayer (≥75% of that observed in the negative control).

V. PROCEDURE NOTE

Alternative staining methods for the detection of *C. trachomatis* inclusions (iodine, Giemsa) are not as sensitive as IF (2, 3).

References

1. **Becker, Y.** 1978. The *Chlamydia:* molecular biology of procaryotic obligate parasites of eucaryocytes. *Microbiol. Rev.* **42:**274–306.
2. **Munday, P. E., A. P. Johnson, B. J. Thomas, and D. Taylor-Robinson.** 1980. A comparison of immunofluorescence and Giemsa for staining *Chlamydia trachomatis* in cycloheximide-treated McCoy cells. *J. Clin. Pathol.* **33:** 177–179.
3. **Schachter, J.** 1985. Immunodiagnosis of sexually transmitted disease. *Yale J. Biol. Med.* **58:** 443–452.
4. **Schachter, J., and W. E. Stamm.** 1995. *Chlamydia,* p. 669–675. *In* P. R. Murray, E. J. Baron, M. A. Pfaller, F. C. Tenover, and R. H. Yolken (ed.), *Manual of Clinical Microbiology,* 6th ed. American Society for Microbiology, Washington, D.C.

Supplemental Reading

Barnes, R. 1989. Laboratory diagnosis of human chlamydial infection. *Clin. Microbiol. Rev.* **2:** 119–136.

Petersen, E. M. 1992. Isolation of *Chlamydia* spp. in cell culture, p. 8.23.1–8.23.12. *In* H. D. Isenberg (ed.), *Clinical Microbiology Procedures Handbook,* vol. 2. American Society for Microbiology, Washington, D.C.

8.6 Reagent Preparation: Viral and Chlamydial Assays

For detailed procedures regarding preparation and QC of media and reagents, see CMPH 8.19. Since most laboratories purchase transport and culture media, this section includes only a brief summary of selected reagents. If sterile disposable cell culture labware is not used, it is important to follow scrupulous washing, rinsing, and sterilization of glassware (Appendix 8.6–1).

I. VIRAL TRANSPORT MEDIUM (VTM)

Include QC information on reagent container and in QC records.

A. Hanks' balanced salt solution (HBSS) with bovine serum albumin

1. Aseptically combine the following sterile reagents:
   Bovine serum albumin, 7.5% sterile solution ....33.00 ml
   HEPES (*N*-2-hydroxyethylpiperazine-*N*′-2-ethanesulfonic acid),
   1 M solution ....10.00 ml
   antimicrobial agents (*see* item II below).. ml
   HBSS, q.s. to ....500.0 ml
2. The final pH should be 7.2 to 7.4 (reddish-orange color). Adjust the pH with 7.5% sodium bicarbonate solution before adding the HEPES.
3. Dispense 2.0- to 3.0-ml volumes into sterile 4-ml screw-cap tubes; dispense 5.0- to 7.0-ml volumes into 15-ml centrifuge tubes. Store frozen at −20°C for up to 6 months. Thaw just before use.

   NOTE: Nystatin is insoluble in water; keep material in suspension during preparation and dispensing.

   NOTE: Include 4 to 5 small (e.g., 5-mm) sterile glass beads in each vial and tube to facilitate specimen processing. Tubes should be able to withstand freezing at −70°C or lower for specimen storage (e.g., NUNC cryotubes; polypropylene centrifuge tubes).

B. 2-SP (0.2 M sucrose–0.02 M phosphate) medium

1. Dissolve the following ingredients separately in approximately 300 ml of deionized water.
   KH_2PO_4..2.01 g
   Na_2HPO_4..1.13 g
   sucrose...68.46 g
2. Combine the solutions and add deionized water to 1,000 ml.
3. Adjust the pH to 7.2 to 7.4.
4. Add HEPES to a final concentration of 20 mM (20.0 ml of 1 M solution per 1,000 ml).
5. Sterilize by using a 0.2-μm-pore-size filter.
6. Aseptically add sterile antimicrobial agents (see item II below).
7. Dispense 2- to 3-ml volumes into sterile 4-ml screw-cap tubes; dispense 5- to 7-ml volumes into 15-ml centrifuge tubes.
8. Store frozen at −20°C for up to 6 months. Thaw just before use. See notes above regarding Nystatin and glass beads.

C. Commercial sources of transport medium are provided in procedure 8.3.

II. ANTIMICROBIAL AGENTS

Include QC information on reagent container and in QC records.

A. Concentrations in viral transport medium

The most commonly used combinations include gentamicin-vancomycin-amphotericin B (or nystatin [Mycostatin]), penicillin-streptomycin-gentamicin-amphotericin B, and penicillin-streptomycin-amphotericin B (or nystatin). The following ranges (i.e., final concentrations per milliliter of transport medium or processed specimen) have been used: amphotericin, 2 to 40 μg/ml; vancomycin, 25 to 1,000 μg/ml; gentamicin, 25 to 150 μg/ml; penicillin, 50 to 55 U/ml; streptomycin, 50 to 500 μg/ml; and nystatin, 50 to 200 U/ml. The higher concentrations may be useful for specimens likely to contain high rates of microbial contamination (e.g., stool); however, high antimicrobial-agent concentrations may be toxic to cell culture monolayers. Monitor toxicity by observing cell cultures that have been mock inoculated with VTM or diluent containing the antimicrobial agents.

B. Concentrations in chlamydial transport medium

The recommended final antimicrobial-agent concentrations are as follows: gentamicin, 25 to 50 μg/ml; vancomycin, 100 to 200 μg/ml; and nystatin, 25 to 50 U/ml. Do not include penicillin in chlamydial transport medium or in other reagents used for chlamydial isolation.

C. Antimicrobial-agent mixture concentrates (e.g., 10×, 20×)

Add these to diluent that is to be used for homogenization of solid samples and to liquid samples not collected with viral or chlamydial transport medium. For example, add 0.1 ml of 20× concentrate to each 2.0 ml of liquid sample (e.g., urine) or homogenizing diluent. Purchase concentrated antimicrobial-agent mixtures or prepare in-house. The types and concentrations of antimicrobial agents vary among laboratories; select desired concentrations for viral or chlamydial specimens by using the guidelines presented above. To prepare mixtures aseptically, combine the necessary volumes of each individual antimicrobial reagent in a sterile container. Add sterile demineralized water or HBSS (pH 7.2 to 7.4) to the desired final volume. Dispense in aliquots convenient for the laboratory (e.g., 2 and 5 ml), and freeze at −20°C. The expiration date is 1 year from the date of preparation.

NOTE: Nystatin is insoluble in water; keep material in suspension during preparation and dispensing.

III. CELL CULTURE MEDIUM

Include QC information on reagent container and in QC records.

Sterile liquid media and components are available at the working dilution (1×) or as concentrates (e.g., 10×) which can be combined aseptically and diluted with sterile, cell culture quality, demineralized water.

Eagle's minimum essential medium (EMEM) supplemented with 2% heat-inactivated fetal bovine serum (FBS) is the most widely used cell culture medium for the maintenance of monolayer cell cultures. EMEM consists of essential amino acids, vitamins, and other nutrients in either HBSS or Earle's balanced salt solution (EBSS) containing glucose, salts, a bicarbonate buffering system, and phenol red. Most laboratories now purchase this medium in a sterile liquid form as a 1× or 10× solution; it can be purchased without glutamine or sodium bicarbonate. Observe manufacturer's expiration dates for reagents.

Use heat-inactivated FBS (2 to 5%) to supplement maintenance medium for monolayer cell cultures. Serum may contain substances that are toxic to monolayers or inhibitory to viral and chlamydial replication; inactivate complement before using the serum by heating serum to 56°C for 30 min (*see* item III.C.9 below). Serum may also be a source of mycoplasmal and viral contaminants and should be certified by the manufacturer to be free of these agents. Serum-free medium is used by many laboratories for influenza and parainfluenza virus cultures; the

III. CELL CULTURE MEDIUM *(continued)*

formulation described below, omitting the FBS, may be used for short-term maintenance of inoculated cultures. Alternatively, supplemented serum-free medium is commercially available; for example, Ultraculture from Whittaker Bioproducts, Walkersville, Md., appears to be useful for a variety of cell lines without preculture adaptation or weaning.

A. Viral cell culture maintenance medium

EMEM in HBSS with 2% FBS and HEPES buffer (*see* CMPH 8.19 for additional information)

NOTE: HBSS is used with tightly capped culture vessels; EBSS is used when caps are loosened and incubation is in a 5% CO_2 atmosphere.

1. To 800 ml of sterile distilled, deionized water, aseptically add the following:

EMEM in HBSS, 10× (without L-glutamine or $NaHCO_3$)	100.0 ml
L-glutamine, 200 mM solution	10.0 ml
HEPES, 1 M solution	20.0 ml
$NaHCO_3$, 7.5% solution	29.3 ml
FBS, heat-inactivated	20.0 ml
gentamicin, 50 mg/ml	0.20 ml
vancomycin, 50 mg/ml	2.0 ml
nystatin (Mycostatin), 10,000 U/ml	2.5 ml
sterile distilled, deionized water	q.s. to 1,000 ml

2. The pH of the medium should be 7.2 to 7.4 (reddish orange to cherry red); adjust if necessary with sterile 1 N NaOH or 1 N HCl.
3. Aseptically dispense the medium into sterile containers.
4. Store at 4°C; use within 3 months of adding FBS.

   NOTE: L-Glutamine is labile; use medium within 3 weeks of adding the glutamine; add additional glutamine beyond 3 weeks as necessary.

B. *Chlamydia* culture medium (*see* CMPH 8.23 for additional information)

Chlamydia culture medium is used to refeed cultures after inoculation; it is not intended for use as a preinoculation monolayer maintenance medium. *Chlamydia* isolation medium is prepared as described above for viral culture medium, with the following modifications:

1. FBS is added to a final concentration of 10%.
2. Cycloheximide concentrate (e.g., 100×) is added at a final concentration of 1 μg/ml; however, this may vary (e.g., 0.5 to 2.0 μg/ml) with each new lot or batch. Use medium within 2 weeks of adding cycloheximide.

   Assay each new batch of cycloheximide at several concentrations; the optimal concentration is that which yields an inclusion count equal to or greater than the count obtained with the positive control by using the current batch of cycloheximide and which is not toxic to the culture monolayer.

   To prepare the cycloheximide concentrate, dissolve the powder in demineralized water and sterilize by using a 0.2-μm-pore-size filter. Aliquot in small volumes, store frozen in the dark; use within 1 year.

   NOTE: Cycloheximide is toxic; avoid inhalation or direct contact.
3. Gentamicin may be added at a final concentration of 50 μg/ml; *see* antimicrobial item B above.
4. A high-glucose EMEM (e.g., 30 μM) is frequently used; however, this requirement was established when staining inclusions with iodine, and its need when cultures are screened by IF has not been determined. EMEM with a high glucose concentration can be purchased, or a stock sucrose may be prepared in-house and added at the time of medium preparation.

C. Culture medium notes

1. Use aseptic techniques when preparing and using medium (*see* Appendix 8.2–2).
2. Assign a lot number to each batch of medium, and record the lot numbers, sources, and expiration dates of all medium components.
3. Use sterile, distilled, deionized water for the preparation of all cell culture

reagents. The removal of contaminating pyrogens and minerals from water is imperative for the success of a given culture system.

4. L-Glutamine will develop a precipitate when frozen and thawed but is readily resolubilized by warming briefly to 37°C upon thawing. Avoid repeated freeze-thaw cycles; aliquot stock material into single-use volumes.
5. FBS may vary extensively from lot to lot; assess each new lot (2). Perform medium toxicity checks with each new lot, and assess for viral/chlamydial inhibitory activity as described below. Obtain samples of different lots of FBS, which can be held on reserve until assessment is completed.
6. Viral/chlamydial inhibitory activity
   Select two or three viruses (e.g., influenza, CMV) or *C. trachomatis* and perform parallel titrations with medium prepared with the current and candidate lots of FBS. Reject lots that show viral inhibitory activity.
7. Culture medium toxicity testing
   a. Feed uninoculated cell cultures with each new lot of medium.
   b. Incubate the cultures and observe for 10 to 14 days for signs of toxicity (rounding, monolayer detachment from the tube surface).
   NOTE: If toxicity is apparent, prepare medium with one new component at a time; water is a frequent cause of toxicity.
8. Sterility check
   Inoculate broth medium (e.g., THIO, BHI, TSB) with 1-ml aliquots of each lot of medium; incubate and observe for 7 days. Do not use medium until sterility is confirmed.
9. Heat inactivation of FBS
   a. Thaw frozen FBS at room temperature just before preparing medium.
   b. Place thawed FBS in a 56°C water bath; the water level should match that of the FBS. Swirl the bottles every 5 minutes so that the entire volume of FBS uniformly reaches 56°C. A thermometer inserted into a bottle containing the same volume of water can be used to assess the time required for the FBS to reach proper temperature.
   c. Inactivate the FBS for 30 min at 56°C, swirling the bottle every 10 min.
   NOTE: Avoid repeated freeze-thaw cycles of FBS.
10. When using MDCK cells for influenza virus isolation, supplement culture medium with TPCK-trypsin (1-tosylamide-2-phenylethyl-chloromethyl ketone) at a final concentration of 1 to 2 μg/ml (1, 3).
11. Cell culture medium reagents are available for numerous commercial sources, including those supplying cell lines (*see* EPCM procedure 8.2).

IV. PHOSPHATE-BUFFERED SALINE (PBS), 0.01 M

A. 10× stock solutions

1. NaH_2PO_4, 0.1 M
   NaH_2PO_4 (anhydrous) .... 13.9 g
   deionized water, q.s. to .... 1,000.0 ml
2. Na_2HPO_4, 0.1 M
   $Na_2HPO_4 \cdot 7H_2O$ .... 26.8 g
   deionized water, q.s. to .... 1,000.0 ml
3. NaCl, 8.5%
   NaCl .... 85.0 g
   deionized water, q.s. to .... 1,000.0 ml
4. Sterilize 10× stock solutions by filtration with a 0.2-μm-pore-size filter. Store at 4°C. Shelf life is 6 months or longer.

IV. PHOSPHATE-BUFFERED SALINE (PBS), 0.01 M *(continued)*

B. Prepare PBS of the desired pH by aseptically combining the following volumes:

pH	10× NaH_2PO_4 (ml)	10× Na_2HPO_4 (ml)
7.0	39.0	61.0
7.2	28.0	72.0
7.4	19.0	81.0
7.6	13.0	87.0
7.8	8.5	91.5
8.0	5.3	94.7

REFERENCES

1. **Frank, A. L., R. B. Couch, C. A. Griffis, and G. D. Baxter.** 1979. Comparison of different tissue cultures for isolation and quantitation of influenza and parainfluenza viruses. *J. Clin. Microbiol.* **10:**32–26.
2. **NCCLS.** 1995. *Fetal Bovine Serum.* Approved guideline M25-A. NCCLS, Wayne, Pa.
3. **Tobita, K., A. Sigiura, C. Enomoto, and M. Furuyama.** 1975. Plaque assay and primary isolation of influenza A viruses in an established line of canine kidney cells (MDCK) in the presence of trypsin. *Med. Microbiol. Immunol.* **162:**9–14.

APPENDIX 8.6–1

Procedure for Washing and Sterilizing Glassware for Cell Culture Reagents

Reprinted from Appendix 4 in CMPH procedure 8.19.

Reusable glassware used for dispensing and storing media and reagents must be meticulously cleaned to eliminate toxic substances that may interfere with the growth of cells in culture.

A. Washing
  1. Rinse all glassware in tap water immediately after use. Do not allow soiled glassware to dry, as this makes cleaning more difficult. If plugged glass pipettes are used, remove the cotton plugs before rinsing.
  2. Soak the glassware overnight in a detergent suitable for cell culture work (e.g., 7X, Liquinox).
  3. Wash the glassware by hand or machine. Use a siphon pipette washer for pipettes.
  4. Rinse bottles and other containers thoroughly in 10 to 12 complete changes of tap water followed by two or three complete changes of distilled, deionized water. Rinse pipettes by siphoning action of the pipette washer for a minimum of 2 h in tap water followed by three changes of distilled, deionized water.
  5. Drain glassware dry, and cap all bottles and other containers with aluminum foil. Sort pipettes by size, and plug them with cotton.

B. Sterilization
  1. Dry (hot-air oven) or moist (autoclave) heat can be used for sterilization. Dry heat is preferred, since toxic residues can be deposited on glassware during steam sterilization.
  2. Place the glassware in a hot-air oven at 160°C for 1 to 2 h. Use a sterility indicator and thermometer to monitor the load.
  3. Allow the glassware to cool, and remove it from the oven.
  4. If ethylene oxide sterilization is used, allow proper aeration as recommended by the manufacturer.

8.7 Direct Specimen Testing: Viral and Chlamydial Infections

I. DIRECT ASSAYS

Direct assays are used by most virology laboratories in conjunction with culture methods. For a summary of viral agents for which these methodologies are frequently used, see EPCM Tables 8.1–1 and 8.1–2.

A. Antigen and nucleic-acid assays (*see* Table 8.7–1)

Table 8.7–1 Antigen and nucleic-acid assays for the laboratory detection of viral and chlamydial infections[a]

Assay and principle	Specimens[b]	Representative agents
Antigen-detection assays		
Immunostaining (e.g., IF, immunoenzymatic staining) IF is the most frequently used immunostaining technique and is adaptable to detecting many different agents (4, 8, 10, 13, 24, 25, 31, 33). Antibodies directed against numerous agents are available, including those listed in this table. In direct IF, specimen cells fixed to a microscope slide are overlaid with a fluorochrome-labeled antiviral or antichlamydial[c] antibody; after incubation and washing, the cells are observed by using a fluorescence microscope to detect areas of antigen localization. With indirect IF, the primary antibody is unlabeled, and a second antibody (i.e., a labeled anti-immunoglobulin antibody) is added, followed by an additional incubation and washing step. MAbs are usually preferred because of their specificity. Limitations imposed by the very high specificity of MAbs and lower binding efficiencies are overcome by using mixtures of MAbs directed against different viral epitopes (28). Mixtures of MAbs (blends, pools) directed against two or more viruses are also available for detecting multiple viruses by using a single reagent (22, 39).	Respiratory-tract specimens (NPA, NPS, BAL, etc.)	Adenoviruses, influenza virus types A and B, measles, parainfluenza virus types 1 to 4, respiratory syncytial virus, *C. trachomatis*; pools of antibodies to several respiratory viruses are available commercially.
	Lesion samples	Herpes simplex virus types 1 and 2 (HSV-1 and -2), varicella-zoster virus (VZV)
	Endocervical swabs	*C. trachomatis*
	Blood leukocytes	CMV antigenemia assay
	Other fluids (e.g., urine, CSF), tissue	Diagnostic kits have not been extensively investigated for staining these specimens.

(continued)

8 *Viruses, Rickettsiae, Chlamydiae, Mycoplasmas*

Table 8.7–1 Antigen and nucleic-acid assays for the laboratory detection of viral and chlamydial infections[a] *(continued)*

Assay and principle	Specimens[b]	Representative agents
EIAs		
These are convenient and relatively easy to introduce into many laboratory settings but may be less sensitive than an optimized culture or IF (3, 7, 17, 40, 42, 43, 44, 45). Solid-phase EIA encompasses a number of formats and variations, including the type of solid phase to which the antigen will be immobilized, the type and format of antiviral or antichlamydial[c] antibodies, and the design of the enzyme-labeled detector system. In general, antigen in the specimen is first captured by an agent-specific antibody and immobilized to the solid phase (e.g., antibody-coated microtiter well, bead, or membrane) followed by binding of enzyme-labeled detector system to the immobilized antigen if present. Bound antigen is then detected by adding a substrate-chromophore mixture, which is converted by the enzyme to yield a colored or fluorescent compound. Confirmatory blocking assays are available for some chlamydial products; in other cases, IF confirmation may be performed on the EIA sample residue. EIAs are available as manual single-use tube or cartridge assays, as microtiter or multiwell assays using plate washers and readers, or as assays in which samples, after preparation, are loaded into reaction vessels which are then placed into automated analyzers.		
Microtiter/multiwell assays	Nasopharyngeal specimens, pharyngeal specimens	Respiratory syncytial virus (RSV), influenza virus type A, *C. trachomatis*
	Serum, plasma	Hepatitis B virus (HBV), human immunodeficiency virus type 1 (HIV-1) (including immune complex dissociation formats)
	Stool	Adenoviruses (group), enteric adenoviruses (types 40 and 41), rotaviruses
	Lesion swabs (genital, oral, dermal)	HSV (negative results may require confirmation by culture)
	Ocular (conjunctival) swabs	HSV (negative results may require confirmation by culture), *C. trachomatis*
	Female endocervical swabs, male urethral swabs, urine	*C. trachomatis*
Single-sample reaction vessels with automated detection and reading	Female endocervical swabs, male urethral swabs, urine	*C. trachomatis*
	Nasopharyngeal specimens (aspirates, washes, swabs)	RSV
	Stool	Rotaviruses

(continued)

Table 8.7–1 Antigen and nucleic-acid assays for the laboratory detection of viral and chlamydial infections[a] *(continued)*

Assay and principle	Specimens[b]	Representative agents
Manual, single-cartridge membrane assays	Nasopharyngeal specimens (aspirates, washes, or swabs)	RSV, influenza virus type A
	Stool	Rotaviruses
	Endocervical swabs	*C. trachomatis*
Manual, tube assays	Nasopharyngeal specimens	RSV
	Stool	Rotaviruses
Latex agglutination	Stool	Rotaviruses
Other antigen assays		
Manual, single-use cartridge Extracted antigen reacts with anti-LPS-coated colored-latex particles; beads carrying the captured antigen become immobilized as they pass an area of the membrane containing antichlamydial antibody; as the colored particles accumulate, a colored line becomes visible (Clearview, Wampole).	Endocervical swabs, male urethral swabs, male urine	*C. trachomatis*
OIA (optical immunoassay, Biostar) Reaction takes place on a silicon wafer and allows for the direct visual detection of the physical change in optical thickness of molecular thin films resulting from the binding reactions between antibodies and antigens. The signal is generated by the change in the reflection of light (color change) through the molecular thin films formed on an optical substrate.	Endocervical swabs	*C. trachomatis*
Nucleic acid detection assays (41, 46)		
Unamplified hybridization assays		
Chemiluminescent hybrid capture RNA probes hybridize with target DNA; hybrids are captured onto a solid phase (tube) coated with antibodies specific for RNA-DNA hybrids. Detector system uses alkaline phosphatase-conjugated antibodies and a chemiluminescent substrate.	Cervical swabs, tissue biopsies	Human papillomavirus (HPV) DNA types 6, 11, 42, 43, and 44 (low-risk types) and types 16, 18, 31, 33, 35, 45, 51, 52, and 56 (high-risk types)
	Blood leukocytes	CMV DNA
	Plasma, serum	HBV DNA
Radiological liquid hybridization After incubation of specimen with ^{125}I-labeled probes, the mixture is loaded into a column from which target-probe hybrids are eluted.	Plasma, serum	HBV DNA
Hybrid protection assay Target rRNA of *C. trachomatis* hybridizes to acridinium ester-labeled probes; unhybridized probes are degraded while hybrids are used to generate signal. Confirmatory probe competition assay available.	Endocervical, male urethral swabs, ocular swabs	*C. trachomatis*

(continued)

Table 8.7–1 Antigen and nucleic-acid assays for the laboratory detection of viral and chlamydial infections[a] *(continued)*

Assay and principle	Specimens[b]	Representative agents
Amplified assays		
Branched DNA (bDNA) Signal amplification system; target nucleic acid is captured by probes immobilized to a solid phase; bDNA amplifier is added followed by the addition of enzyme-labeled probes specific for the bDNA and a chemiluminescent substrate.	Plasma, serum	HBV DNA, HCV RNA (may be useful [16] to address difficulties in predicting and monitoring therapeutic response [5])
	Plasma	HIV-1 viral load assay (prognostic marker of disease progression and indicator of response to therapy; measures genomic RNA of virions shed into plasma [35, 37]).
Ligase chain reaction (LCR) Probe amplification assay in which product is formed by ligation of two oligonucleotide probes after their binding to adjacent complementary target sequences; product subsequently serves as target during additional cycles. Product detection is accomplished by microparticle EIA.	Endocervical swabs, male urthral swabs, male urine	*C. trachomatis* plasmid DNA
Nucleic-acid sequence-based amplification (NASBA) Isothermal target amplification assay in which a DNA intermediate is transcribed from target RNA; large amounts of RNA are produced by using the DNA intermediates as template. Product detection is based on a hybridization step and electrochemiluminescence.	Plasma and other samples	HIV-1 viral load assay (prognostic marker of disease progression and indicator of response to therapy; measures genomic RNA of virions shed into plasma [35, 37])
PCR DNA target amplification system involving repeated cycles of denaturation–primer annealing–primer extension. PCR is the most commonly applied DNA amplification procedure, and many laboratories have developed in-house assays. Commercial PCR assays are available for HIV-1, *C. trachomatis*, and enteroviruses with biotinylated primers. The biotinylated product is captured by complementary probes immobilized to wells of a microtiter plate; detection uses an enzyme-based system.	Plasma (heparin is inhibitory)	HIV-1 DNA; detects proviral DNA in leukocytes. Useful for detecting infection in infants born to HIV-1-infected mothers (32) and for resolution of infection status in individuals with consistent indeterminate Western blots.
	Endocervical swabs, male urethral swabs, urine	*C. trachomatis* plasmid DNA
	CSF, bodily fluids, etc.	Primers and reagents are commercially available for in-house development and verification. Assays are available for several viruses through reference laboratories (e.g., herpesviruses, parvovirus B-19, hantavirus). PCR is useful for viral detection in materials and clinical situations elusive to other diagnostic methods (e.g., CMV in vitreous humor fluid with retinitis [11]; CMV in CSF in peripheral [34] and central [19] nervous system disease).

(continued)

Table 8.7–1 Antigen and nucleic-acid assays for the laboratory detection of viral and chlamydial infections[a] *(continued)*

Assay and principle	Specimens[b]	Representative agents
Reverse transcription PCR (RT-PCR) Similar to PCR as described above; reverse transcription of target RNA to yield DNA templates for the PCR reaction	Plasma (heparin is inhibitory)	HIV-1 viral load assay (prognostic marker of disease progression and indicator of response to therapy; measures genomic RNA of virions shed into plasma [35, 37])
	CSF	Enteroviruses
Transcription-mediated amplification (TMA) An isothermal target amplification assay in which primers anneal to specific target sequences and, in the case of *C. trachomatis*, a 23S rRNA target is amplified via DNA intermediates. A chemiluminescent DNA probe hybridizes to amplicon to form a stable RNA-DNA hybrid which is detected by using the hybrid protection assay.	Endocervical swabs, male urethral swabs, male urine	*C. trachomatis*
In situ hybridization Useful for detecting viral infection as well as for investigating virus-cell interactions	Cells, tissues	Probes and reagents for several viruses (e.g., HPV, HSV, CMV) are available for in-house development and verification

[a] See Appendix 8.7–1 for commercial reagent sources.

[b] Refer to kit manufacturer regarding specimens for which the reagent can be used and specific instructions regarding collection, transport, and storage of specimens. NPA, nasopharyngeal aspirate; NPS, nasopharyngeal swab.

[c] MAbs are either genus- (anti-LPS) or species-specific (anti-MOMP). Anti-LPS antibodies may cross-react with bacteria present in vaginal, rectal, and oral specimens, and false-positive results have been reported with EIA testing of vaginal specimens from children (2, 15, 18, 29) and of rectal specimens from adults (30) and children (29) and with direct IF staining of rectal smears from adults and children (18, 29).

I. DIRECT ASSAYS *(continued)*

B. Electron microscopy (EM)

EM (6, 12, 23) can be used to visualize viral particles in specimens and can be very useful for studying virus-cell relationships and for attempting to assign a viral origin to a disease of unknown etiology. However, EM is generally less sensitive than other methods and is of limited availability. Negative staining, usually with phosphotungstic acid, can be used to visualize viral particles in specimen fluids and extracts containing at least 10^5 to 10^6 particles per ml; thin sectioning is useful for detecting viruses in tissues. Enhancement methods such as immunoelectron microscopy and/or concentration by ultracentrifugation increase the sensitivity of this method.

EM is useful in investigating suspected outbreaks of viral gastroenteritis caused by viruses for which other methods are not readily available (e.g., Norwalk agent, astroviruses, and caliciviruses). Collect stool within 2 to 3 days of onset of symptoms. Store and ship at 4°C; do not freeze. Contact the local health department or the Enteric Diseases Branch, CDC, Atlanta, GA 30333 (telephone 404-639-3653).

C. Cytohistopathology

Examination of exfoliated cells, biopsies, and autopsy tissues may reveal nonspecific changes suggestive of viral infections, and several viruses are associated with the development of characteristic inclusion bodies (20, 38; Table 8.7–2). Cytohistopathology is generally less sensitive than other viral and chlamydial detection methods. However, this method is useful for detecting agents for which other diagnostic assays are not available (e.g., molluscum contagio-

Table 8.7–2 Inclusion morphology

Agent	Inclusion
CMV (Fig. 8.7–1A)	Enlarged (cytomegalic) cells with a large single basophilic intranuclear inclusion that nearly fills the nucleus and is surrounded by a clear halo (owl's eye appearance); associated with small Pap-positive cytoplasmic inclusions
C. trachomatis (Fig. 8.7–1B)	Giemsa- or Papanicolaou-stained smears may reveal typical intracytoplasmic inclusions. Sensitive method for examination of ocular specimens, for which the sensitivity has been estimated to approach 95% compared with culture, but insensitive for urogenital samples
HSV, VZV	Large eosinophilic Cowdry type A intranuclear inclusion with a halo of nuclear chromatin (mature form); associated with multinuclearity with indented (molded) nuclei. The Tzanck smear (Fig. 8.7–1C) is a classic method involving Giemsa staining of epithelial cells scraped from the lesion base.
HPV	Squamous cells appear swollen and contain an enlarged nucleus and perinuclear halo with poor cytoplasmic keratinization resulting in irregular staining (koilocytic change).
Measles	Large multinucleated giant cells with eosinophilic nuclear and cytoplasmic inclusions; high sensitivity for giant-cell pneumonia in immunocompromised patients (Warthin-Finkelday inclusions)
Molluscum contagiosum (Fig. 8.7–1D)	Large dense intracytoplasmic eosinophilic inclusions (molluscum bodies) occupying the entire squamous cell and resulting in peripheral displacement of a flattened nucleus
Rabies	Seller's stain reveals large eosinophilic cytoplasmic inclusions (Negri body) often with blue-staining granules or inner bodies are often arranged in concentric layers; several inclusion bodies, usually of variable size, may be present in one neuron.

I. DIRECT ASSAYS *(continued)*

sum) and to demonstrate organ or tissue involvement. For example, cytomegalovirus (CMV) inclusions in alveolar epithelial cells may be a better indicator of pulmonary CMV disease than positive culture or polymerase chain reaction (PCR) results on BAL samples.

II. ASSAY SELECTION

Factors important in the selection of direct assays include cost, turnaround time, ease of performance, suitability for work flow patterns, analytical sensitivity and specificity, and clinical sensitivity and specificity. In selecting assays it is essential that the laboratorian not rely solely on the manufacturer's marketing data regarding these performance characteristics (9, 36). Assessment of assay performance should include a critical review of published studies and consultation with full-service virology laboratories having extensive experience with these assays. Familiarity with gold-standard methods and appreciation of how variables may affect evaluations are essential.

III. SPECIMENS

A. Inadequate or inappropriate specimens compromise assay sensitivity and specificity and increase the number of uninterpretable specimens. Select, collect, transport, and store specimens as instructed by the manufacturer. If specific instructions are not provided or if it is recommended that specimens be collected in the usual manner, collect specimens as described for viral culture (*see* procedure 8.3 for details).

B. Direct assays do not require viable or infectious organisms, and thus cold, rapid transport to the laboratory is not always necessary. However, many direct assays specify the need for rapid and/or cold transport.

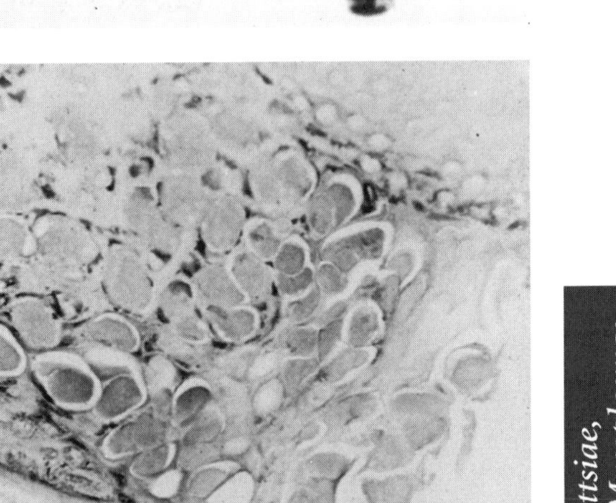

Figure 8.7–1 (A) Kidney tubule with characteristic "owl eye" inclusion of CMV (arrow). Courtesy of W. L. Drew. (B) Giemsa stain of conjunctival cells showing cytoplasmic inclusion (arrow) of *C. trachomatis.* Courtesy of W. L. Drew. (C) Cytologic examination (Tzanck smear) of scrapings from the base of an ulcerative HSV lesion showing multinucleated giant cells. Courtesy of W. L. Drew. (D) Large eosinophilic inclusions fill the cytoplasm of several infected basal cells (molluscum bodies). Courtesy of W. L. Drew.

IV. QUALITY CONTROL, VERIFICATION, AND VALIDATION OF DIRECT ASSAYS

A. Include a positive and negative control in each run; record control results and reagent lot number(s) and expiration date(s) for each run.

B. Do not mix reagents from different lots, follow recommended storage guidelines, and do not use materials beyond their expiration dates.

C. In addition to the manufacturer's controls, it is recommended that a well-characterized positive and negative external control be included in each run or at least weekly. Prepare external controls by using specimens or proficiency samples. Establish acceptable limits (26) by performing replicate testing; aliquot and store samples frozen in single-use aliquots. A weakly reactive positive control is more effective than a strong positive in monitoring assay performance and product deterioration.

D. Perform lot-to-lot testing of kits by overlapping controls using the old and new lots.

E. Perform verification of new kits or test systems of moderate or high complexity before introducing them into the laboratory (9), and address other regulatory agency requirements as necessary. Assay verification can be accomplished by comparing the candidate assay with the reference or gold-standard assay with a panel of positive (e.g., 20) and negative (e.g., 50) specimens (9). If the perfor-

IV. QUALITY CONTROL, VERIFICATION, AND VALIDATION OF DIRECT ASSAYS *(continued)*

mance characteristics stated by the manufacturer are not achieved, troubleshooting and corrective action must be taken by the user and/or the manufacturer, and the verification process must be repeated. More extensive verification is recommended for assays developed in-house, for modified procedures, and for methods that have not been cleared by the U.S. Food and Drug Administration.

F. Perform validation (9) of ongoing assays to monitor quality consistency. For example:

1. Plot results obtained with the weakly reactive and negative controls; periodically test panels of positive and negative samples, particularly if a problem is suspected.
2. Enroll in proficiency-testing programs and develop in-house programs with blinded surveillance samples.

G. Determine that individuals performing assays are and remain competent in performing the assay (27).

H. Perform equipment maintenance and calibrations as recommended by the manufacturer.

V. ASSAY PERFORMANCE

A. EIAs, nucleic-acid-hybridization assays, and nucleic-acid-amplification assays
Kits include manufacturer-directed protocols that should be followed without modification; thus, procedures for these assays are not presented here.

B. Immunofluorescence (IF) assays
Perform based on the manufacturer's protocol, if applicable. When using a reagent without a manufacturer's protocol, follow the procedure presented below.

C. Modification of any aspect of the manufacturer's protocol, including specimen selection, collection, assessment, or storage, requires the user to perform appropriate assay verification (9).

VI. REPORTS

A. Follow manufacturer's recommendations regarding reporting criteria. Develop reporting criteria for IF testing (*see* item VII.C below).

B. Include the test methodology in the report and provide an interpretation, if applicable.

C. Include a qualifier if appropriate, e.g., "for investigational use only."

VII. IF PROCEDURE

A. Supplies, reagents, and equipment

1. Sterile individually wrapped pipettes (1, 2, 5, and 10 ml) and safety pipetting device
2. Waste containers
   Discard specimens, cell cultures, and other infectious waste based on local regulations.
3. Acetone-cleaned, Teflon-coated microscope slides with 5- to 8-mm wells and no. 1 coverslips (e.g., Cel-Line Associates, Inc., Newfield, N.J.)
4. Coplin jars or staining dishes and racks
5. Slide storage boxes and desiccant
6. Scalpels, tongue depressors, or spatulas (for tissue impression smears)
7. Personal protective gear
8. Most antibody reagents used for IF assays are monoclonal antibodies (MAbs); because of their high specificity and lower binding efficiencies, a mixture of MAbs directed against different viral epitopes is generally more useful than individual MAb preparations. Mixtures of MAbs (blends or pools) directed against two or more viruses are also available for detecting

multiple viruses by using a single reagent. Recently, mixed-antibody preparations labeled with different fluorescing reagents have been introduced for the simultaneous identification of common virus pairs (e.g., influenza virus types A and B, herpes simplex virus [HSV], and CMV).

9. Phosphate-buffered saline (PBS) (pH 7.1 to 7.6)
10. Acetone (reagent grade) or fixative recommended by reagent manufacturer (e.g., methanol, ethanol, paraformaldehyde)
11. Buffered-glycerol mounting medium, pH 8.0 for IF
12. Cytofuge and funnels, optional, for cytospin preparations
13. Centrifuge, low speed with carriers to hold 15-ml centrifuge tubes
14. Rotator (15 to 20 rpm) or rocker
15. Fluorescence microscope equipped with filters suitable for fluorescein isothiocyanate (FITC) or fluorochrome of choice
16. Refrigerator
17. Incubator, 37°C

B. QC

1. Include positive and negative cell spots in each run.
2. Unless predetermined and stated by the manufacturer, determine the optimal working dilution to use for antibodies, and conjugate and assess cross-reactivity with other viruses (*see* CMPH procedure 8.9, Appendix 1, for details).

C. Cell spot preparation

1. Aspirates and fluids containing cells (other than blood)
   Follow the procedure described in Fig. 8.7–2. Alternatively, cytospin samples may be prepared following the cytofuge manufacturer's instructions.
2. Swab specimens or scrapings
   Follow the procedure described in Fig. 8.7–3.
   NOTE: Swab specimens or scrapings may be placed into 1 to 2 ml of viral transport medium or PBS. After vortexing to release cells from the swab, the specimen may be processed in the same manner as aspirates and fluids; cell yield may be low for specimens processed in this manner.
3. Blood
   Kits are commercially available for performance of the CMV antigenemia assay for detection of pp65, a CMV lower-matrix phosphoprotein, in the nuclei of peripheral blood PMNs (Fig. 8.7–4).
   a. Collect blood with anticoagulant (e.g., EDTA, heparin) as recommended by manufacturer. Store blood at room temperature before cell separation; for optimal results prepare slides within several hours of specimen collection.
   b. There are four basic steps to the assay.
      (1) Isolation of the leukocytes from blood by dextran sedimentation
      (2) Preparation of a slide by using either a cell spotting or cytospin method
      (3) Immunostaining
      (4) Microscopic examination of cells
   c. Though acetone is frequently used for fixation of cells in viral staining procedures, studies have compared the use of other fixatives. Fixation with paraformaldehyde was reported to yield a greater number of positive cells and greater staining intensity than fixation with methanol-acetone (13) or acetone (1). In another study, acetone fixation yielded a greater staining intensity of positive cells, but cell morphology was better after fixation with paraformaldehyde (21).
   d. The test can be applied quantitatively by applying a known number of leukocytes to the slides; however, it is important to determine whether

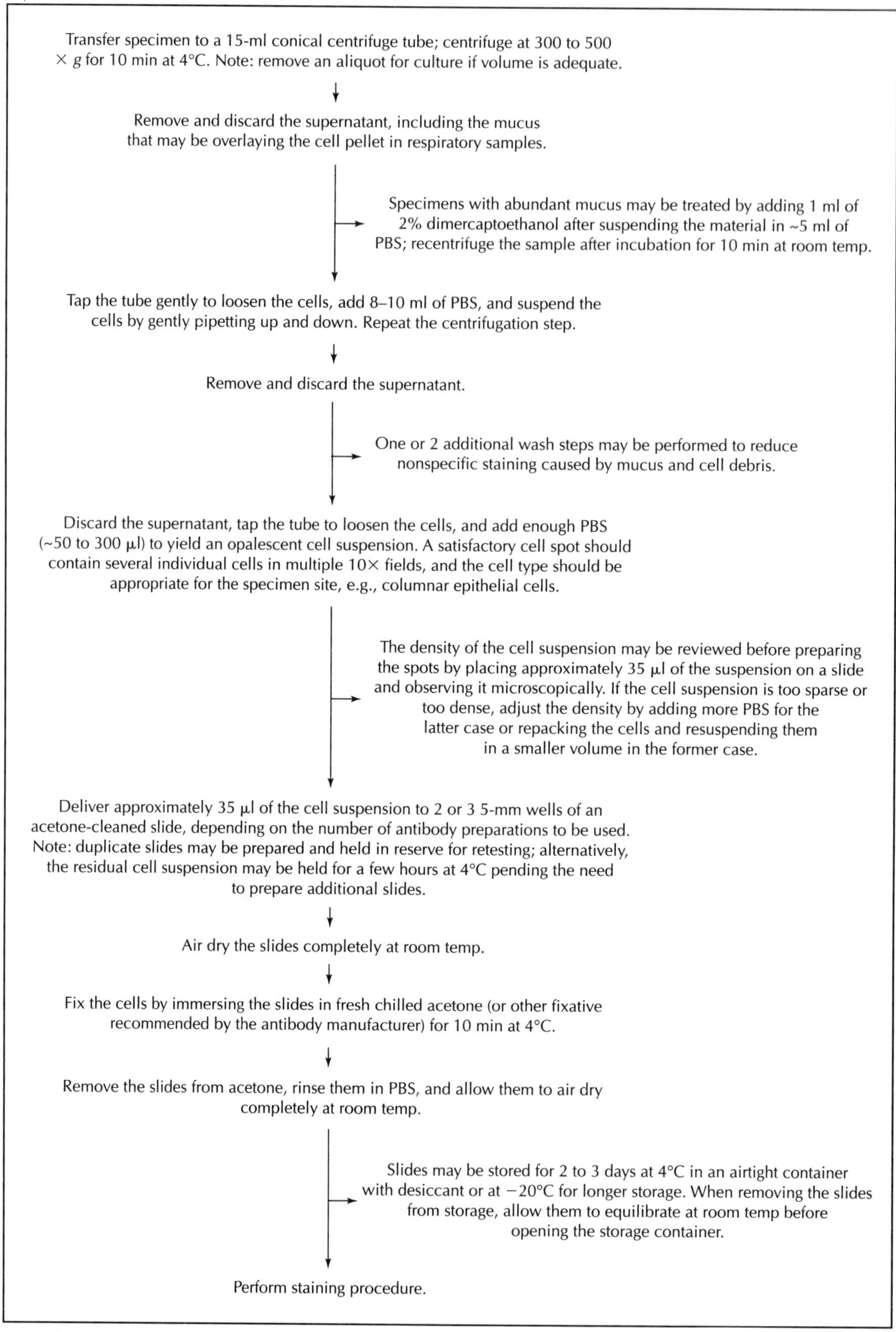

Transfer specimen to a 15-ml conical centrifuge tube; centrifuge at 300 to 500 × *g* for 10 min at 4°C. Note: remove an aliquot for culture if volume is adequate.

↓

Remove and discard the supernatant, including the mucus that may be overlaying the cell pellet in respiratory samples.

↓ → Specimens with abundant mucus may be treated by adding 1 ml of 2% dimercaptoethanol after suspending the material in ~5 ml of PBS; recentrifuge the sample after incubation for 10 min at room temp.

Tap the tube gently to loosen the cells, add 8–10 ml of PBS, and suspend the cells by gently pipetting up and down. Repeat the centrifugation step.

↓

Remove and discard the supernatant.

↓ → One or 2 additional wash steps may be performed to reduce nonspecific staining caused by mucus and cell debris.

Discard the supernatant, tap the tube to loosen the cells, and add enough PBS (~50 to 300 μl) to yield an opalescent cell suspension. A satisfactory cell spot should contain several individual cells in multiple 10× fields, and the cell type should be appropriate for the specimen site, e.g., columnar epithelial cells.

↓ → The density of the cell suspension may be reviewed before preparing the spots by placing approximately 35 μl of the suspension on a slide and observing it microscopically. If the cell suspension is too sparse or too dense, adjust the density by adding more PBS for the latter case or repacking the cells and resuspending them in a smaller volume in the former case.

Deliver approximately 35 μl of the cell suspension to 2 or 3 5-mm wells of an acetone-cleaned slide, depending on the number of antibody preparations to be used. Note: duplicate slides may be prepared and held in reserve for retesting; alternatively, the residual cell suspension may be held for a few hours at 4°C pending the need to prepare additional slides.

↓

Air dry the slides completely at room temp.

↓

Fix the cells by immersing the slides in fresh chilled acetone (or other fixative recommended by the antibody manufacturer) for 10 min at 4°C.

↓

Remove the slides from acetone, rinse them in PBS, and allow them to air dry completely at room temp.

↓ → Slides may be stored for 2 to 3 days at 4°C in an airtight container with desiccant or at −20°C for longer storage. When removing the slides from storage, allow them to equilibrate at room temp before opening the storage container.

Perform staining procedure.

Figure 8.7–2 Preparation of aspirates and fluids for IF.

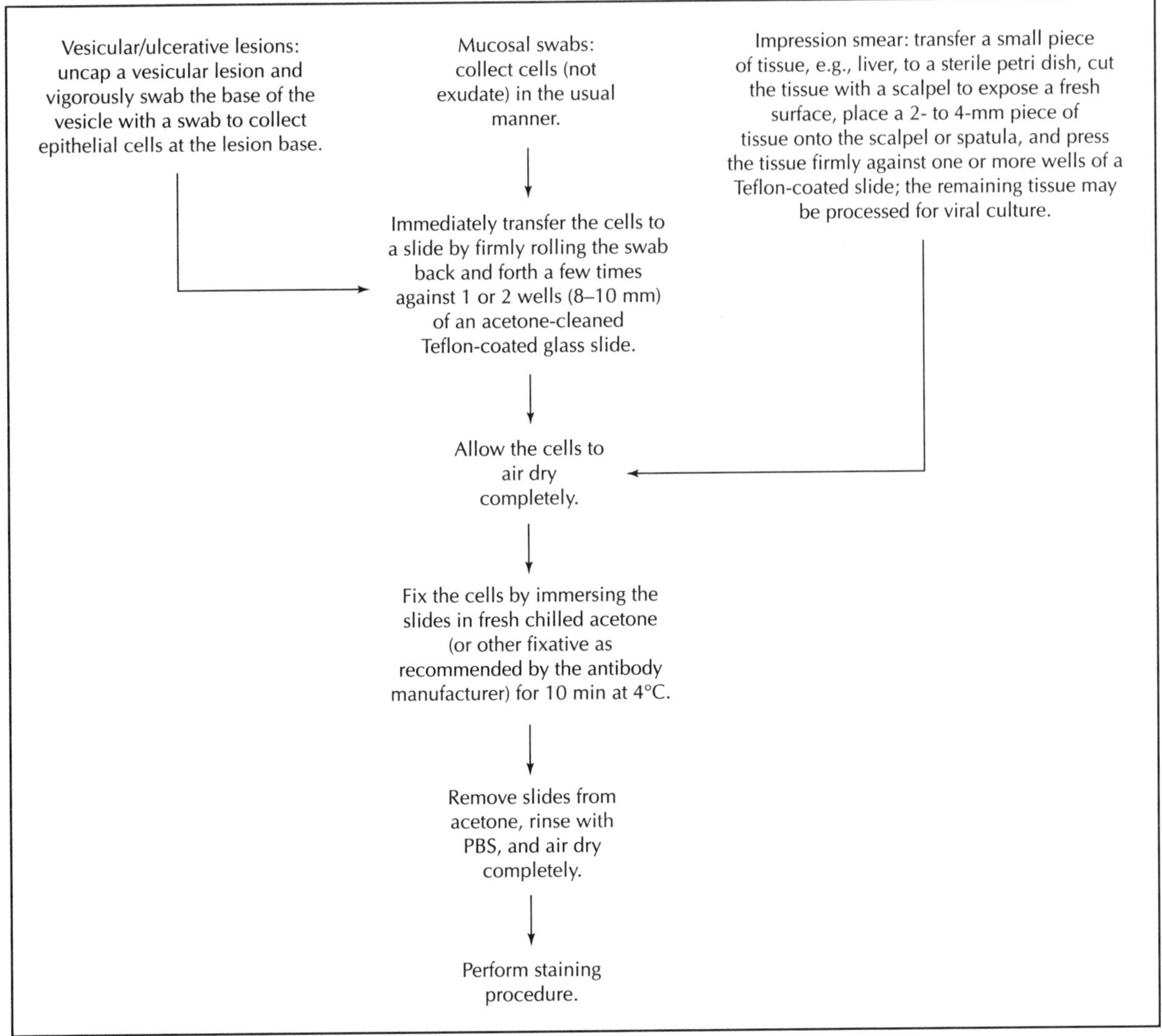

Figure 8.7–3 Preparation of swabs, scrapings, and impression smears for IF.

VII. IF PROCEDURE *(continued)*

an adequate number of cells are actually present at the time of reading. The number of cells can be estimated visually with experience; the number can also be estimated by inserting a micrometer disk in the ocular lens of the microscope and counting the cells in several representative fields. The average number of cells per field multiplied by the number of fields per well will give an estimate of the number of cells.

4. Tissues

a. Prepare impression smears as described in Fig. 8.7–3.

b. Prepare cell spots from tissue homogenates as described for aspirates in Fig. 8.7–2. Homogenates may be prepared by using a tissue grinder (Fig. 8.4–3) or a Tekmar Stomacher tissue grinder (*see* CMPH procedure 8.3).

c. Acetone- or formalin-fixed frozen sections 4 to 6 μm thick are suitable for staining.

d. Paraffin-embedded tissues may be used after deparaffinization.

D. IF staining

Follow the procedure described in Fig. 8.5–7.

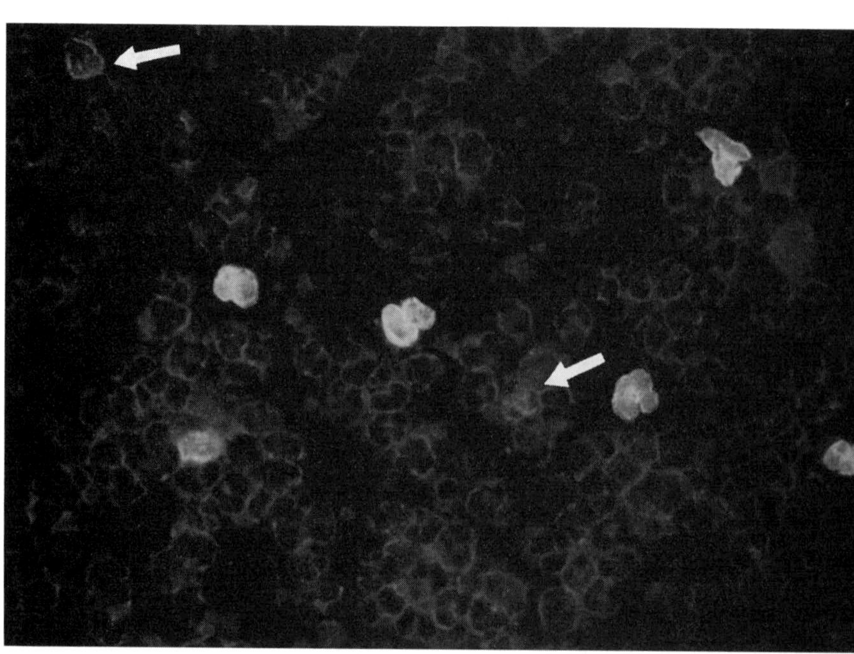

Figure 8.7–4 Leukocyte preparation with several cells showing intense CMV pp65 staining and a few cells (arrows) showing weak speckled nuclear staining (×400). Provided by Chemicon International.

VII. IF PROCEDURE *(continued)*

E. Observation of viral antigens

Use a 10× objective to focus the slide. Then examine the entire cell spot at 20×. Cell morphology should be clear, with nuclear and cytoplasmic details defined. Use a higher magnification (×40 or ×63) for closer examination of cellular detail and for examining questionable areas.

1. Assess control slides before examining test slides; repeat the assay if the controls do not yield the expected results. Refer to Table 8.7–3 for troubleshooting guidelines.
2. The positive control slide (virus-infected cells with homologous antibody) should have numerous cells with characteristic and intense (3+ to 4+) apple-green (for FITC) fluorescence. The number or percentage of fluorescing cells can also be determined, if desired.
3. The negative control should exhibit no (0) or minimal dim (<1+) fluorescence.
4. A positive specimen is one in which one or more intact cells exhibit a characteristic staining pattern with a 2+ or greater staining intensity.

   ◪ **NOTE:** Slides with weak staining and/or a single positive cell might be considered equivocal.
5. A negative specimen is one that contains an adequate number of cells typical of the sampling site and in which all cells exhibit minimal (<1) or no (0) fluorescence.
6. An unsatisfactory specimen is one in which strong nonspecific staining interferes with reading, or the slide contains too few intact cells typical of the site. Clinical samples should contain several cells in several 10× fields. Prepare new cell spots from the residual sample, and repeat the assay or request a new sample.

Table 8.7–3 Troubleshooting IF problems (reprinted from CMPH 8.9)

Problem	Causes	Solution
Weak or no specific fluorescence including positive controls	Wrong (too weak) concn of immunoreagent used	Retitrate reagent.
	Deterioration of reagent	Retitrate and/or replace. Store properly, and aliquot if necessary.
	Counterstain too strong	Review concn and counterstaining time.
	Rapid fading of fluorescence	pH of mounting medium must be greater than that of wash buffer; use mounting medium that contains photobleaching inhibitor.
	Wrong filters or inadequate light source	Review use and maintenance of equipment; replace light bulb.
	Cannot achieve sharp focus because of dirty or improperly focused optics	Review procedures for maintenance and use of optics.
	Nonspecific glare or haze masking specific fluorescence	Allow cell spots to air dry completely before mounting; use fresh acetone.
	Antigens destroyed by fixative	Use alternative fixative
Weak or no fluorescence with test slides but positive controls acceptable	Controls do not adequately reflect actual test conditions	Review procedure. Determine whether same or different reagents or methods were used for test and control slides.
	Improper slide cleaning, preparation, or fixation	Review procedures.
Nonspecific fluorescence and/or false positives with test slides and negative controls	Wrong concn (too strong) of immunoreagent used	Retitrate reagent.
	Cross-reactive immunoreagent	Identify non-cross-reactive concn of reagent, or obtain a better product.
	Binding of antibody via Fc receptors	Use conjugated $F(ab')_2$ fragments.
	Trapping of immunoreagent on heavy or raised cell preparations	Avoid heavy cell preparations, which obscure cellular morphology and complicate reading. Increase number of washes during staining.
	Inappropriate immersion oil	Use immersion oil designated for fluorescence microscopy.
	Inadequate removal of fixative	Air dry cell spots completely after fixing. Rinse slides with PBS and/or water just prior to staining.
	Water in acetone	Replace acetone.
	Slides dried during staining	Use properly sealed moist chamber during incubation. Apply adequate reagent volumes to cell spots.
	Mounting medium not appropriate for IF, or pH too low	Obtain proper reagents. pH of mounting medium must be greater than that of wash buffer.
	Mounting medium applied to wet cells	Allow cell spots to air dry completely before mounting.
	Wrong filters or equipment	Review use and maintenance of equipment.
	Reading error	Must be able to differentiate specific patterns of fluorescence from nonspecific staining.
	Precipitated material in immunoreagent	Microcentrifuge reagent for 1 min, or filter through a 0.45-μm-pore-size filter.
Nonspecific fluorescence and/or false positives with test slides but negative controls acceptable	Controls do not adequately reflect test conditions	Review procedure. Determine whether same or different reagents or methods were used for test and control slides.
	Reading error	More reading experience needed, since controls may be easier to read and interpret than test slides.
	Reagent cross-contamination	Take care to confine reagents to wells.
	Inoculum debris or microorganisms on test slide	Increase number of washes.

VII. IF PROCEDURE *(continued)*

☑ **NOTE:** The specific pattern of fluorescence depends on several variables, including the viral protein(s) being targeted and the type of antibody preparation (e.g., single MAb, mixed MAbs, or polyclonal antiserum). The expected pattern with a particular antibody preparation should be provided by the manufacturer. Positive controls, provided that the cells represent those in the test slides, are the best tool for determining characteristic staining. Refer to Fig. 8.5–9 in procedure 8.5 for examples of IF staining.

F. Observation for *C. trachomatis* elementary bodies (EBs) (staining for other chlamydial species may be performed in the same manner)

☑ **NOTE:** Fix slides for *C. pneumonia* by using acetone, not methanol.

Extracellular EBs are approximately 300 nm in diameter and appear as bright apple-green punctate structures; EBs show a smooth edge when stained with anti-major outer membrane protein (MOMP) antibodies. Chlamydial reticulate bodies or intermediate forms of the organism may stain with a peripheral halo and are larger than EBs; do not include these structures in the EB counts.

1. Examine control slides and determine whether they are acceptable before observing specimen slides. Examine slides at 40×; use 63× or 100× to facilitate confirmation of suspected EBs (glycerol or oil objectives will facilitate reading).
2. A positive specimen is one that contains a quantity of EBs equal to or greater than the established cutoff number (e.g., 1, 2, 5).
3. A negative specimen is one that contains an adequate number of columnar epithelial cells and no EBs.

   ☑ **NOTE:** Report as positive for *C. trachomatis* if an anti-MOMP MAb is used; report as positive for *Chlamydia* spp. if an anti-lipopolysaccharide (LPS) antibody is used.
4. An equivocal sample is one that contains fewer EBs than the established positive cutoff number.
5. An unsatisfactory sample is one in which EBs were not observed but which exhibits strong nonspecific staining that interferes with reading or that contains an insufficient number of cells. Specimen slides should contain two to three columnar epithelial cells per high dry field.

G. IF procedure notes

1. Wetting cell spots with deionized water or buffer just before staining them may facilitate distribution of reagent and improve staining quality.
2. Counterstain is frequently incorporated into commercial conjugates. For in-house preparation, Evans blue diluted to 0.01% may be incorporated in the final wash buffer; negative cells will appear dull red. Evans blue is a potential carcinogen. If skin contact occurs, flush with water immediately and dispose of reagent based on local regulations.
3. A quantity (e.g., 0.1%) of sodium azide may be included in reagents as a preservative. Sodium azide may produce explosive metal azides in lead or copper plumbing. For disposal, the reagents should be flushed with large volumes of water. Occasional decontamination of the drain system with 10% sodium hydroxide is recommended; sodium hydroxide is a caustic solution. Flush with large volumes of water if contact with skin or eyes occurs, and contact a physician immediately.
4. Thimerosal may be used as a preservative and is toxic. Avoid ingestion and dispose of based on local regulations.
5. Paraformaldehyde is a highly toxic, allergenic, and potentially carcinogenic reagent which should be handled by using a fume hood; avoid skin or eye contact.
6. Handle staining reagents aseptically since microbial contamination will affect performance.

7. Acetone fixation may not completely inactivate viral infectivity. Thus, handle slides as potentially infectious materials.
8. Follow the manufacturer's directions for the selection of barrier and excitation filters.
9. An epifluorescence illumination system is simpler to use than a transmitted-light fluorescence system, since the need for a dark-field condenser is eliminated with epifluorescence. Since the objective functions as the condenser, objectives with the highest N.A. can be used, so that the images are brighter. In addition, oil or glycerol objectives facilitate reading by providing heightened fluorescence.

REFERENCES

1. **Boeckh, M., P. M. Woogerd, T. Stevens-Ayers, C. G. Ray, and R. A. Bowden.** 1994. Factors influencing detection of quantitative cytomegalovirus antigenemia. *J. Clin. Microbiol.* **32:**832–834.
2. **CDC.** 1991. False-positive results with the use of chlamydia tests in the evaluation of suspected sexual abuse. *Morbid. Mortal. Weekly Rep.* **39:**932–935.
3. **Clarke, L. M., M. F. Sierra, B. Daidone, J. Covino, N. Lopez, and W. M. McCormack.** 1993. Comparison of Syva Micro Trak EIA and Gen-Probe PACE 2 to cell culture for detection of *Chlamydia trachomatis* in cervical specimens. *J. Clin. Microbiol.* **31:**968–971.
4. **Coffin, S. E., and R. L. Hodinka.** 1995. Utility of direct immunofluorescence and virus culture for detection of varicella-zoster virus in skin lesions. *J. Clin. Microbiol.* **33:** 2792–2795.
5. **Cuthbert, J. A.** 1994. Hepatitis C: progress and problems. *Clin. Microbiol. Rev.* **7:** 505–532.
6. **Doane, F. W., and N. Anderson.** 1987. *Electron Microscopy in Diagnostic Virology: Practical Guide and Atlas.* Cambridge University Press, New York.
7. **Dominguez, E. A., L. H. Taber, and R. B. Couch.** 1993. Comparison of rapid diagnostic techniques for respiratory syncytial and influenza A virus respiratory infections in young children. *J. Clin. Microbiol.* **31:**2286–2290.
8. **Drew, W. L., and L. Mintz.** 1980. Rapid diagnosis of varicella-zoster virus infection by direct immunofluorescence. *Am. J. Clin. Pathol.* **73:**699–701.
9. **Elder, B. L., S. A. Hansen, J. A. Kellogg, R. J. Marsik, and R. J. Zabransky.** 1997. *Cumitech 31. Verification and Validation of Procedures in the Clinical Microbiology Laboratory.* Coordinating ed., B. W. McCurdy. American Society for Microbiology, Washington, D.C.
10. **Emmons, R. W., and J. L. Riggs.** 1977. Application of immunofluorescence to diagnosis of viral infections. *Methods Virol.* **6:**1–28.
11. **Fenner, T. E., J. Garweg, F. T. Hufert, M. Boehnke, and H. Schmitz.** 1991. Diagnosis of human cytomegalovirus-induced retinitis in human immunodeficiency virus type 1-infected subjects by using the polymerase chain reaction. *J. Clin. Microbiol.* **29:**2621–2622.
12. **Fong, C. K. Y.** 1994. Electron microscopy and immunoelectron microscopy, p. 99–107. *In* G. D. Hsiung, C. K. Y. Fong, and M. L. Landry (ed.), *Hsiung's Diagnostic Virology,* 4th ed. Yale University Press, New Haven, Conn.
13. **Gerna, G., M. G. Revello, E. Percivalle, and F. Morini.** 1992. Comparison of different immunostaining techniques and monoclonal antibodies to the lower matrix phosphoprotein (pp65) for optimal quantitation of cytomegalovirus antigenemia. *J. Clin. Microbiol.* **30:** 1045–1048.
14. **Gardner, P. S.** 1986. Immunofluorescence, p. 95–109. *In* S. Specter and G. Lancz (ed.), *Clinical Virology Manual.* Elsevier, New York.
15. **Goudswaard, F., L. Sabbe, and C. van Belzen.** 1989. Interference by gram-negative bacteria in the enzyme immunoassay for detecting *Chlamydia trachomatis. J. Infect. Dis.* **18:** 94–96.
16. **Gretch, D. R., C. dela Rosa, R. L. Carithers, Jr., R. A. Wilson, B. Williams, and L. Corey.** 1995. Assessment of hepatitis C viremia using molecular amplification technologies: correlations and clinical implications. *Ann. Intern. Med.* **123:**321–329.
17. **Halstead, D. C., S. Todd, and G. Fritch.** 1990. Evaluation of five methods for respiratory syncytial virus detection. *J. Clin. Microbiol.* **28:**1021–1025.
18. **Hammerschlag, M. R., P. J. Rettig, and M. E. Shields.** 1988. False positive results with the use of chlamydia antigen detection tests in the evaluation of suspected sexual abuse in children. *Pediatr. Infect. Dis. J.* **7:**11–14.
19. **Holland, N. R., C. Power, V. P. Mathews, J. D. Glass, M. Forman, and J. C. McArthur.** 1994. Cytomegalovirus encephalitis in acquired immunodeficiency syndrome (AIDS). *Neurology* **44:**507–514.
20. **Hsiung, G. D.** 1994. Histocytochemical staining, p. 83–90. *In* G. D. Hsiung, C. K. Y. Fong, and M. L. Landry (ed.), *Hsiung's Diagnostic Virology,* 4th ed. Yale University Press, New Haven, Conn.
21. **Landry, M. L., and D. Ferguson.** 1993. Comparison of quantitative cytomegalovirus antigenemia assay with culture methods and correlation with clinical disease. *J. Clin. Microbiol.* **31:**2851–2856.

REFERENCES *(continued)*

22. **Matthey, S., D. Nicholson, S. Ruhs, B. Alden, M. Knock, K. Schultz, and A. Schmuecker.** 1992. Rapid detection of respiratory viruses by shell vial culture and direct staining by using pooled and individual monoclonal antibodies. *J. Clin. Microbiol.* **30:** 540–544.
23. **Miller, S. E.** 1986. Detection and identification of viruses by electron microscopy. *J. Electron. Microsc. Technol.* **4:**265–301.
24. **Minnich, L. L., F. Goodenough, and C. G. Ray.** 1991. Use of immunofluorescence to identify measles virus infections. *J. Clin. Microbiol.* **29:**1148–1150.
25. **Minnich, L. L., T. F. Smith, and C. G. Ray.** 1988. *Cumitech 24. Rapid Detection of Viruses by Immunofluorescence.* Coordinating ed., S. Specter. American Society for Microbiology, Washington, D.C.
26. **NCCLS.** 1994. *Molecular Diagnostic Methods for Infectious Diseases.* Proposed guideline MM3-P. NCCLS, Wayne, Pa.
27. **NCCLS.** 1995. *Training Verification of Laboratory Personnel.* Approved guideline GP10-A. NCCLS, Wayne, Pa.
28. **Payne, W. J., Jr., D. L. Marshall, R. K. Shockley, and W. J. Martin.** 1988. Clinical laboratory applications of monoclonal antibodies. *Clin. Microbiol. Rev.* **1:**313–329.
29. **Poder, K., N. Sanchez, P. M. Robin, M. McHugh, and M. R. Hammerschlag.** 1989. Lack of specificity of Chlamydiazyme for detection of vaginal chlamydial infection in prepubertal girls. *Pediatr. Infect. Dis. J.* **8:** 358–360.
30. **Pratt, B. C., I. A. Tait, and W. I. Anyaegbunam.** 1989. Rectal carriage of *Chlamydia trachomatis* in women. *J. Clin. Pathol.* **42:** 1309–1310.
31. **Ray, C. G., and L. L. Minnich.** 1987. Efficiency of immunofluorescence for rapid detection of common respiratory viruses. *J. Clin. Microbiol.* **25:**355–357.
32. **Rogers, M. F., C.-Y. Ou, M. Rayfield, P. A. Thomas, E. E. Schoenbaum, E. Abrams, K. Krasinski, P. A. Selwyn, J. Moore, A. Kaul, K. T. Grimm, M. Bamji, G. Schochetman, and the New York City Collaborative Study of Maternal HIV Transmission and Montefiore Medical Center HIV Perinatal Transmission Study Group.** 1989. Use of the polymerase chain reaction for early detection of the proviral sequences of human immunodeficiency virus in infants born to seropositive mothers. *N. Engl. J. Med.* **320:** 1649–1654.
33. **Rossier, E., H. R. Miller, and P. H. Phipps.** 1989. *Rapid Viral Diagnosis by Immunofluorescence: an Atlas and Practical Guide.* University of Ottawa Press, Ottawa.
34. **Roullet, E., V. Assuerus, J. Gozlan, A. Ropert, G. Said, M. Baudrimont, M. El Amrani, C. Jacomet, C. Duvivier, G. Gonzales-Canali, M. Kirstetter, M.-C. Meyohas, O. Picard, and W. Rozenbaum.** 1994. Cytomegalovirus multifocal neuropathy in AIDS: analysis of 15 consecutive cases. *Neurology* **44:**2174–2182.
35. **Saag, M. S., M. Holodniy, D. R. Kuritzkes, W. A. O'Brien, R. Coombs, M. E. Poscher, D. M. Jacobsen, G. M. Shaw, D. D. Richman, and P. A. Volberding.** 1996. HIV viral load markers in clinical practice. *Nature Med.* **2:**625–629.
36. **Schachter, J., W. E. Stamm, M. A. Chernesky, K. W. Hook III, R. B. Jones, F. N. Judson, J. A. Kellogg, B. LeBar, P.-A. Mardh, W. M. McCormack, T. C. Quinn, G. L. Ridgway, and D. Taylor-Robinson.** 1992. Nonculture tests for genital tract chlamydial infection. What does the package insert mean, and will it mean the same thing tomorrow? *Sex. Transm. Dis.* **19:**243–245.
37. **Shearer, W. T., T. C. Quinn, P. LaRussa, J. F. Lew, L. Mofenson, S. Almy, K. Rich, E. Handelsman, C. Diaz, M. Pagano, V. Smeriglio, and L. A. Kalish.** 1997. Viral load and disease progression in infants infected with human immunodeficiency virus type 1. *N. Engl. J. Med.* **336:**1337–1342.
38. **Smith, R. D., and K. Sutherland.** 1992. The cytopathology of viral infections, p. 71–88. *In* S. Specter and G. Lancz (ed.), *Clinical Virology Manual,* 2nd ed. Elsevier, New York.
39. **Stout, C., M. D. Murphy, S. Lawrence, and S. Julian.** 1989. Evaluation of a monoclonal antibody pool for rapid diagnosis of respiratory viral infections. *J. Clin. Microbiol.* **27:** 448–452.
40. **Swierkosz, E. M., R. Flanders, L. Melvin, J. D. Miller, and M. K. Kline.** 1989. Evaluation of the Abbott TESTPACK RSV enzyme immunoassay for detection of respiratory syncytial virus in nasopharyngeal swab specimens. *J. Clin. Microbiol.* **27:**1151–1154.
41. **Tenover, F. C.** 1989. *DNA Probes for Infectious Diseases.* CRC Press, Boca Raton, Fla.
42. **Thomas, E. E., and L. E. Book.** 1991. Comparison of two rapid methods for detection of respiratory syncytial virus (RSV) (TestPack RSV and Ortho RSV ELISA) with direct immunofluorescence and virus isolation of the diagnosis of pediatric RSV infection. *J. Clin. Microbiol.* **29:**632–635.
43. **Todd, S. J., L. Minnich, and J. L. Waner.** 1995. Comparison of rapid immunofluorescence procedure with TestPack RSV and Directigen FLU-A for diagnosis of respiratory syncytial virus and influenza A virus. *J. Clin. Microbiol.* **33:**1650–1651.
44. **Verano, L., and F. J. Michalski.** 1995. Comparison of direct antigen enzyme immunoassay, Herpcheck, with cell culture for detection of herpes simplex virus from clinical specimens. *J. Clin. Microbiol.* **33:**1378–1379.
45. **Waner, J. L., N. J. Whitehurst, S. J. Todd, H. Shalaby, and L. V. Wall.** 1990. Comparison of Directigen RSV with viral isolation and direct immunofluorescence for the identification of respiratory syncytial virus. *J. Clin. Microbiol.* **28:**480–483.
46. **Wolcott, M. J.** 1992. Advances in nucleic acid-based detection methods. *Clin. Microbiol. Rev.* **5:**370–386.

SUPPLEMENTAL READING

Gleaves, C. A., R. L. Hodinka, S. L. G. Johnston, and E. M. Swierkosz. 1994. *Cumitech 15A. Laboratory Diagnosis of Viral Infections.* Coordinating ed., E. J. Baron. American Society for Microbiology, Washington, D.C.

Keller, W. W. 1992. Detection and identification of viruses by immunofluorescence, p. 8.9.1–8.9.10. *In* H. D. Isenberg (ed.), *Clinical Microbiology Procedures Handbook,* vol. 2. American Society for Microbiology, Washington, D.C.

Keller, W. W. 1992. Preparation of cell spots for immunofluorescence, p. 8.10.1–8.10.9. *In* H. D. Isenberg (ed.), *Clinical Microbiology Procedures Handbook,* vol. 2. American Society for Microbiology, Washington, D.C.

Murray, P. R., E. J. Baron, M. A. Pfaller, F. C. Tenover, and R. H. Yolken (ed.). 1995. *Manual of Clinical Microbiology,* 6th ed. American Society for Microbiology, Washington, D. C.

Persing, D. H., T. F. Smith, F. C. Tenover, and T. J. White (ed.). 1993. *Diagnostic Molecular Biology: Principles and Applications.* American Society for Microbiology. Washington, D.C.

APPENDIX 8.7–1

Commercial sources of reagents

Antigen detection kits and reagents[a]

Abbot Laboratories, Inc. (2, 3)
100 Abbott Park Rd.
Abbott Park, IL 60064
(800) 323-9100

Becton Dickinson Microbiology Systems (3)
P.O. Box 243
Cockeysville, MD 21030
(800) 638-8663

Behring Diagnostics, Inc. (2)
3403 Yerba Buena Rd.
San Jose, CA 95135
(800) 227-9948

bioMerieux Vitek, Inc. (2)
595 Anglum Dr.
Hazelwood, MO 63042
(800) 638-4835

BioStar, Inc. (4)
6655 Lookout Road
Boulder, CO 80301
(800) 637-3717

Biotest Diagnostics Corp. (1)
66 Ford Rd., Suite 131
Denville, NJ 07834
(800) 522-0090

BioWhittaker, Inc. (1)
8830 Biggs Ford Rd.
Walkersville, MD 21793
(800) 638-8174

Chemicon International, Inc. (1)
28835 Single Oak Dr.
Temecula, CA 92590
(800) 437-7500

Dako Corp. (1)
6392 Via Real
Carpinteria, CA 93013
(800) 235-5763

Difco Laboratories (1, 2)
P.O. Box 1058
Detroit, MI 48232

Incstar Corp. (1)
1990 Industrial Blvd.
Stillwater, MN 55082
(800) 328-1482

Meridian Diagnostics Inc. (1, 2, 3)
3471 River Hill Dr.
Cincinnati, OH 45244
(800) 343-3858

Murex Diagnostics, Inc. (3)
3075 Northwoods Circle
Norcross, GA 30071
(800) 334-8570

NEN Life Sciences (2)
(DuPont Co.)
549 Albany St.
Boston, MA 02118
(800) 551-2121

Organon Teknika Corp. (2)
800 Capitola Dr.
Durham, NC 27713
(800) 682-2666

Sanofi Diagnostics Pasteur, Inc. (1, 2, 3)
1000 Lake Hazeltine Dr.
Chaska, MN 55318
(800) 666-5111

Vector Laboratories, Inc. (1)
(Novocastra Laboratories Ltd.)
30 Ingold Rd.
Burlingame, CA 94010
(800) 227-6666

Wampole Laboratories (4)
P.O. Box 1001
Cranbury, NJ 08512
(800) 257-9525

APPENDIX 8.7–1 *(continued)*

In situ nucleic acid hybridization reagents

Dako Corp.
6392 Via Real
Carpinteria, CA 93013
(800) 235-5763

Enzo Biochem Corp.
60 Executive Blvd.
Farmingdale, NY 11735
(800) 221-7705

Vector Laboratories, Inc.
(Novocastra Laboratories Ltd.)
30 Ingold Rd.
Burlingame, CA 94010
(800) 227-6666

Nucleic acid hybridization kits

Abbott Laboratories, Inc. (Liquid Hybridization Assay)
100 Abbott Park Rd.
Abbott Park, IL 60064
(800) 323-9100

Digene (Hybrid Capture Assay)
2301-B Broadbirch Dr.
Silver Spring, MD 20904
(800) 344-3631

[a] Symbols: 1, IF/immunoenzymatic staining reagents; 2, semiautomated/automated EIA; 3, manual EIA; or 4, manual antigen assay other than EIA.

Gen-Probe Inc. (Hybrid Protection Assay)
9880 Campus Point Dr.
San Diego, CA 92121
(800) 523-5001

Amplified nucleic acid assays

Abbott Laboratories, Inc. (LCR)
100 Abbott Park Rd.
Abbott Park, IL 60064
(800) 323-9100

Chiron Diagnostics Corp. (bDNA)
115 Norwood Park South
Norwood, MA 02062
(800) 255-3232

Gen-Probe Inc. (TMA)
9880 Campus Point Dr.
San Diego, CA 92121
(800) 523-5001

Organon Teknika Corp. (NASBA)
800 Capitola Dr.
Durham, NC 27713

Roche Molecular Systems, Inc. (PCR, RT-PCR)
1080 U.S. Highway 202
Branchburg, NJ 08876
(800) 526-1247

8.8 Laboratory Diagnosis of Rickettsial Infections

Rickettsiae are gram-negative obligate intracellular bacteria that multiply in an arthropod vector as part of their life cycle and are associated with severe infections in humans (Table 8.8–1) (2–3). Replication is limited in cell cultures, and these organisms are best isolated in guinea pigs or mice. This, together with the need for BSL 3 containment that is required for many rickettsiae (4), limits the isolation of these organisms to reference laboratories. Similarly, PCR and other detection methods are generally limited to highly specialized reference laboratories and public health laboratories. Specimen selection and collection are summarized in Table 8.8–2. See EPCM procedure 8.10 for details regarding serologic testing.

Table 8.8–1 Human rickettsial infections[a]

Organism and transmission	Clinical manifestations
Spotted fever group (e.g., Rocky Mountain spotted fever [RMSF], *Rickettsia rickettsii*, tick bite; rickettsialpox, *R. akari*, mite bite)	Fever, headache, nausea and vomiting, abdominal pain; maculopapular, petechial rash with RMSF (rash frequently absent in elderly or black individuals [1]). Rash with rickettsialpox is papulovesicular or maculopapular.
Typhus group (e.g., epidemic typhus, *R. prowazekii*, close contact with louse feces; murine typhus, *R. typhi*, flea feces)	Headache, fever, maculopapular or petechial rash; Brill Zinsser disease (recrudescence of latent *R. prowazekii* infection)
Q fever (*Coxiella burnetti*, airborne exposure to excreta of infected animals, including cattle, sheep, goats, rabbits, parturient cats)	Undifferentiated febrile illness, pneumonia, endocarditis, hepatitis, meningoencephalitis (rare)
Ehrlichiosis (e.g., *Ehrlichia chaffeensis*, in U.S., tick bite)	Fever, chills, headache, myalgia, malaise, nausea, anorexia, weight loss; maculopapular or petechial rash in less than half of patients

[a] Members of the genus *Bartonella*, formerly *Rochalimaea*, are more closely related to *Brucella* and *Agrobacterium* spp. than to members of the *Rickettsiaceae*; these organisms can be cultured on bacteriologic media and are discussed elsewhere.

Table 8.8–2 Collection of specimens for rickettsial culture (reprinted from CMPH 8.2)

Specimen	Means of collection[a]	Comments
Blood[b]	10–12 ml, heparinized, or clotted	Snap-freeze immediately in acetone-dry ice slurry. Store at −70°C, and ship on dry ice. Contact reference laboratory[c].
	10–12 ml, clotted	Remove serum for serologic tests, and snap-freeze clot for culture as described above.
Sputum, urine	Collect in sterile leakproof container.	Useful for *Coxiella burnetti*
Tissue: brain, liver, spleen, etc.	Place in sterile leakproof container appropriate for snap-freezing.	Snap-freeze as described above.

[a] Do not include antimicrobial agents in transport medium.

[b] Collect blood for culture as soon after onset of infection as possible and before antibiotic therapy. Serum samples (acute and convalescent phase) should also be obtained for serologic testing.

[c] Most rickettsiae are BSL 3 organisms. Obtain specific handling and shipping information from the CDC, Viral and Rickettsial Branch, (404) 639-3574.

REFERENCES

1. **Helmick, C. G., K. W. Bernard, and L. J. D'Angelo.** 1984. Rocky Mountain spotted fever: clinical, laboratory, and epidemiological features of 262 cases. *J. Infect. Dis.* **150:** 480–486.
2. **Olson, J. G., and J. E. McDade.** 1995. Rickettsia and Coxiella, p. 678–685. *In* P. R. Murray, E. J. Baron, M. A. Pfaller, F. C. Tenover, and R. H. Yolken (ed.), *Manual of Clinical Microbiology,* 6th ed. American Society for Microbiology, Washington, D.C.
3. **Olson, J. G., and J. E. Dawson.** 1995. *Ehrlichia*, p, 686–689. *In* P. R. Murray, E. J. Baron, M. A. Pfaller, F. C. Tenover, and R. H. Yolken (ed.), *Manual of Clinical Microbiology,* 6th ed. American Society for Microbiology, Washington, D.C.
4. **Public Health Service, CDC, and NIH.** 1988. *Biosafety in Microbiological and Biomedical Laboratories.* U. S. Department of Health and Human Services publication no. (NIH) 88-8395, 2nd ed. U. S. Department of Health and Human Services, U.S. Government Printing Office, Washington, D.C.

8.9 Laboratory Diagnosis of *Mycoplasma* and *Ureaplasma* Infections

Mycoplasmas are small pleomorphic bacteria 0.3 to 0.8 μm in diameter; they lack a cell wall and are bounded by only a plasma membrane, and several species have been associated with human disease (1–4; Table 8.9–1). The ability of members of the genus *Ureaplasma* to hydrolyze urea sets them apart from the other mycoplasmas. Mycoplasma cultures are available through specialized reference laboratories; culture procedures for *M. pneumoniae, M. hominis*, and *U. urealyticum* cultures are described in CMPH 8.24. See EPCM 8.10 for serologic testing recommendations.

Table 8.9–1 *Mycoplasma* spp.: respiratory and genitourinary tract infections

Agent	Characteristics
M. pneumoniae	Year-round cause of respiratory disease in older children, adolescents, and young adults and also of bronchitis, laryngitis, pharyngitis, and otitis media. While it is the major cause of primary atypical pneumonia, accounting for 10 to 20% of X-ray-documented pneumonia cases, this organism causes mild upper respiratory disease more often than pneumonia.
M. hominis	Postpartum fever, opportunistic infection
U. urealyticum	Nongonococcal urethritis, chorioamnionitis, low birth weight

REFERENCES

1. **Clyde, W. A., Jr., G. E. Kenny, and J. Schachter.** 1984. *Cumitech 19, Laboratory Diagnosis of Chlamydial and Mycoplasmal Infections.* Coordinating ed., W. L. Drew. American Society for Microbiology, Washington, D.C.
2. **Cummings, M.** 1992. Mycoplasmas, p. 8.24.1–8.24. *In* H. D. Isenberg (ed.), *Clinical Microbiology Procedures Handbook*, vol. 2. American Society for Microbiology, Washington, D.C.
3. **Madoff, S., and D. C. Hooper.** 1988. Nongenitourinary infections caused by *Mycoplasma hominis* in adults. *Rev. Infect. Dis.* **10:**602–613.
4. **Taylor-Robinson, D.** 1995. Mycoplasma and ureaplasma, p. 652–662. *In* P. R. Murray, E. J. Baron, M. A. Pfaller, F. C. Tenover, and R. H. Yolken (ed.), *Manual of Clinical Microbiology,* 6th ed. American Society for Microbiology, Washington, D.C.

8.10 Serologic Diagnosis of Viral, Chlamydial, Rickettsial, and Mycoplasmal Infections

I. INTRODUCTION

Serologic assays designed to detect antibodies in serum samples after exposure to an infectious agent are useful for laboratory diagnosis of several infections (4). See Tables 8.1–1 and 8.1–2 for a summary of viruses for which serologic diagnosis is useful. Antibody assays are also frequently used to determine immune status in individuals not suspected of having active infections (e.g., donor, prenatal, and pre-employment screening). EIA, immunofluorescence (IF), immunoblot, and latex agglutination assays represent the most commonly used serologic techniques. In addition, reference and public health laboratories perform other assays that are less widely available (e.g., complement fixation [CF], neutralization). Serologic reagents are available from numerous sources, including those listed in Appendix 8.10–1.

II. ANTIBODY RESPONSES

Viral, chlamydial, rickettsial, and mycoplasmal antibody assays are generally designed to detect immunoglobulin G (IgG) and/or IgM antibodies, and antibody response patterns are somewhat similar for these different types of organisms. A primary antibody response is characterized by a rapid and early IgM response followed by an IgG response. IgG antibodies typically plateau several weeks after infection and generally persist for many years or for life. By contrast, the IgM response is transient; it declines rapidly after several weeks, but low levels may persist for several months. Reinfection with the same or antigenically related agent is characterized by an anamnestic (recall) response showing a rapid IgG rise and variable IgM production; thus, repeated rounds of viral reactivation (e.g., cytomegalovirus [CMV]) frequently yield variable responses and thus present complicated situations for serologic diagnosis.

NOTE: Antibodies detected by the CF assay are frequently not detectable many years after infection.

III. SPECIMENS

A. Refer to manufacturer's package insert or reference laboratory for specific guidelines.

B. Contaminated, viscous, severely lipemic, or grossly hemolyzed samples are unacceptable for testing; in some cases heat-inactivated samples are not acceptable.

C. Serum

Collect 8 to 10 ml of blood without anticoagulant. Store at ambient temperature or at 2 to 8°C; do not freeze blood. Centrifuge the sample briefly (e.g., 900 to 1,500 × *g* for 10 to 15 min), and separate the serum from the clot, preferably within several hours of collection; avoid microbial contamination. Serum can be stored briefly at 2 to 8°C or at −20°C or lower for storage beyond a few days. Avoid repeated freeze-thaw cycles.

D. Plasma
Collect 8 to 10 ml of blood with heparin or other anticoagulant that is compatible with the assay to be performed; plasma is suitable for many, but not all, serologic assays. Centrifuge and store as described for serum.

E. CSF
Store as described for serum; bloody CSF is unsuitable for testing, even after centrifugation. Examination of CSF is of limited value and does not appear to be useful for making therapeutic decisions in suspected cases of herpes simplex virus (HSV) infections of the central nervous system (CNS [11]). Local antibody response may be slow and difficult to distinguish from transport across a damaged blood-brain barrier, and, there have not been extensively characterized or standardized applications. If CSF is to be examined, collect and test CSF and peripheral blood in parallel to establish serum/CSF ratios.

NOTE: The integrity of the blood-brain barrier can be assessed by the method described by Tourtelotte (9) in which an immunoglobulin G (IgG) index is calculated by dividing the CSF/serum ratio of albumin by the CSF/serum ratio of IgG.

F. Other specimens
Oral fluids and dried blood spots are used in special settings (HIV-1). Collect and store as per manufacturer's instructions.

'V. SPECIMEN NUMBER AND TIMING

Guidelines regarding the use of single or paired specimens and time points for collection are summarized in Table 8.10–1.

V. ASSAY VERIFICATION, VALIDATION, AND QUALITY CONTROL

See item IV in procedure 8.7 for details.

VI. SPECIAL CONSIDERATIONS

A. Epstein-Barr virus (EBV) (5) and hepatitis B virus (HBV) serology
Antibody response patterns to different antigens are useful when interpreted in the context of clinical presentation for EBV (Table 8.10–2) and HBV (Table 8.10–3) infections.

B. Herpes simplex virus (HSV) serology
Significant cross-reactivity exists between HSV types 1 and 2, and it is impossible to determine whether infection has occurred with type 1 or type 2 based on serology performed by many current assays. Serologic assays based on detection of antibodies to glycoprotein G, which differs significantly between HSV-1 and HSV-2, have been recently introduced and appear useful for detecting type-specific antibodies (1, 7).

C. Respiratory syncytial virus (RSV) serology
CF is insensitive for diagnosis in young infants (10); immunofluorescence assay (IFA) and enzyme immunoassay (EIA) are preferred.

D. Antibody assays are less reliable in immunocompromised individuals, and persistent infection may exist in the setting of minimal antibody responses. For example, in immunocompromised patients, CMV IgM serology may not be reliable (2) and may be misleading in the diagnosis of acute or reactivation CMV disease.

E. Chlamydial serology
CF and micro-immunofluorescence (IF) methods are available, with micro-IF being more sensitive than CF (8). These assays are generally limited to large centers or reference laboratories. Insufficient data are available to determine

Table 8.10–1 Utility of single and paired serum samples for serologic testing[a]

Clinical situation	Using single serum sample	Using paired acute and convalescent serum samples[b]
Screening for evidence of infection or immunization (e.g., HIV serostatus, immune status, blood/organ donor screening, prenatal testing, pretransplant)	Presence of IgG antibody is consistent with recent or past infection or immunization. Absence of IgG and IgM antibodies is consistent with no prior infection or immunization. A negative result following immunization may indicate that the specimen was collected too soon after immunization and a follow-up specimen in 2 to 3 weeks is generally recommended.	Paired specimens not necessary.
Diagnosing a current or recent infection[c]	Presence of IgM antibody is consistent with a current or recent infection. However, low levels of IgM infection may persist for several months after primary infection and may appear after reinfection or reactivation of latent infection. NOTE: IgM testing is available for a limited number of viruses (e.g., CMV, EBV, HAV, measles, mumps, parvovirus B-19, rubella) Presence of high levels of antibodies, particularly by CF or IFA, may be suggestive of recent infection. Absence of antibody (particularly IgG) in an acute-phase sample collected within the first few days or first week after onset does not rule out acute infection and may indicate that the specimen was collected too soon after onset; a follow-up specimen collected 2 to 3 weeks after the first specimen is necessary (see paired serum testing). NOTE: A single acute-phase sample tested for IgG antibody is generally without diagnostic merit.	Seroconversion or a fourfold or greater rise in antibody titer or equivalent, between acute- and convalescent-phase samples, is consistent with a current or very recent infection. It is advised that paired specimens be tested in parallel to avoid complications regarding interpretation owing to run-to-run variation. If the acute-phase sample is collected beyond 2 weeks after onset, the antibody levels may be approaching an extended plateau phase, and thus a diagnostic rise may not be demonstrable. Presence of high levels of antibodies, particularly by CF or IFA, may be suggestive of recent infection.
Congenital infection[c]	Detection of viral IgM in a neonate is consistent with in utero infection, since IgM antibody does not cross the placenta. Detection of IgG antibody in a neonate is consistent with the presence of passively acquired maternal antibodies. Testing at monthly intervals will show a steady decline of maternally derived antibodies and disappearance of detectable antibody at about 6 months of age, whereas congenital infection is supported by stationary or increasing titers during the first several months of life. Failure to detect IgG antibodies at or soon after birth is consistent with lack of maternal exposure to the agent.	

[a] HIV, human immunodeficiency virus; HAV, hepatitis A virus.

[b] Acute phase, collect within the first week to 10 days after onset; convalescent phase, collect 2 to 3 weeks after the acute-phase specimen.

[c] Submit adequate clinical history to the testing laboratory. The onset date of symptoms and other pertinent history are critical in determining whether adequate specimens have been collected and whether other testing should be performed (e.g., other serologic assays, culture, polymerase chain reaction [PCR], etc.).

Table 8.10–2 EBV antibody response patterns[a]

Antibody	Pattern	Antibody profile at different stages of infection				
		Very early	Acute phase	Recent	Past	Reactivated
VCA-IgM	Early and of relatively short duration	±	+	±	–	–
VCA-IgG	Lifetime	–	+	+	+	+
EBNA-IgG						
EBNA-1	Elevated in NPC	–	–	±	+	+
EBNA-2		–	+	±	±	+
EA-IgG	Elevated levels consistent with protracted or severe infections					
EA/D	Elevated in NPC	–	±	+	–	+
EA/R	Elevated in Burkitt's lymphoma	–	±	±	–	±

[a] VCA, viral capsid antigen; EA, early antigen; EBNA, Epstein-Barr nuclear antigen; D, diffuse; NPC, nasopharyngeal carcinoma; R, restricted. Note: heterophile antibody response appears early and is of relatively short duration.

VI. SPECIAL CONSIDERATIONS *(continued)*

the performance characteristics of EIA reagents. CF is a genus-specific assay and is useful for systemic infections with all three species (e.g., psittacosis, lymphogranuloma venereum [LGV]). Micro-IF is more sensitive than CF for superficial infections (e.g., conjunctivitis, sexually transmitted diseases [STDs]) in which a good CF response may not be developed; however, the utility of this approach in detection of STDs is limited because chronic and repeat infections make rising titers difficult to detect. Serology is useful in diagnosing chlamydial pneumonia in neonates since micro-IF is useful for detecting IgM antichlamydial antibodies; IgG testing is not helpful because of the presence of maternal antibody.

F. Rickettsial serology

Antibody assays (e.g., IFA, CF) are available through reference laboratories and public health laboratories; ELISA assays (and their reagents) are available through specialized research labs. Testing includes assays for typhus fever group, the Rocky Mountain spotted fever (RMSF) group, and Q fever (6).

G. Mycoplasmal serology

Laboratory diagnosis of mycoplasmal respiratory disease is generally limited to serology (3), although the sensitivity and specificity of available methods

Table 8.10–3 HBV antibody response patterns[a]

Antigen or antibody	Pattern	Stage of infection		
		Acute	Recent	Past
HBsAg	First serologic marker to appear	+	–	–
HBeAg	Appears early and of short duration; persistence is consistent with chronic carrier state	+	±	–
Anti-HBs	Last serologic marker to appear (~ 4 mo); consistent with recovery; marker of immunity following immunization	–	+	+
Anti-HBc	Appears early and remains detectable for several years	±	+	+
Anti-HBe	Consistent with recovery	–	±	+

[a] Hepatitis B virus serologic testing profiles include a combination of antigen and antibody assays. Abbreviations: HBsAg, hepatitis B surface antigen; HBc, hepatitis B core antigen.

VI. SPECIAL CONSIDERATIONS (continued)

may be less than desired. Assays are available for detection of antibodies to *Mycoplasma pneumoniae*, including EIA, high-density particle agglutination, IFA, and CF. Serology does not appear suitable for genital-tract infections at this time. Cold agglutinins can be detected in individuals with primary atypical pneumonia and in certain hemolytic anemias. A fourfold rise usually begins to appear late in the first 2 weeks after onset and begins to decline between 4 and 6 weeks after onset.

NOTE: For the cold agglutinin assay, do not refrigerate blood before removing serum.

REFERENCES

1. **Ashley, R. L., J. Militoni, F. Lee, A. Nahmias, and L. Corey.** 1988. Comparison of Western blot (immunoblot) and glycoprotein G-specific immunodot enzyme assay for detecting antibodies to herpes simplex virus types 1 and 2 in human sera. *J. Clin. Microbiol.* **26:**662–667.
2. **Chou, S., D. Y. Kim, K. M. Scott, and D. L. Sewell.** 1987. Immunoglobulin M to cytomegalovirus in primary and reactivation infections in renal transplant recipients. *J. Clin. Microbiol.* **25:**52–55.
3. **Clyde, W. A., Jr., G. E. Kenny, and J. Schachter.** 1984. *Cumitech 19, Laboratory Diagnosis of Chlamydial and Mycoplasmal Infections.* Coordinating ed., W. L. Drew. American Society for Microbiology, Washington, D.C.
4. **James, K.** 1990. Immunoserology of infectious diseases. *Clin. Microbiol. Rev.* **3:** 132–152.
5. **Okano, M., G. M. Thiele, J. R. Davis, H. L. Grierson, and D. T. Purtilo.** 1988. Epstein-Barr virus and human diseases: recent advances in diagnosis. *Clin. Microbiol. Rev.* **1:** 300–312.
6. **Olson, J. G., and J. E. McDade.** 1995. Rickettsia and Coxiella, p. 678–685. *In* P. R. Murray, E. J. Baron, M. A. Pfaller, F. C. Tenover, and R. H. Yolken (ed.), *Manual of Clinical Microbiology,* 6th ed. American Society for Microbiology, Washington, D.C.
7. **Safrin, S., A. Arvin, J. Mills, and R. Ashley.** 1992. Comparison of the Western immunoblot and a glycoprotein B enzyme immunoassay for detection of serum antibodies to herpes simplex virus type 2 in patients with AIDS. *J. Clin. Microbiol.* **30:**1312–1314.
8. **Schachter, J., and W. E. Stamm.** 1995. *Chlamydia,* p. 669–675. *In* P. R. Murray, E. J. Baron, M. A. Pfaller, F. C. Tenover, and R. H. Yolken (ed.), *Manual of Clinical Microbiology,* 6th ed. American Society for Microbiology, Washington, D.C.
9. **Tourtellotte, W.** 1970. On cerebrospinal fluid immunoglobulin-G (IgG) quotients in multiple sclerosis and other diseases: a review and a new formula to estimate the amount of IgG synthesized per day by the central nervous system. *J. Neurol. Sci.* **10:**279–304.
10. **Welliver, R. C.** 1988. Detection, pathogenesis, and therapy of respiratory syncytial virus infections. *Clin. Microbiol. Rev.* **1:**27–39.
11. **Whitley, R. J., and F. Lakeman.** 1995. Herpes simplex virus infections of the central nervous system: therapeutic and diagnostic considerations. *Clin. Infect. Dis.* **20:**414–420.

SUPPLEMENTAL READING

Lennette, E. T. 1995. Epstein Barr Virus (EBV), p. 299–312. *In* E. H. Lennette, D. A. Lennette, and E. T. Lennette (ed.), *Diagnostic Procedures for Viral, Rickettsial, and Chlamydial Infections,* 7th ed. American Public Health Association, Washington, D.C.

Murray, P. R., E. J. Baron, M. A. Pfaller, F. C. Tenover, and R. H. Yolken (ed.). 1995. *Manual of Clinical Microbiology*, 6th ed. American Society for Microbiology, Washington, D.C.

Rose, N. R., E. Conway de Macario, J. D. Folds, H. C. Lane, and R. M. Nakamura (ed.). 1997. *Manual of Clinical Laboratory Immunology,* 5th ed. American Society for Microbiology, Washington, D.C.

·PPENDIX 8.10–1

Sources of serologic reagents

Abbott Laboratories, Inc.
100 Abbott Park Rd.
Abbott Park, IL 60064
(800) 323-9100

Bartels, Inc.
2005 NW Sammamish Road, Suite 107
Issaquah, WA 98027
(800) BARTELS

Becton Dickinson Microbiology Systems
P.O. Box 243
Cockeysville, MD 21030
(800) 638-8663

Behring Diagnostics, Inc.
3403 Yerba Buena Rd.
San Jose, CA 95135
(800) 227-9948

bioMerieux Vitek, Inc.
595 Anglum Dr.
Hazelwood, MO 63042
(800) 638-4835

Diagnostic Technology
240 Vanderbilt Motor Parkway
Hauppauge, NY 11788
(516) 582-4949

Gull Laboratories, Inc.
1011 East Murray Holladay Rd.
Salt Lake City, Utah 84117
(800) 448-4855

Incstar Corp.
1990 Industrial Blvd.
Stillwater, MN 55082
(800) 328-1482

Integrated Diagnostics, Inc.
1756 Sulphur Spring Rd.
Baltimore, MD 21227
(800) TEC-INDX

Murex Diagnostics, Inc.
3075 Northwoods Circle
Norcross, GA 30071
(800) 334-8570

Organon Teknika Corp.
800 Capitola Dr.
Durham, NC 27713
(800) 682-2666

Sanofi Diagnostics Pasteur, Inc.
1000 Lake Hazeltine Dr.
Chaska, MN 55318
(800) 666-5111

Tri-Delta Diagnostics, Inc.
2 Ridgedale Ave.
Cedar Knolls, NJ 07927
(800) 241-1202

Virion (U.S.), Inc.
4 Upperfield Rd.
Morristown, NJ 07960
(800) 524-2689

Wampole Laboratories
P.O. Box 1001
Cranbury, NJ 08512
(800) 257-9525

SECTION 9

Immunology

SECTION EDITOR: *William M. Janda*

The following table is a summary of various tests that can be used in clinical microbiology laboratories for the serologic diagnosis of infectious diseases. For further details, see section 9 in CMPH.

Table 9–1 Generic description of immunologic assays used in the serologic diagnosis of infectious diseases[a]

Test	Basic principle	Specimen type evaluated	Basic materials	Basic QC
Latex agglutination for antigen detection	Latex spheres coated with specific immunoglobulins agglutinate in the presence of antigen.	CSF, urine, or serum	1. Latex beads coated with specific antibodies; 2. Positive and negative antigen controls; 3. Mixing sticks; 4. Reaction slides; 5. Heat-block or boiling water bath; 6. Centrifuge	Positive and negative controls should be performed with each batch of patient specimens.
Latex agglutination for antibody detection	Latex spheres coated with specific antigens agglutinate in the presence of serum containing the corresponding specific antibody.	Serum or plasma	1. Latex beads coated with specific antibodies; 2. Mixing sticks; 3. Glass agglutination slides; 4. Appropriate positive and negative controls	Inclusion of positive and negative controls each time test is performed
Staphylococcal coagglutination	Staphylococcal cells are sensitized with antibody against specific antigens by the ability of staphylococcal protein A to bind immunoglobulin to the bacterial cell surface by the Fc region of the antibody molecules.	Suspension of the organism to be identified prepared in PBS, buffered saline, etc., depending on the kit (usually used for bacterial identification)	1. Staphylococcal cells sensitized with specific antibody; 2. Nonsensitized staphylococcal cells (negative control reagent); 3. Reaction slides or cards; 4. Mixing sticks	Inclusion of positive and negative controls each time the test is performed

Basic procedure	Expected results	Basic limitations	Generic examples
Mix specimen with reagent on a glass slide and inspect for specific agglutination.	1. Positive control must yield a strong agglutination reaction within 10 min; 2. Negative control must show no agglutination; 3. Specimens showing agglutination are positive; 4. Specimens showing no agglutination are negative.	1. Tests should not be used to replace culture (CSF antigen detection); 2. Urine and serum specimens require additional processing to prevent uninterpretable results; 3. Low antigen levels may result in false-negative tests	Several commercial kits for detection of bacterial antigens (group B streptococcus, *Haemophilus influenzae* type b, *Streptococcus pneumoniae*, and *Neisseria meningitidis* serogroups A, C, Y, and W135) in CSF; also kits for detection of *Cryptococcus neoformans* capsular polysaccharide in CSF
1. Mix serum specimen on a glass slide; 2. Rock or rotate slide for specified period; 3. Inspect for agglutination.	1. Positive control must produce appropriate specific agglutination reaction; 2. Negative control must not show agglutination; 3. Patient specimens showing agglutination contain specific antibody; 4. Negative specimens show no agglutination.	1. Low levels of antibody may not be reliably detected; 2. Some kits may not be able to produce results that correlate with traditional antibody titers.	Latex agglutination tests for detection of rubella and CMV antibodies
1. Standardized, properly processed bacterial cell suspension (1 drop) is reacted with antibody-sensitized staphylococcal cells (1 drop) and cells that are not sensitized in circles or walls of a reaction slide; 2. Mix and rock or rotate the reaction slide for specified period; 3. Inspect for specific agglutination.	1. Positive control suspension must produce at least 2+ agglutination with the sensitized cells, and no agglutination with the control cells; 2. Negative control suspension must show no agglutination with either the sensitized or the control staphylococcal cells; 3. Agglutination reactions of 2+ or greater with the sensitized cells and no agglutination with the non-sensitized cells on patient isolates are positive.	1. Isolates that show no agglutination must be identified by other methods; 2. Not all strains of an organism may be identified by a particular test. For example, not all strains of *Neisseria gonorrhoeae* are identified by commercially available coagglutination tests due to lack of appropriate monoclonal antibodies on the surface of the sensitized staphylococcal cells.	Coagglutination tests for the identification of *Neisseria gonorrhoeae* (e.g., Phadebact OMNI, Gonogen, Meritec)

(continued)

Table 9–1 Generic description of immunologic assays used in the serologic diagnosis of infectious diseases[a] *(continued)*

Test	Basic principle	Specimen type evaluated	Basic materials	Basic QC
Enzyme immunoassay for antibody detection	Antigen is bound to a solid phase (bead or microtiter well). Antibody present in the specimen binds to the antigen on the solid phase. This complex is then reacted with a second antibody that is enzyme-labeled and also directed against the antigen (the conjugate is often an anti-human immunoglobulin raised in goats). Subsequent exposure of this complex to the enzyme substrate results in the generation of a colored end-product that is detected spectrophotometrically.	Serum or plasma	1. Microtiter walls/beads coated with antigen; 2. PBS with Tween 80 for serum dilution; 3. Detection antibody-enzyme conjugate; 4. Chromogenic enzyme substrate; 5. Spectrophotometer; 6. Incubator; 7. Pipettes	Inclusion of sera known to be positive and negative for antibody being detected; diluent buffer control
Enzyme immunoassay for antigen detection	In this case, antibody (e.g., ''capture'' antibody) instead of antigen is bound to a solid phase. Antigen present in a specimen binds to the capture antibody on the solid phase. Following wash steps, this complex is reacted with a second enzyme-labeled antibody that is directed against another epitope of the ''captured'' antigen. After washing, enzyme substrate is added, resulting in the generation of a colored end-product that is detected visually or spectrophotometrically.	Serum, plasma, other body fluids (e.g., stool for detection of *Giardia lamblia* antigen; respiratory tract secretions for detection of respiratory syncytial virus, etc.)	1. Antibody-sensitized microtiter wells, beads, or membrane filters; 2. Second enzyme-conjugated antibody directed against other epitopes of the antigen; 3. Positive and negative controls that contain antigen or lack the antigen, respectively; 4. Some antigen detection EIAs contain calibrators and standards to allow semiquantitation of antigen levels present in certain specimens; 5. Spectrophotometric detection of positive results is used with some kits, although many rely on visual reading for detection of positive and negative tests.	Known positive and negative control samples

Basic procedure	Expected results	Basic limitations	Generic examples
1. Diluted serum specimen is incubated in antigen-sensitized wall or with antigen-sensitized bead; 2. Incubation and wash stage; 3. Addition of enzyme-labeled conjugate directed against antibody (e.g., anti-human IgG raised in goats and linked to horseradish peroxidase); 4. Incubation and wash steps; 5. Addition of enzyme substrate and generation of signal; 6. Addition of stop solution; 7. Generated color assessed spectrophotometrically	Positive results are determined by comparison with negative control; generally absorbance values 2- to 3-fold greater than the mean absorbance of a group of negative controls are considered positive. Some kits may include low- and high-titer positive controls, calibrations, etc.	1. Sera obtained during the acute phase of infection may contain only antibodies of the IgM class and will not be detected in IgG-specific assays; the converse is also true; 2. Contaminated, icteric, lipemic, heat-inactivated or hemolyzed sera may produce erroneous results; 3. Departure from specific procedure may affect test results.	EIA test kits for detection of antibodies are available for a wide variety of agents, particularly certain bacteria (e.g., *Borrelia burgdorferi*) and viruses (HIV-1, CMV, rubella, HSV, measles, mumps, HIV-1/2)
1. The specimen is incubated in a microtiter well or tube containing the capture antibody; 2. Incubate for specified period; 3. Wash steps; 4. A second enzyme-labeled antibody that is also directed against epitopes of the antigen is added; 3. Incubate for a specified period; 4. Wash steps; 5. Enzyme substrate is added and allowed to incubate for a specified period; 6. Addition of a ''stop'' solution; 7. Color reaction is read visually or with a spectrophotometer; 8. In semiquantitative assay, standard curves are prepared from calibrators run in the same assay.	1. Positive results determined visually by the presence of a colored endpoint. This may also be done with a spectrophotometer. 2. Semiquantitative results may be determined by standard curves with cutoffs determined by negative controls and regression analysis of standards.	1. Tests must be used exactly as described in package inserts; 2. Low levels of antigen may not be detected; 3. Antigens already present in serum specimens as antigen-antibody complexes will not be detected unless immune complexes are first dissociated.	Immunoassays for detection of group A streptococci directly in throat swabs; immunoassays for detection of RSV and influenza A in respiratory tract specimens; assays for detection of antigen in stool of patients with *Giardia lamblia* and *Cryptosporidium parvum*; assays for detection of HIV p24 antigen in serum

(continued)

Table 9–1 Generic description of immunologic assays used in the serologic diagnosis of infectious diseases[a] *(continued)*

Test	Basic principle	Specimen type evaluated	Basic materials	Basic QC
Direct fluorescent antibody test for antigen detection	Specific antibody conjugated to a fluorescent "tag" (e.g., FITC) is reacted directly with a specimen on a glass slide or with a smear of an organism prepared from a culture plate. After incubation and washing, the slide is examined under a fluorescence microscope and examined for specific immunofluorescence.	Specimens may include smears made directly from clinical specimens (e.g., endocervical smears for detection of *Chlamydia trachomatis*) or from cultures, where this procedure may be used for identification (e.g., identification of *Legionella* spp. from cultures grown on BCYE medium, identification of respiratory viruses growing in shell vials, etc.)	1. Appropriate clinical specimens; 2. Teflon-coated slides used for immunofluorescence (i.e., with specimen wells); 3. Specific antibodies conjugated directly to a fluorophore (e.g., FITC); 4. Humidor for slide incubation; 5. Pipettes	Positive and negative controls performed along with clinical specimens
IFA test for antigen detection	A specimen fixed on a slide is overlaid with an excess of unlabeled immune serum directed against the antigen and incubated. After a wash step, the specimen-antibody complex is reacted with a fluorochrome-labeled antibody directed against the species of the first unlabeled antibody (e.g., goat anti-human immunoglobulin conjugated to FITC). After a wash step, the specimen is examined under a fluorescence microscope fitted with the appropriate barrier filters for detection of specific immunofluorescence.	IFA tests can be used for detection of antigens in various clinical specimens (e.g., IFA for *Pneumocystis carinii* in bronchoalveolar lavage specimens).	1. Appropriate clinical specimens; 2. Unlabeled specific antibodies directed against the antigen of interest; 3. Fluorochrome-conjugated antibodies directed against the unlabeled antibodies (e.g., FITC-labeled anti-goat antibodies raised in mice); 4. Teflon-coated immunofluorescence slides with multiple wells; 5. Humidor for incubation; 6. Pipettes	Positive and negative control slides prepared from previously tested patient specimens are processed through the procedure along with unknown patient specimens.

Basic procedure	Expected results	Basic limitations	Generic examples
1. Specimen is applied to a well on a Teflon-coated immunofluorescence slide and fixed; 2. Specimen is overlaid with labeled conjugate and incubated in humidor for 15- to 30-min period; 3. Slide washed to remove unbound conjugate; 4. Air-dry specimen and apply mounting fluid and cover slip; 5. Examine under a microscope outfitted with a UV light source and appropriate barrier filters for assay under consideration.	Positive control and positive patient specimens show specific apple-green immunofluorescence (FITC); negative control and negative patient samples show no fluorescence.	Usually used only for antigen detection; while more rapid and less nonspecific, direct fluorescence procedures may be less sensitive and produce duller fluorescence than indirect immunofluorescence antigen detection methods.	Direct fluorescent antibody test for *Chlamydia trachomatis*
1. Appropriate clinical specimen and positive and negative control material are placed in walls of Teflon-coated immunofluorescence slide; 2. Slides are dried and fixed; 3. The wells are overlaid with unlabeled antibodies directed against the antigen of interest; 4. Incubate slides for 15–30 min in a humidor; 5. Wash and blot slide dry; 6. Overlay wells with fluorochrome-conjugated antibody; 7. Incubate in humidor for 15–30 min; 8. Wash and blot slide dry; 9. Add mounting medium and place a cover slip over wells; 10. Read wells for specific immunofluorescence under a fluorescence microscope with appropriate filter system for fluorochrome detection.	1. Positive control material reveals presence of specific immunofluorescence of the antigen; 2. No immunofluorescence should be observed with negative control material; 3. Positive and negative patient specimens are determined by comparison of unknowns with the immunofluorescence observed in the positive control.	1. Positive and negative controls of human origin should be tested to verify reagent performance; 2. Care must be taken to minimize nonspecific immunofluorescence caused by ''trapping'' of the fluorochrome (e.g., treatment of sputa with mucolytic agents when looking for *Pneumocystis carinii* by IFA); 3. Test may not detect antigens present in low concentrations or in inappropriately collected specimens.	*Pneumocystis carinii* IFA test

(continued)

Table 9–1 Generic description of immunologic assays used in the serologic diagnosis of infectious diseases[a] *(continued)*

Test	Basic principle	Specimen type evaluated	Basic materials	Basic QC
IFA test for antibody detection	This test is performed as described for the indirect essay; however, the first unlabeled antibody used in the test is that of the patient. Patient serum is serially diluted and placed on a series of immunofluorescence slide wells containing the same antigen. After an incubation and wash step, fluorochrome-labeled goat anti-human antibody is applied to each well. After another incubation/wash step, the slide is observed with a fluorescence microscope. The reciprocal of the highest dilution of the patient's serum that shows specific immunofluorescence is called the titer. Titers may be determined for both IgG and IgM. In the latter determination, the fluorochrome-labeled conjugate is directed against human IgM.	An antibody titer can be determined with a single serum. To document recent infection, an acute-phase specimen (collected early in the illness) and a convalescent-phase specimen (collected 2–3 weeks later) is needed to demonstrate an increase in titer against the antigen being tested.	1. Teflon-coated immunofluorescence slides, each well of which contains the antigen of interest (e.g., cells infected with CMV and expressing CMV-specific antigens); 2. Patient serum (acute and convalescent); 3. Low- and high-titer positive and negative control sera; 4. Fluorochrome-labeled anti-human immunoglobulin raised in another animal species (e.g., FITC-labeled anti-human immunoglobulins raised in goats); 5. Humidor for incubations; 6. Pipettes	1. Negative, low-titer positive, and high-titer positive control sera are performed with each test run performed on patient specimens. 2. Negative control serum is usually tested at titers of 4 and 16, while the low- and high-titer positive controls are run at four dilutions, usually one twofold dilution below and two twofold dilutions above the expected titer (e.g., if the low positive control has an expected titer of 64, it should be tested at titers of 32, 64, 128, and 256). 3. A conjugate control using PBS is also included.

Basic procedure	Expected results	Basic limitations	Generic examples
1. Prepare twofold serial dilutions of the control sera and patients' sera. Depending on individual lab procedures, patient serum may be screened at one dilution (e.g., 1:16) or at two dilutions (e.g., 1:16 and 1:64). 2. Place 0.10 ml of the diluted control and patient sera on appropriately labeled wells of the immunofluorescence slide. 3. Incubate slide in a humidor for 30 min at 37°C. 4. Rinse off sera and gently blot slide dry. 5. Place a drop of fluorochrome-tagged, appropriately diluted goat anti-human immunoglobulin on each of the slide wells. 6. Incubate in a humidor for 30 min at 37°C. 7. Rinse slides, blot dry. 8. Place small drops of buffered glycerol on each well and place a cover slip on the slide. 9. Read for specific immunofluorescence under the high-dry objective of the fluorescence microscope.	1. No immunofluorescence should be observed with either dilution of the negative control or with the PBS conjugate control. 2. Low- and high-titer control sera should agree with the predetermined titers, or neither titer should be $> \pm 1$ dilution off the expected titer. 3. If the controls are correct, the antibody titer is defined as the reciprocal of the highest dilution giving characteristic immunofluorescence.	Rheumatoid factor present in patient sera may cause false positive results in tests for both IgG and IgM.	IFA tests are available for the detection of antibodies against several viral agents (e.g., CMV, HSV, HIV-1) and toxoplasma.

(continued)

Table 9–1 Generic description of immunologic assays used in the serologic diagnosis of infectious diseases[a] *(continued)*

Test	Basic principle	Specimen type evaluated	Basic materials	Basic QC
Immunoblot procedure	In this modified enzyme immunoassay, a mixture of different antigens (e.g., a lysate prepared from a virus growing in culture) is separated into its component proteins by molecular weight with electrophoresis. The separated components are then electrophoretically transblotted onto a sheet of nitrocellulose paper. The sheet is cut into strips that are used as the solid phase for reaction with patient sera. Subsequently, the strip is incubated with enzyme-labeled anti-human immunoglobulin. After wash steps, enzyme substrate is added. The presence of antibody in patient serum that is reactive with antigens separated on the nitrocellulose strip is indicated by the appearance of a colored band at the location of that particular antigen on the strip.	Serum or plasma	1. Nitrocellulose strips containing electrophoresed antigen (these may be prepared in-house or purchased from a vendor if available); 2. Plastic container with troughs for holding and incubating nitrocellulose blot strips; 3. Pipettes; 4. Positive and negative control sera; 5. HRP-labeled goat anti-human antibodies (alternatively, biotinylated goat anti-human conjugate and avidin-HRP can be used in place of the HRP-labeled goat anti-human immunoglobulin); 6. HRP substrate	Negative and low- and high-titer controls should be performed with each test.

Basic procedure	Expected results	Basic limitations	Generic examples
1. The nitrocellulose strip with the electrophoresed antigens is placed in an incubation trough; 2. A dilution of the serum to be tested is placed in the trough along with the antigen strip; 3. After incubation and wash steps, the strip is incubated with goat anti-human IgG labeled with an enzyme. Alternatively, goat anti-human immunoglobulin conjugated to biotin may be used as the conjugate. 4. After incubation and wash steps, the enzyme substrate is added to the trough. The appearance of colored bands at the locations of the various antigens indicates the presence of antibody to that antigen in the original serum specimen. If a biotinylated conjugate is used, then avidin-enzyme is added and, after incubation and wash steps, the enzyme substrate is added for color development.	The appearance of colored bands on the strip indicates that antibody against that particular antigen was present in the original serum specimen.	1. Test must be performed exactly as described in the package insert; 2. Western blot is as sensitive as and more specific than standard colorimetric EIA.	HIV-1 Western immunoblot supplemental test for detection of specific anti-HIV-1 antibodies

(continued)

Table 9–1 Generic description of immunologic assays used in the serologic diagnosis of infectious diseases[a] *(continued)*

Test	Basic principle	Specimen type evaluated	Basic materials	Basic QC
Complement fixation	Terminal components of the complement cascade (C7, 8, 9) damage cell membranes in the presence of specific antibody, which fixes complement to the cell surface. In the CF test, erythrocytes are used as the target cells and complement-induced ''leakiness'' can be detected colorimetrically or visually as an increase in free (rather than cell-bound) hemoglobin. In the presence of specific antibodies to an infectious agent, any complement added to the test system becomes bound, leaving no residual complement for reaction with antibodies to the ''indicator'' target erythrocytes. Therefore, the presence of specific antibody is indicated by the absence of hemolysis.	Serum	1. Sheep erythrocyte suspension; 2. Hemolysin (commercially available)	Known antibody-positive and antibody-negative controls; serum control to detect anticomplementary activity; antigen controls without serum to detect anticomplementary activity; tissue control (the cells or tissue in which the antigen was prepared); buffer control; back titration of commercial complement to ensure use of 5 CH_{50} units in each well; cell control to demonstrate absence of hemolysis; reference hemolysis standards

Basic procedure	Expected results	Basic limitations	Generic examples
Sensitization of erythrocytes: 1. Serial dilutions of hemolysin incubated with 2.8% solution of sheep RBCs in buffer; 2. As hemolysin concentration increases, degree of hemolysis increases, to achieve a maximal "plateau." Quantitation of complement: 1. Varying amounts of a 1:400 dilution of cold, reconstituted guinea pig complement are incubated with sensitized RBCs; 2. Volume of complement producing 50% lysis of RBCs (CH_{50} unit) is calculated from graphed results of the hemolysis. CF test: 1. Block titration of serial dilutions of both antigen and antibody is performed in presence of 5 CH_{50} units of complement at 4°C for 16 h; 2. Patient sera are heated at 56°C for 30 min to inactivate endogenous complement; 3. Subsequent addition of RBCs that has been sensitized with the optimal dilution of hemolysin is incubated at 37°C for 30 min; 4. Plates are centrifuged to sediment unlysed erythrocytes.	Highest dilution of antibody providing 3+ to 4+ fixation of complement (<30% hemolysis) is the endpoint.	All reagents must be free of anticomplementary activity, the correct quantity of complement must be present, and controls must produce expected reactions.	CF test is considered the reference method for detection of antibodies against a wide variety of viral agents and *Mycoplasma pneumoniae*. The CF test has been largely supplanted by EIA methodology.

(continued)

Table 9–1 Generic description of immunologic assays used in the serologic diagnosis of infectious diseases[a] *(continued)*

Test	Basic principle	Specimen type evaluated	Basic materials	Basic QC
Hemagglutination inhibition	Some viruses produce surface proteins that agglutinate erythrocytes from various species. Presence of antibodies to such viruses in patient serum can be determined by specific inhibition of that hemagglutination.	Serum or plasma	1. Erythrocytes from appropriate species (usually chicken) collected in Alsever's solution or heparin; 2. Diluent of appropriate pH; 3. Solutions to remove nonspecific agglutinins from serum; 4. Standardized viral antigen	Known positive and negative sera; back titration of antigen to ensure that the correct concentration of hemagglutinating units was tested.

Basic procedure	Expected results	Basic limitations	Generic examples
Determination of hemagglutinating titer: 1. Diluted suspension of RBCs is incubated of 4°C or at room temperature with serial dilutions of viral antigen until RBCs in tubes lacking virus settle as a button; 2. Highest antigen dilution that produces partial or complete agglutination equals 1 HAU. Treatment of sera: Treat serum by physical means (kaolin), enzymes (neuraminidase), etc., to remove nonspecific agglutinins. Inhibition test: Dilutions of antigen containing 4 HAU are mixed with erythrocytes and twofold dilutions of pretreated patient sera and incubated as for the determination of HAU titer.	Endpoint is the last well in which partial or complete inhibition of viral agglutination of RBCs occurs.	Use of test is limited to detection of antibodies against those viruses that possess surface hemagglutinins (e.g., rubella).	

(continued)

9 Immunology

Table 9–1 Generic description of immunologic assays used in the serologic diagnosis of infectious diseases[a] *(continued)*

Test	Basic principle	Specimen type evaluated	Basic materials	Basic QC
Immunodiffusion test	Double immunodiffusion is used to detect antigen or antibody. Known and unknown reactants are placed in adjoining wells of an agarose matrix and are allowed to passively diffuse towards one another for 12 to 24 h. A precipitin line forms where optimal levels of antigen and antibody are present. Double diffusion is usually used for the detection of antibodies. CIE is similar to double diffusion, except that the migration of one or both reactants is directed by an electrical field. (Cells to the right describe materials and procedures for fungal immunodiffusion testing.)	Patient serum or plasma for detection of antibody. CIE has been used with other body fluids (e.g., CSF) primarily for the detection of bacterial antigens.	Materials listed are those needed for fungal immunodiffusion. 1. *Blastomyces dermatitidis* immunodiffusion antigen; 2. *Blastomyces dermatitidis* immunodiffusion rabbit antisera; 3. *Coccidioides immitis* immunodiffusion antigen; 4. *Coccidioides immitis* immunodiffusion rabbit antisera; 5. *Histoplasma capsulatum* immunodiffusion antigen; 6. *Histoplasma capsulatum* immunodiffusion rabbit antisera; 7. 1% purified agar; 8. Immunodiffusion matrix pattern containing a single central well and 6 peripheral wells located equidistant from the central well and from adjacent wells; 9. Reading box containing an incandescent light source with a flat, black background	For fungal immunodiffusion, control serum containing the antibodies being sought is required to discern lines of identity that may form between the antigen, the known antisera, and the patient's specimen. In addition to homologous testing, each test serum and antigen should also be tested for immunodiffusion reactivity with heterologous sera and antigens.

[a] Abbreviations: CMV, cytomegalovirus; PBS, phosphate-buffered saline; HIV, human immunodeficiency virus; HSV, herpes simplex virus; IgM, immunoglobulin M; IgG, immunoglobulin G; RSV, respiratory syncytial virus; BCYE, buffered charcoal-yeast extract; IFA, indirect fluorescent antibody; HRP, horseradish peroxidase; CF, complement fixation; HAU, hemagglutinating unit(s); CIE, countercurrent immunoelectrophoresis; FITC, fluorescein isothiocyanate.

Basic procedure	Expected results	Basic limitations	Generic examples
1. Each immunodiffusion grid can be used to detect antibodies against a given antigen for four patients. 2. Using *B. dermatitidis* immunodiffusion testing, as an example, *B. dermatitidis* antigen is placed in the central well (1), and reference antisera are placed in wells 2 and 3. 3. Patients' sera to be tested for *B. dermatitidis* antibodies are placed in wells 4, 5, 6, and 7. 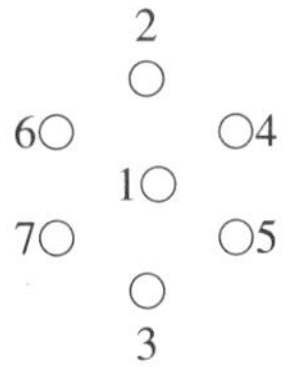 2 ○ / 6○ ○4 / 1○ / 7○ ○5 / ○ 3 4. Similar plates are set up for *C. immitis* and *H. capsulatum* antigens and reference antisera. The patient sera are again placed in the four wells as stated in step 3 above. 5. Incubate plates at room temperature and read for lines of precipitation after 24 and 48 h of incubation.	The location and identity of precipitin lines should be noted. Reactions of identity (i.e., precipitin lines formed by the test reagents which are continuous with precipitin lines formed by reference reagents) indicate that the patient serum specimen contains antibodies directed against the antigen in the central well. 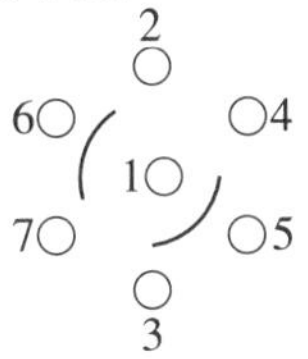 2 ○ / 6○ ○4 / 1○ / 7○ ○5 / ○ 3	Low levels of antibodies may not be detected by this method.	Fungal immunodiffusion for detection of antibodies to the systemic molds and *Aspergillus* spp.

SECTION 10 Molecular Biology

SECTION EDITOR: *Michael A. Pfaller*

(continued)

10.1 Introduction

The techniques of molecular biology have contributed tremendously to our understanding of the pathogenesis and epidemiology of infectious diseases. The molecular diagnosis of infectious diseases requires the isolation of nucleic acids from microorganisms and clinical material and the use of restriction endonuclease enzymes, gel electrophoresis, and nucleic acid hybridization techniques. Increasingly, newer techniques for amplification of nucleic acids are applied to clinical material. DNA sequence analysis, although not practical at the present time for use in the clinical laboratory, provides the ultimate means of characterizing organisms (1, 6–8).

The use of molecular methods for identification and direct detection of microorganisms in clinical microbiology is evolving gradually. Several products are now available commercially and are becoming assimilated into the routine practice of diagnostic laboratories of all sizes. Practical applications include the use of nucleic acid probes and amplification-based techniques for direct detection of organisms in clinical material (Table 10.1–1) or for identification of previously isolated organisms (Table 10.1–2). In addition, molecular methods provide a powerful means of characterizing organisms to the subspecies level in epidemiologic investigations (Table 10.1–3). Finally, molecular techniques may be used to define selected characteristics of organisms such as anti-

Table 10.1–1 Commercial nucleic acid probes and amplification systems for direct pathogen detection and monitoring[a,b]

Technology	Organism	Format	Detection/reporter system
Nucleic acid probe detection	Cytomegalovirus	Solution phase	DNA/RNA hybrid capture and chemiluminescence[c]
	Human papillomavirus	Solution phase	DNA/RNA hybrid capture and chemiluminescence[c]
	Hepatitis B virus	Solid-phase microtiter tray	Branched chain DNA[d]
		Solution phase	DNA/RNA hybrid capture and chemiluminescence[c]
	Hepatitis C virus	Solid-phase microtiter tray	Branched chain DNA[d]
	Chlamydia trachomatis	Solution phase	Acridinium ester[e]
	Neisseria gonorrhoeae	Solution phase	Acridinium ester[e]
	Streptococcus pyogenes	Solution phase	Acridinium ester[e]
Nucleic acid amplification	Hepatitis C virus	PCR	Microtiter capture probe[f]
	Human immunodeficiency virus	PCR	Microtiter capture probe[f]
	Mycobacterium tuberculosis	PCR, strand displacement amplification	Microtiter capture probe[f]; ethidium bromide-stained gel[g]
		Transcription-mediated amplification	Acridium ester probe[e]
	Chlamydia trachomatis	PCR	Microtiter capture probe[f]
		LCR	Antibody capture of labeled product[h]
	Neisseria gonorrhoeae	PCR	Microtiter capture probe[f]
		LCR	Antibody capture of labeled product[h]

[a] The table contains examples of available systems and is not all-inclusive.
[b] Abbreviations: PCR, polymerase chain reaction; LCR, ligase chain reaction.
[c] Digene, Silver Spring, Md.
[d] Chiron, Emeryville, Calif.
[e] Gen-Probe, Inc., San Diego, Calif.
[f] Roche, Branchburg, N.J.
[g] Becton Dickinson, Cockeysville, Md.
[h] Abbott Laboratories, North Chicago, Ill.

Table 10.1–2 Commercial nucleic acid probes for culture identification[a,b]

Organism	Hybridization format	Reporter system
Mycobacterium tuberculosis	Solution phase	Acridinium ester
M. avium-M. intracellulare	Solution phase	Acridinium ester
M. gordonae	Solution phase	Acridinium ester
M. kansasii	Solution phase	Acridinium ester
Histoplasma capsulatum	Solution phase	Acridinium ester
Coccidioides immitis	Solution phase	Acridinium ester
Blastomyces dermatitidis	Solution phase	Acridinium ester
Cryptococcus neoformans	Solution phase	Acridinium ester
Listeria monocytogenes	Solution phase	Acridinium ester
Staphylococcus aureus	Solution phase	Acridinium ester
Streptococcus pneumoniae	Solution phase	Acridinium ester
Streptococcus agalactiae	Solution phase	Acridinium ester
Streptococcus pyogenes	Solution phase	Acridinium ester
Enterococcus spp.	Solution phase	Acridinium ester
Escherichia coli	Solution phase	Acridinium ester
Haemophilus influenzae	Solution phase	Acridinium ester
Neisseria gonorrhoeae	Solution phase	Acridinium ester
Campylobacter spp.	Solution phase	Acridinium ester

[a] Adapted from Cormican and Pfaller (2).

[b] The probes listed have been developed by Gen-Probe, Inc., San Diego, Calif. The table contains examples of described probes; it is not all-inclusive.

Table 10.1–3 Molecular methods for epidemiologic typing of microorganisms

Method	Substrate	Principal characteristics	Examples
Plasmid fingerprinting	Plasmid DNA	Potentially unstable owing to loss of plasmids. May be augmented by restriction endonuclease digestion	*Staphylococcus aureus*; coagulase-negative staphylococci; *Klebsiella, Serratia*, and *Enterobacter* spp.
Restriction endonuclease analysis of chromosomal DNA with conventional electrophoresis	Chromosomal DNA	Broadly applicable. Large numbers of bands. Often difficult to interpret	*Clostridium difficile, Enterococcus faecium, S. aureus*
Restriction fragment length polymorphism analysis with DNA probes	Chromosomal DNA	Broadly applicable. Multistep process. Limited power to discriminate (ribotyping). Includes IS*6110* analysis of *Mycobacterium tuberculosis* and ribotyping	*M. tuberculosis, S. aureus, Candida albicans*, members of the family *Enterobacteriaceae*
Pulsed-field-gel electrophoresis	Chromosomal DNA	Broadly applicable. Uses infrequent cutting restriction enzymes to generate large DNA fragments (10–800 kb). Fewer bands. Excellent reproducibility and discriminatory power. Expensive equipment	*S. aureus, Enterococcus* spp., *Enterobacteriaceae, Pseudomonas* spp., *Candida* spp.
Random amplification of polymorphic DNA	Chromosomal DNA	Broadly applicable. Rapid and moderately easy to perform. Difficult to standardize. Only moderate reproducibility and discriminatory power	*S. aureus, C. difficile, Enterobacteriaceae*

Table 10.1–4 Application of molecular methods to detect antimicrobial-agent-resistance genes[a]

Organism	Target gene	Antimicrobial-agent resistance	Detected by[b]
Staphylococcus spp.	*mecA*	Methicillin/oxacillin	Probe, PCR
	ant	Aminoglycosides	Probe
	erm, msr	Macrolides	Probe, PCR
	gyrA	Fluoroquinolones	Probe, PCR
Streptococcus pneumoniae	*pbp1A, 2B, 2X*	β-Lactams	PCR
Enterococcus spp.	*vanA, B, C*	Glycopeptides	Probe, PCR
	aac, aph, ant	Aminoglycosides	Probe, PCR
Enterobacteriaceae	*tem, shv*, other β-lactamase genes	β-Lactams	Probe, PCR
	aph, ant, aac	Aminoglycosides	Probe, PCR
	cat	Chloramphenicol	Probe
	tet	Tetracycline	Probe, PCR
	erm, ere	Macrolides	Probe, PCR
	dhfr	Trimethoprim	Probe
Haemophilus influenzae	*tem*	β-Lactams	Probe, PCR
Mycobacterium tuberculosis	*rpoB*	Rifampin	PCR
	katG	Isoniazid	PCR
	rpsL, rrs	Streptomycin	PCR
	rrl	Clarithromycin	PCR
	gyrA	Fluoroquinolones	PCR

[a] List not all-inclusive.

[b] Polymerase chain reaction (PCR) is frequently used to amplify a gene relevant to antimicrobial-agent resistance, which must be characterized by other methods (DNA sequence analysis, single-strand conformational polymorphism) to determine if a mutation in the gene which confers resistance is present.

microbial-agent-resistance genes (Table 10.1–4).

As molecular techniques become routine, the critical questions of cost and potential for contribution to patient care need to be addressed. In the area of diagnosis of infectious diseases, improved patient outcomes and reduced costs of antimicrobial agents and duration of hospital stays may outweigh increases in laboratory costs which are sure to accompany the use of these more sophisticated testing methods. For communicable diseases, the cost may be justified by improved infection control measures to protect patients and health care workers. Molecular epidemiology may make a real contribution to the detection and control of the nosocomial spread of infection, with the potential to reduce infection rates and the requirement for expensive antimicrobial agents to treat resistant organisms. Although the use of this new technology is attractive, one must remember that the value of any of these molecular methods is a function of the degree to which it addresses the limitations of current methods. Presently, much of the justification for expenditures on molecular diagnostics and epidemiology is speculative, and we must continue to critically examine the role of these methods.

The molecular methods and applications presented in this section of the manual are those that are either commercially available for individual laboratories or are accessible through reference laboratories. In most instances the procedures presented address infectious agents or characteristics of microorganisms that are difficult to detect and identify in a timely fashion by conventional methods. The protocols provided are merely representative of the ever-increasing array of molecular techniques that can be applied to the diagnosis of infectious diseases. The interested reader is referred to several excellent manuals that provide additional protocols for use in both diagnostic and research laboratories (3–5).

REFERENCES

1. **Armann, R., W. Ludwig, and K. H. Schleifer.** 1994. Identification of uncultured bacteria: a challenging task for molecular taxonomists. *ASM News* **60:**360–365.
2. **Cormican, M. G., and M. A. Pfaller.** 1996. Molecular pathology of infectious diseases, p. 1390–1399. *In* J. B. Henry (ed.), *Clinical Diagnosis and Management by Laboratory Methods,* 19th ed. The W. B. Saunders Co., Philadelphia.
3. **Erlich, H. A.** 1992. *PCR Technology: Principles and Applications for DNA Amplification.* W. H. Freeman & Co., New York.
4. **Persing, D. H. (ed.)** 1996. *PCR Protocols for Emerging Infectious Diseases.* ASM Press, Washington, D.C.
5. **Persing, D. H., T. F. Smith, F. C. Tenover, and T. J. White.** 1993. *Diagnostic Molecular Microbiology: Principles and Applications.* ASM Press, Washington, D.C.
6. **Relman, D. A., and D. H. Persing.** 1996. Genotypic methods for microbial identification, p. 3–31. *In* D. H. Persing (ed.), *PCR Protocols for Emerging Infectious Diseases.* ASM Press, Washington, D.C.

10.2.1 Introduction

The detection of pathogenic microorganisms directly in clinical specimens by molecular methods has been investigated extensively by using a variety of nucleic acid probe hybridization, target amplification, and signal-generating formats. Commercial product development has focused on direct diagnosis of blood-borne and sexually transmitted diseases and respiratory pathogens by using solid- and solution-phase hybridization with nonisotopic nucleic acid probes and several different target amplification methods including the polymerase chain reaction (PCR), ligase chain reaction (LCR), strand displacement amplification, and transcription-mediated amplification (Table 10.1–1).

10.2.2 Nucleic Acid Probe-Based Methods

Direct diagnosis by nucleic acid probe hybridization is simple, rapid, and relatively free of the contamination and inhibition problems associated with target amplification methods. The sensitivity of probe hybridization methods is limited by the relatively large number of copies (~10^4) of the target sequence required to generate a positive signal. Efforts to improve the sensitivity of the nucleic acid probe-based methods include the use of RNA targets, various signal amplification formats, and reduction of the background signal (1–3).

REFERENCES

1. **Enns, R. K.** 1988. DNA probes: an overview on comparisons with current methods. *Lab. Med.* **19:**295–300.
2. **Lorincz, A. T.** 1992. Diagnosis of human papilloma virus infection by the new generation of molecular DNA assays. *Clin. Immunol. Newsl.* **12:**8.
3. **Urdea, M. S., T. Horn, and T. Fultz.** 1992. Branched DNA amplification multimers for the sensitive direct detection of human hepatitis viruses. *Nucleic Acids Res. Symp. Ser.* **24:** 197–200.

PART 1

Gen-Probe PACE 2 Nucleic Acid Hybridization Test for Detecting *Chlamydia trachomatis* and *Neisseria gonorrhoeae*

I. PRINCIPLE

The Gen-Probe PACE 2 Systems (Gen-Probe, Inc., San Diego, Calif.) for *Chlamydia trachomatis* and *Neisseria gonorrhoeae* detection are rapid DNA probe tests that utilize the technique of nucleic acid hybridization for detection of *C. trachomatis* and *N. gonorrhoeae*, respectively, in endocervical and male urethral-swab specimens. The test for *C. trachomatis* may also be applied to conjunctival specimens. The Gen-Probe PACE 2 systems use a chemiluminescently labeled DNA probe that is complimentary to the rRNA of the target organism. The labeled DNA:RNA hybrid is separated from nonhybridized probe, and the luminescence of the hybrid is measured in a luminometer.

NOTE: Procedure adapted from that of Cintron (1).

II. SPECIMEN

A. Specimens from the urogenital tract (endocervical or male urethral) and conjunctiva must be obtained by using the Gen-Probe specimen collection kit (containing a transport tube and swab).

B. Collection tubes containing the appropriate swab specimens must be transported to the laboratory at 2 to 25°C and may be stored at 2 to 25°C for up to 7 days before testing. If longer storage is necessary, process the specimen as described in item IV.A below, and freeze at −20 to −70°C.

C. Grossly bloody specimens (greater than 80 μl of whole blood in 1 ml of transport medium) could interfere with the assay.

III. MATERIALS

Include QC information on reagent container and in QC records.

■ **NOTE:** Do not freeze reagents contained in PACE 2 kits.

A. Reagents

1. Gen-Probe PACE 2 System for *C. trachomatis*
2. Gen-Probe PACE 2 System for *N. gonorrhoeae*
3. Gen-Probe PACE 2 detection reagent kit

B. Supplies

1. Gen-Probe PACE 2 specimen collection kit
   a. Urethral-conjunctival collection kit
   b. Endocervical collection kit
2. Wash bottle and cap assembly (provided)
3. Sealing cards (provided)
   ■ **NOTE:** Items 4 through 7 below required but not provided.
4. Disposable polystyrene tubes (12 by 75 mm)
5. Pipette tip (1 to 5 ml)
6. Pipette tip (100 μl)
7. Absorbent paper

C. Equipment (required but not provided)

1. Leader 1 luminometer
2. Covered water bath (60 ± 1°C)
3. Vortex mixer
4. Magnetic rack
5. Certified thermometer
6. Micropipettor (100 μl)
7. Pipettes capable of delivering 1 to 25 ml

IV. PROCEDURE

A. Sample preparation

1. Allow the transport tubes containing the swab specimens to come to room temperature and then vortex for 5 to 10 s.
2. Express all liquid from the swab by pressing the swab against the wall of the tube. Discard the swab.
3. Vortex the tubes for 5 to 10 s to ensure homogeneity.

B. Reagent preparation

1. Probe reagent
   a. Remove hybridization buffer from the kit, vortex for 10 s, warm in a water bath (60°C) for 3 to 4 min and then vortex for 10 s to ensure homogeneity.
   b. Pipette 6.0 ml of the buffer into lyophilized probe reagent, let stand at room temperature for 2 min and vortex for 10 s. Visually inspect the mixture to ensure complete rehydration and homogeneity.
   c. Record the expiration date of the preparation on the label. Reagent is stable for 3 weeks at 2 to 8°C.
2. Separation suspension
   a. Determine number of tests to be performed.
   b. Calculate volume of selection reagent and separation reagent needed as follows:
      (1) Volume (ml) of selection reagent
         (a) When using an Eppendorf pipettor = number of tests + two extra tests.
         (b) When using a bottle top dispenser = number of tests + 10 extra tests.
      (2) Volume (ml) of separation reagent = volume of selection reagent divided by 20.
   c. Pour required volume of selection reagent into a clean, dry container.
   d. Add the required volume of separation reagent to the selection reagent and mix well. The separation suspension is stable at room temperature for 6 h.

C. Hybridization

1. Label tubes with sample identification numbers and insert into magnetic rack. Include three tubes for negative reference and one for the positive control.

2. Vortex each specimen for 5 s.
3. Pipette 100 μl of each control and each specimen into the bottoms of the appropriate tubes.
4. Pipette 100 μl of the probe reagent to the bottom of each tube, taking care not to touch the top or sides of the tube.
5. Seal the tubes with sealing cards and shake the rack three to five times to mix the contents.
6. Incubate the rack with tubes in a water bath (60 ± 1°C) for 1 h.

D. Separation

1. Remove the tube rack from the water bath, and pipette 1 ml of the separation solution into each tube.
2. Cover the tubes with sealing cards, shake three to five times, and incubate in a water bath (60 ± 1°C) for 10 min.
3. Place tube rack on the base of the Gen-Probe magnetic separation unit for 5 min at room temperature.
4. Holding the tube rack and base of magnetic separation unit together, invert to decant the supernatants, shake the unit two or three times, and then blot the inverted tubes on absorbent paper.
5. *Do not remove the tube rack from the magnetic separation base.* Fill each tube to the rim with wash solution and let sit on the magnetic separation base for 20 min at room temperature.
6. Holding the tube rack and base together, decant wash fluid. *Do not blot.* Approximately 50 to 100 μl of wash solution should remain in each tube.
7. Separate the tube rack from the base, and shake to suspend the pellets.

E. Equipment preparation

1. Prepare the Gen-Probe Leader luminometer for operation. Make sure there are sufficient volumes of detection reagents 1 and 2 to complete the tests.
2. Select the appropriate protocol from the instrument software.

F. Detection

1. Wipe each tube to ensure that no residue is present on the outside of the tube.
2. Mix each tube to suspend the pellets, and insert into the instrument based on the prompts provided with the software.
3. Read the tubes in the following order:
   a. Negative reference: three tubes
   b. Positive control: one tube
   c. Specimen tubes
4. When the analysis is completed, remove the last tube, replace it with an empty tube, and follow the instrument prompts.

V. RESULTS

A. Calculation

1. Results are calculated on the basis of the difference between the response in relative light units (RLU) of the specimen and the mean of the negative reference.
2. The mean of the negative reference equals the sum of the three negative reference replicates divided by 3.

B. Interpretation

1. The luminometer prints the specimen response (in RLU) and compares this response with an assigned assay cutoff. A positive or negative interpretation resulting from this comparison is printed.
2. A positive result indicates that the target organism is present in the specimen and strongly supports a diagnosis of chlamydial or gonococcal infection.

V. RESULTS *(continued)*

◩ **NOTE:** A positive direct test for *N. gonorrhoeae* should prompt the collection of an additional specimen for culture and antimicrobial-agent susceptibility testing.

  a. Positive for *C. trachomatis*. The response difference is ≥350 RLU.
  b. Positive for *N. gonorrhoeae*. The response difference is ≥300 RLU.

3. A negative result indicates that the target organism is not present in the specimen.

  a. Negative for *C. trachomatis*. The response difference is <350 RLU.
  b. Negative for *N. gonorrhoeae*. The response difference is <300 RLU.

C. QC

1. Negative reference

  a. The response of each negative reference should be <200 but >20 RLU.
  b. All negative reference values should fall within 30% of the mean response of the negative reference.
  c. If one value falls outside of these ranges, it may be deleted.
  d. If two values fall outside these ranges, the results should not be reported, and the assay must be repeated.

2. Positive control

  a. The difference in the response of the positive control and the mean response of the negative reference should be >600 RLU.
  b. If the positive control value is not in the required range, the test results are not acceptable, and the assay must be repeated.

VI. PROCEDURE NOTES

A. Hybridization buffer and probe reagent

Heating and swirling of the buffer and reconstituted probe solution at 60°C are essential to prevent gel formation and ensure a homogeneous solution.

B. Specimens

1. Occasionally a specimen may be too viscous to pipette.
2. Ensure that specimens are at room temperature and vortex to liquefy.
3. Gen-Probe Fast Express reagent may be used to simplify specimen preparation.

C. Pipetting

1. A repeating pipettor may be used for adding probe solution, separation solution, and wash solution.
2. Pipettors with disposable tips are recommended to avoid sample carryover and cross-contamination.

D. Blotting

1. Discard absorbent paper after each blotting to avoid contamination.
2. Do not blot after the wash step.

E. Temperature

1. The hybridization and separation reactions are very temperature dependent. It is imperative that the water bath and reaction tubes be equilibrated uniformly during these steps.
2. Use a covered water bath capable of maintaining 60 ± 1°C.

F. Washing

1. The washing solution must be forcefully injected into each tube.
2. Angle the wash reagent toward the front sides of the tubes, not straight to the bottoms, to avoid splashback.
3. The appearance of a 1.0-cm-deep foam head at the top of a reaction tube indicates that the wash was forceful enough.
4. After adding washing solution to all tubes, top off by adding additional solution so that no foam remains.
5. Failure to deliver wash reagent in the proper manner may lead to spurious results.

REFERENCE

1. **Cintron, F.** 1994. DNA hybridization for detecting *Chlamydia trachomatis* and *Neisseria gonorrhoeae*, p. 10.5.b.1–10.5.b.6. *In* H. D. Isenberg (ed.), *Clinical Microbiology Procedures Handbook,* Supplement 1. American Society for Microbiology, Washington, D.C.

SUPPLEMENTAL READING

Clarke, L. M., M. F. Sierra, B. J. Daidone, N. Lopez, J. M. Covino, and W. M. McCormack. 1993. Comparison of the Syva Micro Trak enzyme immunoassay and Gen-Probe PACE 2 with cell culture for diagnosis of cervical *Chlamydia trachomatis* infection in a high-prevalence female population. *J. Clin Microbiol.* **31:**968–971.

Hale, Y. M., M. E. Melton, J. S. Lewis, and D. E. Willis. 1992. Evaluation of the PACE 2 *Neisseria gonorrhoeae* assay by three public health laboratories. *J. Clin. Microbiol.* **31:**451–453.

Harper, M., and R. Johnson. 1990. The predictive value of culture for the diagnosis of gonorrhoea and chlamydia infections. *Clin. Microbiol. Newsl.* **12:**54–56.

Hosein, I. K., A. M. Kaunitz, and S. J. Craft. 1992. Detection of cervical *Chlamydia trachomatis* and *Neisseria gonorrhoeae* with deoxyribonucleic acid probe assay in obstetric patients. *Am. J. Obstet. Gynecol.* **167:**588–591.

Iwen, P. C., T. M. H. Blair, and G. L. Woods. 1991. Comparison of the Gen-Probe PACE 2 system, direct fluorescent-antibody, and cell culture for detecting *Chlamydia trachomatis* in cervical specimens. *Am. J. Clin. Pathol.* **95:**578–582.

Kluytmans, J. A. J. W., H. G. M. Niesters, J. W. Mouton, W. G. V. Quint, J. A. J. Ijeplaar, J. H. Van Rijsoort-Vos, L. Habbema, E. Stolz, M. F. Michel, and J. H. T. Wagenvoort. 1991. Performance of a nonisotopic DNA probe for detection of *Chlamydia trachomatis* in urogenital specimens. *J. Clin. Microbiol.* **29:**2685–2689.

Lees, M. I., D. M. Newman, and S. M. Garland. 1991. Comparison of a DNA probe assay with culture for the detection of *Chlamydia trachomatis. J. Med. Microbiol.* **35:**159–161.

Limberger, R. J., R. Biega, A. Evanco, L. McCarthy, L. Slivienski, and M. Kirkwood. 1992. Evaluation of culture and the Gen-Probe PACE 2 assay for detection of *Neisseria gonorrhoeae* and *Chlamydia trachomatis* in endocervical specimens transported to a state health laboratory. *J. Clin. Microbiol.* **30:**1162–1166.

Ossewaarde, J. M., M. Rieffe, M. Rozenberg-Arska, P. M. Ossenkoppele, R. P. Nawrocki, and A. M. van Loon. 1992. Development and clinical evaluation of a polymerase chain reaction test for detection of *Chlamydia trachomatis. J. Clin. Microbiol.* **30:**2122–2128.

Panke, E. S., L. J. Yang, P. A. Leist, P. Magevney, R. J. Fry, and R. F. Lee. 1991. Comparison of Gen-Probe DNA probe test and culture for the detection of *Neisseria gonorrhoeae* in endocervical specimens. *J. Clin. Microbiol.* **29:**883–888.

Vlaspolder, F., J. A. E. M. Mutsaers, F. Blog, and A. Notowicz. 1991. Value of a DNA probe assay (Gen-Probe) compared with that of culture for diagnosis of gonococcal infection. *J. Clin. Microbiol.* **31:**107–110.

Warren, R., B. Dwyer, M. Plackett, K. Pettit, N. Rizvi, and A. Baker. 1993. Comparative evaluation of detection assay for *Chlamydia trachomatis. J. Clin. Microbiol.* **31:**1663–1666.

Yang, L. I., E. S. Panke, P. A. Leist, R. J. Fry, and R. F. Lee. 1991. Detection of *Chlamydia trachomatis* endocervical infection in asymptomatic and symptomatic women: comparison of Gen-Probe PACE 2 DNA probe test and tissue culture. *Am. J. Obstet. Gynecol.* **165:**1444–1453.

PART 2

Solution Hybridization Antibody Capture Assay for the Chemiluminescent Detection and Quantitation of Human Cytomegalovirus DNA in WBCs

I. PRINCIPLE

The Digene Hybrid Capture CMV DNA Assay (Digene Diagnostics, Inc., Silver Spring, Md.) is a solution hybridization antibody capture assay for the chemiluminescent detection and quantitation of human cytomegalovirus (CMV) in WBCs. Specimens containing the target DNA hybridize with a specific CMV RNA probe cocktail. The resultant RNA:DNA hybrids are captured onto the surface of a tube coated with antibodies specific for RNA-DNA hybrids. Immobilized hybrids are then reacted with alkaline phosphatase-conjugated antibodies specific for the RNA-DNA hybrids and detected with a chemiluminescent substrate. As the substrate is cleaved by the bound alkaline phosphatase, light is emitted and is measured as relative light units (RLU) on a luminometer. The intensity of the light emitted is proportional to the amount of target DNA in the specimen.

II. SPECIMEN

Whole blood (4 to 7 ml) collected in an EDTA tube must be processed within 48 h from the time of collection. After 48 h, adequate cell recovery may not occur. Immediately after collection, whole blood may be stored for up to 24 h at 20 to 25°C. Thereafter, the blood should be stored at 2 to 8°C until processing.

III. MATERIALS

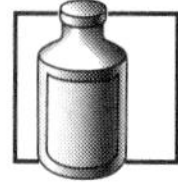

Include QC information on reagent container and in QC records.

A. Reagents (items 1 through 11 provided)

1. Lysis buffer concentrate
2. Negative control
3. Positive standards 1 to 3
4. Positive control
5. Sample diluent
6. Denaturation reagent
7. Indicator dye
8. CMV probe
9. Probe diluent
10. Detection reagents 1 and 2
11. Wash buffer pack
12. Sodium hypochlorite solution (household bleach [required but not provided]).

B. Supplies (items 3 through 13 required but not provided; items 8,12, and 13 available from Digene Diagnostics, Inc.)

1. Hybridization tubes and caps (provided)
2. Capture tubes (provided)
3. Graduated polypropylene conical vials (15 ml) with screw caps
4. Disposable transfer pipettes (5 ml) with standard tips
5. Disposable pipettes (5 ml) with extra fine tips
6. Sterile, RNase-free disposable micropipettor tips
7. Disposable pipette tips (25, 50, 75, and 250 μl)
8. Disposable 1-ml plastic transfer pipettes
9. Parafilm or equivalent
10. Polystyrene tubes (12 × 75 mm)
11. Specimen test tube racks
12. Hybridization racks
13. Decanting racks

C. Equipment (required but not provided; items 5 and 6 available from Digene Diagnostics, Inc.)

1. Clinical centrifuge with swinging-bucket rotor
2. 70 ± 2°C water bath
3. Vortex mixer
4. Rotary shaker with adjustable speed setting
5. Wash apparatus
6. DCR-1 luminometer or equivalent
7. Micropipettor with variable settings (20 through 200 μl)
8. Repeating positive displacement pipettor

IV. PROCEDURE

A. Sample preparation

1. Add 3.5 ml of whole blood to 10 ml of 1× lysis buffer. Mix and incubate for 15 min at 20 to 25°C. Centrifuge at 1,000 × *g* for 15 min. Discard supernatant.
2. Label hybridization tubes.
3. Suspend cell pellet in 1.5 ml of 1× lysis buffer. Transfer to hybridization tubes and incubate for 10 min at 20 to 25°C.
4. Centrifuge hybridization tubes to pellet cells. Discard supernatant.
   NOTE: At this point the specimens (cell pellets) may be tested immediately in the CMV DNA assay or may be stored at −20°C for future testing.

B. Denaturation

1. Place specimen hybridization tubes (containing specimen) in a hybridization rack.
2. Pipette 75 μl of sample diluent and 50 μl of denaturation reagent into each specimen hybridization tube. Cap and vortex each specimen.
3. Pipette 100 μl of controls or standards into the bottoms of control and standard hybridization tubes.
4. Pipette 50 μl of denaturation reagent into control and standard hybridization tubes.
5. Tighten caps on hybridization tubes and shake rack to mix on rotary shaker set at 1,100 ± 100 rpm for 5 ± 2 min.
   Standards will be dark purple: specimens will be purple to greenish-purple.

6. Incubate in a 70 ± 2°C water bath for 25 ± 5 min (label a new set of hybridization tubes for each specimen during this incubation).
7. Remove hybridization rack from water bath. Visually ensure that pellets have dissolved. Vortex individual tubes to suspend any remaining material.
8. Transfer specimens to fresh hybridization tubes.
9. Return hybridization rack to 70 ± 2°C water bath and incubate for 25 ± 5 min (prepare CMV probe mix).

C. Hybridization

1. Loosen hybridization tube caps. Pipette 50 μl of CMV probe mix into each tube. Recap hybridization tubes with same caps and shake on rotary shaker at 1,100 ± 100 rpm for 5 ± 2 min.
   ◩ Controls, standards, and specimens should turn yellow.
2. Incubate in a 70 ± 2°C water bath for 120 ± 5 min (label capture tubes).

D. Hybrid capture

1. Transfer contents from each hybridization tube to corresponding capture tube. Cover with Parafilm or equivalent. Shake at 1,100 ± 100 rpm at 20 to 25°C for 60 ± 5 min (prepare wash buffer).
2. Decant and blot capture tubes.

E. Hybrid detection

1. Pipette 250 μl of detection reagent 1 into each capture tube. Cover capture tubes and shake side-to-side and back-to-front by hand. Incubate at 20 to 25°C for 30 ± 3 min.
2. Decant and blot capture tubes.
3. Wash five times with wash buffer. Drain 5 min on absorbent paper.

F. Signal generation and reading

1. Pipette 250 μl of detection reagent 2 into each capture tube and incubate at 20 to 25°C for 30 ± 3 min.
2. Wipe and read capture tubes on luminometer.
3. Validate assay and interpret specimen results.

V. RESULTS

A. QC

◩ **NOTE:** As a validation procedure, users must test Positive Standards (PS) in triplicate until satisfactory results have been obtained in several consecutive assays.

1. For a qualitative assay, the Negative Control (NC) must be tested in triplicate and Positive Standard 1 (PS1) must be tested in duplicate for each test run.
2. For a quantitative assay, the NC and PS1, 2, and 3 must be tested in duplicate for each test run.
3. The detection reagent 2 blank should have an RLU value <5,000 and less than all control values. Values above 5,000 RLU suggest detection reagent 2 contamination. Specimen results cannot be interpreted in this case.
4. All control results should demonstrate a percent variability (%VAR) ≤30. If the %VAR of any triplicate result is >30, discard the control value with a RLU value furthest from the mean as an outlier and recalculate the mean by using the remaining two control values. If the difference between the mean and each of the two values is ≤30%, proceed to the next step; otherwise, the assay is invalid and must be repeated.
5. The mean of the NC results should be ≤12,000 RLU. If the mean of the NC is >12,000 RLU, the assay is invalid and must be repeated.
6. The PS mean (PS_{mean}) and NC mean (NC_{mean}) results are used to calculate the ratios that validate the assay. These results must be within the following acceptable ranges to validate the assay before the specimen results can be interpreted:

V. **RESULTS** *(continued)*

Assay	Calculation	Acceptable range
Qualitative	$PS1_{mean}/NC_{mean}$	1.5–10.0
Quantitative	$PS1_{mean}/NC_{mean}$	1.5–10.0
	$PS2_{mean}/NC_{mean}$	10.0–200
	$PS3_{mean}/NC_{mean}$	100–1,200

7. Calculate the appropriate ratios shown above by using the mean values on the luminometer printout tape. If all of the results are within their respective ranges, proceed to the next step. If any of the calculated values fall outside its expected range, the assay is invalid and must be repeated.
8. The Positive Control (PC) contains a defined concentration of CMV DNA that is detectable with the CMV DNA assay. This sample may be tested to meet the QC requirements of the testing laboratory. Acceptable values for the PC are as follows:

Assay	Calculation	Acceptable range
Qualitative	PC/NC_{mean}	1.8–15
Quantitative	—	10.5–42 pg/ml

B. Interpretation

1. Once the test run has been validated as described in item V.A above, the positive cutoff value (PCV) is calculated by the formula PCV = $[NC_{mean}] \times 2$.
2. Specimens with RLU values ≥PCV are considered ''positive'' for CMV DNA.
3. Specimens with RLU values <PCV and ≥90% of the PCV are considered ''equivocal'' for CMV DNA. A second sample should be obtained and tested.
4. Specimens with RLU values <90% of the PCV contain CMV DNA levels that are below the detection limit of the assay or do not contain CMV DNA. These should be interpreted as ''no CMV DNA detected.''
5. When the quantitative assay is performed, only specimens with RLU values between PS1 and PS3 can be quantified accurately since this is in the linear range of the assay. Samples giving RLU values greater than PS3 should be reported as ''high positive'' and positive samples giving RLU values below PS1 but greater than the PCV should be reported as ''low positive.''
6. To determine the actual concentration of CMV in the specimen, the number of genome equivalents detected must be calculated from the concentration of complementary DNA (in picograms per milliliter) reported by the luminometer from the following formula:

$$\frac{\text{CMV genomes}}{\text{Assay}} = \text{Reported CMV DNA concentration (pg/ml)} \times 2{,}410\ \frac{\text{genomes-ml}}{\text{assay-pg}}$$

The number of CMV genomes per milliliter of specimen can be calculated as follows:

$$\frac{\text{CMV genomes}}{\text{Assay}} \times \text{Initial specimen volume (ml)} = \frac{\text{CMV genomes}}{\text{ml of specimen}}$$

SUPPLEMENTAL READING

Boeckh, M., P. M. Woogerd, T. Stevens-Ayers, C. G. Ray, and R. A. Bowden. 1994. Factors influencing detection of quantitative cytomegalovirus antigenemia. *J. Clin. Microbiol.* **32:** 832–834.

Digene Diagnostic, Inc. 1994. Digene Hybrid Capture™ System CMV DNA Assay, package insert. Digene Diagnostic, Inc., Silver Spring, Md.

Erice, A., M. A. Holm, P. C. Gill, S, Henry, C. L. Dirksen, D. L. Dunn, R. P. Hillam, and H. H. Balfour, Jr. 1992. Cytomegalovirus (CMV) antigenemia assay is more sensitive than shell vial cultures for rapid detection of CMV in polymorphonuclear blood leukocytes. *J. Clin. Microbiol.* **30:**2822–2825.

Gerna, G., D. Zipeto, M. Parea, M. G. Revello, E. Silini, E. Percivalle, M. Zavattoni, P. Grossi, and G. Milanesi. 1991. Monitoring of human cytomegalovirus infections and ganciclovir treatment in heart transplant recipients by determination of viremia, antigenemia, and DNA emia. *J. Infect. Dis.* **164:**488–498.

Ho, M. 1991. *Cytomegalovirus: Biology and Infection*, 2nd ed. Plenum Publishing Corp., New York.

Munderli, W., M. K., Kagi, E. Gruter, and J. D. Auracher. 1989. Detection of cytomegalovirus in peripheral leukocytes by different methods. *J. Clin. Microbiol.* **27:**1916–1917.

PART 3

Solution Hybridization Antibody Capture Chemiluminescent Assay for the Detection of Human Papillomavirus Types in Cervical Specimens

I. PRINCIPLE

The Digene Hybrid Capture HPV DNA Assay is a hybridization antibody capture assay that uses chemiluminescence to qualitatively detect the presence of 14 human papillomavirus (HPV) types: 6, 11, 16, 18, 31, 33, 35, 42, 43, 44, 45, 51, 52, and 56. The HPV DNA Assay can differentiate between two HPV DNA groups based on their risk of cervical neoplasm (low-risk HPV types 6, 11, 42, 43, and 44 and high- or intermediate-risk HPV types 16, 18, 31, 33, 35, 45, 51, 52, and 56) in cervical specimens (cervical swabs and fresh cervical biopsies) collected by using the Digene Specimen Collection Kit or Digene Specimen Transport Medium (Digene Diagnostic, Inc., Silver Spring, Md.). Specimens containing the target DNA hybridize with a specific HPV RNA probe cocktail. The resultant RNA:DNA hybrids are captured onto the surface of a tube coated with antibodies specific for RNA: DNA hybrids. Immobilized hybrids are then reacted with alkaline phosphatase-conjugated antibodies specific for the hybrids and detected with a chemiluminescent substrate.

II. SPECIMEN

A. Cervical specimens (cervical swabs and cervical biopsies) must be collected by using the Digene Specimen Collection Kit and transported in Digene Specimen Transport Medium. Specimens taken with other sampling devices or transported in other transport media are not suitable for this assay.

B. Appropriately collected cervical swab specimens may be held for up to 2 weeks at room temperature and shipped without refrigeration to the testing laboratory. At the testing laboratory, specimens should be stored at 2 to 8°C if the assay is to be performed within 1 week of collection. If the assay will be performed later than 1 week after collection, store specimens at −20°C.

C. Freshly collected cervical biopsies up to 5 mm in cross section may also be analyzed with the HPV DNA Assay. The biopsy specimen must be placed immediately into 1 ml of Digene Specimen Transport Medium and stored at −20°C. Biopsy specimens may be shipped at 20 to 25°C for overnight delivery to the testing laboratory and stored at −20°C until processed. Biopsies less than 2 mm in diameter should not be used.

III. MATERIALS

Include QC information on reagent container and in QC records.

A. Reagents (items 1 through 11 provided)

1. Negative control (NC)
2. Positive control A (PCA)
3. Positive control B (PCB)
4. Denaturation reagent
5. Indicator dye
6. Probe diluent
7. HPV probe A (HPV 6, 11, 42, 43, and 44 RNA probe cocktail)
8. HPV probe B (HPV 16, 18, 31, 33, 35, 45, 51, 52, and 56 RNA probe cocktail)
9. Detection reagent 1

III. MATERIALS *(continued)*

10. Detection reagent 2
11. Wash buffer pack
12. Sodium hypochlorite solution (household bleach [required but not provided])

B. **Supplies** (items 5 through 13 required but not provided; items 6, 7, and 13 available from Digene Diagnostics, Inc., Silver Spring, Md.)

1. Specimen transportation tube vent caps (provided)
2. Hybridization tubes (provided)
3. Hybridization tube caps (provided)
4. Capture tubes (provided)
5. Specimen test tube racks
6. Hybridization racks
7. Decanting racks
8. Disposable bench cover, paper towels, powder-free gloves, Kimwipes, or equivalent
9. Parafilm or equivalent
10. Polystyrene tubes (12 × 75 mm)
11. Sterile RNase-free disposable pipette tips (20 to 200 μl)
12. Disposable pipette tips (50, 250, and 500 μl)
13. Disposable 1-ml transfer pipettes

C. **Equipment** (required but not provided; items 4 and 5 available from Digene Diagnostics)

1. 65 ± 2°C water bath
2. Vortex mixer
3. Rotary shaker with adjustable speed setting
4. Wash apparatus
5. DCR-1 luminometer or equivalent
6. Micropipettor with variable settings for 20- to 200-μl delivery volumes
7. Repeating positive displacement pipettor

IV. PROCEDURE

A. **Setup**

1. Specimens
   a. Remove specimens and all required reagents from refrigerator before beginning assay.
   b. Allow specimens to reach 20 to 25°C for at least 15 to 30 min.
   c. Mix specimens thoroughly by vortexing.
   d. Specimens may be tested once with the combined probe cocktail or once with HPV probe A cocktail and once with HPV probe B cocktail if using the two-probe method.
2. Controls
   a. If using the combined probe cocktail method, PCA, PCB, and NC should each be tested in triplicate with the combined probe cocktail.
   b. If using the two-probe method, PCA and NC should be tested in triplicate with HPV probe A cocktail. PCB and NC should be tested in triplicate with HPV probe B cocktail.

B. **Denaturation and hybridization: combined probe cocktail and two-probe methods**

1. Pipette 500 μl of denaturation reagent into bottoms of control and specimen tubes. Cap tubes with vent caps and vortex to mix thoroughly. Check that all tubes show a purple color.

   NOTE: Some cervical specimens may contain blood or other material which may mask the color changes upon addition of denaturation reagent and probe. In these cases, failure to exhibit the proper color change will not affect the results of the assay.
2. Incubate at 65 ± 2°C for 45 ± 5 min.
3. Label hybridization tubes and prepare probe cocktails.
4. Pipette 50 μl of appropriate probe cocktail (combined, probe A, or probe B) into the bottom of the hybridization tubes.
5. Mix denatured specimen well, remove swab, vortex, and pipette 150 μl into the bottoms of hybridization tubes.
6. Recap tubes and shake on rotary shaker at 1,100 ± 100 rpm for 3 ± 2 min. Check that all tubes show yellow color.
7. Incubate at 65 ± 2°C for 60 ± 5 min.
8. Label capture tubes.

C. Hybrid capture

1. Transfer contents from each hybridization tube to corresponding capture tube.
2. Cover tubes with Parafilm or equivalent and shake at 1,100 ± 100 rpm at 20 to 25°C for 60 ± 5 min. Prepare wash buffer.
3. Decant and blot tubes.

D. Hybrid detection

1. Pipette 250 μl of detection reagent 1 into each capture tube. Cover tubes and shake by hand several times side-to-side and back-to-front.
2. Incubate at 20 to 25°C for 30 ± 3 min.
3. Decant and blot tubes.
4. Wash five times with wash buffer and drain 5 min on absorbent paper.

E. Signal generation

1. Pipette 250 μl of detection reagent 2 into each capture tube. Include detection reagent 2 blank at this time.
2. Incubate in dark at 20 to 25°C for 20 ± 3 min.
3. Wipe and read tubes in luminometer.
4. Validate assay (QC) and interpret specimen results.

V. RESULTS

A. QC

1. Combined probe cocktail method
   - **a.** Three replicates of NC, PCA, and PCB must be tested with the combined probe cocktail for each test run.
   - **b.** PCA and PCB should be assayed separately.
2. Two-probe method
   - **a.** Three replicates of PCA and NC must be tested with HPV probe A for each test run.
   - **b.** Three replicates of PCB and NC must be tested with HPV probe B for each test run.
3. Both methods
   - **a.** The detection reagent 2 blank should have an RLU value <5,000 and less than all control values.
   - **b.** All control results should have a demonstrated percent variability (%VAR) ≤30.
   - **c.** The positive control mean (PC_{mean}) and negative control mean (NC_{mean}) results are used to calculate the PC_{mean}/NC_{mean} ratio for each probe. These ratios must all be ≥1.5 to validate the assay and before the specimen results can be interpreted.

B. Interpretation

1. Once the assay has been validated, the cutoff values for determining positive specimens are as follows:
   - **a.** Combined probe cocktail method

$$\frac{0.8 \times (PCA_{mean} + PCB_{mean})}{2}$$

   - **b.** Two-probe method

     Probe A cutoff $= 1 \times PCA_{mean}$

     Probe B cutoff $= 1 \times PCB_{mean}$
2. All specimen RLU values should be converted into a ratio to the appropriate cutoff value. For example, all assays tested with HPV probe A should be expressed as specimen RLU divided by cutoff value A.
3. Specimens with RLU/cutoff value ratios ≥1.0 with the combined probe cocktail are considered ''positive'' for one or more of HPV types 6, 11, 16, 18, 31, 33, 35, 42, 43, 44, 45, 51, 52, and 56.

V. RESULTS *(continued)*

4. Specimens with RLU/cutoff value ratios ≥1.0 with HPV probe A only are considered "positive" for one or more of HPV types 6, 11, 42, 43, and 44.
5. Specimens with RLU/cutoff value ratios ≥1.0 with HPV probe B only are considered "positive" for one or more of HPV types 16, 18, 31, 33, 35, 45, 51, 52, and 56.
6. Specimens with RLU/cutoff value ratios of ≥1.0 with both probe A and probe B are considered "positive" for one or more HPV types from each group of probes.
7. Specimens with RLU/cutoff value ratios <1.0 for combined probe cocktail of both HPV probe A and probe B are considered "negative" or "none detected" for the 14 HPV types tested. Either HPV DNA sequences are absent or the HPV DNA levels are below the detection limit of the assay.

SUPPLEMENTAL READING

Cox, J. T., A. T. Lorincz, M. H. Schiffman, M. E. Sherman, A. Cullen, and R. J. Kurman. 1995. Human papillomavirus testing by hybrid capture appears to be useful in triaging women with a cytologic diagnosis of atypical squamous cells of undetermined significance. *Am. J. Obstet. Gynecol.* **172:**946–954.

Digene Diagnostics, Inc. 1995. Digene Hybrid Capture™ System HPV DNA Assay, package insert. Digene Diagnostics, Inc., Silver Spring, Md.

Ferenczy, A. 1995. Viral testing for genital human papillomavirus infections: recent progress and clinical potentials. *Int. J. Gynecol. Cancer* **5:** 321–328.

Koutsky, L. A., K. K. Holmes, C. W. Critchlow, C. E. Stevens, J. Paavonen, A. M. Beckmann, T. A. DeRouen, D. A. Galloway, D. Vernon, and N. B. Kiviat. 1992. A cohort study of the risk of cervical intraepithelial neoplasia grade 2 or 3 in relation to papillomavirus infection. *N. Engl. J. Med.* **327:**1272–1278.

Morrison, E. A. B., G. Y. F. Ho, S. H. Vermund, G. L. Goldberg, A. S. Kadish, K. F. Kelley, and R. D. Burk. 1991. Human papillomavirus infection and other risk factors for cervical neoplasia: a case-control study. *Int. J. Cancer* **49:**6–13.

Schiffman, M. H., N. B. Kiviat, R. D. Burk, K. V. Shah, R. W. Daniel, R. Lewis, J. Kuypers, M. M. Manos, D. R. Scott, M. E. Sherman, R. J. Kurman, M. H. Stoler, A. G. Glass, B. B. Rush, I. Mielzynska, and A. T. Lorincz. 1995. Accuracy and interlaboratory reliability of human papillomavirus DNA testing by hybrid capture. *J. Clin. Microbiol.* **33:**545–550.

Sherman, M. E., M. H. Schiffman, A. T. Lorincz, M. M. Manos, D. R. Scott, R. J. Kurman, N. B. Kiviat, M. Stoler, A. G. Glass, and B. B. Rush. 1994. Toward objective quality assurance in cervical cytopathology: correlation of cytopathologic diagnoses with detection of high-risk human papillomavirus types. *Am. J. Clin. Pathol.* **102:**182–187.

10.2.3 Amplification-Based Methods

INTRODUCTION

Nucleic acid target amplification has the potential to overcome the lack of sensitivity that is the primary limitation of probe hybridization methods when used for direct detection of infectious agents in clinical material. As the first amplification technique to be developed, polymerase chain reaction (PCR) remains the most widely applied molecular method for diagnosis of infectious diseases. Commercial kits employing PCR technology for detection of *Mycobacterium tuberculosis*, hepatitis C virus (HCV), human immunodeficiency virus (HIV), *Neisseria gonorrhoeae*, and *Chlamydia trachomatis* have been developed and are or will soon be available. The availability of commercial products with built-in contamination controls, standardized reagents, and potential for automation will almost certainly increase the reproducibility and ease of application of this technology. The newer amplification technologies such as ligase chain reaction (LCR), transcription-mediated amplification, and strand displacement amplification will form the basis of additional diagnostic systems that are, or are soon likely to be, available for use in the clinical laboratory (Table 10.1–1).

PART 1 Detection of *Chlamydia trachomatis* in Genitourinary Specimens by Using the Roche Amplicor PCR Kit

I. PRINCIPLE

The Amplicor *Chlamydia trachomatis* test is a direct DNA probe test that uses the polymerase chain reaction (PCR) and nucleic acid hybridization for detection of *C. trachomatis* plasmid DNA in endocervical, urethral (male), and urine (male) samples. *C. trachomatis* contains, in addition to its chromosomal DNA, a cryptic plasmid (ca. 7,500 base pairs) that is common to all serovars of *C. trachomatis*. For the detection of *C. trachomatis* DNA in the Amplicor *C. trachomatis* test, the specific target sequence DNA is located on the cryptic plasmid and is 207 base pairs in length. Each elementary body of *C. trachomatis* contains about 10 copies of this plasmid and the reticulate body stage contains at least 10 copies of the plasmid, depending on its state of replication. The Amplicor *C. trachomatis* test is based on three major processes: PCR amplification, hybridization of the amplified product to a specific nucleic acid probe, and detection of the amplified product by color formation.

Because of the sensitivity of the PCR method, it is necessary to limit the potential for contamination by performing each step of the process in a *separate area* (preferably separate rooms) of the laboratory and to dedicate equipment to each of the areas. Workflow in the laboratory must proceed in a unidirectional manner, beginning in the Reagent Preparation Area and moving to the Specimen Preparation Area and then to the Amplification/Detection Area. Preamplification activities must begin with reagent preparation and proceed to specimen preparation. Reagent preparation activities and specimen preparation activities must be performed in separate, segregated areas. Supplies and equipment must be dedicated to each activity and not used for other activities or moved between areas. Lab coats and gloves must be worn in each area and must be changed before leaving that area. Equipment and supplies used for reagent preparation must not be used for specimen preparation activities or for pipetting or processing amplified DNA or other sources of target DNA. Amplification and detection supplies and equipment must be confined to the Amplification/Detection Area at all times.

10 Molecular Biology

II. MATERIALS

(Unless otherwise indicated, materials listed are required but not provided.)

Include QC information on reagent container and in QC records.

A. Area 1: reagent preparation

1. Amplicor *Chlamydia trachomatis* Amplification Kit (provided)
2. Biological safety cabinet or dead-air box equipped for UV irradiation.
3. Dedicated lab coat (store in area 1)
4. Powder-free gloves
5. 10% bleach
6. 70% ethanol
7. Pipette with plugged pipette tips (100 μl)
8. MicroAmp tray, base, tray, and retainer

II. MATERIALS *(continued)*

9. Repeat pipettor with 1.25 μl of sterile Combitips
10. Plastic baggie

B. Area 2: specimen preparation

1. Amplicor STD Swab Specimen Preparation Kit (provided)
2. Amplicor STD Urine Preparation Kit (provided)
3. Amplicor PCR controls (positive and negative) (provided)
4. Dedicated lab coat (store in area 2)
5. Powder-free gloves
6. 10% bleach
7. 70% ethanol
8. Pipetaide
9. Sterile 2-ml serologic pipettes
10. Sterile transfer pipettes
11. 37°C (±2°C) incubator
12. Vortex mixer
13. Test tube racks
14. 15 ml conical polypropylene centrifuge tubes
15. Centrifuge
16. 4 × 4 gauze
17. Pipette with plugged pipette tips (50 and 1,000 μl)
18. MicroAmp tube caps and capping tool

C. Area 3: amplification and detection

1. Amplicor *Chlamydia trachomatis* Detection Kit (provided)
2. Dedicated lab coat (store in area 3)
3. Powder-free gloves
4. 10% bleach
5. Thermal cycler with printer
6. Spatula or tool to remove tube caps
7. Multichannel pipette with plugged pipette tips (100 μl)
8. Pipette with plugged pipette tips (25 μl)
9. Disposable reagent reservoirs
10. Erlenmeyer flask (1 liter)
11. Graduated cylinder (100 ml)
12. 37°C (±2°C) incubator
13. Disposable 96-well plate with lid
14. Microwell plate washer
15. Sterile 15-ml polypropylene centrifuge tubes
16. ELISA plate reader

III. SPECIMENS

A. Use Amplicor STD Specimen Collection and Transport Kit (provided).

B. Female endocervical swabs

C. Male urethral swabs

D. Male first-catch urine collected in clean *plastic* container

IV. SPECIMEN TRANSPORT AND STORAGE

A. Transport the specimen at room temperature. It is stable for 24 h at room temperature. Upon receipt the specimen should be stored at 2 to 8°C. Process swab specimen within 10 days of collection. Process urine specimens within 4 days of collection.

◪ Specimens that are received with the swab still in the transport medium are unacceptable for testing.

B. Specimens that require shipment to off-site test centers must be stored at 2 to 8°C until shipment. Specimens should be shipped with guaranteed arrival within 24 h.

◪ Urine specimens that have been frozen are unacceptable for testing.

V. PROCEDURE

A. Area 1: reagent preparation

◪ **NOTE:** To minimize contamination, the use of a biological safety cabinet (hood) or dead-air box equipped for UV irradiation is suggested. All reagent preparation should be performed in this cabinet.

1. Clean out dead-air box or hood with 10% bleach. Wipe out with 70% ethanol.
2. Add 100 μl of Amperase to each tube of master mix (one tube is sufficient for 32 amplifications).
3. Recap and mix by inverting 10 to 15 times. Discard Amperase tube when empty. Record date of preparation on master mix tube.

4. Determine the number of PCR tubes needed and place in MicroAmp sample tray and lock in place with retainer. (Run one positive and three negative controls with each run. Well locations for controls are as follows: positive, A1; negatives, B1, H1, and A2.)
5. Pipet 50 μl of working master mix into each PCR tube by using repeat pipettor with sterile Combitip or a micropipette with plugged tips.
6. Place tray in a baggie and transport to area 2.

B. Area 2: specimen preparation

1. Control preparation
   a. Add 0.75 ml of control diluent to the tube containing the negative control. Recap the tube and vortex well to mix. Incubate at room temperature for 10 min.
   b. Add 0.75 ml of control diluent to the tube containing the positive control. Recap the tube and vortex well to mix. Incubate at room temperature for 10 min.
2. Urine preparation
   a. Check urine for precipitates. If present, warm the specimen for up to 30 min at 37°C and mix to dissolve. If some precipitate remains after heating, it will not interfere with the test.
   b. Vortex specimen for 3 to 10 s.
   c. Transfer 7 to 8 ml to a conical *polypropylene* 15-ml centrifuge tube.
   d. Centrifuge at 1,500 × *g* for 10 min at room temperature.
   e. Discard supernatant. Tap tube on clean 4 × 4 gauze. (Use new gauze for each specimen.)
   f. Add 2 ml of urine resuspension buffer. Cap and vortex for 3 to 10 s to suspend the pellet. (A multitube vortexer may be used.)
   g. Let stand for 1 h at room temperature.
   h. Add 2 ml of urine diluent to each tube. Cap and vortex for 3 to 10 s.
   i. Let stand at room temperature for 10 min.
3. Swab preparation
   a. Check the transport tube for swab. If a specimen has been transported containing a swab, contact the lead scientist or section manager. Verify that the seal has been broken on transported specimens.
   b. Pipet 1 ml of specimen diluent to each tube with a sterile serologic pipette.
   c. Recap the tube and vortex for 5 to 10 seconds.
   d. Incubate at room temperature for 10 min.
4. Tray preparation
   a. Using a clean plugged tip for each sample, pipette 50 μl of processed sample into appropriate MicroAmp tubes. Be careful not to pipette any precipitated material that may not have been resuspended.
   b. Cap the tubes tightly with the capping tool.
   c. Move the prepared samples in tray to area 3.

C. Area 3: Amplification and detection

1. Amplification
   a. Remove MicroAmp tray from base.
   b. Place MicroAmp tray into the thermal cycler block. Check to make sure the notch in the sample tray is at the left of the block and that the rim of the tray is seated in the channel around the block.
   c. Slide the cover forward.
   d. Turn the knob clockwise until hand tight. (The white mark on the cover knob should line up with the white mark on the cover.)
   e. Program the thermal cycler as follows:
      Program 1 CYCLE Program (1 cycle): 5 min, 95°C; 1 min, 60°C
      Program 2 CYCLE Program (29 cycles): 30 s, 95°C; 1 min, 60°C
      Program 3 HOLD Program: 5 min, 72°C

V. **PROCEDURE** *(continued)*

Program 4 HOLD Program: 72°C constant
In CYCLE programs, the ramp times are left at 0:00. Link the four programs together into METHOD Program 5.

f. Start the METHOD program (program runs about 1.5 h).
g. Remove the completed PCR sample tray from the thermal cycler and place in tray base. *Do not remove from area 3.*
h. Remove caps carefully to avoid aerosolizing PCR products.
i. *Immediately* add 100 μl of denaturation solution to each PCR tube with a multichannel pipette (program 1 on Amplicor pipettor) with plugged tips.
j. Incubate for 10 min at room temperature. Store the denatured, amplified samples at room temperature only if the detection test will be performed within 1 to 2 h. If not, store the samples at 2 to 8°C for up to 1 week.
k. Review the thermal cycler run parameters for "HISTORY FILE" and record on "RUN QUALITY CONTROL LOG," or print run parameters on printer during run.

2. Detection

a. Prepare working wash solution by adding 1 vol of wash concentrate (10×) to 9 vol of distilled, deionized water. Mix well.
b. Allow microwell plate to warm to room temperature before removing from the foil pouch. Remove the appropriate number of 8-well strips and set into microwell plate frame. Return unused strips to foil pouch and reseal bag, making sure desiccant pillow remains in the pouch.
c. Add 100 μl of hybridization buffer to each well (program 2 on Amplicor pipettor). If the amplified samples were stored at 2 to 8°C, it may be necessary to incubate them at 37°C for 2 to 4 min to reduce viscosity.
d. Using plugged pipette tips, pipette 25 μl of denatured amplified samples to the appropriate microtiter well. (A multichannel pipettor can be used.) Place lid on tray and gently tap the plate 10 to 15 times until color changes from blue to light yellow.
e. Cover plate and place in 37°C (±2°C) incubator for 1 h.
f. Wash plate five times manually or by microwell plate washer, using the prepared 1× washing solution.
  (1) Manually (do not use squirt bottle):
    (a) Empty contents of plate and tap on paper towel.
    (b) Pipette working wash solution to fill each well (400 to 450 μl).
    (c) Soak 30 s. Empty contents and tap dry on a paper towel.
    (d) Repeat steps 2 and 3 four additional times.
  (2) Automated (program washer):
    (a) Aspirate contents of well.
    (b) Fill each well to top with working wash solution (350 to 450 μl), soak for 30 s and aspirate dry.
    (c) Repeat step 2 four additional times.
    (d) Tap plate dry.
g. Add 100 ml of avidin-horseradish peroxidase (HRP) conjugate to each well. Cover plate and incubate 15 min at 37°C (±2°C).
h. Wash plate as described in step f.
i. Prepare working substrate by mixing 2 ml of substrate A and 0.5 ml of substrate B into a polypropylene tube for each pair of 8-well strips. Prepare substrate no more than 3 h before use and protect from the light.
j. Pipette 100 μl of prepared working substrate into each well. Check wells for uneven blue color and note these wells on the worksheet.
k. Allow color to develop for 10 min at room temperature *in the dark.*
l. Add 100 μl of stop reagent to each well.
m. Measure the optical density at 450 nm within 1 h of adding stop reagent.

VI. RESULTS

The presence of *C. trachomatis* in the sample is determined by relating the absorbance of the unknown specimen to that of the cutoff value.

A. Assay validation

1. All three negative control values *must* be <0.25 optical density units (ODU). If not, the test (entire batch) must be repeated.
2. Each of the three negative control values *should* be within ±25% of the negative control mean.
3. The positive control *must* be >2.000 ODU. If not, the test must be repeated.
4. The thermal cycler run QC log *must* show that there was a successful run.
5. The test must be repeated if any of the following conditions apply:
   a. A negative control of >0.25 ODU
   b. A positive control of <2.000 ODU
   c. A failed thermal cycler run
6. A sample processing control should be run monthly. McCoy cells (10^3–10^4) infected with *C. trachomatis* should be added to a fresh tube of specimen transport medium (provided in Collection and Transport Kit) for 1 h and treated as a normal clinical specimen. A positive signal of >0.25 ODU at 450 nm should be obtained.

B. Protocol for repeat testing

Store specimens at 4°C and retest within 4 days of initial processing. Warm specimens to room temperature and vortex each specimen 10 to 15 s to mix.

C. Interpretation of results

1. A clinical specimen with an A_{450} reading of <0.2 ODU should be reported as negative for *C. trachomatis*.
2. A clinical specimen with an A_{450} reading of >0.5 ODU should be reported as positive for *C. trachomatis*.
3. A clinical specimen with an A_{450} between 0.2 and 0.5 ODU should be considered equivocal. The specimen should be repeated in duplicate. The final test result should be determined as follows:
   a. Compile all three results (initial and duplicate repeat) of the sample.
   b. If two of the three results are <0.250 ODU, sample is negative.
   c. If two of the three results are >0.250 ODU, sample is positive.
4. The analytical sensitivity of this assay is 1 inclusion forming unit (approximately 10 plasmid copies) of any *C. trachomatis* serovar (including serovars responsible for lymphogranuloma venereum).

VII. LIMITATIONS

A. The Amplicor *C. trachomatis* test will not detect plasmid-free variants of *C. trachomatis*.

B. *Only standard chlamydial cell culture methods should be used for evaluation of suspected sexual abuse and for other medico-legal indications.*

C. Additional testing is recommended in any circumstance when false positive results could lead to adverse medical, social, or psychological consequences. Test results should be interpreted with consideration of clinical and laboratory findings.

D. The presence of PCR inhibitors may cause false-negative results.

E. The presence of spermicides in excess of 1%, or surgical lubricants in excess of 10%, in the specimen for testing may have an inhibitory effect.

F. The presence of mucus and blood in cervical specimens has not been demonstrated to have any direct biochemical effect.

G. A negative result does not exclude the possibility of infection because reliable results are dependent on adequate specimen collection.

H. Therapeutic success or failure cannot be determined, because chlamydial plasmid DNA may persist after appropriate antimicrobial-agent therapy.

SUPPLEMENTAL READING

Bauwens, J. E., A. M. Clark, M. J. Loeffelholz, and S. A. Herman. 1993. Diagnosis of *Chlamydia trachomatis* urethritis in men by polymerase chain reaction of first-catch urine. *J. Clin. Microbiol.* **31:** 3013–3016.

Bauwens, J. E., A. M. Clark, and W. E. Stamm. 1993. Diagnosis of *Chlamydia trachomatis* endocervical infections by a commercial polymerase chain reaction assay. *J. Clin. Microbiol.* **31:** 3023–3027.

Centers for Disease Control and Prevention. 1993. Recommendations for the prevention and management of *Chlamydia trachomatis* infections. *Morbid. Mortal. Weekly Rep.* **42:**1–39.

Jaschek, G., C. A. Gaydos, L. E. Welsh, and T. C. Quinn. 1993. Direct detection of *Chlamydia trachomatis* in urine specimens from symptomatic and asymptomatic men by using a rapid polymerase chain reaction assay. *J. Clin. Microbiol.* **31:** 1209–1212.

Ossewaarde, J. M., M. Rieffe, M. Rozenberg-Arska, P. M. Ossenkoppele, R. Nawrocki, and A. M. VanLoon. 1992. Development and clinical evaluation of a polymerase chain reaction test for detection of *Chlamydia trachomatis*. *J. Clin. Microbiol.* **30:**2122–2128.

Roche Diagnostic Systems. 1993. Amplicor™ 1993. *Chlamydia trachomatis* test, package insert. Roche Diagnostic Systems, Nutley, N.J.

Vogels, W. H. M., P. C. vanVoorst Vader, and F. P. Schroder. 1993. *Chlamydia trachomatis* infection in a high-risk population: comparison of polymerase chain reaction and cell culture for diagnosis and follow-up. *J. Clin. Microbiol.* **31:** 1103–1107.

PART 2

Qualitative Detection of Hepatitis C Virus RNA by Using the Roche Amplicor Reverse Transcriptase-PCR Kit

I. PRINCIPLE

The major cause of posttransfusion non-A, non-B hepatitis is the hepatitis C virus (HCV). The entire genome of HCV has been cloned and sequenced. The 5′ untranslated region of the virus is remarkably well conserved and serves as the target sequence for several molecular detection assays including the Roche Amplicor HCV test. Detection of HCV RNA in serum or plasma has been applied to the acute diagnosis of the disease, as a supplement to results obtained by immunoassays, and as a monitor of therapeutic protocols. The Amplicor HCV test is based on four major processes: reverse transcription of RNA to generate a complementary DNA (cDNA), polymerase chain reaction (PCR) target amplification, hybridization of the amplified product to a specific oligonucleotide probe, and detection of the amplified product by color formation.

Because of the sensitivity of the PCR method, it is necessary to limit the potential for contamination by performing each step of the process in a separate area (preferably separate rooms) of the laboratory and to dedicate equipment to each of the areas. Workflow in the laboratory must proceed in a unidirectional manner, beginning in the Reagent Preparation Area and moving to the Specimen Preparation Area and then to the Amplification/Detection Area. Preamplification activities must begin with reagent preparation and proceed to specimen preparation. Reagent preparation activities and specimen preparation activities must be performed in separate, segregated areas. Supplies and equipment must be dedicated to each activity and not used for other activities or moved between areas. Lab coats and gloves must be worn in each area and must be changed before leaving that area. Equipment and supplies used for reagent preparation must not be used for specimen preparation activities or for pipetting or processing amplified DNA or other sources of target DNA. Amplification and detection supplies and equipment must be confined to the Amplification/Detection Area at all times.

II. MATERIALS

(Unless indicated, materials listed are required but not provided.)

Include QC information on reagent container and in QC records.

A. Area 1: reagent preparation

1. Amplicor HCV Amplification Kit (provided)
2. Biological safety cabinet or dead-air box equipped for UV irradiation.
3. Dedicated lab coat (store in area 1)
4. Powder-free gloves
5. 10% bleach
6. 70% ethanol
7. Pipette with plugged pipette tips (100 μl)
8. MicroAmp tray, base, tray, and retainer
9. Repeat pipettor with 1.25-μl sterile Combitips
10. Plastic baggie

B. Area 2: specimen preparation

1. Amplicor HCV Specimen Preparation Kit (provided)
2. Amplicor HCV PCR controls (positive and negative; provided)
3. Amplicor HCV Serum Control Kit (provided)
4. Dedicated lab coat (store in area 2)
5. Powder-free gloves
6. 10% bleach
7. 70% ethanol
8. Pipetaide
9. Attached screw-cap microcentrifuge tubes (1.5 μl), conical bottom
10. 60°C ± 2°C dry-heat block

11. 95% ethyl alcohol
12. Isopropyl alcohol
13. Sterile transfer pipettes (graduated)
14. Vortex mixer
15. Test tube racks
16. Microcentrifuge
17. Pipettes with plugged pipette tips (50, 100, and 1,000 μl)
18. MicroAmp tube caps and capping tool

C. Area 3: amplification and detection

1. Amplicor HCV Detection Kit (provided)
2. Dedicated lab coat (store in area 3)
3. Powder-free gloves
4. 10% bleach
5. 70% ethanol
6. Thermal cycler with printer
7. Spatula or tool to remove tube caps
8. Multichannel pipette with plugged pipette tips (100 μl)
9. Pipette with plugged pipette tips (25 μl)
10. Disposable reagent reservoirs
11. Erlenmeyer flask (1 liter)
12. Graduated cylinder (100 μl)
13. 37°C (±2°C) incubator
14. Disposable 96-well plate with lid
15. Microwell plate washer
16. Sterile 15-μl polypropylene centrifuge tubes
17. ELISA plate reader

III. SPECIMENS

Serum or plasma collected in acid citrate dextrose (ACD) or EDTA

IV. SPECIMEN TRANSPORT AND STORAGE

A. Transport the specimen at room temperature.

B. Within 3 h of blood draw, blood collection tubes must be centrifuged at 1,500 $\times$ g for 20 min at room temperature.

C. Collect serum or plasma and store in 150-μl aliquots at 2 to 8°C for 72 h. For longer storage, store at −20 to −80°C.

V. QUALITY CONTROL

A. For each batch of specimens processed, include three replicates of the positive serum control and one replicate of the negative serum control. Process the controls in the exact manner as the patient specimens.

B. Specimens and controls from separate preparation batches may be amplified and detected at the same time. Each separate specimen batch is validated individually by the set of serum controls for that batch.

C. All test specimens and controls prepared in the same batch should be amplified and detected in adjacent positions in the thermal cycler and on the detection plate. The exact order of the placement is not critical.

D. For each PCR amplification, run one positive reagent control and one negative reagent control.

VI. PROCEDURE

A. Area 1: reagent preparation

NOTE: To minimize contamination, the use of a biological safety cabinet (hood) or dead-air box equipped with UV irradiation is suggested. All reagent preparation should be performed in this cabinet.

1. Warm all reagents to room temperature before beginning.
2. Clean out dead-air box or hood with 10% bleach. Wipe out with 70% ethanol.
3. Add 100 μl of Amperase to one tube of master mix (one tube is sufficient for 32 amplifications).
4. Recap and mix by inverting 10 to 15 times. Discard Amperase tube when empty. Record date of preparation on master mix tube.

VI. PROCEDURE *(continued)*

5. Determine the number of PCR tubes needed and place in MicroAmp sample tray and lock in place with retainer. (See QC section for number of controls.)
6. Pipette 50 μl of working master mix into each PCR tube by using a repeat pipettor with a sterile Combitip or a micropipette with plugged tips.
7. Place tray in a baggie and transport to area 2. Store at 2 to 8°C until use

B. Area 2: specimen preparation

◩ **NOTE:** Prewarm the HCV lysis reagent by placing it in a 37°C incubator. Ensure that crystals have gone into solution.

1. Clean out hood with 10% bleach. Wipe out with Sanicloth disinfectant.
2. Reagent control preparation

   ◩ **NOTE:** Do not extract the reagent controls with lysis reagent. These reagent controls are added directly to the PCR tubes.

   a. Add 50 μl of HCV(−) control to a tube containing 200 μl of control diluent. Recap and vortex *well.* Incubate at room temperature for 10 min. Negative reagent control must be amplified within 5 h of preparation. Do not reuse.

   b. Add 50 μl of HCV(+) control to a tube containing 200 μl of control diluent. Recap and vortex *well.* Incubate at room temperature for 10 min. Positive reagent control must be amplified within 5 h of preparation. Do not reuse.
3. Serum control preparation

   a. Label three 1.5-μl microcentrifuge tubes as positive serum controls and one as a negative serum control. Draw a vertical line on the tubes for pellet orientation during centrifugation. Add 400 μl of lysis reagent to each tube.

   b. Add 100 μl of normal serum and mix well by vortexing for 10 s.

   ◩ **NOTE:** The normal serum is not included with the kit. Use a serum which has previously been tested as negative by PCR.

   c. Vortex controls and add 50 μl of the appropriate control to each tube. Vortex well.
4. For each patient specimen, label a 1.5-μl microcentrifuge tube and draw a vertical line for orientation of the pellet after centrifugation. Add 400 μl of lysis reagent to each tube.
5. Thaw specimens at room temperature and vortex. Add 100 μl to the appropriate tube and mix well by vortexing.
6. Incubate all tubes for 10 min at 60 ± 2°C; then mix well by vortexing.
7. Using a new tip for each specimen, add 500 μl of 100% isopropyl alcohol to each tube and mix well by vortexing.
8. Incubate all tubes for 2 min at room temperature.
9. Centrifuge tubes for 15 min at 13,000 to 16,000 × *g* at room temperature. Position tubes so that all are in the same orientation in the rotor.
10. Using a different transfer pipette for each specimen, remove and discard the supernatant into a container containing 10% bleach. Let bleach stand for 24 h and discard down the drain. Flush sink with running water after discarding bleach.
11. Add 1.0 ml of 70% ethanol (made fresh) and vortex.
12. Centrifuge for 5 min at 13,000 to 16,000 × *g* at room temperature. Position tubes so that all are in the same orientation in the rotor.
13. Remove the supernatant from each tube with a different transfer pipette. Discard into a container with 10% bleach. Let sit for 24 h before discarding down the drain. Flush sink with running water after discarding bleach.
14. Suspend the pellet in 1.0 ml of specimen diluent by carefully, but thoroughly, dislodging the pellet with a pipette tip from the *side* of the tube.

◪ NOTE: The specimens and controls should be held at room temperature until ready to use. RNA preparations should be amplified within 3 h of preparation. If this is not possible, the RNA preparations can be stored at −20°C for up to 1 month. Thaw frozen RNA preparations at room temperature before proceeding to tray preparation.

15. Tray preparation
   a. Using a clean plugged tip for each sample, pipette 50 μl of processed sample into appropriate MicroAmp tubes. Be careful not to pipette any precipitated material that may not have been resuspended.
   b. Cap the tubes tightly with the capping tool.
   c. Move the prepared samples in tray to area 3.

C. Area 3: amplification and detection

1. Amplification
   a. Remove MicroAmp tray from base.
   b. Place MicroAmp tray into the thermal cycler block. Check to make sure the notch in the sample tray is at the left of the block and that the rim of the tray is seated in the channel around the block.
   c. Slide the cover forward.
   d. Turn the knob clockwise until hand tight. (The white mark on the cover knob should line up with the white mark on the cover.)
   e. Program the thermal cycler as follows:

Program 13 HOLD Program:	2 min, 50°C
Program 14 HOLD Program:	30 min, 60°C
Program 15 HOLD Program:	1 min, 95°C
Program 16 CYCLE Program (2 cycles):	15 s, 95°C; 20 s, 60°C
Program 17 CYCLE Program (38 cycles):	15 s, 90°C; 20 s, 60°C
Program 18 HOLD Program:	72°C (*at least 5 min; do not exceed 15 min*)

   In CYCLE programs, the ramp times are left at 0:00.
   Link the programs together into METHOD Program 19.
   f. Start the METHOD program (program runs about 1.75 h).
   ◪ NOTE: Specimens must be removed within 15 min of the start of the final hold program.
   g. Remove the completed PCR sample tray from the thermal cycler and place in tray base. *Do not remove from area 3.*
   h. Remove caps carefully to avoid aerosolizing PCR products.
   i. *Immediately* add 100 μl of denaturation solution to each PCR reaction tube with a multichannel pipette (program 1 on Amplicor pipettor) with plugged tips.
   j. Incubate for 10 min at room temperature. Store the denatured, amplified samples at room temperature only if the detection test will be performed within 1 to 2 h. If not, store the samples at 2 to 8°C for up to 1 week.
   k. Review the thermal cycler run parameters for "HISTORY FILE" and record on "RUN QUALITY CONTROL LOG," or print run parameters on printer during run.

2. Detection
   a. Prepare working wash solution by adding 1 vol of wash concentrate (10×) to 9 vol of distilled, deionized water. Mix well.
   b. Allow microwell plate to warm to room temperature before removing from the foil pouch. Remove the appropriate number of 8-well strips and set into microwell plate frame. Return unused strips to foil pouch and reseal bag, making sure desiccant pillow remains in the pouch.

VI. PROCEDURE *(continued)*

c. Add 100 μl of hybridization buffer to each well (program 2 on Amplicor pipettor). If the amplified samples were stored at 2 to 8°C, it may be necessary to incubate them at 37°C for 2 to 4 min to reduce viscosity.
d. Using plugged pipette tips, pipette 25 μl of denatured amplified samples to the appropriate microtiter well. (A multichannel pipetter may be used.) Place lid on tray and gently tap the plate 10 to 15 times until color changes from blue to light yellow.
e. Cover plate and place in 37°C (±2°C) incubator for 1 h.
f. Wash plate 5 times manually or by microwell plate washer, using the prepared 1× washing solution.
    (1) Manually (do not use squirt bottle):
        (a) Empty contents of plate and tap on paper towel.
        (b) Pipette working wash solution to fill each well (400 to 450 μl).
        (c) Soak 30 s. Empty contents and tap dry on a paper towel.
        (d) Repeat steps 2 and 3 four additional times.
    (2) Automated (program washer):
        (a) Aspirate contents of well.
        (b) Fill each well to top with working wash solution (350 to 450 μl), soak for 30 s, and aspirate dry.
        (c) Repeat step 2 five additional times.
        (d) Tap plate dry.
g. Add 100 μl of avidin-HRP conjugate to each well. Cover plate and incubate 15 min at 37°C (±2°C).
h. Wash plate as described in step f.
i. Prepare working substrate by mixing 2 ml of substrate A and 0.5 ml of substrate B into a polypropylene tube for each pair of 8-well strips. Prepare substrate no more than 3 h before use and protect from the light.
j. Pipette 100 μl of prepared working substrate into each well. Check wells for uneven blue color and note these wells on the worksheet.
k. Allow color to develop for 10 min, at room temperature, *in the dark*.
l. Add 100 μl of stop reagent to each well.
m. Measure the optical density at 450 nm within 1 h of adding stop reagent.

VII. EVALUATION OF RESULTS

The presence of HCV RNA in the sample is determined by relating the absorbance of the unknown specimen to that of the cutoff value.

A. Assay validation

1. The absorbance of the negative serum control should be less than or equal to 0.25. If the negative serum control in a specimen batch is greater than 0.25, that batch of specimens should be invalidated and the entire test procedure for that batch repeated.
2. The absorbance of at least two of the three positive serum controls in each batch must be greater than or equal to 1.50. If this criterion is not met, the entire test procedure for that batch should be invalidated and repeated.
3. The absorbance of the negative reagent control should be less than or equal to 0.25. If this criterion is not met, the entire amplification and detection run should be invalidated and repeated.
4. The absorbance of the positive reagent control should be greater than or equal to 2.00. If this criterion is not met, the entire amplification and detection run should be invalidated and repeated.
5. The thermal cycler run QC log *must* show that there was a successful run.

6. The test must be repeated if any of the following conditions apply:
   a. A negative control of >0.25 ODU
   b. Two of three positive serum controls of <1.50 ODU
   c. A positive reagent control of <2.00 ODU
   d. A failed thermal cycler run

B. Interpretation of results

1. A clinical specimen with an A_{450} reading of <0.25 ODU should be reported as negative for HCV RNA.
2. A clinical specimen with an A_{450} reading of >0.60 ODU should be reported as positive for HCV RNA.
3. A clinical specimen with an A_{450} between 0.25 and 0.60 ODU should be considered equivocal. The specimen should be repeated in duplicate. The final test result should be determined as follows.
   a. Compile all three results (initial and duplicate repeat) of the sample.
   b. If two of the three results are <0.40, sample is negative.
   c. If two of the three results are >0.40, sample is positive.

SUPPLEMENTAL READING

Busch, M. P., J. C. Wilber, P. Johnson, L. Tobler, and C. S. Evans. 1992. Impact of specimen handling and storage on detection of hepatitis C virus RNA. *Transfusion* **32:**420–425.

Cha, T. A., J. Kolberg, B. Irvine, M. Stempen, E. Beall, M. Yano, Q.-L. Choo, M. Houghton, G. Kuo, J. H. Han, and M. S. Urdea. 1991. Use of signature nucleotide sequence of the hepatitis C virus for the detection of viral RNA in human serum and plasma. *J. Clin. Microbiol.* **29:** 2528–2534.

Nolte, F. S., C. Thurmond, and M. W. Fried. 1995. Preclinical evaluation of AMPLICOR hepatitis C virus test for hepatitis C virus RNA. *J. Clin. Microbiol.* **33:**1775–1778.

Roche Diagnostic Systems, 1995. Amplicor™ Hepatitis C Virus RNA Reverse Transcriptase-PCR Kit, package insert. Roche Diagnostic Systems, Nutley, N.J.

Young, K. K. Y., R. M. Resnick, and T. W. Myers. 1993. Detection of hepatitis C virus RNA by a combined reverse transcription-polymerase chain reaction assay. *J. Clin. Microbiol.* **31:** 882–886.

PART 3

Quantitative Measurement of Hepatitis C Virus RNA by Using the Roche Amplicor HCV Monitor Reverse Transcriptase-PCR Kit

I. PRINCIPLE

The Amplicor HCV Monitor test is a polymerase chain reaction (PCR) assay for the quantitative measurement of hepatitis C virus (HCV) RNA in serum or plasma. The test includes an RNA quantitation standard of known copy number that is coamplified with the target and is used to assign the copy level to the specimen. The quantitation standard (QS) is a synthetic RNA molecule with primer binding sites identical to the HCV target RNA and with a unique internal sequence specific for the QS probe. The test is based on four major processes: reverse transcription of RNA to generate a complementary DNA (cDNA), PCR target amplification, hybridization of the amplified product to a specific oligonucleotide probe, and detection of the amplified product by a color formation.

Because of the sensitivity of the PCR method, it is necessary to limit the potential for contamination by performing each step of the process in a separate area (preferably a separate room) of the laboratory and to dedicate equipment to each of the areas. Workflow in the laboratory must proceed in a unidirectional manner, beginning in the Reagent Preparation Area and moving to the Specimen Preparation Area and then to the Amplification/Detection Area. Preamplification activities must begin with reagent preparation and proceed to specimen preparation. Reagent and specimen preparation activities must be performed in separate, segregated areas. Supplies and equipment must be dedicated to each activity and not used for other activities or moved between areas. Lab coats and gloves must be worn in each area and must be changed before leaving that area. Equipment and supplies used for reagent preparation must not be used for specimen preparation activities or for pipetting or processing amplified DNA or other sources of target DNA. Amplification and detection supplies and equipment must be confined to the Amplification/Detection Area at all times.

II. MATERIALS

(Unless indicated, materials listed are required but not provided.)

Include QC information on reagent container and in QC records.

A. Area 1: reagent preparation

1. Amplicor HCV Amplification Kit (provided)
2. Biological safety cabinet or dead-air box equipped for UV irradiation
3. Dedicated lab coat (store in area 1)
4. Powder-free gloves
5. 10% bleach
6. 70% ethanol
7. Pipette with plugged pipette tips (100 μl)
8. MicroAmp tray, base, tray, and retainer
9. Repeat pipettor with 1.25 μl of sterile Combitips
10. Plastic baggie
11. 95% ethyl alcohol
12. Isopropyl alcohol
13. Sterile transfer pipettes (graduated)
14. Vortex mixer
15. Test tube racks
16. Microcentrifuge
17. Pipettes with plugged pipette tips (50, 100, and 1,000 μl)
18. MicroAmp tube caps and capping tool

B. Area 2: specimen preparation

1. Amplicor HCV Specimen Preparation Kit (provided)
2. Amplicor HCV PCR Controls (positive and negative; provided)
3. Amplicor HCV Serum Control Kit (provided)
4. Dedicated lab coat (store in area 2)
5. Powder-free gloves
6. 10% bleach
7. 70% ethanol
8. Pipetaide
9. Attached screw-cap microcentrifuge tubes (1.5 μl), conical bottom
10. 60°C ± 2°C dry-heat block

C. Area 3: amplification and detection

1. Amplicor HCV Detection Kit (provided)
2. Dedicated lab coat (store in area 3)
3. Powder-free gloves
4. 10% bleach
5. 70% ethanol
6. Thermal cycler with printer
7. Spatula or tool to remove tube caps
8. Multichannel pipette with plugged pipette tips (100 μl)
9. Pipette with plugged pipette tips (25 μl)
10. Disposable reagent reservoirs
11. Erlenmeyer flask (1 liter)
12. Graduated cylinder (100 μl)
13. 37°C (±2°C) incubator
14. Disposable 96-well plate with lid
15. Microwell plate washer
16. Sterile 15-μl polypropylene centrifuge tubes
17. ELISA plate reader

III. SPECIMENS

Serum or plasma collected in ACD or EDTA

IV. SPECIMEN TRANSPORT AND STORAGE

A. Transport the specimen at room temperature.

B. Within 3 h of blood draw, blood collection tubes must be centrifuged at 1,500 × *g* for 20 min at room temperature.

C. Collect serum or plasma and store in 150-μl aliquots at 2 to 8°C for 72 h. For longer storage, store at −20 to −80°C.

V. QUALITY CONTROL

A. For each batch of specimens processed, include one replicate of the positive serum control and one replicate of the negative serum control. Process the controls in the same manner as the patient specimens.

B. Specimens and controls from separate preparation batches may be amplified and detected at the same time. Each separate specimen batch is validated individually by the set of serum controls for that batch.

C. All test specimens and controls prepared in the same batch should be amplified and detected in adjacent positions in the thermal cycler and on the detection plate. The exact order of the placement is not critical.

VI. PROCEDURE

A. Area 1: reagent preparation

◪ **NOTE:** To minimize contamination, the use of a biological safety cabinet (hood) or dead-air box equipped for UV irradiation is suggested. All reagent preparation should be performed in this cabinet.

1. Clean out dead-air box or hood with 10% bleach. Wipe out with 70% ethanol.
2. Remove HCV lysis reagent from storage at 4°C. Mix by inversion to dissolve any precipitate. Vortex the Internal QS and add the appropriate volume. The copy number of Internal QS varies with the kit lot. Check the kit insert for the amount of Internal QS specified.
3. Determine the number of PCR tubes needed for patient samples and controls, place in MicroAmp sample tray, and lock in place with retainer.
4. Pipet 50 μl of working master mix into each PCR tube by using repeat pipettor with sterile Combitip or a micropipette with plugged tips. Inspect the reaction tubes after pipetting to ensure the same volume was pipetted into each tube.
5. Place the tray in a baggie and transport to area 2. Store at 2 to 8°C until used.

B. Area 2: specimen preparation

1. Prepare aliquots of isopropanol and 70% ethanol, made fresh daily.
2. Label 1.5-ml microcentrifuge tubes for one positive serum control, one negative serum control, and patient samples. Draw a vertical line on the tubes for pellet orientation during centrifugation. Add 400 μl of lysis reagent to each tube.
3. Control preparation
   a. Mix normal serum well by vortexing and add 100 μl to the lysis reagent in the control tubes. Vortex well after mixing.
   ◪ **NOTE:** The normal serum is not included with the kit. Use a serum which has previously been tested as negative by PCR.
   b. Add 100 μl of the appropriate control to each tube.
4. Vortex the patient serum or plasma after thawing, add 100 μl to the appropriate tube, and mix well by vortexing.
5. Incubate all tubes for 10 min at 60°C ($\pm$2°C), then mix well by vortexing.
6. Add 500 μl of 100% isopropyl alcohol to each tube and mix well by vortexing.
7. Incubate all tubes for 2 min at room temperature.
8. Centrifuge tubes for 15 min at 13,000 to 16,000 $\times$ g at room temperature. Position tubes so that all are in the same orientation in the rotor. The RNA pellet may not be visible at this point.
9. Remove and discard the supernatant from each tube with a different transfer pipette.
10. Add 1.0 ml of 70% ethanol (made fresh) and vortex.
11. Centrifuge for 5 min at 13,000 to 16,000 $\times$ g at room temperature. Position tubes so that all are in the same orientation in the rotor.
12. Remove the supernatant from each tube with a different transfer pipette. Discard into a container with 10% bleach. Let discarded supernatant sit for 24 h before discarding down the drain. Flush sink with running water after discarding bleach.
13. Suspend the pellet in 1.0 ml of specimen diluent by carefully dislodging the pellet with a pipette tip from the side of the tube. Break it apart as much as possible by scraping the tube wall with a P200 pipettor fitted with a P200 plugged pipette tip. Scrape the top third, the middle third, the bottom third, and then the entire area of the tube wall. Vortex for 5 s. Let the

VI. PROCEDURE *(continued)*

undissolved particles settle to the bottom of the tube. Do not vortex just before adding to the master mix. Pipette samples from near the top of the preparation and avoid pipetting the chunks of precipitate.

NOTE: The specimens and controls should be held at room temperature until ready to use. RNA preparations should be amplified within 3 h of preparation. If this is not possible, the RNA preparations can be stored at −20°C for up to 1 month.

14. Tray preparation
- **a.** Using a clean plugged tip for each sample, pipette 50 μl of processed sample into appropriate MicroAmp tubes. Be careful not to pipette any precipitated material that may not have been resuspended.
- **b.** Cap the tubes tightly with the capping tool.
- **c.** Move the prepared samples in tray to area 3.

C. Area 3: amplification and detection

1. Amplification
- **a.** Remove MicroAmp tray from base.
- **b.** Place MicroAmp tray into the thermal cycler block. Check to make sure the notch in the sample tray is at the left of the block and that the rim of the tray is seated in the channel around the block.
- **c.** Slide the cover forward.
- **d.** Turn the knob clockwise until hand tight. (The white mark on the cover knob should line up with the white mark on the cover).
- **e.** Program the thermal cycler as follows:

Program 13 HOLD Program:	2 min, 50°C
Program 14 HOLD Program:	30 min, 60°C
Program 15 HOLD Program:	1 min, 95°C
Program 16 CYCLE Program (2 cycles):	15 s, 95°C; 20 s, 60°C
Program 17 CYCLE Program (38 cycles):	15 s, 90°C; 20 s, 60°C
Program 18 HOLD Program:	72°C (*at least 5 min; do not exceed 15 min*)

  In CYCLE programs, the ramp times are left at 0:00.
  Link the programs together into METHOD Program 19.
- **f.** Start the METHOD program (program runs about 1.75 h).

  NOTE: Specimens must be removed within 15 min of the start of the final hold program.
- **g.** Remove the completed PCR sample tray from the thermal cycler and place in tray base. *Do not remove from area 3.*
- **h.** Remove caps carefully to avoid aerosolizing PCR products.
- **i.** *Immediately* add 100 μl of denaturation solution to each PCR reaction tube by using a multichannel pipette (program 1 on Amplicor pipettor) with plugged tips.
- **j.** Incubate for 10 min at room temperature. Store the denatured, amplified samples at room temperature only if the detection test will be performed within 1 to 2 h. If not, store the samples at 2 to 8°C for up to 1 week.
- **k.** Review the thermal cycler run parameters for "HISTORY FILE" and record on "RUN QUALITY CONTROL LOG," or print run parameters on printer during run.

2. Detection
- **a.** Prepare working wash solution by adding 1 vol of wash concentrate (10×) to 9 vol of distilled, deionized water. Mix well.

b. Allow microwell plate to warm to room temperature before removing from the foil pouch. Remove the appropriate number of 8-well strips and set into microwell plate frame.

c. Add 100 μl of hybridization buffer to each well (program 2 on Amplicor pipettor).

d. Using plugged pipette tips, pipette 25 μl of denatured amplified samples to the appropriate microtiter well. (A multichannel pipettor may be used.) Place lid on tray and gently tap the plate 10 to 15 times until color changes from blue to light yellow.

e. Cover plate and place in a 37°C (±2°C) incubator for 1 h.

f. Wash plate 5 times manually or with a microwell plate washer, using the prepared 1× washing solution.

(1) Manually (do not use squirt bottle):

(a) Empty contents of plate and tap on paper towel.

(b) Pipette working wash solution to fill each well (400 to 450 μl).

(c) Soak 30 s. Empty contents and tap dry on a paper towel.

(d) Repeat steps 2 and 3 four additional times.

(2) Automated (program washer):

(a) Aspirate contents of well.

(b) Fill each well to top with working wash solution (350 to 450 μl), soak for 30 s, and aspirate dry.

(c) Repeat step (b) five additional times.

(d) Tap plate dry.

g. Add 100 μl of avidin-HRP conjugate to each well. Cover plate and incubate 15 min at 37°C (±2°C).

h. Wash plate as described in step f.

i. Prepare working substrate by mixing 2 ml of substrate A and 0.5 ml of substrate B into a polypropylene tube for each pair of 8-well strips. Prepare working substrate no more than 3 h before use and protect from the light.

j. Pipette 100 μl of prepared working substrate into each well. Check wells for uneven blue color and note these wells on the worksheet.

k. Allow color to develop for 10 min at room temperature, *in the dark.*

l. Add 100 μl of stop reagent to each well.

m. Measure the optical density (OD) at 450 nm within 1 h of adding stop reagent.

VII. EVALUATION OF RESULTS

The presence of HCV RNA in the sample is determined by relating the absorbance of the unknown specimen to that of the cutoff value.

A. Assay validation

1. The absorbance of the negative serum control should be less than or equal to 0.25. If the negative serum control in a specimen batch is greater than 0.25, that batch of specimens should be invalidated and the entire test procedure for that batch repeated.
2. The absorbance of at least two of the three positive serum controls in each batch must be greater than or equal to 1.50. If this criterion is not met, the entire test procedure for that batch should be invalidated and repeated.
3. The absorbance of the negative reagent control should be less than or equal to 0.25. If this criterion is not met, the entire amplification and detection run should be invalidated and repeated.
4. The absorbance of the positive reagent control should be greater than or equal to 2.00. If this criterion is not met, the entire amplification and detection run should be invalidated and repeated.

VII. EVALUATION OF RESULTS *(continued)*

5. The thermal cycler run QC log *must* show that there was a successful run.
6. The test must be repeated if any of the following conditions apply:
   - **a.** A negative control of >0.25 ODU
   - **b.** Two of three positive serum controls of <1.50 ODU
   - **c.** A positive reagent control of <2.00 ODU
   - **d.** A failed thermal cycler run

B. Interpretation of results

1. For each sample, select the highest dilution of the amplicon that gives an OD between 0.2 and 2.0 on the HCV microwells (rows A through E). Repeat the same determination for the QS microwells (rows F through H). Enter the optical density and the dilution factors for the HCV and QS microwells in the worksheet. These values will be used to calculate the number of HCV copies per milliliter.

   NOTE: For each sample, the OD for the Internal QS must be above 0.5 in at least one of the wells for the determination to be valid. If it is not, then both the specimen preparation and amplification/detection must be repeated for that sample.
2. To manually calculate the results, follow the instructions below:
   - **a.** Choose the appropriate HCV well that has an OD value in the range of 0.2 to 2.0 ODU. If more than one well is in this range, choose the well with the larger dilution factor (i.e., the smaller OD).
   - **b.** Subtract the background from the selected HCV OD value. Background = 0.07 ODU.
   - **c.** Calculate the total HCV OD by multiplying (OD − background value of the selected HCV well) times the dilution factor associated with that well).
   - **d.** Choose the appropriate QS well that has an OD value in the range of 0.2 to 2.0 ODU. If more than one well is in range, choose the well with the larger dilution factor (i.e., the smaller OD). The OD value of the QS in row F must be greater than or equal to 0.5 for the test result to be valid. If it is not, then both the specimen preparation and amplification/detection must be repeated for that sample.
   - **e.** Subtract the background from the selected QS OD value. Background = 0.07 ODU.
   - **f.** Calculate the total QS OD by multiplying (OD − background of the selected QS well) times the dilution factor associated with that well.
   - **g.** Calculate the HCV RNA copies per milliliter as follows:

     $$\left(\frac{\text{Total HCV OD}}{\text{Total QS OD}} \times \text{input copies QS}\right) \times 200$$
   - **h.** Unexpected results
     - (1) All HCV values are <0.20 ODU. If all HCV wells have OD values less than 0.2, but the QS wells have the expected values, the result should be reported as undetectable.
     - (2) All HCV wells are >2.0 ODU. If all HCV wells have OD values greater than 2.0 but the QS wells have the expected values, then either an error occurred in the test or the HCV copy number is above the dynamic range of the assay. Report the result as Not Determined. If it is suspected that the sample has a very high titer, repeat the entire test using 10 μl of serum instead of 100 μl. Calculate the results as above, then multiply the final result by 10.
     - (3) All QS values are <0.2 ODU. If all QS wells are less than 0.2, then either the processed sample was inhibitory to the amplification or the RNA was not recovered during sample preparation. Report

the result as Not Determined and repeat the entire test, including sample preparation and amplification/detection.

(4) All QS values are >2.0 ODU. If this is the case, then an error occurred. Report the result as Not Determined and repeat the assay.

SUPPLEMENTAL READING

Gretch, D., L. Corey, and J. Wilson. 1994. Assessment of hepatitis C virus levels by quantitative competitive RNA polymerase chain reaction: high titer viremia correlates with advanced stage of disease. *J. Infect. Dis.* **169:**1219–1225.

Marzin, A., P. Bagnarelli, S., Menzo, F. Giostra, M. Brugia, R. Francesconi, F. B. Bianchi, and M. Clementi. 1994. Quantitation of hepatitis C virus genome molecules in plasma samples. *J. Clin. Microbiol.* **32:**1939–1944.

Nolte, F. S., C. Thurmond, and M. W. Fried. 1995. Preclinical evaluation of AMPLICOR hepatitis C virus test for detection of hepatitis C virus RNA. *J. Clin. Microbiol.* **33:**1775–1778.

Roche Diagnostic Systems. 1995. Amplicor™ HCV Monitor Reverse Transcriptase-PCR Kit, package insert. Roche Diagnostic Systems, Nutley, N.J.

Roth, W. K., J. H. Lee, B. Ruster, and S. Zeuzem. 1996. Comparison of two quantitative hepatitis C virus reverse transcriptase PCR assays. *J. Clin. Microbiol.* **34:**261–264.

PART 4

Quantitative Measurement of Human Immunodeficiency Virus Type 1 RNA by Using the Amplicor HIV-1 MONITOR Test Kit

I. PRINCIPLE

The Amplicor HIV-1 MONITOR Test is an in vitro nucleic acid amplification test for the quantitation of human immunodeficiency virus type 1 (HIV-1) RNA in human plasma. The Amplicor HIV-1 MONITOR Test is based on five major processes: specimen preparation; reverse transcription (RT) of target RNA to generate cDNA; PCR amplification of target cDNA by using HIV-1 specific complementary primers; hybridization of the amplified products to oligonucleotide probes specific to the target(s); and detection of the probe-bound amplified products by colorimetric determination. The Amplicor HIV-1 MONITOR Test amplifies and detects a 142-base target sequence located in a highly conserved region of the HIV-1 *gag* gene, defined by the primers SK431 and SK462. The test quantitates viral load by utilizing a second target sequence quantitation standard (QS) that is added to the amplification mixture at a known concentration. The QS is a noninfectious 219-base, in vitro-transcribed RNA molecule that contains SK431 and SK462 primer binding sites and generates a target of the same length (142 base pairs) as the HIV-1 target. The probe-binding region of the QS has been modified to differentiate QS-specific amplicons from HIV-1 target amplicons. HIV-1 levels in the test specimen are determined by comparing the absorbance of the specimen with that obtained for the QS.

Workflow in the laboratory must proceed in a unidirectional manner, beginning in the Reagent Preparation Area and moving to the Specimen Preparation Area and then to the Amplification/Detection Area. Preamplification activities must begin with reagent preparation and proceed to specimen preparation. Reagent preparation activities and specimen preparation activities must be performed in separate, segregated areas. Supplies and equipment must be dedicated to each activity and not used for other activities or moved between areas. Lab coats and gloves must be worn in each area and must be changed before leaving that area. Equipment and supplies used for reagent preparation must not be used for specimen preparation activities or for pipetting or processing amplified DNA or other sources of target DNA. Amplification and detection supplies and equipment must be confined to the Amplification/Detection Area at all times.

10 Molecular Biology

II. MATERIALS

Include QC information on reagent container and in QC records.

A. Materials provided

1. Specimen preparation reagents
   a. Lysis reagent
   b. QS
   c. Specimen diluent
2. Controls
   a. Negative plasma (human)
   b. Negative control
   c. Positive control—low
   d. Positive control—high
3. Amplification reagents
   a. Master mix
   b. Manganese solution
4. Detection reagents
   a. Denaturation solution
   b. Hybridization buffer
   c. Avidin-HRP conjugate
   d. Substrate A
   e. Substrate B
   f. Stop reagent
   g. 10× wash concentrate
   h. Oligonucleotide probe-coated microwell plate

II. MATERIALS *(continued)*

B. Materials required but not provided

1. Preamplification reagent preparation
   a. Consumables for thermal cycler
      (1) Reaction tubes
      (2) Caps
      (3) Tray/retainers
      (4) Base
   b. Plastic resealable bag
   c. Repeating pipettor with 1.25-ml reservoir
   d. Micropipettes (adjustable volume, 20 to 200 μl)
   e. Plugged (aerosol barrier) pipette tips (50 and 200 μl)
   f. Latex gloves
   g. Vortex mixer
2. Preamplification specimen preparation
   a. Microcentrifuge
   b. Sterile screw-cap tubes (2.0 ml)
   c. Tube racks
   d. Ethyl alcohol, absolute (freshly diluted to 70% with deionized water)
   e. Isopropanol, reagent grade (100%)
   f. Fine-tip, sterile transfer pipettes
   g. Vortex mixer
   h. Latex gloves
   i. Sterile disposable, polystyrene pipettes (5, 10, and 25 ml)
   j. Micropipettes with plugged (aerosol barrier) or positive displacement tips (50, 100, 200, 400, 600, 800, and 1,000 μl)
3. Postamplification area—amplification/detection
   a. Multichannel pipettor (25 and 100 μl)
   b. Plugged (aerosol barrier) micropipette tips (25 and 100 μl) and unplugged tips (100 μl)
   c. Thermal cycler (Perkin-Elmer GeneAmp PCR System 9600 or GeneAmp PCR System 2400)
   d. MicroAmp Base (Perkin-Elmer GeneAmp PCR System 9600 or GeneAmp PCR System 2400)
   e. Microwell plate washer
   f. Microwell plate reader
   g. Disposable reagent reservoirs
   h. Microwell plate lid
   i. Incubator, 37°C (±2°C)
   j. Graduated vessels (25 to 100 ml)
   k. Distilled or deionized H_2O

III. SPECIMEN

A. Specimen collection and storage

1. The Amplicor HIV-1 MONITOR Test is for use with plasma samples only.
2. Blood should be collected in sterile tubes with EDTA or ACD as the anticoagulant.
   ◩ Specimens anticoagulated with heparin are unsuitable for this test.
3. Store blood at 20 to 25°C for no longer than 6 h. Do not refrigerate whole blood.
4. Separate plasma within 6 h by centrifugation at 800 to 1,600 × *g* for 20 min at room temperature.
5. Plasma samples may be stored at room temperature for 24 h, at 2 to 8°C for up to 5 days, or at −20°C to −80°C.
6. Plasma specimens may be frozen and thawed up to three times.

B. Specimen transport

1. Whole blood must be transported at 2 to 25°C and processed within 6 h of collection.
2. Plasma may be transported at 2 to 8°C or frozen.

IV. PROCEDURE

A. Reagent preparation

1. Performed in Preamplification-Reagent Preparation Area
2. Prepare working master mix by adding 100 μl of HIV-1 MONITOR Test manganese solution to one tube of HIV-1 MONITOR Test master mix.
3. Place reaction tubes in tray and lock in place with tube retainer.
4. Pipette 50 μl of master mix into each tube.
5. Place tray in plastic resealable bag, and move to Specimen Preparation Area.

6. Store the tray at 2 to 8°C until specimen preparation is completed. Amplification must begin within 4 h of preparation of working master mix.

B. Specimen and control preparation

1. Performed in Preamplification-Specimen Preparation Area
2. Prepare 70% ethanol.
3. Prepare working lysis reagent. Add 100 μl of HIV-1 MONITOR Test QS to one bottle of lysis reagent and mix (enough for 12 specimens).
4. Label 2.0-ml screw-cap microcentrifuge tube for each specimen and control.
5. Thaw plasma specimens at room temperature and vortex 3 to 5 s.
6. Spin tube briefly to collect specimen in base of tube.
7. Dispense 600 μl of working lysis reagent into each tube.
8. Add 200 μl of each patient specimen to appropriate tube, cap, and vortex 3 to 5 s.
9. For each negative and positive control, add 200 μl of negative plasma (human) to the appropriate tubes. Cap the tubes and vortex 3 to 5 s, then add 50 μl of HIV-1 MONITOR Test (−) control, HIV-1 MONITOR Test low (+) control, and HIV-1 MONITOR Test high (+) control to the appropriate tubes. Cap the tubes and mix.
10. Incubate tubes 10 min at room temperature.
11. Add 800 μl of 100% isopropanol to each tube, cap, and vortex 3 to 5 s.
12. Put an orientation mark on each tube and place tubes into microcentrifuge with orientation mark facing outward, so that the pellet will align with the orientation mark. Centrifuge specimens at maximum speed (at least 12,500 $\times$ *g*) for 15 min at room temperature (begin timing from the moment of reaching maximum speed).
13. Using a new, fine-tip disposable transfer pipette for each tube, carefully remove and discard supernatant from each tube, being careful not to disturb the pellet.
14. Add 1.0 ml of 70% ethanol to each tube, cap, and vortex 3 to 5 s.
15. Centrifuge tubes for 5 min at maximum speed (at least 12,500 $\times$ *g*) at room temperature.
16. Carefully remove supernatant without disturbing the pellet. Remove as much of the supernatant as possible. Residual ethanol can inhibit amplification.
17. Add 400 μl of HIV-1 MONITOR Test specimen diluent to each tube, cap, and vortex for 10 s to suspend extracted RNA.
18. Amplify the processed specimens within 2 h of preparation (step 19) or store frozen at −20°C or colder for up to 1 week.
19. Pipette 50 μl of each prepared control and patient specimen to appropriate reaction tube by using a micropipettor with plugged tips. Cap and seal the tubes.
20. Transfer the tray with sealed tubes containing the processed specimens and controls in working master mix to the amplification/detection area.

C. Reverse transcription and amplification

1. Performed in Amplification/Detection Area
2. Turn on the thermal cycler at least 30 min before amplification.
3. Place the tray/retainer assembly into the thermal cycler block.
4. Program the thermal cycler as follows:

Hold Program:	2 min, 50°C
Hold Program:	30 min, 60°C
Cycle Program (4 cycles):	10 s, 95°C; 10 s, 55°C; 10 s, 72°C
Cycle Program (26 cycles):	10 s, 90°C; 10 s, 60°C; 10 s, 72°C
Hold Program:	15 min, 72°C

IV. PROCEDURE *(continued)*

5. Link the five programs into a METHOD Program.
6. Start the METHOD program. The program runs approximately 30 min.
7. Remove the tray from the thermal cycler at any time during the final Hold program, place in MicroAmp base and continue immediately with step 8. Do not allow the reaction tubes to remain in the thermal cycler past the end of the final Hold program and do not extend the final Hold program beyond 15 min.

   ◪ **NOTE:** Do not bring amplified samples into the preamplification area. Amplified controls and specimens should be considered a major source of contamination.
8. Remove caps from the reaction tubes, immediately pipette 100 μl of MONITOR Denaturation Solution into each tube, and mix by pipetting up and down five times.
9. The denatured amplicon can be held at room temperature no more than 2 h before proceeding to the detection reaction. If the detection reaction cannot be performed within 2 h, recap the tubes and store at 2 to 8°C for up to 1 week.

D. Detection

1. Performed in Amplification/Detection Area
2. Warm all reagents to room temperature before use.
3. Prepare working wash solution as follows: add 1 vol of 10× wash concentrate to 9 vol of distilled, deionized water and mix well. Store in a clean, closed plastic container at 2 to 25°C for up to 2 weeks.
4. Allow the HIV-1 MONITOR microwell plate (MWP) to warm to room temperature before removing from the foil pouch. Add 100 μl of hybridization buffer to each well. Rows A through F of the MWP are coated with the HIV-specific oligonucleotide probe; rows G and H are coated with the QS-specific oligonucleotide probe.
5. Add 25 μl of the denatured amplicon to the HIV wells in row A of the MWP, and mix up and down 10 times with a 12-channel pipettor with plugged tips. Make serial fivefold dilutions in the HIV wells in rows B through F as follows: transfer 25 μl from row A to row B and mix as before; continue through row F. Mix row F and remove and discard 25 μl. Discard pipette tips.
6. Add 25 μl of the denatured amplicon to the QS wells in row G of the MWP and mix. Transfer 25 μl from row G to row H. Mix as before, then remove 25 μl from row H and discard.
7. Cover the MWP and incubate 1 h at 37°C (±2°C).
8. Wash the MWP five times with the working wash solution, using an automated MWP washer.
   a. Fill each well to top (400 to 450 μl). Let soak 30 s. Aspirate.
   b. Repeat step a four additional times.
   c. Tap the plate dry.
9. Add 100 μl of avidin-HRP conjugate to each well. Cover the MWP and incubate for 15 min at 37°C (±2°C).
10. Wash the MWP as described in step 8.
11. Prepare the working substrate solution. For each MWP, mix 12 ml of substrate A with 3 ml of substrate B. Protect working substrate from direct light. Working substrate must be kept at room temperature and used within 3 h of preparation.
12. Pipette 100 μl of working substrate solution into each well.
13. Allow color to develop for 10 min at room temperature in the dark.
14. Add 100 μl of stop reagent to each well.
15. Measure the OD at 450 nm (single wavelength) within 10 min of adding stop reagent.

V. QUALITY CONTROL

A. It is recommended that one replicate of the HIV-1 MONITOR Test (−) control, one replicate of the HIV-1 MONITOR Test low (+) control, and one replicate of the HIV-1 MONITOR Test high (+) control be included in each test run.

B. All controls and patient specimens should yield OD values for the QS that meet the criteria described in the Results section, demonstrating that the specimen processing, reverse transcription, amplification, and detection steps were performed correctly.

   1. If any specimen has a QS OD value that does not meet the criteria described above, the result for that specimen is invalid, but the run is still acceptable.

   2. If any control has a QS OD value that does not meet the criteria described above, *the entire run is invalid.*

C. The expected range for each of the HIV-1 MONITOR Test (+) controls is specific for each lot of control and is provided on the HIV-1 MONITOR Test data card supplied with the kit.

   1. The HIV-1 RNA copy number per milliliter for both the low (+) and high (+) controls should fall within the range indicated on the control data card.

   2. The HIV-1 MONITOR Test (−) control should yield a ''Not Detected'' result; i.e., all HIV OD values are <0.20.

VI. RESULTS

A. For each specimen and control, calculate the HIV-1 RNA level.

B. Choose the appropriate HIV-1 well.

   1. The HIV wells in rows A through F represent undiluted and 1:5, 1:25, 1:125, 1:625, and 1:3,125 serial dilutions, respectively, of the HIV-1 amplicon. The absorbance values should decrease with the serial dilutions, with the highest OD at 450 nm (OD_{450}) for each specimen or control in row A and the lowest OD_{450} in row F.

   2. Choose the well with the lowest OD_{450} that is ≥0.20 and ≤2.00 ODU.

   3. If any of the following conditions exist, see ''Unexpected results'' (step I, below).

      a. All HIV OD values are <0.20.

      b. All HIV OD values are >2.00.

      c. HIV OD values are not in sequence (OD values do not decrease from well A to well F).

C. Subtract background from selected HIV OD value (background = 0.07 ODU).

D. Calculate the total HIV OD by multiplying the background-corrected OD value of the selected HIV well by the dilution factor associated with that well.

E. Choose the appropriate QS well.

   1. The QS wells in rows G and H represent neat and 1:5 dilutions, respectively, of the QS amplicon. The absorbance value in row G should be greater than the value in row H.

   2. Choose the well with the lowest OD_{450} that is ≥0.30 and ≤2.00 ODU.

   3. If the following conditions exist, see ''Unexpected results'' (step I, below).

      a. Both QS OD values are <0.30.

      b. Both QS OD values are >2.00.

      c. QS OD values are not in sequence (well H has a higher OD than well G).

F. Subtract background from the selected QS OD values (background = 0.07 ODU).

G. Calculate the total QS OD by multiplying the background-corrected OD value of the selected QS well by the dilution factor associated with that well.

H. Calculate HIV-1 RNA copies per milliliter of plasma as follows:

$$\text{HIV RNA copies/ml} = \left(\frac{\text{Total HIV OD}}{\text{Total QS OD}} \times [\text{Input QS copies/PCR}] \times 40\right)$$

VI. RESULTS *(continued)*

where input QS copies per PCR is specific to each lot of QS (*see* the Amplicor HIV-1 MONITOR Test data card for the input QS copies per PCR. Verify that the QS lot number matches the lot number on the Amplicor HIV-1 MONITOR Test data card), and 40 is a factor to convert copies per PCR to copies per milliliter of plasma.

I. Unexpected results

1. All HIV OD values are <0.20. If QS wells have the expected values, use 0.20 as the raw HIV-1 OD and 1 as the dilution factor. Calculate the result as above and report the result as ''No HIV-1 RNA detected, less than . . . . .'' (the calculated value).
2. All HIV OD values are >2.0. This means that the HIV-1 copy number is above the linear range of the assay. Report the result as ''Not Determined.'' Prepare a 1 : 50 dilution of the original specimen with HIV-negative human plasma and repeat the test. Calculate the HIV-1 result as above, then multiply the final result by 50.
3. HIV OD values are out of sequence. The OD values for HIV wells should follow a pattern of decreasing OD values with increasing dilution factor, except for wells that are saturated (OD >2.3) and wells with background (OD <0.1) OD values. If this pattern is not observed, a dilution error may have occurred, and the test procedure should be repeated.
4. Both the QS OD values are <0.30. This suggests that either the specimen was inhibitory to the amplification or the RNA was not recovered during specimen processing. Repeat the entire test procedure.
5. Both QS OD values are >2.0. This indicates that an error has occurred and the result for that specimen is invalid.
6. QS OD values are out of sequence. An error has occurred, and the result of that specimen is invalid.

VII. PROCEDURAL PRECAUTIONS

A. Workflow in the laboratory must proceed in a unidirectional manner, beginning in the Reagent Preparation Area and moving to the Specimen Preparation Area and then to the Amplification/Detection Area. Reagent preparation activities and specimen preparation activities must be performed in separate, segregated areas. Supplies and equipment must be dedicated to each activity.

B. Owing to the analytical sensitivity of this test and the potential for contamination, extreme care should be taken to preserve the purity of kit reagents or amplification mixtures. All reagents should be closely monitored for purity. Discard any reagents that may be suspect.

C. All pipettors, pipettes, bulbs, pipette tips, etc., should be dedicated to, and used only for, each PCR laboratory activity.

VIII. PROCEDURE LIMITATIONS

A. This test has been validated for use only with human plasma anticoagulated with EDTA or ACD. Heparin inhibits PCR; specimens collected with heparin as the anticoagulant should *not* be used with this test.

B. The presence of AmpErase in the master mix reduces the risk of amplicon contamination. However, contamination from HIV-positive controls and HIV-positive clinical specimens can be avoided only by good laboratory practices and careful adherence to the procedures specified by the manufacturer.

C. Use of this product should be limited to personnel trained in the techniques of PCR.

D. As with any diagnostic test, results from the Amplicor HIV-1 MONITOR Test should be interpreted with consideration of all clinical and laboratory findings.

E. Interfering substances
   1. Elevated levels of lipids, bilirubin, and hemoglobin in specimens have been shown *not* to interfere with the quantitation of HIV-1 RNA by this test.
   2. The following drug compounds have been shown not to interfere with the quantitation of HIV-1 RNA by this test: AZT, ddI, ddC, d4T, HBY097, nevirapine, saquinavir, isoniazid, foscarnet, and ganciclovir.
   3. Heparin inhibits PCR.

IX. PERFORMANCE CHARACTERISTICS

A. Limit of detection
Studies demonstrate that the Amplicor HIV-1 MONITOR Test can detect less than 2 copies of HIV RNA per reaction and that 7.5 or more copies were detected 100% of the time. Five copies of HIV RNA per PCR are equivalent to 200 HIV RNA copies per ml of sample.

B. Limit of quantitation
Studies demonstrate that the HIV-1 MONITOR Test can quantitate virion-associated HIV-1 RNA in plasma at concentrations as low as 400 RNA copies/ml of plasma provided that the OD of the selected microwell is within the specified OK range (0.2 to 2.0).

C. Linear range
The HIV-1 MONITOR Test was found to give a linear response between 400 and 750,000 HIV-1 RNA copies per ml. Samples with results greater than 750,000 HIV-1 RNA copies per ml must be diluted with HIV-negative human plasma and tested.

SUPPLEMENTAL READING

Roche Diagnostic Systems. 1996. Amplicor HIV-1 Monitor™ Test, package insert. Roche Diagnostic Systems, Nutley, N.J.

Coffin, J. M. 1995. HIV-1 population dynamics in vivo: implications for genetic variation, pathogenesis, and therapy. *Science* **267:**483–489.

Ho, D. D. 1996. Viral counts in HIV infection. *Science* **272:**1124–1125.

Mellors, J. W., C. R. Rinaldo, Jr., P. Gupta, R. M. White, J. A. Todd, and L. A. Kingsley. 1996. Prognosis of HIV-1 infection predicted by the quantity of virus in plasma. *Science* **272:** 1167–1170.

Pantaleo, G., C. Graziosi, and A. S. Fauci. 1993. New concepts in the immunopathogenesis of human immunodeficiency virus (HIV-1) infection. *New Engl. J. Med.* **328:**327–335.

Saag, M. S., M. Holodniy, D. R. Kuritzkes, W. A. O'Brien, R. Coombs, M. E. Poscher, D. M. Jacobsen, G. M. Shaw, D. D. Richman, and P. A. Volberding. 1996. HIV viral load markers in clinical practice. *Nature Med.* **2:**625–629.

PART 5

Detection of *Mycobacterium tuberculosis* in Respiratory Specimens by Using the Roche Amplicor PCR Kit

I. PRINCIPLE

The Amplicor *Mycobacterium tuberculosis* Test is a direct DNA probe test that utilizes PCR and nucleic acid hybridization for the detection of the *M. tuberculosis* complex in digested, decontaminated sputum and bronchial alveolar lavage (BAL) samples. The test is based on three major processes, PCR target amplification, hybridization of the amplified product to a specific nucleic acid probe, and detection of the amplified product by color formation. Genus-specific primers located in a highly conserved region of the 16S rRNA gene of mycobacteria are biotinylated and used to amplify a 584-base-pair sequence. After PCR, the amplicons are denatured and captured by oligonucleotide probes bound to the wells of a microwell plate. The capture probes were selected from the hypervariable region of the 16S rRNA gene and are specific for the *M. tuberculosis* complex. The bound amplicons are then detected by a color reaction with an avidin-HRP conjugate.

Workflow in the laboratory must proceed in a unidirectional manner, beginning in the Reagent Preparation Area and moving to the Specimen Preparation Area and then to the Amplification/Detection Area. Preamplification activities must begin with reagent preparation and proceed to specimen preparation. Reagent preparation and specimen preparation activities must be performed in separate, segregated areas. Supplies and equipment must be dedicated to each activity and not used for other activities or moved between areas. Lab coats and gloves must be worn in each area and must be changed before leaving that area. Equipment and supplies used for reagent preparation must not be used for specimen preparation activities or for pipetting or processing amplified DNA or other sources of target DNA. Amplification and detection supplies and equipment must be confined to the Amplification/Detection Area at all times.

II. MATERIALS

Include QC information on reagent container and in QC records.

A. **Materials provided** (*see* Amplicor *Mycobacterium tuberculosis* Test package insert)
   1. Sputum preparation kit
      a. Wash solution
      b. Lysis reagent
      c. Neutralization reagent
   2. Amplification kit
      a. Master mix
      b. AmpErase
      c. Tuberculosis (TB) positive control
      d. Negative control
   3. Detection kit
      a. Denaturation solution
      b. Hybridization buffer
      c. Avidin-HRP conjugate
      d. Substrate A
      e. Substrate B
      f. Stop reagent
      g. 10× wash concentrate
      h. *M. tuberculosis* DNA probe-coated microwell plate

B. **Materials required but not provided** (*see* Amplicor *Mycobacterium tuberculosis* Test package insert)
   1. Area 1: reagent preparation
      a. Consumables for thermal cycler
         (1) Reaction tubes
         (2) Caps
         (3) Base
         (4) Tray/retainers
      b. Repeat pipettor
      c. Individually-wrapped Combitips (1.25 ml)
      d. Micropipettes
      e. Plugged (aerosol barrier) or positive displacement pipette tips (50, 100, and 1,000 μl)
   2. Area 2: specimen preparation
      a. Microcentrifuge
      b. Sterile screw-cap tubes
      c. Tube racks
      d. Sterile fine-tip transfer pipettes
      e. Repeat pipettor with 12.5-ml individually wrapped Combitips
      f. Vortex mixer
      g. Thermal cycler base and capping tool
      h. Micropipettes with plugged (aerosol barrier) or positive displacement tips (50 and 100 μl)
      i. Dry-heat blocks, 60°C (±2°C)
   3. Area 3: amplification and detection
      a. Micropipettes with plugged (aerosol barrier) tips (25 and 100 μl) and unplugged tips (100 μl).
      b. Thermal cycler
      c. Multichannel pipettor (25 and 100 μl) with plugged (aerosol barrier) tips (25 and 100 μl) and unplugged tips (100 μl)
      d. Disposable reagent reservoirs
      e. Microwell plate lid
      f. ELISA well key for strip removal
      g. Incubator 37°C (±2°C)
      h. Distilled or deionized water
      i. Microwell plate washer
      j. Microwell plate reader and printer

III. SPECIMEN

Sputum or BAL fluid is obtained from the patient and concentrated by using either *N*-acetyl-L-cysteine (NALC)–NaOH or NaOH liquefaction-decontamination, as recommended by the CDC, before inoculation in a growth medium.

IV. PROCEDURE

NOTE: This procedure should be performed in three areas of the laboratory as dictated by the instructions below. Before starting this procedure, turn on the thermal cycler to allow it warm up. All reagents must be at ambient temperature before use. Use micropipettes with plugged (aerosol barrier) or positive displacement tips.

■ *Use extreme care to ensure selective amplification.*

A. **Reagent preparation**
   1. Performed in Reagent Preparation Area (area 1)
   2. If reagent preparation (part A), specimen preparation (part B), and amplification (part C) can not be completed within 1 day, complete specimen preparation on Day 1. Reagent preparation (part A) and amplification and detection (part C) should then be performed on Day 2.
   3. Prepare master mix with AmpErase by adding 100 μl of AmpErase to one tube of master mix (sufficient for 32 amplifications).

4. Determine appropriate number of PCR tubes needed for patient specimen and control testing. It is recommended that one positive and three negative controls be run with each amplification. Place tubes in sample tray and lock in place with retainer.
5. Pipette 50 μl of master mix with AmpErase into each tube by using a repeat pipettor and 1.25-ml Combitip or a micropipette with a plugged tip.
6. Cover tubes loosely with caps and move to Specimen Preparation Area (area 2). Store sample tray at 2 to 8°C until specimen preparation is complete.

B. Specimen preparation

1. Performed in Specimen Preparation Area (area 2).
2. Add 100 μl of decontaminated sputum or BAL to 500 μl of sputum wash solution in a 1.5-ml screw-cap tube. Vortex.
3. Centrifuge at 12,500 × *g* for 10 min.
4. Aspirate supernatant and add 100 μl of lysis reagent to the cell pellet. Vortex to suspend pellet.
5. Prepare positive and negative control stocks.
   - **a.** Pipette 100 μl of negative control into a tube by using a micropipette with a plugged tip. Add 400 μl of lysis reagent. Vortex. This is the negative control stock.
   - **b.** Pipette 100 μl of positive control into a tube by using a micropipette with a plugged tip. Add 400 μl of lysis reagent. Vortex. This is the positive control stock.
   - **c.** Pipette 100 μl from each control stock and place into a 1.5-ml screw-cap tube to be processed.
6. Incubate specimens and controls in 60°C (±2°C) dry-heat block for 45 min.
7. Remove tubes from heat block and pulse centrifuge for 5 s.
8. Add 100 μl of neutralization reagent. Vortex.
9. Pipette 50 μl of prepared patient specimens and prepared controls (one positive and three negative) to appropriate PCR tubes by using a micropipette with plugged tips. Record positions of the tubes in the tray. Cap the tubes.
10. Move the prepared specimens in the sample tray to the Amplification and Detection Area (area 3).

C. Amplification and detection

1. Performed in amplification and detection area (area 3).
2. Amplification
   - **a.** Place sample tray into the thermal cycler sample block.
   - **b.** Make certain that the cover knob of the thermal cycler is turned completely counter-clockwise. Slide the cover forward.
   - **c.** Turn the thermal cycler cover knob clockwise until hand tight.
   - **d.** Program the thermal cycler for amplification of the Amplicor *Mycobacterium tuberculosis* Test as follows:

HOLD Program:	2 min, 50°C
CYCLE Program (2 cycles):	20 s, 98°C; 20 s, 62°C; 45 s, 72°C
CYCLE Program (35 cycles):	20 s, 94°C; 20 s, 62°C; 45 s, 72°C
HOLD Program:	5 min, 72°C
HOLD Program:	72°C constant

In the CYCLE Programs, the ramp times should be left at the default setting (0:00), which is the maximum rate, and the allowed setpoint error should be at the default setting (2°C). Link the 5 programs together into a METHOD Program.

IV. PROCEDURE *(continued)*

e. Start the METHOD Program. The program runs approximately 1.5 h. Specimens may be removed at any time during the final HOLD program, but must be removed within 24 h.
f. Remove completed PCR amplification specimen from the thermal cycler.

   ■ *Do not bring amplified DNA into area 1 or area 2. The amplified specimens and controls should be strictly confined to area 3.*
g. Immediately pipette 100 μl of denaturation solution to the first column (or row) of PCR tubes by using a multichannel pipettor with plugged tips, and mix by pipetting up and down. For each column (or row), repeat this procedure with a fresh set of tips. Incubate for 10 min at room temperature to allow complete denaturation.
h. Store denatured, amplified specimens at room temperature only if the detection will be performed within 1 to 2 h. If not, store the specimens at 2 to 8°C until the detection assay is performed. Amplicons may be stored for up to 1 week at 2 to 8°C.

3. Detection
   a. Warm all reagents to room temperature.
   b. Prepare working wash solution by adding 1 vol of 10× wash concentrate to 9 vol of distilled or deionized water.
   c. Allow the microwell plates to warm to room temperature before removing from the foil pouch.
   d. Add 100 μl of hybridization buffer to each well to be tested on the microwell plate.
   e. With plugged tips, pipette 25 μl of denatured amplification specimen to the appropriate well. Gently tap the plate approximately 10 to 15 times until the color changes from blue to light yellow (indicates sufficient mixing has occurred).
   f. Cover the plate and incubate for 1.5 h at 37°C (±2°C).
   g. Wash plate five times manually or by using a microwell plate washer. Use the prepared washing solution for washing the plate.
   h. Add 100 μl of avidin-HRP conjugate to each well. Cover plate and incubate for 15 min at 37°C (±2°C).
   i. Wash plate as in step g above.
   j. Prepare working substrate by mixing 2.0 ml of substrate A and 0.5 ml of substrate B for each multiple of two 8-well microwell plate strips (16 tests). Prepare this reagent no more than 3 h before use and protect from exposure to direct light.
   k. Pipette 100 μl of prepared working substrate reagent into each well being tested.
   l. Allow color to develop for 10 min at room temperature (20 to 25°C) in the dark.
   m. Add 100 μl of stop reagent to each well.
   n. Measure the A_{450} within 1 h of adding the stop reagent. Record the absorbance value for each patient specimen and control tested. Calculate the results.

V. RESULTS

A. The presence of *M. tuberculosis* in the specimen is determined by relating the absorbance of the unknown specimen to that of the cutoff value.
B. An A_{450} of 0.35 has been selected as the cutoff value for this assay.
C. A clinical specimen with an A_{450} reading equal to or greater than 0.35 is positive for the presence of *M. tuberculosis*.
D. A clinical specimen with an A_{450} reading less than 0.35 is considered negative for *M. tuberculosis*.

VI. QUALITY CONTROL

A. It is recommended that at least one positive control and three negative controls be run each time the test is performed.

B. Negative control
   1. The assay result of each negative control should be less than 0.25 A_{450} units.
   2. If one or more of the negative control values are greater than 0.25 A_{450} units, the entire run should be discarded and the entire assay, including amplification, should be repeated.

C. Positive control
   1. The response of the positive control should be greater than 3.0 A_{450} units.
   2. If the value of the positive control falls below 2.0 A_{450} units, the entire run should be discarded and the entire assay, including amplification, should be repeated.

D. Sample processing control
   1. To test the effectiveness of sample processing (recommended on a monthly basis), 10^4 *M. tuberculosis* cells (determined by quantitative plating or turbidimetric methods) should be processed as described for specimen preparation (item IV.B above) and then treated as a normal clinical specimen.
   2. A positive signal above 3.0 A_{450} should be obtained on the microwell plate if the sample is properly processed.

VII. PROCEDURAL LIMITATIONS

A. The Amplicor *Mycobacterium tuberculosis* Test has been validated by using sputum or BAL specimens that have been liquefied, concentrated, and decontaminated with either NALC-NaOH or NaOH. Performance with other specimens has not been evaluated and may result in false-negative or -positive results.

B. The addition of AmpErase to master mix enables selective amplification of target DNA; however, reagent purity is maintained only by good laboratory practices and careful adherence to the recommended procedure.

C. Detection of *M. tuberculosis* is dependent on the number of organisms present in the specimen. This may be affected by specimen collection methods and patient factors such as age, history of respiratory disease, presence of symptoms, etc.

D. A negative test does not exclude the possibility of infection.

E. Therapeutic success or failure cannot be determined by using this assay.

SUPPLEMENTAL READING

Beavis, K. G., M. B. Lichty, D. J. Jungkind, and O. Giger. 1995. Evaluation of Amplicor PCR for direct detection of *Mycobacterium tuberculosis* from sputum specimens. *J. Clin. Microbiol.* **33:** 2582–2586.

D'Amato, R. F., A. A. Wallman, L. H. Hochstein, P. M. Colaninno, M. Scardamaglia, E. Ardila, M. Ghouri, K. Kim, R. C. Patel, and A. Miller. 1995. Rapid diagnosis of pulmonary tuberculosis by using Roche AMPLICOR *Mycobacterium tuberculosis* PCR test. *J. Clin. Microbiol.* **33:**1832–1834.

Moore, D. F., and J. I. Curry. 1995. Detection and identification of *Mycobacterium tuberculosis* directly from sputum sediments by Amplicor PCR. *J. Clin. Microbiol.* **33:**2686–2691.

Nolte, F. S., B. Metchock, J. E. McGowan, Jr., A. Edwards, O. Okwumabua, C. Thurmond, P. S. Mitchell, B. Plikaytis, and T. Shinnick. 1993. Direct detection of *Mycobacterium tuberculosis* in sputum by polymerase chain reaction and DNA hybridization. *J. Clin. Microbiol.* **31:**1777–1782.

Roche Diagnostic Systems. 1994. Amplicor™ Mycobacterium Tuberculosis Test, package insert. Roche Diagnostic Systems, Nutley, N.J.

Vuorinen, P., A. Miettinen, R. Vuento, and O. Hallstrom. 1995. Direct detection of *Mycobacterium tuberculosis* complex in respiratory specimens by Gen-Probe Amplified Mycobacterium Tuberculosis Direct Test and Roche Amplicor Mycobacterium Tuberculosis Test. *J. Clin. Microbiol.* **33:**1856–1859.

PART 6

Detection of *Mycobacterium tuberculosis* in Respiratory Specimens by Using the Gen-Probe Amplified *Mycobacterium tuberculosis* Direct Test

I. PRINCIPLE

The Gen-Probe Amplified *Mycobacterium tuberculosis* Direct (MTD) Test is a target-amplified nucleic acid probe test for the in vitro diagnostic detection of *Mycobacterium tuberculosis* complex rRNA in acid-fast bacillus (AFB) smear-positive concentrated sediments prepared from sputum, bronchial specimens (e.g., bronchial lavages or bronchial aspirates), or tracheal aspirates.

The MTD test is a two-part test in which amplification and detection take place in a single tube. After nucleic acid extraction, the MTD test utilizes Transcription-Mediated Amplification (TMA) and the Hybridization Protection Assay (HPA) to detect *M. tuberculosis* complex rRNA. In contrast to PCR, which requires thermal cycling and uses a thermostable polymerase to produce multiple copies (amplicons) of DNA, TMA uses a constant 42°C temperature and amplifies a specific rRNA target by using RNA polymerase and reverse transcriptase enzymes to produce multiple RNA amplicons. Detection of the amplicons is accomplished with a chemiluminescence-labeled DNA probe that is complementary to the RNA amplicons. When stable RNA:DNA hybrids are formed, separation of hybridized from unhybridized probe is done by adding a selection reagent that hydrolyzes the chemiluminescent label on the unhybridized probe. The label on the hybridized probe is *protected* (hybridization protection) and may be detected in a luminometer. The entire assay is performed in a single tube.

II. MATERIALS

Include QC information on reagent container and in QC records.

A. **Materials provided** (*see* package insert)
   1. Specimen dilution buffer
   2. Amplification reagent
   3. Reconstitution buffer
   4. Lysing tubes
   5. Oil reagent
   6. Enzyme reagent
      a. Reverse transcriptase
      b. RNA polymerase
   7. Enzyme dilution buffer
   8. Termination reagent
   9. Hybridization positive control
   10. Hybridization negative control
   11. Probe reagent
   12. Hybridization buffer
   13. Selection reagent
   14. Sealing cards

B. **Materials required but not provided** (*see* package insert)
   1. Luminometer
   2. Sonicator
   3. Detection reagent kit
   4. Dry-heat bath (95°C [±5°C])
   5. Sonicator rack
   6. Pipette tips with hydrophobic plugs
   7. Polypropylene tubes (12 × 75 mm)
   8. Micropipettes capable of dispensing 20, 25, 50, 100, 200, and 300 μl
   9. Snap-top polypropylene caps for 12 × 75-mm tubes
   10. Water bath and/or dry-heat bath (42°C [±1°C] and 60°C [±1°C])
   11. Vortex mixer
   12. Sterile water
   13. Culture tubes
   14. Glass beads (3 mm)
   15. Screw-cap microcentrifuge tubes
   16. Positive cell controls (e.g., *M. tuberculosis* ATCC 25177 or ATCC 27294)
   17. Negative cell controls (e.g., *Mycobacterium gordonae* ATCC 14470 or *Mycobacterium terrae* ATCC 15755).
   18. Household bleach (5.25% hypochlorite solution)
   19. Plastic-backed bench covers

III. SPECIMEN

A. **Specimen type**

Acceptable specimens include sputum (induced or expectorated), bronchial (e.g., bronchial lavages or aspirates), or tracheal aspirates.

NOTE: The efficacy of this test has not been demonstrated (and it is not FDA approved) for the direct detection of *M. tuberculosis* rRNA with other clinical specimens (e.g., blood, urine, stool, or tissue).

B. **Specimen collection and storage**
   1. Specimens must be collected in sterile, plastic containers and stored at 2 to 8°C until transported or processed.
   2. Specimens should be processed (decontaminated and concentrated) within 24 h of collection (including transport time) as recommended by the CDC.

C. Processing (decontamination and concentration)

1. Specimens that are grossly bloody should not be tested with the MTD test.
2. The MTD test is designed to detect rRNA from members of the *M. tuberculosis* complex by using sediments prepared from generally accepted current adaptations of the *N*-acetyl-L-cysteine–NaOH or NaOH decontamination protocols, described by the CDC, with 1 to 1.5% NaOH for 15 to 20 min and centrifugation at $\geq$3,000 $\times$ *g*.
3. Processed sediments may be stored at 2 to 8°C for up to 3 days before testing.

IV. PROCEDURE

A. Equipment preparation

1. Degas water in sonicator.
2. Adjust one dry-heat bath to 95°C, one dry-heat or water bath to 60°C, and another dry-heat bath or water bath to 42°C ($\pm$2°C).
3. Wipe down work surfaces, equipment, and pipettors with a 1:1 dilution of household bleach (2.6% NaOCl) before starting. Cover work surfaces with plastic-backed laboratory bench covers.
4. Prepare luminometer for operation. Make sure there are sufficient volumes of detection reagents I and II to complete the tests.

B. Reagent preparation

1. Reconstitute one vial (25 tests) of lyophilized *M. tuberculosis* amplification reagent with 750 μl of *Mycobacterium* reconstitution buffer. Vortex and let sit at room temperature until clear.
2. The reconstituted amplification reagent may be stored at 2 to 8°C for 2 months.
3. The reconstituted amplification reagent should be allowed to come to room temperature before use.

C. Controls

1. Lysis and amplification controls
   a. Cells used for the cell positive control should be a member of the *M. tuberculosis* complex, such as *M. tuberculosis* ATCC 25177 or 27294, suspended in sterile water. Cells used for the cell negative control should be mycobacteria other than tuberculosis complex, such as *M. gordonae* ATCC 14470 or *M. terrae* ATCC 15755.
   b. Preparation and storage of amplification controls.
      (1) Place three to five sterile 3-mm glass beads in a clean culture tube.
      (2) Add several 1-μl loopfuls of growth from the appropriate culture to 1 to 2 ml of sterile water, cap, and vortex several times.
      (3) Allow the suspension to settle and transfer the supernatant to a clean culture tube. Adjust turbidity to that of a number 1 McFarland standard.
      (4) Make a 1:100 dilution (dilution 1) of the suspension into sterile water. Cap and vortex.
      (5) Make a 1:100 dilution (dilution 2) of dilution 1.
      (6) Take 100 μl of dilution 2 and place into 6 ml of sterile water (dilution 3). This contains approximately 25 CFU per 50 μl.
      (7) Plate 50 μl of each dilution onto Lowenstein-Jensen culture medium and freeze the remaining stock at -20 or -70°C. Test all dilutions by using the MTD test.
      (8) The dilutions that give between 25 and 150 CFU per 50 μl on plated culture medium and that perform as expected with the MTD test should be thawed, aliquoted, and used as controls.

IV. PROCEDURE *(continued)*

(9) The dilutions must be aliquoted into clean 1.5-ml screw-cap microcentrifuge tubes as single use aliquots (100 μl). The tubes may be stored frozen at −20°C for 6 months or −70°C for 1 year. Frost-free freezers must *not* be used.

c. A single replicate of the cell controls must be tested with each run.

d. Each laboratory should determine target values and means for the controls.

2. Hybridization controls

a. Hybridization positive controls and hybridization negative controls are provided in the MTD test kit.

b. A single replicate of each of the hybridization controls should be tested with each run for QC purposes.

3. Specimen inhibition controls

a. When the AFB smear is positive and the MTD test is negative for un treated patients, one must consider the following possibilities:

(1) The specimen is inhibitory.

(2) The specimen contains mycobacteria other than *M. tuberculosis.*

(3) The specimen contains a mixture of large number of non-tuberculosis mycobacteria and a low number of *M. tuberculosis.*

b. Testing patient sediments for inhibition in the MTD test

(1) Place 200 μl of specimen dilution buffer into two lysing tubes.

(2) Add 50 μl of amplification positive cell control and 50 μl of sediment from patient specimen to one tube (seeded). Add 50 μl of sediment to the second tube (unseeded). Proceed with the test as usual.

(3) If the luminometer reading of the seeded tube is ≥30,000 relative light units (RLU), then the sample is not inhibitory to amplification, and there apparently was no target available for amplification.

(4) If the RLU value of the seeded tube is below 30,000, then the sample is inhibitory to amplification, and another sample should be evaluated.

4. Laboratory contamination monitoring control

a. Place 2 ml of sterile water in a clean tube.

b. Wipe the area of bench or equipment to be tested by using a premoistened sterile polyester or Dacron swab.

c. Place the swab in the water and swirl gently. Remove the swab and express fluid along the side of the tube.

d. Add 25 μl of the water to an amplification tube containing 25 μl of amplification reagent and 200 μl of oil reagent.

e. Follow the test procedure for amplification and detection.

f. If the results are ≥30,000 RLU, the surface is contaminated and should be decontaminated with bleach.

D. Sample preparation

1. Pipette 200 μl of specimen dilution buffer into labeled lysing tubes.

2. Transfer 50 μl of decontaminated, well-vortexed specimen or cell control from its container to the correspondingly labeled lysing tube.

3. Cap the tubes and vortex for 3 s.

E. Sample lysis

1. Push the lysing tubes through the sonicator rack so that the reaction mixture in the bottom of the tubes is submerged but the caps are above water. Place sonicator rack on water bath sonicator. Do not allow the tubes to touch the bottom or sides of the sonicator.

2. Sonicate for 15 min but no more than 20 min. Samples and controls that have been sonicated are referred to as lysates.

F. Amplification

1. Add 25 μl of reconstituted amplification reagent to each appropriately labeled amplification tube by using a repeat pipettor. Add 200 μl of oil reagent to each tube by using a repeat pipettor.
2. Transfer 50 μl of lysate to the bottom of the appropriately labeled amplification tube by using a separate extended-length, hydrophobically plugged pipette tip for each transfer.
3. Incubate tubes at 95°C for 15 min, but for no more than 20 min, in the dry-heat bath.
4. Prepare the enzyme mix by adding 1.4 ml of enzyme dilution buffer to the lyophilized enzyme reagent. Swirl to mix. Do not vortex.
5. Transfer the tubes to the 42°C (± 1°C) dry-heat bath or water bath and incubate for 5 min. Do not allow tubes to cool at room temperature. Do not cover the water bath.
6. Add 25 μl of enzyme mix to each amplification tube while the tubes are at 42°C (± 1°C). Shake to mix. Incubate at 42°C for 2 h, but no more than 3 h. Sealing cards or snap caps should be used during this step. Do not cover the water bath.
7. After the 2-h incubation, tubes may be kept at 2 to 8°C for up to 2 h or at −20°C overnight.

G. Termination

1. Add 20 μl of termination reagent to each tube. Cover tubes and shake to mix. Incubate at 42°C for 10 min.
2. Tubes may be covered and placed at 2 to 8°C for up to 2 h or at −20°C overnight.

H. Hybridization

1. Reconstitute lyophilized probe reagent with hybridization buffer. Vortex the solution until clear. The reconstituted probe reagent is stable for 1 month at 2 to 8°C.
2. Place 100 μl of hybridization positive control and 100 μl of hybridization negative control into correspondingly labeled 12 × 75-mm polypropylene tubes.
3. Add 100 μl of reconstituted probe reagent to each tube. Cover the tubes and vortex eight times until the reaction mixture is uniformly yellow.
4. Incubate 60°C for 15 min, but no more than 20 min, in a dry-heat or water bath.

I. Selection

1. Remove tubes from the 60°C water or dry-heat bath and add 300 μl of selection reagent. Cover tubes and vortex three times until the reaction mixture is uniformly pink.
2. Incubate tubes at 60°C for 10 min, but no more than 11 min, in a dry-heat or water bath.
3. Remove tubes from water or dry-heat bath and cool at room temperature for at least 5 min, but not more than 1 h.

J. Detection

1. Select the appropriate protocol from the menu of the luminometer software. Use a 2 read time.
2. Wipe each tube and insert into the luminometer. Tubes must be read within 1 h of selection step 3 above.
3. When the analysis is complete, remove the tubes from the luminometer.
4. Decontaminate tubes and work surfaces with bleach.

V. PROCEDURAL NOTES

A. Reagents

1. Enzyme reagent should not be held at room temperature for more than 15 min after it is reconstituted.
2. Hybridization buffer may precipitate. Warming and mixing at 60°C will dissolve the precipitate.

B. Temperature

1. The amplification, hybridization, and selection reactions are temperature dependent; ensure that the water bath or dry-heat bath is maintained within the specified temperature range.
2. The tubes must be cooled at 42°C for 5 min before addition of enzyme mix for optimal performance.
3. The temperature is critical for the amplification (42°C ± 1°C).

C. Water bath

1. The level of water in the water bath must be maintained so that the entire liquid volume in the tubes is submerged.
2. During the amplification step, water bath covers should not be used to ensure that condensation cannot drip into or onto the tubes.

VI. TEST INTERPRETATION

A. The results of the MTD test are based on a 30,000 RLU cutoff value. Samples producing signals greater than or equal to the cutoff value are considered positive. Samples producing signals less than the cutoff value are considered negative.

B. QC results and acceptability

1. The amplification cell negative control and amplification cell positive control should produce the following values:
   a. Amplification cell negative control of <20,000 RLU
   b. Amplification cell positive control of ≥500,000 RLU
2. The hybridization negative control and hybridization positive control should produce the following values:
   a. Hybridization negative control of <5,000 RLU
   b. Hybridization positive control of ≥15,000 RLU
3. Patient test results must not be reported if the MTD test control values do not meet the criteria above.
4. Target values for amplification cell positive and cell negative controls should be determined in each laboratory by using test results for each batch of prepared controls.

C. Patient test results

1. If the controls do not yield the expected results, test results on patient specimens in the same run must not be reported.
2. A value of ≥30,000 RLU is considered positive for *M. tuberculosis*.
3. A value of <30,000 RLU is considered negative for *M. tuberculosis*.

VII. REPORTING OF RESULTS

A. If the AFB smear and MTD test result are positive, report the following: "AFB smear positive and *Mycobacterium tuberculosis* complex rRNA detected. AFB culture pending. Specimen may contain *M. tuberculosis* alone or in combination with a non-tuberculosis mycobacterium."

B. If the AFB smear is positive and the MTD test is negative, then report the following: "AFB smear positive; no *Mycobacterium tuberculosis* complex rRNA detected. AFB culture pending. Specimen may not contain *M. tuberculosis*, or result may be falsely negative owing to low numbers of *M. tuberculosis* in the presence of other mycobacterial species, or interference with assay detection by specimen inhibitors."

SUPPLEMENTAL READING

Abe, C., K. Hirano, M. Wada, Y. Kazumi, M. Takahashi, Y. Fukosawa, T. Yoshimura, C. Miyagi, and S. Goto. 1993. Detection of *Mycobacterium tuberculosis* in clinical specimens by polymerase chain reaction and Gen-Probe Amplified Mycobacterium Tuberculosis Direct Test. *J. Clin. Microbiol.* **31:**3270–3274.

Gen-Probe Inc. 1995. Gen-Probe Amplified Mycobacterium Tuberculosis Direct Test, package insert. Gen-Probe Inc., San Diego, Calif.

Jonas, V., M. A. Alden, J. J. Curry, K. Kamisango, C. A. Knott, R. Lankford, J. M. Wolfe, and D. F. Moore. 1993. Detection and identification of *Mycobacterium tuberculosis* directly from sputum sediments by amplification of rRNA. *J. Clin. Microbiol.* **31:**2410–2416.

Miller, N., S. G. Hernandez, and T. J. Cleary. 1994. Evaluation of Gen-Probe Amplified Mycobacterium Tuberculosis Direct Test and PCR for direct detection of *Mycobacterium tuberculosis* in clinical specimens. *J. Clin. Microbiol.* **32:** 393–397.

Vuorinen, P., A. Miettinen, R. Vuento, and O. Hallstrom. 1995. Direct detection of *Mycobacterium tuberculosis* complex in respiratory specimens by Gen-Probe Amplified Mycobacterium Tuberculosis Direct Test and Roche Amplicor Mycobacterium Tuberculosis Test. *J. Clin. Microbiol.* **33:**1856–1859.

10.3 Molecular Methods for Identification of Cultured Microorganisms

10.3.1 Introduction

Bacteria and fungi can be identified by using nucleic acid hybridization techniques. Nucleic acid probes can be used to confirm the identification of culture isolates that have been tested by presumptive identification methods. Alternatively, they can be used as the primary method for identifying isolated organisms. Culture identification by probe hybridization is not dependent on the ability to detect minute quantities of nucleic acid, and thus sensitivity is not a limiting factor in this application of molecular technology. The advantage of probe-based identification is greatest for slow-growing organisms like the mycobacteria or for organisms for which convenient commercial identification systems are not available. Although the specificity of the available commercial probes is high and they facilitate rapid identification of a number of pathogens (Table 10.1–2), misidentifications do occur and serve to emphasize the need for caution in using any single characteristic in identification of a species. It should also be emphasized that at this point identification by probe hybridization is more expensive than conventional techniques for many organisms.

10.3.2 Identification of Bacteria and Fungi by Using Nucleic Acid Probes

I. PRINCIPLE

Bacteria and fungi can be identified by using nucleic acid probes. Nucleic acid probes can be used to confirm the identification of isolates that have been tested by presumptive identification methods. Alternatively, they can be used as the primary method for identifying isolated organisms. The AccuProbe system (Gen-Probe, Inc., San Diego, Calif.) is currently the only DNA probe system available commercially. DNA probes are available for the organisms listed in Table 10.1–2. The sample preparation procedure for mycobacterial and fungal probes differs from that for bacterial probes and is described in procedure 10.3.3.

NOTE: Sample preparation methods are identical for mycobacteria and fungi.

The AccuProbe system is based on nucleic hybridization of a DNA probe (oligomer) that is complementary to the rRNA of the target organism. The probes are labeled with an acridinium ester. The organism is treated to release the rRNA and then incubated with the labeled single-stranded DNA probe. If the rRNA and DNA probes are complementary, a stable DNA-RNA hybrid forms. A hydrolyzing reagent is added to selectively remove unbound DNA probe. The chemiluminescence of the DNA-RNA hybrids is quantitated in relative light units (RLU) with a luminometer.

II. SPECIMEN

Test growth from appropriate solid media with morphology suggestive of the organism in question. Samples may be tested as soon as growth is visible, but they should be less than 48 h old. If a single colony is to be tested, it should be at least 1 mm in diameter. Alternatively, several (three or four) smaller colonies can be tested. Do not use confluent growth.

III. MATERIALS

Include QC information on reagent container and in QC records.

A. Reagents

1. Identification reagent kit (provided)
   - **a.** Reagent 1 (lysis reagent)
   - **b.** Reagent 2 (hybridization buffer)
   - **c.** Reagent 3 (selection reagent)
2. Detection reagent kit (provided)
   - **a.** Detection reagent 1
   - **b.** Detection reagent 2
3. Organism-specific probe kit (provided)

B. Materials required but not provided

1. Plastic sterile inoculating loops (1 μl), applicator sticks
2. Reference strains for controls
3. Incubator or water bath (35 to 37°C; for gram-positive bacteria)
4. Water bath or heating block (60°C)
5. Micropipettes (50 and 300 μl)
6. Repipettors (50 and 300 μl)
7. Vortex mixer
8. Leader 1 luminometer or AccuLDR luminometer

IV. PROCEDURE

A. Equipment preparation

1. Adjust the incubator or water bath to 35 to 37°C if testing gram-positive bacteria.
2. Adjust the water bath or heating block to 60°C (±1°C).
3. Prepare the luminometer for operation. Make sure there are sufficient volumes of detection reagents 1 and 2.

IV. PROCEDURE *(continued)*

B. Sample preparation

1. Label the probe reagent tubes. Remove and retain the caps.
2. Pipette 50 μl of reagent 1 (lysis reagent) into each probe reagent tube.
3. Transfer one or several colonies from solid medium into probe reagent tubes. Avoid taking any of the solid medium with the cells. Twirl the loop or stick in reagent 1 to remove the cells, and mix thoroughly.
4. For gram-positive organisms, recap the probe reagent tubes and incubate them at 35 to 37°C for 5 min. Remove the tubes from the water or dry bath, and then remove and retain the caps.

C. Hybridization

1. Pipette 50 μl of reagent 2 (hybridization buffer) into each probe reagent tube.
2. Recap the tubes, and mix the contents by shaking or vortexing.
3. Incubate for 15 min at 60°C in a water bath or heating block.

D. Selection

1. Remove the probe reagent tubes from the water or dry bath. Remove and retain the caps. Pipette 300 μl of reagent 3 (selection reagent) into each tube. Recap the tubes, and vortex them to mix the contents completely.
2. Incubate the probe reagent tubes for 5 min at 60°C.
3. Remove the tubes from the water or dry bath and leave them at room temperature for at least 5 min. Remove and discard the caps. Read results in the luminometer within 30 min (within 1 h for *Streptococcus pneumoniae*) after removing the tubes from the water bath or heating block.

E. Detection

1. Select the appropriate protocol from the menu of the luminometer software.
2. Using a damp tissue or paper towel, wipe each tube to ensure that no residue is present on the outside of the tube, and insert the tube into the luminometer based on the instrument instructions.
3. When the analysis is complete, remove the tube from the luminometer. The sample cannot be read again.

F. QC

Include in each run appropriate reference strains representing a positive and negative control for each organism being tested.

V. PROCEDURE NOTES

A. A precipitate may form in the hybridization buffer. Warming and mixing the solution at 35 to 60°C will dissolve the precipitate.

B. The hybridization and selection reactions are temperature dependent. Therefore, it is imperative that the incubator, water bath, or heating block be maintained within the specified temperature. The entire liquid reaction volume in the probe reagent tube must be exposed to the required temperature.

C. With gram-positive bacteria, the hybridization reaction should be started within 30 min of adding cells and reagent 1 to the probe reagent tubes.

D. The hybridization and selection reactions are time dependent. Hybridize for at least 20 min. Incubate the probe reaction tubes during the selection step for at least 5 min but no more than 6 min.

E. It is critical to have a homogenous mixture during the selection (vortex) step, specifically after the addition of reagent 3.

F. Troubleshooting

1. Elevated negative control values of more than 20,000 RLU in the Leader 1 luminometer or more than 600 photometric light units (PLU) in the AccuLDR luminometer (formerly PAL) can occur. If they do, streak a portion of the growth onto the appropriate agar medium and incubate it to check for multiple colony types.

2. Low positive control values (less than 50,000 RLU in the leader or less than 1,500 PLU in the AccuLDR) can be caused by insufficient cell numbers, by testing mixed or aged cultures, or, for gram-positive organisms, by leaving cells exposed to the lysis reagent for more than 30 min before hybridization reagents are added.

VI. RESULTS

A. Interpretation

The results of the Accuprobe Culture Identification Test are based on the following cutoff values. Samples producing signals higher than or equal to the cutoff values are considered positive. Samples with signals lower than these values are considered negative. Results from samples in the indeterminate range must be retested.

	AccuLDR (PLU)	Leader (RLU)
Cutoff value	1,500	50,000
Repeat range	1,200–1,499	40,000–49,000

B. QC

Negative- and positive-control strains should yield the following values.

	AccuLDR (RLU)	Leader (RLU)
Negative control	<600	<20,000
Positive control	>1,500	>50,000

VII. PERFORMANCE CHARACTERISTICS

A. Sensitivity, 100%

B. Specificity, 100%

C. Percent agreement, 100%

SUPPLEMENTAL READING

Daily, J. A., N. L. Clifton, K. C. Seskin, and W. M. Gooch III. 1991. Use of rapid, nonradioactive DNA probes in culture confirmation tests to detect *Streptococcus agalactiae, Haemophilus influenzae*, and *Enterococcus* spp. from pediatric patients with significant infections. *J. Clin. Microbiol.* **29:** 80–82.

Denys, G. A., and R. B. Carey. 1992. Identification of *Streptococcus pneumoniae* with a DNA probe. *J. Clin. Microbiol.* **30:**2725–2727.

Gen-Probe, Inc. Accuprobe Culture Identification Test, package insert. Gen-Probe, Inc., San Diego, Calif.

Kohne, D. E. 1990. The use of DNA probes to detect and identify microorganisms, p. 11–35. *In* B. Kleger et al. (ed.), *Rapid Methods in Clinical Microbiology.* Plenum Press, New York.

Lewis, J. S., D. Kranig-Brown, and D. A. Trainor. 1990. DNA probe confirmatory test for *Neisseria gonorrhoeae. J. Clin. Microbiol.* **28:** 2349–2350.

Okwumabua, O., et al. 1992. Evaluation of chemiluminescent DNA probe assay for the rapid confirmation of *Listeria monocytogenes. Res. Microbiol.* **143:**183–189.

Popovic-Uroic, T., et al. 1991. Evaluation of an oligonucleotide probe for identification of *Campylobacter* spp. *Lab. Med.* **22:**533–539.

Stockman, L., K. A. Clark, J. M. Hunt, and G. D. Roberts. 1993. Evaluation of commercially available acridinium ester-labeled chemiluminescent DNA probes for culture identification of *Blastomyces dermatitidis, Coccidioides immitis, Cryptococcus neoformans*, and *Histoplasma capsulatum. J. Clin. Microbiol.* **31:**845–850.

10.3.3 DNA Probes for the Identification of Mycobacteria

I. PRINCIPLE

A commercially manufactured system, the AccuProbe Culture Confirmation Test, is available for identification of isolates of several species of mycobacteria. The system is based on the use of DNA probes that are complementary to species-specific rRNA. The mycobacterial cells are lysed by sonication, heat killed, and exposed to DNA that has been labeled with a chemiluminescent tag. The labeled DNA probe combines with the organism's rRNA to form a stable DNA-RNA hybrid. A selection reagent "kills" the signal on all unbound DNA. Chemiluminescence produced by DNA-RNA hybrids is measured in a luminometer.

Probes for the identification of *Mycobacterium tuberculosis* complex, *Mycobacterium avium, Mycobacterium intracellulare, Mycobacterium gordonae, M. avium* complex, and *Mycobacterium kansasii* are available. Nucleic acid probes provide the most rapid identification of mycobacteria of any identification systems.

II. SPECIMEN

Any actively growing culture less than 1 month old recovered on any solid or broth medium such as Middlebrook 7H9, 7H10, 7H11, or 7H11 Selective or Lowenstein-Jensen can be used.

III. MATERIALS

Include QC information on reagent container and in QC records.

A. Reagents (Table 10.3.3–1)
Indicate the expiration date on the label and in the work record or on the manufacturer's label.

B. Supplies
1. Micropipettes (100 μl)
2. Micropipettor (300 μl)
3. Repipettor (300 μl)
4. Repipettor syringes
5. Plastic sterile inoculating loops (1 μl)

C. Equipment
1. Luminometer (Leader 1, Leader 250, or PAL)
2. Vortex mixer
3. Water or dry bath, 90°C (±5°C)
4. Water or dry bath, 60°C (±1°C)
5. Water bath sonicator

Table 10.3.3–1 Reagents for AccuProbe culture confirmation

Reagent[a]	Description	Storage temp (°C)
Lysing tubes	Lyophilized glass beads and buffer	23–25
Probe reagent	Tubed lyophilized species-specific DNA (in foil pouches)[b]	4
Reagent 1	Specimen diluent	23–25
Reagent 2	Probe diluent	23–25
Reagent 3	Selection reagent	23–25
Detection I	H_2O_2 in nitric acid solution containing 1 N NaOH	23–25
Detection II	H_2O_2 in nitric acid solution containing 1 N NaOH	23–25
Ultrasonic enhancer	Detergent-like solution to increase sonication power	23–25

[a] All reagents are stable for approximately 1 year from date of manufacture.
[b] After pouch is opened, stable for 2 months.

Table 10.3.3–2 Control organisms for AccuProbe mycobacterial culture confirmation tests

Probe to be tested	Control	
	Positive	Negative
M. avium	*M. avium* ATCC 25291	*M. intracellulare* ATCC 13950
M. intracellulare	*M. intracellulare* ATCC 13950	*M. avium* ATCC 25291
M. tuberculosis complex	*M. tuberculosis* ATCC 25177	*M. avium* ATCC 25291
M. gordonae	*M. gordonae* ATCC 14470	*M. scrofulaceum* ATCC 19073 type
M. avium complex	*M. intracellulare* ATCC 13950	*M. gordonae* ATCC 14470
M. kansasii	*M. kansasii* ATCC 12478	*M. tuberculosis* ATCC 25177

IV. QUALITY CONTROL

A. Frequency and tolerance of controls

1. Run a positive and negative control with each batch of organisms tested.
2. If controls are not within limits, check age of culture (should be less than 1 month old), level of water in sonicator, quality of sonicator (by observing waving water patterns while sonicator is on; water should ''dance''), and temperature of hybridization; if necessary, run a tritium standard.

B. Control organisms (Table 10.3.3–2)

V. PROCEDURE

A. Degas the water for 15 min. Turn on 95°C water bath (or heat block) and 60°C water bath (or heat block).

B. Add 100 μl of reagent 1 (specimen diluent) and 100 μl of reagent 2 (probe diluent (Table 10.3.3–1) to each lysing tube.

C. Working in a biological safety cabinet, transfer a loopful (1 μl) of test organism into the lysing tube. Twirl the loop against the side of the tube to remove the entire inoculum.

D. Place the lysing tubes in the sonicator. Ensure that the water level is high enough to cover the contents of the tube. Do not allow the tubes to touch the sides of the sonicator. Sonicate at room temperature for 15 min.

E. Place the lysing tubes in the 95°C water bath (or heat block) for 15 min.

F. Allow the tubes to cool at room temperature for 5 min.

G. Pipette 100 μl of the killed lysate into the probe reaction tube.

H. Incubate the tubes for 15 min at 60°C in the water bath (or heat block).

I. Pipette 300 μl of reagent 3 (selection reagent) (Table 10.3.3–1) into each tube. Recap and vortex the tubes, and immediately place them back into the 60°C water bath (or heat block). Incubate 5 min for *M. avium* complex and *M. gordonae*, 8 min for *M. kansasii*, and 10 min for *M. tuberculosis* complex.

J. Prepare the luminometer for operation by completing a wash cycle. Wipe each tube with a damp tissue before inserting it into the luminometer. Read each tube and controls. The luminometer records relative light units (RLU).

VI. RESULTS

A. A positive result is >30,000 RLU.

B. Signals below 30,000 RLU are considered negative.

C. Repeat any test with a result between 20,000 and 29,999 RLU.

VII. PROCEDURE NOTES

A. It is important to maintain the 60°C (± 1°C) dry bath, because annealing of the DNA-RNA hybrid is temperature sensitive.

B. The water level of the sonicator should be high enough to cover the hybridization buffer, beads, and organisms.

C. Other sources of error include failure to vortex after addition of the organism-buffer mixture to the probe tube and after addition of reagent 3, which inactivates the acridinium ester not bound to rRNA.

VIII. LIMITATIONS OF THE PROCEDURE

A. This method can not be used directly on fresh clinical specimens. It is for culture identification only.

B. A small number of biochemically determined *M. avium* complex isolates will not produce a positive result with the *M. avium* complex probe. The taxonomic status of these strains is currently uncertain.

C. The *M. tuberculosis* complex culture confirmation test does not differentiate between *M. tuberculosis, Mycobacterium bovis, M. bovis* BCG, *Mycobacterium africanum*, and *Mycobacterium microti*. Additional tests such as niacin and nitrate reductase are required to differentiate *M. tuberculosis* from other members of the group (*see* section 4).

D. The *M. avium* complex culture confirmation test does not differentiate between *M. avium* and *M. intracellulare*.

E. The sample preparation for the identification of fungal pathogens *Histoplasma capsulatum, Blastomyces dermatitidis*, and *Coccidioides immitis* is identical to that of the mycobacteria. These organisms may then be identified by using the appropriate Gen-Probe kits and reagents.

SUPPLEMENTAL READING

Ellner, P. D., T. E. Kiehn, R. Cammarata, and M. Hosmer. 1988. Rapid detection and identification of pathogenic mycobacteria by combining radiometric and nucleic acid probe methods. *J. Clin. Microbiol.* **26:**1349–1352.

Evans, K., A. Nakasone, P. Sutherland, L. de la Maza, and E. Peterson. 1992. Identification of *Mycobacterium tuberculosis* and *Mycobacterium avium-M. intracellulare* directly from primary BACTEC cultures by using acridinium ester-labeled DNA probes. *J. Clin. Microbiol.* **30:** 2427–2431.

Gen-Probe, Inc. AccuProbe Culture Confirmation Test for Mycobacteria, package insert. Gen-Probe, Inc., San Diego, Calif.

Goto, M., S. Oka, K. Okuzumi, S. Kimura, and K. Shimada. 1991. Evaluation of acridinium ester-labeled DNA probes for identification of *Mycobacterium tuberculosis* and *Mycobacterium avium-Mycobacterium intracellulare* complex in culture. *J. Clin. Microbiol.* **29:**2473–2476.

Musial, C. E., L. S. Tice, L. Stockman, and G. P. Roberts. 1988. Identification of mycobacteria from culture by using the Gen-Probe rapid diagnostic system for *Mycobacterium avium* complex and *Mycobacterium tuberculosis* complex. *J. Clin. Microbiol.* **26:**2120–2123.

10.4.1 Introduction

The ability to identify specific strains within a species of pathogen is an important aid in the rational development of effective measures to prevent and control nosocomial infections. The efforts of both microbiologists and hospital epidemiologists are facilitated greatly by the availability of the newer molecular epidemiologic typing techniques. The variety of molecular epidemiologic tools available at present is considerable, and based on current experience the methods that appear to be the most practical and useful for both large- and small-scale epidemiologic studies are the DNA-based methods such as pulsed-field gel electrophoresis (PFGE) (Table 10.1–3). Although these methods clearly have limitations, they generally are a significant improvement over the more conventional typing methods, many of which are too cumbersome, insensitive, and time-consuming to be of practical value for epidemiologic evaluations. It is important that no single technique is universally applicable and that the choice for a particular application is related to the species studied, the scope of the question posed, and the convenience of the technique. The techniques of molecular epidemiology are useful in answering real clinical and infection control questions and are not limited to research uses. Examples include distinguishing between relapse and reinfection in an individual patient and tracking the spread of an individual strain of a bacterium or fungus within the hospital environment (1–3).

REFERENCES

1. **Arbeit, R. D.** 1995. Laboratory procedures for the epidemiologic analysis of microorganisms, p. 190–208. *In* P. R. Murray, E. J. Baron, M. A. Pfaller, F. C. Tenover, and R. H. Yolken (ed.), *Manual of Clinical Microbiology,* 6th ed. American Society for Microbiology, Washington, D.C.
2. **Maslow, J. N., M. E. Mulligan, and R. D. Arbeit.** 1993. Molecular epidemiology: application of contemporary techniques to the typing of microorganisms. *Clin. Infect. Dis.* **17:** 153–164.
3. **Sader, H. S., R. J. Hollis, and M. A. Pfaller.** 1995. The use of molecular techniques in the epidemiology and control of infectious diseases. *Clin. Lab. Med.* **15:**407–431.

10.4.2 Plasmid Fingerprinting of Gram-Negative Organisms

I. PRINCIPLE

Plasmids are extrachromosomal circles of DNA present in many but not all species of bacteria. Many organisms, particularly those that are resistant to multiple antimicrobial agents, may harbor two or more plasmids. The presence of similarly sized plasmids in a series of isolates collected as part of an epidemiological investigation can be used in conjunction with other typing schemes to indicate the presence of an epidemic strain in a hospital or community. However, two or more isolates of the same species that are devoid of plasmid DNA can not be assumed to be the same strain. Isolates of the same species each showing a single plasmid of similar molecular size should be investigated further by cleaving the plasmid DNA with restriction endonucleases to confirm the similarity of the plasmids.

Plasmid fingerprinting techniques focus on the recovery of the supercoiled, covalently closed, circular form of the molecule. Lysis of the cell wall with detergent at high pH allows denaturation of chromosomal DNA without irreparable harm to plasmid DNA. Precipitation of plasmid DNA with high NaCl and isopropanol results in preparations relatively free of chromosomal DNA. Although plasmids normally have a molecular size ranging from 1 to 100 kb, some *Pseudomonas* plasmids may be as large as 400 kb. It is important to run molecular size standards consisting of supercoiled plasmids and not restriction endonuclease fragments for accurate determinations of plasmid size.

II. SPECIMENS

Most enteric gram-negative bacilli and *Pseudomonas* species grown on a nonselective agar medium for 18 to 24 h will yield plasmid DNA with this procedure.

III. MATERIALS

NOTE: Autoclave all centrifuge tubes and pipette tips before use to destroy any nucleases that may be present.

A. Sterile microcentrifuge tubes (1.5 ml) with caps
B. Pipettor, variable volume (0 to 20 μl)
C. Pipettor, variable volume (20 to 200 μl)
D. Sterile pipette tips
E. Disposable 5-ml pipettes
F. Microcentrifuge (maximum speed, 12,000 $\times$ *g*)
G. Wet ice in bucket
H. Vacuum line with trap

IV. PREPARATION OF REAGENTS

(Unless otherwise stated, all reagents are available from Sigma, St. Louis, Mo.)

Include QC information on reagent container and in QC records.

A. TE buffer (50:10), pH 8.0 (50 mM Tris–10 mM EDTA)
B. Lysis buffer, pH 12.4 (stable for 1 month): Trizma base–Na_2EDTA–sodium dodecyl surfate (SDS)–NaOH
C. 2 M Tris, pH 7.0 (stable for 6 months)
D. 5 M NaCl (indefinite stability)
E. 100% ethanol
F. Stop mix (running dye) (stable for 1 year): bromophenol blue–SDS–Ficoll type 400

V. PROCEDURE

A. Pipette 40 μl of TE buffer (50:10) into a 1.5-ml microcentrifuge tube.
B. Suspend one-fourth loopful (0.01-ml inoculating loop) of bacteria taken from 18- to 24-h growth on a nonselective agar plate in the buffer.
C. Add 0.6 ml of lysis buffer, and mix by inverting tube 20 times. Do not vortex.
D. Incubate at 37°C in a water bath for 20 min.
E. Add 30 μl of 2 M Tris (pH 7.0), and invert tube 10 times to mix.
F. Add 160 μl of 5 M NaCl, and invert tube 10 times.
G. Place tubes on wet ice for at least 20 min. Incubation may continue for up to 1 h.
H. Centrifuge tubes at 12,000 $\times$ g in a microcentrifuge for 5 min. Pour supernatant into a fresh (sterile) 1.5-ml centrifuge tube.
I. Add 0.55 ml of cold absolute isopropanol (store at -20°C) and mix by inverting tube.
J. Freeze at -20°C for at least 20 min. Tubes can be left in the freezer overnight.
K. Remove tubes from freezer and centrifuge at 12,000 $\times$ g for 3 min. Discard supernatant, and cautiously remove remaining alcohol with a Pasteur pipette connected to a vacuum line. Do *not* remove pellet at bottom of tube. Place inverted tubes on a paper towel for 30 min. Make sure that all alcohol is removed.
L. Suspend DNA in 30 μl of TE buffer (10:1). Add 6 μl of stop mix. Allow at least 1 h for DNA to suspend.
M. Electrophorese 15-μl samples through a 0.7% agarose gel at 200 V/h.
N. Stain gel in ethidium bromide (0.2 μg/ml of water).
NOTE: Wear gloves when handling ethidium bromide.
O. Photograph gel with orange filter and midrange UV transilluminator.

VI. INTERPRETATION OF RESULTS

Some chromosomal-DNA fragments will be present in plasmid preparations and will be seen on agarose gels, indicating that the bacteria have lysed. The chromosomal-DNA band, which tends to be diffuse, will migrate to the 12- to 15-kb area of the gel. Plasmids produce sharp, well-delineated bands on agarose gels. Plasmids >15 kb in size will migrate more slowly than the chromosomal band, while smaller plasmids will migrate more quickly than the chromosomal band. Isolates with multiple plasmids producing identical or very similar plasmid migration profiles after electrophoresis are likely to represent multiple isolates of the same strain. Plasmids placed in the outside lanes of a gel may migrate slightly more slowly (smile effect). This problem can be reduced by running stop mix in the two outside lanes of the gel. Two isolates determined to be devoid of plasmids cannot be assumed to be the same strain on this basis. If several isolates each contain only a single plasmid, all similar in size, other typing procedures or restriction enzyme analysis of the plasmids should be performed. Typing of isolates should be done only within a defined epidemiological situation or to assess the relatedness of isolates from a single patient.

A. Organisms such as *Stenotrophomonas maltophilia, Serratia marcescens*, and some species of *Acinetobacter*, all of which are frequently involved in outbreaks of nosocomial infections, are often devoid of plasmid DNA. The presence of a chromosomal-DNA band in the absence of plasmid DNA indicates that the technique is working and that the isolates did undergo lysis. In this situation, alternative typing methods must be used.
B. The presence of a large bright band in the pocket of the gel without indications of DNA elsewhere in the gel suggests that the isolates did not lyse well and that the DNA has not migrated out of the pocket of the gel. In this situation, the pH of the lysis buffer should be checked.

VII. QUALITY CONTROL

A standard strain containing well-characterized plasmids should be lysed in parallel with each plasmid screen to ensure the quality of the lysis buffer. *Escherichia coli* V517, which contains eight plasmids, is a good QC strain and also serves as an excellent source of molecular size standards (it is available from the American Type Culture Collection [#37514]). Supercoiled standards also can be purchased from several commercial sources.

VIII. LIMITATIONS OF THE PROCEDURE

Not all organisms have plasmid DNA. Large plasmids tend to cluster in the upper portions of the gel, and it is difficult to distinguish the sizes of plasmids that are >100 kb. Exercise caution when determining the relatedness of large plasmids. Because plasmids of identical size can be totally unrelated to one another, always exercise caution when isolates with a single plasmid each are encountered.

IX. REPORTING RESULTS

The manner in which the results of plasmid fingerprinting investigations are reported depends on three factors: why the tests were ordered, who ordered them, and the expertise and comfort level of the microbiologist performing the test. Plasmid fingerprinting is a novel procedure. Thus, physicians and many infection control practitioners who request such tests will need to be given both the results and an interpretation of the results. The following guidelines are suggested.

A. Provide a duplicate picture of the gel for the record. If you are not comfortable releasing a copy of the gel, you probably are not comfortable with the results, and the procedure should be deemed inconclusive. Some isolates of *Pseudomonas* spp. are difficult to lyse, and a report of ''Inconclusive'' is better than a guess.

B. Include in your report the sizes of the plasmids, if possible, and a comment such as ''Plasmid patterns of isolates A, B, and C suggest that they are the same strain'' or ''Plasmid patterns suggest the presence of three unrelated isolates.''

C. Report a series of isolates that each have a single plasmid, all of approximately the same size, with a caution that further studies may be warranted. Whether further studies are done depends on the strength of the other data collected as part of the epidemiological investigation. For example, if the isolates were collected within a short time from patients on the same ward with a common health care practitioner and if the antibiograms for all of the isolates are identical and have an unusual resistance pattern, this is strong supporting evidence of strain identity, and further information is probably not required. The strength of the data should be reflected in the report. If, on the other hand, the isolates were collected from patients on different wards, have antibiograms showing nothing but susceptibility, were collected over the course of 2 months, and do not have a substantial epidemiological link, additional typing data should be sought, and the report should indicate this is being done.

D. Results generated for infection control investigations do not necessarily have to be entered on a patient's chart. Provide a copy of the gel to the infection control practitioner, and discuss the results in the context of what is known about the outbreak.

SUPPLEMENTAL READING

Portnoy, D. A., S. L. Moseley, and S. Falkow. 1981. Characterization of plasmid and plasmid-associated determinants of *Yersinia enterocolitica* pathogenesis. *Infect. Immun.* **31:**775–782.

Tenover, F. C. 1985. Plasmid fingerprinting: a tool for bacterial strain identification and surveillance of nosocomial and community acquired infections. *Clin. Lab. Med.* **5:**413–436.

Tenover, F. C., K. Phillips, and L. G. Carlson. 1994. Plasmid fingerprinting of gram-negative organisms, p. 10.4.1–10.4.4. *In* H. D. Isenberg (ed.), *Clinical Microbiology Procedures Handbook, Supplement 1.* American Society for Microbiology, Washington, D.C.

10.4.3 Plasmid Fingerprinting of Staphylococci

I. PRINCIPLE

Plasmid analysis is based on the fact that different bacterial strains often carry different types or numbers of plasmids. This is a useful epidemiological tool for investigations of strain relatedness. Approximately 80 to 90% of the clinical isolates of staphylococci carry one or more plasmids. Plasmid DNA is isolated from the bacterial cells, digested with a restriction endonuclease, and electrophoresed through agarose. Different plasmids display different patterns of fragment sizes. For initial analysis, the restriction endonucleases *Eco*RI and *Hin*dIII are recommended.

II. MATERIALS

(Unless otherwise stated, all reagents are available from Sigma Chemical Co., St. Louis, Mo.)

Include QC information on reagent container and in QC records.

A. Stocks and purchased chemicals

Indicate the expiration date on the label and in the work record or on the manufacturer's label. All materials are stable for 2 years at room temperature unless otherwise noted. Stock solutions that must be autoclaved or filter sterilized are noted.

1. 1 M Tris, pH 8.0 (autoclave)
2. 1 M Tris, pH 7.0 (autoclave)
3. Sodium dodecyl sulfate (SDS), 20% (filter sterilize)
4. 0.5 M EDTA, pH 8.0 (autoclave)
5. 4 M NaCl (autoclave)
6. 2 N NaOH
7. Sucrose (50%) (filter sterilize)
8. Ethanol (95%)
9. Agarose
10. Mueller-Hinton broth for coagulase-negative staphylococci
11. TSB for *Staphylococcus aureus*
12. Restriction endonucleases *Eco*RI and *Hin*dIII and buffers (New England BioLabs, Inc., Beverly, Mass.)
13. Molecular size standards (New England BioLabs, Inc.); use *Eco*RI or *Hin*dIII digests of bacteriophage lambda
14. 3 M sodium acetate (autoclave)
15. Lysostaphin stock (1 mg/ml) (Applied Microbiology, New York, N.Y.)
16. Lysozyme stock (20 mg/ml)
17. RNase A stock solution (10 mg/ml [New England BioLabs, Inc.])
18. RNase dilution buffer (New England BioLabs, Inc.): 1 M Tris (pH 8.0)–1 M Tris (pH 7.0)–4 M NaCl
19. Dye mixture: 20 mM EDTA, 50% glycerol, 0.1% bromphenol blue, 1% xylene cyanol
20. 10× TBE buffer: Tris base, boric acid, Na_2EDTA

B. Working solutions

Indicate the expiration date on the label and in the work record or on the manufacturer's label. Working solutions are made from stock solutions and are stable for 1 year at room temperature.

1. TES–SDS: 100 mM Tris–70 mM EDTA–50 mM NaCl–1.5% SDS
2. TES-sucrose: 100 mM Tris–70 mM EDTA–50 mM NaCl–5% sucrose
3. 0.375 M EDTA, pH 8.0
4. TE buffer (10:0.1): 10 mM Tris–0.1 mM EDTA
5. 0.4% SDS–0.4 N NaOH
6. RNase A working solution: dilute RNase A stock solution 1:100 in 10 mM Tris (pH 7.5)–15 mM NaCl. Boil in water bath for 15 min. Cool gradually to room temperature. Aliquot and store at −20°C.

7. Loading dye, 1 : 1 mixture. Mix the dye and 1% SDS in a 1 : 1 ratio.
8. Loading dye, 1 : 1 : 4 mixture. Mix 1 part dye mixture, 1 part 1% SDS, and 4 parts distilled water.
9. DNA wash solution: 10 mM Tris–100 mM EDTA–75 mM sodium acetate

C. **Supplies and equipment**
1. Power supply
2. Gel box for electrophoresis
3. Microcentrifuge
4. Microcentrifuge tube (1.5 ml) (supplier not critical)
5. Shaking water bath or shaking platform in incubator
6. Vacuum line or vacuum unit (for aspiration of liquids)
7. Transilluminator (midrange UV)
8. Camera unit (e.g., Polaroid MP-4 or hand-held Polaroid-type Fotodyne FCR-10)

III. PROCEDURE

A. **DNA preparation**
1. Using three to five well-isolated colonies, inoculate one tube of growth medium for each test and control strain to be screened. For *S. aureus*, use 10 ml of TSB; for coagulase-negative staphylococci, use 20 ml of Mueller-Hinton broth divided into two 10-ml cultures (for convenience in processing).
2. Incubate with shaking at 37°C for 16 to 24 h.
3. Centrifuge broth for 15 min at 7,000 × *g*, and remove supernatant by aspiration.
4. Suspend cells in 1 ml of sterile saline, and transfer them to a sterile 1.5-ml microcentrifuge tube (supplier not critical).
5. Centrifuge at top speed (12,000 × *g*) for 15 s. Carefully aspirate supernatant.
6. Suspend cells in 100 μl of TES-sucrose. For *S. aureus*, add 50 μl of dilute lysostaphin. Make dilute lysostaphin by mixing 5 μl of lysostaphin stock with 45 μl of TES-sucrose. For coagulase-negative staphylococci, add 15 μl of lysostaphin stock and 50 μl of lysozyme stock. Mix well.
7. Incubate tubes in water bath at 37°C for 20 min for *S. aureus* and 60 min for coagulase-negative staphylococci.
8. Add 50 μl of 0.375 M EDTA (pH 8.0) to each tube. Mix by inverting tube 10 to 15 times. Solution may become very viscous.
9. Incubate on ice for 15 min.
10. Add 100 μl of TES-SDS (SDS must be in solution with no precipitate). Mix by inverting tube 10 to 15 times.
11. Place on ice for at least 30 min. Hold for additional time if necessary.
12. Centrifuge at 12,000 × *g* for 15 min. Use of a refrigerated microcentrifuge or a centrifuge located in a cold room is optimal but not mandatory. Centrifugation at room temperature will also work.
13. Remove 300 μl of supernatant, and place it in a new microcentrifuge tube. Avoid the pellet and any viscous material. If less than 300 μl is available, make up difference with TES-sucrose.
14. Add 150 μl of 0.4% SDS–0.4 N NaOH. Mix by inverting tube 10 times. Quickly place on ice for 5 min.
15. Add 150 μl of 3 M sodium acetate (pH 5.4), and invert tube 10 times. Return to ice quickly, and incubate for 60 min or longer.
16. Centrifuge at 12,000 × *g* for 15 min. Remove supernatant to a fresh tube. Avoid white pellet.
17. Add 1 ml of 95% cold ethanol to supernatant. Mix by inverting tube 20 times until contents are well mixed. This can be stored overnight at −20°C.
18. Centrifuge at 12,000 × *g* for 30 min. (Shorter centrifugation times will reduce yield.) Remove supernatant by aspiration with drawn-glass pipette, taking care to save pellet (may be difficult to see).
19. Suspend pellet in 500 μl of DNA wash solution. Incubate at 37°C for 20 min. Vortex gently (low setting) at 10 min and again at 20 min.

III. PROCEDURE *(continued)*

20. Centrifuge at 12,000 × *g* for 10 min.
21. Transfer supernatant to a new microcentrifuge tube containing 1 ml of cold 95% ethanol. Mix by inverting tube 20 times. This mixture can be stored overnight at −20°C.
22. Centrifuge at 12,000 × *g* for 30 min, and remove supernatant, taking care to save the pellet. Centrifuge an additional 5 min, and aspirate remaining ethanol.
23. Dry under vacuum for 30 min. This is sufficient time for 20 to 30 tubes.
24. Remove 3 μl of plasmid preparation, and combine it with 10 μl of the 1:1:4 dye mixture. Electrophorese through an agarose gel to screen for whole plasmid DNA. Electrophoresis times will vary depending on the size of the gel and the apparatus used. Alternatively, the samples can be stored at −20°C. DNA is stable in a −20°C non-frost-free freezer (defrosting cycle is hard on DNA samples) for approximately 6 months.

B. Restriction analysis of plasmid DNA

1. Place 3 μl of the plasmid preparation in each of two fresh tubes for restriction enzyme digestion (one with *Eco*RI, the other with *Hin*dIII), and store the remainder at −20°C.
2. To the 3-μl DNA sample add the following: 4.5 μl of sterile water; 1.0 μl of 10× restriction buffer, per manufacturer; 1.0 μl of RNase A working solution; and 0.5 μl of restriction enzyme (*Eco*RI or *Hin*dIII).
3. Centrifuge at 12,000 × *g* for 15 s. Place in 37°C water bath for 2 h. Add 3 μl of 1:1 mixture of SDS and dye.
4. Prepare 0.7% agarose gel by using 1× TBE buffer. Load entire 13-μl sample into a well of the gel. Use *Hin*dIII or *Eco*RI digest of phage lambda loaded onto a separate well as molecular size standards for restriction digests.
5. Electrophorese for 4 h at 100 V or for 1.5 h at 130 V (slower is better). The dark blue dye should move >3 cm. Photograph by using midrange (260 nm) transilluminator and orange filter (Kodak 22A Wrattan or equivalent).

IV. INTERPRETATION OF RESULTS

One of the most difficult aspects of interpreting plasmid patterns is determining whether there was adequate lysis of the cells during the protocol to provide DNA for analysis. In preparations with no detectable plasmid DNA, small amounts of chromosomal DNA must be present for the preparation to be considered adequate for analysis. If no chromosomal DNA is present (it usually appears as a faint smear in the molecular size range of 12 to 15 kb) and no plasmid DNA is detectable, repeat the procedure.

Two or more isolates are presumed to be the same strain if all of the bands in both the *Hin*dIII and *Eco*RI restriction digests are identical. If multiple bands do not match between isolates, they are presumed to represent different strains.

V. REPORTING

A report of the results of plasmid analysis, detailing which isolates are related or distinct, should be issued to the service that generated the request for typing (see legend to Fig. 10.4.3–1 for example). Photographs of the results may accompany the report at the discretion of the laboratory.

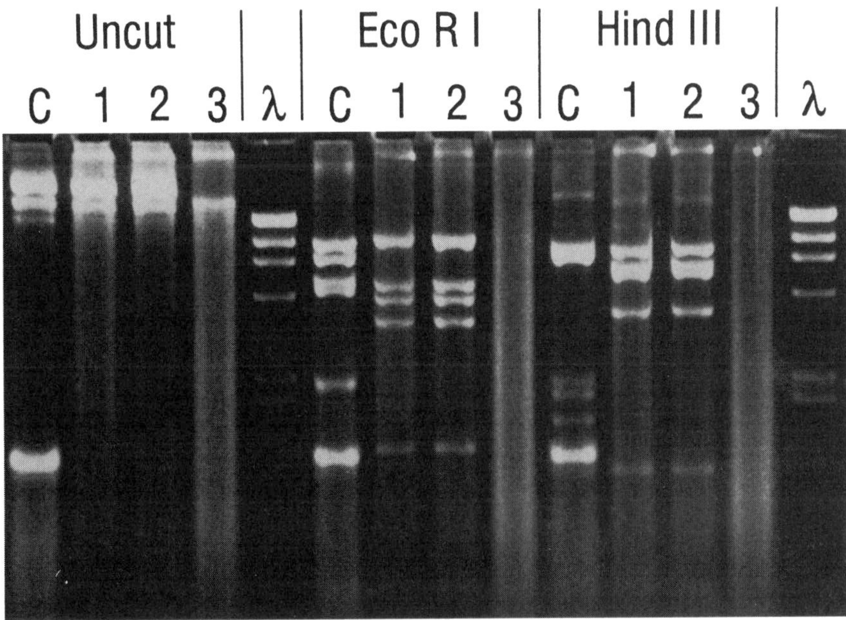

Figure 10.4.3–1 *S. aureus* plasmid DNA preparations. Uncut, undigested plasmid DNA; *Eco*RI, DNA digested with *Eco*RI; *Hin*dIII, plasmid DNA digested with *Hin*dIII. Lanes: C, control strain SM818-73; 1, 2, and 3, isolates collected from a single patient. Isolate 3 does not have detectable plasmid DNA. The report for this group of isolates stated: ''Based on restriction endonuclease analysis of plasmid DNA, these strains were grouped as follows: isolates 1 and 2 are identical to each other; isolate 3 is not related to isolates 1 and 2.''

VI. QUALITY CONTROL

A staphylococcal strain of known plasmid content should be run in parallel with each screening procedure. *S. aureus* SM818-73 and coagulase-negative strain SM734-41 are available from the authors of this procedure as controls.

SUPPLEMENTAL READING

Mayer, L. W. 1988. Use of plasmid profiles in epidemiologic surveillance of disease outbreaks and in tracing the transmission of antibiotic resistance. *Clin. Microbiol. Rev.* **1:**228–243.

Nahaie, M. E., M. Goodfellow, and G. Harwood. 1984. A rapid screening procedure for staphylococcal plasmids. *J. Microbiol. Methods* **2:** 73–81.

Pfaller, M. A., D. S. Wakefield, R. Hollis, M. Fredrickson, E. Evans, and R. M. Massanari. 1991. The clinical microbiology laboratory as an aid in infection control: the application of molecular techniques in epidemiologic studies of methicillin-resistant *Staphylococcus aureus*. *Diagn. Microbiol. Infect. Dis.* **14:**209–217.

10.4.4 Method for Ribotyping by Using a Chemiluminescent Probe

I. PRINCIPLE

Ribotyping was first described in 1986 by Grimont and Grimont (1), and since that time it has emerged as one of the most powerful tools of molecular epidemiology. Ribotyping involves the use of Southern blot analysis to detect polymorphisms in the chromosomal regions containing the rRNA genes. Classic ribotyping uses labeled probes containing *Escherichia coli* 16S and 23S rRNA sequences. Both the nature of the probe and the mode of labeling, however, may vary. Ribotyping has three main advantages: (i) the genes coding for rRNA are highly conserved, allowing for the use of a single probe to subtype all eubacteria; (ii) because most bacteria contain multiple rRNA genes, a reasonable number of bands are obtained after probing; and (iii) all strains have rRNA genes (rDNA). The availability of chemiluminescent-labeling techniques and the recent development of an automated ribotyping system (3) make ribotyping more feasible for the investigation of nosocomial outbreaks by the clinical laboratory.

NOTE: The following procedure has been adapted from that of Gustaferro (2) and Stull and LiPuma (4). The method described is optimized for *Staphylococcus aureus* but can also be used for ribotyping other gram-positive and gram-negative bacteria (1–5).

II. MATERIALS

(Unless otherwise stated, all reagents are available from Sigma Chemical Co., St. Louis, Mo.)

Include QC information on reagent container and in QC records.

A. Reagents

1. Lysostaphin (500 U/ml)
2. 10% bleach
3. Denhardt's solution
4. Chondroitin sulfate A
5. Bromophenol blue
6. Proteinase K (20 mg/ml)
7. RNasin RNase inhibitor
8. High-pressure liquid chromatography (HPLC) water
9. Enhanced chemiluminescence (ECL) kit (Amersham)
10. 16S and 23S rRNA from *E. coli* MRE600
11. Restriction endonucleases (individualized for each organism [available from New England BioLabs Inc., Beverly, Mass.])
12. BHI broth or another appropriate liquid medium
13. TES buffer: 0.01 M Tris (pH 8)–0.1 M NaCl–0.001 M EDTA
14. 50× TAE buffer: 242.2 g of Tris base, 57.15 ml of 17.5 M glacial acetic acid, 100 ml of 0.5 M EDTA, water to make up 1 liter (pH 8.5)
15. Denaturation solution: 1.5 M NaCl–0.5 M NaOH
16. Depurination solution: 250 mM HCl
17. Neutralization solution: 1 M Tris HCl (pH 8.0)–1.5 M NaCl
18. Phenol
19. Chloroform
20. Isoamyl alcohol
21. Prehybridization buffer: 25 mM KPO_4 (pH 7.4), 5× SSC (1× SSC is 0.15 M NaCl plus 0.015 M sodium citrate), 5× Denhardt's solution, 50 μg of salmon sperm DNA per ml, 6 M urea, diethylpyrocarbonate (DEPC)-treated water
22. Hybridization buffer: 6 M urea, 1.1% sodium dodecyl sulfate (SDS), 0.5× SSC, DEPC-treated water
23. Secondary wash buffer: 2× SSC, DEPC-treated water
24. Electrophoresis loading buffer: 27 ml of glycerol, 3 ml of 10× TAE, 0.1% (wt/vol) SDS, 10 ml of 0.5 M EDTA (pH 8.0), 10 mg of bromophenol blue, water to make up 100 ml
25. Glutaraldehyde
26. Agarose
27. Ethidium bromide
28. 5 M NaCl
29. 70% and 100% ethanol

B. Supplies

1. Cellophane
2. Nitrocellulose membrane

3. Gel blot paper
4. Nylon mesh
5. X-ray film
6. Scissors
7. Forceps
8. Eppendorf tubes (1.5 ml)

C. **Equipment**
1. Agarose gel electrophoresis apparatus
2. Power supply
3. UV transilluminator and photodocumentation setup
4. Autoradiograph cassettes and intensifying screens
5. Vacuum desiccator
6. Vacuum oven
7. Microcentrifuge
8. Block heater (60°C)

III. PROCEDURE

A. DNA preparation

1. Methods for extraction of genomic DNA from *Staphylococcus aureus* are outlined below.
2. Inoculate one colony into 4 ml of broth and incubate overnight at 37°C.
3. Pipette 1 ml of the sample into each of three 1.5-ml Eppendorf tubes, and centrifuge for 5 min at 14,000 × *g* in a microcentrifuge.
4. Wash pellet once with TES, vortex, and pellet for 5 min at 14,000 × *g*. Discard supernatant into 10% bleach. Suspend pellet in 200 μl of TES by vortexing.
5. Add 10 μl of lysostaphin to the cell suspension, vortex, and incubate at 37°C for 30 min.
6. Add 20 μl of SDS and 1.2 μl of proteinase K, and incubate in a 60°C block heater for 1 to 2 h.
7. Extract genomic DNA with two phenol-chloroform-isoamyl alcohol (25:24:1) extractions followed by one chloroform-isoamyl alcohol extraction.
8. Precipitate the DNA by adding 1/25 vol of 5 M NaCl and 2.5 vol of cold 100% ethanol to the aqueous layer of the final extraction. Mix by inverting the tube four or five times. Store at −20°C for 30 min.
9. Centrifuge at 14,000 × *g* for 10 min, wash with 70% ethanol, and centrifuge again (14,000 × *g* for 10 min).
10. Dry the pellet in a vacuum oven at 60°C. Suspend it in 20 μl of HPLC water. After suspension, consolidate all three samples from the same isolate.

B. DNA restriction

1. Restriction enzymes must be individualized for each organism. The enzymes *Xba*I, *Hin*dIII, *Kpn*I, and *Eco*RI have been used successfully to identify strains of *S. aureus*.
2. Calculate the A_{260}/A_{280} ratio to determine the purity of the DNA sample. This ratio should be 1.8 or greater. If it is less than 1.8, the DNA should be reextracted and precipitated.
3. Determine the amount of DNA present by using the A_{250}.
4. Dilute 4 μg of DNA to a total volume of 18 μl. Add 2 μl of 10× restriction buffer. Add restriction enzyme, mix, and digest as recommended by manufacturer.
5. Prepare 1% agarose gel in 1× TAE buffer. Electrophoretically separate the restricted DNA in TAE buffer for 4 h at 100 V.
6. Stain the gel with ethidium bromide and visualize the restriction products by using a UV light box.
7. Depurinate, denature, and neutralize the gel by soaking with gentle shaking in depurination and denaturation buffer for 15 min each and neutralization buffer for 30 min. Rinse the gel with distilled water twice between each buffer. Transfer the DNA to a nitrocellulose membrane by Southern blotting overnight.

III. PROCEDURE *(continued)*

8. Soak the blot in 6 × SSC for 5 min. Allow it to air dry. Sandwich it between sheets of fresh 3MM Whatman paper and bake it at 80°C under vacuum for 2 h.

C. Labeling of probe

◩ **NOTE:** The ECL kit is designed to label DNA. To label RNA, the following protocol should be followed in place of the kit instructions.

1. Dilute 1 μl of 16S and 23S rRNA from *E. coli* MRE 600 (4 μg/ml) with 399 μl of kit water in a 1.5-ml microcentrifuge tube (final concentration, 10 ng/μl).
2. Boil the mixture for 2 min in a water bath and then snap-cool it in an ice water bath for 5 min. Centrifuge briefly to settle contents to bottom.
3. Add 400 μl of DNA-labeling reagent and mix gently. Add 400 μl of glutaraldehyde solution and mix gently.
4. Incubate at 37°C for 10 min. Add 30 U of RNasin RNase inhibitor, mix well, and immediately place on ice.
5. Store the labeled probe in 50% glycerol in a 1.5-ml microcentrifuge tube at −20°C.

D. Hybridization

◩ **NOTE:** The hybridization buffer supplied in the ECL kit must not be used. All solutions must be prepared with DEPC-treated water.

1. Roll the blot into a nylon mesh with the DNA surface facing inward. Place it in a hybridization tube. Prehybridize the blots at 40°C in prehybridization buffer (0.125 ml per cm^2 of nylon mesh) in a hybridization oven.
2. After 3 h, discard the prehybridization buffer and replace it with hybridization buffer. Add 20 ng of probe per ml of hybridization buffer. Do not pipette the probe directly onto the blot.
3. Hybridize for 5 h at 40°C.
4. Wash the blot twice in primary wash buffer for 20 min each at 40°C.
5. Remove the blot from the tube and place it in a clean container. Wash it twice in secondary wash buffer for 5 min at room temperature with gentle agitation.

E. Detection

1. Prepare a tray containing the following items: scissors, forceps, blotting paper, detection reagents 1 and 2 (at 4°C and separate until use), washed blot in secondary wash buffer, cassette and X-ray film, cellophane, timer.
2. Working as quickly as possible in a darkroom, with the lights on, remove the blot from the wash buffer with forceps. Drain off excess reagents. Place the blot on the tray. Mix equal amounts (~10 ml) of each detection reagent, and pour the mixture onto the blot. Incubate for exactly 1 min.
3. Drain off excess reagents, and lay the blot on a piece of blot paper for 10 s to absorb excess moisture. Cover the blot with cellophane and place it in the cassette with the DNA side up. Smooth out all air bubbles.
4. Turn off the lights. Place the film in the cassette (clip off a corner to mark the orientation). Obtain a 1-min exposure.

IV. RESULTS

A. Interpretation

1. Fig. 10.4.4–1 shows examples of ribotype patterns of *Staphylococcus aureus*.
2. Strains with identical ribotype patterns are considered to be clonal.
3. Strains with one or two band shifts are also considered to be clonally related subtypes.
4. Strains that differ by three or more bands are considered to represent independent strains.

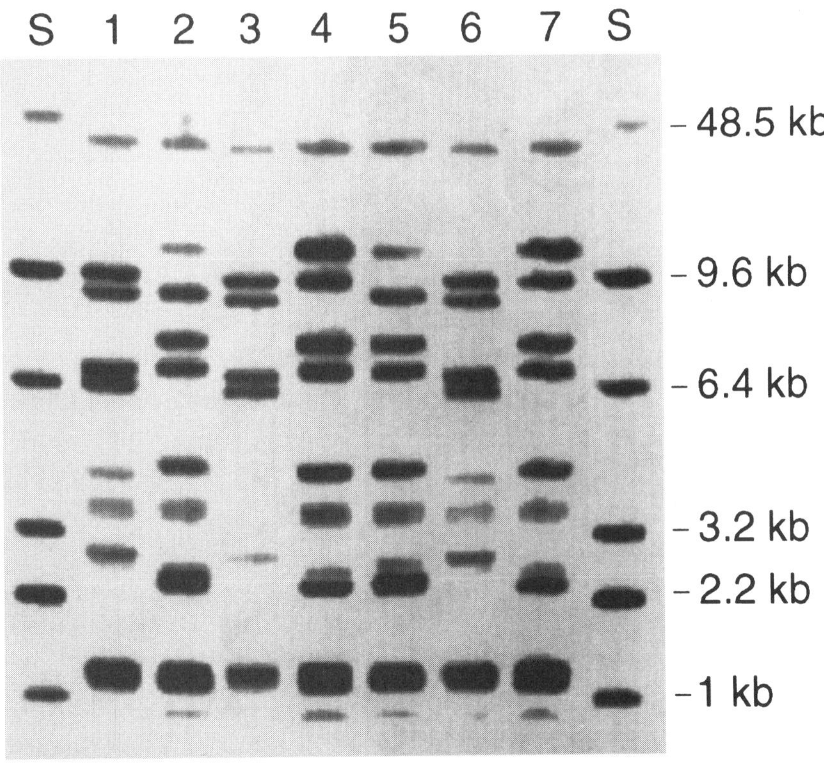

Figure 10.4.4–1 Ribotype profile of *Staphylococcus aureus* isolates. Lanes: S, molecular size standards (in kilobases); 1 and 6, ribotype A; 2 and 5, ribotype B; 3, ribotype C; 4 and 7, ribotype D.

B. Reporting

A report of the results of ribotyping that details which isolates are indistinguishable (clonal), similar (subtypes), or different should be issued to the service that generated the request for typing.

C. QC

1. A standard control strain of the same species, if available, should be prepared along with every preparation.
2. The use of molecular typing tests to identify related strains is most effective if a group of appropriate epidemiologically unrelated control organisms can be prepared and compared with the isolates in question. Appropriate control strains include the following:
   a. Identical species isolated in the same unit at different dates
   b. Identical species isolated on the same date in different units

V. COMMENTS

A. The sensitivity of chemiluminescent ribotyping is directly related to the amount of DNA analyzed in each reaction. Underloaded lanes will require longer exposure times. Luminescence persists for 3 to 7 h.

B. Labeled probe has been used successfully for up to 3 months when stored at −20°C in 50% glycerol.

C. Molecular size standards can also be labeled with the ECL kit to provide convenient determination of molecular weights.

D. The chemiluminescent ribotyping procedure has been completely automated. The instrumentation and ribotyping kits are available commercially from Qualicon, Wilmington, Del.

REFERENCES

1. **Grimont, F., and P. A. D. Grimont.** 1986. Ribosomal ribonucleic acid gene restriction patterns as potential taxonomic tools. *Ann. Inst. Pasteur/Microbiol.* **137B:**165–175.

2. **Gustaferro, C. A.** 1993. Chemiluminescent ribotyping, p. 584–589. *In* D. H. Persing, T. F. Smith, F. C. Tenover, and T. J. White (ed.), *Diagnostic Molecular Microbiology: Principles and Applications.* American Society for Microbiology, Washington, D.C.

3. **Pfaller, M. A., C. Wendt, R. J. Hollis, R. P. Wenzel, S. J. Fritschel, J. J. Neubauer, and L. A. Herwaldt.** 1996. Comparative evaluation of an automated ribotyping system versus pulsed-field gel electrophoresis for epidemiological typing of clinical isolates of *Escherichia coli* and *Pseudomonas aeruginosa* from patients with recurrent gram-negative bacteremia. *Diagn. Microbiol. Infect. Dis.* **25:**1–8.

4. **Stull, T. L. and J. J. LiPuma.** 1994. Methods for ribotyping, p. 10.5.1–10.5.8. *In* H. D. Isenberg (ed.), *Clinical Microbiology Procedures Handbook, Supplement 1.* American Society for Microbiology, Washington, D.C.

5. **Stull, T. L., J. J. LiPuma, and T. D. Edlind.** 1988. A broad-spectrum probe for molecular epidemiology of bacteria: ribosomal RNA. *J. Infect. Dis.* **157:**280–286.

10.4.5 Chromosomal Restriction Fragment Analysis by Pulsed-Field Gel Electrophoresis—Application to Molecular Epidemiology

I. PRINCIPLE

Chromosomal restriction fragment analysis is based on the fact that chromosomes are not static and can undergo both rearrangements and point mutations. Changes in nucleotide sequence are reflected in the restriction endonuclease patterns of chromosomal DNA when the fragments are separated on agarose gels. Physical characterization of bacterial and fungal DNA at the genomic level has been limited by (i) the difficulty in obtaining suitable fragments of chromosomal DNA because of the shearing of large DNA molecules in solution and (ii) the inability of conventional (constant-field) agarose gel electrophoresis to resolve DNA molecules in an appropriate size range (40 to 4,000 kbp). These constraints have been overcome by the development of techniques for (i) preparing unit-length chromosomal DNA by the in situ lysis of organisms embedded in agarose and subsequent digestion of the DNA directly in the agarose with restriction enzymes that have infrequent recognition sites and (ii) resolving the resulting large DNA fragments in agarose gels by alternation of electric fields by a technique known as pulsed-field gel electrophoresis (PFGE). Thus, total genomic DNA can be resolved into a limited number of restriction fragments with distinct electrophoretic mobilities. PFGE has been shown to be highly effective in molecular epidemiologic studies of bacterial and yeast isolates.

NOTE: The following procedure addresses the use of PFGE for typing bacterial isolates and has been adapted from Maslow et al. (1993) and Pfaller et al. (1994).

II. MATERIALS

Unless otherwise stated, all reagents are available from Sigma Chemical Co., St. Louis, Mo.)

Include QC information on reagent container and in QC records.

A. Reagents

1. InCert agarose (low melting temperature) (FMC BioProducts, Rockland, Maine)
2. SeaKem agarose (high melting temperature) (FMC BioProducts)
3. Ethidium bromide
4. Restriction endonucleases (individualized for each organism, available from New England BioLabs)
5. 10× restriction buffers as provided by manufacturer (New England BioLabs)
6. Molecular size standards, 48.5 lambda ladder molecular weight standards (New England BioLabs)
7. Brij 58 (polyoxyethylene 20 cetyl ether)
8. Sodium deoxycholate
9. Sodium lauroyl sarcosine
10. RNase A (molecular biology grade)
11. Lysostaphin
12. Lysozyme
13. Proteinase K
14. PIV buffer: 10 mM Tris (pH 7.6)–1 M NaCl (total volume, 500 ml; autoclave and store at 4°C)
15. Lysis buffer: 6 mM Tris (pH 7.6)–1 M NaCl–100 mM EDTA (pH 7.6)–0.5% Brij 58–0.2% sodium deoxycholate–0.5% sodium lauroyl sarcosine (total volume, 100 ml; filter sterilize and store at 4°C)
16. Lysis solution: 20 μg of RNase per ml, 1 mg of lysozyme per ml, 5 U of lysostaphin per ml (add enzymes to lysis buffer just before use)
17. ES buffer: 0.5 M EDTA (pH 8.0)–10% sodium lauroyl sarcosine (total volume, 500 ml; filter sterilize and store at room temperature)
18. ESP solution: 100 μg of proteinase K per ml of ES buffer (total volume, 100 ml; store at 4°C) (20× proteinase stock solution: 100 mg of proteinase K in 50 ml of ES buffer; incubate at 50°C for 1 h; store at 4°C [this solution is highly stable])

II. MATERIALS *(continued)*

19. 1× TE buffer: 10 mM Tris (pH 7.6)–0.1 mM EDTA (pH 7.6) (total volume, 500 ml; autoclave and store at room temperature)
20. Lysostaphin buffer: 50 mM Tris (pH 7.6)–0.15 M NaCl (total volume, 50 ml; autoclave and store at room temperature)

B. Supplies

1. Snap-top tubes (15 ml)
2. Snap-top tubes (5 ml)
3. Round-bottom tubes, screw cap (15 ml)
4. Six-well tissue culture plate
5. Insert mold
6. Glass tray (20 by 30 cm), metal tray (15 by 25 cm)

C. Equipment

1. Orbital shaker
2. Incubators at a variety of tempera tures (25, 37, and 50°C)
3. Vacuum line for aspiration of lic uids
4. Camera unit for photodocumenta tion
5. Transilluminator, midrange UV
6. Test tube roller
7. Heating platform
8. Pulsed-field gel box, pump, ge molds, cooling water bath, powe supply, pulse wave switches

NOTE: Electrophoresis con ditions specified in this procedur require the CHEF-DRII syster from Bio-Rad with their Mini Chiller. However, other forms o PFGE apparatus may be used.

III. PROCEDURE

A. Sample preparation

1. Isolate single colony and prepare overnight (ON) broth.
2. For each strain, inoculate a single colony into 0.5 ml of broth (typicall TSB).
3. Grow for 2 h or until turbid.
4. Streak out onto an agar plate, and incubate ON.
5. Pick a single colony, inoculate 5 ml of broth, and incubate ON.
6. Prepare the plugs on day 1.
   - **a.** Mix InCert agarose (1.3% [wt/vol]) in PIV buffer in a 15-ml tube Prepare 1 ml per strain (plus 1 ml extra), vortex, and place into a beake of boiling water to dissolve. Revortex before aliquoting.
   - **b.** For each strain, label one 15-ml snap-top tube, one 5-ml snap-top tube and one 15-ml round-bottom tube.
   - **c.** For each strain, label one plug mold. Then tape the bottom of the plug molds, place them in a metal tray, put this tray in a glass tray fille with ice, and place it in the refrigerator for at least 30 min.
   - **d.** Put 5 ml of cold PIV buffer into each 15-ml snap-top tube and plac on ice.
   - **e.** Carefully dispense 1 ml of dissolved agarose into each 5-ml snap-top tube, and keep the tubes in a 50°C heat block.
   - **f.** Dispense 1.5 ml of the ON culture into each 15-ml snap-top tube wit cold PIV.
   - **g.** Centrifuge the 15-ml snap-top tubes at 1,100 × *g* for 15 min at 4°C.
   - **h.** Decant the PIV buffer from the cell pellet, suspend the cells thoroughl in 1.5 ml of cold PIV, and place on ice.
   - **i.** Add 1 ml of cells in PIV to the 5-ml snap-top tube with 1 ml of molte InCert agarose, and vortex the tube lightly.
   - **j.** Working quickly, dispense 105 μl of the mixture into each well of the plug molds.

     NOTE: The final concentration of organisms is ~1 × 10^9 CFU/ml Since 1 CFU ≈ 1 × 10^{-14} g of DNA, each plug has ~1 μg of DNA
   - **k.** Keep the molds in the tray on ice, and place at 4°C for 30 min to solidify.
7. Lyse the plugs on days 2, 3, and 4.
   - **a.** Make fresh lysis solution by adding RNase and lysozyme (for gram negative bacteria) or lysostaphin and lysozyme (for gram-positive bacte ria) to stock lysis buffer.

(1) RNase stock solution: dissolve RNase at 10 mg/ml in sterile water; heat in boiling water for 20 to 30 min; dispense 100 μl per microcentrifuge tube and store at −20°C.
(2) Lysozyme stock solution: dissolve 0.5 g of lysozyme in 10 ml of sterile water (final concentration, 50 mg/ml); dispense 200 μl per microcentrifuge tube and store at −20°C.
(3) Lysostaphin stock solution: dissolve 5,000 U of lysostaphin in 10 ml of lysostaphin buffer; dispense 200 μl per microcentrifuge tube and store at −20°C.

b. Dispense 4 ml of lysis solution into each labeled 15-ml round-bottom tube.
c. For each strain, push out the 12 plugs into the tube with lysis solution.
d. Incubate overnight at 37°C with gentle shaking.
e. Chill the tubes on ice for at least 15 min to harden the plugs.
f. Carefully aspirate the lysis solution.
g. Dispense 4 ml of ESP solution into each tube and incubate ON at 50°C, shaking gently.
h. Chill the tubes. Change to fresh ESP solution and again incubate ON at 50°C.
i. The plugs can now be used directly for restriction enzyme digestion as outlined below or can be placed in fresh ESP solution and stored at 4°C (stable for up to 2 years).

B. Restriction enzyme digestion

◩ Perform steps 1 and 2 on day 1 of restriction enzyme digestion.

1. Turn on a water bath at the temperature appropriate for the restriction enzyme.
2. Wash the plugs in 10 ml of 1× TE at 37°C on a roller. Four washes are required, at least two of which should be a minimum of 2 h; the others should be at least 1 h. For convenience, one of the washes can continue ON.

◩ Perform steps 3 through 6 on day 2 of restriction enzyme digestion.

3. Complete the TE washes, if necessary.
4. For each plug, prepare a labeled microcentrifuge tube containing 10× restriction enzyme buffer and bovine serum albumin (final concentration, 100 μg/ml), and add water to a final volume of 250 μl.
5. Add a washed plug to the above tube.
6. Add restriction enzyme, mix gently, and incubate ON at the appropriate temperature. See Table 10.4.5–1 for restriction endonucleases appropriate for some species.

◩ Perform steps 7 through 11 on day 3.

7. Chill the tubes before handling the inserts.
8. Pour out a plug into a labeled well of a six-well tissue culture plate. Save the plate.
9. Add 1 ml of ES buffer to a microcentrifuge tube and put the plugs into the tube.
10. Incubate the tube at 50°C for 2 h.
11. Pour out the plug into the same tissue culture plate well as above (step 8). Add 300 μl of ESP solution to each tube. Replace the plug into the tube and store at 4°C.

C. Gel preparation, loading, and electrophoresis

1. Make the gel with SeaKem high-melting-temperature agarose (1% [wt/vol]) dissolved in 0.5× TBE. Wash the gel ON in 0.5× TBE.
2. By using a glass coverslip or surgical blade, cut a slice ~1 mm thick off the end of the plug and load it into a well of the gel.
3. Fill all the wells with agarose, and place the gel in a PFGE box with 0.5× TBE.

III. **PROCEDURE** *(continued)*

Table 10.4.5–1 Suggested enzymes and running conditions for some species

Organism	Enzyme	Fragment size range (kb)[a]	Running conditions[b]
Acinetobacter spp.	*Sma*I	50–250	C
Citrobacter spp.	*Spe*I	50–300	B/C
Escherichia coli	*Spe*I	50–300	B/C
	*Xba*I	50–400	B
Enterobacter spp.	*Spe*I	50–500	B
	*Not*I	100–700	A
Enterococcus spp.	*Sma*I	<50–250	C
Klebsiella spp.	*Spe*I	50–400	B
Legionella pneumophila	*Sfi*I	100–500	B
	*Asc*I	100–600	A
Moraxella catarrhalis	*Bgl*II	<50–100	C
	*Pme*I	50–250	C
Neisseria meningitidis	*Spe*I	50–250	C
	*Sfi*I	100–350	B/C
Providencia spp.	*Sfi*I	50–450	B
Pseudomonas spp.	*Spe*I	50–500	A
	*Dra*I	50–300	B/C
	*Xba*I	50–250	C
Serratia spp.	*Spe*I	50–450	B
	*Dra*I	<50–100	C
Staphylococcus spp.	*Sma*I	50–400	B
	*Eag*I	50–400	B
Stenotrophomonas spp.	*Xba*I	50–500	B
	*Spe*I	50–400	B
Candida spp.	*Sfi*I	300–800	A
	*Bss*HII	100–600	A
	*Eag*I	50–500	A
	Karyotype (uncut)	500–2,000	D

[a] Size range includes the majority of fragments created after digestion with the corresponding enzyme. Some species will have several fragments under 48.5 kb; however, it is difficult to analyze them when the 48.5-kb lambda ladder molecular weight standard is used. To analyze those fragments, use *Hin*dIII-cut lambda as a molecular weight standard and/or a lower switch time, such as 5 to 15 s.

[b] See Table 10.4.5–2.

4. Run the gel in accordance with the manufacturer's recommendations for the equipment being used. See Table 10.4.5–2 for CHEF-DRII running conditions.
5. Store the remainder of the plug in ESP at 4°C in a microcentrifuge tube.

D. Gel visualization

1. Stain the gel for 30 min with ethidium bromide solution (0.5 μg/ml in water).
2. Rinse the gel with deionized water and destain in water for at least 2 h.
3. Photograph under UV illumination.

Table 10.4.5–2 Examples of running conditions[a]

Running condition	Agarose concn (%)	Switch time (s) Initial	Switch time (s) Final	Time (h)	Run (V/cm)	Range of band sizes best separated (kb)	Largest band resolved (kb)
A	1.0	10.0	90	24	6.0	100–650	776.0
B	1.0	5.0	60	23	6.0	50–500	630.5
C	1.0	5.0	30	23	6.0	<50–350	436.5
D	0.7	120	280	48	4.5	500–2,000	2,000

[a] All electrophoretic conditions are run at 13°C.

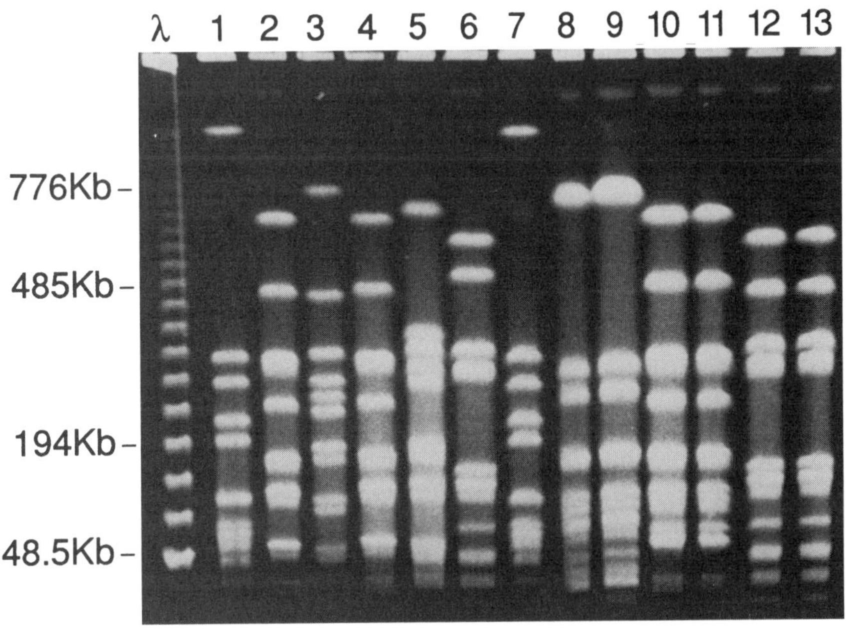

Figure 10.4.5–1 Restriction fragment length polymorphism of *S. aureus* chromosomal DNA digested with *Sma*I. Switch time ramped from 10 to 90 s (running condition A, Table 10.4.5–2). Lanes 1 through 7 show different patterns. Lanes 8 and 9, 10 and 11, and 12 and 13 represent three pairs of indistinguishable patterns. This photo was kindly provided by Andreas Widmer.

IV. RESULTS

A. Interpretation

1. Examples of chromosomal-DNA patterns of *Staphylococcus aureus* and *Stenotrophomonas maltophilia* are shown in Fig. 10.4.5–1 and 10.4.5–2.
2. Strains with identical PFGE patterns are considered to be clonal.
3. Strains with one to three band shifts consistent with a single genetic event (e.g., a point mutation resulting in the loss or gain of a restriction site, an insertion, a deletion, or a chromosomal inversion) are also considered to be clonally related subtypes (Fig. 10.4.5–2).
4. Strains that differ at four or more bands are considered to represent independent strains, although strains with multiple similarities may have a common ancestry.

B. Reporting

A report of the results of PFGE typing that details which isolates are indistinguishable (clonal), similar (subtypes), or different should be issued to the service that generated the request for typing.

V. QUALITY CONTROL

A. A standard control strain of the same species, if available, should be prepared along with every preparation as a control for the DNA isolation procedures.

B. The use of molecular typing tests to identify related strains is most effective if a group of appropriate epidemiologically unrelated control organisms can be prepared and compared to the isolates in question. Appropriate control strains would be as follows:

1. Identical species isolated in the same unit on different dates.
2. Identical species isolated on the same date in different units.

C. The many points to be checked are as follows:

1. All lanes should have electrophoresed more or less straight.

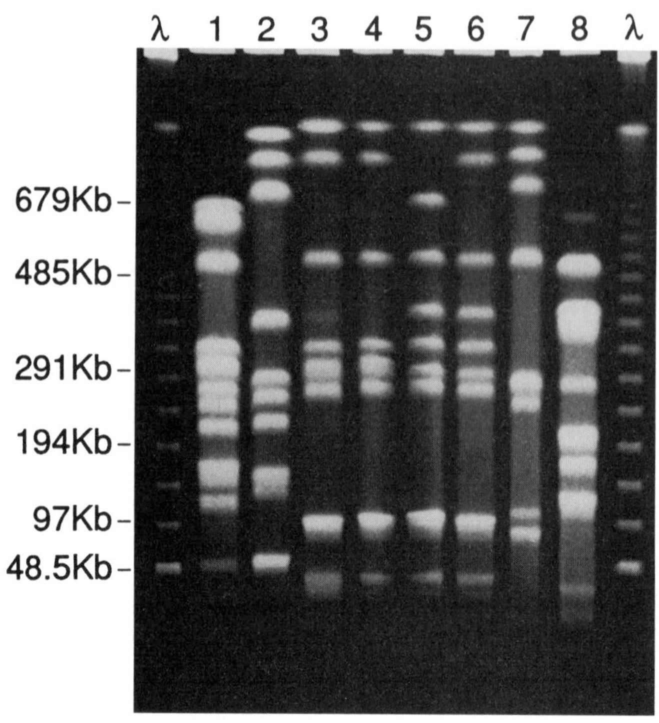

Figure 10.4.5–2 RFLP of *S. maltophilia* chromosomal DNA digested with *Spe*I (running condition A, Table 10.4.5–2). Lanes 3 and 4 show identical patterns, while the pattern represented in lanes 5 and 6 are similar but not identical and represent subtypes of a single strain.

V. QUALITY CONTROL *(continued)*

2. The restriction digests should be complete. A complete digest will have few if any dim bands located above or between bright bands in the same lane. Bright and dim are relative terms and apply only within the context of the same region of one lane. An incomplete (or partial) digest will have one or more dim bands located above or between bright bands in the same lane.
3. The DNA in all lanes should have digested and have visible bands. Usually all strains of the same species will have about the same number of bands with the same size distribution for any given restriction enzyme and running conditions.
4. If any of the above conditions are not met, then some or all of the samples must be repeated. Sometimes the plugs do not digest, and the failed samples need to be prepared again.

D. DNA size standards

1. *S. aureus* ATCC 8325 digested with *Sma*I
2. The chromosomes of *Saccharomyces cerevisiae* (Bio-Rad Laboratories)
3. Lambda ladder (48.5 kbp)

SUPPLEMENTAL READING

Arbeit, R. D. 1995. Laboratory procedures for the epidemiologic analysis of microorganisms, p. 190–208. *In* P. R. Murray, E. J. Baron, M. A. Pfaller, F. C. Tenover, and R. H. Yolken (ed.), *Manual of Clinical Microbiology,* 6th ed. American Society for Microbiology, Washington, D.C.

Birren, B., and E. Lai (ed.). 1993. *Pulsed-Field Gel Electrophoresis: a Practical Guide.* Academic Press, Inc., San Diego, Calif.

Goering, R. V. 1993. Molecular epidemiology of nosocomial infection: analysis of chromosomal restriction fragment patterns by pulsed-field gel electrophoresis. *Infect. Control Hosp. Epidemiol.* **14:**595–600.

Maslow, J. N., M. E. Mulligan, and R. D. Arbeit 1993. Molecular epidemiology: the application of contemporary techniques to typing bacteria. *Clin. Infect. Dis.* **17:**153–164.

Maslow, J. N., A. M. Slutsky, and R. D. Arbeit. 1993. The application of pulsed-field gel electrophoresis to molecular epidemiology, p. 563–572. *In* D. H. Persing, T. F. Smith, F. C. Tenover, and T. J. White (ed.), *Diagnostic Molecular Microbiology: Principles and Applications.* American Society for Microbiology, Washington, D.C.

Pfaller, M. A., R. J. Hollis, and H. S. Sader. 1994. Chromosomal restriction fragment analysis by pulsed-field gel electrophoresis, p. 10.5.C.1–10.5.C.12. *In* H. D. Isenberg (ed.), *Clinical Microbiology Procedures Handbook, Supplement 1.* American Society for Microbiology, Washington, D.C.

Sader, H. S., R. J. Hollis, and M. A. Pfaller. 1995. The use of molecular techniques in the epidemiology and control of infectious diseases. *Clin. Lab. Med.* **15:**407–431.

10.4.6 Characterization of Pathogenic Microorganisms by Genomic Fingerprinting with Arbitrarily Primed PCR

I. PRINCIPLE

Genomic fingerprinting by the arbitrarily primed PCR (AP-PCR) is a method for sampling homologous sequences of related pathogens in such a way that small sequence differences are observed. Primers are chosen arbitrarily and annealed (hybridized) to the genome at low stringency (low temperature and high salt concentration). Subsequent PCR results in the amplification of sequences bounded by these low-stringency annealing events. The products are then arrayed by electrophoresis to yield "fingerprints," which differ depending on the relatedness of the genomic templates. The more closely related two genomic templates are, the more similar the resultant fingerprints will be. For very closely related but distinct templates, additional arbitrary primers can be used until a distinction is revealed.

◪ **NOTE:** This procedure is adapted from that of Welsh and McClelland (1).

II. MATERIALS

(Unless otherwise stated, all materials are available from Sigma Chemical Co., St. Louis, Mo.)

Include QC information on reagent container and in QC records.

A. Reagents

1. TE buffer: 10 mM Tris (pH 8.0)–1 mM EDTA
2. PK solution: 1 mg of proteinase K per ml, 0.2 M EDTA, 1% N-lauroyl sarcosine, and 0.1 M Tris (pH 8.0)
3. 10× *Taq* (Mg) buffer: 100 mM Tris (pH 8.3)–500 mM KCl–40 mM $MgCl_2$
4. AP-PCR master: 500 μl of 10× *Taq* (Mg) buffer, 20 μl of 200 μM oligonucleotide primer, 200 μl of deoxyribonucleoside triphosphates (dNTP) (5 mM each dNTP), 3,730 μl of water, and 25 μl of Ampli*Taq* (5 U/ml) (total volume, 4,500 μl) (New England BioLabs and others)
5. Oligonucleotide primers: The MB universal-sequencing and reverse-sequencing primers work well, but almost any primer ranging from 10 to 50 nucleotides in length will work (New England BioLabs and others).
6. Formamide dye: 10 ml of 98% deionized formamide, 10 mg of xylene cyanol, 10 mg of bromophenol blue, and 200 μl of 0.5 M EDTA (pH 8.0)
7. Agarose (FMC BioProducts)
8. Ethidium bromide (10 mg/ml)

B. Equipment

1. Power supply
2. Electrophoresis chamber
3. UV transilluminator and photodocumentation equipment
4. 96-well thermal cycler with disposable tubes and racks
5. Multichannel Pipetman

III. PROCEDURE

A. Specimen

Use a fresh colony of the organism to be typed grown on suitable agar medium.

B. DNA sample preparation

1. Grow a 1- to 5-ml culture to late log phase or saturation in a 10-ml culture tube.
2. Pellet cells by low-speed centrifugation (3,000 × *g* for 10 min).
3. Suspend the cells in 5 ml of TE and repellet them.
4. Resuspend the cells in 0.2 ml of TE and add 0.2 ml of PK solution.

5. Incubate the mixture at 50°C for 1 h to overnight, until the solution is clear.
6. Add 3 ml of TE and transfer 0.4 ml to a 1.5-ml centrifuge tube. Add 0.4 ml of phenol, vortex, and centrifuge briefly to separate the phases. Recover and save the aqueous phase. Add 0.4 ml of chloroform, vortex, and centrifuge briefly to separate phases. Recover and save the aqueous phase.
7. Add 1/10 vol of 3 M sodium acetate and 2 vol of ethanol. Chill the mixture at −20°C. Pellet the DNA by centrifugation in a microcentrifuge.
8. Dry the pellet, and dissolve the DNA in 100 μl of TE.
9. Check the concentration of DNA by running several twofold serial dilutions on a 1% agarose gel and comparing the ethidium bromide fluorescence of the DNA with a molecular mass standard (New England BioLabs).

C. Genomic fingerprinting

NOTE: To select the optimum range of DNA concentrations to be used for fingerprinting, an initial experiment with multiple dilutions of the template DNA should be performed. Once the optimum range of DNA concentration has been determined, as few as two dilutions of each DNA can be used.

1. Array the purified DNAs in a "master" 96-well microtiter tray at about 10 ng/μl.
2. Using a multichannel pipette, perform a series of four fivefold dilutions on a representative subset of DNAs in additional 96-well dishes. The final desired DNA concentration for bacteria in the AP-PCR experiment is generally about 1 ng/μl, but this can vary depending on the DNA quality. For this reason, it is usually wise to perform the initial amplification on more than one dilution at a time.
3. In the 96-well array, combine 45 μl of AP-PCR mix with 5 μl of DNA dilution stocks of 10, 2, 0.4, 0.08, and 0.016 μg/μl from step 2.
4. Perform the amplification as follows: two cycles of 5 min at 94°C for template denaturation, 5 min at 40°C for primer annealing, and 5 min at 72°C for primer extension, followed by 40 cycles of 1 min at 94°C for template denaturation, 1 min at 40°C for primer annealing, and 2 min at 72°C for primer extension.
5. Separate the amplification products by electrophoresis in 1.5% agarose at 100 V for 4 h. Stain the gel with ethidium bromide and visualize under UV light.
6. After this initial experiment, choose two dilutions that give the most robust and informative fingerprints and use them for all subsequent experiments, allowing 48 strains to be examined at one time.

IV. INTERPRETATION

A. Standard guidelines for interpretation of AP-PCR are not yet available, and the interpretation of fingerprinting results obtained with this technique is empiric.

B. Under conditions in which variability of amplicon sizes can be demonstrated among epidemiologically unrelated isolates, those showing no differences or changes only in band intensity can be considered epidemiologically related.

C. AP-PCR patterns with three or more fragment (amplicon) differences may be considered different.

D. AP-PCR patterns differing by one or two bands remain difficult to interpret but may be considered subtypes of one another and possibly epidemiologically related.

V. COMMENTS

A. In AP-PCR even slight variations in the pH or ionic strength of the buffers used, the temperature of the reaction, or the source of the DNA polymerase may result in wide variations in the intensities of the individual amplicons.

B. The reproducibility and discriminatory power of each primer and amplification protocol must be validated by analyzing sets of isolates that have been previously well defined by epidemiologic data and/or independent typing studies.

C. The most reliable epidemiologic results are obtained when a set of isolates is tested in a single amplification reaction and analyzed on a single electrophoretic gel.

REFERENCE

1. **Welsh J., and M. McClelland.** 1993. Characterization of pathogenic microorganisms by genomic fingerprinting using arbitrarily primed PCR, p. 595–602. *In* D. H. Persing, T. F. Smith, F. C. Tenover, and T. J. White (ed.), *Diagnostic Molecular Microbiology: Principles and Applications.* American Society for Microbiology, Washington, D.C.

SUPPLEMENTAL READING

van Belkum, A. 1994. DNA fingerprinting of medically important microorganisms by use of PCR. *Clin. Microbiol. Rev.* **7:**174–184.

Welsh, J., and M. McClelland. 1990. Fingerprinting genomes using PCR with arbitrary primers. *Nucleic Acids Res.* **18:**7213–7218.

10.5.1 Introduction

Antimicrobial-agent resistance is an increasing problem worldwide, particularly among critically ill hospitalized patients. For this reason, there is a renewed interest in monitoring the development and spread of antimicrobial-agent resistance and a recognition of the need for effective interventions to limit the spread of resistance to prolong the therapeutic life of the available antimicrobial agents. Unfortunately, conventional methods to perform antimicrobial-agent susceptibility testing may be too slow and insensitive in detecting antimicrobial-agent resistance to be of much use clinically. The techniques of molecular biology have been used to characterize resistance at the DNA level and may provide rapid, sensitive, and specific information to the clinician for use in therapeutic decision making (1–3). Genetic material that confers antimicrobial-agent resistance may be carried on the bacterial chromosome or on transposons or plasmids and has been detected by probe hybridization or by DNA amplification with the PCR (Table 10.1–4). Molecular detection of resistance has potential value for decisions directly related to patient care and is useful for calibration of conventional susceptibility tests and for precise definition of the mechanisms of resistance to selected antimicrobial agents. Molecular techniques have been used to detect genes encoding several different mechanisms of resistance, e.g., resistance to β-lactam agents, aminoglycosides, macrolides, and fluoroquinolones, after isolation of a clinical isolate (Table 10.1–4).

REFERENCES

1. **Arlet, G., and A. Philippon.** 1992. PCR-based approaches for the detection of bacterial resistance, p. 665–687. *In* H. A. Erlich (ed.), *PCR Technology: Principles and Applications for DNA Amplification.* W. H. Freeman & Co., New York.
2. **Persing, D. H., D. A. Relman, and F. C. Tenover.** 1996. Genotypic detection of antimicrobial resistance, p. 33–57. *In* D. H. Persing (ed.), *PCR Protocols for Emerging Infectious Diseases.* ASM Press, Washington, D.C.
3. **Tenover, F. C., T. Popovic, and O. Olsvik.** 1995. Genetic methods for detecting antibacterial resistance genes, p. 1368–1378. *In* P. R. Murray, E. J. Baron, M. A. Pfaller, F. C. Tenover, and R. H. Yolken (ed.), *Manual of Clinical Microbiology,* 6th ed. American Society for Microbiology, Washington, D.C.

10.5.2 Detection of Enterococcal Vancomycin Resistance by Multiplex PCR

I. PRINCIPLE

The emergence of vancomycin resistance among enterococci is considered one of the major antimicrobial-resistance threats of the 1990s. The expression of vancomycin resistance involves the complex interaction of several genes, but classification of resistance is based on the altered ligase (an enzyme involved in the D-alanine branch of peptidoglycan synthesis), which is central to the production of cell wall precursors with decreased affinity for vancomycin (Table 10.5.2–1). Currently, the ligase designations include Van A (*vanA*), Van B (*vanB*), and Van C1 to 3 (*vanC-1, vanC-2, vanC-3*). All five Van types may be encountered clinically, but resistance encoded by *vanA* or *vanB* presents the greatest infection control and therapeutic threat because it is genetically transferable, results in high-level vancomycin resistance (MICs, ≥32 μg/ml), and occurs in the two most common enterococcal species (*Enterococcus faecalis* and *E. faecium*), both of which are frequently resistant to multiple agents including penicillins and aminoglycosides. Although presumptive identification of the mechanism of vancomycin resistance may be inferred on the basis of phenotypic criteria, the only definitive method for establishing the mechanism of resistance is detection of the ligase gene(s). PCR provides an ideal molecular tool for definitively establishing the mechanism of vancomycin resistance in clinical isolates of enterococci. In the present protocol, a PCR procedure that uses primers internal to each of the known *van* genes is described. This method may be used to arbitrate equivocal susceptibility results obtained by phenotypic methods, to characterize "outbreak" strains, and to evaluate the accuracy of various phenotypic susceptibility testing methods.

◪ **NOTE:** Procedure adapted from that of Free and Sahm (2).

Workflow in the laboratory must proceed in a unidirectional manner, beginning in the Reagent Preparation Area and moving to the Specimen Preparation Area and then to the Amplification/Detection Area. Preamplification activities must begin with reagent preparation and proceed to specimen preparation. Reagent preparation activities and specimen preparation activities must be performed in separate, segregated areas. Supplies and equipment must be dedicated to each activity and not used for other activities or moved between areas. Lab coats and gloves must be worn in each area and must be changed before leaving that area. Equipment and supplies used for reagent preparation must not be used for specimen preparation activities or for pipetting or processing amplified DNA or other sources of target DNA. Amplification and detection supplies and equipment must be confined to the Amplification/Detection Area at all times.

Table 10.5.2–1 PCR primers for detection of *vanA, vanB, vanC1*, and *vanC2* in enterococci

Target gene[a]	Primer designation	Nucleotide sequence, 5′ to 3′[b]	Product size (bp)
vanA	Van A1	GCT ATT CAG CTG TAC TC	783
	Van A2	CAG CGG CCA TCA TAC GG	
vanB	Van B1	CAT CGC CGT CCC CGA ATT TCA AA	297
	Van B2	GAT GCG GAA GAT ACC GTG GCT	
vanC-1	Van C1-1	GGT ATC AAG GAA ACC TC	822
	Van C1-2	CTT CCG CCA TCA TAG CT	
vanC-2	Van C2-1	CTC CTA CGA TTC TCT TG	439
	Van C2-2	CGA GCA AGA CCT TTA AG	

[a] GenBank accession numbers: *vanA*, X56895; *vanB*, U00456; *vanC-1*, M75132; *vanC-2*, L29638.

[b] Primer sequence for Van C1-1, Van C1-2, Van C2-1, and Van C2-2 from Dutka-Malen et al. (1).

II. SPECIMEN

Test growth of *Enterococcus* spp. with elevated vancomycin MIC ($\geq$8.0 μg/ml) or growth on an agar screen plate from appropriate solid medium. Select two to four well-isolated colonies from pure culture. The optimal colony age is 18 to 48 h; substantially older colonies have been used successfully.

III. MATERIALS

(Unless otherwise stated, all materials are available from Sigma Chemical Co., St. Louis, Mo.)

Include QC information on reagent container and in QC records.

A. Reagents

1. Agarose (type 1-A, low electroendosmosis)
2. Autoclaved Tris-borate-EDTA (TBE) buffer (pH 8.3 to 8.5)
3. 6× gel electrophoresis loading buffer (0.25% bromophenol blue, 0.25% xylene cyanol, and 30% glycerol in H_2O)
4. Sterile distilled H_2O
5. Ethidium bromide (0.5 μg/ml of H_2O)
6. Autoclaved light mineral oil
7. PCR master mix
   - **a.** 10× PCR buffer (100 mM Tris-HCl [pH 9]–500 mM KCl–1% Triton X-100 [Promega, Madison, Wis.])
   - **b.** 10 mM each deoxyribonucleoside triphosphate
   - **c.** 25 μM each primer (primer designations, primer sequences, and amplicon sizes given in Table 10.5.2–1)
   - **d.** 25 mM $MgCl_2$
   - **e.** 5,000 U of *Taq* (Promega) per ml
8. Size marker, ϕX174 phage DNA-*Hae*III digest (Promega), diluted 1:40 with 6× electrophoresis buffer

B. Supplies

1. Sterile 0.5-ml snap-cap PCR thermal reactor tubes
2. Sterile, plugged micropipette tips (10, 200, and 1,000 μl)
3. Sterile bacteriologic transfer needles or sterile toothpicks

C. Equipment

1. Vortex mixer
2. Thermal cycler
3. Electrophoresis gel box
4. Electrophoresis power supply
5. UV transilluminator and photodocumentation equipment

IV. PROCEDURE

A. Preparation of PCR master mix for *vanA* and *vanB* detection (prepare in Area 1: Reagent Preparation Area)

1. Volumes based on 99-μl reaction mixture

   NOTE: Overestimating the final required volume by one or two reactions allows for volume loss and pipetting errors and is recommended.
2. Using sterile technique, add the following volumes of each reaction component to each PCR thermal reactor tube:
   - **a.** 52.5 μl of distilled H_2O
   - **b.** 10 μl of 10× PCR buffer
   - **c.** 2 μl of each of the following:
     - (1) Deoxynucleoside triphosphate stock (10 mM) (total volume, 8 μl)
     - (2) *vanA-1* primer stock (25 μM)
     - (3) *vanA-2* primer stock (25 μM)
     - (4) *vanB-1* primer stock (25 μM)
     - (5) *vanB-2* primer stock (25 μM)
   - **d.** 20 μl of 25 mM $MgCl_2$ stock
   - **e.** 0.5 μl of 5,000-U/ml *Taq* stock

     NOTE: If *vanC-1* and *vanC-2* genes are of interest, their respective primers would replace those of *vanA* and *vanB* at the same volume and concentration.
3. Use the vortex mixer to gently, but thoroughly mix the contents.
4. Transfer 99 μl of mixture to each reaction tube.

B. Sample preparation—preparation of DNA template (prepare in Area 2: Specimen Preparation Area)

IV. PROCEDURE *(continued)*

1. Use a sterile toothpick or inoculating needle to select two to four well-isolated colonies from pure culture (18 to 48 h old).
2. Emulsify colonies in PCR mixture.
3. Using aseptic technique, gently overlay each tube with 75 μl of sterile mineral oil (not necessary if self-sealing PCR tubes are used).

C. PCR amplification (Area 3: Amplification and Detection Area)

NOTE: Have the thermocycler on so that cycling can begin as soon as the tubes are ready.

Amplify the target region of *van* genes in a programmed thermocycler as follows:

1. 3-min hot start at 95°C (bacterial lysis and release of target DNA).
2. 30 cycles of 94°C for 1 min, 56°C for 1 min, and 72°C for 1 min, with a final extension at 72°C for 1 min

   NOTE: If not able to perform detection of PCR product soon after completion of this cycle, program the machine to hold at 4°C or place reaction mixture in a refrigerator or freezer until needed for electrophoretic analysis.

D. Electrophoretic detection of PCR products (Area 3: Amplification and Detection Area)

1. Prepare a 1.5% agarose gel in 1× TBE.
2. Remove 8.3 μl of reaction mixture from below the mineral oil layer, combine with 1.7 μl of 6× loading buffer, and mix by pipetting.
3. Prepare size marker by adding 0.5 μl of marker to 7.5 μl of H_2O and 2 μl of 6× loading buffer.
4. Apply a total of 10 μl of sample to each well, and perform electrophoresis (70 V for 50 min) in 1× TBE containing ethidium bromide (0.5 μg/ml).
5. Visualize the gel under UV light and photograph with Polaroid type 667 film.

   NOTE: The gel may be destained in H_2O (ca. 30 min) to optimize visualization of results.

V. RESULTS

A. Interpretation

1. Amplification of a 783- or 297-bp product indicates the presence of *vanA* or *vanB*, respectively (Fig. 10.5.2–1). Either result establishes the presence of acquired and transferable vancomycin resistance and confirms the need to initiate appropriate infection control and patient management measures.
2. If no PCR product is observed with a strain that appeared nonsusceptible by a phenotype-based method and if appropriate results are obtained with control strains (see below), the isolate probably contains a *vanC* gene. This can be confirmed by using the *vanC-1* and *vanC-2* primers, which yield 822- and 439-bp amplification products, respectively (Fig. 10.5.2–1).

B. QC

1. Positive controls
   a. *vanA, E. faecium* 228
   b. *vanB, E. faecalis* V583 or *E. faecalis* ATCC 51299
   c. *vanC-1, E. galllinarum* AIB39
   d. *vanC-2, E. casseliflavus* ATCC 25788
   e. Strains 228, V583, and AIB39 are available from D. F. Sahm (MRL Pharmaceutical Services, 11921 Freedom Drive, Suite 400, Reston, VA 20190).
2. Negative control, *E. faecalis* ATCC 29212
3. Run appropriate positive and negative controls with each test procedure.

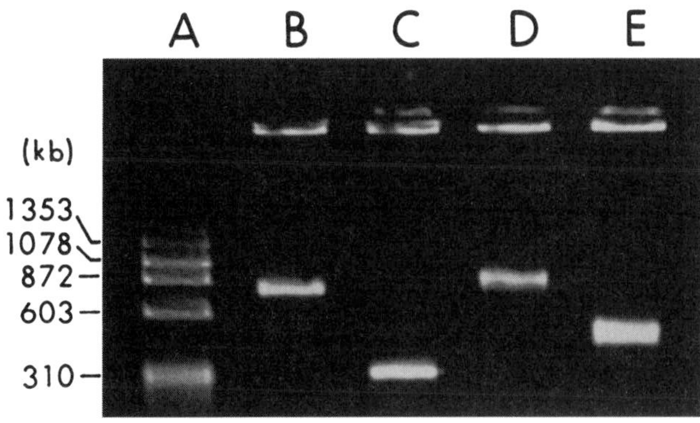

Figure 10.5.2–1 Agarose gel electrophoresis pattern of *van* PCR products. Lanes: A, size marker ϕX174 phage DNA-*Hae*III digest; B, *vanA* product (783 bp); C, *vanB* product (297 bp); D, *vanC-1* product (822 bp); E, *vanC-2* product (439 bp). PCR products for *vanA* and *vanB* resulted from the reaction described above; *vanC-1* and *vanC-2* products were obtained with single primer sets per reaction mixture. Figure from Free and Sahm (2).

REFERENCES

1. **Dutka-Malen, S., S. Evers, and P. Courvalin.** 1995. Detection of glycopeptide resistance genotypes and identification to the species level of clinically relevant enterococci by PCR. *J. Clin. Microbiol.* **33:**24–27.

2. **Free, L., and D. F. Sahm.** 1996. Detection of enterococcal vancomycin resistance by multiplex PCR, p. 150–155. In D. H. Persing (ed.), *PCR Protocols for Emerging Infectious Diseases.* ASM Press, Washington, D.C.

SUPPLEMENTAL READING

Arthur, M., and P. Courvalin. 1993. Genetics and mechanisms of glycopeptide resistance in enterococci. *Antimicrob. Agents Chemother.* **37:** 1563–1571.

Clark, N. C., R. C. Cooksey, B. C. Hill, J. M. Swenson, and F. C. Tenover. 1993. Characterization of glycopeptide-resistant enterococci from U.S. hospitals. *Antimicrob. Agents Chemother.* **37:** 2311–2317.

Federal Register. 1994. Preventing the spread of vancomycin resistance—report from the Hospital Infection Control Practices Advisory Committee. *Fed. Regist.* **59:**25758–25763.

Free, L., and D. F. Sahm. 1995. Investigation of the reformulated Remel Synergy Quad plate for detection of high-level aminoglycoside and vancomycin resistance among enterococci. *J. Clin. Microbiol.* **33:**1643–1654.

Miele, A., M. Bandera, and B. P. Goldstein. 1995. Use of primers selective for vancomycin resistance genes to determine *van* genotype in enterococci and to study gene organization in *van* A isolates. *Antimicrob. Agents Chemother.* **39:** 1772–1778.

Navarro, F., and P. Courvalin. 1994. Analysis of genes encoding D-alanine-D-alanine ligase-related enzymes in *Enterococcus casseliflavus* and *Enterococcus flavescens. Antimicrob. Agents Chemother.* **38:**1788–1793.

Sahm, D. F., L. Free and S. Handwerger. 1995. Inducible and constitutive expression of *vanC-1*-encoded resistance to vancomycin in *Enterococcus gallinarum. Antimicrob. Agents Chemother.* **39:**1480–1484.

Swenson, J. M., N. C. Clark, M. J. Ferraro, D. F. Sahm, G. Doern, M. A. Pfaller, L. B. Reller, M. P. Weinstein, R. J. Zabransky, and F. C. Tenover. 1994. Development of a standard screening method for detection of vancomycin-resistant enterococci. *J. Clin. Microbiol.* **32:** 1700–1704.

Tenover, F. C., J. M. Swenson, C. M. O'Hara, and S. A. Stocker. 1995. Ability of commercial and reference antimicrobial susceptibility testing methods to detect vancomycin resistance in enterococci. *J. Clin. Microbiol.* **33:**1524–1527.

10.5.3 Detection of Methicillin Resistance in Staphylococci by PCR

I. PRINCIPLE

The primary mechanism of resistance to methicillin and related compounds (oxacillin, nafcillin) in both *Staphylococcus aureus* and coagulase-negative staphylococci is the production of a unique penicillin-binding protein, termed PBP2a or PBP2′, that has a low affinity for β-lactam antibiotics. The gene encoding PBP2a, known as *mecA*, has been cloned and sequenced. The phenotypic expression of *mecA* is often heterogeneous and difficult to detect in many antimicrobial-agent-susceptibility testing systems. An alternative method for identifying methicillin-resistant strains of staphylococci is to detect the *mecA* gene, which is absent in susceptible strains. Detection of *mecA* in staphylococci has been accomplished by both PCR and DNA probe methods. In the present protocol, a PCR procedure that uses oligonucleotide primers to amplify a 533-bp region of *mecA* is described. This method is applicable to both *S. aureus* and coagulase-negative staphylococci and may be used to rapidly identify methicillin-resistant strains for clinical purposes and to aid in the development of phenotypic methods that can more accurately predict methicillin resistance.

Workflow in the laboratory must proceed in a unidirectional manner, beginning in the Reagent Preparation Area and moving to the Specimen Preparation Area and then to the Amplification/Detection Area. Preamplification activities must begin with reagent preparation and proceed to specimen preparation. Reagent preparation activities and specimen preparation activities must be performed in separate, segregated areas. Supplies and equipment must be dedicated to each activity and not used for other activities or moved between areas. Lab coats and gloves must be worn in each area and must be changed before leaving that area. Equipment and supplies used for reagent preparation must not be used for specimen preparation activities or for pipetting or processing amplified DNA or other sources of target DNA. Amplification and detection supplies and equipment must be confined to the Amplification/Detection Area at all times.

NOTE: Procedure adapted from that of Murakami and Minamide (1).

II. SPECIMEN

Test growth of *S. aureus* or coagulase-negative staphylococci from Mueller-Hinton agar. Select several colonies from pure culture. The optimal colony age is 18 to 24 h.

III. MATERIALS

(Unless otherwise specified, all materials are available from Sigma Chemical Co., St. Louis, Mo.)

Include QC information on reagent container and in QC records.

A. Reagents

1. Agarose, standard low M_r (Bio-Rad Laboratories, Richmond, Calif.)
2. Autoclaved Tris buffer (TE): 10 mM Tris-HCl (pH 8.0)–1 mM EDTA
3. Electrophoresis loading buffer (0.25% bromphenol blue, 0.25% xylene cyanol, and 30% glycerol in 1× TAE buffer).
4. Sterile distilled H_2O
5. Ethidium bromide (0.5 μg/ml of H_2O)
6. Autoclaved light mineral oil
7. PCR master mix (Gene Amp PCR Reagent Kit [Perkin-Elmer Corp., Norwalk, Conn.])
   - **a.** PCR buffer: 50 mM KCl–10 mM Tris-HCl (pH 8.3)
   - **b.** $MgCl_2$ (1.5 mM)
   - **c.** 0.01% gelatin
   - **d.** Deoxyribonucleoside triphosphates (200 mM each)
   - **e.** AmpliTaq DNA polymerase (0.025 U per μl)
   - **f.** Primers RSM-2647 and RSM-2648 (0.25 μM [final concentration each]) (primer sequences given in Table 10.5.3–1)
   - **g.** Size marker, ϕX174 phage DNA-*Hae*III digest (Nippon

10 Molecular Biology

Table 10.5.3–1 PCR primers for detection of *mecA* in staphylococci

Function	Name	Nucleotide sequence, 5′ to 3′
Primer	RSM-2647	AAA ATC GAT GGT AAA GGT TGG C
Primer	RSM-2648	AGT TCT GCA GTA CCG GAT TTG C

Gene, Toyama, Japan); add 1 vol of electrophoresis loading buffer to 5 vol of the digest to give a DNA concentration of 40 ng/μl.

8. Achromopeptidase solution: 20,000 U of purified achromopeptidase (Wako BioProducts, Richmond, Va.) per ml of TE

B. Supplies

1. Sterile 0.5-ml snap-cap PCR thermal reactor tubes
2. Sterile, plugged micropipette tips (10, 200, and 1,000 μl)
3. Sterile bacteriologic transfer needles or sterile toothpicks
4. 0.5 McFarland turbidity standard
5. Microcentrifuge tubes (1.5 ml)
6. Glass test tubes (12 × 75 mm)

C. Equipment

1. Vortex mixer
2. Thermal cycler
3. Electrophoresis gel box
4. Electrophoresis power supply
5. Microcentrifuge
6. Heating block (55°C and 100°C)
7. UV transilluminator and photodocumentation equipment

IV. PROCEDURE

A. Preparation of template DNA

1. Use a sterile toothpick or inoculating needle to select several well-isolated colonies from pure culture.
2. Emulsify the bacteria in 1 ml of TE by vigorous vortexing to give a turbidity equal to a 0.5 McFarland standard (approx. 1.5×10^8 CFU/ml). Make sure that the bacteria are dispersed well in the TE.

   NOTE: Too dense a bacterial suspension may give a false-negative result.
3. Transfer 100 μl of the bacterial suspension to a 1.5-ml microcentrifuge tube containing 2 μl of achromopeptidase solution. Vortex and place the tube in a 55°C heating block for 30 min.
4. Add 2.5 μl of 4% SDS (Polyscience Inc., Warrington, Pa.) to the digest.

   NOTE: Some brands of SDS inhibit PCR.

   Vortex and place tube in the 100°C heating block for 5 min. Centrifuge at 10,000 × *g* for 5 min, and use 5 μl of the bacterial lysate for PCR.

B. PCR amplification

1. Transfer 95 μl of PCR master mix to a 0.5-ml microcentrifuge tube. Add 5 μl of template DNA to the master mix and mix well by vortexing.
2. Add 2 or 3 drops (about 100 μl) of mineral oil to the reaction mixture to prevent evaporation.
3. Add a drop of mineral oil to the tube holder of the thermal cycler for efficient heat transmission, and place the tube in the tube holder.
4. Amplify a target region of *mecA* for 40 cycles in a programmed thermal cycler as follows: denaturation at 94°C for 30 s, annealing at 55°C for 30 s, and extension at 72°C for 1 min, with a final extension at 72°C for 5 min.

C. Electrophoretic detection of the 533-bp PCR product

1. Prepare a 2% agarose minigel in 1× TAE buffer for electrophoresis.
2. Mix 10 μl of PCR product with 2 μl of loading buffer in a 1.5-ml microcentrifuge tube. Vortex. Apply 6 μl of the sample or size marker to each well, and perform electrophoresis in 1× TAE buffer (100V for 20 to 30 min). Positive and negative controls are also applied to each gel.
3. Stain the gel by gentle shaking in ethidium bromide solution until the 310-bp size marker is visible.
4. Visualize the gel under UV light and photograph with Polaroid type 667 film.

V. QUALITY CONTROL

A. Positive control
S. aureus ATCC 43300

B. Negative control
S. aureus ATCC 25923

C. Run positive and negative controls with each test procedure.

VI. RESULTS

A. Amplification of a 533-bp region of *mecA* indicates that the strain tested is methicillin resistant.

B. A few *mecA*-positive strains appear phenotypically susceptible to β-lactam antibiotics because of their inability to express PBP2′. Since these strains have the potential to become truly resistant, they should be regarded as cryptically methicillin-resistant strains.

C. Use of bacterial suspension of too high a density for preparation of template DNA may lead to false-negative results. A bacterial suspension of as few as 2×10^5 CFU/ml for preparation of template DNA is enough to detect *mecA*.

D. Methicillin-resistant strains of coagulase-negative staphylococci can also be detected by this method. Species in which *mecA* has been detected by using this procedure include *S. epidermidis, S. haemolyticus, S. sciuri, S. saprophyticus, S. hominis, S. capitis*, and *S. caprae*.

REFERENCE

1. **Murakami, K., and W. Minamide.** 1993. PCR identification of methicillin-resistant *Staphylococcus aureus*, p. 539–542. *In* D. H. Persing, T. F. Smith, F. C. Tenover, and T. J. White (ed.), *Diagnostic Molecular Microbiology: Principles and Applications.* American Society for Microbiology, Washington, D.C.

SUPPLEMENTAL READING

Archer, G. L., and E. Pennell. 1990. Detection of methicillin resistance in staphylococci by using a DNA probe. *Antimicrob. Agents Chemother.* **34:** 1720–1724.

Brakstad, O. G., J. A. Maeland, and Y. Tveten. 1993. Multiplex polymerase chain reaction for detection of genes for *Staphylococcus aureus* thermonuclease and methicillin resistance and correlation with oxacillin. *APMIS* **101:**681–688.

Chambers, H. F. 1993. Detection of methicillin-resistant staphylococci. *Infect. Dis. Clin. N. Am.* **7:**425–433.

Geha, D. J., J. R. Uhl, C. A. Gustaferro, and D. H. Persing. 1994. Multiplex PCR for identification of methicillin-resistant staphylococci in the clinical laboratory. *J. Clin. Microbiol.* **32:** 1768–1772.

SECTION 11

Epidemiologic and Infection Control Microbiology

SECTION EDITOR: *Henry D. Isenberg*

11.1 Introduction

The clinical microbiology laboratory plays a pivotal role in infection control and institutional epidemiological efforts. CMPH lists many activities that may be required on occasion to help pinpoint cluster epidemics or establish relationships between isolates, provide means of detecting resistant organisms, and identify the culprits in foodborne outbreaks. These procedures do require experienced personnel who are assigned to perform these analyses. Most clinical laboratories cannot provide complete support for epidemiological activities and must rely on specialized reference laboratories for such services. Note, however, that CMPH provides directions for performing cultures of hospital water for members of the legionellae, cultures and endotoxin assays of hemodialysis water, surveillance cultures of immunocompromised hosts, air cultures for fungi, cultures of environmental and medical-device surfaces, cultures of blood bank products and of orthopedic surgery sites, and quantitative cultures of small bowel contents, as well as various epidemiological tools such as serological typing. CMPH should be consulted by all who wish to practice these approaches. Interested parties must also be aware that many advances in molecular epidemiology are fairly recent, and they should consult pertinent sections of the *Manual of Clinical Microbiology*, 6th edition (1).

The procedures outlined in this section address analyses requested with some frequency in most medical facilities. Infection control committees carry the responsibility for all investigations of an epidemiologic nature and should establish the policies that lead to the involvement of microbiologic investigations. The everyday activity of the microbiology laboratory supplies much useful information for the epidemiologist, who usually works closely with members of laboratory staff. Policies established in concert with infectious-disease specialists and others will guide the efforts of the microbiology service with respect to the epidemiological examinations required. This section provides a few procedures commonly performed by microbiologists.

REFERENCE

1. **Murray, P. R., E. J. Baron, M. A. Pfaller, F. C. Tenover, and R. H. Yolken (ed.).** 1995. *Manual of Clinical Microbiology*, 6th ed. American Society for Microbiology, Washington, D.C.

11.2 Culture of Continuous Ambulatory Peritoneal Dialysis Fluid

I. PRINCIPLE

A. Patient population
Continuous ambulatory peritoneal dialysis (CAPD) has become an increasingly popular therapy for patients with end-stage renal disease.

B. Method
Dialysis fluid is infused into the peritoneum through a permanent (Tenckhoff) catheter. A collapsible bag containing dialysis fluid is attached to the catheter, and the fluid is allowed to flow into the peritoneal cavity. The bag is rolled up and placed in a pouch worn under the clothing. After a diffusion period of approximately 5 to 6 h, the delivery bag becomes the drainage bag. When the drainage bag is full, it is replaced with a fresh full bag, and the cycle is repeated.

C. Precautions
All bag change manipulations are performed by aseptic techniques to reduce the danger of external contamination.

D. Outcome
Even with this precaution, infections are not uncommon, averaging 1.07 to 1.47 episodes per patient year (10). See Appendix 11.1–1 for further discussion.

II. SPECIMEN

A. Collection and transport

1. Safety precautions
   Enclose dialysate bag in a larger plastic bag. Place this bag into a disposable plastic pan, and transport it to the laboratory.
2. Conditions
   a. For immediate delivery, transport at room temperature.
   b. For delayed delivery (>1 h after collection), refrigerate but do not freeze.

B. Processing

1. Mix contents of bag by inverting it 10 times.
2. Disinfect entry port with povidone-iodine, and withdraw 100 ml of fluid.
3. Perform cell count. If there are >50 WBCs per ml, culture may be indicated, depending on the dialysis schedule (4). Cell count may be performed in appropriate laboratory if microbiology division is not equipped to do it.
4. Place fluid into two 50-ml sterile, plastic centrifuge tubes, and centrifuge as for acid-fast cultures, using sufficient time and *g* force (*see* CMPH section 3 for further details).
5. Carefully decant approximately 45 to 48 ml of supernatant from each tube, and suspend the pellet in the balance of the fluid.
6. Wash the pellet by adding approximately 45 ml of sterile distilled water and repeating steps 4 and 5 above (3).
7. Use the resuspended pellet from step 6 above to inoculate media (*see* item IV.B below).

III. MATERIALS

Media, supplies, and instruments are those commonly utilized in the relevant laboratory and are employed selectively, as dictated by the choice of culture protocol (*see* item IV.B below).

IV. PROCEDURE

A. Microscopic examination

1. Gram stain

a. Concentrate the specimen by centrifugation or filtration as described above.

b. Prepare a Gram-stained smear, and review microscopically.

c. If organisms are detected by microscopic observation, report these immediately.

d. Although of low yield if applied to unconcentrated specimen, the Gram stain may promote early therapy if it is positive, and thus it is clinically useful.

2. Acid-fast smear

a. If cell levels are elevated (>100/ml) but no bacteria are identified on review of the Gram-stained smear, an acid-fast stain is indicated (*see* CMPH section 3).

b. Conduct a careful microscopic review for acid-fast bacilli, and report immediately if positive.

3. Fungal KOH preparation

This preparation is electively done on concentrated specimen, preferably employing Calcofluor or phase-contrast microscopy to enhance detection (*see* CMPH section 6).

B. Culture

1. Bacteria

a. Aerobic

Inoculate aerobic bacteriologic culture plates and broth with the concentrated specimen. Blood agar, CHOC, and broth (either THIO or brain heart infusion) media are recommended. Incubate in carbon dioxide (5%) at 35°C for at least 5 days, during which time most bacteria, yeasts, and rapidly growing mycobacteria will appear.

b. Anaerobic

(1) Inoculate nonselective anaerobic bacteriologic culture media (*see* CMPH section 2), and incubate them at 35°C under an anaerobic atmosphere for 7 days.

(2) Ideally, the broth should be incubated for an additional week for the delayed appearance of slowly growing anaerobes.

(3) Though rare in CAPD-related peritonitis, anaerobic bacteria are associated with severe peritonitis, and thus it is useful to specifically document them.

c. Alternative

Inoculate the standard blood culture system in use in the laboratory (*see* CMPH section 1 for specific instructions). In steps 2 to 5 below, the specimen is concentrated by centrifugation, and 5 ml is inoculated into the blood culture bottle.

(1) Lysis-centrifugation (Isolator) gives a good yield of aerobic organisms if 10 ml of unconcentrated peritoneal fluid is used (1, 2, 11).

(2) Bactec gives a yield equivalent to that of Isolator, but a longer time to positivity has been noted in some studies (2, 5, 9, 11).

IV. PROCEDURE *(continued)*

(3) Oxoid Signal gives a high yield with concentrated peritoneal fluid but requires blind subculture (9).
(4) Septi-Chek gives a good yield and rapid recovery of organisms on solid media (8).
(5) Conventional 2-bottle system

2. Mycobacteria and fungi
One of two scenarios is possible.

a. Parallel cultures
(1) Concentrated dialysate is inoculated onto the mycobacterial and fungal media on the day the bacterial cultures are set up.
(2) This approach is definitely indicated if the acid-fast smear or KOH preparation is positive but is optional if either preparation is negative.

b. Tandem cultures
(1) Concentrated dialysate is reserved in the refrigerator for 5 days. If the Isolator system is used, 10 ml of unconcentrated fluid is reserved in the refrigerator.
(2) If the bacterial cultures are negative and the patient has not responded to antibiotic therapy, fungal and mycobacterial cultures are set up. With the extended incubation period advocated initially, many of the cultures will yield bacteria or rapidly growing yeasts and mycobacteria, eliminating the need for many cultures of acid-fast bacteria and fungi.

V. LIMITATIONS OF THE PROCEDURE

A. Biopsy, supplemental
Biopsy may be indicated when bacterial cultures are negative and tuberculous peritonitis is suspected (6).

B. Other supplemental methods
Limulus amebocyte lysate (7) and other tests for the detection of infectious agents may not be sufficiently sensitive or specific to be of value unless used rigidly for specific therapeutic protocols.

VI. INTERPRETATION

A. Contamination
◪ Since patient skin microorganisms are the predominant cause of CAPD-caused peritonitis and since this disease is characterized by low levels of these microorganisms, the primary difficulty in interpretation of culture results will lie in distinguishing exogenous contaminants from those actively involved in an infection. Some guidelines are provided here.

1. Consider the following positive.
a. One or more colonies or types of colonies, regardless of genus or species, that grow on streak lines or in inoculated areas
b. One or more colonies or types of colonies that grow in inoculum or on streak lines if the colonies differ in appearance from colonies that grow outside of streak lines or inoculum

2. Consider the following contaminants.
a. Colonies that are found exclusively outside the inoculum or streak lines
b. Colonies with the same appearance that grow on streak lines and outside of inoculum or streak lines

B. Quantitation
Even very low numbers of organisms are considered potentially significant, but confounding organisms are frequently isolated from patients without peritonitis. Thus, use clinical judgment.

C. Clinical correlates

1. Cell count

a. Normal peritoneal fluid (6-h dwell time)

(1) Contains 50 to 100 cells per ml, mostly mononuclear

(2) Visibly clear

b. Peritonitis dialysis fluid

(1) Contains >100 cells per ml

(2) Visibly cloudy

2. Differential

a. Bacterial peritonitis: neutrophils predominant

b. Mycobacterial peritonitis (6)

(1) Neutrophils are sometimes associated with mycobacterial peritonitis.

(2) Mononuclear cells are often associated with mycobacterial peritonitis.

3. Signs and symptoms

a. Fever

(1) Relatively constant feature with tuberculous peritonitis (6)

(2) Observed in fewer than 50% of CAPD patients with bacterial peritonitis

b. Abdominal tenderness, 70% of patients

c. Abdominal pain, 79% of patients

d. Rebound tenderness, 50% of patients

4. Clinical response

Absence of a clinical response to antibacterial therapy suggests a fungal or mycobacterial infection.

VII. QUALITY ASSURANCE

A. QC

1. Routine QC of media and reagents as described in CMPH section 13

2. Parallel culture of the sterile distilled water used to resuspend the cells in step II.B.6 above

B. QA

Report all positive culture results to the Infection Control Committee for review. Upon establishing that there is an excess infection rate, a repetitive microorganism, or an apparent series of pseudoinfections, the Infection Control Committee should initiate an investigation and consider the institution of appropriate intervention or monitoring strategies.

REFERENCES

1. **Forbes, B. A., P. A. Frymover, R. T. Kopecky, J. M. Wejtaazak, and D. J. Pettit.** 1988. Evaluation of the lysis-centrifugation system for culturing dialysates from continuous ambulatory peritoneal dialysis patients with peritonitis. *Am. J. Kidney Dis.* **11:** 176–179.
2. **Holley, J. L., and A. H. Moss.** 1989. A prospective evaluation of blood culture versus standard plate techniques for diagnosing peritonitis in continuous ambulatory peritoneal dialysis. *Am. J. Kidney Dis.* **13:**184–188.
3. **Ludlam, H., A. Dickens, A. Simpson, and I. Phillips.** 1990. A comparison of four culture methods for diagnosing infection in continuous ambulatory peritoneal dialysis. *J. Hosp. Infect.* **16:**263–269.
4. **Males, B. M., J. J. Walsbe, and D. Amsterdam.** 1987. Laboratory indices of clinical peritonitis: total leukocyte count, microscopy, and microbiologic culture of peritoneal dialysis effluent. *J. Clin. Microbiol.* **25:** 2367–2371.
5. **Males, B. M., J. J. Walsbe, L. Garringer, D. Koscinaki, and D. Amsterdam.** 1986. Addi-Chek filtration, BACTEC, and 10-ml culture methods for recovery of microorganisms from dialysis effluent during episodes of peritonitis. *J. Clin. Microbiol.* **23:**350–353.

REFERENCES *(continued)*

6. **Mallat, S. G., and J. M. Brensilver.** 1989. Tuberculous peritonitis in a CAPD patient cured without catheter removal: case report, review of the literature, and guidelines for treatment and diagnosis. *Am. J. Kidney Dis.* **13:**154–157.
7. **McCartney, A. C., J. N. Brunton, and G. L. Warwick.** 1989. Limulus amoebocyte lysate (LAL) assay and rapid detection of Gram negative bacterial peritonitis in patients receiving CAPD. *J. Clin. Pathol.* **42:**1115.
8. **Ryan, S., and S. Fessia.** 1987. Improved method for recovery of peritonitis-causing microorganisms from peritoneal dialysate. *J. Clin. Microbiol.* **25:**383–384.
9. **Soubulie, M. A., D. L. Sewell, M. D. Holland, and T. A. Golper.** 1989. Comparison of two commercial broth-culture systems for microbial detection in dialysates of patients on continuous ambulatory peritoneal dialysis. *Diagn. Microbiol. Infect. Dis.* **12:**457–461.
10. **Vas, S. I.** 1989. Infections associated with peritoneal and hemodialysis, p. 215–248. *In* A. L. Bisno and F. A. Waldvogel (ed.), *Infections Associated with Indwelling Medical Devices.* American Society for Microbiology, Washington, D.C.
11. **Woods, G. L., and J. A. Washington II.** 1987. Comparison of methods for processing dialysate in suspected continuous ambulatory peritoneal dialysis-associated peritonitis. *Diagn. Microbiol. Infect. Dis.* **7:**155–157.

APPENDIX 11.1–1

CAPD-Associated Infections

A. CAPD-associated infections
   1. Types
      a. Infection of the catheter exit site or subcutaneous tunnel
      b. Infection of the peritoneum
      c. Infection of the catheter exit site or subcutaneous tunnel is often a precursor to infection of the peritoneum.
   2. Etiologic agents
      Although bacteria predominate, fungal and mycobacterial infections are reported and are probably more common than the data suggest. Some reports indicate a possible viral etiology in CAPD-associated peritonitis (3).
   3. Origin
      a. Common skin microorganisms: access via catheter lumen or periluminal space
      b. Intestinal organisms, transmural access
      c. Hematogenous (streptococci and *Mycobacterium tuberculosis*) seeding
      d. Vaginal, ascending (yeasts)
      e. Environmental (e.g., tap water, *Mycobacterium chelonae*)

B. Culture negativity
   A substantial percentage of CAPD patients who experience signs and symptoms of peritonitis are culture negative. There are several possible explanations for this.
   1. Dilution effect
      The large volume of dialysate may lower the concentration of bacteria below the detection limits of the culture method.
   2. Inhibitory substances
      Organisms may be present but have their growth inhibited; coagulase-negative staphylococci do not survive well in dialysate from patients clinically ill with peritonitis. While not all antibacterial factors are known, suppression by the presence of antibiotics and sequestration in phagocytic cells are two documented sources of false-negative cultures.
   3. Exacting growth requirements
      False-negative cultures are also caused by a lack of suitable growth conditions for the spectrum of etiologic agents.
   4. Avoidance
      To minimize the number of false-negative cultures, the laboratory should provide for the following.
      a. Concentration of microorganisms
      b. Dilution or removal of any antibiotics present
      c. Lysis of host cells to release intracellular microorganisms
      d. Parallel or tandem cultures for bacteria, fungi, and mycobacteria

C. Optimal culture protocol

A plethora (1) of different methods have been advocated for the culture of CAPD dialysate, with no one method proven superior to all others. However, a clearly inferior method is the simple culture of a small volume of unconcentrated dialysate. Many methods described in the literature call for special media containing lysing reagents (4, 7), sonication (8), antibiotic-removal resins (9), or a special filtration apparatus for concentration (5) or large-volume culture (2, 6). Laboratories already employing such methods need not discontinue them. For the balance of laboratories, the procedure described here provides a comprehensive approach that utilizes the technology available in the average laboratory.

References

1. **Buggy, B. P.** 1986. Culture methods for continuous ambulatory peritoneal dialysis-associated peritonitis. *Clin. Microbiol. News.* **3**:12–14.
2. **Dawson, M. S., A. M. Harford, B. K. Garner, D. A. Sica, D. M. Landwehr, and H. P. Dution.** 1985. Total volume culture technique for the isolation of microorganisms from continuous ambulatory peritoneal dialysis patients with peritonitis. *J. Clin. Microbiol.* **22:** 391–394.
3. **Lewis, S. L.** 1991. Recurrent peritonitis: evidence for possible viral etiology. *Am. J. Kidney Dis.* **17**:343–345.
4. **Ludlam, H. A., T. N. C. Price, A. J. Berry, and I. Phillips.** 1988. Laboratory diagnosis of peritonitis in patients on continuous ambulatory peritoneal dialysis. *J. Clin. Microbiol.* **26:** 1757–1762.
5. **Males, B. M., J. J. Walshe, L. Garringer, D. Kosciasid, and D. Amsterdam.** 1986. Addi-Chek filtration, BACTEC, and 10-ml culture methods for recovery of microorganisms from dialysis effluent during episodes of peritonitis. *J. Clin. Microbiol.* **23:**350–353.
6. **Sewell, D. L., T. A. Golper, P. B. Hulman, C. M. Thomas, L. M. West, W. Y. Kubey, and L. Holmes.** 1990. Comparison of large volume culture to other methods for isolation of microorganisms from dialysate. *Peritoneal Dialysis Int.* **10:**49–52.
7. **Spancer, R. C., and W. K. Alunad.** 1986. Laboratory diagnosis of peritonitis in continuous ambulatory peritoneal dialysis by lysis and centrifugation. *J. Clin. Pathol.* **39:**925–926.
8. **Taylor, P. C., L. A. Foole-Warren, and R. E. Grundy.** 1987. Increased microbial yield from continuous ambulatory peritoneal dialysis peritonitis effluent after chemical or physical disruption of phagocytes. *J. Clin. Microbiol.* **25:** 580–583.
9. **Vas, S. I., and L. Law.** 1985. Microbiological diagnosis of peritonitis in patients on continuous ambulatory peritoneal dialysis. *J. Clin. Microbiol.* **21:**522–523.

11.3 Culture of Intravascular Devices

I. PRINCIPLE

Intravascular catheters are used to provide continuous vascular access to permit blood sampling; to administer blood products, medications, total parenteral nutrition, and other fluids; and, in the case of pulmonary artery catheters, to permit hemodynamic monitoring of cardiac function. Because these devices penetrate the integument, they put the patient at significant risk for development of device-related infection. The insertion site becomes colonized by bacteria from the patient's own skin or by microorganisms carried on the hands of medical personnel (5). Organisms can also gain access through the lumen of the catheter following contamination of the hub (4) or infusion of contaminated fluids. Invading organisms can then colonize the intravascular catheter surfaces and produce local infection and, in a significant number of cases, bacteremia, fungemia, suppurative phlebitis, or septic thrombosis (5). (*See* Appendix 11.3–1 for further discussion.)

II. SPECIMEN

A. Types

1. Long catheters

These catheters are surgically inserted into the central vein to provide access for administration of therapeutic agents, total parenteral nutrition, and blood products (6).

a. Long term

(1) Percutaneous catheters: Broviac, Hickman, Raaf, Groshong, Quinton (Fig. 11.3–1). These catheters are surgically inserted and subcutaneously tunnelled to a chest wall exit site. A Dacron cuff facilitates growth of fibrous tissue to the catheter and inhibits migration of microorganisms along the surface of the catheter (6).

(2) Subcutaneous ports: Infus-A-Port, Port-A-Cath, Med-i-Port (Fig. 11.3–2). These systems can be entered by inserting a needle through the septum of the reservoir (6).

b. Short-term percutaneous catheters: Swan-Ganz (plus introducer), Intracath, Cordis, multilumen, Udall (for hemodialysis) (Fig. 11.3–1)

2. Short catheters

These catheters are inserted into various sites and are used for short-term vascular access.

a. Peripheral

b. Steel needles

c. Umbilical

d. Arterial

B. Collection and transport

1. Collection

To prevent contamination by skin microorganisms and antibiotic ointment, clean the skin insertion site with an iodophore and alcohol prior to removal of the cannula. Remove the cannula in an aseptic manner once the alcohol has dried. If purulence of the catheter exit site is evident, send pus for culture and Gram stain (5).

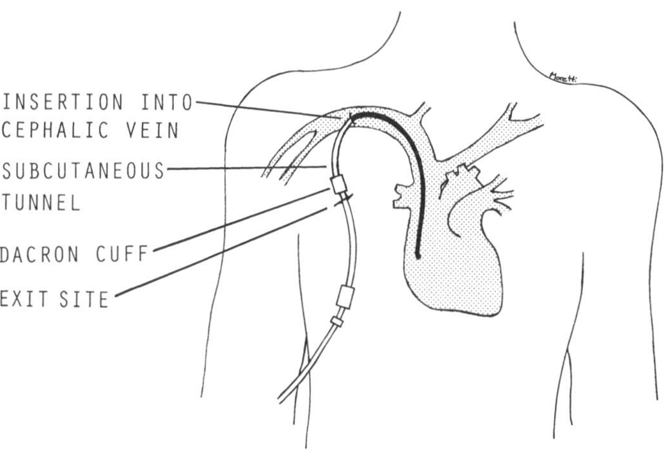

Figure 11.3–1 Placement of surgically implanted long percutaneous catheter (illustration by Carol Moretti)

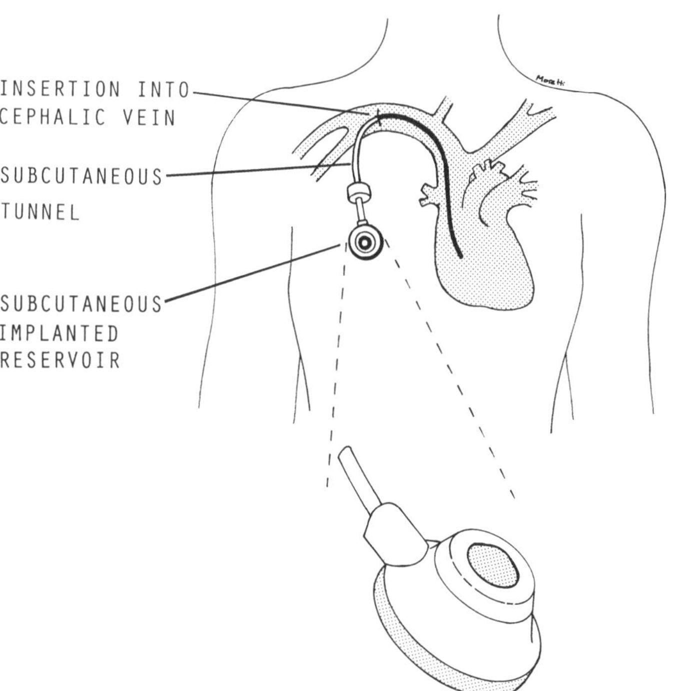

Figure 11.3–2 Placement of surgically implanted subcutaneous port (illustration by Carol Moretti)

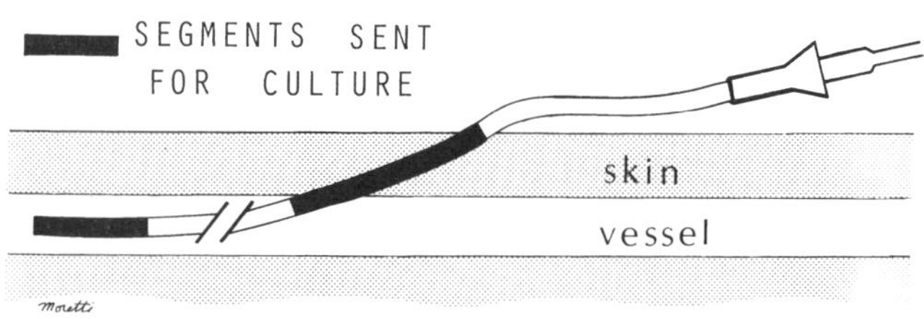

Figure 11.3–3 Segments of long catheter for culture (illustration by Carol Moretti)

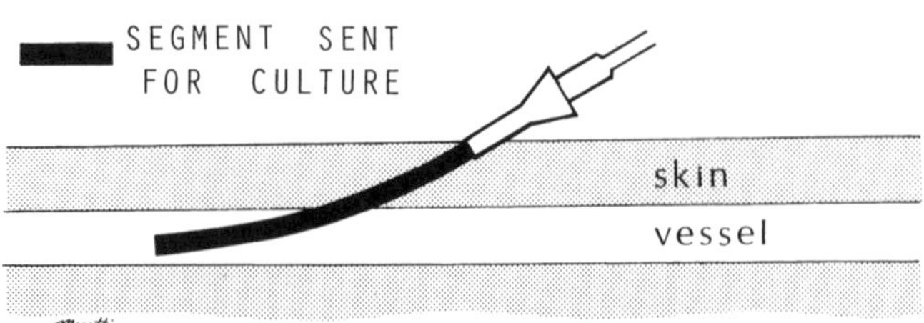

Figure 11.3–4 Segment of short catheter for culture (illustration by Carol Moretti)

II. SPECIMEN *(continued)*

a. Long catheters

Two portions of these catheters should be sent for culture: the distal intravascular tip and the proximal transcutaneous segment. Each segment should be approximately 2 to 3 cm long (Fig. 11.3–3).

b. Short catheters

The cannula is cultured in its entirety following removal of the hub. To remove the hub, use sterile scissors, or snap off steel needles with a sterile hemostat (5) (Fig. 11.3–4).

2. Transport

Transport catheter tips in a sterile container. If tips are cut to proper length, there is no need to bend them for insertion into transport container. Culture tips within 2 h of collection to prevent desiccation of microorganisms.

III. EQUIPMENT, SUPPLIES, AND REAGENTS

Include QC information on reagent container and in QC records.

A. 100-mm 5% BAP
B. Forceps
C. 95% alcohol
D. Sterile TSB
E. 100-μl pipettor and sterile tips
F. Sterile 1-ml syringe and needle
G. Sterile test tubes (12 by 75 mm) with caps

IV. PROCEDURE

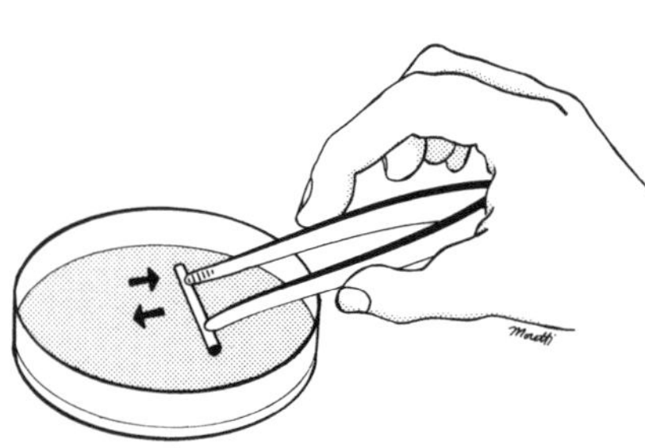

Figure 11.3–5 Maki roll technique (illustration by Carol Moretti)

A. Semiquantitative culture, Maki method

1. Dip forceps in 95% alcohol, and flame to sterilize. Allow to cool.
2. If the catheter segment is too long to be rolled on the plate, cut it in half with sterile scissors, and culture each segment separately.
3. Use sterile forceps to transfer catheter tip from transport container to BAP.
4. Using slight pressure, roll catheter tip back and forth across agar surface at least four times. It is essential that the catheter tip have good contact with the surface of the plate. If tip is bent and hard to roll, use forceps to pick up tip, and rub all surfaces onto plate (Fig. 11.3–5).
5. Dip forceps in 95% alcohol, and flame sterilize.
6. Incubate plate for at least 72 h at 35°C in a CO_2 incubator. It is not necessary to routinely culture for anaerobic bacteria, fungi, or mycobacteria.

B. Culture of implanted subcutaneous ports

At this time, there is no standardized method available for culturing these devices. If sepsis is suspected, quantitative blood cultures may be done without necessitating removal of the device. If the port is subsequently removed, the distal catheter tip is also cultured (7). When a subcutaneous infection is suspected, culture an aspirate or tissue sample of the wound site. There is no evidence that suggests culturing the port itself is of any value.

C. **Culture of the lumen**

1. Aseptically transfer catheter segment from transport container to a sterile test tube (12 by 75 mm).
2. Use a sterile syringe to draw up 1 ml of sterile TSB.
3. Place needle into lumen of catheter, and dispense broth through catheter segment.
4. Cap tube, and vortex to dislodge adherent bacteria.
5. Serially dilute sample 100-fold.
6. Inoculate 100 μl of each dilution to 100-mm BAP. Cross-streak for well-isolated colonies.
7. Incubate plate for at least 72 h at 35°C in a CO_2 incubator. It is not necessary to routinely culture for anaerobic bacteria, fungi, or mycobacteria.

V. INTERPRETATION OF CULTURE

A. **Semiquantitative culture, Maki method**

Identify and perform susceptibility testing on each organism that produces $\geq$15 colonies.

◪ **NOTE:** Some authors use five colonies as the cutoff (2, 3).

B. **Culture of implanted subcutaneous ports**

Identify and perform susceptibility testing on all organisms isolated.

C. **Culture of the lumen**

Calculate the number of CFU per milliliter by multiplying the number of colonies by 10 times the dilution factor and then dividing by the volume of broth used (1 ml). Identify and perform susceptibility testing on all organisms present at >1,000 CFU/ml (1).

VI. LIMITATIONS OF THE PROCEDURE

Fastidious organisms will not be detected. CHOC may help detect these organisms.

VII. QUALITY CONTROL

Perform QC on media based on your laboratory procedure. Report catheter-associated infections to the appropriate committee (e.g., Infection Control, Catheter Care) for QA review.

VIII. REPORTING RESULTS

A. **Semiquantitative culture, Maki method**

For each organism that produces >15 CFU, include actual colony count with the organism identification and susceptibility report. If there is confluent growth, report as such.

B. **Culture of implanted subcutaneous ports**

Report qualitatively (e.g., few, moderate, many) when organism identification and susceptibility are reported.

C. **Culture of lumen**

For each organism that produces >1,000 CFU/ml, include dilution amount with the identification and susceptibility report.

REFERENCES

1. **Cleri, D. J., M. L. Corrade, and S. J. Seligman.** 1980. Quantitative culture of intravenous catheters and other intravascular inserts. *J. Infect. Dis.* **141:**781–786.
2. **Collignon, P. J., R. Chan, and R. Munro.** 1987. Rapid diagnosis of intravascular catheter-related sepsis. *Arch. Intern. Med.* **147:** 1609–1612.
3. **Collignon, P. J., N. Soal, I. Y. Pearson, W. F. Weeds, R. Munro, and T. C. Serrell.** 1986. Is semiquantitative culture of central vein catheter tips useful in the diagnosis of catheter-associated bacteremia? *J. Clin. Microbiol.* **24:** 532–535.
4. **Liñares, J., A. Sitges-Serra, J. Gerau, J. L. Perez, and R. Martin.** 1985. Pathogenesis of catheter sepsis: a prospective study with quantitative and semiquantitative cultures of catheter hub and segments. *J. Clin. Microbiol.* **21:** 357–360.
5. **Maki, D. G.** 1980. Sepsis associated with infusion therapy, p. 207–253. *In* S. Karran (ed.), *Controversies in Surgical Sepsis.* Praeger, New York.
6. **Miller, K. B.** 1988. Cardiovascular system update, p. 1168–1181. *In* R. Berg (ed.), *The APIC Curriculum for Infection Control Practices,* vol. 3. Kendall/Hunt, Dubuque, Iowa.
7. **Moore, C. L., K. A. Erickson, L. B. Yanes, M. Franklin, and L. Gonsalves.** 1986. Nursing care and management of venous access ports. *Oncol. Nurs. Forum* **13:**35–39.

SUPPLEMENTAL READING

Collignea, P. J., and R. Munro. 1989. Laboratory diagnosis of intravascular catheter associated sepsis. *Eur. J. Clin. Microbiol. Infect. Dis.* **8:** 807–814.

Coutlée, F., C. Lenieux, and J. F. Paradis. 1988. Value of direct catheter staining in the diagnosis of intravascular-catheter-related infection. *J. Clin. Microbiol.* **26:**1088–1090.

Hamory, B. H. 1987. Nosocomial bloodstream and intravascular device-related infections, p. 283–319. *In* R. P. Wenzel (ed.), *Prevention and Control of Nosocomial Infections*, vol. 1. The Williams & Wilkins Co., Baltimore.

Kristinsson, K. G., I. A. Burnett, and R. C. Spencer. 1989. Evaluation of three methods for culturing long intravascular catheters. *J. Hosp. Infect.* **14:**183–191.

APPENDIX 11.3–1

Device-Related Infections

Intravascular-device-related infections are most often caused by endogenous skin flora (Table 11.3–A1). Some organisms, such as coagulase-negative staphylococci and *Corynebacterium jeikeium*, that were previously considered contaminants are now recognized as major pathogens in device-related infections. Accurate diagnosis of device-related infection can be difficult, since true infection must be distinguished from contamination of the catheter during its removal from the patient. When device-related infection is suspected, the entire catheter (short catheters) or 2- to 3-cm segments from both the distal tip and the transcutaneous portions (long catheter) should be submitted for culture (5). The clinical microbiology laboratory must be able to identify and quantitate the microorganisms present.

Qualitative culture of the catheter, in which the segment is incubated in liquid medium, is a sensitive technique for diagnosis of infection but provides no information about the number of microorganisms present (7); i.e., the presence of a single contaminant can lead to a positive culture. In an effort to improve the specificity of catheter cultures, quantitative techniques that demonstrate a sensitivity comparable to that of broth culture but with superior specificity and predictive value have been developed. Detection of catheter-related infections is also possible by microscopic examination of the catheter surfaces following Gram (2) or acridine orange (10) staining. This allows rapid detection, preliminary characterization, and quantitation of the organisms present. Although these staining techniques show promise, they are not feasible for all types of catheters and are technically quite demanding and labor intensive.

Maki et al. (7) developed a semiquantitative technique in which the catheter segment is rolled back and forth across the surface of a BAP. Colonies are counted after incubation, and the presence of ≥15 indicates local infection. Because the technique samples only the exterior portion of the catheter, it may not detect infections originating in the lumen of the

Table 11.3–A1 Bloodstream pathogens suggestive of intravenous-infusion-related septicemia[a]

Probable source	Pathogens
Peripheral intravenous catheters	Coagulase-negative staphylococci, *Staphylococcus aureus,* enterococci, *Candida* spp.
Central venous catheters	*Candida* spp., coagulase-negative staphylococci, *S. aureus, Corynebacterium* spp. (especially *C. jeikeium*), *Klebsiella* spp., *Acinetobacter* spp., *Fusarium* spp., *Malassezia furfur*

[a] Modified from reference 6. Reprinted with permission.

catheter. It has been suggested that colonization of the lumen may be an important reservoir of infection in some catheter infections (1, 3, 4, 8). Cleri et al. (1) developed a quantitative technique in which the catheter lumen is flushed with liquid medium and agitated vigorously to dislodge adherent microorganisms. The suspension is then serially diluted and plated. The presence of >1,000 CFU/ml suggests the presence of bacteremia. The theoretical advantage of this approach is that both the interior and exterior surfaces of the catheter are sampled. Thus, a modified procedure for such culture is also included. Rello et al. suggest that using both the semiquantitative culture of the exterior catheter surface and the quantitative culture of the lumen increases the detection of catheter colonization (9).

References

1. **Cleri, D. J., M. L. Corrado, and S. J. Seligman.** 1980. Quantitative culture of intravenous catheters and other intravascular inserts. *J. Infect. Dis.* **141:**781–786.
2. **Cooper, G. L., and C. C. Hopkins.** 1985. Rapid diagnosis of intravascular catheter-associated infection by direct Gram staining of catheter segments. *N. Engl. J. Med.* **312:** 1142–1147.
3. **Haslett, T. M., H. D. Isenberg, K. Hilton, V. Tucci, B. G. Kay, and E. M. Vellozzi.** 1988. Microbiology of indwelling central intravascular catheters. *J. Clin. Microbiol.* **26:** 696–701.
4. **Liñares, J., A. Sitges-Serra, J. Garau, J. L. Perez, and R. Martin.** 1985. Pathogenesis of catheter sepsis: a prospective study with quantitative and semiquantitative cultures of catheter hub and segments. *J. Clin. Microbiol.* **21:** 357–360.
5. **Maki, D. G.** 1980. Sepsis associated with infusion therapy, p. 207–253. *In* S. Karran (ed.), *Controversies in Surgical Sepsis.* Praeger, New York.
6. **Maki, D. G.** 1989. Pathogenesis, prevention, and management of infections due to intravascular devices used for infusion therapy, p. 161–177. *In* A. L. Bisno and F. A. Waldvogel (ed.), *Infections Associated with Indwelling Medical Devices.* American Society for Microbiology, Washington, D.C.
7. **Maki, D. G., C. E. Weise, and H. W. Sarafin.** 1977. A semiquantitative culture method for identifying intravenous-catheter-related infection. *N. Engl. J. Med.* **296:**1305–1309.
8. **Needham, P. M.** 1990. Intraluminal colonization of a Groshong catheter. *Clin. Microbiol. Newsl.* **12:**189–190.
9. **Rello, J., J. M. Gatell, J. Almirall, J. M. Campistol, J. Gonzalez, and J. Puig de la Bellacasa.** 1989. Evaluation of culture techniques for identification of catheter-related infection in hemodialysis patients. *Eur. J. Clin. Microbiol. Infect. Dis.* **8:**620–622.
10. **Zufferey, J., B. Rime, P. Francioli, and J. Bille.** 1988. Simple method for rapid diagnosis of catheter-associated infection by direct acridine orange staining of catheter tips. *J. Clin. Microbiol.* **16:**175–176.

11.4 Culture of Biological Indicators of Sterilization Processes

I. PRINCIPLE

Hospital central-service departments have learned, during their many years of experience with steam sterilization (i.e., autoclaving) and more recently with ethylene oxide gas sterilization and low-temperature hydrogen peroxide plasma, which set of conditions is likely to achieve sterility. However, packaging materials may interfere with steam, heat, or chemical penetration. Moreover, the density of the pack and its relative position in the load as well as normal variations in time, temperature, steam, chemical concentration, and operator performance are additional concerns. To provide some independent means of sterility assurance, physical, chemical, and biological indicators (BIs) are routinely used. As new sterilization technologies are approved, appropriate biological indicators will need to be defined and validated. (*See* Appendix 11.4–1 for further discussion.)

II. SPECIMEN

A. Use of the BI

BIs must be used in a way that maximally challenges the sterilizer. An operator should place the BI within the densest pack that is to be sterilized. This pack should be placed in the sterilizer chamber at the point where optimal conditions are most difficult to achieve. For steam sterilizers, this position is usually in the bottom front over the drain (Table 11.4–1). Occasional challenges in other areas, e.g., rear of autoclave, may also be informative if an air pocket is suspected. For ethylene oxide sterilizers, several packs may be necessary, depending on the design and size of the sterilizer.

Since retrieval of a BI from within a pack often means that the goods must be contaminated, much waste of time and overprocessing of goods could occur. Additionally, different materials may not provide a consistent challenge to the sterilizer. To remedy this situation and to provide some consistency, standard test packs have been developed. The contents and uses of these test packs are prescribed by many accreditating agencies (Table 11.4–1).

Several manufacturers have produced disposable test packs that reportedly present the same challenge but require less labor and fewer supplies (3). Several configurations are available. The laboratory may on occasion have to retrieve a BI from one of these packs. For autoclaving media, other liquids, or microbiological wastes, place BI in the center among flasks or tubes or within autoclave waste bag in the center of the device.

B. BI transport

It is imperative that spore tests be processed promptly and properly. Hospitals do not have the supplies available to allow a quarantine until spore tests are negative. The sooner incubation begins, the more likely it is that if a positive test occurs, goods can be recalled before they are used for a patient. It is also known that sublethally damaged spores have a better chance of being recovered if promptly cultured and incubated (4).

II. MATERIALS, EQUIPMENT, AND SUPPLIES

Include QC information on reagent container and in QC records.

A. Medium

TSB has been shown to be the best recovery medium for both *Bacillus subtilis* and *Bacillus stearothermophilus*. Soybean casein digest broth is another alternative (9). Do not use THIO; it has been documented as toxic for some spore preparations (2).

B. BIs

1. Medium and spore suspensions

These preparations are packaged in sealed glass vials. Since the spores are not exposed to steam or chemicals, they are heat indicators only. Chemspor (AMSCO) is an example of such a BI. (*See* Appendix 11.4–2 for addresses of suppliers.) Each unit is intended to evaluate hard goods in liquids (e.g., soft contact lenses in vials) or liquids that are used in a sterile state (e.g., small-volume parenteral fluids). *B. stearothermophilus* spores are suspended in broth. Phenol red indicates growth by a change in color from red to yellow. A heat-sensitive pellet melts on exposure to a temperature of 250°F (ca. 121°C).

2. Dry-spore paper carrier systems

Two types of paper carrier dry-spore systems are available. One type consists of spore-impregnated strips in sealed envelopes, usually of glassine. After being processed, the strips are sent to the laboratory for inoculation, incubation, and interpretation. This method requires sterile transfer to a culture medium and an incubation period of up to 7 days. Spordi (AMSCO) is an example of such a system. Each spore-impregnated filter paper strip contains 10^4 spores of *B. stearothermophilus* and 10^6 spores of *B. subtilis* sealed in glassine envelopes. Each packet has two compartments. One side contains two spore strips for testing, and the other side contains an additional spore strip to be used as a viability control. This strip is omitted from sterilization but submitted as a control to the laboratory. Media and incubators for use with this system are not supplied by the manufacturer. The Tec-Test Biological Indicator System (Castle/Sybron Medical Products) consists of UniSpore spore strips in peel-open glassine envelopes, tubes containing 5 ml of modified soybean casein digest broth with phenol red, and two preset (37 or 56°C) dry-block incubators.

The second type of system consists of a dry-spore strip and a crushable glass vial containing culture medium within a plastic vial closed with either a ventable or a filter paper friction fit cap. Proof Plus (AMSCO) is such a system. This BI can be used with both steam and ethylene oxide, as each unit contains spores of both *B. stearothermophilus* and *B. subtilis* var. *globigii*. A polycarbonate vial contains one filter paper strip with 10^3 *B. stearothermophilus* and 10^5 *B. subtilis* spores and a glass ampoule containing 2 ml of TSB with phenol red. A chemical indicator strip is attached to the outside of the vial. Proof Flash (AMSCO) is intended to be used in flash sterilization cycles only. A filter paper disk rather than a strip is impregnated with spores of *B. stearothermophilus*. The medium contains bromcresol purple. The outer cap is purple. In the Attest Indicator System (Medical Products Division, 3M Co.), three separate BIs contain approximately 10^5 spores of *B. stearothermophilus* for steam testing (brown caps), 10 spores of *B. stearothermophilus* for flash testing (purple caps), or 10^6 spores of *B. subtilis* var. *niger* for ethylene oxide testing (green caps). A crushable ampoule contains approximately 1 ml of modified TSB. Bromcresol purple is used as the pH indicator in the steam indicators, and bromthymol blue is used with ethylene oxide. Both the spore strips and the medium ampoule are placed inside a flexible polypropylene vial. The system is closed with a color-coded cap containing a hydrophobic filter. Dry-block incubators are available in 56 or 37°C or dual-temperature models.

Table 11.4–1 Test packs[a]

Agency and publication	Steam BI		
	Contents	Placement	Frequency of use
American Hospital Association, *Guidelines for the Hospital Central Service Department,* 1978			Once a day
Army, *Army Regulations* (AR-40-19), 1984			Minimum of once a week
Veterans Administration, *VA Manual GI, MP-2,* 1985, and *MP-2,* subchapter E, change 159, June 22, 1983	3 muslin surgical gowns, 12 towels, 30 gauze sponges, 5 lap sponges, 1 drape or equivalent linen, 10–12 lb, 12 by 12 by 20 in., 2 BIs in center	On edge in area least favorable to sterilization	No less often than weekly; each load of implantables or intravascular materials; following major sterilizer repairs; for evaluating new products or packaging material
Association for the Advancement of Medical Instrumentation (AAMI)			
Good Hospital Practice: Steam Sterilization and Sterility Assurance, 1993	16 towels folded lengthwise into thirds and width-wise in half along new long dimension; place on top of each other with folds opposite each other, 9 by 9 by 6 in.; 2 BIs between 8th and 9th towels in center; CIs placed adjacent to BIs; 2 pieces of tape around entire pack	Flat (layer of towels horizontal), front bottom near drain in full load	At least weekly, preferably daily; each load that contains implantable; installation: three consecutive cycles in empty sterilizer; periodic QA testing; after major repairs
Good Hospital Practice: Flash Sterilization of Patient Care Items for Immediate Use, 1992	Perforated or mesh-bottom instrument tray, organizing case, or single wrapped tray; one or more BIs and a CI	Bottom shelf of empty sterilizer	At least once a week, daily preferable
Good Hospital Practice: Ethylene Oxide Sterilization and Sterility Assurance, 1992			

[a] Abbreviations: CI, chemical indicator; RT, room temperature; RH, relative humidity.

Ethylene oxide BI		
Contents	Placement	Frequency of use
Routine pack: 1 BI in a plastic syringe large enough that plunger does not touch BI when inserted into the barrel. Tip guard is removed. The syringe is placed in the folds of a clean 100% cotton huckaback towel. These items are then placed in a peel pouch of appropriate size.	In center of full load of routine testing	Each cycle

NOTE: Assembled test packs should be stored at room temp (18 to 24°C) at a minimum relative humidity of 35%. CI may be used in conjunction with BIs in test packs. Aeration of test packs prior to retrieval of BI varies. Contents of test packs (e.g., towels, tubing) should be aerated completely before reprocessing.

IV. PROCEDURES

A. Spordi

1. Obtain cooled, processed spore strips from the test pack. Note time, date, sterilizer used, and load information.
2. Obtain one tube containing 5 ml of TSB or soybean casein digest broth for each strip to be cultured.
3. In a laminar-flow hood or in a low-traffic area of the laboratory, use sterile scissors to open the glassine envelope.
4. With sterile forceps, transfer each spore strip to a tube of medium.
5. Incubate the strips for 7 days at 56°C for steam cycles and 37°C for ethylene oxide and dry-heat cycles.
6. Observe at regular intervals (at least daily) for turbidity, which indicates a positive test.
7. Gram stain all positives. A positive should not be reported until gram-positive bacilli resembling *Bacillus* species have been seen. Report all positives as they occur. Record all results.
8. QC

   Incubate one unprocessed strip as a positive control. Incubate a tube of medium as a sterility control. If the positive control does not grow, the test is invalid. Check the temperature of the incubator, the growth medium (for efficacy), and the storage conditions of the strips. Also, check for a mix-up in labeling or inoculating tubes. Keep permanent longitudinal records on growth in the negative control tube; these serve as a record of percent contamination but do not invalidate other tests unless a *Bacillus* sp. is grown.

B. Tec-Test

1. As soon as possible after completion of the sterilization cycle, transport spore strips to processing area.
2. Carefully peel open glassine envelope to 1/4 in. (1 in. = 2.54 cm) above spore strip.
3. Turn opened glassine envelope upside down above prepared medium tube, and slowly peel apart until the spore strip drops into the medium.
4. Replace medium tube cap, and tighten. Incubate spores at 56°C for steam and 37°C for ethylene oxide and dry-heat processes.
5. Observe all tubes daily for 5 days. A positive results is shown by medium turbidity with a color change from red to yellow.

C. Proof Plus and Proof Flash

1. Obtain cooled, processed BI, and check the chemical indicator on the label for color change. (Steam changes the indicator to dark brown or black; ethylene oxide changes the indicator to gold.) If the indicator has not changed color, sterilization conditions have not been met. Check cycle parameters.
2. Activate the BI by fully depressing cap so the enclosed ampoule is crushed.
3. Invert vial, and shake vigorously to mix contents. Return to upright position.
4. Place activated BI in the appropriate manufacturer-supplied incubator (56°C for steam or 37°C for ethylene oxide).
5. Activate positive-control BI by depressing cap on an unprocessed unit and incubating as described above.
6. Incubate for 48 h. Check every day.
7. A positive test is shown by a change of the medium's color from red to yellow for Proof Plus. A positive test for Proof Flash is indicated by a change from purple to yellow. No color change indicates a negative result.

8. QC
   If the positive control does not turn positive, no results can be reported. Check for a mix-up in labeling or in interchanging a challenge vial with the control vial. Also, check for an overheated incubator or inappropriate storage of the spore vials. *The use of incorrect thermometers for confirmation of incubator temperature is a known cause of problems with this test.* The temperature of the incubator should be checked daily with a correctly calibrated thermometer. Records of these thermometer readings are part of the QC system of the laboratory.
9. Record all results.
10. Precautions: Notify by telephone the hospital department submitting the BIs of all positives.
    a. Be certain that caps are fully depressed to crush vials and prevent medium evaporation.
    b. Subculture of positive BIs is not recommended because of the great likelihood of introducing contamination during the procedure.
    c. Check incubator temperature daily.

D. Attest

1. After the processed BI has been retrieved from an appropriate test pack and cooled, the medium ampoule must be crushed. Avoid crushing by hand, as this method may overcrush the BI and cause the medium to spill over the sides of the polypropylene vial. When the dry-block incubator is used, the indicator is automatically crushed when properly inserted into the metal heating block.
2. Incubate at the appropriate temperature.
3. Examine at regular intervals for up to 48 h for any color change. A positive test result is a color change from purple or green-blue to yellow.
4. As a positive control, place a crushed, nonsterilized Attest indicator in each incubator daily.
5. Report by telephone all positives as they occur. Record all results.

V. SPECIAL PRECAUTIONS

A. Transfer in contaminated and/or busy work areas

Since *B. subtilis* is ubiquitous in nature, care must be taken not to introduce it into spore cultures, thereby possibly producing false-positive results. *Use of a laminar-flow hood for transfer of spore strips is preferable, but such a hood may not be available.* Use of an isolated area that is routinely disinfected and kept free of dust may help reduce contamination problems.

B. Improper handling of test materials

Storing or dipping transfer forceps in a disinfectant solution will not ensure that all spores are destroyed. Alcohol flaming has also been shown not to be totally effective (5). Sterilized forceps and/or scissors should be used routinely.

C. Sterilization of positive-control strip

On occasion, the department monitoring the sterilization process will inadvertently submit a sterilized spore strip as the positive control. Some manufacturers (Spordi, AMSCO) have attempted to prevent this problem by placing a spore strip in a sealed envelope for transport to the laboratory. An alternative approach is to keep a supply of nonprocessed spore strips of the same lot on hand in the laboratory.

D. Use of water bath incubators

Covered water baths will become contaminated in time, usually with test organisms from improperly handled control cultures. Although not common, contamination due to tubes being closed with cotton plugs or with loose screw caps has been reported (12).

V. SPECIAL PRECAUTIONS (continued)

E. Other contaminants

Although contaminants would be expected to be more common at 37°C (8), they can occur even at 55°C (7). Gurevich and colleagues (7) recovered *Bacillus coagulans* from the TSB used to culture spore strips. Proper sterility controls should always be performed.

F. Alterations in manufactured BIs

Despite published standards from the U.S. Pharmacopeia (11) and the Association for the Advancement of Medical Instrumentation (1) for the manufacture of spore strips, a number of reports have documented considerable variability in spore concentration and resistance. Numerous false-positive results have created difficult, expensive situations for some facilities. Although no positive spore test should be dismissed, the problem may lie with the spores themselves and not with the sterilizer. Conversely, even negative spore test results do not guarantee that sterility has been achieved. The personnel operating the sterilizers are advised to check other monitors in addition to the BI for each load.

G. Storage conditions

Depending on storage conditions, spores may become more or less resistant. Desiccation, such as occurs during storage in frost-free refrigerators or in overheated laboratories, greatly increases spore resistance (10). Conversely, BIs stored next to sterilizer exhaust such as ethylene oxide have been inactivated over time (6).

REFERENCES

1. **Association for the Advancement of Medical Instrumentation.** 1985. *Biological Indicators for Saturated Steam Sterilization Processes in Health Care Facilities.* Association for the Advancement of Medical Instrumentation, Arlington, Va.
2. **Boris, C., and G. S. Graham.** 1985. The effect of recovery medium and test methodology on biological indicators. *Med. Device Diagn. Ind.* **7:**43–48.
3. **Caporino, P.** 1988. Test packs. Making your own vs. buying commercially. *J. Health Mater. Manage.* **6:**32–35.
4. **Caputo, R. A., K. J. Rohn, and C. C. Mascoli.** 1980. Recovery of biological indicator organisms after sublethal sterilization treatment. *J. Parenter. Drug Assoc.* **34:**394–397.
5. **Doyle, J. E., and B. R. Ernst.** 1968. A possible source of contamination in sterility testing—alcohol flaming. *Tech. Bull. Reg. Med. Technol.* **39:**29–30.
6. **Doyle, J. E., and R. E. Ernst.** 1968. Influence of various pretreatment on the destruction of *Bacillus subtilis* var. *niger* spores with ethylene oxide. *J. Pharm. Sci.* **57:**433–436.
7. **Gurevich, L., J. E. Holmes, and B. A. Cunha.** 1982. Presumed autoclave failure due to false positive spore strip tests. *Infect. Control* **3:**388–391.
8. **Kotilainen, H. R., and N. M. Gantz.** 1985. Biological sterilization monitors: a four year evaluation of two systems. *Infect. Control* **6:** 451–455.
9. **Pflug, I. J., G. M. Smith, and R. Christensen.** 1981. Effect of soybean casein digest media on the number of *Bacillus stearothermophilus* spores. *Appl. Environ. Microbiol.* **42:**226–279.
10. **Reich, R. R.** 1982. Influence of environmental storage and relative humidity on biological indicator resistance, validity, and moisture content. *Appl. Environ. Microbiol.* **43:** 609–614.
11. **U.S. Pharmacopeia.** 1980. *The Pharmacopeia of the United States of America,* vol. XIX. *U.S. Pharmacopeial Convention,* p. 711–713. Mack Publishing Co., Rockville, Md.
12. **Whitbourne, J., and K. West.** 1976. Sterility testing: how appropriate for the hospital. *Med. Instrument.* **10:**291–292.

APPENDIX 11.4–1

Use of Indicators in Sterilization Processes

Physical monitors refer to time–temperature–pressure-recording charts. These charts indicate that the sterilization cycle has occurred. Chemical indicators consist of dye-impregnated tapes or papers that change color on exposure to the desired temperature or agent in combination with other cycle conditions. These monitors indicate that an item was at least processed through a sterilizing cycle. The BI is believed by many to be the best monitor of the sterilization process, since it most directly measures the desired outcome, i.e., the destruction of all microbial life. No one monitor alone can ensure users that goods are sterile; all monitors are part of a comprehensive QA program. All major accrediting agencies require the use of

Table 11.4–A1 Organizations' recommendations on use of BIs

Agency (reference)	Frequency of use		Quarantine until BI results available?	Subculture positive BIs?
	Steam	Ethylene oxide		
Association for the Advancement of Medical Instrumentation (1, 2)	Not less than weekly, preferably daily	Each load	Ethylene oxide-sterilized implantables	Yes
Association of Operating Room Nurses (3)	Daily and with each load of implantables	Each cycle		
Department of the Army (*Regulations AR40-19,* 1972)	Once a week	Each cycle		
CDC (4)	At least weekly and for each load containing implantable objects	At least weekly and for each load containing implantable objects	Implantable objects (ethylene oxide or steam sterilized)	
JCAHO (5)	The organization should establish a policy for consistent sterilization monitoring.			
Veterans Administration (*VA Manual MP-2,* Change 107, July 21, 1975)	Once a week	Each cycle		

BIs as well as physical and chemical monitors. The specifications of these agencies for frequency of use of BIs are listed in Table 11.4–A1.

Variations on the traditional gravity displacement steam sterilization cycle of 121°C for 15 min at 15 lb/in^2 have caused changes in practice. To increase processing speed, the prevacuum sterilizer is used in many central sterile service departments. This sterilizer allows a small but steady influx of steam into the chamber during evacuation, with free air withdrawn by a vacuum pump. Usual prevacuum cycle parameters are 3 to 5 min at 132 to 135°C and 30 to 33 lb/in^2. To rapidly sterilize needed instruments in surgery, the flash sterilizer, which omits the vacuum portion of the cycle and maintains a steady pressure in the jacket to inject steam rapidly, is employed. These units are often referred to as emergency sterilizers and should be considered such. Because items are unwrapped when placed in the chamber to maximize steam contact, recontamination may occur when items are removed.

BI systems cannot be interchanged among different sterilization processes. Specific spore preparations have been developed for steam, ethylene oxide, and other systems as well as for flash sterilization. Product labels should be checked to ensure that the proper BI is being used for each application. Product inserts will provide information on spore resistance and density. Recommendations for plasma and chemo-sterilants are under development at this time.

3M Co. has introduced a biological indicator system that utilizes the ability of *B. stearothermophilus*-associated enzymes to convert a substrate in the growth medium to a fluorescent by-product (the 1291 Attest Rapid Readout Biological Indicator for 270°F [132°C] gravity displacement steam sterilization). It is a self-contained system that contains at least 10^5 spores of *B. stearothermophilus*. Processed BIs may be read after a 1-h incubation at 140°F (60°C) by use of the ''auto-reader'' provided by the manufacturer. A green light indicates sterilization, while a red light indicates a failure. The growth medium also contains a pH indicator (bromcresol purple) that demonstrates inadequate sterilization with a yellow color change within 24 h of incubation. The manufacturer recommends that all BIs that are negative at 1 h be incubated until a positive color change is noted or 24 h elapses.

APPENDIX 11.4–1 *(continued)*

References

1. **Association for the Advancement of Medical Instrumentation.** 1993. *Good Hospital Practice: Steam Sterilization and Sterility Assurance.* Association for the Advancement of Medical Instrumentation, Arlington, Va.
2. **Association for the Advancement of Medical Instrumentation.** 1992. *Good Hospital Practice: Ethylene Oxide Sterilization and Sterility Assurance.* Association for the Advancement of Medical Instrumentation, Arlington, Va.
3. **Association of Operating Room Nurses.** 1996. Recommended practices for sterilization and disinfection. *Assoc. Oper. Room Nurs. J.* **45:**271–282.
4. **Garner, J. S., and M. S. Favero.** 1985. Guide line for handwashing and hospital environmen tal control, p. 10–13. *In Guidelines for the Pre vention and Control of Nosocomial Infections* CDC, Atlanta.
5. **JCAHO.** 1997. *Accreditation Manual for Hos pitals,* 1997 ed. American Hospital Associa tion, Chicago.

APPENDIX 11.4–2

Addresses of Suppliers*

Attest, 1291 Attest:
Medical Products Division
3M Co.
St. Paul, Minn.

Chemspor, Proof, Proof Plus:
American Sterilizer Corp.
Medical Products Division
Erie, Pa.

Tec-Test:
Castle/Sybron
Division of Sybron Corp.
Rochester, N.Y.

* Other products are available; these products may be available through your hospital supply distributor.

SECTION 12

Instrument Maintenance and Quality Control

SECTION EDITORS: *Phyllis Della-Latta and Ernestine M. Vellozzi*

12.1 General Guidelines

I. PRINCIPLE

All instruments located in clinical microbiology laboratories must undergo appropriate preventative maintenance and QC as dictated by the manufacturer's instructions and established standards. This section will address instruments common to most laboratories. For information on other equipment, see the comprehensive coverage in CMPH section 12 and the instrument manual published by CAP (1).

II. MAINTENANCE PROGRAM

A. A preventative maintenance program is essential to ensure the accuracy and longevity of instruments. Scheduled periodic performance checks as recommended by the manufacturer are important to minimize instrument breakdown and the need for service and repair.

B. The laboratory director is responsible for the program. However, service contracts place some of the responsibility back with manufacturers, and in-house biomedical departments and clinical microbiology personnel can be assigned certain levels of maintenance responsibility. This delegation of responsibility depends on the reliability of the hospital biomedical facility, the complexity of the instruments involved, and the affordability of the service agreement.

III. STANDARD OPERATING PROCEDURE MANUAL

A. The Standard Operating Procedure Manual (SOPM) for equipment is best organized in sections (e.g., Incubators, Refrigerators, etc.). A line listing of all equipment must specify the serial numbers, hospital identification numbers, and specific locations in the laboratory.

B. New equipment requires inspection by the biomedical department to ascertain electrical safety; this inspection includes leakage tests, ground test (Ohms), and assessment of the necessity for dedicated lines.

C. Follow NCCLS guidelines (2) to format the manual, including annual review by the laboratory director (signature and date).

D. All records of preventative maintenance checks and instrument failures are included in the manual. Each incident must be documented and all records retained for the life of the instrument.

E. The manual must be kept in a location where it is easily retrievable.

IV. VALIDATION PROCEDURES

A. New equipment most critical to patient care (e.g., automated antibiotic susceptibility testing and blood culture instruments) requires validation studies to determine whether its performance is at least comparable to that of existing equipment or to that of a reference facility.

IV. VALIDATION PROCEDURES *(continued)*

B. Regulatory agencies may vary in their requirements for a minimum number of patient comparisons (e.g., New York State expects at least 25; CAP recommends 40) [1]. The new test is adopted only if results are equal to or better than the existing or ''gold'' standard.

C. Validation records must be kept for the life of the instrument. Should there be more than one identical instrument used for the same assay (e.g., more than one Parallel Processing Center [Abbott Instruments]), there are requirements in some states (e.g., New York) that parallel tests be performed twice a year.

V. INSTRUCTIONS FOR INSTRUMENT USE

A. Clearly written instructions for the operation of equipment must be incorporated in the SOPM (manufacturers' manuals and product inserts are for reference use only).

B. In-service instruction by the manufacturer, either off- or on-site, is recommended for operation of critical instruments. Documentation of personnel training is required.

C. Each protocol must include safety precautions when appropriate along with procedures for cleaning and care of the instrument.

D. Preprinted forms for QC data management are critical for an organized program. Documentation required includes the date and time of incident, evaluation of the problem, corrective action taken, resolution, and follow-up results. Appropriate signatures and dates are necessary, including that of the laboratory director, who is responsible for the final review. Periodic or yearly summaries of all equipment failures are advised to consider the replacement of faulty instruments that may be too costly to maintain. Appendix 12.1–1 has an example of format for a report to be completed when there is an error on the part of an employee who is using an instrument. Similar documentation is required.

VI. CALIBRATION OF INSTRUMENTS

Equipment that requires an exact level of precision to obtain accurate test results requires periodic calibration. The calibration, its frequency, and results should be documented by the person performing the analysis, and records should be kept in the laboratory.

VII. CLEANING PROCEDURES

A. Follow manufacturers' instructions for cleaning equipment precisely, and incorporate these procedures into the SOPM.

B. External and internal surfaces of equipment composed of stainless steel or another metal can be disinfected with 10% household bleach (4–5% NaOCl). Rubber gloves must be worn, and towels may be used. Should the manufacturer indicate that the instrument is susceptible to corrosion, rinse it with water after disinfection.

C. Terminal cleaning must be performed before an instrument leaves the microbiology laboratory, by thoroughly disinfecting with an appropriate disinfectant such as 0.5% amphyl (a phenolic).

VIII. QUALITY CONTROL SCHEDULE FOR INSTRUMENTS

See Appendix 12–1 for details.

EFERENCES

1. **CAP.** 1991. *Laboratory Instrumentation, Verification, and Maintenance,* 4th ed. CAP, Northfield, Ill.
2. **NCCLS.** 1984. *Clinical Laboratory Procedure Manuals. Approved Guidelines.* NCCLS document GP2-A. NCCLS, Wayne, Pa.

UPPLEMENTAL READING

August, M. J., J. A. Hindler, T. W. Huber, and D. L. Sewell. 1990. *Cumitech 3A, Quality Control and Quality Assurance Practices in Clinical Microbiology.* Coordinating ed., A. S. Weissfeld. American Society for Microbiology, Washington, D.C.

Association for the Advancement of Medical Instrumentation. 1986. *AAMI Standards and Recommended Practices: 1987 Reference Book.* Association for the Advancement of Medical Instrumentation, Arlington, Va.

PPENDIX 12.1–1

Example of a record-keeping form to report a laboratory problem or error (from CMPH 12.1.8).

REPORT OF LABORATORY PROBLEM/ERROR

Problem:
Steps taken to solve the problem:
Corrective action to be taken to avoid this problem in the future:
Additional comments:

Prepared by:	Date:
Reviewed by:	Date:

12.2 Instruments Requiring Temperature Monitoring: Freezers, Heat Blocks, Refrigerators, Slide Warmers, and Water Baths

I. PRINCIPLE

Some equipment, such as freezers, heat blocks, refrigerators, slide warmers, and water baths, is simple to operate and requires little maintenance with the exception of temperature monitoring. See CMPH procedures 12.10, 12.18, and 12.21 for additional information on the instrumentation discussed in this section.

II. CHARTS

A. An action and preventative maintenance chart example is provided in Appendix 12.2–1.

B. Temperatures are recorded on other preprinted forms daily or on each day the instrument is used, and this record is placed on the instrument in a plastic envelope with magnetic or adhesive backing. Information on the form must include the acceptable temperature range and the type of instrument and instrument number, and there must be boxes to record the exact degrees and the initials of the recorder and reviewer. Space for explanations of QC failures and corrective action must be provided. Since the corrective action for an instrument is often standard, the form might contain a selection of the usual steps taken for resolution of such problems, to save the technologist writing time and eliminate problems caused by illegible handwriting.

SUPPLEMENTAL READING

CAP. 1991. *Laboratory Instrumentation, Verification, and Maintenance,* 4th ed. CAP, Northfield, Ill.

APPENDIX 12.2–1

Example of form to record preventative maintenance and corrective action (*see* p. 699) (adapted from CMPH 12.1.6).

12 Instrument Maintenance and Quality Control

PREVENTIVE MAINTENANCE/CORRECTIVE ACTION

PREVENTIVE MAINTENANCE FUNCTIONS FREQUENCY*

Instrument:

ID code:

Location:

JANUARY		FEBRUARY		MARCH		APRIL	
Date/Initial	Procedure	Date/Initial	Procedure	Date/Initial	Procedure	Date/Initial	Procedure
MAY		JUNE		JULY		AUGUST	
Date/Initial	Procedure	Date/Initial	Procedure	Date/Initial	Procedure	Date/Initial	Procedure
SEPTEMBER		OCTOBER		NOVEMBER		DECEMBER	
Date/Initial	Procedure	Date/Initial	Procedure	Date/Initial	Procedure	Date/Initial	Procedure

Problem/Corrective action/Follow-up:

*D, daily; W, weekly; BW, biweekly; M, monthly; Q, quarterly; Y, yearly.

12.3 Equipment Requiring Calibration: Balances, pH Meters, Centrifuges, Pipettes, Rotators, and Thermometers

I. PRINCIPLE

Equipment whose precision is critical to the accuracy of a diagnostic test, such as balances, pH meters, centrifuges, pipettes, rotators, and thermometers, must be calibrated immediately upon delivery, before use, and periodically at intervals dictated by the manufacturer, regulatory agencies, or peer recommendations.

II. PROTOCOL CHARTS

A. Table 12.3–1 lists commonly used equipment requiring calibration. It summarizes the principle, procedure, and frequency of testing for each type. For details on these and other instruments, see CMPH procedures 12.6, 12.15, 12.16, 12.17 and 12.20.

B. Calibration can be performed via service contracts or by hospital-based biomedical teams or laboratory personnel. The choice depends on the nature of the instrument, the skill and reliability of staff, and financial considerations.

C. The calibration results must be provided to the laboratory for their permanent record, documenting optimal functioning of instruments essential to quality diagnostic testing (*see* Appendix 12.3–1).

Table 12.3–1 Principle and frequency of testing of common equipment requiring calibration

Instrument	Principle	Protocol
Balances	Single-pan balances are based on built-in weights and balance beams. Analytical balances are more precise and reserved for assays requiring exact measurement, e.g., testing of antibiotics for macrodilution broth susceptibility test.	1. Place in location free of vibration, temperature fluctuations, and humidity. 2. Clean with mild soap solution daily. 3. Protective plastic covers are recommended. 4. Schedule preventative maintenance yearly.
Centrifuges	Centrifugal force is used to separate microorganisms from fluid. Refrigeration is required if high speeds are used. The number of revolutions per minute is not equivalent to the relative centrifugal force, which is expressed in gravities (g) since the rotor radii of centrifuges can differ.	1. Verify temperature of refrigerated centrifuge; check brushes semiannually. 2. Calibrate speed annually.
Pipettes	Air displacement or positive pressure pipetters are used when precise volumes of fluid are required. This is especially critical when performing molecular assays.	Calibration can be performed by the manufacturer or an outside contractor.
Thermometers	Temperatures for equipment used to perform assays or store reagent samples and organisms have optimal ranges for required functions. The accuracy can vary among thermometers; therefore, each must be calibrated with a standard.	1. Use a standard reference thermometer to establish accuracy when each new thermometer is put in place. 2. Thereafter, testing is done periodically and as dictated by regulatory agencies.

UPPLEMENTAL READING

CAP. 1991. *Laboratory Instrumentation, Verification, and Maintenance,* 4th ed. CAP, Northfield, Ill.

PPENDIX 12.3–1

Example of record sheet for centrifuge quality control (adapted from CMPH 12.6.13).

ROUTINE CENTRIFUGE QUALITY CONTROL RECORD

YEAR__________________

Instrument: Centrifuge	Model # Serial #	Ident. #	Date in use:	Section: Room #:

Note: Sign initials in appropriate box after maintenance procedure is completed.

Date	Check temp.	Check speed calibration	Problem and corrective action

12.4 Incinerator Burner

I. PRINCIPLE

The incinerator burner sterilizes through infrared heat. A wire loop or needle with organic material on the surface is inserted deep into the ceramic funnel tube housed in a heat-dissipating metal container. After 5 to 7 s at the optimum sterilizing temperature of 815°C, the material is incinerated. Spattering of infectious microorganisms and the risk of cross-contamination are both prevented. Electric incinerator burners are particularly useful for working in laminar-flow biosafety cabinets and in anaerobic chambers. The absence of an open flame provides a safer alternative to the Bunsen burner. See CMPH Procedure 12.11 for additional information on the incinerator burner.

II. ROUTINE OPERATION

A. Plug electrical cord into the appropriate outlet. A two-prong adapter may be used, but the unit must be properly grounded.

B. The indicator light will come on when the toggle switch is in the ''ON'' position.

C. Gently insert the loop or needle, attached to an insulated loop holder, into the heater element. Avoid scraping the sides of the element, to ensure the longevity of the loop and heater element. Insert the loop deep into the heater element to avoid spattering. Allow the loop to remain within the heater element for a minimum of 5 but no longer than 30 s. It is not necessary to obtain a glowing loop to ensure sterility. Microorganisms (bacteria, fungi, and mycobacteria) will be destroyed within 5 s.

D. The optimum sterilizing temperature (815°C) will be reached 10 min after the unit is plugged in. Turn the incinerator on at the beginning of the workday and leave it on throughout the day. The unit may be left on indefinitely without posing an electrical or fire hazard. If the unit is not going to be used for extended periods, such as for more than 4 h or overnight, turn it off to conserve the life of the heater element, as recommended in CMPH procedure 12.11.

III. QUALITY CONTROL

A. Initial calibration

1. Plug the incinerator into a grounded electrical circuit.
2. Look for a red glow inside the heater element.
3. Be sure that the unit does not generate any smoke or persistent odor suggestive of burning rubber.

B. Routine quality control

1. Inspect the heater element daily to determine whether the core is worn.
2. Inspect for small cracks in the ceramic casing and residue buildup during both cool and heated conditions. In the heated condition, *cracks can be seen as small, intensely yellow-orange fissures.*
3. Document all findings on a recording sheet as described in CMPH section 12.
4. Replace the heater element if any defects are noted.

   ◪ **NOTE:** Cracks do not inhibit the sterilization ability of the unit, but can create an electrical safety hazard.

V. TROUBLESHOOTING

A. Failure to heat

1. Check to see that the instrument is switched on.
2. Make sure the instrument is plugged in.
3. Check for loose or disconnected wires after unplugging the unit.
4. Check for cracks or fissures in the heating element that indicate that the heating element needs to be replaced.

B. Instrument smoking

1. Check for loose or disconnected wires after unplugging and cooling the instrument.
2. Check for cracks or fissures in the heating element. If present, change the heating element.
3. Look inside the tube of the heating element to see whether any debris is lodged inside. If so, let the debris burn off.

12.5 Incubators

I. PRINCIPLE

An incubator is an insulated cabinet that is essential for maintaining a temperature- and moisture-controlled environment to allow the growth of microorganisms. The temperature must be carefully monitored. Some incubators are equipped to maintai a desired level of CO_2 for capnophilic mi croorganisms. The reader is referred to th comprehensive coverage of incubators i CMPH procedure 12.12.

II. SPECIMENS

Inoculated culture media.

III. PROCEDURE

A. Non-CO_2 incubators

1. Set thermoregulator to desired temperature.
2. Once the desired temperature is achieved, record temperature on each da of use on a QC sheet (*see* Appendix 12.5–1 of this section). As per CA guidelines, continuous monitoring is not required (1).
3. Place samples securely on shelves or trays.
4. A humid environment can be maintained by placing a water-filled pan a the bottom of the incubators, reflecting the size of the chamber.

B. CO_2 incubators

Temperature and CO_2 levels are recorded on a QC sheet, on each day of use

IV. RESULTS

After appropriate incubation time, media are observed for adequate growth of mi croorganisms, e.g., *Neisseria gonorrhoeae* in the CO_2 incubator.

V. MAINTENANCE

A. All incubators are cleaned monthly with a mild soap solution.

B. Preventative maintenance should be performed approximately four times pe year by a service company.

C. For safety purposes, CO_2 tanks must be secured to a wall in an upright position with heavy chains. When cylinders are not in use, keep both the valves an caps tightly closed. Empty cylinders can be stored by securely chaining then on a movable gas cylinder cart. Never store gas cylinders at temperatures o >125°F (52°C). Never place a gas cylinder in a horizontal position.

VI. QUALITY CONTROL

A. Non-CO_2 incubators

1. A supervisor must be notified when an incubator temperature falls outsid the acceptable range posted for the unit.

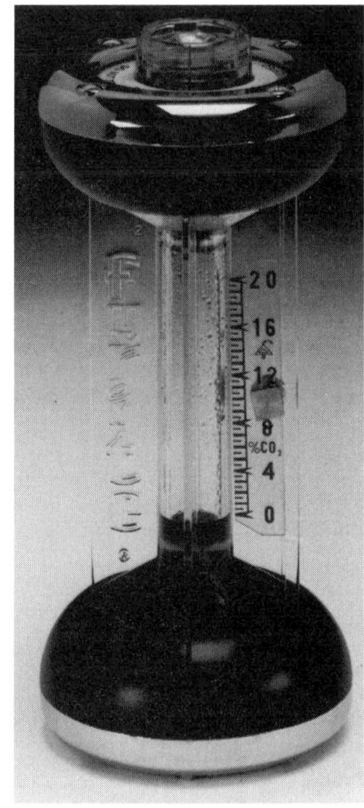

Figure 12.5–1 Fyrite gas analyzer for measuring CO_2 concentration (courtesy of Bacharach Instruments, Pittsburgh, Pa.; reprinted from CMPH procedure 12.12).

2. Corrective action should proceed as follows:
   a. Check power source, unit plug, and circuit panel.
   b. Check temperature set point.
   c. If the unit still malfunctions, call service representative.

B. CO_2 incubators

1. A culture of *Neisseria gonorrhoeae* ATCC 43069 is placed in the incubator. Every day it is subcultured and checked for growth. This organism has an absolute CO_2 requirement.
2. Weekly, the CO_2 level in the incubator is checked with the Fyrite Test Kit (*see* Fig. 12.5–1). The procedure is as follows:
   a. Calibrate Fyrite (*see* CMPH section 12) and adjust the zero on the scale to the top of the fluid column by loosening the locknut at the back of the scale, sliding the scale into place, and retightening the locknut.
   b. Remove the cover of the atmospheric sampling port on the outside of the incubator door, and insert the brass tube at the end of the Fyrite sampling assembly.
   c. Hold the Fyrite upright and away from the face, and vent it by pressing and then releasing the plunger valve (button at top).
   d. Stand the Fyrite upright on a level surface.
   e. Use the heel of your hand to press the round rubber end of the black tubing firmly down onto the top of the Fyrite plunger valve. At the same time, squeeze the black rubber bulb 18 times. Remove your hand from the plunger valve at the same time that the bulb is squeezed for the 18th time.
   f. Invert the Fyrite slowly so that the fluid will run to the opposite end and then slowly turn it upright.
   g. Repeat step ‘‘f’’; hold the Fyrite at a 45° angle for a few seconds, stand upright for a few seconds, then read the percent CO_2 from the scale. Each division is 0.5% CO_2.
   h. Be sure to record the Fyrite reading on a QC sheet.

REFERENCE

1. **CAP.** 1991. *Laboratory Instrumentation, Verification and Maintenance,* 4th ed. CAP, Northfield, Ill.

APPENDIX 12.5–1

Example of record-keeping form for comprehensive incubator QC (from CMPH 12.12.13).

COMPREHENSIVE INCUBATOR QUALITY CONTROL RECORD FORM

Month______________ Year________ Laboratory_____________

Date	CO_2 incubator			Air inc. temp.	42° inc. temp.	30° inc. temp.	Other inc. temp.	Other inc. temp.	Init. of reader
	Temp	% CO_2	% Humid.						
1									
2									
3									
4									
5									
6									
7									
8									
9									
10									
11									
12									
13									
14									
15									
16									
17									
18									
19									
20									
21									
22									
23									
24									
25									
26									
27									
28									
29									
30									
31									
Corrective action:									

12.6 Biological Safety Cabinets

I. PRINCIPLE

Biological safety cabinets are utilized for operator protection when working with class II pathogens, such as *Mycobacterium tuberculosis*. Safety cabinets used in clinical microbiology laboratories are either class I or class II, types A or B. These cabinets can come equipped with UV light to serve as an adjunct aid to surface disinfection.

II. PROTOCOL

A. Switch on fluorescent light.

B. Turn on blower and allow unit to purge for approximately 15 min.

C. Do not block air intake grills.

D. Spills: An overt spill confined to the interior of the cabinet should be disinfected with 0.5% amphyl or other appropriate disinfectant while the system is still operating.

■ *The UV light in the cabinet must be turned off before use.*

E. Avoid use of an open flame within the cabinet (use a microincinerator).

III. MAINTENANCE AND QUALITY CONTROL

A. Service contracts check the following:

1. UV lamp and replacement of bulbs
2. Adjustment and record of air flow
3. Filter and cabinet leakage
4. Adjustment of pressure gauge
5. Sound level and illuminator function
6. Cleaning or replacement of filters
7. Smoke pattern test
8. Alarm system function

B. The interior of the cabinet is thoroughly cleaned weekly, including back walls and plastic shield. Tasks must be documented (*see* Appendix 12.6–1). The surface work area is cleaned with 0.5% amphyl solution daily. The interior of the cabinet may be cultured weekly before cleaning by swabbing and placing the swab in TSB. ''Satisfactory'' means no fungi or large numbers of other microorganisms are present. See CMPH procedure 12.19 for additional information on biological safety cabinets.

SUPPLEMENTAL READING

Kruse, R. H., W. H. Puckett, and J. H. Richardson. 1991. Biological safety cabinetry. *Clin. Microbiol. Rev.* **4:**207–241.

APPENDIX 12.6–1

Example of record-keeping form for safety cabinet QC and maintenance (from CMPH procedure 12.19.16).

Month____________ Year________

BIOLOGICAL SAFETY CABINET QC AND MAINTENANCE RECORD

CABINET NO.________________________

LOCATION________________________

CLASS__________

Day	Visual air flow check	Blower/ gauge reading	Cleaning	Other
1				
2				
3				
4				
5				
6				
7				
8				
9				
10				
11				
12				
13				
14				
15				
16				
17				
18				
19				
20				
21				
22				
23				
24				
25				
26				
27				
28				
29				
30				
31				

12.7 GasPak Systems

I. PRINCIPLE

GasPak systems are self-contained systems used to produce atmospheres required to support the primary isolation and cultivation of anaerobic or microaerophilic bacteria with gas-generating envelopes. The jar and lid used in each of these systems must be checked periodically for proper performance.

II. DESCRIPTIONS OF SELECTED SYSTEMS

A. Self-contained atmosphere-generating systems

1. Generator envelopes (anaerobic, microaerophilic, or carbon dioxide) are commercially available in rigid jars for specific use (Fig. 12.7–1).
2. An anaerobic gas generator envelope contains one tablet of sodium borohydride, one tablet of sodium bicarbonate, and citric acid. A cold catalyst (palladium-coated alumina pellets) is required for the reaction to occur. Addition of 10 ml of water causes the resulting acid solution and other components to mix, generating hydrogen, water vapor, and CO_2. Heat is generated by the catalyst. Some newer (more expensive) systems incorporate the catalyst on the envelope, so that separate handling of the catalyst is not required.
3. CO_2-generating envelopes contain one tablet of sodium bicarbonate and citric acid. The addition of 10 ml of water generates 5 to 10% CO_2 in a small GasPak jar (2.5-liter volume, 12-cm inside diameter; Becton Dickinson Microbiology Products) that accommodates 12 100-mm-diameter petri dishes.
4. Alka-Seltzer (one-half tablet; Miles Laboratories, Elkhart, Ind.) with 5 ml of water generates approximately 5% CO_2 in a small GasPak jar.

B. Oxygen removal and CO_2 generation

1. The candle jar uses combustion to remove oxygen from the atmosphere, forming water and CO_2 in a closed jar. It is used to create a CO_2-rich atmosphere (approximately 3 to 5% CO_2) and oxygen tension lower than that of room air. One side benefit is increased humidity in the system. Although old anaerobic jars are most often used for candle jars, any glass or plastic jar with a tight-fitting lid and a mouth wide enough to admit a petri plate (such as a commercial mayonnaise jar) can be used.

   ◪ **NOTE:** Do not use scented candles because they may create toxic vapors.
2. An iron-filing system relies on formation of iron oxide (which removes oxygen from the atmosphere) after addition of water. It is used primarily to generate an anaerobic environment in small, sealed plastic bags.

II. DESCRIPTIONS OF SELECTED SYSTEMS *(continued)*

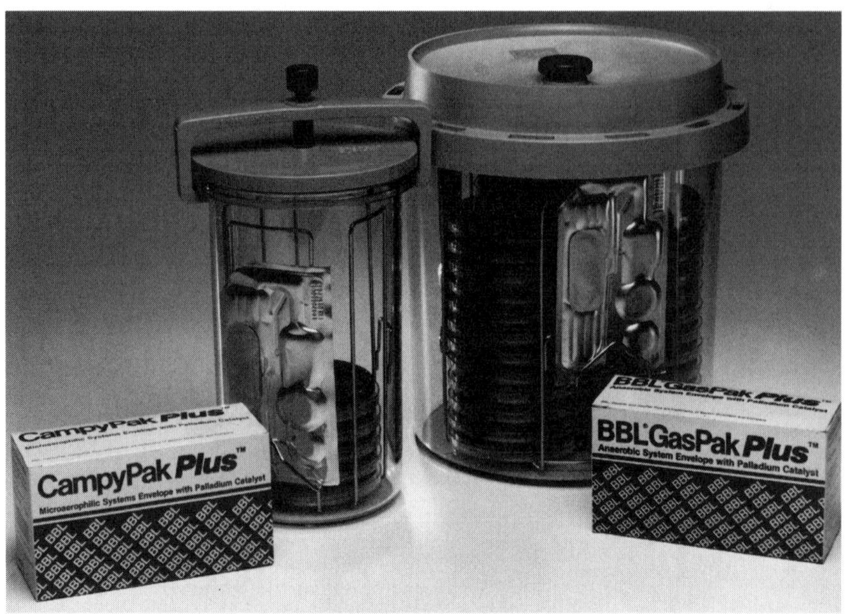

Figure 12.7–1 Two sizes of rigid plastic jars, showing placement of gas-generating envelopes with self-contained catalysts (photograph courtesy of Becton-Dickinson Microbiology Products, Becton-Dickinson Co., Cockeysville, Md.; reprinted from CMPH 12.13.2).

C. Systems using external gas source

Evacuation-replacement systems use a venting apparatus with a rigid container or a pipette held into the squeezed opening of a plastic bag to replace air with the atmosphere of choice. With rigid containers, a vacuum must be drawn to remove air; simply pushing out the air is sufficient with plastic bags. Three atmosphere exchanges are usually used.

D. Uses in microbiology

1. Primary plate incubation (all atmospheres)
   a. Microaerophilic atmosphere for *Campylobacter* species
   b. Anaerobic atmosphere primarily for strict anaerobes
   c. Carbon dioxide-enriched atmosphere primarily for *Neisseria* and *Haemophilus* species, certain microaerophilic streptococci, and other capnophiles
2. Maintenance of specimens (such as tissues) under anaerobic atmosphere

III. PROTOCOL

A. Place inoculated plates in GasPak rack.

B. Peel open the corner of the GasPak envelope along the dotted line and place it in the clip of the rack. Also place an indicator strip in the clip provided in the rack.

C. Add 10 ml of tap or distilled water through the open corner of the envelope.

D. Place the GasPak inside a jar and securely tighten the jar lid.

E. Indicator strip should change color from blue to clear within 6 h with GasPak Plus generator envelopes, but with CampyPak Plus envelopes, visible condensation within 60 min indicates proper functioning.

F. Some microbiologists use an agar plate inoculated with *Clostridium novyi* B (ATCC 19402) as an indicator organism with each use of the jars; bacteria can be maintained in anaerobic broth.

IV. MAINTENANCE AND QUALITY CONTROL

A. Periodically, plastic lids and clamps of jars should be checked for cracks, chips, split O-ring grooves, and other irregularities. Each O-ring gasket should be examined for cracks, bumps, dirt, and improper seating in the O-ring groove. The O-ring must be resilient, clean, and snugly retained within the O-ring groove. Replace any part as necessary.

B. QC organisms are added to the jar with the expected results:

QC strain	ATCC number	Result
Clostridium perfringens	13124	Growth
Micrococcus luteus	9341	No growth
Campylobacter jejuni	33291	Growth

See CMPH 12.13 for additional information on GasPak Systems.

12.8 Autoclave (Steam Sterilizer)

I. PRINCIPLE

The basic function of a steam sterilizer is to provide a means of treating the whole of the material to be sterilized with saturated steam at a given temperature for a specified length of time. Heating uses saturated steam under pressure of approximately 20 psi to achieve a chamber temperature of at least 121°C for a specified time.

II. PROCEDURE

A. Loading the autoclave

1. Load the autoclave to facilitate flow of air down the drain and maximum circulation of steam around and penetration into contents.
2. Place empty or dry containers on their sides or upside down.
3. Stack several items side by side vertically rather than horizontally, so that steam can circulate freely around each item.
4. If horizontal stacking is necessary, use a wire rack to create a shelf that allows free flow of steam.
5. Leave at least 2 in. (ca. 5 cm) of space between objects in the chamber and between objects and the chamber walls and doors.

B. Sterilization of media and solutions

1. Use only flasks and self-venting, automatic-sealing stoppers recommended by the manufacturer for processing liquids in a steam sterilizer.
2. Use a chemical indicator to monitor each load.
3. Start measuring sterilization time for medium solutions from the moment the temperature reaches 121°C. Suggested times are 500 ml, 18 min; 1,000 ml, 21 min; and 1,500 ml, 24 min. As a general rule, add 3 min more for each 500 ml of medium.
4. Follow the instructions provided by the supplier of each medium, since most media have specific time and temperature requirements. Do not exceed the sterilization time and temperature recommended by the manufacturer.

C. Sterilization of liquids (agars and broths)

1. Check that all caps or lids are loose.
2. Never fill a vessel more than two-thirds full.
3. Close the autoclave, and set time and temperature; 15 min at 121°C with 15 min of exhaust time is recommended, but follow manufacturer's instructions.
4. Start the cycle. Wait until temperature and pressure indicators are functioning before leaving the area.
5. After the cycle has ended, chamber pressure reaches zero, and temperature falls below 60°C, put on heat-resistant gloves and eye protection and open the door slowly, standing behind the door to avoid escaping steam.

6. Wearing gloves, gently remove containers of liquid after autoclaving. Rapid changes in temperature may cause glass or contents to boil up.

D. Sterilization of contaminated materials

1. Transport contaminated materials to the autoclave room in double autoclavable bags marked with the biohazard label.
2. Before bags are autoclaved, loosen their ties to ensure penetration of steam to all parts of the load, or add 1 cup (ca. 0.3 liter) of water to sealed bags.
3. Place bags to be autoclaved in trays to prevent melted agar from clogging the chamber drain.
4. Place a piece of autoclave chemical tape on each item to indicate that the contents have been sterilized.
5. When running the weekly biological-sterilizer indicator, place the test ampoule or strip where it can be retrieved from deep inside the bag after the cycle.

E. Sterilization of dry, wrapped materials

1. Arrange packs to allow maximum steam circulation. Do not allow any item to touch the wall of the chamber.
2. Close autoclave and set time and temperature; 25 min at 121°C with fast exhaust and 30-min drying cycle are recommended.
3. Start cycle. To confirm that the door gasket does not leak and that operations are progressing, wait until temperature and pressure begin to rise before leaving the area.
4. After the chamber pressure has reached zero and the temperature has fallen below 60°C, put on heat-resistant gloves and eye protection and open the door slowly, standing behind the door to avoid the escaping steam.

III. QUALITY CONTROL AND MAINTENANCE

See Table 12.8–1 for a list of routine QC and maintenance steps for the autoclave.

Table 12.8–1 Quality control and maintenance[a] schedule for the autoclave

Procedure	Per run	Daily	Weekly	Monthly	As required
1. Clean plug screen		X			
2. Examine recorder plot	X				
3. Use chemical indicator	X				
4. Use biological indicator			X		
5. Record date, time, and temperature on recorder chart paper		X			
6. Clean recorder pan (recording thermometer)				X	
7. Check generator water level		X			
8. Check safety valve			X		
9. Clean interior and exterior					X[b]
10. Check/replace door gasket					X[b]
11. Clean out wastewater drain					X[b]
12. Check recording thermometer accuracy			X		
13. Clean drain and seals			X		

[a] It is best to leave major maintenance tasks to experts. A maintenance contract with the manufacturer or an industrial firm familiar with autoclaves is suggested (1). Certain maintenance functions on a daily or more frequent basis must be carried out by the user.

[b] At least every three months (1).

IV. TROUBLESHOOTING

See Table 12.8–2 for a list of potential problems and suggested solutions.

Table 12.8–2 Trouble analysis chart[a]

Problem	Possible causes	Solution
1. Door does not latch when fully closed.	a. Object restricting b. Cables too loose c. Door latch out of adjustment	a. Remove object. b. Adjust cables. c. Adjust latch.
2. Recorder motor is not operating.	a. Blown control circuit fuse b. Defective recorder motor	a. Replace fuse. b. Replace motor.
3. Jacket is not heating	a. Steam supply and shut off valves not open b. Strainer plugged c. Regulator valve not working	a. Open valves; clean or repair. b. Clean strainer. c. Adjust or clean regulator.
4. Steam does not pressurize chamber.	a. Pressure regulator not working b. Chamber steam trap not functioning c. Door gasket leak	a. Clean or repair regulator. b. Clean or repair trap. c. Clean, lubricate, or replace gasket.

[a] For complete information, always consult the owner's manual or call a repair service.

REFERENCE

1. **CAP.** 1991. *Laboratory Instrumentation, Verification and Maintenance,* 4th ed. CAP, Northfield, Ill.

12.9 Anaerobic Chamber

PRINCIPLE

Anaerobic chambers are used in clinical laboratories to process specimens being investigated for the presence of anaerobic bacteria, to incubate and maintain cultures, and to store media for anaerobic bacteria. These chambers are not absolutely required for anaerobic culturing, but they are recommended for the isolation of anaerobic bacteria, especially fastidious organisms, and for the processing of specimens that require grinding. See CMPH procedure 12.2 for further information on this instrument.

12.10 Automated Instruments

I. MICROBIAL IDENTIFICATION AND SUSCEPTIBILITY TESTING SYSTEMS

While a full description of each instrument is beyond the scope of this chapter, Table 12.10.1 lists currently available instrument systems and briefly summarizes the salient characteristics of each. Performance features among systems and even within a given system may vary considerably, and performance is directly proportional to the investment in mechanized devices and attendant computer hardware and software. See CMPH procedure 12.4 for a more detailed discussion of these instruments.

Table 12.10–1 Automated and semiautomated systems for antimicrobial-agent-susceptibility testing and bacterial identification[a,b]

System (company)	Test unit	Reagent format and storage	MIC panel formats	Breakpoint panel formats (S, I, R results)	Inoculation process		Assisted visual reading	
					Manual	Mechanized	Device	Test data entry
MicroScan (Dade International)	Microdilution tray	Dry, RT	GN, GP, HM, NS, Combo	GN, GP; 1 organism/tray	Renok device		Touchscan-SR	Ring pen
Pasco (Becton Dickinson)	Microdilution tray	Frozen, −20 or −70°C	GN, GP, Combo	GN as Combo; 1 organism/tray	Multiprong disposable tray lid		Semiautomated reader	Light pen
Sceptor (Becton-Dickinson)	Microdilution tray	Dry, RT	GN, GP, ANA, Combo	GN, GP as Combo; 1 organism/tray	Electronic repeating pipette	Autopreparation station	Sceptor reader-recorder	Touch pad
Sensititre (Accumed)	Microdilution tray	Dry, RT	GN, GP	GN, GP; 3 organisms/tray	Multitip pipette	Autoinoculator	Sensi-touch	Membrane keypad
Vitek (bioMerieux Vitek Inc. USA)	Multiwell card	Dry, 4°C	GN, GP	S, I, R derived from MIC calculation		Filling module		

[a] Abbreviations: ANA, anaerobes; ARIS, automated reader incubation system; AST, antimicrobial susceptibility testing; Combo, identification and antimicrobial susceptibility testing on same tray; GN, gram negative; GP, gram positive; HM, *Haemophilus* spp.; ID, identification; NF, nonfermenters; NS, *Neisseria* spp.; RT, room temperature; YS, yeast cells; S, sensitive; I, intermediate; R, resistant; WA, WalkAway. Adapted from CMPH 12.4.2–12.4.3.

[b] All systems offer custom panels.

[c] Rapid and conventional panels available.

II. BLOOD CULTURE INSTRUMENT

Blood cultures are among the most important specimens for the detection of etiologic agents of infection. The presence of microorganisms in blood usually has serious consequences for the patient. The time- and labor-intensive methods used classically for the detection of bacteria or fungi in blood, for visual inspections, and for blind subcultures can now be replaced with commercially available culture systems. Three such systems have gained widespread acceptance by microbiologists: BACTEC (Becton Dickinson Diagnostic Instrument Systems), BacT/Alert (Organon Teknika), and ESP (Difco). See CMPH procedure 12.5 for a detailed discussion of blood culturing instruments.

Table 12.10–1 *(continued)*

Automated reading of AST				Microbial identification		Incubation intervals			Software options for data management		
Device(s)	Operational reading mode	Principle(s)	Incubation mode(s)	Panel formats	Isolates/ panel	Organism group	AST (h)	ID (h)	Antibiograms	Epidemiology data base	Nondedicated software
autoSCAN 4 autoSCAN = Walk-Away[c] WalkAway 96 WalkAway 40	Attended (autoSCAN 4) Nonattended (Walk-Away systems)	Turbidity, colorimetry (all systems) Plus fluorescence in Walk-Away systems	External (autoSCAN 4), internal (autoSCAN WA)	GN, NF, GP, NS, YS, ANA	1	*Enterobacteriaceae* (WA) *Pseudomonas* spp. (WA) *S. aureus* (WA)	3.5–18 3.5–18 3.5–24	2/18[d] 2/18[d] 2/18[d]	Yes	Yes	IBM compatible
Autosceptor	Attended (reader-recorder)	Visual (reader-recorder), turbidity, colorimetry (Autosceptor)	External	GN, GP, ANA	1	*Enterobacteriaceae* *Pseudomonas* spp. *S. aureus*	16–20 16–20 24	16–20 16–20 16–20	Yes	Yes	IBM compatible
Sensititre autoreader with ARIS module	Attended (autoreader), nonattended (with ARIS)	Fluorescence	External (autoreader), internal (with ARIS)	GN, NF, GP	3	*Enterobacteriaceae* *Pseudomonas* spp. *S. aureus*	5 or 18 18 5 or 18	5 or 18 5 or 18 5 or 18	Yes	Yes	IBM compatible
Reader incubator module	Nonattended	Turbidity, colorimetry	Internal incubation module	GN, NF, GP, YS, HM & NS, ANA	1	*Enterobacteriaceae* *Pseudomonas* spp. *S. aureus* Yeast cells	4–15 6–15 6–15	2–18 4–18 2–15 24–48		Yes	Yes

12.11 ELISA Plate Reader

PRINCIPLE

An ELISA plate reader is a photometer designed to measure the light absorbance of samples in microdilution plates. Kinetic ELISA readers, which measure the rate of color development in microtiter wells by taking readings at specified intervals, are now available. The simplest ELISA plate readers provide printed data via a heat-sensitive tape output. Other readers perform tasks such as calculating standard curves and can be connected to a printer to generate a hard copy of results. More sophisticated readers can be connected to computers that perform various calculations and to a printer or a mainframe computer system. For a more comprehensive discussion of ELISA plate readers, see CMPH procedure 12.7.

12.12 Packed-Column Gas-Liquid Chromatograph

PRINCIPLE

Chromatography is a method used to separate components within a mixture based on their size and electrical charge. Clinical microbiology laboratories have traditionally used GLC for identification of anaerobic bacteria. New technological developments in GLC have made possible many more sophisticated uses for microbiology. Identification of bacteria and fungi on the basis of whole-cell fatty-acid content, detection of metabolites in body fluids, and quantitation of antimicrobial agents in body fluids are only a few of these exciting new applications. See CMPH procedure 12.8 for a detailed discussion of packed-column GLC.

12.13 Compressed-Gas Regulator and Handling of Gas Cylinders

I. PRINCIPLE

Gas-pressure regulators are used to control the flow of gas from a pressurized storage tank (usually in the form of a cylinder) to an end-stage instrument that requires gas flow at a pressure less than that in the storage cylinder. Diaphragms made up of interlocking leaves of metal (much like the condenser diaphragm on a microscope) are used to vary the size of the opening through which the gas flows. The smaller the opening, the less gas is allowed to flow through. Flow is controlled precisely, because relatively small movements in the interlocking leaves are achieved with relatively large turns of the threaded control knob or handle. See CMPH procedure 12.9 for a more detailed discussion of compressed-gas regulators and handling of gas cylinders.

II. SPECIAL PRECAUTIONS

A. *The Compressed Gas Association standard numbers for the gas cylinder and the regulator must match!*

B. Be certain that the screw-on valve cap that covers the outlet-valve knob is securely tightened before moving a gas cylinder. Do not move a cylinder with a regulator attached. If the cylinder should fall with the valve open (or if it jars open or breaks), the cylinder will be propelled, severely compromising the safety of employees. In this event, lie flat on the floor and stay out of the way until all compressed gas is exhausted and the cylinder stops moving. Try to keep a very substantial object (like a laboratory bench) between you and the cylinder.

C. Use only cylinders that are secured upright. If a valve should malfunction, the cylinder will be propelled into the floor instead of horizontally across it.

D. For certain cylinder types (identified in the catalog and the product insert and received with the washer in place), you must place a new washer or O-ring between the gas cylinder and the inside of the regulator hexagonal nut before connecting the regulator and cylinder. *Never use oil or grease to lubricate joints; the equipment works without lubricant, which can be flammable.*

E. Open and close the gas cylinder valve briefly to clear the valve of dust and particles before attaching the regulator.

F. Do not allow the pressure on a 35-liter or larger cylinder (diameter, >21 cm) to go below 200 psi before changing the cylinder.

G. Scoring on the outer hexagonal nut of a regulator indicates that it has a left-handed thread and opens by counterclockwise turning.

H. If the temperature of storage is substantially below room temperature (less than 15°C), you must allow the cylinders to equilibrate inside before use. Otherwise, pressure cannot be regulated with consistency. To be safe, when cylinders are stored outside in the cold, bring them indoors 1 day before you plan to use them. If this cannot be anticipated, allow them to equilibrate at room temperature for 8 h before opening.

NOTE: *Take account of partial pressure of individual gases within a mixture when determining the amount of gas allowed for storage. For example, in a* *full cylinder (2,000 psi) of anaerobic gas mixture consisting of 90% nitrogen, 5% carbon dioxide, and 5% hydrogen, only approximately 10 ft^3 (~0.9 m^3) of hydrogen gas is present in the cylinder. This may be useful if regulations allow certain minimum amounts of volatile gases to be stored in the laboratory area.*

III. QUALITY CONTROL

A. Daily

1. Check and record pressure in the cylinder, on a recording form. Change to a new cylinder if pressure goes below 200 psi.
2. Check and record the pressure of the second-stage regulator.

B. As needed

1. Record date and time each fresh cylinder is placed into use.
2. If the regulator requires it, change the washer each time the cylinder is changed.
3. Check tubing and connections for leaks.
4. Clean gauge lens with soapy water, rinse, and wipe dry with soft cloth. Do not use solvents on plastic lenses.

SUPPLEMENTAL READING

Compressed Gas Association. Pamphlet E-1. Standard connections for regulator outlets. Compressed Gas Association, Arlington, Va.

Compressed Gas Association. Standard V-1. Compressed cylinder valve inlet and outlet connections. Compressed Gas Association, Arlington, Va.

12.14 Bright-Field Microscope

I. PRINCIPLE

The term ''bright field'' refers to the dark appearance of a stained or naturally colored, transparent or translucent specimen against a bright white background or field. A microscope consists of a compound magnifying system in which the observer looks at the first (primary) image with the lens that produces an enlarged secondary (virtual) image (Fig. 12.14.1). The two convex lenses are aligned in such a way that the observer sees an enlarged image of the object. The resolving power (also called resolution) of a microscope is the smallest distance between two separate dots that allows the viewer to differentiate two dots from a single dot. See CMPH procedure 12.14 for a more detailed discussion on the theories and function of the illuminating microscope.

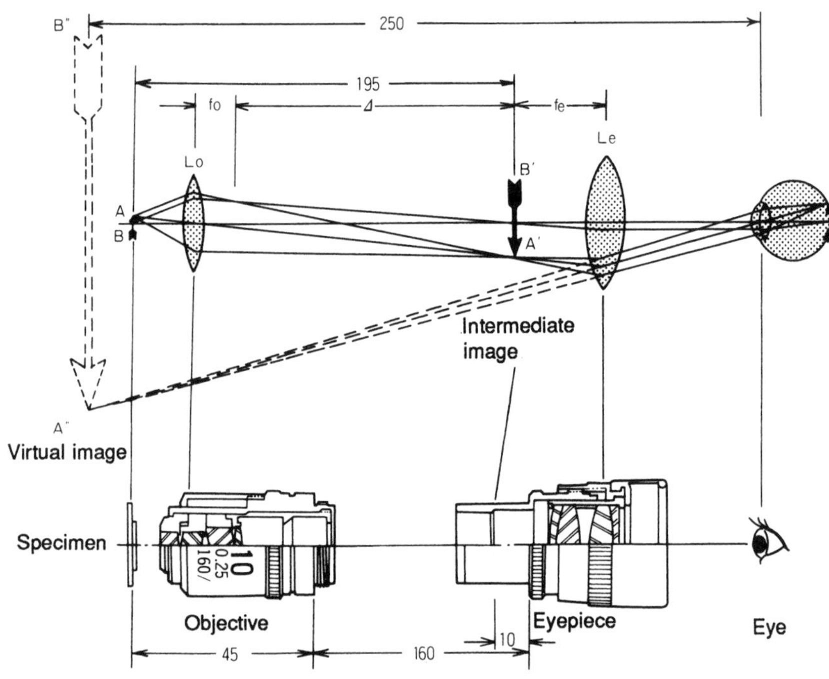

Figure 12.14–1 Basic principle of microscopy (top) and microscope parts (bottom). The objective lens near the specimen (Lo) creates a magnified image of the specimen (small black arrow A′B′). The eyepiece lens (Le) magnifies the image again to create a virtual image (large clear arrow A″B″) (courtesy of A. G. Heinze Co., Inc., Irvine, Calif., and Nikon Inc., Garden City, N.Y.; reprinted from CMPH 12.14).

A. Use in clinical microbiology laboratories

Observation and description of the microscopic morphology of bacteria, fungi, parasites, and host cells in various stained and unstained preparations

B. Basic requirements for optimal performance

1. Koehler illumination: sharply focused, clearly resolved, evenly illuminated field of view
2. Proper maintenance and care of the instrument
3. Good working knowledge of the instrument
4. Consideration when purchasing the microscope

II. MICROSCOPE FUNCTION VERIFICATION

A. Microscope eyepieces (*see* Fig. 12.14–2)

1. Function
   To enlarge primary image to form virtual image
2. Magnification
   Magnifying power of eyepiece; range, 5× to 20× (most common, 10× or 12.5×).
3. Position
   Eyepieces are moved closer or farther apart by a pulling-pushing motion or by moving a ridged dial to adjust for interpupillary distance.

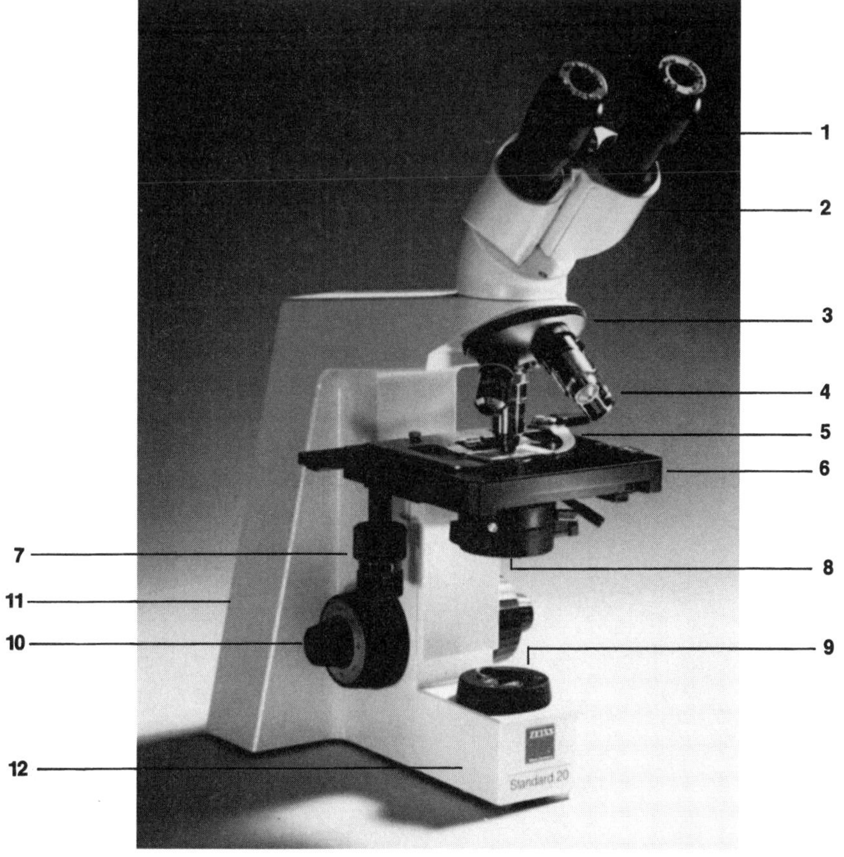

Figure 12.14–2 Photograph of a compound microscope (courtesy of Carl Zeiss Microscopes, Max Erb Instrument Co., Burbank, Calif., and Carl Zeiss Inc., Thornwood, N.Y.). Numbers: 1, eyepieces; 2, binocular tube; 3, nosepiece (4×); 4, objective; 5, specimen holder with spring clip to the right; 6, mechanical stage; 7, coaxial stage drive for movement on *x* and *y* axes; 8, achromatic condenser with 0.9 aperture; 9, luminous-field diaphragm; 10, coaxial coarse- and fine-focus control; 11, on-off switch; 12, 6-V, 20-W integral illuminator. Reprinted from CMPH 12.14.3.

II. MICROSCOPE FUNCTION VERIFICATION *(continued)*

4. Focus
   Some eyepieces can be focused independently of the fine- and coarse-focus adjustment knobs. At least one eyepiece capable of being focused is recommended to compensate for differences in the strengths of users' eyes.
5. Reticle
   An engraved glass disk or glass-mounted microphotograph placed on the eyepiece diaphragm to assist in the following:
   a. Calibration: A micrometer scale is superimposed on the object image so that measurements can be made.
   b. Photomicroscopy: Cross hairs are superimposed on the object image to allow focusing of the image at the photographic film plane.
6. Designations on the eyepiece as follows:
   a. ''Plan'' or ''Plano''
      Correction for field curvature (e.g., ''periplan''). The image is in focus from the center to the periphery of the viewing field at the same time. These lenses are more expensive than standard lenses.
   b. Wide field
      Eyepieces encompass a large viewing area and are more comfortable for long periods of viewing.
   c. High eyepoint
      Sometimes designated with a pair of ''spectacles,'' these lenses are designated for eyeglass wearers. The image is projected some distance from the eyepiece. Certain eye defects, e.g., astigmatism, are not corrected for within the optics of the microscope. Therefore, one can use the microscope while wearing glasses. The high clearance from the top of the eyepiece reduces fatigue when viewing for long periods.

B. Microscope mechanical tube length (Fig. 12.14–2)
This is the distance between the insertion position of the objective and the top of the draw-tube into which the eyepiece fits; 160 mm is considered universal for standard microscopes, although infinity-corrected systems are now available. If an objective with a specified mechanical tube length is interchanged with one that has a different specified length (i.e., is designed for a different microscope), image quality may suffer. *Do not interchange objectives on different microscopes unless you know the specifications are equivalent.*

C. Microscope nosepiece (Fig. 12.14–2)
1. Function
   Rotates and houses the objectives
2. Housing
   Forward or reverse-facing objectives (reverse-facing most common)
3. Arrangement of objectives
   The sequence of low power to intermediate power(s) to high power objectives reduces the risk of getting immersion fluid on high dry objectives.

D. Microscope objectives (Fig. 12.14–2)
1. Function
   Formation of a primary image
2. Magnification (magnifying power of objective)
   Range is 1.25× to 100× (most common, 10×, 40×, and 100×). Magnification is marked on objective, usually with a characteristic colored ring for easy identification.
3. Numerical aperture (N.A.)
   Is the measure of light-gathering power of a lens (the widest angle of the cone of light that can enter the objective lens).
   a. The higher the N.A., the better the resolving power (the ability of the system to allow the observer to distinguish individual tiny objects).

b. When the N.A. is >1.00, a liquid immersion medium (e.g., immersion oil) must be placed between the lens and the specimen slide. N.A. is a function of the refractive index of the material through which light passes from the object to the lens; using a material with a higher refractive index than air results in a higher N.A. because of increased concentration of the light entering the lens.

4. Degree of optical correction for chromatic aberration

Scattering by the lens of white light into colored light (prism effect) around the object, causing halos and decreasing resolution and spherical aberration (distortion in shape of image due to bending of light rays)

a. Achromat

Corrected for brightest part of the spectrum; no spherical correction unless designated "plan."

b. Fluorite (calcium fluorite)

More highly corrected for aberrations; commonly used in fluorescence microscopy; transmits better than crown glass at shorter wavelengths.

c. Apochromat

Least common, most expensive, highest correction for aberrations

E. Microscope mechanical stage (Fig. 12.14–2)

1. Function

Holds the specimen slide. Adjusting knobs move the slide left, right, forward, and backward.

2. The mechanical stage may have calibrated graduated lines. Readings are taken from the x and y axes of the stage grids to facilitate relocation of a specific field of view.

F. Microscope condenser (Fig. 12.14–2)

1. Function

Illumination of specimen and production of contrast

a. Light from the condenser converges, forming a cone of light on the specimen plane. The light then diverges from the specimen plane to form an inverted cone of light angled upward through the objective front lens.

b. The aperture diaphragm located beneath or within the condenser controls the angle of the illuminating cone of light.

2. Degree of optical correction (for chromatic and spherical aberration)

a. Abbe: simplest and least expensive; no correction for optical aberrations

b. Aplanatic: corrected for spherical aberrations

c. Achromatic: corrected for chromatic aberrations

d. Achromatic-aplanatic: most expensive; corrected for both kinds of optical aberrations

3. N.A. of the condenser

a. Resolving power (ability of the system to allow the observer to distinguish between two small objects) is limited by the N.A. *Should be equal to or greater than the N.A. of the highest objective.*

b. Condensers with 0.9 N.A. are most common and do not accept immersion media. N.A. of the optical system of >1.0 cannot be achieved with this condenser.

c. Condensers with N.A. of >1.0 accept immersion media to achieve an N.A. of >1.0 for better resolution. Immersion media must be placed between the condenser and specimen plane.

NOTE: *To optimize the resolving power of an objective with an N.A. >1.0, use a condenser with an N.A. higher than the N.A. of the objective. Also use*

II. MICROSCOPE FUNCTION VERIFICATION *(continued)*

immersion liquid between the objective and the specimen plane and between the condenser and the specimen plane.

G. Other substage components

1. Condenser centering screws are used to correctly align the condenser on the same optical axis as the objective
2. Two-sided mirror (one side concave and the other side flat). The flat side of the mirror is always used with the condenser. The concave side of the mirror is used without the condenser for objectives with very low N.A.
3. Built-in light sources (bulbs)
   a. Tungsten: Turn light source down for standby. Turning it completely off and on allows heat buildup.
   b. Tungsten-halogen: Turn light source off for standby.
4. Filters
   a. Blue changes yellow light source into a more natural ''white'' light source; good for some color photography.
   b. Neutral density reduces brightness without changing color of the background; two can be used together to further reduce brightness and are especially good for photography.
   c. Other filters with light-modifying characteristics are used for different purposes. Consult technical representatives for filter information related to an individual application.

H. Coverslip

In biomedical applications, objectives are corrected for use with a coverslip. Objectives are corrected for spherical aberration only when used with a coverslip of the proper thickness. A no. $1\frac{1}{2}$ coverslip is approximately 0.16 to 0.19 mm thick.

NOTE: *Thickness of the coverslip must not vary more than ±0.05 mm from the thickness indicated on the objective.*

1. A coverslip is mounted on top of the specimen to enhance resolution and protect the specimen.
2. *The appropriate thickness of the coverslip is specified by the number engraved on the objective (0.17 or 0.18 mm).*
3. No coverslip is necessary when the objective has a dash (–) on it.

I. Immersion liquid

With objectives having an N.A. of >1.0, use immersion liquid, usually oil, to expand the angle of light entering the lens and increase the resolution. ''Oil'' printed on an objective indicates that immersion liquid must be used. Medium-viscosity immersion oil is adequate, but for best resolution, use the oil recommended by the manufacturer.

NOTE: *Certain special objectives with N.A. of <1.0 require an immersion medium. The objective will be labeled to indicate this.*

J. Other designations, features, and definitions

1. Flat field, the correction for field curvature, is indicated by the prefix ''plan'' or ''plano'' with the type of objective, e.g., planachromat.
2. An objective may have a built-in adjustable correction system, also called a ''correction ring,'' that allows the operator to modify the objective to accommodate a coverslip thickness other than that designated on the objective.
3. An immersion oil object may have a locking mechanism to lock the objective up out of the way to protect it and to prevent immersion oil from spreading to other objectives.
4. Objectives are usually marked with differently colored rings or different numbers of rings for quick magnification reference.

III. CALIBRATION WITH OCULAR MICROMETER

See EPCM procedure 7.3 for this procedure.

IV. MAINTENANCE AND QUALITY CONTROL

See Table 12.14–1 for a listing of routine maintenance and QC of the bright-field microscope.

Table 12.14–1 Routine maintenance and quality control of the microscope

Procedure	Daily	Monthly	Semiannually
1. Wipe oil from objective, condenser, and stage.	X		
2. Turn off light source.	X		
3. Replace microscope cover.	X		
4. Thoroughly dust all optical surfaces.		X	
5. Clean objectives, eyepieces, condenser, and field diaphragm with lens-cleaning fluid.[a]		X	
6. Remove and clean slide holder.		X	
7. Wipe down the body of the microscope and window of the illuminator in the base.[b]		X	
8. Complete general overhaul of the microscope.[c]			X

[a] Apply fluid with lens paper. *Do not* put fluid directly onto the part.

[b] Include blue filter if used.

[c] This includes a full disassembly with cleaning, oiling, and reassembly and readjustment. This should be performed by factory-trained, authorized personnel.

SUPPLEMENTAL READING

CAP. 1991. *Laboratory Instrumentation, Verification and Maintenance*, 4th ed. CAP, Northfield, Ill.

APPENDIX

Instrument Maintenance and Quality Control

APPENDIX 12–1 QC Activity Schedule for Instruments

QC activity schedule for instruments (*see* CMPH 12.1.3)[a]

Instrument and CMPH procedure	Daily or per test	Weekly	Monthly	Quarterly	Semiannually	Annually[b]
Anaerobic chamber	Check anaerobic indicator; check gas pressure.	Change catalyst and desiccant (twice weekly).	Clean interior; check humidity.		Lubricate airlock; check for leaks; check filters in vacuum pump.	Change vacuum pump oil; change pump air filter.
Autoclave (steam sterilizer) (12.3)	Record time and temp of run; check indicator; clean drain hole.	Check safety valve; test biological indicator.	Clean drain and seals.			
Automated bacterial identification and susceptibility testing instrument (12.4)	Run control organisms; run self-test.				Clean instrument.	
Blood culture instrument (BACTEC 9000 series) (12.5)	Station QC occurs automatically every 10 min when a bottle is entered into the system; record temp.	Check/clean air filter.				
Blood culture instrument (BacT/Alert)	Record temp.	Clean tape drive.				
Blood culture instrument (ESP) (12.5)	Sensor QC occurs automatically when a bottle is entered into the system.	Read calibrated thermometer and record temperature.	Print sensor QC report for records; monthly temperature report available.			Routine maintenance by Difco Field service
Centrifuge (12.6)	Balance load; record temp; wipe interior with disinfectant.	Check vents; wipe all parts with disinfectant.	Vacuum the condenser coils, fan, filters; check brushes; check electrical connections.		Verify temp of refrigerated centrifuge; check brushes.	Calibrate speed.

QC activity schedule for instruments (*see* CMPH 12.1.3)[a] *(continued)*

Instrument and CMPH procedure	Daily or per test	Weekly	Monthly	Quarterly	Semiannually	Annually[b]
ELISA reader (12.7)	Clean instrument; run self-check; run QC samples.		Lubricate plate guide bar.	Clean air filter.	Run linearity, repeatability, diode, and accuracy checks; check mechanical alignment.	
GLC (12.8)	Change septum; run stock samples; clean syringes; check gas pressure.			Clean or change column as needed.		
Gas tank regulator (12.9)	Check tank pressure; check second-stage regulator pressure.			Check for leaks; change washer as needed when changing gas tanks.		
Heating block (12.10)	Record temp; check level of fluid in thermometer tube.		Clean out all holes.			
Incinerator burner (12.11)	Inspect heater element.			As needed, replace heating element.		
Incubator (12.12)	Record temp.	Subculture stock organisms; check humidity; check CO_2 if needed.	Check CO_2 if needed.			Clean thoroughly inside and outside; check tubing for leaks; check blower; check that unit is level.
Jars and pouches for incubation (12.13)	Check atmosphere indicator; rejuvenate catalysts; check jars and lids for damage.	Clean jars.	Check and lubricate seal rings as needed; weigh catalyst baskets; test stock organisms.			
Microscope (12.14)	Clean off oil; cover to protect from dust.		Remove dust and dirt; clean all lenses with lens cleaner; clean slide holder.		Clean, lubricate, and inspect all parts; realign.	
Ocular micrometer (12.14)						Calibrate annually or when any component is changed.

(continued)

QC activity schedule for instruments (*see* CMPH 12.1.3)[a] *(continued)*

Instrument and CMPH procedure	Daily or per test	Weekly	Monthly	Quarterly	Semiannually	Annually[b]
Oven (hot air) (12.15)	Record temp.			Verify thermometers.	Clean interior.	
pH meter (12.16)	Calibrate with standards; check electrode solutions; rinse electrode.		Clean entire instrument; check electrical connections.	As needed: replace or recondition electrodes.		
Pipetters and calibrated loops (12.17)	Check for damage; calibrate disposable loops with each new lot no.		Calibrate pipetters with four samples; calibrate quantitative loops.	Calibrate pipetters with 10 samples.		
Refrigerator or freezer (12.18)	Record temp; check blower.				Defrost freezer; clean interior and exterior; check gaskets.	Check that unit is level; check external alarm system function.
Safety cabinet (12.19)	Check airflow; check air pressure if available; wipe all surfaces with disinfectant.	Clean UV lights.	Check air velocity; clean and disinfectant gutter area; measure UV light output.			Perform scheduled certified maintenance.
Thermometers (12.20)	Check for damage.				Calibrate using a standard reference thermometer.	
Water bath (12.21)	Check temp; check for leaks, level of water, contamination, deposits.		Drain and clean.		Drain and remove mineral deposit buildup.	

[a] Reprinted from CMPH 12.1.4.

[b] All instruments need an annual electrical check.

SECTION 13

Quality Assessment and Control

SECTION EDITOR: *David L. Sewell*

13.1 Introduction

QA is a concept that has become integrated into the outcome-based management of patients and has impacted laboratory practices. In general, QA attempts to continuously monitor, evaluate, and improve the reliability and efficiency of practices related to patient care (1, 3). Benchmarking is an important tool of QA and produces two types of information: (a) comparative measures of performance among departments within an organization or among institutions of similar size and complexity, and (b) practices that produce exceptional performance (2). *If benchmarking is applied without regard to performance quality, it becomes a tool to justify inappropriate reductions in staff and resources.* As a tool for QA in the laboratory, benchmarking has been used to improve utilization of cultures, reduce contamination of cultures, reduce turnaround times, etc. (2). After comparative data are gathered and analyzed, improvements are implemented based on the methods associated with better performance. QA activities in microbiology include determining the clinical value of test information and services to the overall patient care process (1, 3). This includes activities that affect patient test utilization, specimen quality, QC, patient and user complaints, appropriateness of specific tests, and efficiency of the overall testing process. QA activities generally require interaction with many other areas of the organization (e.g., medical center, clinics, physicians' offices, etc.).

QC programs are designed so that test information produced by the microbiology laboratory is accurate, reliable, and reproducible. This is accomplished by assessing the quality of the specimens received; monitoring test procedures, reagents, media, instruments, and personnel; reviewing test results; and assessing the clinical usefulness of the information produced. Efficient operation of a microbiology laboratory depends on both QA and QC practices, such that timely, cost-effective, clinically relevant laboratory data are provided.

REFERENCES

1. **Schifman, R. B.** 1992. Quality assessment and improvement (quality assurance), p. 13.1.1–13.1.29. *In* H. D. Isenberg (ed.), *Clinical Microbiology Procedures Handbook*, vol. 2. American Society for Microbiology, Washington, D.C.
2. **Schifman, R. B.** 1995. Strategies for quality management in clinical microbiology. *Clin. Lab. Med.* 15:437–446.
3. **Sewell, D. L., and R. B. Schifman.** 1995. Quality assurance: quality improvement, quality control, and test validation, p. 55–66. *In* P. R. Murray, E. J. Baron, M. A. Pfaller, F. C. Tenover, and R. H. Yolken (ed.), *Manual of Clinical Microbiology*, 6th ed. American Society for Microbiology, Washington, D.C.

SUPPLEMENTAL READING

JCAHO. 1992. *Accreditation Manual for Hospitals*, p. 137–143. JCAHO, Chicago.

13.2 Quality Assessment Monitors

QA practices are based on examination of the entire patient care experience to find processes that can be improved through increased efficacy, appropriateness, availability, timeliness, effectiveness, continuity, safety, efficiency, and respect and caring (13). The QA subjects most often monitored in a microbiology laboratory are listed in Table 13.2–1. Many of these processes are not under the direct control of the laboratory. Therefore, the laboratory must work closely with other involved health care workers and solicit their input for improvement.

I. SPECIMENS

A critical element affecting test information from a microbiology laboratory is specimen quality. Poor-quality specimens guarantee poor results and poor patient management. The laboratory must provide collectors of specimens with effective procedures for both collection and transport (EPCM section 1, 9, 14) and continuously monitor these processes. For example, CSF, genital, eye, or internal ear specimens should never be refrigerated, because *Neisseria gonorrhoeae*, *N. meningitidis*, and *Haemophilus influenzae* are susceptible to cold temperatures (8). In addition to collection procedures, the laboratory must establish guidelines for appropriate specimen volumes and numbers of specimens required for testing. Insufficient volume compromises the recovery of potential pathogens from CSF (when multiple tests are requested) and from blood. Mermel and Maki (7) have shown that the yield of positive blood cultures in adults increases approximately 3% per ml of blood cultured. Insufficient numbers of specimens compromise the diagnostic reliability of a test (e.g., blood cultures, tuberculosis workup), whereas excessive numbers of specimens are not cost effective (e.g., greater than two specimens for an acute diarrheal workup) and waste laboratory resources. The quality of specimens can be monitored by cytological examination of expectorated sputum and endotrachial aspirate for large numbers of squamous epithelial cells, which suggest excessive contamination with oropharyngeal secretions (10, 11), and by contamination rates of blood, body fluids, and urine specimens (1, 3). Monitoring these factors helps indicate when a problem exists and whether it is hospital-wide or restricted to a particular location or ward. Resolution of the problem may require feedback to the identified area, increased educational activities with the staff, changes in procedures, and follow-up checks for sustained improvement.

Table 13.2–1 QA monitors

Monitors	Goal
Specimens	Monitor the quality, quantity, volume, and acceptability of the specimen.
Test utilization	Monitor test request for appropriateness based on clinical indications, diagnostic yield, and other clinical data.
Reports	Monitor the accuracy, clarity, and timeliness of reports.
User perceptions	Actively query users regarding needs and satisfaction. Monitor resolution of complaints.

. TEST UTILIZATION

Test utilization by health care providers is receiving increasing recognition as a critical element in patient management. The goal is to ensure the availability of clinically relevant test data and eliminate inappropriate and clinically useless testing. This is accomplished by establishing guidelines with the medical staff that define clinical indications for testing (e.g., clinical-practice guidelines). Examples of practices that must be restricted include collecting stool specimens for bacterial culture and ova and parasite examinations from patients hospitalized for greater than 3 days (15); collecting CSF specimens for mycobacterial culture with normal cell counts, glucose, and protein (4); and performing bacterial culturing of poor-quality specimens (e.g., mouth lesions, decubiti, aerobic swabs for anaerobic culture, and multiple specimens collected from the same site [*see* EPCM section 1]. As important, the laboratory should work with other health care providers to implement procedures that ensure notification of the physician or health care team of laboratory results (e.g., CSF smears, positive blood cultures, acid-fast bacillus [AFB] smears, etc.) which are deemed critical for patient management (*see* EPCM section 1) and recognition of discrepancies between laboratory and clinical data. For example, the patient may not be receiving therapy or may be receiving inappropriate therapy, although he or she has a significant culture result. Intervention by the microbiology laboratory may require computerized access to clinical and pharmacological data.

II. REPORTS

Microbiology test information must be accurate and timely to positively impact patient management. The format should clearly identify the test result and, when appropriate, include additional information related to the reference range and critical value and interpretive comments. Therefore, the laboratory must establish medically acceptable turnaround times (from specimen collection to result utilization) for critical tests such as body fluid smears, AFB smears, and cultures. For instance, the median turnaround time for CSF smears in a survey of over 400 laboratories was 45 min (6). Recently, CDC has recommended that laboratories provide AFB smears within 24 h of specimen collection; isolation and identification of *Mycobacterium tuberculosis* within 10 to 14 days; and susceptibility test results within 15 to 30 days (5). The average mycobacteriology laboratory cannot meet the 10- to 14-day guidelines for recovery and identification of *M. tuberculosis* unless the laboratory uses a direct nucleic acid-based probe method. Because of staffing patterns (less than 7-day-per-week mycobacteriology coverage) and current technology costs, it is doubtful that most laboratories will be able to provide mycobacteriology services if these guidelines become standards. QA monitors developed in individual medical centers require a consensus of all health care workers involved in patient management and vary in complexity depending on staffing and technology. After development of the turnaround times, continuous monitoring of the process will identify areas that need improvement.

The accuracy of reports can be monitored for clerical and computer errors by comparing results reported in patient charts or computer records with those recorded on laboratory worksheets. When errors are detected, they should be classified by clinical significance and procedures and policies modified to prevent recurrences. The correction of a final report must not conceal the initial error but should identify the erroneous result and provide the correct data. For example, the report may state "Corrected Report: *S. aureus* was incorrectly reported as *S. epidermidis.*" If the corrected result is clinically significant, the physician should be notified immediately.

IV. USER PERCEPTIONS

Measurement of user satisfaction with questionnaires can provide some insight int whether the microbiology laboratory is meeting its QA goals (2, 12). In addition the daily recording of complaints and how they were resolved can identify recurring problems and effective procedures to eliminate them.

REFERENCES

1. **Baerheim, A., A. Digranes, and S. Hunskaar.** 1992. Evaluation of urine sampling technique: bacterial contamination of samples from women students. *Br. J. Gen. Pract.* **42:** 241–243.
2. **Baron, E. J., D. Francis, and K. M. Peddecord.** 1996. Infectious disease physicians rate microbiology services and practices. *J. Clin. Microbiol.* **34:**496–500.
3. **Bates, D. W., L. Goldman, and T. H. Lee.** 1991. Contaminant blood cultures and resource utilization. The true consequences of false-positive results. *JAMA* **265:**365–369.
4. **Crowson, T. W., E. C. Rich, B. F. Woolfrey, and D. P. Connelly.** 1984. Over-utilization of cultures of CSF for mycobacteria. *JAMA* **251:**70–72.
5. **Doern, G. V.** 1996. Diagnostic mycobacteriology: where are we today? *J. Clin. Microbiol.* **34:**1873–1876.
6. **Howanitz, P. J., and S. Steindel.** 1991. Intralaboratory performance and laboratorians' expectations for stat turnaround times. *Arch. Pathol. Lab. Med.* **115:**977–983.
7. **Mermel, L. A., and D. G. Maki.** 1993. Detection of bacteremia in adults: consequences of culturing an inadequate volume of blood. *Ann. Intern. Med.* **119:**270–272.
8. **Miller, J. M.** 1985. *Handbook of Specimen Collection and Handling in Microbiology. CDC Laboratory Manual.* U.S. Department of Health and Human Services, Atlanta, Ga.
9. **Miller, J. M., and H. T. Holmes.** 1995. Specimen collection, transport, and storage, p. 19–32. *In* P. R. Murray, E. J. Baron, M. A. Pfaller, F. C. Tenover, and R. H. Yolken (ed.), *Manual of Clinical Microbiology*, 6th ed. American Society for Microbiology, Washington, D.C.
10. **Morris, A. J., D. C. Tanner, and L. B. Reller.** 1993. Rejection criteria for endotracheal aspirates from adults. *J. Clin. Microbiol.* **31:** 1027–1029.
11. **Murray, P. R., and J. A. Washington II** 1975. Microscopic and bacteriologic analysis of expectorated sputum. *Mayo Clin. Proc.* **50:** 339–344.
12. **Peddecord, K. M. E. J. Baron, D. F. Francis, and J. A. Drew.** 1996. Quality perceptions of microbiology services: a survey of infectious diseases specialists. *Am. J. Clin. Pathol.* **105:**58–64.
13. **Sewell, D. L., and R. B. Schifman.** 1995. Quality assurance: quality improvement, quality control, and test validation, p. 55–66. *In* P. R. Murray, E. J. Baron, M. A. Pfaller, F. C. Tenover, and R. H. Yolken (ed.). *Manual of Clinical Microbiology,* 6th ed. American Society for Microbiology, Washington, D.C.
14. **Shea, Y. R.** 1992. Specimen collection and transport, p. 1.1.1–1.1.30. *In* H. D. Isenberg (ed.), *Clinical Microbiology Procedures Handbook*, vol. 1. American Society for Microbiology, Washington, D.C.
15. **Valenstein, P., M. Pfaller, and M. Yungbluth.** 1996. The use and abuse of routine stool microbiology. *Arch. Pathol. Lab. Med.* **120:**206–211.

13.3 Quality Control Guidelines

The accuracy and clinical usefulness of test information provided by the microbiology laboratory is dependent on specimen quality; validity of the test method; performance of test procedures, reagents, media, instruments, and personnel; and reporting of results. A QC program continuously monitors these parameters and identifies areas for improvement. The QC guidelines listed in Table 13.3–1 are derived from the Clinical Laboratory Improvement Amendments of 1988 (CLIA '88) (3), the accreditation and licensure guidelines issued by the College of American Pathologists (CAP) (1), and the guidelines of the Joint Commission on Accreditation of Healthcare Organizations (4). Additional specific quality control information can be found in the other sections of this manual. Appendix 13–1 lists some of the common deficiencies noted during CAP laboratory inspections, as well as selected new checklist questions with a discussion of their importance in a laboratory's overall QC program.

I. PERSONNEL

The personnel requirements for microbiology laboratories are based on the complexity test model defined in CLIA '88 (3). All personnel must document annually that they are competent to perform the specified procedures, including preanalytic, analytic, and postanalytic phases of testing. Written training methods should be available to be used for orienting new employees, retraining other laboratory employees, and verifying the competency of current employees (2, 6). Generally, these training methods include reading the procedure manual and performing supervised workup, interpretation, and reporting of cultures. Verification of competency may include a written examination, interpretation of unknowns, and review of culture workup. In addition, employee personnel folders should document continuing education activities and annual performance appraisals.

II. POLICY AND PROCEDURE MANUAL

The laboratory's policy and procedure manual must contain procedures for all aspects of laboratory work, beginning with test order and collection of specimens and ending with the reporting of results. When appropriate, the procedure format should substantially follow the recommendations specified in CLIA '88 or as outlined in NCCLS publication GP2-A2 (5) and the recently published update, GP2-A3 (7). The procedures listed in this manual or CMPH can be easily modified to meet the needs of various sized laboratories. The original manual and all changes must be signed by the current laboratory director. Manuals must be reviewed annually, and discontinued procedures are retained by the laboratory for 2 years.

III. RECORDS AND REPORTS

Sufficient records and reports must be retained to document all the activities related to a specific patient specimen. The test requisition should contain at least the patient's name or identifier, the dates specimens were collected and received in the laboratory, the name of the requester, and the test requested. The procedures performed, by whom, and any communication between the laboratory and the requester should be documented. Laboratory reports must include the name and address of

Table 13.3–1 QC parameters and guidelines[a]

QC parameter	Guidelines
Personnel	Use sufficient qualified personnel for volume and complexity of work. Document competency and training. Document continuing education activities. Provide employees with written performance standards. Evaluate employees annually.
Policy and procedure manual	Write in NCCLS format. Define test performance, tolerance limits, specimen acceptability, reagent preparation, QC, calculations, and reporting. Review and initial annually. Approve and date all changes (laboratory director or designee). Make available in work area (laboratory director or designee). Retain discontinued procedure for 2 years. Use only accredited or licensed reference laboratory.
Records and reports	Record all QC results on QC form. Report out-of-control results to supervisor and note corrective action on QC form. QC records are reviewed weekly by supervisor and monthly by the laboratory director or designee. Report results only to authorized personnel. Notify test requester of critical values immediately. Provide normal ranges where appropriate. Correct errors in patient reports in timely fashion. Include name of reference laboratory on patient's report. Retain records for at least 2 years.
Equipment performance	Document function checks of equipment. Perform as frequently as necessary to ensure proper function or as specified by manufacturer. Document routine preventive maintenance. Retain maintenance records for life of instrument.
Commercially prepared media exempt from QC	Certain primary plating media are exempt form QC testing by user. Retain manufacturer's QC protocol. Obtain written assurance that manufacturer follows NCCLS standards. Inspect each shipment for cracked media or petri dishes, hemolysis, freezing, unequal filling, excessive bubbles, and contamination. Document medium deficiencies and corrective action, and inform manufacturer. Perform in-house QC testing until deficiency is resolved.
User-prepared and nonexempt media	Record amount prepared, source, lot number, sterilization method, preparation date, pH, expiration date, and name of preparer. Check medium for proper color, consistency, depth, smoothness, hemolysis, excessive bubbles, and contamination. Test media with QC microorganisms of known physiologic and biochemical properties.
Stains, reagents, and antisera	Label containers as to contents; concentration; storage requirements; date prepared, received, and/or placed in service; and expiration date or shelf life. If user prepared, record volume, source, and lot number. Store based on manufacturer's recommendation. Test with positive and negative controls before use. Discard outdated material and reagents that fail performance standards.
Commercial systems	Test each new lot and/or shipment. Follow manufacturers' recommendation for QC testing.
Proficiency testing	Participate in appropriate level of external proficiency testing program.
Test verification	Perform prior to test implementation. Determine test sensitivity, specificity, predictive values, accuracy, and precision. Compare with reference method.
Test validation	Document ongoing performance of a verified test. Use QA and QC elements.

[a] Adapted from CMPH 13.2.2–3.

I. RECORDS AND REPORTS (continued)

the laboratory performing the test, reference ranges when appropriate, and reasons for rejection if the specimen is unacceptable. All records and reports (paper or computer) are retained for 2 years. Generally, computer records can be backed up on disks and archived in a safe storage area.

V. MEDIA, REAGENTS, STAINS, AND EQUIPMENT

Every aspect of the performance of laboratory tests is monitored, documented, and evaluated. Negative and positive controls are used for qualitative tests, and, for quantitative tests, two controls of different strengths are included in each batch of patient specimens. Control material and patient specimens are tested by identical procedures. Reagents and supplies are labeled to indicate identity, concentration, storage requirements, and dates of preparation and expiration. The reliability of the reagent or test must be verified before it is used for patient testing. The QC strains, frequency of testing, expected results, and recording forms for a variety of microbiological tests are found in specific sections of this manual and other references (9).

Maintenance and function checks are performed and recorded for each instrument or piece of equipment and retained for the life of the equipment. The laboratory should perform maintenance and function checks as specified by the manufacturer.

V. PROFICIENCY TESTING

All laboratories must participate in an external proficiency program approved by the U.S. Department of Health and Human Services that reflects the specialty and level of expertise of the laboratory. Laboratories must analyze proficiency samples by routine methods and submit correct responses for at least 80% of the samples. Currently approved 1997 proficiency programs for CLIA are listed in Appendix 13–2.

VI. TEST VERIFICATION AND VALIDATION

Laboratories must use the test method that provides the most accurate, reproducible, and clinically relevant result. New tests and reagents must be verified for accuracy and precision before they are made available for patient care or to replace a current test. Also, laboratories must validate existing methodologies for consistency of results and document that personnel remain competent to perform tests and report results (2). Therefore, test verification and validation are integral parts of QA and QC.

Verification is performed before the test is used for patient care and is required for all highly complex tests. Generally, tests of moderate complexity do not require verification unless modified, developed by the user, or not U.S. Food and Drug Administration (FDA) approved. The verification process determines whether the laboratory can reproduce the results obtained by the manufacturer or, for a noncommercial test, whether the test method compares favorably with an accepted reference test. This process requires determination of the test sensitivity, specificity, predictive values, accuracy, and precision when appropriate, and is accomplished by performing the new test or using the reagent in parallel with an accepted reference method (2, 8). Many microbiology items (e.g., media, individual reagents) are adequately monitored by QC checks and do not require verification. The difficulty for smaller laboratories is in defining the appropriate method for verification, obtaining adequate samples for testing in a reasonable timeframe, and finding the time and financial resources to verify new test procedures. Suggested methods for verification of antimicrobial susceptibility systems, blood cultures, nonculture test systems, and commercial tests have been discussed by Elder et al. (2). These authors

VI. TEST VERIFICATION AND VALIDATION *(continued)*

suggest that for commercial tests performed based on the manufacturer's produc insert, the verification process should include at least 20 samples containing th target analyte (positive samples) and 20 to 50 samples lacking the analyte (negativ samples). A minimum of 50 strains (12 to 15 different species) should be teste for verification of test systems used for identification of microorganisms. In general a test is verified if the sensitivity and specificity of the new test are no lower tha 5% below the values of the reference method. Verification data must be kept fo at least 2 years.

Validation documents whether a previously verified test continues to perform satisfactorily and may be documented by QC data, personnel competency, profi ciency testing, equipment calibration, and correlation with clinical data (2). There fore, test validation provides ongoing quality assessment of the procedure, combin ing the elements of QA and QC.

REFERENCES

1. **College of American Pathologists.** 1996. *Commission on Laboratory Accreditation Inspection Checklist.* College of American Pathologists, Northfield, Ill.
2. **Elder, B. L., S. A. Hansen, J. A. Kellogg, F. J. Marsik, and R. J. Zabransky.** 1997. *Cumitech 31, Verification and Validation of Procedures in the Clinical Microbiology Laboratory.* Coordinating ed., B. W. McCurdy. American Society for Microbiology, Washington, D.C.
3. **Health Care Financing Administration.** 1992. Clinical Laboratory Improvement Amendments of 1988; final rule. *Fed. Regist.* **57:**7137–7186.
4. **JCAHO.** 1992. *Accreditation Manual for Hospitals*, p. 137–143. JCAHO, Chicago.
5. **NCCLS.** 1984. *Clinical Laboratory Procedur Manuals.* Approved Guideline GP2-A.2 NCCLS, Wayne, Pa.
6. **NCCLS.** 1995. *Training Verification for Laboratory Personnel GP21-A.* NCCLS, Wayne Pa.
7. **NCCLS.** 1996. *Clinical Laboratory Procedur Manuals.* Approved Guideline GP2-A.3 NCCLS, Wayne, Pa.
8. **Sewell, D. L.** 1992. Quality control, p 13.2.1–13.2.35. *In* H. D. Isenberg (ed.), *Clini cal Microbiology Procedures Handbook*, vol 2. American Society for Microbiology, Washington, D.C.

.3.4 Frequency of Quality Control Performance

Table 13.4–1 has been adapted from H. D. Isenberg (ed.), *Guide to Regulatory Requirements,* American Society for Microbiology, Washington, D.C., 1995. This table lists the recommended frequency for performing QC checks. Refer to other sections in this manual for the specific methods and endpoints.

'able 13.4–1 Frequency of QC performance

Standard for:	CAP	JCAHO	CLIA '88
⁄ledia			
Commercial	Lot or batch; manufacturer's control checks are acceptable if NCCLS standards are met except for Thayer-Martin, Martin-Lewis, chocolate, and Campylobacter agars	Lot or batch; manufacturer's control checks are acceptable if NCCLS standards are met except for Thayer-Martin, Martin-Lewis, chocolate, and Campylobacter agars	Lot or batch; manufacturer's control checks are acceptable if NCCLS standards are met except for Thayer-Martin, Martin-Lewis, chocolate, and Campylobacter agars
User prepared	Lot or batch	Lot or batch	Lot or batch
›tains			
Gram stain	Lot or batch and weekly	Lot or batch and weekly	Lot or batch and weekly
Special stains (spore, capsule, flagella, modified acid fast)	Lot or batch and weekly or day of use	Day of use[a]	Lot or batch and weekly
Ziehl-Neelsen	Lot or batch and weekly	Lot or batch and weekly	Lot or batch and weekly
Fluorochrome	Lot or batch and with each use	Day of use	Lot or batch and weekly
Fluorescent antibody	Lot or batch and with specimen tests	NS[b]	Lot or batch and simultaneous/concurrent[c]
Permanent (parasitology)	Lot or batch and day of use	NS	Lot or batch and monthly
Commercial biochemical identification kits	NS	NS	Lot or batch and as specified by manufacturer

(continued)

13 Quality Assessment and Control

Table 13.4–1 Frequency of QC performance *(continued)*

Standard for:	CAP	JCAHO	CLIA '88
Reagents			
Coagulase (plasma)	Lot or batch and day of use	Lot or batch and day of use	Lot or batch and day of use
Catalase	Lot or batch and day of use	Lot or batch and day of use	Lot or batch and day of use
Oxidase	Lot or batch and day of use	Lot or batch and day of use	Lot or batch and day of use
Bacitracin	Lot or batch and weekly	Lot or batch and weekly	Lot or batch and weekly
β-Lactamase	Lot or batch and day of use	Lot or batch and day of use	Lot or batch and day of use
Optochin	Lot or batch and weekly	Lot or batch and weekly	Lot or batch and weekly
XV disks/strips	Lot or batch and weekly	Lot or batch and weekly	Lot or batch and weekly
ONPG	Lot or batch and weekly	Lot or batch and weekly	Lot or batch and weekly
AFB reagents	Lot or batch and day of use	NS	Lot or batch and day of use
Fungal nitrate reagents	Lot or batch and day of use with peptone control	NS	Lot or batch and day of use
Fungal identification	NS	NS	Lot or batch and weekly
Serum for germ tube	NS	Lot or batch and weekly	NS
Antisera (identification)	Initial use and monthly	Lot or batch and monthly	Lot or batch and monthly
Type 1 water	As specified in NCCLS standard	NS	NS
DNA probes	Lot or batch and day of use	Lot or batch and day of use	Lot or batch and day of use
Yeast morphology media	NS	Lot or batch and weekly	NS
Diluents for tissue culture	Checked for sterility and pH	NS	NS
Bacterial susceptibility tests			
Media	Lot or batch	Lot or batch	Lot or batch
Disks	Lot and weekly	Lot or batch	Lot or batch
Diffusion or broth or agar dilution tests	Day of use or weekly if NCCLS standards are met	Day of use or weekly if NCCLS standards are met	Day of use or weekly if NCCLS standards are met
AFB susceptibility tests	Day of use (Mtb only, NS for other AFB)	NS	Weekly (Mtb only, NS for other AFB)
Fungal susceptibility tests	Day of use, establish control limits	NS	Day of use; must establish own control limits
Cells and cell substrates	Lot or batch with uninoculated control and for correct CPE	NS	Simultaneous/concurrent
Procedures and test systems			
All sections, general	Establish and follow written QC procedures; establish, verify, and document performance characteristics of all procedures; perform and document remedial action when problems or errors are identified	NS	Establish and follow written QC procedures; establish, verify, and document performance characteristics of all procedures; perform and document remedial action when problems or errors are identified
Parasitology	Availability of reference collection of slides or photos and gross specimens	Availability of reference collection of slides or photos and gross specimens	Availability of reference collection of slides or photos and gross specimens
Virology	Availability of appropriate cell lines and test systems for isolation and identification of viruses etiologically related to diagnoses being offered; cultures of uninoculated cell lines as negative controls; documentation of maintenance of cell lines and reactions	NS	Availability of appropriate cell lines and test systems for isolation and identification of viruses etiologically related to diagnoses being offered; cultures of uninoculated cell lines as negative controls; documentation of maintenance of cell lines and reactions

able 13.4–1 Frequency of QC performance *(continued)*

Standard for:	CAP	JCAHO	CLIA '88
quipment and space			
General			
FDA approved	Document that all function checks are performed and that instruments are on regular maintenance schedule; perform, document, and verify calibration method and compliance	NS	Document that all function checks are performed in accordance with manufacturer's recommendations; perform and document calibration; and verify the established reportable ranges of test results every 6 months and when reagents are changed
Not FDA approved	Document that all function checks are performed and that instruments are on regular maintenance schedule; perform, document, and verify calibration method and compliance	NS	Establish function checks, calibration standards, and verification of test results within laboratory; document compliance as above
Temperatures	Daily	Daily	NS
CO_2	NS	Daily	NS
Ocular micrometer	Available and calibrated each time eyepieces or objectives are changed	Available and calibrated, no other specifications	Available and calibrated; no other specifications
Airflow (room)	Adequate for types of work	NS	NS
Biological safety cabinet	Certified annually	NS	NS
Microliter pipettes	Check for accuracy before being placed into service and every 6 months	NS	NS
Thermometers	Checked against NIST standard before being put into use	NS	NS
Automatic diluters	Check for accuracy before being placed into service and every 6 months	NS	NS

''Day of use'' and ''daily'' are used interchangeably.

NS, not specified.

May also be interpreted as day of use.

APPENDIXES

Quality Assessment and Control

APPENDIX 13-1

College of American Pathologists Microbiology Inspection Checklist: Common Deficiencies and Selected New Checklist Questions

Proficiency Testing

Question 04.0025: Does the laboratory integrate the external Surveys testing samples withi the routine laboratory workload, and are those samples analyzed by personnel who rou tinely test patient samples, using the same methods as for patient samples?

Question 04.0035: Are proficiency testing specimens identified to the same level as patien samples?

Answer: The value of proficiency surveys is dependent on the samples being integrate within the routine workload and not assigned to one individual. This means that the sampl should be given a laboratory accession and handled as if it were a patient specimen Eventually, all technologists will be involved in working up and reporting results of th proficiency samples using the same approach and procedures used for patient samples. Document the individuals who worked-up and reported the proficiency sample result.

Procedure Manual

Question 04.1205: Is a complete procedure manual written substantially in compliance with NCCLS GP2-A2 available at the workbench or in the work area?

Question 04.1211: Is there documentation of at least annual review of all procedures in the entire microbiology laboratory by the current laboratory director or designee?

Question 04.1212: If there is a change in directorship, does the new director ensure (over a reasonable period of time) that laboratory procedures are well documented and undergo at least annual review?

Question 04.1213: When a procedure is discontinued, is a copy maintained for at least two years, recording initial date of use and retirement date?

Question 04.1214: Does the laboratory have a system documenting that all personnel are knowledgeable about the contents of procedure manuals relevant to the scope of their testing activities?

Answer: Procedure manuals that are incomplete, unreviewed, or have inadequate content areas continue to account for inspection deficiencies. The NCCLS GP2-2A manual (1992) is still cited in the 1997 CAP inspection checklist and describes the necessary content areas that should be contained in a procedure. Procedures in both CMPH and this manual follow the NCCLS content format. The NCCLS has recently published GP2-A3 (1996), which includes information on electronic procedure manuals. Keep in mind that your procedure manual does not have to be identical to the NCCLS format but must be substantially in compliance. All individual procedure manuals must be reviewed annually by the laboratory director or designee and after there is a change in directorship. In larger laboratories with numerous procedures, this review process must be performed throughout the year in order to complete the task. Because a laboratory must be able to identify the method used to produce a reported result, discontinued procedures are archived for two years. Often this can be accomplished on a computer disk. The requirement for documentation of personnel knowledge about the contents of the procedure manuals is problematic due to the number of procedures but can be accomplished by various methods such as documentation of review of the appropriate manuals during the course of the year, written tests, workup of unknown samples, or supervised workup of specific samples.

Quality Assessment and Control

Question 04.1101: Is there a document for the design and evaluation of the laboratory quality control (QC) program?

Question 04.1105: Is there documentation of corrective action taken when controls, etc., exceed defined tolerance limits?

Answer: The QC program must document that it is an active, ongoing program with daily attention to QC results by personnel performing the tests, at least weekly review by the section supervisor, and monthly review by the laboratory director or designee. There should be a written description of the operation of the QC program. When controls are unacceptable, action must be taken and documented. The intent of QA and QC is to identify areas that need improvement and to develop and implement a plan that improves the quality of work performed and ultimately the quality of patient care.

Question 04.1450: Are critical limits established for the results of certain tests important for prompt patient management decisions?

Answer: The laboratory should identify critical limits or "panic values" that result in immediate notification of the health care provider such as positive smears and cultures from normally sterile body sites, positive malarial smears, positive cryptococcal antigen tests, positive AFB smears, etc. The intent of this guideline is to ensure the prompt communication of information that significantly impacts patient management.

Questions 04.2147, 04.3162, 04.4131: Are control specimens tested in the same manner as patient samples?

Questions 04.2148, 04.3163, 04.4133: Are the results of controls verified for acceptability before reporting patient results?

Answer: These two guidelines are the heart of the laboratory's QC program. In order to ensure that all steps of the testing process are controlled, the control specimens must be tested in the same manner as patient specimens. When control values are unacceptable, it follows that patient results can't be reported. Reporting an erroneous test result is worse than not reporting a result.

Reagents

Questions 04.2158, 04.3130, 04.4110, 04.5110: Are all reagents and biologicals properly labeled as to contents, concentration, date prepared or received and/or lot number, date placed in service, and date of expiration or shelf life as appropriate?

Answer: It is amazing that laboratories continue to acquire inspection deficiencies because reagents are not properly labeled. Inspection guidelines require that all reagent labels contain the information listed in the question.

Questions 04.2161, 04.3133, 04.4112, 04.5119: Are results of reagent quality control testing recorded?

Question 04.2181: If there are multiple components of a reagent kit, does the laboratory use components of reagent kits only within the kit lot unless otherwise specified by the manufacturer?

Questions 04.2160, 04.3132, 04.4111, 04.5118: Are all reagents being used within their indicated expiration date?

Answer: Reagent QC results must be recorded. Otherwise, the inspector will assume that the QC hasn't been performed. The use of appropriate QC sheets can make the task easier. The use of reagents beyond the expiration date continues to be a problem for some laboratories. Nobody wants to discard reagents, but manufacturers assign the expiration date based on available data and it is inappropriate to extend the expiration date without valid data. To minimize this type of reagent loss, the laboratory should determine test volumes, stock quantities that will be used within the expiration date, and procure reagents with a long shelf life.

Components of reagent kits can only be used within the kit lot (unless specified b the manufacturer) to ensure proper function of the test. Therefore, the laboratory shoul purchase kits within the same kit lot to minimize wastage.

Questions 04.2162, 04.3134, 04.4113, 04.5120: Are new reagent lots checked against o reagent lots or with suitable reference material before or concurrently with being place in service?

Answer: The reagent performance must be verified before placing in service or concurre with use. The laboratory may meet this requirement by direct analysis of the reagen testing suitable reference material, or parallel testing with the old reagent lot. The inter is to ensure that the new and previous reagents yield a similar result. Whether a patier sample or reference material is used, it is always preferable to use a weakly positiv instead of a strongly positive, sample to ensure the sensitivity of the new reagent lot.

Mycobacteriology

Question 04.3120: When clinically indicated, are results of acid-fast stains reported withi 24 hours of specimen receipt by the laboratory?

Question 04.3122: Are susceptibility test results for *M. tuberculosis* available in a timel manner?

Question 04.3210: Fluorochrome staining for direct smears?

Question 04.3212: Liquid medium such as BACTEC, Septi-Chek, or MGIT, or thin aga plates with microscopic examination, when appropriate?

Question 04.3214: DNA probes, chromatography, and/or the NAP test for identification o isolates?

Answer: Many of the new mycobacteriology guidelines ''encourage'' laboratories to provide AFB smears (using fluorochrome staining) on a 7 day/week basis and to utilize the mos rapid methods for detection, identification, and susceptibility testing of *M. tuberculosis.* A CDC work group recommends that susceptibility testing be completed within 28 days of specimen receipt (*see* procedure 13.2, reference 5). Although it is doubtful that most mycobacteriology laboratories can meet this recommendation, it is a worthy goal. To improve time to detection, identification, and susceptibility testing, laboratories will have to use liquid media (e.g., BACTEC, Septi-Chek, MGIT, etc.) for cultures; chromatography, nucleic acid probes/PCR, or NAP test for identification; and liquid-based tests for susceptibility testing. If these changes are not possible, the laboratory should consider a reference laboratory that can ensure rapid results.

APPENDIX 13–2 Currently Approved 1997 Proficiency Testing Programs for CLIA

Accutest
P.O. Box 999
Westford, MA 01886-0031
(800) 356-6788

American Academy of Family Physicians
8880 Ward Parkway
Kansas City, MO 64114-2797
(800) 274-7911

American Academy of Pediatrics
AAP-PT
P.O. Box 927
Elk Grove Village, IL 60009-0927
(847) 981-7662
(800) 433-9016 Ext. 7662

American Association of Bioanalysts
Proficiency Testing Service
205 West Levee Street
Brownsville, TX 78520-5596
(800) 234-5315

American Proficiency Institute
1159 Business Park Drive
Traverse City, MI 49686
(800) 333-0958

American Thoracic Society
1740 Broadway
New York, NY 10019-4374
(212) 315-8700

California Thoracic Society (offers only to California resident laboratories)
202 Fashion Lane, Suite 219
Tustin, CA 92680
(714) 730-1944

The College of American Pathologists-Surveys
College of American Pathologists
325 Waukegan Road
Northfield, IL 60093-2750
(847) 832-7000

External Comparative Evaluation for Laboratories-Excel
College of American Pathologists
325 Waukegan Road
Northfield, IL 60093-2750
(847) 832-7000

Idaho Bureau of Laboratories
Proficiency Testing Program
2220 Old Penitentiary Rd.
Boise, ID 83712
(208) 334-2235

Medical Laboratory Evaluation (MLE)
2011 Pennsylvania Avenue, NW, Suite 800
Washington, DC 20008-1808
(202) 835-2746
(800) 338-2746

State of Maryland (available to cytology laboratories possessing Maryland state license)
Office of Licensing and Certification Programs
Division of Laboratory Licensure
4201 Patterson Avenue, 4th Floor
Baltimore, MD 21215
(410) 764-4688

New Jersey Department of Health
Proficiency Testing Program, CN 360
Trenton, NJ 08625-0360
(609) 530-6172

New York State Department of Health
The Governor Nelson A. Rockefeller State Plaza
P.O. Box 509
Albany, NY 12201-0509
(518) 474-8739
(A laboratory located outside the State of New York that holds a valid New York State permit may use the New York State proficiency testing [PT] program to fulfill federal requirements for PT enrollment)

Ohio Department of Health
1571 Perry Street
P.O. Box 2568
Columbus, OH 43216-2568
(614) 466-2278

Pacific Biometrics
110 Eastlake Avenue East
Seattle, WA 98109
(206) 233-9151

Commonwealth of Pennsylvania
Department of Health
Bureau of Laboratories
P.O. Box 500
Exton, PA 19341-0500
(215) 363-8500

Puerto Rico Department of Health
Laboratory Services Program
Department of Health of Puerto Rico, Building A
Call Box 70184
San Juan, Puerto Rico 00936
(809) 274-7735

Solomon Park Research Institute
12815 NE 124th Street, Suite 1
Kirkland, WA 98034
(800) 769-7774

Wisconsin State Laboratory of Hygiene
465 Henry Mall
Madison, WI 53706-1578
(800) 462-5261

ECTION 14

Safety in the Microbiology Laboratory

SECTION EDITOR AND AUTHOR: *Gerald A. Denys*

4.1 Introduction

ms 14.2 and 14.3 deal with the essential ocedures for biohazard prevention and ntainment in the microbiology labora-ry. They provide basic information on tivities which may transmit infection d on safe work practices to prevent the nsmission of infectious agents. Also esented are general control measures for icrobial containment.

Items 14.4 through 14.12 describe common operational procedures performed in the clinical microbiology laboratory. They present proper safety practices for working with equipment and handling infectious material. Procedures for dealing with spills, decontamination, and treatment and disposal of infectious waste are also delineated.

Microbiology laboratories should establish a safety management program that addresses safe work practices and the management of all types of generated waste. The program should be based on regulatory requirements and institutional policy. Items 14.13 through 14.15 outline the essential components of a safety management program and include procedures for compliance with the requirements of most regulatory agencies.

14.2 Biohazard Prevention

I. PRINCIPLE

Many different categories of microbiological hazards are encountered from the time a specimen is collected until it is disposed of permanently. The greatest potential risks of infection for clinical microbiologists are associated with processing primary clinical specimens and manipulating the pathogens isolated from these materials. The management of risks engendered by working with pathogens is accomplished by the development and implementation of standard procedures ar practices for handling infectious materia and will prevent microbial transmissio "Infection" as used in this section impli overt clinical manifestation of disease.

II. ROUTE OF INFECTION AND LABORATORY ACTIVITIES

Exposure to infectious agents can occur by several routes. The actual manifestatic of an infection depends on the concentration and virulence of the infecting agen the route of exposure, and the susceptibility of the host.

A. Inhalation: activities that generate aerosols
- **1.** Manipulating needles and syringes
  - **a.** Expelling air from tubes or bottles
  - **b.** Withdrawing needles from stoppers
  - **c.** Separating needles from syringes
- **2.** Manipulating inoculation needles or loops
  - **a.** Flaming loops
  - **b.** Cooling loops in culture media
  - **c.** Subculturing and streaking culture media
- **3.** Manipulating pipettes
  - **a.** Mixing microbial suspensions
  - **b.** Spilling microbial suspensions on hard surfaces
- **4.** Manipulating specimens and cultures
  - **a.** Centrifugation
  - **b.** Mixing, blending, grinding, shaking, sonication, and vortexing of speci mens or cultures
  - **c.** Pouring or decanting fluids
  - **d.** Removing caps or swabs from culture containers
  - **e.** Spilling infectious material
  - **f.** Filtering specimens under vacuum

B. Ingestion: activities related to oral transmission
- **1.** Pipetting by mouth
- **2.** Splashing contaminated material into the mouth
- **3.** Placing contaminated material or fingers in the mouth
- **4.** Eating, drinking, using lipstick, and smoking in the workplace

C. Inoculation: activities related to direct intravenous and subcutaneou transmission
- **1.** Manipulating needles and syringes
- **2.** Handling broken glass, scalpels, and other sharp objects

D. Inoculation: activities related to contaminated skin and mucus membranes

1. Splashing or spilling material into eyes, mouth, and nose and onto skin
2. Exposing nonintact skin to contaminated material
3. Working on contaminated surfaces
4. Handling contaminated equipment
5. Inappropriate handling of loops, inoculating needles, or swabs containing specimen or culture material

II. SAFE WORK PRACTICES

A. Handling of specimens

1. Gloves and gowns
   - **a.** Wear gloves and gowns (impervious to liquids) at all times when handling and processing patient specimens and decontaminating instruments and countertops and cleaning spills. Coats with snug-fitted sleeves and aprons are optional.
   - **b.** Bandage open cuts and scratches on hands and then wear gloves.
   - **c.** Wear gloves (e.g., latex-free or vinyl) when performing phlebotomy and when handling actual blood specimens.
   - **d.** Wash hands immediately after gloves are removed, after a task that involves heavily contaminated matter, and before leaving the laboratory.
2. Specimen transport
   - **a.** Place specimens in plastic bags and transport in leak-proof containers with the biohazard symbol affixed (*see* Fig. 14.12–3).
   - **b.** Do not accept grossly soiled or contaminated specimens. Notify an individual responsible for submitting such a specimen immediately, and follow the laboratory's specimen rejection policy.
3. Needles and syringes
   - **a.** Use needle-locking syringes or disposable syringe-needle units.
   - **b.** Never recap or bend needles or remove them from syringes.
   - **c.** Do not accept specimens received in syringes with needles attached. Notify the individual responsible for submitting such a specimen immediately and follow the laboratory's specimen rejection policy.
   - **d.** Discard in a puncture-resistant container that has the biohazard symbol affixed.
   - **e.** Use needleless systems for blood samples if available.
   - **f.** Use mechanical devices or one-handed techniques when handling sharp objects.
   - **g.** Secure blood culture bottles before inserting needles into the bottles (e.g., place bottles in support racks).
4. Tubes
   - **a.** Carry tubes in racks.
   - **b.** Use plastic tubes when possible.
   - **c.** Uncap tubes behind a clear plastic shield to contain splashes or sprays (e.g., when removing tops from vacuum tubes).
5. Centrifuges
   - **a.** Centrifuge tubes must be intact and properly balanced when centrifuged.
   - **b.** Centrifuge tubes used in mycobacteriologic areas must be enclosed in sealed safety cups.
   - **c.** Use aerosol-free centrifuges when possible (e.g., in the mycobacteriology laboratory).
   - **d.** Centrifuge safety cups must be opened in a Biological Safety Cabinet (BSC) after centrifugation.
   - **e.** Do not place table top centrifuges in a BSC because air turbulence within the cabinet can allow aerosols to escape.

III. SAFE WORK PRACTICES *(continued)*

6. Hand washing
   a. Perform frequent hand washing after removing gloves, before leavin the laboratory, and before eating, drinking, or applying cosmetics.
   b. Use nonirritating soap for routine washing.
   c. Use antiseptic soap or alcohol planchet followed by thorough hand wash ing for accidental skin contamination.

It is imperative that these cultures be handled in a biosafety hood.

B. Processing of specimens

1. Process all specimens in a BSC. See procedure 14.5 for BSC operatin procedures.
2. Sterilize bacteriological wire needles and loops to avoid spattering of mate rial on heating.
   a. Use electric incinerators if available.
   b. Use Bunsen burners equipped with safety sleeves.
   c. Cool needle and loop tips enough to avoid searing the surface of th medium.
   d. Alternatives to wire needles and loops which require no heating includ plastic disposable loops and spreaders for streaking plates and spreadin material onto slides.
3. Mix or transfer liquids by using disposable plastic pipettes and a rubbe bulb. Alternatively, use mechanical pipetting devices.
4. Cover tubes when mixing, blending, vortexing, etc. (e.g., cap tubes or cove them with parafilm)
5. Work over an absorbent covering or disinfectant-moistened mat (e.g., phenolic-soaked pad for TB).
6. Plan tasks to minimize exposure to known hazards.
7. Follow Universal Precautions when performing nonculture techniques (e.g. antigen detection, polymerase chain reaction, or DNA probes).

Observe universal (standard) precautions.

C. Housekeeping and miscellaneous safe practices

1. General
   a. Avoid or minimize activities outlined in procedure 14.2, item II above
   b. Designate clean and contaminated work areas.
      (1) Wear gloves in contaminated areas.
      (2) Clean and disinfect all surfaces after spills and at the end of eac work shift.
   c. Keep all work areas neat and uncluttered.
      (1) Do not store personal items in the work areas.
      (2) Do not store large quantities of disposable items in the work areas
   d. Remove coats and gowns before leaving the laboratory. Place contami nated laundry (e.g., reusable lab coats and gowns) in designated bag for cleaning by the institution.
   e. Dispose of all contaminated material in containers for treatment (e.g. by autoclaving).
2. Compressed gases
   a. Secure cylinders in an upright position with wall mounts.
   b. Store cylinders away from open flames and sources of heat.
   c. Use the correct pressure regulators.
   d. Verify the contents of the cylinder before gas is used.
   e. Transport cylinders in secured hand trucks or carts.
3. Chemicals
   a. Wear appropriate PPE when handling hazardous chemicals (*see* proce dure 14.10).
   b. Label all reagents with their chemical names and appropriate hazar warnings provided from the MSDS information.
   c. Keep MSDSs for all chemicals, either in the laboratory or in a nearb office.

d. Store flammable and combustible liquids in fire-rated storage cabinets and explosion-proof refrigerators.

e. Store all hazardous chemicals including reagents and dyes below eye level.

(1) Use plastic bottles when appropriate.

(2) Store on the bench volumes necessary for daily work.

f. Place chemical waste in a flume hood until final disposal.

SUPPLEMENTAL READING

Gerson, R., and T. Stimpfel. 1992. Physical ergonomic safety, p. 14.4.1–14.4.3. *In* H. D. Isenberg (ed.), *Clinical Microbiology Procedures Handbook*, vol. 2. American Society for Microbiology, Washington, D.C.

NCCLS. 1991. *Protection of Laboratory Workers from Instrument Biohazards.* Document 117-P. NCCLS, Wayne, Pa.

NCCLS. 1994. *Clinical Laboratory Safety.* Document GP17-T. NCCLS, Wayne, Pa.

Sewell, D. L. 1995. Laboratory-associated infections and biosafety. *Clin. Microbiol. Rev.* **8:** 389–405.

Yablonsky, T. 1996. How safe is your laboratory? *Lab. Med.* **27:**92–98.

14.3 Biohazard Containment

I. PRINCIPLE

In addition to implementing standard microbiological procedures and practices, management of the biohazards associated with working with pathogens includes physical and administrative controls. The BSL is based on the risk associated with the pathogenicity of the microorganisms encountered. Most clinical microbiology laboratories follow BSL 2 practices. When working with highly infectious agents for which the risk of aerosol transmission is greater (e.g., *Brucella* species, *Francisella* species, *Mycobacterium tuberculosis*, and systemic fungi), clinical microbiology laboratories should follow BSL 3 practices. This procedure outlines general physica and administrative controls for microbia containment. Refer to Table 14.3–1 for a sample risk assessment and exposure con- trol plan. Individual laboratories shoulc modify this table to conform with the re- quirements in their own institutions.

Table 14.3–1 Risk assessment and exposure control plan for the clinical microbiology laboratory[a,b]

Laboratory section and task	Exposure risk from		PPE[c]			Engineering control[d]			
	Blood and body fluids	Cultured biological agents	Gloves	Lab coat/ gown	Face shield	Splash shield	BSC	Sharps containers readily assessible[e]	Safety centrifuge[f]
General									
Inventory: media and supplies	Low			Coat					
Clerical: computer entry, telephones, records, reports, calculations, writing	Low		P	Coat					
Instrument maintenance:									
Parts contaminated with blood or body fluids	High	Variable	R	Gown					
Parts not contaminated with blood or body fluids	Low	Variable	D	Coat					
Surface decontamination	Low	Variable	R	Coat					
Infectious-waste disposal	High	Variable	R	Gown	A(D)			Sharps	
Bacteriology									
Primary specimen processing	High	BSL 2	R	Coat			R	Sharps	
Subculture blood culture bottles	High	BSL 2	R	Coat	A[g]	A[g]	A[g]	Needles, vents	
Subculture colonies or broth tubes	Low	BSL 2[h]		Coat				Sharps	
Identification tests and AST	Low	BSL 2		Coat				Sharps	
Prepare smears and fix slides	Low	BSL 2		Coat				Sharps	
Stain fixed slides and read	Low			Coat				Slides	
Mycobacteriology and mycology									
Primary specimen processing	High	BSL 2/3	R	Gown			R[i]	Sharps	R[i]
Prepare smears, wet mounts, India ink preps; fix slides	High	BSL 2/3	R	Gown			R	Sharps	
Read wet mounts, India ink preps from specimens	High	BSL 2	R	Coat				Slides/covers	
Read wet mounts, India ink preps from cultures	Low	BSL 2/3	R	Coat				Slides/covers	
Examine sealed cultures	Low	BSL 2/3[j]		Coat					
Stain fixed slides and read	Low			Coat				Slides	

Table 14.3–1 Risk assessment and exposure control plan for the clinical microbiology laboratory[a,b] (*continued*)

Laboratory section and task	Exposure risk from: Blood and body fluids	Exposure risk from: Cultured biological agents	PPE[c]: Gloves	PPE[c]: Lab coat/ gown	PPE[c]: Face shield	Engineering control[d]: Splash shield	Engineering control[d]: BSC	Engineering control[d]: Sharps containers readily assessible[e]	Engineering control[d]: Safety centrifuge[f]
Handling yeast cultures, smears and fixed slides	Low	BSL 2		Coat			D	Slides	
Handling molds and mycobacteria cultures	Low	BSL 3[k]	R	Gown			R[l]	Sharps	R[l]
Virology[m]									
Primary specimen processing	High	BSL 2	R	Coat			R	Sharps	
Feed and manipulate uninoculated cells	Low			Coat				Pipettes	
Read inoculated cells, tubes or vials for CPE	Low	BSL 2		Coat					
Feed and manipulate inoculated cells	High	BSL 2	R	Coat			R	Pipettes	
Perform tests to identify viruses	High	BSL 2	R	Coat			D	Sharps	
Stain fixed slides and read	Low			Coat				Slides	
Parasitology									
Concentrate fecal specimens, smear, wet mounts	Low	BSL 2	R	Coat				Pipettes/sticks	
Read fecal wet mounts	Low	BSL 2	R	Coat				Slides/covers	
Prepare thick and thin blood smears; fix slides	High	BSL 2	R	Coat	A	A		Slides/covers	
Stain and read slides (focal and fixed blood)	Low			Coat				Slides/covers	
Antigen detection/PCR/DNA probes									
Primary specimen processing	High	BSL 2/3	R	Coat	A[n]	A[n]	A[n]	Pipettes	MTB (R)
Cultured microorganisms	Low	BSL 2	D	Coat			R	Sharps	
	Low	BSL 3	R	Gown				R[o]	Sharps
Antibiotic levels (Schlichter test)									
Manipulate organisms (without serum)	Low	BSL 2		Coat				Pipettes/loops	
Manipulate serum	High	BSL 2	R	Coat	A[n]	A[n]	A[n]	Pipettes	
Serology									
Manipulate serum	High	BSL 2	R	Gown	A	A		Pipettes	
Arrange tubes; prepare and dispense reagents	Low			Gown					
Mix serum and reagents; read and discard tests	High	BSL 2	R	Gown	A[n]	A[n]	A[n]	Pipettes	

[a] Adapted from CMPH (Introduction, p. xxxiii–xxxv).

[b] P, prohibited; R, required; D, discretionary; A, one of the required alternatives; AST, antimicrobial susceptibility testing; CPE, cytopathic effect; MTB, *Mycobacterium tuberculosis*.

[c] Remove PPE when leaving the laboratory. Gowns must have solid fronts and be impervious to liquid. Many employers provide and launder gowns, thus eliminating the need for lab coats.

[d] Recapping of needles should be prohibited. Carry tubes in racks, or use plastic tubes. Plan each task to minimize known hazards. Wash hands when leaving the laboratory.

[e] Sharps include scalpel blades, pipettes, plastic loops, sticks, needles, syringes, slides, and coverslips.

[f] Open sealed cups only in a BSC.

[g] Use a BSC or acrylic splash shield.

[h] Requires surveillance and action plan for occasional isolation of BSL 3 organisms (e.g., *Brucella* species, *Francisella* species, *Mycobacterium* species, and systemic fungi), especially if plates are held for ≥3 days.

[i] MTB requires a BSC and safety centrifuge.

[j] Requires a contingency plan for breakage of culture containers.

[k] Mycobacteria other than tuberculosis (MOTT group) may be handled at BSL 2; however, use BSL 3 practices since most manipulations precede organism identification.

[l] Use a HEPA-filtered mask or respirator in addition to a BSC for culture of *Mycobacterium tuberculosis*.

[m] Special precautions for BSL 4 agents (e.g., hemorrhagic fever virus) should be arranged (e.g., call CDC).

[n] Vortexing or other splatter-generating steps require use of a BSC or safety shield.

[o] Requires BSL 3 practices if there is potential for aerosols.

II. BIOSAFETY LEVELS

A. General laboratory design (BSL 1)

1. Restrict access to authorized personnel.
2. Make sinks for hand washing readily accessible.
3. Make eyewash stations readily accessible.
4. Make appropriate PPE available and ensure their use.
5. Ensure that laboratory benchtops are impervious to liquids and resistant to chemicals.
6. Ensure that laboratory surfaces and equipment are easily cleaned and disinfected and that these procedures are done on a regular basis or whenever the surfaces or equipment are contaminated.
7. Decontaminate solid waste within the laboratory (e.g., by autoclaving) or package the waste to be transported off site.

B. BSL 2

Observe universal (standard) precautions.

1. Follow BSL 1 practices plus the following.
2. Display universal biohazard signs outside of the laboratory (*see* Fig 14.12–3).
3. Perform specimen processing in a BSC (*see* procedure 14.5).
4. Perform centrifugation of mycobacteriologic specimens by using centrifuge safety cups (*see* procedure 14.6).
5. Ensure that an autoclave or other decontamination equipment is available and used for treatment of infectious waste (*see* procedure 14.11).
6. Use the appropriate PPE (e.g., gowns, gloves, and facial barriers).
7. Place all sharps carefully in conveniently located, puncture-resistant containers.
8. Train personnel to observe good microbiological practices and techniques.

C. BSL 3

It is imperative that these cultures be handled in a biosafety hood.

1. Follow BSL 1 and BSL 2 practices plus the following.
2. Control access to the laboratory.
3. Perform all manipulations of cultures and clinical material in a BSC (class II).
4. Maintain a negative-pressure airflow in the laboratory.
5. Include double doors and an anteroom in the laboratory design.
6. Discharge HEPA-filtered exhaust air from BSCs outside the facility.
7. Use all appropriate PPE and containment devices.
8. Use HEPA-filtered respirators or masks when aerosols may be generated.
9. Collect baseline serum samples from all personnel for serological determination of immune status.

III. PPE AND CONTAINMENT DEVICES

A. OSHA Bloodborne Pathogen Standards require laboratory to provide PPE for its employees.

1. Protective clothing
    a. Coats, gowns, or aprons impervious to liquids
    b. Garments fitted snugly around the wrists
2. Gloves
    a. Disposable and powderless gloves
    b. Glove liners
    c. Hypoallergenic or latex-free gloves
3. Face and eye protection
    a. Masks
    b. Goggles or safety glasses
    c. Face shields
    d. Acrylic splash shields

B. Engineering controls should be in place within the laboratory.
  1. Automatic or mechanical pipetting devices
  2. Biohazard bags
  3. Biohazard labels are required on:
    a. Refrigerators and freezers containing blood or infectious materials
    b. Containers used to store, transport, or ship regulated waste, blood, or infectious materials
    c. Sharps containers
    d. Contaminated equipment
  4. BSC
  5. Centrifuge safety cups
  6. Containers
    a. Secondary containers (transport of specimens)
    b. Sharps (puncture-resistant)
    c. Waste (including stainless-steel buckets)
  7. Eyewash stations
  8. Hand washing sinks
  9. HEPA-filtered respirators or masks
  10. Waste collection carts

C. Commercial safety products
  1. Needle stick prevention devices (e.g., those from Becton Dickinson Microbiology Systems, Cockeysville, Md.)
  2. Biogel Revel glove (Regent Hospital Products, Greenville, S.C.)
  3. Automated petri plate spreader (e.g., Isoplater 80, Vista Laboratories Ltd., Edmonton, Alberta, Canada)
  4. Automated tissue processor (e.g., Stomacher, Seward Medical Ltd., London, United Kingdom).

SUPPLEMENTAL READING

Denys, G. A. 1992. Safety in work with bloodborne pathogens: the OSHA bloodborne pathogen standard, p. 14.5.1–14.5.6. *In* H. D. Isenberg (ed.), *Clinical Microbiology Procedures Handbook*, vol. 2, supplement 1. American Society for Microbiology, Washington, D.C.

Fleming, D. O. 1995. Laboratory biosafety practices, p. 203–218. *In* D. O. Fleming, J. H. Richardson, J. J. Tulis, and D. Vesley (ed.), *Laboratory Safety: Principles and Practices*, 2nd ed. American Society for Microbiology, Washington, D.C.

Gershon, R., and I. F. Salin. 1992. Biological safety, p. 14.1.1–14.1.6. *In* H. D. Isenberg (ed.), *Clinical Microbiology Procedures Handbook*, vol. 2. American Society for Microbiology, Washington, D.C.

Gilchrist, M. J. R., J. Hindler, and D. O. Fleming. 1992. Laboratory safety management, p. xxix–xxxvii. *In* H. D. Isenberg (ed.), *Clinical Microbiology Procedures Handbook*, vol. 1. American Society for Microbiology, Washington, D.C.

NCCLS. 1994. *Clinical Laboratory Safety.* Document GP17-T. NCCLS, Wayne, Pa.

Richardson, J. H., and R. R. M. Gershon. 1994. Safety in the clinical microbiology laboratory, p. 37–45. *In* B. J. Howard, J. F. Keiser, T. F. Smith, A. S. Weissfeld, and R. C. Tilton (ed.), *Clinical and Pathogenic Microbiology*, 2nd ed. Mosby-Year Book, Inc., St. Louis, Mo.

14.4 Autoclave

I. PRINCIPLE

An autoclave or steam sterilizer is an insulated pressure chamber in which saturated steam is used to elevate temperature above the normal boiling point of water. A gravity displacement autoclave, in which lighter steam is fed into the chamber to displace heavier air, is the most common type used in the laboratory. The higher the temperature and pressure, the shorter the time required to achieve killing of microorganisms. In addition to time, temperature, and direct steam contact, other factors that can affect killing time include density physical state and size, and organic content of the material treated. Autoclaves are used for sterilization of both culture media and heat-stable supplies and the treatment of infectious waste.

II. MATERIALS

A. Physical monitors (recorder chart)
1. Exposure time
2. Exposure temperature
3. Chamber pressure

B. Biological indicators (*Bacillus stearothermophilus*)
1. Ampoule
2. Spore strip

C. Chemical indicators
1. Chemical-impregnated strip
2. Autoclave tape

D. Containers and packaging
1. Autoclave bags, trays, or pans
2. Metal pails
3. Linen packs

E. PPE
1. Splash goggles
2. Heat-resistant gloves

III. PROCEDURES

A. Media, supplies, and linen packs
1. Initial preparation
   a. Loosen all caps or lids.
   b. Do not fill vessels more than two-thirds full.
   c. Place autoclave tape on items to be sterilized.
2. Load the autoclave with material to be sterilized.
   a. Allow adequate space between objects (at least 2 in. apart).
   b. Use test tube or other wire racks to allow the free flow of steam.
3. Close the autoclave and set the sterilization time and temperature (*see* manufacturer's instructions).
   a. Autoclave culture media for 15 min at 121°C with a 15-min exhaust time. Add 3 min for each 500 ml of medium.
   b. Autoclave dry-wrapped material for 25 min at 121°C with fast exhaust and a 30-min dry cycle.
4. Start the cycle and confirm that temperature and pressure monitors are functioning.
5. After the cycle has ended and the chamber pressure has reached 0 lb/in.^{2}, open the door slowly.
   a. Stand behind the door to avoid steam.
   b. Wear eye protection and heat-resistant gloves.

c. Allow 20 min for contents to cool.
d. Handle all liquid material gently to prevent splattering of hot fluid.

6. Carefully remove material from the autoclave and place on designated area (e.g., benchtop or cart) labeled ''autoclaved.''
7. Tighten caps.

B. Infectious waste

1. For initial preparation, infectious waste should be clearly marked and separated from the microbiological media and supplies to be sterilized.
2. Autoclave bags
   - **a.** Place infectious waste in double autoclave bags with biohazard labels.
   - **b.** Fill bags no more than three-fourths full.
   - **c.** Tie bags loosely or add 1 cup of water to bags before sealing (optional).
   - **d.** Place autoclave tape on the outside of the bags.
   - **f.** Place the bags on autoclave pans or trays.
   - **g.** Place liquid waste in vessels (not to exceed two-thirds full) in pails.
3. Load autoclave bags on pans or trays or in pails to be sterilized.
4. For routine waste, set the time and temperature for 15 min at 121°C with a 15-min exhaust time.
5. For mycobacteriology waste (e.g., stock cultures), set the time and temperature at 30 to 45 min and 121°C with a 15-min exhaust time.
6. Proceed as outlined in procedure 14.4, item III.A.
7. Drain excess fluid from the pans and clean them with soap and water. Allow molten agar to solidify and then discard it in normal trash.

V. LIMITATIONS

A. Steam under pressure and at high temperature presents the risks of scalding and explosion to the operator.

B. Antineoplastic agents, toxic chemicals, or radioisotopes in waste may not be destroyed. In the microbiology laboratory, the most likely types of radiation are ^{14}C and ^{3}H, which can be autoclaved without any radiation hazard.

C. Highly volatile chemicals could become vaporized and disseminated by heat and ignited by a spark from operating equipment.

/. QUALITY CONTROL

A. Each run

1. Check the data recorder to verify correct time and temperature.
2. Check the autoclave tape or chemical indicator strip to verify operating conditions.

B. Daily

1. Remove the sediment screen from the chamber drain hole and clean thoroughly.
2. Periodically clean accessories such as shelves, racks, and trays with soap and water.

C. Weekly

Include a biological indicator ampoule or strip in a routine run based on manufacturer's instructions.

D. Troubleshooting and maintenance

Contact a technical service representative.

SUPPLEMENTAL READING

Joslyn, L. J. 1991. Sterilization by heat, p. 495–526. *In* S. S. Block (ed.), *Disinfection, Sterilization, and Preservation*, 4th ed. Lea & Febiger, Philadelphia.

Nelson, E. A., and E. Molitoris. 1992. Autoclave (steam sterilizer), p. 12.3.1–13.3.7. *In* H. D. Isenberg (ed.), *Clinical Microbiology Procedures Handbook*, vol. 2. American Society for Microbiology, Washington, D.C.

14.5 Biological Safety Cabinet

I. PRINCIPLE

The BSC is the most important primary containment equipment in the clinical microbiology laboratory. Class II laminar flow cabinets are the most commonly used for BSL 2 and 3 practices. Air is drawn into the cabinet by negative air pressure and passes through a HEPA filter. The air flows in a vertical sheet that serves as a barrier between the outside and inside of the cabinet. The cabinet exhaust air is also passed through HEPA filters. Aerosols are contained within the BSC, and the work area is protected from outside contamination when the cabinet is operating under the manufacturer's recommended conditions.

II. MATERIALS

Include QC information on reagent container and in QC records.

A. Disinfectants

1. 10% household bleach solution (0.5% sodium hypochlorite)
   Metal surfaces must be rinsed to avoid corrosion.
2. 70% ethanol
   Store outside of the cabinet.
3. See Appendix 14:8–1 for EPA-registered disinfectants that are tuberculocidal and virucidal. Follow the manufacturer's directions for preparation.

B. Common supplies

1. Sterile disposable loops and spreaders
2. Bacteriological wire needles and loops
3. Sterile swabs
4. Sterile disposable pipettes
5. Sterile forceps
6. Glass slides
7. Absorbent mats or towels

C. Common equipment

1. Incinerator
2. Heat block
3. Vortex
4. Sonicator
5. Tissue grinders
6. Filtering apparatus
7. Pipetting devices
8. Test tube rack
9. Discard container

III. PROCEDURES

A. Start-up

1. Turn on blowers for at least 10 min before QC checks and specimen processing.
2. Open the viewscreen to operating height (8 in.).
3. Turn off UV lights (if present).
4. Turn on a fluorescent light.
5. Perform daily QC as outlined in item IVA, below.
6. Clean the cabinet working surfaces with appropriate disinfectant.

B. Operation

1. Avoid outside sources of air currents (e.g., personnel walking by, doors being opened or closed, etc.).
2. Wear long-sleeved coats or cuffed gowns and gloves during specimen processing.
3. Place all items that will be used in a planned activity inside the cabinet before starting work.

III. PROCEDURES *(continued)*

a. Segregate clean from contaminated materials.
b. Place the minimum number of large devices and supplies inside the cabinet.
   (1) *Do not* block intake or rear grills.
   (2) *Do not* place sterile material or specimens near the sides, front, or back of the cabinet.
   (3) *Do not* place or tape paper notes and procedures on the window.
   (4) *Do not* use a flame in the cabinet.
   (5) *Do not* operate centrifuges in the cabinet.

4. Plan work flow to minimize movements.
   a. Work at least 6 in. inside the front grill intake.
   b. Avoid rapid arm movement in and out of the cabinet while a procedure is in progress.
   c. If necessary, slowly move arms in and out of the cabinet.

C. Shutdown

1. Allow the cabinet to continue running for 15 to 20 min after work is completed and before removing materials.
2. Allow the cabinet to continue running for at least 3 h after processing acid-fast or fungal specimens.
3. Place contaminated materials in covered containers or closed bags, or immerse them in disinfectant, before removal from the cabinet.
4. Disinfect the surfaces for any contaminated materials before removal from the cabinet.
5. Clean up spills by following the steps outlined in procedure 14.9, item III.C.
6. Clean interior surfaces with disinfectant (e.g., 70% alcohol).
7. Turn off the blower and lights.
8. Turn on UV light (if present).
9. Close the viewscreen (if present).
10. Do not shut down cabinets that function 24 h a day, unless a malfunction occurs or routine maintenance is required.

IV. QUALITY CONTROL

A. Daily

1. Disinfect all cabinet surfaces while the cabinet is running.
2. Check air velocity (recommended for BSCs not equipped with velocity gauges and airflow indicators in the work area).
   a. Place a vaneometer (Lab Safety Supply, Janesville, Wis.) at the opening of the cabinet toward the workspace area, but not on the airflow grill.
   b. Level the vaneometer and record face airflow velocity.
      (1) The air velocity should be greater than 100 ft/min.
      (2) Do not use the BSC if velocity remains less than 100 ft/min. Contact a technical service consultant.
3. Check the blower function.
   a. Record the manometer gauge reading (if provided).
   b. Perform a visible-smoke test.
      (1) Pass a smoke source (e.g., smoke bottle or smoke stick; Lab Safety Supply, Janeville, Wis.) from one end of the cabinet opening to the other.
      (2) The smoke should show a smooth downward flow with no dead spots or reflux.
      (3) Record results.
   c. Observe airflow check strips (optional).
      (1) Strips should be drawn toward the work space area.
      (2) Record results.

B. Weekly
Clean UV lights (if present) with 70% ethanol.

C. Monthly
Clean the gutter area with disinfectant.

D. Semiannually

1. Have mycobacteriology and mycology BSCs recertified by certified personnel
2. Have certified personnel measure UV light output (if present).

E. Annually (or every 1,000 h)
Recertify routine and virology BSC by certified personnel.

V. LIMITATIONS

A. BSCs should not be used as fume hoods.
   1. Toxic, radioactive, or flammable vapors or gases are not removed by HEPA filters.
   2. Potentially hazardous amounts of volatile material may build up in class II-type cabinets.
   3. Exhausted vapors can be vented into the laboratory air.

B. The use of gloves when performing routine manipulations should not substitute for proper hand washing practices. Wash hands with soap and water between tasks and when they are heavily soiled with clinical material.

SUPPLEMENTAL READING

Kruse, R. H., W. H. Puckett, and J. H. Richardson. 1991. Biological safety cabinet. *Clin. Microbiol. Rev.* **4**:207–241.

Mann, M. B. 1992. Biological safety cabinet, p. 12.19.1–12.19.16. *In* H. D. Isenberg (ed.), *Clinical Microbiology Procedures Handbook*, vol. 2, supplement 1. American Society for Microbiology, Washington, D.C.

National Sanitation Foundation. 1988. *Standard 49 for Class II (Laminar Flow) Biohazard Cabinetry.* National Sanitation Foundation, Ann Arbor, Mich.

APPENDIX 14.5–1

Biological Safety Cabinet Quality Control and Maintenance Record

Month ________ Year ________ Cabinet no./Class ________ Location ________

Date	Visual airflow reading	Blower/gauge reading (manometer)	Vaneometer reading (optional) ≥100 ft/min	Cleaning	Other
1					
2					
3					
4					
5					
6					
7					
8					
9					
10					
11					
12					
13					
14					
15					
16					
17					
18					
19					
20					
21					
22					
23					
24					
25					
26					
27					
28					
29					
30					
31					

Corrective action Date: Reviewed by:

Note: Sign initials in appropriate box after each task is completed.

14.6 Centrifuge

I. PRINCIPLE

A centrifuge produces centrifugal force that is used to separate particles of various densities. In the microbiology laboratory, centrifugation is used for concentrating microorganisms and cells, separating components of body fluids or mixed suspensions, and removing particulate matter. Cytocentrifugation is also used in the laboratory as a rapid means of concentrating small volumes of body fluid specimens for staining. This procedure focuses on general safety issues related to centrifugation. The laboratory should follow the manufacturers' recommendations for the operation and maintenance of specific instruments.

II. MATERIALS

A. Supplies
   1. Disinfectants and cleaning solutions
   2. Paper towels
   3. Gloves, gowns, masks (for spills)

B. Equipment
   1. Centrifuge
      a. Horizontal rotor, fixed-angle rotor, microtube, etc.
      b. Cytocentrifuge (Shandon, Inc., Pittsburgh, Pa.)
   2. Centrifuge tubes, bottles
   3. Blank containers
   4. Safety cups with O-rings
   5. Balance (two-pan)
   6. Calibrated thermometer
   7. Tachometer (e.g., photoelectric)

III. PROCEDURE

A. Balance load
   1. Weigh similar items and balance to within 0.5 g.
   2. Use blank tubes or containers filled with water for balancing.
   3. Weigh buckets with tubes.
      a. Add balance tubes until both buckets are equal in weight.
      b. Distribute tubes and adapters in rotor or bucket holder in a symmetrical arrangement.
      c. Place all buckets (empty or full) in place.

B. Containment of material
   1. All specimens must be capped or covered.
   2. Use nonbreakable plastic screw-cap tubes.
   3. Place tubes in tightly covered safety cups or rotors (e.g., tubes containing mycobacteria and systemic fungi specimens)
   4. Use aerosol-free centrifuges if available.
   5. Cytocentrifuge (double-enclosed system)
      a. *Do not* overfill specimen containers.
      b. All cytospin funnels must be capped.
      c. Open sealed heads in a BSC.
   6. Open sealed safety cups in a BSC after centrifugation

C. Operation
   1. Set appropriate time, rotation speed, temperature, and brake speed (if necessary).

III. PROCEDURE *(continued)*

2. Close and lock the top.
3. Turn the centrifuge on. For manual speed adjustment, slowly increase the speed until the desired rpm is reached.
4. Once the desired speed is reached, check for undue vibration.
   a. Turn the centrifuge off if excess vibration occurs.
   b. *Do not* apply the brake.
5. *Do not* open the top of the centrifuge until it has come to a complete stop.

IV. QUALITY CONTROL

A. Each run

1. Check tubes for cracks.
2. Check inside of carrier cups for rough walls or glass and other debris, and remove any of these from the base of the cushion.
3. Check that each load is properly balanced.
4. Check the operating temperature.
5. Clean up any spills immediately (*see* procedure 14.8).
   a. Use a 10% bleach solution, and rinse surfaces thoroughly after cleaning.
   b. Use alternative disinfectants (*see* Appendix 14.8–1).

B. Daily

1. Clean the inside of the bowl with disinfectant and rinse thoroughly.
2. Refrigerated centrifuges
   a. Allow the bowl to dry if it is turned off at night.
   b. Leave the top closed when the unit is under refrigeration.

C. Weekly

1. Clean and open all vents.
2. Clean and disinfect rotors, carriers, and the inside of the bowl. Rinse and dry thoroughly.

D. Monthly

1. Vacuum clean the condenser coils, fan, screens, and filters (refrigerated centrifuges).
2. Check brushes.
3. Check electrical connections.

E. Semiannually

1. Check the internal temperature of refrigerated units with a calibrated thermometer.
2. Check brushes.
3. Calibrate speed with a photoelectric tachometer.

F. Annually

Have speed and temperature control recertified by qualified personnel.

V. TROUBLESHOOTING

A. Vibration
1. Check for proper balance.
2. Check all tube holders and buckets.
3. Check that the centrifuge is on a level surface.

B. Breakage
1. Check balance.
2. Check tubes for proper size or use.
3. Check tube holders for proper cushion.

C. Corks or stoppers pop off.
1. Use screw-cap tops.
2. Use parafilm over the stopper.

3. Use clean tube holders, free of glass.
4. Use lower temperatures in (refrigerated) centrifuges.

D. Fine gray dust in rotor chamber
1. Thoroughly vacuum clean the chamber, and run the centrifuge empty several times.
2. Clean between operations until dust is gone.

'I. LIMITATIONS

A. *Do not* centrifuge large volumes of flammable liquids which could become vaporized and then be ignited by a spark from operating equipment.

B. Concentration of infectious material for culture, such as bronchoalveolar lavage specimens, by cytocentrifugation may be performed in the routine culture area. Preparations should not produce spatters or aerosols if specimen containers are properly filled.

C. Sample preparation of infectious material for smears by cytocentrifugation for detection of acid-fast bacilli should be pretreated with equal amounts of fresh 5% sodium hypochlorite (household bleach).

UPPLEMENTAL READING

Baron, J. E., and T. Ricker. 1992. Centrifuge, p. 12.6.1–12.6.13. *In* H. D. Isenberg (ed.), *Clinical Microbiology Procedures Handbook*, vol. 2. American Society for Microbiology, Washington, D.C.

Saceanu, C. A., N. C. Pfeiffer, and T. McLean. 1993. Evaluation of sputum smears concentrated by cytocentrifugation for detection of acid-fast bacilli. *J. Clin. Microbiol.* **31**:2371–2374.

APPENDIX 14.6–1

Centrifuge Quality Control and Maintenance Record

Year ____________ Location: ________________ Date in use: ________________

Centrifuge: Model no. ________________ Serial no. ________________ Identification no. ________________

Date	Temperature check	Speed calibration check	Cleaning	Other

Corrective action	Date:	Reviewed by:

Note: Sign initials in appropriate box after each task is completed.

14.7 Gas Cylinders

POLICY

Laboratory personnel should be cognizant of hazards from compressed gases. A falling cylinder can cause physical injury to personnel. A gas cylinder with ruptured pressure-reducing fittings or broken valve heads can become an ''unguided missile,'' destroying everything in its path. Some gasses may be toxic and/or flammable. If heated, cylinders may explode.

I. PROCEDURE

A. Transportation

1. *Do not* move a cylinder with a regulator attached. Valve safety covers must be tightened securely during transport.
2. *Do not* drag or roll cylinders.
   a. Use hand trucks, dollies, carts, etc., to move cylinders.
   b. Properly secure cylinders to hand trucks, etc.
3. *Do not* place cylinders in a horizontal position.

B. Storage

1. *Do not* accept cylinders that are marked with wired-on tags or are color coded. All cylinders must be labeled with the name of the contents and appropriate warning labels (e.g., U.S. Department of Transportation labels).
2. Secure all gas cylinders in an upright position in secure racks or to wall mounts.
3. *Do not* remove valve safety covers until pressure regulators are attached.
4. *Do not* store cylinders with or near flammable materials.
5. *Do not* store cylinders with highly combustible material in the laboratory. A separate room or enclosure should be reserved for stored cylinders.
6. *Do not* store cylinders at temperatures of >125°F (55°C).
7. *Do not* discard empty cylinders.
   a. Mark cylinders as empty and keep them secured in an upright position with safety covers in place.
   b. Return empty or unused cylinders to the manufacturer.

C. Pressure regulators and needle valves

1. Use only designated fittings for gases indicated.
2. *Do not* use oil, grease, or lubricants on valves, regulators, or fittings. Threads and surfaces should be clean and tightly fitted.
3. *Do not* attempt to repair damaged cylinders or to force open frozen cylinder valves.
4. Use a proper size of wrench to tighten regulators.

II. PROCEDURE *(continued)*

5. Open the diaphragm control knob completely before the control valve i opened.
6. Open all valves slowly.
   a. Stand to the side of gauges.
   b. *Do not* force open valves that stick.
7. Check all connections for leaks.
   a. Test before and after regulators or gauges are attached.
   b. Test piped lines before use and periodically thereafter.
   c. Use a "snoop" or soap solution.
8. Leave valve handles attached to cylinders.
9. Regulation of gas flow rate
   a. Use high-pressure valve on the cylinder to set the maximum rate of ga flow.
   b. Use needle valve for fine tuning of gas flow.
10. Shut off valves on gas cylinders when not in use.

D. Cryogenic liquid nitrogen container

1. Handle liquid nitrogen with care.
   a. Use protective clothing.
      (1) Insulated gloves must be worn when handling any object that ha been in contact with liquid nitrogen.
      (2) Safety glasses are recommended during transfer of liquid nitroger
   b. *Do not* allow objects cooled by liquid nitrogen to touch bare skin.
2. Cryobiological storage container
   a. Storage
      (1) Always store in an upright position.
      (2) Store only in areas that are fully ventilated.
      (3) Store in clean dry areas.
   b. *Do not* roll, drag, or tip containers.
   c. *Do not* vibrate, jolt, or drop a container (which might damage its vacuur insulation system).
   d. *Do not* overfill containers.
   e. *Do not* seal containers tightly.
   f. *Do not* tamper with relief valves.
   g. Use solid metal or wooden dipsticks.
   h. Avoid moisture, caustic cleaners, chemicals, or other substances tha might cause corrosion.

III. QUALITY CONTROL

Complete and initial the Gas Cylinder Quality Control Record after each task is com pleted (*see* Appendix 14.7–1).

SUPPLEMENTAL READING

Mulholland, P. A., and D. Citron. 1992. Compressed-gas regulator and handling of gas cylinders, p. 12.9.1–12.9.8. *In* H. D. Isenberg (ed.), *Clinical Microbiology Procedures Handbook,* vol. 2. American Society for Microbiology, Washington, D.C.

NCCLS. 1994. *Clinical Laboratory Safety.* Docu ment GP17-T, p. 20–22. NCCLS, Wayne, Pa.

APPENDIX 14.7–1 Gas Cylinder Quality Control Record

Year ______________ Location ______________________________

Date cylinder received in lab	Cylinder idenification code	Date placed into use	Pressure when placed into use	Date empty and replaced

Corrective action	Date:	Reviewed by:

Note: Complete information and sign initials in appropriate box after each task is completed.

14.8 Decontamination

I. PRINCIPLE

Routine disinfection and cleaning of the work environment is the responsibility of all laboratory workers at both the basic and highly sophisticated laboratory levels. To accomplish this effectively, the work areas should be uncluttered, with clean and dirty materials separated and clearly identified. The properties of chemical disinfectants and their effectiveness and application in the clinical microbiology laboratory are given in reference 1. Appendix 14.8–1 is a partial list of EPA-registered hospital disinfectants. The manufacturer's instructions for preparation and use should be carefully followed. This procedure presents common decontamination practices used in the microbiology laboratory.

II. MATERIALS

(See Appendix 14.8–1.)

A. Alcohol (e.g., 70% ethanol)
Use to decontaminate work surfaces only.

B. Glutaraldehyde
1. Dirty glassware
2. Equipment decontamination

C. Chlorine, iodophors, phenolic compounds, quaternary ammonium compounds (e.g., T.B.Q. germicidal detergent, Calgon Vestal Labs, St. Louis, Mo.)
1. Work surfaces
2. Dirty glassware
3. Equipment decontamination

III. PROCEDURES

A. Preparation of 10% household bleach (0.5% sodium hypochlorite) working solution
1. Prepare fresh daily.
2. Add one part household bleach to nine parts tap water (5,000 mg/liter of available chlorine).
3. Dispense in wash bottles.
4. Record the date prepared on the bottles.
5. Leave wash bottles in the work areas.
6. After 24 h, pour unused bleach solution down the drain and flush the drain with running water to prevent corrosion of pipes.
7. Allow bottles to air dry.

B. Decontamination of work surfaces
1. Follow manufacturer's instructions for cleaning and appropriate disinfecting solutions. Allow bleach solutions to air dry (minimum contact time should be 10 min).
2. Use a paper towel or soft cloth soaked with the recommended disinfectant solution.
3. Decontaminate
   a. Before and at the end of the work shift
   b. Upon completion of a procedure
   c. When surfaces become overtly contaminated (*see* procedure 14.9)

C. Decontamination of equipment (including vortex, centrifuge, and telephone)

1. Follow manufacturer's instructions for cleaning and suitable disinfecting solutions.
2. Use a soft cloth moistened with recommended disinfecting solution.
3. Decontaminate
   a. Upon completion of a procedure
   b. When surfaces become overtly contaminated (*see* procedure 14.9)
   c. Rinse equipment with tap water (except when alcohol is used).
4. If the equipment needs to be replaced or serviced, it must be decontaminated before removal.

D. Decontamination of reusable biohazard pails

1. Remove biohazard lining bags and tape shut with white tape.
   a. If a bag is ripped or torn, place it inside a second bag.
   b. Place red biohazard tape on each bag.
   c. Place each bag on a biohazard cart or in a waste dumpster for disposal.
2. Wipe the entire inside surface of the pails with soft cloths soaked with a fresh solution of 10% bleach.
3. Allow a pail to air dry before placing a new biohazard lining bag into it.
4. Inspect each pail for holes or leakage.
5. Decontaminate
   a. Whenever the biohazard lining bag is changed or at least once per week
   b. Whenever a biohazard lining bag is torn

E. Decontamination of computer keyboards

1. Clean and disinfect keyboards at the end of every shift and when the keyboards are visibly contaminated.
2. Unplug the keyboard from the computer.
3. Clean the keyboard surface with a soft cloth moistened with 70% alcohol or use an alcohol prep pad.
   a. If the keyboard has a fitted cover, inspect the cover for holes or tears.
   b. If the keyboard is uncovered, avoid spilling alcohol underneath the keys.
4. Plug the keyboard back into the computer.
5. If the computer is overtly contaminated with a specimen, clean with 10% bleach solution and rinse with water.

IV. LIMITATIONS

A. Disposable gloves, impervious gowns, eye protection, and masks should be worn when preparing disinfectants and when cleaning and decontaminating soiled equipment.

B. Overtly contaminated reusable equipment and surfaces should be cleaned with an aqueous detergent solution before application of the disinfectant for optimum effectiveness.

C. *Do not* use alcohol solutions for equipment decontamination in poorly ventilated areas or near open flames.

D. *Do not* use sodium hypochlorite solutions on metal parts because they can cause rusting and pitting.

E. *Do not* use glutaraldehyde solutions as general surface disinfectants because of their irritating vapors and prolonged contact time.

V. QUALITY CONTROL

Sign initials in the appropriate box on the decontamination worksheet after completion of each task. See Appendix 14.8–2 for a sample QC daily recording worksheet.

REFERENCE

1. **Marsik, F. J., and G. A. Denys.** 1995. Sterilization, decontamination, and disinfection procedures for the microbiology laboratory, p. 86–98. *In* P. R. Murray, E. J. Baron, M. A. Pfaller, F. C. Tenover, and R. H. Yolken (ed.), *Manual of Clinical Microbiology,* 6th ed. American Society for Microbiology, Washington, D.C.

SUPPLEMENTAL READING

NCCLS. 1991. *Protection of Laboratory Workers from Instrument Biohazards.* Document 117-P. NCCLS, Wayne, Pa.

Vesley, D., and J. Lauer. 1995. Decontamination, sterilization, disinfection, and antisepsis, p. 219–237. *In* D. O. Fleming, J. H. Richardson, J. J. Tulis, and D. Vesley (ed.), *Laboratory Safety: Principles and Practices,* 2nd ed. American Society for Microbiology, Washington, D.C.

APPENDIX 14.8–1

EPA-registered disinfectants that are tuberculocidal and viricidal[a]

Production (formulation)	Supplier and location
Chlorine	
CLOROX (5.25% sodium hypochlorite)	Chlorox, Oakland, CA
ALCIDE[b] (1.4–2.7% sodium chlorite)	Alcide, Norwalk, CT
Glutaraldehyde	
CIDEX (2–12%)	Johnson and Johnson, New Brunswick, NJ
METRICIDE (2%)	Metex, Parker, CO
OMNICIDE[b] (2%)	Omnitec, Laguna Hills, CA
SONACIDE[b] (2%)	Ayerst, New York, NY
SPORICIDIN[b] (2%)	Schattner, Washington, DC
STERIL-IZE[b] (2%)	Larson Labs, Erie, PA
TOTACIDE[b] (2%)	Amsco Medical, Erie, PA
UCARCIDE[b] (2%)	Union Carbine, Danbury, CT
WAVICIDE (2%)	Wave Energy, New York, NY
3M GLUTAREX[b] (2%)	3M Company, Saint Paul, MN
Iodophor	
BAB-O-DYNE	Babson, Oak Brook, IL
BIOPAL NR	GAF Corporation, Wayne, NJ
DETERGDYNE	Continental, Redwood City, CA
LIFLEX 1	Wiz Chemical, Bala Cynwyd, PA
MIKROLENE	Economic Lab, Saint Paul, MN
PURINA I-O	Continental, Redwood City, CA
STARDYNE	Star Chemical, Miami, FL
SUPER-1	Northern, Pipestone, MN
WESCODYNE	West Chemical, Long Island City, NY
Phenolic	
AMPHYL	Lehn and Fink, Montvale, NJ
CHEMI-CAP	Chemical Pack, Ft. Lauderdale, FL
DI PHEN	Dow Chemical, Midland, MI
DI-CROBE	Huntington, Huntington, IN
ENVIRON-H	Vestal, St. Louis, MO
LYSOL	Lehn and Fink, Montvale, NJ
NIPACIDE	Napa Labs, Wilmington, DE
PHENOLA DETERGENT-GERMICIDE	West Chemical, Long Island City, NY
RESIST	Hysan, Chicago, IL
VESPHENE II	Vestal, St. Louis, MO
WEX-CIDE	Wexford, Kirkwood, MO

[a] These products represent only a partial list and their inclusion does not constitute endorsement by the author.
[b] Also registered as a sterilant.

APPENDIX 14.8–2 Decontamination Worksheet

Month ______________ **Year** ______________ **Location** ______________

Date	Equipment	Biohazard pail	Work surface
1			
2			
3			
4			
5			
6			
7			
8			
9			
10			
11			
12			
13			
14			
15			
16			
17			
18			
19			
20			
21			
22			
23			
24			
25			
26			
27			
28			
29			
30			
31			

Corrective action	Date:	Reviewed by:
Note: Sign initials in appropriate box after completion of decontamination tasks.		

14.9 Biohazardous Spills

I. PRINCIPLE

Management of accidental biohazardous spills depends on the infectious agent, the quantity of material spilled, and whether an aerosol was generated. A number of effective disinfectants have been described for decontamination; however, a tuberculocidal disinfectant is appropriate for most spills. The following procedure is for the management of accidental biohazardous spills of infectious material or release of infectious microorganisms into the laboratory.

II. MATERIALS

(Keep readily available)

Include QC information on reagent container and in QC records.

A. Disinfectants
1. 10% household bleach
2. See Appendix 14.8–1.

B. Equipment and supplies
1. Paper towels
2. PPE
3. Autoclavable squeegee and dust pan
4. Autoclavable plastic bags
5. Biohazardous-waste container
6. BioZorb (Biological Absorbent and Disinfectant, Ulster Scientific, Inc., Highland, N.Y.)

III. PROCEDURE

A. Cleanup of major spills (possible aerosol formation)

1. Evacuate the area or room, taking care not to breathe in aerosolized material.
2. Alert personnel in the laboratory to evacuate the area.
3. Close the doors to the affected area.
4. After 30 min, when aerosols have settled, enter the area to clean.
   a. Cleanup should be performed by the individual who committed the spill or by assigned personnel.
   b. PPE should include gloves (e.g., heavy-weight, utility), disposable booties or water-impermeable shoe covers, long-sleeved gowns, and masks.
   c. For high-risk agents, a full-faced respirator or HEPA-filtered mask should also be used.
5. Remove and discard any broken glass or other objects.
   a. *Do not* allow contact with hands.
   b. Use rigid cardboard or squeegee and dust pan.
   c. Discard these items into a plastic biohazardous-waste container.
6. Cover the spill with disposable absorbent material (e.g., paper towels).
7. After absorption of liquid, discard all contaminated material in a biohazardous-waste container.
8. Carefully clean the spill site of any visible material, from the edges of the spill into the center, with an aqueous detergent solution.
9. Pour fresh disinfectant on the spill site or wipe down the site with disinfectant-soaked disposable towels.

10. Allow disinfectant to remain on the site for 20 min.
11. Absorb disinfectant solution with disposable material or allow the disinfectant to dry.
12. Rinse the spill site with water, and air dry the site to prevent slipping.
13. Discard all paper towels, gloves, and other disposable items into an autoclavable plastic bag or biohazardous-waste container.
14. Place gowns in a container for autoclaving after the cleanup process is completed.
15. Wash hands with soap and water.

B. Cleanup of minor spills

1. PPE should include gloves and gowns.
2. Wipe up contaminated or spilled material with disinfectant-soaked paper towels.
3. Rinse materials with water if necessary.
4. Discard all materials in a biohazardous-waste container.
5. Wash hands with soap and water.

C. Cleanup of spills in BSC

1. See procedure 14.5 for operation of the BSC.
2. Do not turn off the cabinet.
3. Pour disinfectant over the spill and cover this area with paper towels. Do not use alcohol.
4. Allow 20 min of contact time with the disinfectant.
5. Discard paper towels in a biohazardous-waste container.
6. Using a soft cloth soaked in disinfectant, wipe down all cabinet surfaces and equipment as needed.
7. If spills leak through the vent cover, remove and clean the gutter area with disinfectant.
8. Place gowns in a container for autoclaving after the cleanup process is completed.
9. Allow the cabinet blower to run 10 min before resuming activity.
10. For major spills of potentially infectious materials, contact a technical service consultant for decontamination.
11. Do not turn off the cabinet.

D. Commercial products (BioZorb)

1. Apply powder liberally over the spill, and wait until the spill has been completely absorbed.
2. Remove the absorbed material with the disposable scoop and scraper provided.
3. Rinse the spill site with water, and air dry to prevent slipping.
4. Discard all materials in a biohazardous-waste container.
5. Wash hands with soap and water.

IV. LIMITATIONS

A. *Do not* pour hypochlorite solutions into pools of urine, blood, or feces. Highly irritating gases may be produced.

B. *Do not* use low-level disinfectants, such as quaternary ammonium compounds, for disinfecting spills.

V. QUALITY ASSURANCE

Complete Biohazardous Spill Report and submit it to a laboratory safety committee representative. See Appendix 14.9–1 for a sample report form.

SUPPLEMENTAL READING

Baron, E. J. 1992. General guidelines to instrument maintenance and quality control, p. 12.1.1–12.1.8. *In* H. D. Isenberg (ed.), *Clinical Microbiology Procedures Handbook*, vol. 2, supplement 1. American Society for Microbiology, Washington, D.C.

Vesley, D., and J. Lauer. 1995. Decontaminatio sterilization, disinfection, and antisepsis, 219–237. *In* D. O. Fleming, J. H. Richardson, I. Tulis, and D. Vesley (ed.), *Laboratory Safet Principles and Practices*, 2nd ed. American Soc ety for Microbiology, Washington, D.C.

APPENDIX 14.9–1

Biohazardous Spill Report

Department_______________

Biohazard name:
Time and date of spill:
Location of spill:
Cleanup procedure:
Names of staff in affected area:
Apparent injuries:
Reported by:
Cleaned by:
Signature: Date:
Note: Return form to department safety officer.

14.10 Chemical Spills

I. PRINCIPLE

Laboratory standards and recommendations have been developed to protect laboratory workers from exposure to hazardous chemicals. Response to small-scale chemical spills must be planned and personnel made aware of handling procedures for such situations. Several commercial products are available for cleaning up spills. A spill control pillow is useful for absorbing most liquid spills including chemical, biohazardous, or radioactive spills. The following procedure is for the management of small chemical spills of disinfectants, solvents, dyes, and reagents commonly used in the microbiology laboratory. If there is a large-scale spill, a trained emergency response team should be immediately notified. The reader should refer to the Chemical Hygiene Plan of one's laboratory facility (*see* procedure 14.13).

II. MATERIALS

Include QC information on reagent container and in QC records.

A. PPE

1. All chemicals and reagent bottles must be clearly labeled with the chemical name and the hazards.
2. Refer to the MSDS information for specific hazards associated with a particular chemical and recommendations for safe handling and disposal.
3. Depending on the nature of the spill, PPE may include one or more of the following:
   a. Heavy-duty gloves
   b. Face shields
   c. Safety glasses or chemical splash goggles
   d. Nose and mouth respirators
   e. Gas masks
   f. Impervious gowns
   g. Shoe covers

B. Sand or kitty litter

C. Commercial products

1. Spill Control Pillow (Science Related Materials, Inc., Janesville, Wis.)
2. Hg Absorb Merc Jar (Lab Safety Supply, Janesville, Wis.)

III. PROCEDURE

A. Small chemical spills

1. Wear appropriate PPE.
2. Pour sand or kitty litter on the spill to isolate and contain the spill.
3. Sweep up the sand or kitty litter with a broom and dust pan, and place material in a designated waste container.
4. Add cold water to small amounts of acid in the spill area, then mop up.
5. Add copious amounts of water to solvents, then mop up.
6. Dispose of material as specified in local, state, and federal regulations.
7. Wash the spill area thoroughly with soap and water.
8. Remove PPE, and wash hands with soap and water.

B. Small chemical and biohazardous spills

1. Wear appropriate PPE.
2. Add absorbent material to the spill area (paper towels, sand or kitty litter, spill pillow).
3. Remove the absorbed material, and discard in a biohazardous-waste container.

III. PROCEDURE *(continued)*

4. Wipe up the contaminated material or spill area with disinfectant-soaked paper towels, and discard towels in a biohazardous-waste container.
5. Wash the spill area thoroughly with soap and water and air dry.
6. Remove PPE, and wash hands with soap and water.

C. Spill control pillow

1. Wear appropriate PPE.
2. Dilute any formaldehyde spill with VYTAC solvent (Baxter Diagnostics Inc., Deerfield, Ill.) to reduce dangerous properties.
3. Estimate the volume of the spill, and carefully place the appropriate number of spill pillows on the spill.
4. Allow spill pillows to absorb the spill, taking care not to spread the spilled material.
5. Remove each pillow, and place it in a designated waste container.
6. Dilute the residue with an appropriate solvent, and use additional pillows to pick up any remains of the spill.
7. Place pillows in a designated waste container.
8. Dispose of material as specified by local, state, and federal regulations.
9. Scrub the surface of the spill area with soap and water, and clean by ordinary means.
10. Remove PPE, and wash hands with soap and water. Clean any contaminated PPE with soap and water before removal and storage for reuse.

D. Hg Absorb Merc Jar

1. Use to remove small drops of mercury from surfaces.
2. Wear appropriate PPE.
3. Unscrew the jar lid with the attached sponge.
4. Add approximately 2 ml of water to the surface of the sponge.
5. Spread the water with a gloved finger to moisten evenly.
6. After 1 min, slowly move the sponge over the surface to be cleaned.
7. When all the mercury is absorbed onto the sponge, screw the lid with the attached sponge back onto the jar.
8. Store the jar under a fume hood.
9. Dispose of the jar as specified by local, state, and federal regulations.

IV. LIMITATIONS

A. *Do not* use spill pillows on hydrofluoric acid spills.
B. *Do not* reuse spill pillows on more than one type of spill (e.g., solvents, dyes, and other reagents).
C. *Do not* inhale fumes.

V. QUALITY ASSURANCE

A. Complete the Chemical Spill Report, and submit it to a laboratory safety committee representative. See Appendix 14.10–1 for a sample report form.
B. If an injury results from a spill, complete an accident report form, and submit it to a supervisor.

SUPPLEMENTAL READING

Gershon, R, and I. F. Salkin. 1992. Chemical safety, p. 14.2.1–14.2.5. *In* H. D. Isenberg (ed.), *Clinical Microbiology Procedures Handbook*, vol. 2. American Society for Microbiology, Washington, D.C.

.PPENDIX 14.10–1

Chemical Spill Report

Department________________

Chemical name:
Time and date of spill:
Location of spill:
Cleanup procedure:
Names of staff in affected area:
Apparent injuries:
Reported by:
Cleaned by:
Signature: Date:
Note: Return form to department safety officer.

14.11 Infectious-Waste Treatment and Disposal

I. PRINCIPLE

The microbiology laboratory is a primary generator of infectious waste. Waste that requires special attention includes cultures and stocks of infectious agents, pathological specimens, blood and blood products, and sharps. The most common treatment and disposal methods are incineration and steam sterilization. Various alternative methods for treatment and disposal of waste have been described. Application of these methods depends on local, state, and federal regulations. These procedures are designed to minimize the exposure of laboratory workers to infectious waste during the period from its generation to its final destruction.

II. MATERIALS

A. PPE
1. Gloves
2. Impervious gowns
3. Facial barriers

B. Containers and packaging with the red biohazard symbol
1. Sharps: rigid, puncture-resistant containers
2. Solids: autoclave bags or heavy plastic bags
3. Liquids and glass: disposable leak-proof containers with secure lids and stainless-steel buckets
4. Broken glass: rigid, puncture-resistant containers

C. Collection carts with red biohazard symbols

D. Disinfectants
1. 10% household bleach
2. See Appendix 14.8–1 for EPA-registered disinfectants that are tuberculocidal and viricidal.

III. INITIAL PREPARATION

A. Identification of infectious waste
1. Human blood and blood products
   a. All human blood, serum, plasma, and blood products
   b. Blood-contaminated tubes, microscope slides, and coverslips
   c. Blood-soaked bandages
2. Contaminated waste from patient care
   Suction canisters, tubing, and hemodialysis waste
3. Cultures and stocks of infectious agents
4. Other contaminated laboratory waste
   a. Specimens and containers (cups, bottles, tubes, and flasks)
   b. Culture plates and devices used to transfer, inoculate, and mix cultures (swabs, pipettes, etc.)
   c. Diagnostic-kit components
   d. PPE grossly contaminated with blood, body secretions, or cultures (gloves, masks, gowns, or coats)
5. Contaminated sharps
   a. Hypodermic needles, syringes, and scalpel blades

   b. Disposable pipettes, capillary tubes, microscope slides, coverslips, and broken glass
6. Unused sharps
7. Pathology waste
   a. Body tissue, organs, and body parts
   b. Body fluids removed during surgery, autopsy, or biopsy
8. Discarded live and attenuated vaccines
9. Contaminated animal waste, including carcasses, body parts, and bedding

B. Handling

1. Use appropriate PPE (*see* Table 14.3–1).
2. Discard wastes directly into designated containers placed as close as possible to the point of generation.
3. Handle waste as little as possible.
4. Avoid spills and accidents.
5. Sharps containment
   a. Place sharps directly into rigid and puncture-resistant containers.
   b. Fill containers no more than three-fourths full.
   c. Tape the lids shut.
6. Solid-waste containment
   a. Place solid waste into biohazard bags.
   b. Repackage or double bag wet or leaking bags immediately.
   c. Seal outer bags with tape.
7. Liquid-waste containment
   a. Place liquid waste in rigid containers with secure lids.
   b. Use glass containers if waste is to be autoclaved (*see* procedure 14.4).
   c. Use plastic disposable containers if waste is to be incinerated.
8. Segregate mixed waste containing infectious and radioactive (e.g., specimens containing ^{125}I) or infectious and toxic chemical (e.g., solvents and carcinogens) components, and manage collectively based on the severity of their hazards.

C. Storage

1. Treat and dispose of waste as soon as possible.
2. Place waste in containers in a designated collection cart.
3. Cover the cart and secure its lid.
4. Clean and disinfect the collection cart immediately after use.
5. Clinical specimens saved for further workup
   a. Place specimens in secondary containers and label with the dates they must be discarded.
   b. Store specimens in a designated and labeled area at 4°C.
   c. Discard specimens in a biohazardous-waste container.
6. Representative culture isolates for further testing
   a. Seal all plates and slants containing mycelial growth (e.g., possible dimorphic fungi).
   b. Place culture plates, tubes, or slants in secondary containers and label them with the dates they must be discarded.
   c. Store specimens in a designated and labeled area at room temperature or 4°C.
   d. Limit access to the storage area.
   e. Discard specimens in a biohazardous-waste container.

IV. TREATMENT AND DISPOSAL

A. Within the laboratory

1. Steam sterilize waste as outlined in procedure 14.4.
2. Transport waste within the facility or off site if additional treatment is re quired (e.g., to render waste unrecognizable).

B. Within the facility

1. Use a collection cart to transport waste to the treatment equipment.
2. Clean and disinfect the collection cart.
3. Incinerate waste or use an alternative waste treatment option that complie with local, state, and federal regulations.

C. Off site

1. Pack plastic bags and other containers in rigid secondary containers suc as plastic barrels or heavy cartons.
2. Seal containers.
3. Transport in closed and leak-proof dumpsters or trucks.

V. QUALITY ASSURANCE

A. Check disposable sharp containers daily to prevent overfilling and possibl needle stick exposure while disposing of sharps.

B. Spot check for proper waste collection, transport, and storage.

C. Document waste treatment effectiveness (see procedure 14.4).

SUPPLEMENTAL READING

Gordon, J. G., and G. A. Denys. 1996. Minimization of waste generation in medical laboratories, p.163–193. *In* P. A. Reinhart, K. L. Leonard, and P. C. Ashbrook (ed.), *Pollution Prevention and Waste Minimization in Laboratories.* CRC Press, Inc., Boca Raton, Fa.

Marsik, F. J., and G. A. Denys. 1995. Sterilization, decontamination, and disinfection procedures for the microbiology laboratory, p. 86–98. *In* P. R. Murray, E. J. Baron, M. A. Pfaller, F. C. Tenover, and R. H. Yolken (ed.), *Manual of Clinical Microbiology*, 6th ed. American Society for Microbiology, Washington, D.C.

Reinhardt, P. A., and J. G. Gordon. 1991. *Infectious and Medical Waste Management.* Lewis Publishers, Chelsea, Mich.

14.12 Packaging and Shipping Infectious Substances

I. PRINCIPLE

Transporting of clinical specimens is broadly regulated based on the assumption that those materials most likely to contain an etiologic agent will be treated like infectious substances. Packaging that includes the Shipper's Declaration for Dangerous Goods (DGR) document is acceptable under all regulations, and this labeling is required by many couriers. Shippers using private commercial carriers should consult them for shipping requirements. The following procedure ensures the viability of an infectious agent, as well as providing maximum protection for those handling the shipment both in transit and after it arrives at the laboratory.

II. SPECIMENS

A. Clinical specimens should be submitted in 2-ml or larger screw-cap glass tubes or vials (not to exceed 50 ml).

B. Bacterial cultures should be submitted as agar slant or stab cultures or on swab transport devices and lyophilized cultures.

III. MATERIALS

A. Leak-proof containers (caps or lids taped)
   1. Primary container (can be transport medium tube with organism)
   2. Secondary container (can be outer canister)

B. Outer shipping container (can be cardboard or Styrofoam)

C. Shock-absorbent material

D. Diamond-shaped label "INFECTIOUS SUBSTANCE" (Fig. 14.12–1)

E. DGR (Fig. 14.12–2)

F. Postal-service label or other shipping document

G. U.S. Public Health Service label (Fig. 14.12–3)

H. Plastic sleeve for placement of DGR

IV. PROCEDURE

A. Fill out two identical DGR forms.
   1. Fill in the shipper's box with the name, address, and phone number of the center shipping the material.
   2. Fill in the airports of departure and destination.
   3. Cross out the box "RADIOACTIVE."
   4. Under "Proper Shipping Name" write "Infectious Substances Affecting Humans" and list each isolate type (genus and species) included in the box.
   5. Write "6.2" under "Class or Division."
   6. Write "UN 2814" under "UN or ID No."
   7. Write "None" under "Subsidiary Risk."
   8. Under "Type of Packing" write "1 Fiberboard Box ____ gms." Use weight of package.
   9. Write "602" under "Packing Instructions."

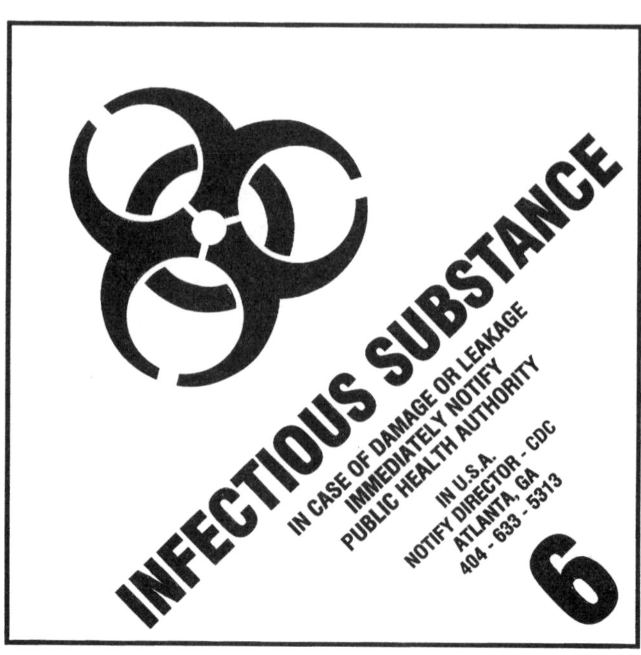

Figure 14.12–1 International label for packages containing infectious substances or etic logic agents.

IV. PROCEDURE *(continued)*

10. Write "CHEMTREC 1-800-424-9300" under "Additional Handling In formation."
11. Each copy must have an original signature.
12. Fold both forms together in half lengthwise and again by width. Plac forms in a clear sleeves and attach the sleeve to the outer container. D not seal the sleeve.

B. Inside packaging

1. Seal the tube or vial with waterproof tape.
2. Pack absorbent material around the tube or vial. If tubes are to be shipped, wrap each tube individually with paper towels
3. Place tube(s) in a watertight secondary container. Add enough absorbent material between primary and secondary container to absorb the total fluid volume of the transport medium.
4. Seal the secondary container, and place it in an outer shipping container
5. Seal the outside shipping container securely.

C. Outside-box labeling

1. Attach a postal-service label or shipping document with the name, address and phone number of the recipient.
2. Label all organisms contained in the box.
3. Attach the DGR.
4. Attach a diamond-shaped label, "INFECTIOUS SUBSTANCES 6."
5. If transported media are liquid, use "up arrows" to show the correct orien tation for shipment.
6. Attach U.S. Public Health Service label (Fig. 14.12–3) as specified by U.S. Postal Service regulation.

Figure 14.12–2 DGR form required by International Air Transport Association (IATA and U.S. Postal Service to accompany air shipments of infectious substances.

SHIPPER'S DECLARATION FOR DANGEROUS GOODS (**Provide at least two copies to the airline.**)

Shipper

Air Waybill No.

Page of Pages

Shipper's Reference Number
(optional)

Consignee

Two completed and signed copies of this Declaration must be handed to the operator

WARNING

Failure to comply in all respects with the applicable Dangerous Goods Regulations may be in breach of the applicable law, subject to legal penalties. This Declaration must not, in any circumstances, be completed and/or signed by a consolidator, a forwarder or an IATA cargo agent.

TRANSPORT DETAILS

This shipment is within the limitations prescribed for:
(delete non-applicable)

PASSENGER AND CARGO AIRCRAFT	CARGO AIRCRAFT ONLY

Airport of Departure

Airport of Destination:

Shipment type: *(delete non-applicable)*

NON-RADIOACTIVE	RADIOACTIVE

NATURE AND QUANTITY OF DANGEROUS GOODS

Dangerous Goods Identification				Quantity and type of packing	Packing Inst.	Authorization
Proper Shipping Name	Class or Division	UN or ID No.	Subsidiary Risk			

Additional Handling Information

I hereby declare that the contents of this consignment are fully and accurately described above by proper shipping name and are classified, packed, marked and labelled, and are in all respects in the proper condition for transport by air according to the applicable International and National Government Regulations.

Name/Title of Signatory

Place and Date

Signature
(see warning above)

FORM 1803 HOBBS & WARREN INC. OP-30 JAN. 1983

Figure 14.12–3 U.S. Public Health Service label required by U.S. Postal Service on packages containing etiologic agents.

V. PROCEDURE NOTES

A. If dry ice is used as a refrigerant, it must be placed outside the secondary container. Affix a dry ice label on the outer container and mark net weight on the package.

B. When possible, ship material so that it will arrive at its destination at the beginning or middle of the work week.

SUPPLEMENTAL READING

McVicar, J. W., and J. Suen. 1995. Packaging and shipping biological materials, p. 239–246. *In* D. O. Fleming, J. H. Richardson, J. I. Tulis, and D. Vesley (ed.), *Laboratory Safety: Principles and Practices,* 2nd ed. American Society for Microbiology, Washington, D.C.

NCCLS. 1994. *Procedures for the Handling and Transport of Diagnostic Specimens and Etiologic Agents.* Document H5-A3. NCCLS, Wayne, Pa.

U.S. Public Health Service. 1996. 42 CFR Part 72, Interstate Shipment of Etiologic Agents. *Fed Regist.* **61**(207):5519 (October 24, 1996).

14.13 Chemical Hygiene Plan

I. POLICY

It is the responsibility of laboratory management to provide a safe working environment for employees. A chemical hygiene plan (CHP) must be developed and include policies and procedures to help the employee be aware of potentially hazardous chemicals in the workplace. The CHP should also give guidelines for developing safe work practices.

It is essential that employers assume responsibility for laboratory safety. All employees should participate in a training program and have access to pertinent safety information through their management staff. When safety concerns arise, employees should be encouraged to contact their managers.

This procedure outlines the essential components of the CHP. Refer to the references provided below for more extensive recommendations for the development and application of a CHP.

II. COMPONENTS OF THE CHP

A. Basic rules and procedures

1. Accidents and spills
   - **a.** Promptly flush eyes with water for a prolonged period (minimum 10 min).
   - **b.** Promptly drink large amounts of water if ingestion occurs.
   - **c.** Promptly flush affected areas of skin contact with a large volume of water and remove contaminated clothing.
   - **d.** Seek medical attention if symptoms persist after washing.
   - **e.** Promptly clean up spills (*see* procedure 14.10).
2. Avoidance of "routine" exposure
   - **a.** Develop safe personal and work habits.
   - **b.** Avoid unnecessary exposure to chemicals by any route.
3. Choice of chemicals

   Do not use chemicals if the ventilation system is inadequate.
4. Eating, smoking, etc.
   - **a.** *Do not* eat, drink, or apply cosmetics in areas where chemicals are present.
   - **b.** *Do not* store, handle, or consume food or beverages in areas of laboratory operation.
5. Equipment and glassware
   - **a.** Avoid practices that might damage laboratory glassware.
   - **b.** *Do not* use damaged glassware.
   - **c.** Use equipment only for its designated purpose.
   - **d.** Centrifuge (*see* procedure 14.6)
   - **e.** Gas cylinders (*see* procedure 14.7)
6. Exiting

   Do not leave the laboratory without washing areas of exposed skin.
7. Horseplay

   Do not play practical jokes or behave in a manner that might confuse, startle, or distract a co-worker.

II. COMPONENTS OF THE CHP
(continued)

8. Mouth suction
Do not mouth pipette.

9. Personal apparel
Do not wear apparel contradictory to the laboratory dress code.

10. Personal housekeeping
- **a.** *Do not* leave the work area soiled or cluttered with chemicals that ar improperly labeled or stored.
- **b.** *Do not* leave the work area at the end of an operation or shift or afte a spill without proper clean up.

11. Personal protection
Do not perform laboratory operations without appropriate PPE.
- **a.** Use eye protection where chemicals are stored or handled (e.g., safet glasses with side shield).
- **b.** Use appropriate gloves that are resistant to toxic chemicals (e.g., natur rubber, nitrile, butyl rubber, polyvinyl chloride).
- **c.** Use appropriate respiratory equipment when air contaminant concentra tions are not sufficiently restricted by engineering controls (e.g., Particu late-Filter Respirator, Technol, Inc., Fort Worth, Tex.).
- **d.** Use other protective and emergency apparel and equipment as appro priate.
- **e.** Avoid wearing contact lenses in the laboratory unless necessary.
- **f.** Remove laboratory coats immediately upon significant contaminatio

12. Planning
- **a.** *Do not* perform a new operation without prior knowledge about hazard PPE, and proper positioning and operation of equipment.
- **b.** Review MSDS before doing any work with hazardous chemicals.

13. Unattended operations
Do not leave operations unattended without providing appropriate precau tions.
- **a.** Leave lights on.
- **b.** Place a sign on the door.
- **c.** Provide for containment of toxic substances should a utility service fai

14. Use of hood
- **a.** *Do not* perform operations that might result in the release of toxi chemical vapors or dust, without the use of a hood.
- **b.** *Do not* use a hood before its adequate performance has been confirmed
  - (1) Keep the hood closed at all times except when adjustments withi the hood are being made.
  - (2) *Do not* block vents or airflow with stored materials.
  - (3) Keep materials stored in hoods to a minimum.
- **c.** *Do not* turn the hood off if toxic substances are stored in it (limite amounts) or if the adequacy of laboratory ventilation is uncertain whe the hood is off.

15. Vigilance
Be alert to unsafe conditions, and see that they are corrected.

16. Waste disposal
- **a.** Deposit chemical waste in appropriately labeled receptacles and follov waste disposal procedures explained in procedure 14.14.
- **b.** *Do not* discharge into the sewer concentrated acids and bases or sub stances that are highly toxic, malodorous, or lachrymatory.
- **c.** *Do not* discharge into the sewer any substance which might interfer with the biological activity of waste treatment plants, create fire o explosion hazards, cause structural damage, or obstruct flow.

17. Working alone
    Do not work alone in the laboratory if the procedures being performed are hazardous.
18. When working with chemicals of moderate chronic or high acute toxicity, refer to reference 1 for supplemental rules to those mentioned above.
19. When working with chemicals of highly chronic toxicity, refer to reference 2 for supplemental rules to those mentioned above.

B. Chemical procurement, distribution, and storage

1. Procurement
   - a. *Do not* receive substances without information on proper handling, storage, and disposal (MSDS sheet).
   - b. *Do not* accept containers without an adequate identifying label.
2. Stockrooms/storerooms
   - a. Segregate toxic substances in a well-identified area with local exhaust ventilation.
   - b. Store in unbreakable secondary containers highly toxic chemicals or chemicals whose containers have been opened.
   - c. Examine stored chemicals periodically for replacement, deterioration, and container integrity.
   - d. *Do not* use stockrooms or storerooms as preparation or repackaging areas.
3. Handling and transport
   - a. *Do not* hand carry chemicals without placing them in an outside container or bucket.
   - b. *Do not* open bottles that might be under pressure, without first covering them with a towel and using appropriate PPE (e.g., face shields, safety glasses, and goggles).
   - c. *Do not* heat flammable liquids with an open flame, hot plate, or uninsulated resistance heater.
   - d. Avoid combinations of chemicals that are explosive, poisonous, or hazardous in some other way.
4. Laboratory storage
   - a. *Do not* store more than a 2-day supply of solvents and toxic and corrosive chemicals in the work area.
     - (1) Use procedures that require small quantities of chemicals.
     - (2) Substitute nonhazardous chemicals (e.g., ethyl acetate for concentrating of stool parasites) for hazardous chemicals (e.g., ether) when possible.
   - b. *Do not* store chemicals on benchtops or in hoods.
   - c. *Do not* expose chemicals to heat or direct sunlight.
   - d. *Do not* store flammables in domestic refrigerators (non-explosion proof).
   - e. *Do not* place heavy containers on shelves above eye level.
   - f. *Do not* place chemicals on the floor.
   - g. Discard or return unneeded items to the storeroom periodically.

C. Environmental monitoring

Do not monitor for airborne concentrations unless a highly toxic substance is stored or used regularly (e.g., monitor formaldehyde and glutaraldehyde levels every 6 months).

D. Housekeeping, maintenance, and inspections

1. Cleaning
   Clean floors regularly.
2. Inspections
   Conduct formal housekeeping and chemical hygiene inspections quarterly or semiannually, depending on the frequency of personnel use.

II. COMPONENTS OF THE CHP *(continued)*

3. Maintenance
   a. Inspect eyewash fountains weekly.
   b. Inspect respirators as required.
   c. Inspect other safety equipment regularly (e.g., every 3 to 6 months).
   d. Discard equipment that is no longer safe for use.
4. Pathways
   a. *Do not* use hallways for storage areas.
   b. *Do not* block access to exits, emergency equipment, and utility controls.

E. Medical program

1. Compliance with regulations
   Provide all employees who work with hazardous chemicals the opportunity to receive medical attention to the extent required by regulations.
2. Routine surveillance
   Provide employees who work with toxicologically significant chemicals the opportunity to consult a qualified physician to determine whether a regular schedule of medical surveillance is desirable.
3. First aid
   Make personnel available during working hours who are trained in first aid, and ensure that an emergency room with medical personnel is nearby.

F. PPE

The following should be available in the laboratory:

1. Protective aprons and gowns
2. Gloves: latex, vinyl (e.g., blood and body fluids), and other specially required gloves (e.g., those for toxic chemicals, autoclaved material, and broken glass)
3. Face shields, goggles, and stationary shields
4. Sand and/or kitty litter in each area for spills
5. Safety showers
6. Permanent eyewash fountains
7. Spill pillows
8. Fire blankets
9. Fire extinguishers
10. Respirators
11. Fire alarm and telephone for emergency use
12. Other items designated by the laboratory supervisor

G. Records

1. Retain written accident records.
2. Retain CHP records that document that facilities and precautions are compatible with current knowledge and regulations.
3. Retain inventory and usage records for highly toxic substances (e.g., carcinogens, toxic or highly toxic agents, and reproductive toxins).
   a. Amounts of materials on hand
   b. Amounts of materials used
   c. Names of workers involved
4. Retain medical records that document compliance with state and federal regulations (30 years).

H. Signs and labels to be posted

1. Telephone numbers of emergency personnel, facility supervisors, and laboratory workers
2. Identity labels showing contents of containers, including waste receptacles and those holding associated hazards
3. Location signs for safety showers, eyewash stations, and other safety and first-aid equipment

4. Location signs for exits and areas where food and beverage consumption and storage are permitted
5. Warnings in areas or at equipment where special or unusual hazards exist

I. **Spills and accidents** (*see* procedure 14.10)

1. Establish an emergency plan that includes procedures for ventilation failure, evacuation, medical care and reporting, and drills.
2. Establish an alarm system to alert people in all parts of the facility.
3. Establish a spill control policy that includes prevention, containment, cleanup, and reporting.
4. Analyze accidents or near accidents, and distribute reports to all who might benefit.

J. **Information and training**

1. Aim
   Instruct all individuals at risk about the nature of their work in the laboratory, its risks, and procedures if an accident occurs.
2. Emergency and personal protection training
   a. Instruct every laboratory worker on the location and proper use of PPE.
   b. Instruct full-time personnel in the proper use of emergency equipment and procedures.
   c. Encourage and make available first-aid instruction.
3. Receiving and stockroom or storeroom personnel
   Insist that personnel be knowledgeable about hazards, equipment handling, protective apparel, and relevant regulations.
4. Frequency of training
   Provide an education program on a regular and continuing basis.
5. Literature and consultation
   Make information regarding chemical hygiene available to all laboratory personnel.

K. **Waste disposal program** (*see* procedure 14.14)

1. Aim
   Assure that minimal harm to people, other organisms, and the environment will result from the disposal of waste laboratory chemicals.
2. Content
   a. Specify how waste is to be collected, segregated, stored, and transported, and identify materials that can be incinerated.
   b. Transport from the institution must be in compliance with Department of Transportation regulations.
3. Discarding chemical stocks
   a. Promptly dispose of unlabeled containers of chemicals and solutions.
   b. Do not open partially used chemicals.
4. Frequency of disposal
   Remove waste from laboratories to a central waste storage area at least once per week and from the central waste storage area at regular intervals.
5. Method of disposal
   a. Incineration is an environmentally acceptable and practical method for disposing of combustible laboratory waste.
   b. Recycling or chemical decontamination should be used when possible.
   c. *Do not* indiscriminately pour waste chemicals down the drain or add them to mixed refuse for landfill burial.
   d. *Do not* use a hood for disposal of volatile chemicals.

REFERENCES

1. **National Research Council, Committee on Hazardous Substances in the Laboratory.** 1981. *Prudent Practices for the Handling of Hazardous Substances in Laboratories*, p. 39–41. National Academy Press, Washington, D.C.

2. **National Research Council, Committee on Hazardous Substances in the Laboratory.** 1981. *Prudent Practices for the Handling of Hazardous Substances in Laboratories*, p. 47–50. National Academy Press, Washington, D.C.

SUPPLEMENTAL READING

Occupational Safety and Health Administration. 1990. Occupational *Exposure to Hazardous Chemicals in Laboratories.* 29 CFR 1910 1450. FRSS 3300–3335, Jan. 31, 1990. U.S. Government Printing Office, Washington, D.C.

U.S. Department of Labor. 1990. Occupational exposures to hazardous chemicals in laboratories, final rule. 29 CFR Part 1910.1450. U.S. Government Printing Office, Washington, D.C.

14.14 Hazardous-Waste Plan

I. POLICY

It is the responsibility of laboratory management to inform employees of significant chemical hazards. It is the responsibility of laboratory personnel to learn of these hazards and to manage hazardous materials, laboratory-generated waste, and other specific wastes in a manner that provides protection for all employees, visitors, and the community. The complex waste management planning and implementation process is illustrated in Figure 14.14–1.

II. DEFINITIONS

A. Hazardous wastes (based on Resource Conservation and Recovery Act [RCRA] regulations)

For details, consult the laws and regulations in each community or contact the Environmental Protection Agency, Office of Solid Waste Management and Emergency Response, Washington, D.C. (202-382-4700).

1. Corrosive (rust removers, acid, alkaline, and battery acids)
2. Toxic (mercury, cadmium, lead, and dioxins)
3. Carcinogenic (benzene and toluene)
4. Ignitable (some paint, acetone, degreasers, ether, ethanol, solvents, and xylene)
5. Reactive (cyanides, bleaches, and oxidizers)

B. General wastes

Paper products, food, and rubbish

C. Special wastes

1. Gaseous by-products of anesthesia and some chemical agents
2. Infectious (*see* procedure 14.15)
3. Cytotoxic: chemotherapy agents
4. Radioactive: in liquid, solid, or gaseous form

III. PROCEDURE (GENERAL)

A. All hazardous materials including wastes must be handled in compliance with applicable federal, state, and local laws and regulations.

B. Laboratory employees must be trained in techniques that minimize exposure to hazardous substances.

C. MSDS must be maintained on all hazardous substances and be readily accessible for employees during all work shifts.

D. Containers of hazardous chemicals

1. Label, tag, or mark the identity of the chemicals in each container.
2. Attach appropriate hazard warnings.

E. Storage of corrosives

1. Store the minimum amount of chemicals consistent with their rate of use.
2. Store materials near the floor.

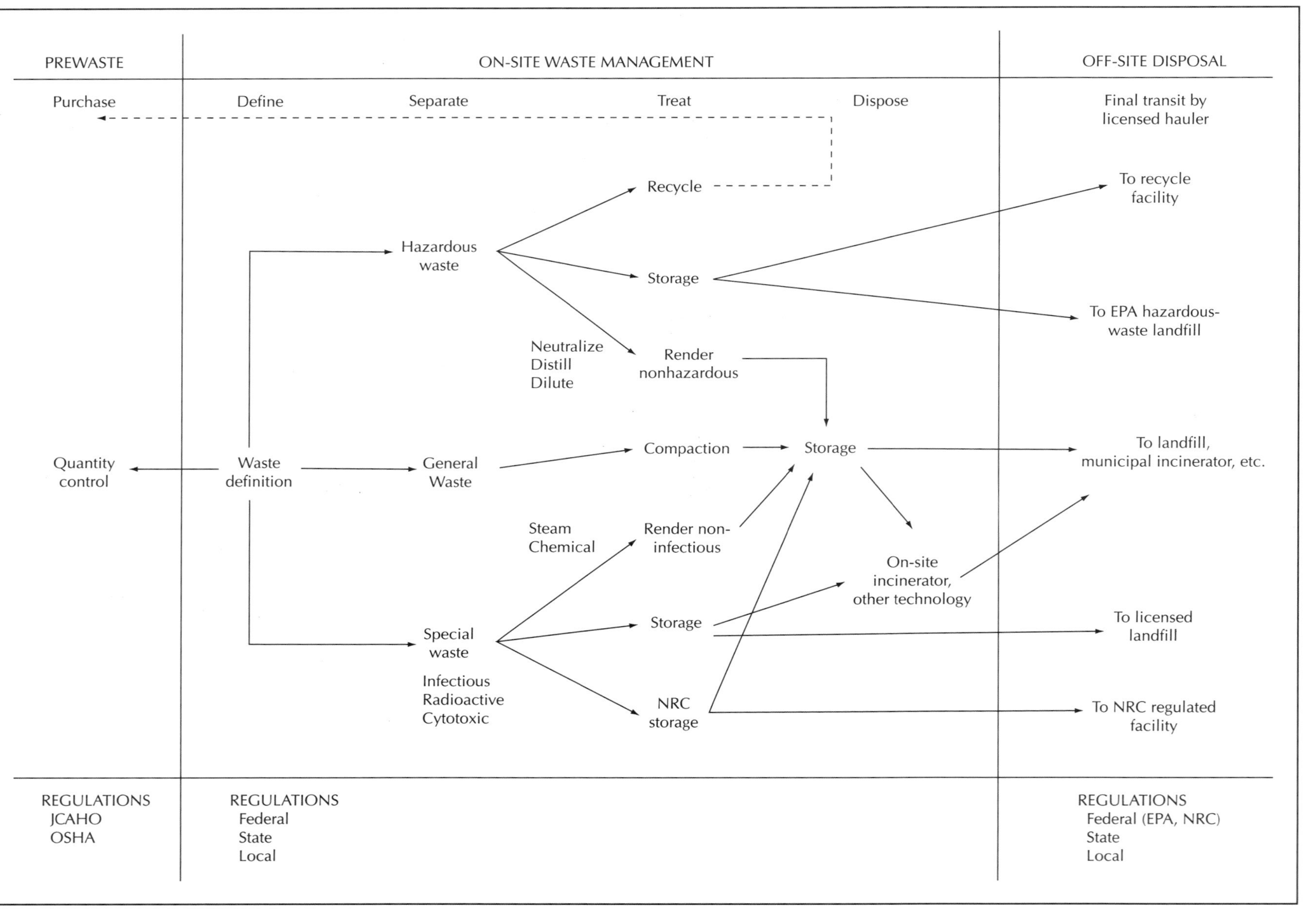

Figure 14.14–1 Waste management planning and implementation process (adapted from reference 1 with permission).

III. PROCEDURE *(continued)*

3. Use acid-bottle carriers for storing containers of concentrated reagents larger than 500 ml.
4. *Do not* store mutually incompatible chemicals in the same area. See reference 1 for a guide to compatible chemicals.
5. *Do not* handle corrosives without appropriate PPE, adequate emergency showers, and eyewashing devices.
6. *Do not* use chemicals without specific arrangement for containing breaks or spills.
7. Handle corrosives under a chemical-fume hood when their vapors present a health hazard.

F. Storage of flammables

1. Store the minimum amount of flammable chemicals consistent with their rate of use.
2. Store flammable and combustible liquids in approved containers when possible.
3. Store flammable or combustible liquids in compliance with National Fire Prevention Association (N.F.P.A.) no. 30 approved storage cabinet.
4. *Do not* store ignitable solvents near open flames or other sources of heat.
5. Store solvents in refrigerators designed to minimize ignition of vapors, and label the doors accordingly.

G. Storage of waste chemicals

1. Store chemicals by compatibility groups (e.g., flammables, oxidizers, corrosives, and poisons) until they are removed from the laboratory by qualified personnel.
2. Specific combinations of waste which are acceptable are listed in item IVB.
3. Waste containers
   a. One-gallon glass jugs (used solvent bottles)
   b. Fill only to the top of the shoulder to leave space for expansion.
   c. 2.5-liter ''ACID'' carboys for concentrated acids and bases

H. Combining of chemical wastes into a container <1,000 ml in volume

1. Activity should take place in a hood.
2. Wear appropriate goggles and gloves.

I. Radioactive waste

1. Most radioisotopes used in the microbiology laboratory have short half lives and can be stored on site for decay before disposal.
2. Long-lived liquid wastes must be transferred to the Radiation Safety Officer for disposal via a commercial waste disposal service.

J. Multihazardous waste

1. Rank priority based on the greatest hazard.
   a. Refer to MSDS information.
   b. Contact the Safety Director.
2. Radioactive-material exposure
   a. If Geiger counter readings are less than two times background readings, waste may be disposed of in accordance with waste disposal policy.
   b. If Geiger counter readings are greater than two times background readings, waste must be disposed of via a commercial waste disposal service.
   c. Contact the Radiation Safety Officer.
3. Waste with biohazardous material (*see* procedure 14.15)
   Add 100 ml of 10% bleach per gallon.

IV. DISPOSAL

A. Labeling requirements

1. Segregate chemicals based on volatility, flammability, and reactivity.
2. Affix a special label to the vessel.
3. List the specific chemicals in the jug and approximate percentage by volume of each.
   If blood or serum is included, list as ''aqueous biohazardous material.''

B. Specific chemicals (Refer to Appendix 14.14–1 for a summary of waste disposal.)

1. Acetone may be combined with other flammable organic solvents.
2. Acetonitrile may be combined with other flammable organic solvents.
3. Acids
   a. Acids (6 N or less) may be poured down the sink with copious amounts of water.
   b. Acids (strong) should be collected in 2.5-liter acid bottles or their original containers.
4. Alkali (*see* bases below)
5. Amylacetate may be combined with other flammable organic solvents (e.g., toluene).
6. Bases
   a. Bases (<6 N) may be poured down the sink with copious amounts of water.
   b. Bases (strong) should be collected in 2.5-liter acid bottles or original containers.
7. Chloroform and other halogenated hydrocarbons must be segregated from flammable waste. Store these in a separate container.
8. Dichloromethane may be combined with chloroform.
9. Ethanol should be combined with other flammable liquids.
10. Ether (diethyl) may be mixed with other flammable liquids. Add alcohol to waste to inhibit the formation of explosive peroxides.
    a. Dispose of small quantities (<55 ml) by evaporation from paper towels in a chemical-fume hood.
    b. Check for peroxides if retained more than 6 months.
    c. *Do not* store unopened for more than 1 year.
11. Empty ether cans should be rinsed with water, and then discarded in general waste.
12. Ether (petroleum) may be combined with other flammable organic solvents.
13. *Do not* store or mix formalin with hydrochloric acid or chloride-containing solutions (e.g., bleach).
14. Hexane may be combined with other flammable organic solvents.
15. Methanol should be combined with other flammable liquids.
16. Methylene chloride (*see* dichloromethane above)
17. Methyl isobutyl ketone may be combined with other flammable solvents.
18. Segregate oxidizers (perchlorates, permanganates, periodates, peroxides, etc.) and call the Safety Director for disposal.
19. Phosphorous pentoxide should be handled only with gloves and while wearing goggles. Contact the Safety Director for disposal.
20. Toluene may be combined with other flammable organic solvents.
21. Metal salts and solutions of these metals should be collected in 1-gal bottles

REFERENCE

1. **Denys, G. A.** 1992. *Encyclopedia of Microbiology*, p. 503. Academic Press, Inc., San Diego.

SUPPLEMENTAL READING

American Association of Bioanalysis and International Society for Clinical Laboratory Technology. 1992. *A Guide to OSHA Requirements for Hospitals, Independent, and Physician Office Laboratories.* American Association of Bioanalysts, St. Louis.

NCCLS. 1994. *Clinical Laboratory Safety.* Document GP17-T, p. 23–29. NCCLS, Wayne, Pa.

Reinhardt, P. A., K. L. Leonard, and P. C. Ashbrook (ed.). 1996. *Pollution Prevention and Waste Minimization in Laboratories.* Lewis Publishers, CRC Press, Inc., Boca Raton. Fla.

U.S. Environmental Protection Agency. 1980. *Hazardous Waste Management Systems.* Code of Federal Regulations, 40 CFR Parts 260–263. U.S. Government Printing Office, Washington, D.C.

APPENDIX 14.14–1

Summary of chemical waste disposal

Flammables	Chlorinated hydrocarbons	Metal wastes	Oxidizers	Dilute acids (<6 N)	Dilute bases (<6 N)
May be mixed. Leave 4-in. head space above liquid.	May be mixed. Leave 3-in. head space above liquid.	Collect in a 1-gal bottle.	Keep separate.	Keep separate. May be poured down drain.	May be mixed. May be poured down drain.
Acetone	Chloroform	Arsenic	Perchlorates	Acetic	NaOH
Amyl acetate	Dichloromethane	Barium	Periodates	Acid-alcohol reagent	KOH
Ethanol		Cadmium	Permanganates	Hydrochloric	NH_4OH
Ethyl acetate		Lead	Peroxides	Nitric	
Ethyl ether		Mercury		Perchloric	
Hexane		Selenium		Sulfuric	
Methanol		Silver			
Methyl isobutyl Ketone					
Petroleum ether					
Toluene					

14.15 Infectious-Waste Plan

I. POLICY

It is the responsibility of laboratory management to develop a comprehensive waste management plan. Such a plan is needed to protect all employees handling infectious waste and all visitors who might be exposed to the waste and to be in compliance with the various regulatory requirements. All employees should receive appropriate training and the equipment needed to properly handle infectious waste.

Several factors should be considered when developing an infectious-waste management plan. These include regulatory requirements, institutional policy, availability of equipment and services, and strategies for waste minimization. This procedure outlines the essential components of the infectious-waste management plan. Please refer to the references provided for more detailed information. Figure 14.14–1 illustrates the complex waste management planning and implementation process.

II. MANAGEMENT PLAN

A. Identification of infectious wastes

Refer to procedure 14.11, item III.A.

B. Waste discard

1. Discard infectious waste into designated containers or red bags.
2. Place all sharp objects directly into impervious, rigid, and puncture-resistant containers.
3. Place glass and liquids in disposable cardboard containers and leak-proof boxes or containers with secure lids.
4. Segregate mixed waste (containing infectious and radioactive or infectious and toxic chemical waste) and direct it to the appropriate treatment procedure based on the relative severity of the hazards.

C. Waste handling and collection

1. Establish a procedure to minimize the potential for exposure.
2. Cover collection carts and disinfect them after use.

D. Waste storage

1. Treat infectious waste as soon as possible.
2. Waste storage area
   - **a.** Establish a separate area with biohazard signs.
   - **b.** Lock the area to limit access only to authorized personnel.
   - **c.** Keep the area free from rodents and vermin.
3. Special considerations
   - **a.** State regulations may require refrigeration of the waste.
   - **b.** Identify types of waste and containers.
   - **c.** Track storage times.

E. Waste treatment

Refer to procedure 14.11, item IV.

F. Waste disposal

1. Handle treated solid waste as general trash.
2. Pour treated liquid waste down the drain.
3. Handle certain treatment residues (e.g., incinerator ash) as hazardous waste.

G. Off-site treatment and disposal
1. Pack plastic bags in rigid containers (e.g., plastic barrels or heavy cartons).
2. Transport in closed and leak-proof dumpsters or trucks.
3. Mark containers with universal biohazard symbol.

H. Contingency plan

Establish a plan for use when equipment fails or maintenance, unscheduled repairs, or staffing issues arise.
1. Waste storage
2. Exchange agreement with another institution
3. Off-site treatment by a licensed contractor

I. Emergency plan

Implement a response plan for an accidental spill or loss of containment.
1. Establish clean-up procedures.
2. Use PPE.
3. Dispose of spill residue.

J. Training

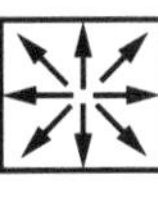

Observe universal (standard) precautions.

1. Implement training programs for all personnel who generate or handle infectious waste.
   a. Training should be based on universal precautions.
   b. Education and training are required by OSHA's hazardous communication and right-to-know regulations and the bloodborne-pathogens rule (1, 2).
2. Implement educational programs.
   a. When the infectious-waste management plan is first developed
   b. When new employees are hired
   c. Whenever infectious-waste management practices are changed

K. Recordkeeping
1. Comply with regulatory requirements
2. Document compliance.
3. Review and analyze the waste management plan.

L. QA/QC procedures
1. Ensure that the infectious-waste management system is working.
2. Document treatment effectiveness.
3. Analyze accident or incident records.

M. Incident and accident reports
1. Identify occupational issues and deficiencies.
2. Modify the waste management plan or practices accordingly.

REFERENCES

1. **OSHA.** 1991. Occupational exposure to bloodborne pathogens; Final Rule. 29 CFR Part 1910.1030. *Fed. Regist.* **56:**64175–64182.
2. **OSHA.** 1987. Hazard communication; Final Rule. 29 CFR Parts 1910, 1915, 1917, 1918, 1926, and 1928. *Fed. Regist.* **52:**31877–31886.

SUPPLEMENTAL READING

Denys, G. A. 1992. Infectious waste management, p. 493–504. *Encyclopedia of Microbiology*, vol. 2. Academic Press, Inc., San Diego, Calif.

Denys, G. A. 1994. Safety in work with bloodborne pathogens: the OSHA bloodborne pathogen standard, p. 14.5.1–14.5.6. *In* H. D. Isenberg (ed.), *Clinical Microbiology Procedures Handbook*, vol. 2, supplement 1. American Society for Microbiology, Washington, D.C.

Gordon, J. G. 1994. Safety in waste management: a comprehensive plan for infectious waste management, p. 14.6.1–14.6.6. *In* H. D. Isenberg (ed.), *Clinical Microbiology Procedures Handbook*, vol. 2, supplement 1. American Society for Microbiology, Washington, D.C.

Reinhardt, P. A., and J. G. Gordon. 1991. *Infectious and Medical Waste Management.* Lewis Publishers, Chelsea, Mich.

APPENDIX A

Selected Commercial Suppliers for Clinical Microbiology

Although numerous suppliers of microbiological products are listed in this appendix, not all manufacturers are listed. We recommend that you consult pertinent procedures in this handbook, regional suppliers, and the major national distributors of scientific products for additional sources.

A. Bacteriology Supplies and Equipment

1. Autoclaves and supplies

Amerex Instruments, Inc.
P.O. Box 787
Lafayette, CA 94549
(510) 937-0182
FAX: (510) 937-0950

Baxter Healthcare Corp.
1430 Waukegan Rd.
McGaw Park, IL 60085
(708) 689-8410
FAX: (708) 473-0804

Becton Dickinson Microbiology Systems
P.O. Box 243
Cockeysville, MD 21030
(410) 771-0100
(800) 638-8663

Biotest Diagnostics Corp.
66 Ford Road, Suite 131
Denville, NJ 07834
(201) 625-1300
FAX: (201) 625-5882

Brinkmann Instruments, Inc.
One Cantiague Rd.
Westbury, NY 11590-0207
(800) 645-3050
FAX: (516) 334-7506

Kulman Technologies, Inc.
10512 NE 68th St.
Kirkland, WA 98033
(206) 822-8282
FAX: (206) 827-9055

Raven Biological Laboratories, Inc.
5017 Leavenworth St.
Omaha, NE 68106-1428
(402) 556-6690
FAX: (402) 556-4722

Steris Corp.
(AMSCO International, Inc.)
5960 Heisley Road
Mentor, OH 44060-1834
(216) 354-2600
FAX: (216) 639-4459

2. Instruments

Abbott Laboratories
100 Abbott Park Rd.
Abbott Park, IL 60064
(800) 323-9100
FAX: (847) 938-3616

AccuMed International Inc.
29299 Clemens Rd. Suite 1-K
Westlake, OH 44145
(800) 871-8909
FAX: (216) 808-0400

Becton Dickinson Microbiology Systems
P.O. Box 243
Cockeysville, MD 21030
(410) 771-0100
(800) 638-8663

bioMérieux Vitek, Inc.
595 Anglum Dr.
Hazlewood, MO 63042-2395
(314) 731-8504
FAX: (314) 731-8700

Chiron Diagnostics
4560 Horton St.
Emeryville, CA 94608
(510) 601-3079
FAX: (510) 601-3307

Dade Microscan, Inc.
1584 Enterprise Blvd.
West Sacramento, CA 95691
(916) 374-2110
FAX: (916) 371-3831

Difco Laboratories
P.O. Box 331058
Detroit, MI 48232-7058
(313) 462-8500
FAX: (313) 462-8517

Digene Diagnostics, Inc.
2301-B Broadbirch Dr.
Silver Spring, MD 20904
(301) 470-6536
FAX: (301) 680-0696

Gen-Probe Inc.
9880 Campus Pont Dr.
San Diego, CA 92121
(800) 523-5001
FAX: (619) 452-5848

Microbial ID, Inc.
115 Barksdale Professional Center
Newark, DE 19711-9918
(302) 737-4297
(302) 737-7781

Organon-Teknika USA
100 Akzo Ave.
Durham, NC 27712
(919) 620-2000
FAX: (919) 620-2107

Roche Diagnostic Systems, Inc.
1080 U.S. Highway 202
Branchburg Township
Somerville, NJ 08876
(800) 526-1247
FAX: (800) 544-6130

Vista Technology, Inc.
8432 45 St.
Edmonton, Alberta T6B 2N6
Canada
(403) 468-0020
FAX: (403) 465-9732

3. Centrifuges

Astel Enterprises, Inc.
110-D Industrial Dr.
Winchester, VA 22602
(800) 556-2323
FAX: (540) 869-5937

Beckman Instruments, Inc.
200 S Kraemer Rd.
Brea, CA 92621
(800) 526-3821

B. Braun Biotech
999 Postal Rd.
Allentown, PA 18103
(610) 266-6262
FAX: (610) 266-9319

Brinkmann Instruments, Inc.
One Cantiague Rd.
Westbury, NY 11590-0207
(800) 645-3050
FAX: (516) 334-7506

International Equipment Co.
300 Second Ave.
Needham Heights, MA 02194
(617) 449-8060
FAX: (617) 455-9799

Jouan, Inc.
110-B Industrial Dr.
Winchester, VA 22602
(800) 662-7477
FAX: (703) 869-8626

New Brunswick Scientific, Inc.
44 Talmadge Rd., P.O. Box 4005
Edison, NJ 08818-4005
(800) 631-5417
FAX: (908) 287-4222

Shandon, Inc.
171 Industry Dr.
Pittsburgh, PA 15275
(412) 788-1133
(800) 245-6212

Sorvall Centrifuges
31 Peck's Lane
Newtown, CT 06470
(203) 270-2207
FAX: (203) 270-2500

4. Gas regulators, gas cylinders, and supplies

Curtin Matheson Scientific
9999 Veterans Memorial Dr.
Houston, TX 77038
(281) 820-9898
(800) 640-0640

Nupro Co.
4800 East 345 St.
Willoughby, OH 44094
(216) 951-7100

5. Gel electrophoresis apparatus and power supplies

Bio-Rad Laboratories, Inc.
4000 Alfred Nobel Dr.
Hercules, CA 94547
(800) 424-6723

C.B.S. Scientific Co., Inc.
420 S. Cedros Ave.
Solana Beach, CA 92075
(619) 755-4959
FAX: (619) 755-0733

E-C Apparatus Corp.
3831 Tyrone Blvd. North
St. Petersburg, FL 33709
(800) 327-2643

Fotodyne, Inc.
950 Walnut Ridge Dr.
Hartland, WI 53029-9388
(414) 369-7000
FAX: (414) 369-7017

GIBCO BRL, Life Technologies, Inc.
P.O. Box 68
Grand Island, NY 14072-0068
(800) 828-6686

Hoeffer Scientific Instruments
654 Minnesota St.
P.O. Box 77387
San Francisco, CA 94107
(800) 227-4750

Owl Scientific, Inc.
10 Commerce Way
Woburn, MA 01801
(617) 935-9499
FAX: (617) 935-8499

Pharmacia/Biotech, Inc.
800 Centennial Ave.
Piscataway, NJ 08855-1327
(800) 526-3593
FAX: (800) 329-3593

Stratagene
11011 North Torrey Pines Rd.
La Jolla, CA 92037
(619) 535-5400
FAX: (619) 535-0071

6. Heating block incubation units

Barnstead/Thermolyne Corp.
2555 Kerper Blvd.,
Dubuque, IA 52004-0797
(319) 556-2241
FAX: (319) 556-0695

Perkin-Elmer Corp.
Applied Biosystems Div.
850 Lincoln Center Dr.
Bldg. 100-3
Foster City, CA 94404
(415) 638-5715
FAX: (415) 638-6604

7. Incinerator burners

Baxter Healthcare Corp.
1430 Waukegan Rd.
McGaw Park, IL 60085-6787
(708) 689-8410
FAX: (708) 473-0804

Matrix Technologies Corp.
44 Stedman St.
Lowell, MA 01851-2734
(508) 454-5690
FAX: (508) 458-9174

Oxford-Sherwood-Davis & Geck
1915 Olive St.
St. Louis, MO 63103
(314) 241-5700
FAX: (314) 241-1673

8. Incubators

Barnstead/Thermolyne Corp.
2555 Kerper Blvd.
Dubuque, IA 52001
(319) 556-2241
FAX: (319) 556-0695

Baxter Healthcare Corp.
1430 Waukegan Rd.
McGaw Park, IL 60085-6787
(708) 689-8410
FAX: (708) 473-0804

Fisher Scientific Co
711 Forbes Ave.
Pittsburgh, PA 15219
(412) 490-8300

Forma Scientific, Inc.
P.O. Box 649
Marietta, OH 45750-0649
(614) 374-1851
FAX: (614) 374-1817

Hotpack Corp.
10940 Dutton Rd.
Philadelphia, PA 19154
(215) 824-1700
FAX: (215) 637-0519

Lab-Line Instruments, Inc.
15th & Bloomingdale Ave.
Melrose Park, IL 60160-1491
(708) 450-2600
FAX: (708) 450-0943

Nor-Lake Scientific
727 Second St.
Hudson, WI 54016
(800) 477-5253
FAX: (715) 386-6149

VWR Scientific Products
1310 Goshen Parkway
West Chester, PA 19380
(610) 429-2728
FAX: (610) 436-1761

9. Loop calibration products

Evergreen Scientific
2300 East 49th St.
Los Angeles, CA 90058-0248
(213) 583-1331
FAX: (213) 581-2503

Remel
12076 Santa Fe Dr.
Lenexa, KS 66215
(800) 255-6730
FAX: (800) 621-8251

10. Microscopes

Nikon Inc., Instrument Group
1300 Walt Whitman Rd.
Melville, NY 11747
(516) 547-8500
FAX: (516) 547-8518

Olympus America, Inc.
Two Corporate Center Dr.
Melville, NY 11747-3157
(516) 844-5055
FAX: (516) 844-5112

Carl Zeiss, Inc.
One Zeiss Dr.
Thornwood, NY 10594
(800) 233-2343
FAX: (914) 681-7445

11. Reagents, supplies, and kits

AB Biodisk North America, Inc.
200 Centennial Ave.
Piscataway, NJ 08854
(908) 457-8408
FAX: (908) 457-8980

Abbott Laboratories
100 Abbott Park Rd.
Abbott Park, IL 60064
(800) 323-9100
FAX: (847) 938-3616

AccuMed International Inc.
29299 Clemens Rd., Suite 1-K
Westlake, OH 44145
(800) 871-8909
FAX: (216) 808-0400

Accurate Chemical and Scientific Corp.
30 Shames Dr.
Westbury, NY 11590
(516) 333-2221
FAX: (516) 997-4948

Alpha-Tec Systems, Inc.
P.O. Box 5435
Vancouver, WA 98682
(360) 260-2779
FAX: (360) 260-3277

Ambion, Inc.
2130 Woodward St., Suite 200
Austin, TX 78744-1832

Anaerobe Systems
2200 Zanker Road, Suite C
San Jose, CA 95131
(408) 432-9103
FAX: (408) 432-9481

Applied Biosystems, Inc.
850 Lincoln Center Dr.
Foster City, CA 94404
(800) 345-5224

Bard Medical Div.
8195 Industrial Blvd.
Covington, GA 30209
(770) 784-6705
FAX: (770) 784-6495

Baxter Healthcare Corp.
1430 Waukegan Rd.
McGaw Park, IL 60085
(708) 689-8410
FAX: (708) 473-0804

Becton Dickinson Microbiology Systems
P.O. Box 243
Cockeysville, MD 21030
(410) 771-0100
(800) 638-8663

Bellco Glass, Inc.
340 Edrudo Road
Vineland, NJ 08360

Bio 101, Inc.
1070 Joshua Way
Vista, CA 92083

Biolog, Inc.
3938 Trust Way
Hayward, CA 94545
(510) 785-2591
FAX: (510) 782-4639

Biomed Diagnostics, Inc.
1430 Koll Circle, Suite 101
San Jose, CA 95112
(408) 451-0400
FAX: (408) 451-0409

Biostar
6655 Lookout Rd.
Boulder, CO 80301
(303) 530-3888
FAX: (303) 530-6601

Carlson Scientific
514 South Third St.
Peotone, IL 60468
(708) 258-6377
FAX: (708) 258-6378

Carr-Scarborough Microbiologicals, Inc.
5342 Panola Industrial Blvd.
Decatur, GA 30035
(404) 987-9300
FAX: (404) 987-9345

Chiron Diagnostics
4560 Horton Street
Emeryville, CA 94608
(510) 601-3079
FAX: (510) 601-3307

Chromagar
267 Rue Lecourbe
Paris, F75015
France

Difco Laboratories
P.O. Box 331058
Detroit, MI 48232-7058
(313) 961-0800
(800) 521-0851

Digene Diagnostics, Inc.
2301-B Broadbirch Dr.
Silver Spring, MD 20904
(301) 470-6536
FAX: (301) 680-0696

Evergreen Scientific
2300 East 49th St.
Los Angeles, CA 90058-0248
(213) 583-1331
FAX: (213) 581-2503

EY Laboratories, Inc.
107 North Amphlett Blvd.
San Mateo, CA 94401
(415) 342-3296
FAX: (415) 342-2648

FMC Bioproducts
191 Thomaston St.
Rockland, ME 04841
(800) 341-1574
FAX: (800) 362-5552

Gelman Sciences
600 South Wagner Rd.
Ann Arbor, MI 48103
(313) 665-0651
FAX: (313) 913-6383

Gen-Probe Inc.
9880 Campus Point Dr.
San Diego, CA 92121
(800) 523-5001
FAX: (619) 452-5848

Hardy Diagnostics
1430 West McCoy Lane
Santa Maria, CA 93455
(800) 266-2222
FAX: (805) 346-2766

IDEXX Laboratories, Inc.
One IDEXX Dr.
Westbrook, ME 04092
(207) 856-0300
FAX: (207) 856-0675

Innovative Diagnostic Systems
2797 Petersen Place
Norcross, GA 30071
(770) 409-0713
FAX: (770) 409-0789

Key Scientific Products
149C Texas Ave.
Round Rock, TX 78664
(512) 218-1913
FAX: (512) 218-8580

Lab Safety Supply, Inc.
P.O. Box 1368
Janesville, WI 53547-1368
(800) 356-0783

Meridian Diagnostics, Inc.
3471 River Hills Dr.
Cincinnati, OH 45244
(513) 271-3700
FAX: (513) 272-5432

Microtech Medical Systems, Inc.
401 Laredo St.
Aurora, CO 80011

Nalge Nunc International
75 Panorama Creek Dr.
P.O. Box 20365
Rochester, NY 14625-0365

New England Biolabs
32 Tozer Rd.
Beverly, MA 01915
(508) 927-5054
FAX: (508) 921-1350

Oncor
209 Perry Parkway
Gaithersburg, MD 20877
(800) 776-6267

Organon Teknika
100 Akzo Ave.
Durham, NC 27704
(919) 620-2000
FAX: (919) 620-2107

Oxoid, Inc.
217 Colonnade Rd.
Nepean, Ontario K2E 7K3
Canada

Perkin-Elmer Corp.
850 Lincoln Center Dr.
Bldg. 100-3
Foster City, CA 94404
(415) 638-5715
FAX: (415) 638-6604

Promega Corp.
2800 Woods Hollow Rd.
Madison, WI 53711
(608) 277-2639
FAX: (608) 277-2601

Reagent Hospital Products, Ltd.
80 International Dr., Suite 300
Greenville, SC 29615
(803) 297-4343
FAX: (803) 234-9516

Remel
12076 Santa Fe Dr.
Lenexa, KS 66215
(800) 255-6730
FAX: (800) 621-8251

Roche Diagnostics Systems, Inc.
1080 U.S. Highway 202
Branchburg Township
Somerville, NJ 08876
(800) 526-1247
FAX: (800) 544-6130

Sarstedt, Inc.
1025 St. James Church Rd.
Newton, NC 28658

Tekmar Co.
P.O. Box 429576
Cincinnati, OH 45249
(800) 543-4461
FAX: (513) 247-7050

Vangard International, Inc.
1111-A Green Grove Rd.
Neptune, NJ 07754-0308
(908) 922-0784
FAX: (908) 922-0557

VWR Scientific Products
1310 Goshen Parkway
West Chester, PA 19380
(610) 429-2728
FAX: (610) 436-1761

Wako Chemicals USA, Inc.
1600 Bellwood Rd.
Richmond, VA 23237
(800) 992-9265
FAX: (804) 271-7791

12. Refrigerators and freezers

Forma Scientific, Inc.
P.O. Box 649
Marietta, OH 45750
(614) 374-1851
FAX: (614) 374-1817

Harris Manufacturing/Queue Systems
275 Aiken Rd.
Ashville, NC 28804
(800) 221-4201
FAX: (704) 645-3368

Hotpack Corp.
10940 Dutton Rd.
Philadelphia, PA 19154
(215) 824-1700
FAX: (215) 637-0519

Jordon Scientific Products
2200 Kennedy St.
Philadelphia, PA 19137
(215) 535-8300
FAX: (215) 289-1597

Nor-Lake Scientific
727 Second St.
Hudson, WI 54016
(800) 477-5253
FAX: (715) 386-6149

Revco Scientific Inc.
275 Aiken Rd.
Asheville, NC 28804
(704) 658-2711
FAX: (704) 645-3368

Sanyo Scientific
900 North Arlington Heights Rd.
Suite 310
Itasca, IL 60143
(800) 858-8442
FAX: (708) 775-0044

13. Safety cabinets

Baker Co.
P.O. Drawer E
Sanford, ME 04073
(207) 324-8773
FAX: (207) 324-3869

Forma Scientific
P.O. Box 649
Marietta, OH 45750
(614) 374-1851
FAX: (614) 374-1817

Labconco Corp.
8811 Prospect
Kansas City, MO 64132
(816) 333-8811
FAX: (816) 363-0130

Nu Aire, Inc.
2100 Fernbrook
Plymouth, MN 55447
(612) 553-1270
FAX: (612) 553-0459

B. Anaerobic Bacteriology Supplies and Equipment

Anaerobe Systems
2200 Zanker Rd., Suite C
San Jose, CA 95131
(408) 432-9103
FAX: (408) 432-9481

Coy Laboratory Products
14500 Coy Dr.
Grass Lake, MI 49240
(313) 475-2200
FAX: (313) 475-1846

Forma Scientific, Inc.
P.O. Box 649
Marietta, OH 45750
(614) 374-1851
FAX: (614) 374-1817

Gelman Sciences
600 South Wagner Rd.
Ann Arbor, MI 48103
(313) 665-0651
FAX: (313) 913-6383

Key Scientific Products
149C Texas Ave.
Round Rock, TX 78664
(512) 218-1913
FAX: (512) 218-8580

Nupro Co.
4800 East 345th St.
Willoughby, OH 44094
(216) 951-7100

Plas Labs, Inc.
917 East Chilson St.
Lansing, MI 48906
(517) 372-7177
FAX: (517) 372-2857

Sheldon Manufacturing, Inc.
300 North 26th Ave.
Cornelius, OR 97113
(503) 640-3000
FAX: (503) 640-1366

Supelco, Inc.
Supelco Park
Bellefonte, PA 16823
(814) 359-5488
FAX: (814) 359-5459

C. Mycology Supplies and Equipment

Becton Dickinson Microbiology Systems
P.O. Box 243
Cockeysville, MD 21030
(410) 771-0100
(800) 638-8663

bioMérieux Vitek, Inc.
595 Anglum Dr.
Hazlewood, MO 63042-2395
(314) 731-8504
FAX: (314) 731-8700

Dade Microscan, Inc.
1584 Enterprise Blvd.
West Sacramento, CA 95691
(916) 374-2110
FAX: (916) 371-3831

Gen-Probe Inc.
9880 Campus Point Dr.
San Diego, CA 92121
(800) 523-5001
FAX: (619) 452-5848

Immuno-Mycologics, Inc.
P.O. Box 1151
Norman, OK 73070
(405) 288-2383
FAX: (405) 288-2228

Innovative Diagnostic Systems
2797 Peterson Place
Norcross, GA 30071
(770) 409-0713
FAX: (770) 409-0789

Meridian Diagnostics, Inc.
3471 River Hills Dr.
Cincinnati, OH 45209
(513) 271-3700
FAX: (513) 272-5432

Remel
12076 Santa Fe Dr.
Lenexa, KS 66215
(800) 255-6730
FAX: (800) 621-8251

Wampole Laboratories
P.O. Box 1001
Cranbury, NY 08512
(800) 257-9525
FAX: (609) 655-6898

D. Mycobacteriology Supplies and Equipment

Becton Dickinson Microbiology Systems
P.O. Box 243
Cockeysville, MD 21030
(410) 771-0100
(800) 638-8663

Difco Laboratories
P.O. Box 331058
Detroit, MI 48232-7058
(313) 462-8500
FAX: (313) 462-8517

Gen-Probe Inc.
9880 Campus Point Dr.
San Diego, CA 92121
(800) 523-5001
FAX: (619) 452-5848

Organon Teknika
100 Akzo Ave.
Durham, NC 27712
(919) 620-2000
FAX: (919) 620-2107

Roche Diagnostic Systems, Inc.
1080 U.S. Highway 202
Branchburg Township
Somerville, NJ 08876
(800) 526-1247
FAX: (800) 544-6130

E. Parasitology Supplies and Equipment

Alexon-Trend, Inc.
14000 Unity St., N.W.
Ramsey, MN 55303
(612) 323-7800
FAX: (612) 323-7858

Alpha-Tec Systems, Inc.
P.O. Box 5435
Vancouver, WA 98668
(360) 260-2779
FAX: (360) 260-3277

American Type Culture Collection
12301 Parklawn Dr.
Rockville, MD 20852
(301) 881-2600
FAX: (301) 816-4367

Antibodies Inc.
P.O. Box 1560
Davis, CA 95617
(916) 758-4400
FAX: (916) 758-5672

Diagnostic Technology, Inc.
240 Vanderbilt Motor Parkway
Hauppauge, NY 11788
(516) 582-4949
FAX: (516) 582-4694

Diamedix Corp.
2140 North Miami Ave.
Miami, FL 33127
(800) 327-4565
FAX: (305) 324-2395

DiaSys Corp.
49 Leavenworth St.
Waterbury, CN 06702

Empyrean Diagnostics, Inc.
2761 Marine Way
Mountain View, CA 94043
(415) 960-0516
FAX: (650) 960-0515

Evergreen Scientific
2300 East 49th St.
Los Angeles, CA 90058
(213) 583-1331
FAX: (213) 581-2503

Gull Laboratories, Inc.
1011 East Murray Holladay Rd.
Salt Lake City, UT 84117
(801) 263-3524
FAX: (801) 265-9268

Hardy Diagnostics
1430 West McCoy Lane
Santa Maria, CA 93455
(800) 266-2222
FAX: (805) 346-2760

IncStar Corp.
1990 Industrial Blvd.
P.O. Box 285
Stillwater, MN 55082

Interfacial Dynamics Corp.
17300 Southwest Upper Boones Ferry Rd., Suite 120
Portland, OR 97224
(503) 256-0076
FAX: (503) 255-0989

International Health Services
2166 Old Middlefield Way
Mountain View, CA 94043
(415) 960-0401
FAX: (415) 960-0402

Medical Chemical Corp.
1909 Centinela Ave.
Santa Monica, CA 90404
(310) 829-4304
FAX: (310) 453-1212

Meridian Diagnostics, Inc.
3471 River Hills Dr.
Cincinnati, OH 45244
(513) 271-3700
FAX: (513) 271-0124

MML Diagnostic Packaging
P.O. Box 458
Troutdale, OR 97060
(503) 666-8398

PML Microbiologicals
15845 Southwest 72nd Ave., Bldg. C
Portland, OR 97224
(503) 639-1500
FAX: (800) 765-4415

Scientific Device Laboratory, Inc.
411 East Jarvis
Des Plaines, IL 60018
(847) 803-9495
FAX: (847) 803-8251

Trend Scientific, Inc.
P.O. Box 120266
St. Paul, MN 55112-0266
(612) 633-0925
FAX: (612) 633-6073

Volu-Sol, Inc.
5095 West 2100 South
Salt Lake City, UT 84120
(801) 974-9474
FAX: (800) 860-4317

F. Virology Supplies and Equipment

ABI/Advanced Biotechnologies, Inc.
9108 Guilford Road
Columbia, MD 21046-2701
(301) 470-3220
FAX: (301) 497-9773

Accurate Chemical & Scientific Corp.
300 Shames Dr.
Westbury, NY 11590
(516) 333-2221
FAX: (516) 997-4948

American Type Culture Collection
12301 Parklawn Dr.
Rockville, MD 20852
(301) 881-2600
FAX: (301) 816-4367

Amicon, Inc.
72 Cherry Hill Dr.
Beverly, MA 01915
(508) 777-3622
FAX: (508) 777-6204

Bartels, Inc.
2005 Northwest Sammamish, Suite 107
Issaquah, WA 98027
(206) 557-7606
FAX: (206) 392-6391

Biowhittaker, Inc.
8830 Biggs Ford Rd.
Walkersville, MD 21793
(301) 898-7025
FAX: (301) 845-4024

Blackhawk Biosystems, Inc.
12945 Alcosta Blvd.
San Ramon, CA 94583
(510) 866-1458
FAX: (510) 866-2941

Chemicon International, Inc.
28835 Single Oak Dr.
Temecula, CA 92592
(909) 676-8080
FAX: (909) 676-9209

Chiron Diagnostics
4560 Horton St.
Emeryville, CA 94608
(510) 601-3079
FAX: (510) 601-3307

Difco Laboratories
P.O. Box 331058
Detroit, MI 48232-7058
(313) 462-8500
FAX: (313) 462-8517

Digene Diagnostics, Inc.
2301-B Broadbirch Dr.
Silver Spring, MD 20904
(301) 470-6536
FAX: (301) 680-0696

Irvine Scientific
2511 Daimler St.
Santa Ana, CA 92705
(714) 261-7800
FAX: (714) 261-6522

Ortho Diagnostic Systems
1001 US Highway 202
Raritan, NJ 08869
(908) 218-8503
FAX: (908) 704-3905

Roche Diagnostic Systems
1080 U.S. Highway 202
Branchburg Township
Somerville, NJ 08876
(800) 526-1247
FAX: (800) 544-6130

Sigma Chemical Co.
P.O. Box 14508
St. Louis, MO 63178
(314) 771-5750
FAX: (314) 652-9930

Viromed Laboratories, Inc.
6101 Blue Circle Dr.
Minneapolis, MN 55343
(800) 582-0077
FAX: (612) 939-4215

G. Serology and Immunology Supplies and Equipment

Abbott Laboratories
100 Abbott Park Rd.
Abbott Park, IL 60064-3500
(800) 323-9100
FAX: (847) 938-3616

Alexon-Trend, Inc.
14000 Unity St., N.W.
Ramsey, MN 55303
(612) 323-7800
FAX: (612) 323-7858

Baxter Healthcare Corp.
1430 Waukegan Rd.
McGaw Park, IL 60085
(708) 689-8410
FAX: (708) 473-0804

Becton Dickinson Microbiology Systems
P.O. Box 243
Cockeysville, MD 21030
(410) 771-0100
(800) 638-8663

Behring
3403 Yerba Buena Rd.
San Jose, CA 95161
(408) 239-2000

Binax, Inc.
217 Read St.
Portland, ME 04103
(207) 772-3988
FAX: (207) 761-2074

Biokit USA Inc.
113 Hartwell Ave.
Lexington, MA 02173
(617) 861-4060
FAX: (617) 861-4065

Biomedical Products Corp.
10 Halstead Rd.
Mendham, NJ 07945
(201) 543-7434
FAX: (201) 543-7497

bioMériux Vitek, Inc.
595 Anglum Dr.
Hazelwood, MO 63042-2395
(314) 731-8504
FAX: (314) 731-8700

Bio-Rad Laboratories
2000 Alfred Nobel Dr.
Hercules, CA 94547
(510) 741-8044
FAX: (510) 234-2642

Bio-Tek Instruments, Inc.
Highland Park, Box 998
Winooski, VT 05404
(802) 655-4040
FAX: (802) 655-7941

Biowhittaker, Inc.
8830 Biggs Ford Rd.
Walkersville, MD 21793
(301) 898-7025
FAX: (301) 845-4024

Cambridge Biotech Corp.
365 Plantation St.
Worcester, MA 01605
(508) 897-5777
FAX: (508) 797-4014

Corning Costar Corp.
One Alewife Center
Cambridge, MA 02140
(617) 868-6200
FAX: (617) 868-2076

Diagnostic Products Corp.
5700 West 96th St.
Los Angeles, CA 90045-5597
(213) 776-0180
FAX: (310) 642-1176

Diamedix Corp.
2140 North Miami Ave.
Miami, FL 33127
(800) 327-4565
FAX: (800) 578-3377

Difco Laboratories, Inc.
P.O. Box 331058
Detroit, MI 48232-7058
(313) 462-8500
FAX: (313) 462-8517

Dynatech Laboratories, Inc.
14340 Sullyfield Circle
Chantilly, VA 22021
(703) 803-1243
FAX: (703) 631-7833

Elias USA, Inc.
373 280th St.
Osceola, WI 54020
(715) 294-2144
FAX: (715) 294-3921

Elkay Products Inc.
800 Boston Turnpike
Shrewsbury, MA 01545-7247
(800) 673-3755
FAX: (508) 842-1338

Genbio
15222 Ave. of Science, Suite A
San Diego, CA 92128
(619) 592-9300
FAX: (610) 592-9400

G.F.M.D., Ltd.
8710 Royal Lane
Irving, TX 75063
(214) 929-1222
FAX: (214) 929-8815

Gull Laboratories
1011 Murray Holiday Rd.
Salt Lake City, UT 84117
(801) 263-3524
FAX: (801) 265-9268

Harlan Bioproducts for Science, Inc.
P.O. Box 29176
Indianapolis, IN 46229-0176
(317) 899-7511
FAX: (317) 899-1766

Hemagen Diagnostics, Inc.
4-40 Bear Hill Rd.
Waltham, MA 02154-1002
(617) 890-3766
FAX: (617) 890-3748

Immunetics, Inc.
63 Rogers St.
Cambridge, MA 02142
(617) 492-5416
FAX: (617) 868-7879

Immuno Concepts Inc.
9779 Business Park Dr.
Sacramento, CA 95827
(916) 363-2649
FAX: (916) 363-2843

Immunostics, Inc.
505 Sunset Ave.
Ocean, NJ 07712
(908) 918-0770
FAX: (908) 918-0618

INCSTAR Corp
P.O. Box 285
1990 Industrial Blvd.
Stillwater, MN 55082
(612) 779-1745
FAX: (612) 779-7847

Inova Diagnostics, Inc.
10451 Roselle St.
San Diego, CA 92121
(619) 455-9495
FAX: (619) 455-0912

Integrated Diagnostics, Inc.
1756 Sulphur Spring Rd.
Baltimore, MD 21227
(410) 737-8500
FAX: (410) 536-1212

International Immuno-Diagnostics
1155 Chess Dr., #121
Foster City, CA 94404
(415) 345-9518
FAX: (415) 578-1810

MarDx Diagnostics, Inc.
5919 Farnsworth Ct.
Carlsbad, CA 92008
(619) 929-0125
FAX: (619) 929-0500

Medix Biotech Inc.
420 Lincoln Center Dr.
Foster City, CA 94404
(415) 573-3315

Meridian Diagnostics, Inc.
3471 River Hills Dr.
Cincinnati, OH 45209
(513) 271-3700
FAX: (513) 272-5432

Murex Diagnostics, Inc.
3075 Northwoods Circle
Norcross, GA 30071
(404) 662-0660
FAX: (404) 449-4018

Organon Teknika Corp.
100 Akzo Ave.
Durham, NC 27712
(919) 620-2000
FAX: (919) 620-2107

Orion Diagnostica Inc.
71 Veronica Ave.
Somerset, NJ 08873
(800) 526-2125
FAX: (908) 246-0570

Ortho Diagnostic Systems, Inc.
1001 U.S. Highway 202
Raritan, NJ 08869
(908) 218-1300
FAX: (908) 704-3905

Roche Diagnostic Systems, Inc.
1080 U.S. Highway 202
Branchburg Township
Somerville, NJ 08876
(800) 526-1247
FAX: (800) 544-6130

Sanofi Diagnostics Pasteur
1000 Lake Hazeltine Dr.
Chaska, MN 55318
(612) 368-1124
FAX: (612) 368-1100

SciMedx
400 Ford Rd., Bldg. 1
Denville, NJ 07834
(201) 625-8822
FAX: (201) 625-8796

Seradyn, Inc.
1200 Madison Ave.
Indianapolis, IN 46225
(317) 266-2000
FAX: (317) 266-2991

Sienna Biotech, Inc.
9115 Guilford Road, Suite 180
Columbia, MD 21046
(301) 497-0007
FAX: (301) 497-8796

Sigma Diagnostics
P.O. Box 14508
St. Louis, MO 63178
(314) 771-5750
FAX: (314) 652-9930

Stellar Bio Systems, Inc.
9075 Guilford Rd.
Columbia, MD 21046
(410) 381-8550
FAX: (410) 381-8984

Tecan Inc.
P.O. Box 13953
Research Triangle Park, NC 27709
(919) 361-5200
FAX: (919) 361-5201

The Binding Site, Inc.
5889 Oberlin Drive, Suite 101
San Diego, CA 92121
(619) 453-9177
FAX: (619) 453-9189

VIRION (U.S.) Inc.
4 Upperfield Rd.
Morristown, NJ 07960
(201) 993-8219
FAX: (201) 993-3683

Virostat
P.O. Box 8522
Portland, ME 04104
(207) 883-1491
FAX: (207) 883-1482

Wampole Laboratories
Half Acre Rd.
P.O. Box 1001
Cranbury, NJ 08512
(800) 257-9525
FAX: (609) 655-6898

Zeus Scientific, Inc.
P.O. Box 38
Raritan, NJ 08869
(800) 526-3874
FAX: (908) 526-2058

H. Laboratory Media

Accurate Chemical & Scientific Corp.
300 Shames Dr.
Westburg, NY 11590
(516) 333-2221
(800) 645-6264
FAX: (516) 997-4948

Adams Scientific
771 Main St.
West Warwick, RI 02893
(401) 828-5250
FAX: (401) 828-5613

American Laboratory Supply, Inc.
P.O. Box 11
Gibsonville, NC 27249
(919) 449-6102

Amersco Inc.
30175 Solon Industrial Parkway
Solon, OH 44139
(216) 349-1313
FAX: (216) 349-1182

Anaerobe Systems
2200 Zanker Rd., Suite C
San Jose, CA 95131
(408) 432-9103
FAX: (408) 432-9481

Becton Dickinson Microbiology Systems
P.O. Box 243
Cockeysville, MD 21030
(410) 771-0100
(800) 638-8663

Carr-Scarborough Microbiologicals Inc.
5342 Pannola Industrial Blvd.
Decatur, GA 30035
(404) 987-9300
FAX: (404) 987-9345

Clinical Standards Laboratory
2011 University Dr.
Rancho Dominquez, CA 90220-6445
(310) 537-6800
FAX: (310) 763-1250

Cultech Diagnostics
1013 Commercial Dr.
Owensville, MO 65066
(800) 325-0167
FAX: (573) 437-5263

Difco Laboratories, Inc.
P.O. Box 331058
Detroit, MI 48232-7058
(313) 462-8500
FAX: (313) 462-8517

Hardy Diagnostics
1430 West McCoy Lane
Santa Maria, CA 93455
(805) 346-2766
(800) 266-2222
FAX: (805) 346-2760

Marcor Development Corp.
108 John St.
Hackensack, NJ 07601
(201) 489-5700
FAX: (201) 489-7357

Medical Wire and Equipment
7 The Boardwalk
Sparta, NJ 07871
(201) 729-5944
FAX: (201) 729-0589

Microbiologics, Inc.
217 Osseo Ave. North
St. Cloud, MN 56303-4455
(612) 253-1640
FAX: (612) 253-6250

Oxoid, Inc.
217 Colonnade Rd.
Nepean, Ontario K2E 7K3
Canada

PML Microbiologicals
15845 Southwest 72nd Ave.
Portland, OR 97224
(503) 639-1500
FAX: (503) 968-4844

Remel
12076 Santa Fe Dr.
Lenexa, KS 66215
(800) 255-6730
FAX: (800) 621-8251

Troy Biologicals, Inc.
1238 Rankin
Troy, MI 48084
(313) 585-9720

Index

B

C

D

I

J

Q

R

S

T